First Aid in Case of an Accident

The occurrence of an accident of any kind in the laboratory should be reported promptly to your instructor, even if it seems relatively minor.

FIRE

Your first consideration is to remove yourself from any danger, *not* to extinguish the fire. *If it is possible to do so without endangering yourself*, turn off any burners and remove containers of flammable solvents from the immediate area to prevent the fire from spreading. For the most effective use of a fire extinguisher, direct its nozzle toward the *base* of the flames. Burning oil may be put out with an extinguisher classified for use on "ABC" type fires.

If your clothing is on fire, DO NOT RUN; rapid movement will only fan the flames. Roll on the floor to smother the fire and to help keep the flames away from your head. Your neighbors can help to extinguish the flames by using fire blankets, laboratory coats, or other items that are immediately available. Do not hestitate to aid your neighbor if he or she is involved in such an emergency; a few seconds delay may result in serious injury. A laboratory shower, *if close by*, can be used to extinguish burning clothing, as can a carbon dioxide extinguisher, which must be used with care until the flames are extinguished *and only if the flames are not near the head*.

If burns are minor, apply a burn ointment. In the case of serious burns, do not apply any ointment; seek professional medical treatment at once.

CHEMICAL BURNS

Areas of the skin with which corrosive chemicals have come in contact should be immediately and thoroughly washed with soap and warm water. If the burns are minor, apply burn ointment; for treatment of more serious burns, see a physician.

Bromine burns can be particularly serious. Such burns should first be washed with soap and warm water and then thoroughly soaked with 0.6 M sodium thiosulfate solution for 3 h. Apply cod liver oil ointment and a dressing; see a physician.

If chemicals, in particular corrosive or hot reagents, come in contact with the eyes, immediately *flood* the eyes with water *from the nearest outlet*. It is best to use a specially designed eyewash fountain if one is available in the laboratory. *Do not touch the eye*. The eyelid as well as the eyeball should be washed with water for several minutes. In all instances where sensitive eye tissue is exposed to corrosive chemicals, an ophthalomologist should be consulted as soon as possible.

CUTS

Minor cuts may be treated by ordinary first-aid procedures; seek professional medical attention for serious cuts. If severe bleeding indicates that an artery has been severed, attempt to stop the bleeding with compresses and pressure; a tourniquet should be applied only by those who have received first-aid training. Arrange for emergency room treatment at once.

A person who is injured severely enough to require a physician's treatment *should be accompanied* to the doctor's office, or infirmary, even if he or she claims to be all right. Persons in shock, particularly after suffering burns, are often more seriously injured than they appear to be.

List of New Experiments

Ch 10   C. Bromination of (*E*)-Cinnamic Acid
Ch 17   B. Transfer Hydrogenation of Cinnamic Acid Derivatives
            A. Tartaric Acid-Mediated Enantioselective Reduction of Methyl Acetoacetate
Ch 19   Preparation of 3-Ethylhex-5-En-3-Ol
            Preparation of 4'-Methyl-(1,1'-Biphenyl)-4-Methanol

List of New Discovery Experiments

Ch 3   C. Who Else Has My Compound?
Ch 10   C. Bromination of (*E*)-Cinnamic Acid
Ch 14   Competing Nucleophiles in S_N Reactions
            Competition between Substitution and Elimination
Ch 15   B. Electrophilic Aromatic Bromination of Monosubstituted Arenes
            C. Nitration of Monosubstituted Arenes
            Azo Dyes and the Chemistry of Dyeing Fabrics
Ch 17   B. Transfer Hydrogenation of Cinnamic Acid Derivatives
            A. Tartaric Acid-Mediated Enantioselective Reduction of Methyl Acetoacetate
Ch 20   Identifying Unknown Esters Produced by Fischer Esterification

List of New Green Experiments

Ch 15   Friedel-Crafts Acylation of Anisole
Ch 16   C. Aerobic Oxidation of Benzylic Alcohols
Ch 18   Synthesis of Ethyl 6-Methyl-2-Oxo-4-Phenyl-1,2,3,4-Tetrahydropyrimidine-5-Carboxylate

Experimental Organic Chemistry

A Miniscale and Microscale Approach

SIXTH EDITION

John C. Gilbert

Santa Clara University

Stephen F. Martin

University of Texas at Austin

 CENGAGE

Australia • Brazil • Mexico • Singapore • United Kingdom • United States

***Experimental Organic Chemistry: A Miniscale
and Microscale Approach*, Sixth Edition**
John C. Gilbert and Stephen F. Martin

Product Director: Mary Finch

Product Manager: Krista Mastroianni

Content Developer: Thomas Martin

Managing Developer: Peter McGahey

Product Assistant: Morgan Carney

Associate Media Developer: Elizabeth Woods

Senior Marketing Manager: Julie Schuster

Art and Cover Direction, Production
Management, and Composition:
Lumina Datamatics, Inc

Manufacturing Planner: Judy Inouye

Cover Image: © Images created using VIDA
[v4.4.0] from OpenEye Scientific Software, Inc.
(www.eyesopen.com)

For product information and technology assistance, contact us at
**Cengage Customer & Sales Support, 1-800-354-9706
or support.cengage.com.**

For permission to use material from this text or product, submit all
requests online at **www.cengage.com/permissions.**

Library of Congress Control Number: 2014948891

ISBN: 978-1-305-08046-1

Cengage
20 Channel Street
Boston, MA 02210
USA

Cengage is a leading provider of customized learning solutions
with employees residing in nearly 40 different countries and sales in more
than 125 countries around the world. Find your local representative at:
www.cengage.com.

Cengage products are represented in Canada by
Nelson Education, Ltd.

To learn more about Cengage platforms and services, register or access
your online learning solution, or purchase materials for your course,
visit **www.cengage.com.**

Printed in Mexico
Print Number: 07   Print Year: 2020

Contents in Brief

Table of Contents

Preface

The management and teaching of an introductory laboratory course in organic chemistry is ever-changing, even though the fundamental chemical principles remain the same. Some of the compelling reasons for innovation and change are linked to the increasing cost associated with purchase and disposal of the chemicals used. There is the added concern of their possible toxicological hazards, both to students and to the environment. These factors dictate that many experiments be performed on reduced scales according to procedures commonly termed as *miniscale* (sometimes called small-scale) and *microscale*. This edition of our textbook maintains our practice of providing both miniscale and microscale procedures for most experiments. This unusual feature gives instructors maximal flexibility in customizing the course for use of apparatus and glassware already on hand and to suit the specific needs of you, the student.

The experiments are thoughtfully selected to introduce you to the common laboratory practices and techniques of organic chemistry and to illustrate the chemistry of the wide range of functional groups that are present in organic molecules. Some experiments are designed to familiarize you with the kinetic and thermodynamic principles underlying chemical reactions. Others allow you to synthesize specific compounds—some of which are found in nature or are of commercial importance—using reactions that are fundamental to organic synthesis. Still others introduce you to *discovery-based* and *green-chemistry* approaches. The discovery-based procedures—there are more than 50 of these in this edition—allow you to develop your own protocols for addressing a particular question experimentally, as you might do in a research laboratory. Discovery experiments are listed inside the front cover and are indicated when they appear in the book with the magnifying glass icon shown in the margin. The five procedures involving green chemistry show you how some chemical transformations may be performed using more environmentally friendly procedures. Green chemistry experiments are indicated when they appear in the book with the leaf icon shown in the margin. Most of the chapters are accompanied by a Historical Highlight, which are only available online. These 21 essays focus on interesting topics in organic chemistry that we believe will broaden your interest in the subject. Overall, our hope is that your experiences in this course will inspire you to take additional laboratory and lecture courses in chemistry, to seize the opportunity to work in a research laboratory as an undergraduate student, and perhaps even to pursue a career in research.

Background Information

Our textbook is distinct from many other laboratory manuals because the focused discussions preceding each Experimental Procedure provide the essential theoretical and "how-to" background, so other sources need not be consulted in order to understand the mechanistic and practical aspects of the specific reactions and procedures being performed. These discussions offer the advantage of making the textbook self-contained, and because they focus on the experiments themselves, they also significantly augment the material found in your lecture textbook.

Experimental Procedures

The *miniscale approach* appeals to instructors who believe in the importance of performing experiments on a scale that allows isolation and characterization of products using conventional laboratory glassware. The quantities of starting materials used are usually in the range of 1–3 g, so the costs associated with purchasing and disposing of the chemicals are modest. The amounts of material may be easily handled, and it is possible to develop the techniques required to purify the products and characterize them by comparing their physical properties with those reported in the scientific literature. You will also be able to characterize the starting materials and products by spectroscopic techniques, so that you can see how their spectral properties differ. In short, you will be able to experience the real world of organic chemistry in which usable quantities of compounds are synthesized.

The *microscale approach* is especially attractive for minimizing the cost of purchasing and disposing of chemicals. The specialized glassware and other apparatus required for performing experiments on such small scales is now readily available. Indeed, many of the components found in a microscale kit are also found in the advanced organic laboratory, where trained researchers often work with minute amounts of material. The amounts of starting materials that are used in these procedures are often only 100–300 mg. Because of the small quantities of materials being handled, you must be meticulous in order to isolate products from microscale reactions. Purifying small quantities of materials by distillation or recrystallization is often tedious, so it will frequently be impractical to characterize pure products. Nevertheless, the experiments performed on the microscale should provide tangible quantities of material so that you can verify that the product was formed using chemical tests as well as some spectroscopic and analytical techniques.

Organization

The experiments we have included are intended to reinforce concepts given in the lecture course in organic chemistry and to familiarize you with the techniques that modern organic chemists routinely use. The basic types of apparatus you will need are described in Chapter 2. In addition, videos illustrating the steps required to assemble many of the set-ups are available online, and we urge you to view these *prior* to going to the laboratory. In subsequent chapters, we provide figures in the margins of the pages to remind you how the assembled apparatus appears. The procedures in Chapters 3–6 are designed to introduce you to the different techniques for distillation, liquid-liquid and liquid-solid extraction, and thin-layer, column, and gas-liquid chromatography; the basic principles for these techniques are also described in their respective chapters. The spectroscopic methods that are fundamental to analyzing organic compounds are described in Chapter 8. Experiments that illustrate concepts such as selectivity of free-radical substitution (Chapter 9), kinetic and thermodynamic control of reactions (Chapter 13), kinetics of nucleophilic substitution reactions (Chapter 14), kinetics and regiochemistry of electrophilic aromatic substitution reactions (Chapter 15), and the stereochemistry and regiochemistry of addition reactions (Chapters 10, 11, 12, and 17) are intended to provide a better understanding of these important subjects. Other experiments illustrate specific chemical transformations such as the generation, reactions, and rearrangements of carbocations (Chapters 10 and 15), electrophilic aromatic and nucleophilic substitution processes (Chapters 15 and 14, respectively), eliminations (Chapters 10 and 11), oxidations and reductions (Chapters 16 and 17, respectively), nucleophilic additions to carbonyl compounds and imines (Chapters 17 and 18, respectively), the generation and reactions of Grignard and organozinc reagents (Chapter 19), and the formation of various carboxylic acid derivatives (Chapter 20). An experiment in Chapter 15 will introduce you to the fun of making your own dyes, and in Chapter 20 you will be able to observe the fascinating phenomenon of chemiluminescence. The value of enzymes for effecting enantioselective reactions is illustrated in Chapter 17. Because the current practice of organic chemistry in industry frequently involves multistep transformations, several examples of multi step synthesis are contained in Chapter 21. Many industrial processes rely on the use of transition-metal catalysts, and you will be able to study one such reaction in Chapter 19. Experiments designed to introduce you to basic concepts of carbohydrate chemistry and polymer chemistry are provided in Chapters 22 and 23, respectively, and the experiments given in Chapter 24 give you an opportunity to explore one aspect of the world of bio-organic chemistry through synthesis of a dipeptide. A rational approach to solving the structures of unknown compounds with and without the aid of spectroscopic data is given in Chapter 25, which is available online.

Spectroscopic Techniques

Spectroscopy may be the single most powerful tool for analyzing organic compounds. Consequently, thorough discussions of the theory and practical techniques for infrared, nuclear magnetic resonance (including ^{1}H and ^{13}C NMR), UV-Vis, and mass spectrometry are presented in Chapter 8. To reinforce the basic spectroscopic principles and to provide an opportunity for interpreting spectroscopic data, the infrared and nuclear magnetic spectra of all of the organic starting materials and products are provided in this textbook and online. It is also possible for you to perform simple manipulations of the ^{1}H NMR and IR spectra that are available at the website. For example, you will be able to measure chemical shifts, integrals, and coupling constants directly on the ^{1}H NMR spectra. You will also be able to determine the position of an absorption in the IR spectrum that is associated with a specific functional group. This "hands-on" experience has proved an invaluable aid in teaching the basics of interpreting ^{1}H NMR and IR spectra and is unique to this laboratory textbook.

Safety and the Environment

Important sections entitled "Safety Alert" and "Wrapping It Up" are included with each experimental procedure. The information in the "Safety Alert" is designed to inform you and your instructor of possible hazards associated with the operations being performed. The abbreviated Material Safety Data Sheets (MSDSs) are available online and provide additional information regarding flammability and toxicological properties of the chemicals being used and produced. Because of the flammable nature of the solvents and the chemicals that are handled in the laboratory, the use of *flameless* heating is emphasized and should be implemented in order to make the laboratory a safe workplace. The guidelines and methods in the "Wrapping It Up" section will familiarize you with the proper procedures for disposing of chemicals and other by-products after you have completed the experiment. Using these recommended methods will help protect the environment and lessen the costs associated with the ultimate disposal of these materials.

Essays

A feature of most of the chapters in the textbook is a Historical Highlight which are only available online. These 21 essays, some of which are biographical in nature, are designed to familiarize you with the lives of some of the chemical pioneers who have advanced science. These accounts will also provide you with a sense of the excitement and insights of individuals whose scientific observations form the basis for some of the experiments you will perform. Other essays are intended to relate organic chemistry to your everyday life. We hope they will whet your appetite for the subject of organic chemistry and enrich your experience as you further develop your scientific expertise.

Pre-Lab and Post-Lab Exercises

Each experiment is accompanied by two sets of questions. The Pre-Lab Exercises are provided online and are designed to test your understanding of potential safety hazards and basic concepts, so you will be able to perform the experiments and techniques safely and successfully. Because these questions will assist you in preparing for work in the laboratory, we strongly recommend that you answer them *before* performing the experiment. The Post-Lab Exercises are found under the heading "Exercises" after each Experimental Procedure. These questions are written to reinforce the principles that are illustrated by the experiments and to determine whether you understand the observations you have made and the operations you have performed. Furthermore, questions on spectroscopy will help you develop the skills required to interpret IR and ^{1}H and ^{13}C NMR spectra.

the Fifth Edition
Significant Changes from

This edition of the textbook includes a total of 15 new experiments. Ten of these are discovery experiments in which you will have an opportunity to identify unknown compounds and to explore reactivity and mechanism. Three new green chemistry experiments have been added that will enable you to learn how chemical reactions can be performed in an environmentally friendly manner. Finally, four experiments have been added so you can explore several modern methods to induce selective transformations.

Chapter 25—*Identifying Organic Compounds* and Chapter 26—*The Literature of Organic Chemistry* and the Historical Highlights are only available online. Moving these documents online provided space to add new experiments in various chapters as needed.

This edition includes a new OWLv2 LabSkills program that can be bundled with the textbook. LabSkills provides students the opportunity to practice skills and prepare for real laboratory classes in a safe environment. Students work through the assignments in advance of the lab so they can actively rehearse common practical techniques by engaging with simulations, videos, and quizzes. Because LabSkills focuses on techniques, instructors can customize assignments to align with the experiments that students will perform in any given lab session. The online course will also offer assessment specific to this edition's experiments, and include a Mind-Tap Reader. The new OWLv2 course with MindTapReader includes assignable pre-lab exercises integrated with LabSkills Safety and Techniques videos. All content is mobile compatible. The eBook contains interactive spectra.

Feedback

As always, we seek your comments, criticisms, and suggestions for improving our textbook. Despite our best efforts, we are certain that there are typographical errors and the like that have escaped our notice, and we would appreciate your bringing them to our attention; our e-mail and snail-mail addresses are provided below. No matter how busy we might be, we shall respond to any messages you send.

Supporting Materials

Please visit http://www.cengage.com/chemistry/gilbert/eoc6e for information about student and instructor resources for this book and about custom versions.

Acknowledgments

As with previous editions, we are grateful to a number of individuals who contributed to making this one a reality. These include Kevin Cannon (Penn State Abington College), Elliot Ennis (Missouri Southern State University), David Flaxbart (The University of Texas, Austin), Deborah Lieberman (University of Cincinnati), Robert Long (Texas Tech University), Barbara Mayer (California State University, Fresno), ArtemMelman (Clarkson University), Cornelia Mihai (Northwestern Oklahoma State University), Felix Ngassa (Grand Valley State University), Christine Pruis (Arizona State University), Narayan Sharma (State University of New York Institute of Technology), and Richard Sheardy (Texas Woman's University). Dr. Warren Martin (Texas Radiology Associates) kindly provided the MRI images and interpretations accompanying the Historical Highlight in Chapter 8. We are also grateful to Professor Scott Bur (GustavusAdolphus College), Professor Shannon Stahl and Dr. Nicholas Hill (The University of Wisconsin, Madison), and Dr. John Colapret and Dr. Conrad Fjetland (The University of Texas, Austin) for developing new experimental procedures contained in this edition. Graduate students, Katherine Ambrose, Alexander Goodnough, Daniel Klosowski, and Rachel Wypych (all of The University of Texas, Austin) also assisted with developing new procedures. We thank OpenEye Scientific Software of Santa Fe, New Mexico (http://www.eyesopen.com) for granting permission to use VIDA v4.4.0 to generate images for the cover, and Zachary White and Kristina Mose-Libon for helping to create the cover design. The capable staff at Cengage Learning provided invaluable support as we prepared this edition. We particularly acknowledge the efforts of Krista Mastroianni, the Product Manager, and Thomas Martin, the Content Developer, for their wise counsel and diligence that helped make this new edition a reality, and Greg Teague for his careful copyediting of the manuscript. We acknowledge the use of ^{13}C NMR spectra data from the Spectral Database for Organic Compounds of the JapaneseNational Institute of Advanced Industrial Science and Technology and also Aldrich Chemical Company, and of mass spectral data from the NIST database. We also thank Bill Vining of SUNY Oneonta and Bill Rohan, Cow Town Productions, for developing the Web-based information.

John C. Gilbert
jgilbert@scu.edu
Department of Chemistry & Biochemistry
Santa Clara University
Santa Clara, CA 95053

Stephen F. Martin
sfmartin@mail.utexas.edu
Department of Chemistry
The University of Texas at Austin
Austin, TX 78712

CHAPTER 1

Introduction, Record Keeping, and Laboratory Safety

This chapter sets the stage as you undertake the adventure of experimental organic chemistry. Although we may be biased, we think that this laboratory experience is one of the most valuable you will have as an undergraduate student. There is much to be learned as you progress from the relatively structured format of your first laboratory course in organic chemistry to the much less defined experimental protocols of a scientific research environment. The laboratory practices described in the following sections should serve you well in the journey.

1.1 INTRODUCTION

The laboratory component of a course in organic chemistry has an important role in developing and augmenting your understanding of the subject matter. The theoretical concepts, functional groups, and reactions presented in the lecture part of the course may seem abstract at times, but they are more understandable as a result of the experiments you perform. The successes, challenges, and, yes, frustrations associated with the "hands-on" experience gained in the laboratory, as you gather and interpret data from a variety of reactions, provide a sense of organic chemistry that is nearly impossible to communicate in formal lectures. For example, it is one thing to be told that the addition of bromine (Br_2) across the π-bond of most alkenes is a rapid process at room temperature. It is quite another to personally observe the *immediate* decoloration of a reddish solution of bromine in dichloromethane (Br_2/CH_2Cl_2) as a few drops of it are added to cyclohexene. The principles developed in the lectures will help you to predict what reaction(s) should occur when various reagents are combined in experimental procedures and to understand the mechanistic course of the process(es). Performing reactions allows you to test and verify the principles presented in lecture. Moreover, careful and thoughtful analysis of the observations and results of the experiments will foster the skills in critical thinking that are a hallmark of successful scientists.

Of course, the laboratory experience in organic chemistry has another important function beyond reinforcing the concepts presented in lecture—to introduce you to the broad range of techniques and procedures that are important to the successful practice of experimental organic chemistry. You will learn how to handle a variety of chemicals safely and how to manipulate apparatus properly, talents that are critical to your success as a student of the chemical sciences. Along with

1

becoming more skilled in the technical aspects of laboratory work, you should also develop a proper scientific approach to executing experiments and interpreting the results. By reading and, more importantly, *understanding* the concepts of this chapter, you will be better able to achieve these valuable goals.

1.2 PREPARING FOR THE LABORATORY

A common misconception students have about performing experiments is that it is much like cooking; that is, you merely follow the directions given—the "recipe"—and the desired product or data will result. Such students enter the laboratory expecting to follow the experimental procedure in a more or less rote manner. This unfortunate attitude can lead to inefficiencies, accidents, and minimal educational benefit and enjoyment from the laboratory experience.

To be sure, cooking is somewhat analogous to performing experiments. The successful scientist, just like a five-star chef, is a careful planner, a diligent worker, a keen observer, and, importantly, is fully prepared for failures! Experiments may not work despite your best efforts, just as a cake may fall even in the hands of a premier pastry chef.

The correct approach to being successful in the laboratory is *never* to begin any experiment until you understand its overall purpose and the reasons for each operation that you are to do. This means that you must *study*, not *just read*, the entire experiment *prior* to arriving at the laboratory. Rarely, if ever, can you complete the necessary preparation in 5 or 10 minutes, which means that you should not wait until just before the laboratory period begins to do the studying, thinking, and writing that are required. *Planning* how to spend your time in the laboratory is the key to efficient completion of the required experiments. Your performance in the laboratory will benefit enormously from proper advance work, and so will your grade!

The specific details of what you should do before coming to the laboratory will be provided by your instructor. However, to help you prepare in advance, we have developed a set of Pre-Lab Exercises for each of the experimental procedures we describe. These exercises are Web-based and are available online.

Your instructor may require you to submit answers to the Pre-Lab Exercises for approval before authorizing you to proceed with the assigned experiments. Even if you are not required to submit the exercises, though, you will find that answering them *prior* to the laboratory period will be a valuable educational tool to self-assess your understanding of the experiments to be performed.

You undoubtedly will be required to maintain a laboratory notebook, which will serve as a complete, accurate, and neat record of the experimental work that you do. Once more, your instructor will provide an outline of what specific information should appear in this notebook, but part of what is prescribed will probably necessitate advance preparation, which will further enhance your ability to complete the experiments successfully. The laboratory notebook is a *permanent record* of your accomplishments in the course, and you should take pride in the quality and completeness of its contents!

1.3 WORKING IN THE LABORATORY

You should be aware that experimental organic chemistry is *potentially* dangerous, because many of the chemicals used are toxic and/or highly flammable, and most of the procedures require the use of glassware that is easily broken. Careless handling of these chemicals and sloppy assembly of apparatus are sources of danger not only to you but also to those working near you. You *should* not be afraid of the chemicals and equipment that you will be using, but you should treat them with the respect and care associated with safe experimental practices. To facilitate this, there is an emphasis on the proper handling of chemicals and apparatus throughout the textbook, and the importance of paying particular attention to these subjects *cannot* be overemphasized. In a sense, laboratory safety is analogous to a chain that is only as strong as its weakest link: The possibility that an accident will occur is only as great as the extent to which unsafe practices are followed. In other words, if you and your labmates adhere to proper laboratory procedures, the risk of an accident will be minimized.

It is important that you follow the experimental procedures in this textbook closely. There is a good reason why each operation should be performed as it is described, although that reason may not be immediately obvious to you. Just as it is risky for a novice chef to be overly innovative when following a recipe, it is *dangerous* for a beginning experimentalist to be "creative" when it comes to modifying the protocol that we've specified. As you gain experience in the organic laboratory, you may wish to develop alternative procedures for performing a reaction or purifying a desired product, but *always* check with your instructor *before* trying any modifications.

Note that rather detailed experimental procedures are given early in the textbook, whereas somewhat less detailed instructions are provided later on. This is because many of the basic laboratory operations will have become familiar to you in time and need not be spelled out explicitly. It is hoped that this approach to the design of procedures will decrease your tendency to think that you are essentially following a recipe in a cookbook. Moreover, many of the experimental procedures given in the literature of organic chemistry are relatively brief and require the chemist to "fill in the blanks," so it is valuable to gain some initial experience in figuring out some details on your own.

Most of your previous experience in a chemistry laboratory has probably required that you measure quantities precisely, using analytical balances, burets, pipets, and other precise measuring devices (Secs. 2.5 and 2.6). Indeed, if you have done quantitative inorganic analysis, you know that it is often necessary to measure weights to the third or fourth decimal place and volumes to at least the first. Experiments in organic chemistry that are performed at the **microscale** level, that is, experiments in which less than about 1 mL of the principal reagents is used and the amounts of solvents are less than 2 or 3 mL, also require relatively precise measuring of quantities. For example, if you are to use 0.1 g of a reagent and your measuring device only allows measuring to the nearest 0.1 g, you could easily have as much as about 0.15 g or as little as 0.05 g of the reagent. Such deviations from the desired quantity represent significant *percentage* errors in measurement and can result in serious errors in the proportions of reagents involved in the reaction. Consequently, weights should be accurate to within about 0.01 g and volumes to within about 0.1 mL. This requires the use of appropriate analytical balances and graduated pipets.

Experiments being performed at the **miniscale** level, which we define as involving 1–5 g of reagents and usually less than about 25 mL of solvent, normally do not require such precise measuring. Weighing reagents to the nearest tenth of a gram is usually satisfactory, as is measuring out liquids in graduated cylinders, which are accurate to ±10%. For example, if you are directed to use 20 mL of diethyl ether as solvent for a reaction, the volume need *not* be 20.0 mL. In fact, it probably will make little difference to the success of the reaction whether anywhere from

15 mL to 25 mL of the solvent is added. This is not to say that care need not be exercised in measuring out the amounts of materials that you use. Rather, it means that valuable time need not be invested in making these measurements highly precise.

We've inserted markers in the form of **stars** (★) in many of the experimental procedures in this textbook. These indicate places where the procedure can be interrupted without affecting the final outcome of the experiment. These markers are designed to help you make the most efficient use of your time in the laboratory. For example, you may be able to start a procedure at a point in the period when there is insufficient time to complete it but enough time to be able to work through to the location of a star; you can then safely store the reaction mixture and finish the sequence during the next laboratory period. We've *not* inserted stars at every possible stopping point but only at those where it is not necessarily obvious that interruption of the procedure will have no effect on the experimental results. Consult your instructor if in doubt about whether a proper stopping point has been reached.

As noted above, a *carefully* written **notebook** and *proper* **safety procedures** are important components of an experimental laboratory course. These aspects are discussed further in the following two sections.

1.4 THE LABORATORY NOTEBOOK

One of the most important characteristics of successful scientists is the habit of keeping a complete and understandable record of the experimental work that has been done. Did a precipitate form? Was there a color change during the course of the reaction? At what temperature was the reaction performed, and for how long did the reaction proceed? Was the reaction mixture homogeneous or heterogeneous? On what date(s) was the work performed? These are observations and data that may seem insignificant at the time but may later prove critical to the interpretation of an experimental result or to the ability of another person to reproduce your work. All of them belong in a properly kept laboratory notebook. We make suggestions for such a document in the following two sections. Your instructor may specify other items to be included, but the list we give is representative of a good notebook.

1.5 GENERAL PROTOCOL FOR THE LABORATORY NOTEBOOK

1. Use a *bound* notebook for your permanent laboratory record to minimize the possibility that pages will be lost. If a number has not been printed on each page, do so manually. Some laboratory notebooks are designed with pairs of identically numbered pages so that a carbon copy of all entries can be made. The duplicate page can then be removed and submitted to your instructor or put in a separate place for safekeeping. Many professional scientists use this type of notebook.

2. Reserve the first page of the notebook for use as a title page, and leave several additional pages blank for a Table of Contents.

3. Use as the main criterion for what should be entered in the notebook the rule that the record should be sufficiently complete so that anyone who reads it will know exactly what you did and will be able to repeat the work in precisely the way you originally did it.

4. Record all experimental observations and data in the notebook *as they are obtained.* Include the date and, if appropriate, the *time* when you did the work. In a legal sense, the information entered into the notebook *at the time of performance* constitutes the primary record of the work, and it is important for you to follow this principle. Many patent cases have been determined on the basis of dates and times recorded in a laboratory notebook. One such example is described in the Historical Highlight *The Importance of Record Keeping,* which is available online.

5. Make all entries in ink, and *do not delete anything you have written* in the notebook. If you make a mistake, cross it out and record the correct information. Using erasers or correction fluid to modify entries in your notebook is unacceptable scientific practice!

   Do not scribble notes on odd bits of paper with the intention of recording the information in your notebook later. Such bad habits only lead to problems, since the scraps of paper are easily lost or mixed up. They are also inefficient, since transcribing the information to your notebook means that you must write it a second time. This procedure can also result in errors if you miscopy the data.

   Finally, do not trust your memory with respect to observations that you have made. When the time comes to write down the information, you may have forgotten a key observation that is critical to the success of the experiment.

6. Unless instructed to do otherwise, do not copy *detailed* experimental procedures that you have already written elsewhere in your notebook; this consumes valuable time. Rather, provide a specific reference to the source of the detailed procedure and enter a *synopsis* of the written procedure that contains enough information that (1) you need not refer to the source while performing the procedure and (2) another chemist will be able to *duplicate* what you did. For example, when performing an experiment from this textbook, give a reference to the page number on which the procedure appears, and detail any *variations* made in the procedure along with the reason(s) for doing so.

7. Start the description of each experiment on a new page titled with the name of the experiment. The recording of data and observations from several different procedures on the same page can lead to confusion, both for yourself and for others who may read your notebook. If you are unable to complete the write-up of an experiment on sequential pages, be certain to specify the page(s) on which the continuation appears.

1.6 TYPES OF ORGANIC EXPERIMENTS AND NOTEBOOK FORMATS

There are two general classes of experiments, **investigative** and **preparative,** in this textbook. Investigative experiments normally involve making observations and learning techniques that are common to laboratory work in organic chemistry but do not entail conversion of one compound into another. Some examples are solubility tests, distillation, recrystallization, and qualitative organic analysis. In contrast, preparative experiments involve interconversion of different compounds. Most of the procedures described in this textbook fall into the latter category.

The format of the laboratory notebook is usually different for these two types of experiments. Once again, your instructor may have a particular style that is recommended, but we provide suggested formats below.

Notebook Format for Investigative Experiments

1. **Heading.** Use a new page of the notebook to start the entries for the experiment. Provide information that includes your name, the date, the title of the experiment, and a reference to the place in the laboratory textbook or other source where the procedure may be found.

2. **Introduction.** Give a brief introduction to the experiment in which you clearly state the purpose(s) of the procedure. This should require no more than one-fourth of a page.

3. **Summary of MSDS Data.** As directed by your instructor, either briefly summarize the Material Safety Data Sheet (MSDS) data (Sec. 1.10) for the solvents, reagents, and products encountered in the experiment or give a reference to where a printout of these data is located.

4. **Synopsis of and Notes on Experimental Procedure—Results.** Enter a one- or two-line statement for each part of an experiment. Reserve sufficient room to record results as they are obtained. As noted in Section 1.5 of "Notebook Format for Preparative Experiments," do *not* copy the experimental procedure from the textbook, but provide a synopsis of it.

   Much of this section of the write-up can be completed before coming to the laboratory, to ensure that you understand the experiment and that you will perform all parts of it.

5. **Interpretation of Instrumental Data.** If instructed to do so, discuss any instrumental data, such as gas-liquid chromatographic analyses and spectral data that you have obtained or are provided in the textbook.

6. **Conclusions.** Record the conclusions that can be reached, based on the results you have obtained in the experiment. If the procedure has involved identifying an unknown compound, summarize your findings in this section.

7. **Answers to Exercises.** Enter answers to any exercises for the experiment that have been assigned from the textbook.

   A sample write-up of an investigative experiment is given in Figure 1.1.

Notebook Format for Preparative Experiments

1. **Heading.** Use a new page of the notebook to start the entries for the experiment. Provide information that includes your name, the date, the title of the experiment, and a reference to the place in the laboratory textbook or other source where the procedure may be found.

2. **Introduction.** Give a brief introduction to the experiment in which you clearly state the purpose(s) of the procedure. This should require no more than one-fourth of a page.

3. **Main Reaction(s) and Mechanism(s).** Write *balanced* equations giving the main reaction(s) for conversion of starting material(s) to product(s). The reason for balancing the equations is discussed in Part **4 Table of Reactants and Products**. Whenever possible, include the detailed mechanisms for the reactions that you have written.

4. **Table of Reactants and Products.** Set up a Table of Reactants and Products as an aid in summarizing the amounts and properties of reagents and catalysts being used and the product(s) being formed. Only those reactants, catalysts, and

1.   **Your Name**
                                                                          Date

Separation of Green Leaf Pigments by TLC

Reference: *Experimental Organic Chemistry: A Miniscale and Microscale Approach,* 6th ed.,
by Gilbert and Martin, Section 6.2.

2. INTRODUCTION

The pigments in green leaves are to be extracted into an organic solvent, and the extract is to be analyzed by thin-layer chromatography (TLC). The presence of multiple spots on the developed TLC plate will indicate that more than a single pigment is contained in the leaves.

3. MSDS DATA

These data are available on the printouts inserted at the back of my lab book.

4. SYNOPSIS OF AND NOTES ON EXPERIMENTAL PROCEDURE—RESULTS

Procedure: Grind five stemless spinach leaves in mortar and pestle with 5 mL of 2:1 pet. ether and EtOH. Swirl soln. with 3×2-mL portions H_2O in sep. funnel; dry org. soln. for few min over anhyd. Na_2SO_4 in Erlenmeyer. Decant and concentrate soln. if not dark-colored. Spot 10-cm $\times$ 2-cm TLC plate about 1.5 mm from end with dried extract; spot should be less than 2 mm diam. Develop plate with $CHCl_3$. *Variances and observations:* Procedure followed exactly as described in reference. Org. soln. was dark green in color; aq. extracts were yellowish. Half of org. layer lost. TLC plate had five spots having colors and R_f-values shown on the drawing below.

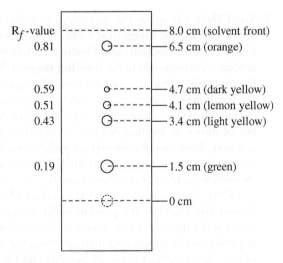

5. INTERPRETATION OF INSTRUMENTAL DATA

No data provided for this experiment.

6. CONCLUSIONS

Based on TLC analysis, the procedure used allows the extraction of at least five different pigments from the spinach leaves. Judging from colors, one of these is a carotene, three are xanthophylls, and the last is chlorophyll *b*.

7. ANSWERS TO EXERCISES

(Answers intentionally omitted.)

Figure 1.1
Sample notebook format for investigative experiments.

products that appear in the main reaction(s) should be listed in the table; many other reagents may be used in the work-up and purification of the reaction mixture, but these should *not* be entered in the table.

Your instructor will have specific recommendations about what should appear in the table, but the following items are illustrative.

a. The name and/or structure of each reactant, catalyst, and product.

b. The molar mass of each compound.

c. The weight used, in grams, of each reactant and the volume of any liquid reactant. We recommend that the weight and/or volume of any catalysts used be entered for purposes of completeness.

d. The molar amount of each reactant used; this can be calculated from the data in Parts **b** and **c.**

e. The theoretical mole ratio, expressed in whole numbers, for the reactants and products; this ratio is determined by the *balanced* equation for the reaction, as given in Part **3**, Main Reaction(s) and Mechanism(s).

f. Physical properties of the reactants and products. This entry might include data such as boiling and/or melting point, density, solubility, color, and odor.

g. As directed by your instructor, either briefly summarize the MSDS data (Sec. 1.10) for the solvents, reagents, and products encountered in the experiment or give a reference to where a printout of these data is located.

5. **Yield Data.** Compute the maximum possible amount of product that can be formed; this is the **theoretical yield.** This can easily be calculated from the data in the Table of Reactants and Products as follows. First determine which of the reactants corresponds to the **limiting reagent.** This is the reagent that is used in the *least* molar amount relative to what is required theoretically. In other words, the reaction will stop once this reactant is consumed, so its molar quantity will define the maximum quantity of product that can be produced. From the number of moles of limiting reagent involved and the balanced equation for the reaction, determine the theoretical yield, in moles (written as "mol" when used as a unit, as in "g/mol"), of product. This value can then be converted into the theoretical yield in grams, based on the molar mass of the product.

Once the isolation of the desired product(s) has been completed, you should also calculate the **percent yield,** which is a convenient way to express the overall efficiency of the reaction. This is done by obtaining the **actual yield** of product(s) in grams, and then applying the expression in Equation 1.1. Generally, the calculated value of percent yield is rounded to the nearest whole number. As points of reference, most organic chemists consider yields of 90% or greater as being "excellent," and those below 20% as "poor."

$$\text{Percent yield} = \frac{\text{Actual yield (g)}}{\text{Theoretical yield (g)}} \times 100 \qquad (1.1)$$

6. **Synopsis of and Notes on Experimental Procedure.** Provide an outline of the experimental procedure that contains enough detail so that you do not have to refer to the textbook repeatedly while performing the experiment. Note any variations that you use, as compared to the referenced procedure, and observations that you make while carrying out the formation and isolation of the product(s).

7. **Observed Properties of Product.** Record the physical properties of the product that you have isolated in the experiment. Appropriate data under this heading might include boiling and/or melting point, odor, color, and crystalline form, if the product is a solid. Compare your observations with those available on the compound in various reference books (e.g., the *CRC Handbook of Chemistry and Physics* or *Lange's Handbook of Chemistry*).

8. **Side Reactions.** If instructed to do so, list possible side reactions (those reactions leading to undesired products) that are likely to occur in the experiment. It is important to consider such processes because the by-products that are formed must be removed by the procedure used to purify the desired product. You may need to consult your lecture notes and textbook to predict what side reactions might be occurring.

9. **Other Methods of Preparation.** If instructed to do so, suggest alternative methods for preparing the desired compound. Such methods may involve using entirely different reagents and reaction conditions. Your lecture notes and textbook can serve as valuable resources for providing possible entries for this section.

10. **Method of Purification.** Develop a flowchart that summarizes the sequence of operations that will be used to purify the desired product. The chart will show at what stages of the work-up procedure unchanged starting materials and unwanted by-products are removed. By understanding the logic of the purification process, you will know why each of the various operations specified in the purification process is performed.

    Purifying the final product of a reaction can be the most challenging part of an experimental procedure. Professional organic chemists are constantly required to develop work-up sequences that allow isolation of a pure product, free from starting materials and other contaminants. They do this by considering the chemical and physical properties of both the desired and undesired substances, and it is important for you to gain experience in devising such schemes as well.

11. **Interpretation of Instrumental Data.** If instructed to do so, discuss any instrumental data, such as gas-liquid chromatographic analyses and spectral data you have obtained or that are provided in the textbook.

12. **Answers to Exercises.** Enter answers to any exercises for the experiment that have been assigned from the textbook.

A detailed example of the write-up for a preparative experiment involving the dehydration of cyclohexanol (Sec. 10.3) is given in Figure 1.2. You may not actually perform this reaction; nevertheless, you should carefully study the example to see how to prepare specific entries for the first eight items listed. The various entries in Figure 1.2 are labeled with circled, **boldface** numbers and are discussed further in the following paragraphs. It is assumed for illustrative purposes that an actual yield of 2.7 g is obtained.

1 Use a new page of the notebook to start the entries for the experiment. Provide information that includes your name, the date, the title of the experiment, and a reference to the place in the laboratory textbook or other source where the procedure can be found.

2 Self-explanatory.

① 1.   **Your Name**
                                                                                                              Date

Dehydration of Cyclohexanol

Reference: *Experimental Organic Chemistry: A Miniscale and Microscale Approach,* 6th ed.,
by Gilbert and Martin, Section 10.3.

② **2. INTRODUCTION**

Cyclohexene is to be prepared by the acid-catalyzed dehydration of cyclohexanol.

③ **3. MAIN REACTION(S) AND MECHANISM(S)**

1   **2**

(Mechanism intentionally omitted.)

④ **4. TABLE OF REACTANTS AND PRODUCTS**

⑤ Compound	⑥ M.M.	⑦a Volume Used (mL)	⑦b Weight Used (g)	⑧ Moles Used	⑨ Moles Required	⑩ Other Data
Cyclohexanol	100.2	5.2	5	0.05	1	bp 161 °C (760 torr), mp 25.1 °C, d 0.962 g/mL, colorless
Sulfuric acid (9 *M*)	98.1	2.5	#	#	0	d 1.84 (18 *M* H$_2$SO$_4$)
Cyclohexene	82.2	*	*	*	1	bp 83 °C (760 torr), mp 103.5 °C, d 0.810, colorless

#Entry left blank because this row is for the catalyst.

*Entry left blank because this row is for the product.

⑪ LIMITING REAGENT: Cyclohexanol

⑫ MSDS DATA

These data are available on the printouts inserted at the back of my lab book.

⑬ **5. YIELD DATA**

Theoretical yield of cyclohexene = moles of limiting reagent (cyclohexanol) × M.W. of cyclohexene

= 0.05 mol × 82.2 g/mol

= 4.1 g

Actual yield = 2.7 g   *(Continued)*

Figure 1.2
Sample notebook format for investigative experiments.

Percent yield = [Actual yield (g)/theoretical yield (g)] × 100

= [2.7/4.1] × 100 = 66%

⑭ **6.** SYNOPSIS OF AND NOTES ON EXPERIMENTAL PROCEDURE—RESULTS

Procedure: Put alcohol in 25-mL rb flask and add H_2SO_4. Mix, add stirbar, attach to fractional dist. apparatus. Heat with oil bath; heating rate such that head temp. stays below 90 °C. Stop when 2.5 mL remain in rxn. flask. Put distillate in 25-mL Erlenmeyer and add 1–2 g K_2CO_3.★ Occasionally swirl mix. for 15 min and transfer liquid to 10-mL rb by decantation or pipet. Add stirbar and do simple distillation (no flames!); receiver must be close to drip tip of adapter to minimize losses by evaporation. Collect product at 80–85 °C (760 torr).

Variances and observations: Procedure followed exactly as described in reference. Distillate cloudy throughout dehydration step; formed two layers in receiver. Head temperature never exceeded 77 °C. Liquid in stillpot darkened as reaction proceeded. Addition of carbonate (1 g) to distillate caused evolution of a few bubbles of gas (CO_2?). Had to add about 0.5 g more of carbonate to get rid of cloudiness. Left solution over drying agent for one week (next lab period). Used pipet to transfer dried liquid to distillation flask. Collected cyclohexene in ice-cooled 10-mL rb flask attached to vacuum adapter protected with $CaCl_2$ tube. Stopped distillation when about 1 mL of yellowish liquid remained in stillpot.

⑮ **7.** OBSERVED PROPERTIES OF PRODUCT

bp 80–84 °C (760 torr); colorless liquid; insoluble in water; decolorizes Br_2/CH_2Cl_2 solution and produces brown precipitate upon treatment with $KMnO_4/H_2O$.

⑯ **8.** SIDE REACTIONS

Figure 1.2 *(Continued)*

⑰ **9. OTHER METHODS OF PREPARATION**

(X = Halogen)

⑱ **10. FLOWCHART FOR PURIFICATION**

⑲ **11. INTERPRETATION OF INSTRUMENTAL DATA**

(*Intentionally omitted.*)

⑳ **12. ANSWERS TO ASSIGNED EXERCISES**

(*Answers intentionally omitted.*)

Figure 1.2 (*Continued*)

3 There is only a single reaction in our example, but in many cases more than one step is involved; write equations for *all* of the main reactions. A mechanism for the reaction is intentionally omitted in our example.

4 Use the illustrated format for the Table of Reactants and Products unless instructed to do otherwise.

5 Enter the name or structure of each reactant catalyst, if any, and desired product.

6 Record the molar mass (M.M.) of each reactant and desired product. For completeness, make an entry for any catalyst used, although this may be optional.

7a Give the volume, in milliliters (mL) or microliters (µL), of each *liquid* reactant and catalyst.

7b Record the weight, in grams (g) or milligrams (mg), of each reactant. This entry is optional for liquid catalysts but should be provided for reference purposes.

8 Calculate the moles used of each reactant. For completeness, a value for the catalyst is computed in our example.

9 Obtain the **theoretical ratio** for reactant(s) and product(s) by referring to the *balanced* main equation(s) for the reaction.

10 List selected physical properties of reactant(s) and product(s). The information needed is generally available in reference books.

11 Determine the **limiting reagent** in the following way. Compare the actual ratio of reactants used to that theoretically required. The reagent that is used in the least molar amount, relative to the theoretical amount, is the limiting reagent. In our example, there is only a single reagent, cyclohexanol, so it obviously must be the limiting reagent.

12 Self-explanatory. However, if you are instructed to provide MSDS data in the form of printouts from the online resources, be sure to read them. Ignorance is *not* bliss when it comes to handling chemicals!

13 Calculate the **theoretical yield** of the desired product both in moles and in grams. Knowing in our case that the limiting reagent is cyclohexanol and, from the main equation, that 1 mole of alcohol yields 1 mole of cyclohexene, it is clear that no more than 0.05 mole of the alkene can be formed.

Assuming that you were able to isolate 2.7 g of pure cyclohexene in the experiment, the **percent yield** would be calculated according to Equation 1.1.

14 Self-explanatory.

15 Self-explanatory.

16 Self-explanatory.

17 Self-explanatory.

18 Develop this diagram by considering what components, in addition to the desired product, may be present in the reaction mixture *after* the main reaction is complete. The chart shows how and where each of the inorganic and organic contaminants of the product is removed by the various steps of the work-up procedure. Ideally, pure cyclohexene results.

19 Self-explanatory.

20 Self-explanatory.

1.7 SAMPLE CALCULATIONS FOR NOTEBOOK RECORDS

Students frequently have difficulty in setting up Tables of Reactants and Products and calculating theoretical yields, so two hypothetical examples are provided for your reference.

Example 1

Problem   Consider the reaction shown in Equation 1.2. Assume that you are to use 5 g (7.8 mL) of 1-pentene and 25 mL of concentrated HBr solution. Prepare a Table of Reactants and Products, determine the limiting reagent, and calculate the theoretical yield for the reaction.

$$CH_2 = CHCH_2CH_2CH_3 + HBr \longrightarrow CH_3CH(Br)CH_2CH_2CH_3 \quad (1.2)$$

Answer   First of all, note that the equation is balanced, because the "1" that signifies that 1 mole of each reactant will react to produce 1 mole of product is omitted by convention. Because an aqueous solution of HBr, rather than the pure acid, is being used, the amount of HBr present must be determined. Concentrated HBr is 47% by weight in the acid, and its density, d, is 1.49 g/mL, a value that would be recorded in the column headed "Other Data." Consequently, 25 mL of this solution contains 17.5 g of HBr (25 mL × 1.49 g/mL × 0.47). The needed data can then be entered into Table 1.1.

The limiting reagent is 1-pentene because theory requires that it and HBr react in a 1:1 molar ratio, yet they have been used in a ratio of 0.07:0.22. This means that no more than 0.07 mole of product can be formed, since theory dictates that the ratio between 2-bromopentane and 1-pentene also be 1:1. The calculation of the theoretical yield is then straightforward.

You may find it convenient to use units of milligrams (mg), microliters (μL), and millimoles (mmol) instead of grams, milliliters, and moles, respectively, in performing measurements and calculations when small quantities of reagents are used, as is the case for microscale reactions. For example, let's consider how Table 1.1 would be modified if 0.1 g of 1-pentene and 0.5 mL of concentrated HBr solution were used. If the calculations were done in grams and moles, the entries

Table 1.1 *Table of Reactants and Products for Preparation of 2-Bromopentane*

Compound	M.M.	Volume Used (mL)	Weight Used (g)	Moles Used	Moles Required	Other Data
1-Pentene	70.14	7.8	5	0.07	1	*
HBr	80.91	25	17.5	0.22	1	*
2-Bromopentane	151.05	†	†	†	1	*

*These entries have been intentionally omitted in this example.
†These entries are left blank because this line is for the product.
Limiting reagent: 1-pentene
Theoretical yield: 151.05 g/mol × 0.07 mol = 10.5 g

Table 1.2 *Table of Reactants and Products for Preparation of 2-Bromopentane*

Compound	M.M.	Volume Used (mL)	Weight Used (g)	Mmols Used	Mmols Required	Other Data
1-Pentene	70.14	0.16	100	1.4	1	*
HBr	80.91	0.5	358	4.4	1	*
2-Bromopentane	151.05	†	†	†	1	*

*These entries have been intentionally omitted in this example.
†These entries are left blank because this line is for the product.
Limiting reagent: 1-pentene
Theoretical yield: 151.05 g/mol × 1.4 mol × 211 mg

under "Moles Used" would be 0.0014 and 0.4, respectively. Errors can arise when making such entries, because a zero may inadvertently be added or dropped.

This potential problem is less likely if you enter the data as milligrams and millimoles. If you recognize that 0.1 g of alkene is 100 mg and 0.5 mL of HBr solution contains 358 mg of HBr (0.5 mL × 1490 mg/mL × 0.48), the entries would be those shown in Table 1.2. You may then determine the limiting reagent and calculate the theoretical yield as in Example 1. Note that the necessary cancellation of units occurs when the molar mass is expressed in mg/mmol.

Although the volume of 1-pentene to be used is expressed in milliliters, you may be measuring out this amount with a device that is calibrated in microliters (μL, 1 μL = 10^{-3} mL). Thus, in the present example, you would be using 160 μL of 1-pentene (0.16 mL × 10^3 μL/mL).

Example 2

Problem

Now consider the transformation illustrated in Equation 1.3. Assume that you are to use 7 mL of ethanol and 0.1 mL of concentrated H_2SO_4 as the catalyst. Prepare a Table of Reactants and Products, determine the limiting reagent, and calculate the theoretical yield for the reaction.

$$2\ CH_3CH_2OH \xrightarrow[\text{(Catalytic amount)}]{H_2SO_4} CH_3CH_2OCH_2CH_3 + H_2O \quad (1.3)$$

Answer

As in the previous example, a volumetric measurement must first be converted to a weight. The density of ethanol is 0.789 g/mL, information that would be entered under the column headed "Other Data," so that means 5.5 g is being used. Table 1.3 can then be completed. Note that the catalyst, although recorded in the table, is not used in any of the calculations because, by definition, it is not consumed during the reaction. Including it should help remind the experimentalist that it is indeed required to make the reaction occur!

Calculation of the theoretical yield is performed as in Example 1, with the important exception that a factor of 0.5 is incorporated to adjust for the fact that only one-half mole of diethyl ether would be produced for each mole of ethanol that is used.

Table 1.3 *Table of Reactants and Products for Preparation of Diethyl Ether*

Compound	M.M.	Volume Used (mL)	Weight Used (g)	Moles Used	Moles Required	Other Data
Ethanol	46.07	7	5.5	0.12	2	*
Sulfuric acid	†	0.1	†	†	†	*
Diethyl ether	74.12	‡	‡	‡	1	*

*These entries have been intentionally omitted in this example.
†These entries are left blank for reactants that serve only as catalysts.
‡These entries are left blank because this line is for the product.
Limiting reagent: ethanol
Theoretical yield: 74.12 g/mol × 0.12 mol · 0.5 = 4.4 g

1.8 SAFE LABORATORY PRACTICE: OVERVIEW

There is little question that one of the most important abilities that you, the aspiring organic chemist, can bring to the laboratory is a sound knowledge of how to perform experimental work safely. But just knowing *how* to work safely is insufficient! You must also make a *serious* commitment to follow standard safety protocols. In other words, having the knowledge about safety is useless if you do not put that knowledge into practice. What you actually do in the laboratory will determine whether you and your labmates are working in a safe environment.

Chemistry laboratories are potentially dangerous because they commonly house flammable liquids, fragile glassware, toxic chemicals, and equipment that may be under vacuum or at pressures above atmospheric. They may also contain gas cylinders that are under high pressure. The gases themselves may or may not be hazardous—for example, nitrogen is not, whereas hydrogen certainly is—but the fact that their containers are under pressure makes them so. Imagine what might happen if a cylinder of nitrogen fell and ruptured: You could have a veritable rocket on your hands, and, if the tank contained hydrogen, the "rocket" might even come equipped with a fiery tail! This is another way of saying *all* substances are hazardous under certain conditions.

Fortunately the laboratory need be no more dangerous than a kitchen or bathroom, but this depends on you *and* your labmates practicing safety as you work. Should you observe others doing anything that is unsafe, let them know about it in a friendly manner. Everyone will benefit from your action. We'll alert you repeatedly to the possible dangers associated with the chemicals and apparatus that you will use so that you can become well trained in safe laboratory practice. Mastery of the proper procedures is just as important in the course as obtaining high yields of pure products, and carefully reading our suggestions will assist you in achieving this goal. Some safety information will be contained in the text describing a particular experiment or in the experimental procedure itself. It will also appear in highlighted sections titled "**Safety Alert.**" These are designed to draw your special attention to aspects of safety that are of particular importance. We urge you to read these sections carefully and follow the guidelines in them carefully. You will then be fully prepared to have the fun and fulfillment of the laboratory experience.

1.9 SAFETY: GENERAL DISCUSSION

We highlight here, in the form of a Safety Alert, some general aspects regarding safe practices in the laboratory. Such alerts will appear throughout the textbook and should be read carefully.

SAFETY ALERT

Personal Attire

1. *Do not wear shorts or sandals in the laboratory; the laboratory is not a beach!* Proper clothing gives protection against chemicals that may be spilled accidentally. It is advisable to wear a laboratory coat, but in any case, the more skin that is protected by clothing the better.

2. *Always wear safety glasses or goggles in the laboratory.* This applies even when you are writing in your laboratory notebook or washing glassware, since nearby workers may have an accident. It is best *not* to wear contact lenses in the laboratory. Even if you are wearing eye protection, chemicals may get into your eyes, and you may not be able to get the contact lenses out before damage has occurred. Should you have to wear corrective glasses while working in the laboratory, make certain that the lenses are shatterproof. Wearing goggles over such glasses is recommended because the goggles give additional protection from chemicals entering your eyes from the sides of the lenses.

3. *Wear suitable protective gloves, which may be made of latex or a plastic such as chloroprene, when working with particularly hazardous chemicals.* Some reagents are especially hazardous if they come into contact with your skin. The ones you are most likely to encounter in the organic laboratory are concentrated acids and bases, and bromine and its solutions. Check with your instructor whenever you are uncertain whether you should be wearing gloves when handling reagents.

General Considerations

1. *Become familiar with the layout of the laboratory room.* Locate the exits from the room and the fire extinguishers, fire blankets, eyewash fountains, safety showers, and first-aid kits in and near your workspace. Consult with your instructor regarding the operation and purpose of each of the safety-related devices.

2. *Find the nearest exits from your laboratory room to the outside of the building.* Should evacuation of the building be necessary, use stairways rather than elevators to exit. Remain calm during the evacuation, and walk rather than run to the exit.

3. *Become knowledgeable about basic first-aid procedures.* The damage from accidents will be minimized if first aid is applied promptly. Read the section "First Aid in Case of an Accident" on the inside front cover of this book.

4. *Never work alone in the laboratory.* In the event of an accident, you may need the immediate help of a coworker. Should you have to work in the laboratory outside of the regularly scheduled periods, do so only with the express permission of your instructor and in the presence of at least one other person.

5. Before performing any experiment, you should perform a hazard analysis by reviewing the set-up for the experiment and the chemicals that will be used and identifying any potential hazards. You should consider how these hazards might be eliminated or mitigated and what you will do in case something goes wrong.

Apparatus and Chemicals

1. *Always check carefully for imperfections in the glassware that you will be using.* This should be done not only when checking into the laboratory for the first time but also when setting up the apparatus needed for each experiment. Look for cracks, chips, or other imperfections in the glass that weaken it. Use care in storing your glassware so that it is not damaged when you open or close the locker or drawer.

   Pay particular attention to the condition of round-bottom flasks and condensers. The flasks often have "star" cracks (multiple cracks emanating from a central point) as a result of being banged against a hard surface. Heating or cooling a flask having this type of flaw may cause the flask to rupture with loss of its contents. This could result in a serious fire, not just loss of the desired product. To detect such cracks, hold the flask up to the light and look at all its surfaces closely. With respect to condensers, their most vulnerable points are the ring seals—the points where the inner tube and the water jacket of the condenser are joined. Special care must be given to examining these seals for defects, because if cracks are present water might leak into your apparatus and contaminate your product or, worse, cause violent reactions.

   If you detect imperfections in your glassware, consult with your teacher immediately regarding replacement. Cracked or seriously chipped apparatus should always be replaced, but glassware with slight chips may still be safe to use.

2. *Dispose of glassware properly.* The laboratory should be equipped with a properly labeled special container for broken glassware and disposable glass items such as Pasteur pipets and melting-point capillaries. It is not appropriate to throw such items in the regular trash containers, because maintenance personnel may injure themselves while removing the trash. Broken thermometers are a special problem because they usually contain mercury, which is toxic and relatively volatile. There should be a separate, closed container for disposal of thermometers. If mercury has spilled as a result of the breakage, it should be cleaned up immediately. Consult with your instructor about appropriate procedures for doing so.

3. *Know the properties of the chemicals used in the experiments.* Understanding these properties helps you to take the proper precautions when handling them and to minimize danger in case of an accident. *Handle all chemicals with care.*

   Refer to MSDSs (Sec. 1.10) to learn about toxicity and other potential hazards associated with the chemicals you use. Most chemicals are at least slightly toxic, and many are *very* toxic and irritating if inhaled or allowed to come in contact with the skin. It is a good laboratory practice to wear suitable protective gloves when handling chemicals, and there may be times when it is imperative to do so. Your instructor will advise you if you do *not* need to use gloves.

   Should chemicals come in contact with your skin, they can usually be removed by a thorough and *immediate* washing of the affected area with soap and water. Do *not* use organic solvents like ethanol or acetone to rinse chemicals from your skin, as these solvents may actually assist the absorption of the substances into your skin.

4. *Avoid the use of flames as much as possible.* Most organic substances are flammable, and some are highly volatile as well, which increases their potential for being ignited accidentally. Examples of these are diethyl ether, commonly used as a solvent in the organic laboratory, and acetone. Occasionally,

open flames may be used for flame-drying an apparatus or distilling a high-boiling liquid. In such cases, a Safety Alert section will give special precautions for their use. Some general guidelines follow.

   a. *Never use an open flame without the permission of your instructor.*

   b. *Never use a flame to heat a flammable liquid in an open container.* Use a water or steam bath, hot plate, aluminum block, or similar electrical heat device instead. If a flammable liquid must be heated with an open flame, equip the container holding the liquid with a *tightly* fitting reflux condenser.

   Information about the flammability of many commonly used organic solvents is provided in Table 3.1. Do *not* assume that a solvent is not flammable just because it is not listed in the table, however. In such cases, refer to the MSDSs (Sec. 1.10) or other sources to determine flammability.

   c. *Do not pour flammable liquids when there are open flames within several feet.* The act of transferring the liquid from one container to another will release vapors into the laboratory, and these could be ignited by a flame some distance away.

   d. *Do not pour flammable water-insoluble organic solvents into drains or sinks.* First of all, this is an environmentally unsound way to dispose of waste solvents, and second, the solvents may be carried to locations where there are open flames that could ignite them. Water-soluble solvents can be flushed down the drain if local regulations permit; consult with your instructor about this.

5. *Avoid inhaling vapors of organic and inorganic compounds.* Although most of the pleasant and unpleasant odors you encounter in everyday life are organic in nature, it is prudent not to expose yourself to such vapors in the laboratory. Work at a fume hood when handling particularly noxious chemicals, such as bromine or acetic anhydride, and, if possible, when performing reactions that produce toxic gases.

6. *Never taste anything in the laboratory unless specifically instructed to do so.* You should also never eat or drink in the laboratory, as your food may become contaminated by the chemicals that are being used.

7. *Minimize the amounts of chemicals you use and dispose of chemicals properly.* This aspect of laboratory practice is so important that we have devoted a portion of Section 1.10 to it. Read the relevant paragraphs *carefully* and consult with your instructor if there are any questions about the procedures.

1.10 SAFETY: MATERIAL SAFETY DATA SHEETS (MSDSs)

The variety and potential danger of chemicals used in the organic chemistry laboratory probably exceed that of any laboratory course you have had. It is imperative to understand the nature of the substances with which you are working. Fortunately, the increased emphasis on the proper handling of chemicals has led to a number of publications containing information about the chemical, physical, and toxicological properties of the majority of organic and inorganic compounds used in the experiments in this textbook. A comprehensive source is *The Sigma-Aldrich Library of Chemical Safety Data* (Reference 6), and it or similar compilations should be available in your library or some other central location. Alternatively, you can usually access MSDS information by typing the name of the chemical of interest into the

Name	Ether	Reviews and standards	OSHA standard-air: TWA 400 ppm.
Other names	Diethyl ether	Health hazards	May be harmful by inhalation, ingestion, or skin absorption. Vapor or mist is irritating to the eyes, mucous membranes, and upper respiratory tract. Causes skin irritation. Exposure can cause coughing, chest pains, difficulty in breathing, and nausea, headache, and vomiting.
CAS Registry No.	60-29-7	First aid	In case of contact, immediately flush eyes or skin with copious amounts of water for at least 15 min while removing contaminated clothing and shoes. If inhaled, remove to fresh air. If not breathing, give artificial respiration; if breathing is difficult, give oxygen. If ingested, wash out mouth with water. Call a physician.
Structure	$(CH_3CH_2)_2O$	Incompatibilities	Oxidizing agents and heat.
MP	$-116\,°C$	Extinguishing media	Carbon dioxide, dry chemical powder, alcohol, or polymer foam.
BP	$34.6\,°C$ (760 torr)	Decomposition products	Toxic fumes of carbon monoxide, carbon dioxide.
FP	$-40\,°C$	Handling and storage	Wear appropriate respirator, chemical-resistant gloves, safety goggles, other protective clothing. Safety shower and eye bath. Do not breathe vapor. Avoid contact with eyes, skin, and clothing. Wash thoroughly after handling. Irritant. Keep tightly closed. Keep away from heat, sparks, and open flame. Forms explosive peroxides on prolonged storage. Refrigerate. Extremely flammable. Vapor may travel considerable distance to source of ignition. Container explosion may occur under fire conditions. *Danger:* Tends to form explosive peroxides, especially when anhydrous. Inhibited with 0.0001% BHT.
Appearance	Colorless liquid	Spillage	Shut off all sources of ignition. Cover with activated carbon adsorbent, place in closed containers, and take outdoors.
Irritation data	Human eye 100 ppm	Disposal	Store in clearly labeled containers until container is given to approved contractor for disposal in accordance with local regulations.
Toxicity data	Man, oral LDL_0 260 mg/kg		

Figure 1.3
Summary of MSDS for diethyl ether.

query box of your browser and using the associated search function. The data provided by these sources are basically summaries of the information contained in the MSDSs published by the supplier of the chemical of interest. Your instructor may be able to provide these sheets because by federal regulation an MSDS must be delivered to the buyer each time a chemical is purchased.

The information in an MSDS can be overwhelming. For example, the official MSDS for sodium bicarbonate is some six pages long. Even the summaries provided in most compilations are quite extensive, as illustrated in Figure 1.3, which contains specific data for diethyl ether. Entries regarding the structure and physical properties of the compound, including melting point (mp), boiling point (bp), and flash point (fp), are included, along with its CAS (Chemical Abstracts Service) Registry Number, which is unique for each different chemical substance, and RTECS (Registry of Toxic Effects of Chemical Substances) number. Further data are provided concerning its toxicity, the permissible levels set by OSHA for exposure to it in the air you breathe (time-weighted average of 400 ppm), and possible health consequences resulting from contact with the compound. For diethyl ether, the entry for "Toxicity Data" represents the *lowest* recorded *lethal* concentration for ingestion of the chemical. Valuable information is also given regarding first-aid procedures, classes of substances with which diethyl ether reacts and thus is "incompatible" with, products of its decomposition, and materials suitable for extinguishing fires involving ether. Finally, protocols for safe handling and storage are included, along with procedures for disposing and cleaning up spills of diethyl ether.

Accessing MSDS information from commercial sources can be very time-consuming, although it is useful to refer to one or more of them if you need more complete MSDS information than is available online (see the online resources associated with this textbook) or if you are to use or produce a chemical that is not listed on it. We've developed the Web-based MSDSs to provide you with a rapid and convenient way to obtain important information on the chemicals you will be using or producing in the experimental procedures performed when using this textbook. The data we've provided online for this textbook are much more abbreviated than those in other sources, as seen in Figure 1.4. In developing our summaries of MSDS data, we've focused on just those data most relevant to your needs in the introductory organic laboratory.

We noted in the discussion of notebook formats (Sec. 1.6) that you may be required to summarize MSDS data in your laboratory book. This could be a daunting assignment, given the amount of information with which you might be faced, as illustrated in Figure 1.3. To assist you in doing this, we have provided one possible format for a summary in Figure 1.5. A summary for a particular chemical has to be provided only once and can be recorded at the end of your laboratory notebook on pages reserved for that purpose. Whenever the chemical is encountered in later experiments, you would only need to refer to the location of the summary of its MSDS information. *However*, you should reread the MSDS information so that you can continue to handle the chemical properly. This same recommendation applies if you have a file of MSDS-related printouts from the online resources for this textbook.

To summarize, you may think that reading about and recording data like those contained in Figures 1.3–1.5 is not a good investment of time. This is absolutely *wrong!* By knowing more about the chemicals that are used in the laboratory, you will be able to work safely and to deal with accidents, should they occur. The end result will be that you accomplish a greater amount of laboratory work and have a more valuable educational experience.

1.11 SAFETY: DISPOSAL OF CHEMICALS

The proper disposal of inorganic and organic chemicals is one of the biggest responsibilities that you have in the organic laboratory. Your actions, and those of your labmates, can minimize the environmental impact and even financial cost to

Diethyl Ether
$C_4H_{10}O$

CAS No.	PS	Color	Odor	FP	BP	MP	d	VP	VD	Sol
60-29-7	Liquid	Colorless	Sweet	40	35	−116	0.7	442 @ 20	2.6	6.9 @ 20

Types of Hazards/Exposures	Acute Hazards/Symptoms	Prevention	First Aid/Fire
Fire	*Severe* fire hazard, *severe* explosion hazard; may form explosive peroxides; vapors or gases may ignite at distant ignition sources.	*No* flames, *no* sparks, *no* contact with hot surfaces.	Alcohol-resistant foam, carbon dioxide, regular dry chemical powder, water.
Inhalation	Central nervous system depression with drowsiness, dizziness, nausea, headache, and lowering of the pulse and body temperature.	Ventilation, local exhaust.	Remove from exposure immediately and seek medical advice.
Skin	Irritation, defatting, and drying of the skin.	Protective gloves and clothing.	Remove contaminated clothes/jewelry; thoroughly wash skin with soap and water; and seek medical advice.
Eyes	Painful inflammation.	Safety goggles.	Thoroughly flush eyes with water for several min, removing contact lenses if possible, and seek medical advice immediately.
Ingestion	Central nervous system depression with nausea, vomiting, drowsiness, dizziness; stomach may become promptly distended, which may hinder breathing.	Do *not* eat or drink in the laboratory.	Seek medical advice immediately.
Carcinogenicity	Not a known carcinogen.	**Mutagenicity**	Possible mutagen.

For more detailed information, consult the Material Safety Data Sheet for this compound.
Abbreviations: CAS No. = Chemical Abstracts Service Registry Number; PS = physical state; FP = flash point (°C); BP = boiling point (°C) @ 760 torr unless otherwise stated; MP = melting point (°C); d = density or specific gravity (g/mL); VP = vapor pressure (torr) at specified temperature (°C); VD = vapor density relative to air (1.0); Sol = solubility in water (g/100 mL) at specified temperature; N/A = not available or not applicable.

Figure 1.4
Example of MSDS data provided online for this textbook.

your school for handling the waste chemicals that are necessarily produced in the experiments you do.

The experimental procedures in this textbook have been designed at a scale that should allow you to isolate an amount of product sufficient to see and manipulate, but they also involve the use of minimal quantities of reactants, solvents, and

Compound	Health Hazards, First Aid, Incompatibilities, Extinguishing Media, and Handling
Diethyl Ether	May be harmful by inhalation, ingestion, or skin absorption. Avoid contact with eyes, skin, and clothing. In case of contact, immediately flush eyes or skin with copious amounts of water. Keep away from hot surfaces, sparks, and open flames. Extremely flammable. Vapor may travel considerable distance to source of ignition. If spilled, shut off all sources of ignition. Extinguish fire with carbon dioxide, dry chemical extinguisher, foam, or water.

Figure 1.5
Abstract of MSDS for diethyl ether.

drying agents. Bear in mind, however, that minimizing the amounts of chemicals that are used is only the *first* part of an experimental design that results in the production of the least possible quantity of waste. The *second* part is to reduce the amounts of materials that you, the experimentalist, *define* as waste, thereby making the material subject to regulations for its disposal. From a legal standpoint, the laboratory worker is empowered to declare material as waste; that is, unneeded materials are not waste until you say they are! Consequently, a part of most of the experimental procedures in this textbook is reduction of the quantity of residual material that eventually must be consigned to waste. This means some additional time will be required for completion of the experiment, but the benefits—educational, environmental, and economic in nature—fully justify your efforts. The recommended procedures that should be followed are described under the heading **Wrapping It Up.**

How do you properly dispose of spent chemicals at the end of an experiment? In some cases this involves simply flushing chemicals down the drain with the aid of large volumes of water. As an example, solutions of sulfuric acid can be neutralized with a base such as sodium hydroxide, and the aqueous solution of sodium sulfate that results can safely be washed into the sanitary sewer system. However, the environmental regulations that apply in your particular community may require use of alternative procedures. *Be certain to check with your instructor before flushing any chemicals down the drain!*

For water-insoluble substances, and even for certain water-soluble ones, this option is not permissible under *any* circumstances, and other procedures must be followed. The laboratory should be equipped with various containers for disposal of liquid and solid chemicals; the latter should not be thrown in a trash can, because this exposes maintenance and cleaning personnel to potential danger, and it is environmentally unsound. The containers must be properly labeled as to what can be put in them, because it is very important for safety and environmental reasons that different categories of spent chemicals be segregated from one another. Thus, you are likely to find the following types of containers in the organic laboratory: hazardous solids, nonhazardous solids, halogenated organic liquids, hydrocarbons, and oxygenated organic liquids. Each student must assume the responsibility for seeing that her or his spent chemicals go into the appropriate container; otherwise dangerous combinations of chemicals might result and/or a much more expensive method of disposal be required.

REFERENCES

1. Lunn, G.; Sansone, E. B. *Destruction of Hazardous Chemicals in the Laboratory*, 3rd ed., John Wiley & Sons, New York, 2012. A handbook providing procedures for decomposition of materials or classes of materials commonly used in the laboratory.

2. Committee on Hazardous Substances in the Laboratory. *Prudent Practices for Disposal of Chemicals from Laboratories*, National Academy Press, Washington, D.C., 1995. An excellent reference containing information for the minimization of waste generated in the laboratory and for the proper handling and disposal of waste chemicals, both organic and inorganic. Available at no cost online at www.nap.edu/catalog.php?record_id=4911#toc.

3. Young, J. A., ed. *Improving Safety in the Chemical Laboratory: A Practical Guide*, 2nd ed., John Wiley & Sons, New York, 1991. A book containing thorough discussions of the full range of safe practices in the laboratory.

4. Lide, D. A., ed. *CRC Handbook of Chemistry and Physics*, annual editions, CRC Press, Boca Raton, FL. Available online to subscribers at www.hbcpnetbase.com.

5. Speight, J. G., ed. *Lange's Handbook of Chemistry*, 16th ed., McGraw-Hill, New York, 2005. Available online to subscribers on McGraw-Hill's site Access Engineering.

6. Lenga, R. E.; Votoupal, K. L., eds. *The Sigma-Aldrich Library of Chemical Safety Data*, 2nd ed., Sigma-Aldrich, Milwaukee, WI, 1988.

7. O'Neil, M. J., ed. *The Merck Index of Chemicals and Drugs*, 15th ed., Merck and Co., Rahway, NJ, 2013. Available free online at www.rsc.org/merck-index.

8. Armour, M. A. *Hazardous Laboratory Chemical Disposal Guide*, 3rd ed., CRC Press, Boca Raton, FL, 2003.

CHAPTER 2

Techniques and Apparatus

In this chapter we introduce the basic experimental techniques and associated glassware and apparatus that are commonly used in the organic chemistry laboratory. In some instances, only the practical aspects of a particular technique are discussed in this chapter, and the theoretical principles underlying it are presented in later chapters.

It is important to skim the contents of this entire chapter, but we do not recommend that you read all parts of it in detail at this time. We make reference to specific techniques and/or apparatus as part of the procedures for many experiments; thus, when you are preparing for an experiment, *carefully* read the appropriate sections of this chapter. This should be done *prior* to entering the laboratory to perform the experimental procedure!

2.1 GLASSWARE: PRECAUTIONS AND CLEANING

Laboratory experiments in organic chemistry are commonly conducted in specialized glassware that is usually expensive. Since you are responsible for maintaining your glassware, you should follow proper procedures for safely handling and cleaning it. Failure to do so is likely to result in injury to yourself, breakage, or dirty glassware that is difficult to clean.

The cardinal rule in handling and using laboratory glassware is *Never apply undue pressure or strain to any piece of glassware*. Strained glassware may break at the moment the strain is induced, when it is heated, or even after standing for a period of time. When setting up a glassware apparatus for a particular experiment, be sure that the glassware is properly positioned and supported to avoid breakage (Sec. 2.4).

Sometimes it is necessary to insert thermometers or glass tubes into rubber stoppers or rubber tubing. *If you have to force it, do not do it!* Either make the hole slightly larger or use a smaller piece of glass. You may also lubricate the glass tube with a little glycerol prior to insertion into stoppers or tubing. When inserting a glass tube into a stopper or tubing, always grasp the glass piece as close as possible to the rubber part. It is also wise to wrap a towel around the glass tube and the rubber stopper while inserting the tube. This usually prevents a serious cut in the event the glass happens to break.

Glassware should be thoroughly cleaned *immediately* after use. Residues from chemical reactions may attack the surface of the glass, and cleaning becomes more difficult the longer you wait. Before washing glassware, it is good practice to wipe off any lubricant or grease from standard-taper ground-glass joints (Sec. 2.2) with

25

a towel or tissue moistened with a solvent such as hexane or dichloromethane. This prevents the grease from being transferred during washing to inner surfaces of the glassware, where it may be difficult to remove. Most chemical residues can be removed by washing the glassware using a brush, special laboratory soap, and water. Acetone, which is miscible with water, dissolves most organic residues and thus is commonly used to clean glassware. To avoid unnecessary waste, use as little solvent as possible to do the job. Acetone should not be used to clean equipment that contains residual amounts of bromine, however, since bromoacetone, a powerful lachrymator, may form. A **lachrymator** is a chemical that adversely affects the eyes and causes crying, and may also affect the lungs and produce a burning sensation. Tear gas is an example of a lachrymator. After use, the spent wash acetone should be transferred to an appropriately marked container for disposal.

Stubborn residues sometimes remain in your glassware. These are often removable by carefully scraping the residue with a bent spatula in the presence of soap and water or acetone. If this technique fails, more powerful cleaning solutions may be required, but these must be used with great care as they are usually highly corrosive. *Do not allow these solutions to come into contact with your skin or clothing; they will cause severe burns and produce holes in your clothing.*

Chromic acid, which is made from concentrated sulfuric acid and chromic anhydride or potassium dichromate, is often an effective cleaning agent, but because it is a strong oxidizing acid, it must be used with great care. When handling chromic acid, always wear suitable protective gloves and pour it *carefully* into the glassware to be cleaned. After the glassware is clean, pour the chromic acid solution into a specially designated bottle, *not* into the sink.

Another powerful cleaning solution is alcoholic potassium hydroxide. This is prepared by putting some ethanol into the flask, adding a few pellets of solid potassium hydroxide, and warming the solution gently with swirling to facilitate dissolution. When the glassware is clean, pour the solution into a specially designated bottle, and rinse the glassware thoroughly with soap and water to complete the process. *Before using any cleaning solutions other than soap and water, consult your instructor for permission and directions concerning their safe handling and disposal.*

Brown stains of manganese dioxide that are left in the glassware can generally be removed by rinsing the apparatus with a 30% (4 M) aqueous solution of sodium bisulfite, $NaHSO_3$. If this fails, wash the equipment with water and then add a small amount of 6 M HCl. This must be done in a hood, since chlorine gas is evolved; try this technique only after obtaining permission from your instructor.

The fastest method for drying your glassware is to rinse the residual droplets of water from the flask with a small volume (<5 mL) of acetone; this acetone may be recovered as wash acetone to remove organic residues in the future. Flasks and beakers should be inverted to allow the last traces of solvent to drain. Final drying is also aided by directing a *gentle* stream of dry compressed air into the flask or beaker or by placing the glassware in a drying oven.

2.2 STANDARD-TAPER GLASSWARE FOR MINISCALE PROCEDURES

Your laboratory kit contains **standard-taper glassware** with **ground-glass joints**. Before the advent of standard-taper glassware, we had to bore corks or rubber stoppers to fit each piece of equipment, and the glass apparatus was assembled

Female ground-glass joint

(a)

Thin, even coating of lubricant   Male ground-glass joint

(b)

Figure 2.1

Standard-taper (ℑ) joints: (a) female; (b) male.

by connecting the individual pieces of equipment with glass tubing. This was not only time-consuming, but many chemicals reacted with and/or were absorbed by the cork or rubber stoppers. Although standard-taper glassware has greatly simplified the task of assembling the apparatus required for numerous routine laboratory operations, it is expensive, so handle it carefully.

A pair of standard-taper joints is depicted in Figure 2.1. Regardless of the manufacturer, a given size of a male standard-taper joint will fit a female joint of the same size. The joints are tapered to ensure a snug fit and a tight seal. Standard-taper joints come in a number of sizes and are designated by the symbol ℑ followed by two sets of numbers separated by a slash, as in 14/20, 19/22, 24/40. The first number is the diameter of the joint in millimeters at its widest point, and the second is the length of the joint in millimeters. A standard-taper joint that is designated as ℑ 19/22 therefore has a widest diameter of 19 mm and a length of 22 mm.

When using glassware with standard-taper ground-glass joints, the joints should be properly lubricated so that they do not freeze and become difficult, if not impossible, to separate. To lubricate a joint properly, first spread a *thin* layer of joint grease around the outside of the upper half of the male joint. Then mate the two joints, and rotate them gently together to cover the surfaces of the joints with a thin coating of lubricant. *Applying the correct amount of grease to the joints is important.* If you use too much, the contents of the flask, including your product, may become contaminated with grease; if you use too little lubricant, the joints may freeze. As soon as you have completed the experiment, disassemble the glassware to lessen the likelihood that the ground-glass joints will stick. If the pieces do not separate easily, the best way to pull them apart is to grasp the two pieces as close to the joint as possible and try to loosen the joint with a *slight* twisting motion.

Sometimes the pieces of glass still cannot be separated. In these cases there are a few other tricks that can be tried. These include the following options: (1) Tap the joint *gently* with the wooden handle of a spatula, and then try pulling it apart as described earlier. (2) Heat the joint in a hot-water or steam bath before attempting to separate the pieces. (3) As a last resort, heat the joint *gently* in the yellow portion of the flame of a Bunsen burner. Heat the joint slowly and carefully until the outer joint breaks away from the inner section. Wrap a cloth towel around the hot joint to avoid burning yourself, and pull the joint apart as described earlier. Consult your instructor before attempting this final option, as it is tricky.

The common pieces of standard-taper glassware with ground-glass joints for miniscale procedures are shown in Figure 2.2, and apparatus involving them are pictured throughout the textbook. The proper techniques for erecting and supporting apparatus are discussed in Section 2.4. They are also illustrated in videos online.

2.3 STANDARD-TAPER GLASSWARE FOR MICROSCALE PROCEDURES

As with miniscale procedures (Sec. 2.2), **standard-taper glassware** with **ground-glass joints** is commonly used for performing reactions at the microscale level. The apparatus is smaller, however, so the joints themselves are shorter, typically ℑ 14/10 (Fig. 2.3a).

When using glassware with standard-taper ground-glass joints, the joints should be properly lubricated so that they do not freeze and become difficult, if not impossible, to separate. To lubricate a joint properly, first spread a *thin* layer of joint

Figure 2.2

Miniscale standard-taper (T) glassware.

West condenser

Hempel column

Round-bottom flask

Claisen adapter

Stillhead

Thermometer adapter with neoprene fitting

Vacuum adapter

Figure 2.3

Standard-taper (T) 14/10 joints and greaseless assembly.

14 mm

10 mm

Screw-cap
O-ring
Male T 14/10 joint

(a)   (b)   (c)

grease around the outside of the upper half of the male joint. Then mate the two joints, and rotate them gently together to cover the surfaces of the joints with a thin coating of lubricant. *Applying the correct amount of grease to the joints is important.* If you use too much, the contents of the flask, including your product, may become contaminated with grease; if you use too little lubricant, the joints may freeze. As soon as you have completed the experiment, disassemble the glassware to lessen the likelihood that the ground-glass joints will stick. If the pieces do not separate easily, the best way to pull them apart is to grasp the two pieces as close to the joint as possible and try to loosen the joint with a *slight* twisting motion.

Sometimes the pieces of glass still cannot be separated. In these cases there are a few other tricks that can be tried. These include the following options: (1) Tap the joint *gently* with the wooden handle of a spatula, and then try pulling it apart as described earlier. (2) Heat the joint in a hot-water or steam bath before attempting to separate the pieces. (3) As a last resort, heat the joint *gently* in the yellow portion of the flame of a Bunsen burner. Heat the joint slowly and carefully until the outer joint breaks away from the inner section. Wrap a cloth towel around the hot joint to avoid burning yourself, and pull the joint apart as described earlier. Consult your instructor before attempting this final option, as it is tricky.

Figure 2.4

Microscale standard-taper (ℑ) glassware.

Conical vial   Round-bottom flask   Claisen adapter   Air condenser   Reflux condenser   Hickman stillhead with port   Drying tube

The design of some microscale apparatus allows for *greaseless* mating of joints, and your kit of glassware may permit this option. As shown in Figure 2.3a, a female joint may have threads on the outside, which permits use of Teflon-lined screw-caps to join the pieces of glassware. The screw-cap contains a silicone rubber O-ring. The male joint is first inserted through the cap and O-ring (Fig. 2.3b) and then mated with the female joint. Tightening the cap compresses the O-ring, producing a gas-tight seal that holds the unit together (Fig. 2.3c). Not using grease has obvious advantages with respect to preventing contamination of reaction mixtures and simplifying cleanup of the glassware. You must disconnect greaseless joints *immediately* after completing use of the apparatus, however; otherwise, the joint will become frozen.

Various pieces of microscale glassware you may be using are depicted in Figure 2.4. A number of different assemblies using this type of glassware are portrayed in this chapter as well as others. The proper techniques for erecting and supporting apparatus are discussed in Section 2.4.

2.4 ASSEMBLING APPARATUS

To perform most of the experiments in the organic chemistry laboratory, you must first assemble the appropriate apparatus. In the sections that follow, you will be introduced to the various techniques. All experiments require certain items of glassware, various heating or cooling devices, and a means to mix the contents of a reaction vessel. The assembly of the various pieces of glassware and equipment according to the diagrams will seem awkward at first, but with practice you will soon be able to set up your own apparatus with ease.

Providing adequate support for your apparatus is paramount to preventing breakage. In general, you will accomplish this by clamping the apparatus to a ring stand or other vertical support rod using a three-finger clamp (Fig. 2.5a) or a jaw clamp (Fig. 2.5b). Plastic Keck clips (Fig. 2.5c) are then used as needed to secure the union of the glass joints. It is best to begin assembling the apparatus with the

Figure 2.5

(a) Three-finger clamp. (b) Jaw clamp. (c) Keck clips.

(a)

(b)

(c)

Figure 2.6
Laboratory jack.

reaction or distillation vessel and to use a single ring stand to minimize the manipulations that are required if the apparatus must be moved.

If the experimental technique requires heating or cooling the apparatus, then you should assemble the apparatus and support it about 15–25 cm above the base of the ring stand or bench top. You can then use a laboratory jack (Fig. 2.6) to raise and lower heating sources or cooling baths without moving the apparatus itself. In addition to providing this convenience, the laboratory jack allows you to lower a heating source quickly if an exothermic reaction begins to get out of control. If such lab jacks are not available, you should use an iron ring clamp to support the heating source or cooling bath; the ring clamp then may be lowered if necessary to remove the heat source or the cooling bath from the flask.

Miniscale Setup

As a rule, when you assemble the apparatus for a miniscale experiment, you should *only* tighten the clamp to the round-bottom flask that contains your reaction or distillation mixture. The proper way to tighten a clamp is to place the neck of the flask or other piece of glassware being secured against the stationary side of the clamp, and then turn the wingnut or screw to bring the movable side of the clamp into contact with the glassware so that the piece of glassware fits snugly in the clamp. To avoid breaking the glassware, *do not overtighten the clamp.* Other clamps should be positioned to support the remainder of the apparatus and should be *loosely* tightened to avoid straining the glassware and risking breakage. Alternatively, plastic Keck clips (Fig. 2.5c) or rubber bands may be used in place of additional clamps. Proper locations of the clamps for the various laboratory setups are provided in the diagrams that accompany the description of the technique.

Microscale Setup

Because the glassware required for *microscale* experiments is substantially smaller than that used in miniscale experiments, a single rigid clamp is usually sufficient to secure the apparatus. To avoid breaking the glassware, *do not overtighten the clamp.* The plastic screw-caps hold the different pieces of glassware together, so additional clamping is unnecessary. Generally you will clamp a reflux condenser or a stillhead by placing the glassware against the stationary side of the clamp and then turning the wingnut or screw to bring the movable side of the clamp into contact with the glassware so that the piece of glassware fits snugly in the clamp.

2.5 MEASURING AND TRANSFERRING LIQUIDS

Proper handling of liquid reagents and products is an important technique. It is particularly critical when no more than 1–2 mL of a liquid are involved. Indeed, when working at the microscale level, quantities of liquids being used often are less than 0.5 mL (500 μL). Even when larger amounts of liquids are being measured, knowing how to transfer them safely and accurately is important.

**Graduated Cylinders, Beakers,
Erlenmeyer Flasks, and Vials**

Graduated cylinders (Fig. 2.7) are calibrated containers, which may be either glass or plastic, that are typically used to measure volumes in the range of 2–500 mL, depending on the size of the cylinder. The markings on the cylinder shaft represent the approximate volume contained when the bottom of the meniscus of the liquid is at the top of the line. The accuracy of calibration is only about ±10%, so graduated cylinders are not used for delivering precise quantities of liquids. As shown in Figure 2.7, a glass cylinder may be fitted with a plastic collar to minimize the

— Funnel

— Plastic collar

— Bottom of
meniscus

Figure 2.7
Graduated cylinder and funnel.

chance of breakage if the cylinder is accidentally knocked over. You should not fill graduated cylinders by pouring liquids directly from very large containers such as 3-L bottles. In such cases, you should first pour the liquid into a beaker, trying to estimate how much you will need so as to minimize waste, and then transfer the liquid from the beaker into the graduated cylinder. You may use a stemmed funnel to help avoid spillage.

Beakers, Erlenmeyer flasks, and **conical vials** may also bear calibration marks. Keep in mind that, just as with graduated cylinders, volumetric measurements using these containers are only approximate, and other apparatus are used when more precise measurements are needed.

Pasteur Pipet

The **Pasteur pipet** is a glass tube, drawn out at one end to form a tip through which liquid is pulled with the aid of a latex suction bulb (Fig. 2.8). These pipets are commonly available in two lengths. Liquid is drawn into the pipet by first compressing the bulb and then inserting the tip of the pipet into the liquid to be transferred. To maximize the efficiency of the transfer, it is best not to compress the bulb more than is needed to draw the desired amount of liquid into the tip and barrel of the pipet. With practice, you will learn how much to compress the bulb to pull up a given volume of liquid.

Pasteur pipets are *not* used for quantitative measurements of volumes, but they can be used for qualitative measurements. This requires calibration of the pipet. Do this by first weighing, to the nearest 0.1 g, a specific amount of water (density 1.0 g/mL) into a test tube. You may also use a graduated cylinder to measure a given volume of liquid. Carefully draw the liquid into the pipet to be calibrated, and use a file to score (scratch) the pipet lightly at the level reached by the liquid (Fig. 2.8b). You may wish to place several calibration marks on the pipet, say at the 0.5-, 1-, and 1.5-mL levels. It is a good idea to calibrate several pipets at once since more than one may be needed in a given experimental procedure and because they are easily broken.

Although Pasteur pipets are often called "disposable pipets," they can be reused many times if given proper care. This involves cleaning (Sec. 2.1) and drying the pipet as soon as possible after use. When the pipet is no longer cleanable or the tip has become chipped or broken, it should be discarded in a special container labeled for broken glassware. Placing broken glassware in a wastebasket could lead to accidental injury to cleaning personnel.

— 1.5 mL
— 1.0 mL
— 0.5 mL

(a)   (b)

Figure 2.8
Pasteur pipets.

Modified Pasteur Pipets

Plug of cotton
or glasswool

(a)   (b)

Figure 2.9
Modified Pasteur pipets:
(a) Pasteur filter-tip pipet;
(b) Pasteur filtering pipet.

The **Pasteur filter-tip pipet** (Fig. 2.9a) and **filtering pipet** (Fig. 2.9b) are two useful modifications of the Pasteur pipet. The filter-tip pipet is helpful for transferring liquids containing particularly volatile components like diethyl ether (bp 35 °C). The plug in the tip of the pipet resists the backpressure that develops when the liquid is drawn into the pipet and volatilizes. This backpressure may force the liquid out of the pipet and onto the bench. The filtering pipet has its plug at the base of the barrel rather than in the tip (Fig. 2.9b). This modification is useful for operations such as removing solids from solutions during microscale recrystallizations (Sec. 2.17).

Cotton is the material preferred for the plug because it packs more tightly than glasswool, making removal of finely divided solids more effective. However, glasswool is used when the particles to be filtered from a liquid are relatively large or when transferring or filtering strongly acidic solutions that may react with cotton.

You may prepare a filter-tip pipet by first inserting a small piece of the packing material in the top of a Pasteur pipet and using a length of heavy-gauge wire, a narrow rod, or the tip of another pipet to push it to the base of the barrel and then into the tip of the pipet. The second step is omitted when making a filtering pipet. For both types of pipets, the plug that results should neither be too large nor packed too tightly, because the flow of liquid into and out of the pipet may then be too slow.

Graduated Pipets

Quantitative volumetric measurements are possible with **graduated pipets**, of which there are a number of styles and sizes. Two examples are shown in Figure 2.10. The graduated "blow-out" pipet (Fig. 2.10a), sometimes labeled as a "to deliver" (TD) pipet, is calibrated so that the full volume of liquid in the pipet is dispensed only when the liquid remaining in the tip is forced out using a pipet bulb. Any drops remaining on the tip must also be added to the delivered liquid, and this is done by touching the drop to the inside of the container or to the surface of the liquid. Of course, you may measure volumes less than the maximum contained in the pipet by difference, in which case the entire contents of the pipet are *not* blown out.

The graduated pipet shown in Figure 2.10b is designed to deliver the full volume of liquid in the pipet without having to be blown out. Thus, a small amount of liquid will remain in the tip after dispensing. Proper use of this pipet requires that the bulb used to draw liquid into the pipet be removed as the contents are being delivered into a container. If there is a drop of liquid remaining on the tip, it too should be added to the dispensed liquid.

Syringes

Plastic or glass **syringes** (Fig. 2.11) are often used to deliver liquids into reaction mixtures. Syringes marked with gradations showing the amount of liquid contained in them normally have an accuracy of volumetric measurement on the order of ±5%, although the syringes originally designed for use in gas chromatography (Sec. 6.4) and those having volumes of 500 μL or less are more accurate.

Figure 2.10
Graduated pipets: (a) to deliver
(TD); (b) full-volume delivery.

Figure 2.11
Fixed and demountable syringes:
(a) fixed needle; (b) demountable,
locking; (c) demountable,
nonlocking.   **(a)**   **(b)**   **(c)**

The needles on syringes are either fixed (Fig. 2.11a) or demountable (Fig. 2.11b, c). With the latter style of syringe, needles having various gauges can be affixed to the barrel. Removing and attaching the needles often require twisting the base of the needle to disengage it from or engage it to the barrel; this type of syringe is shown in Figure 2.11b. Otherwise, the needle simply slides on or off the plastic or ground-glass tip of the barrel (Fig. 2.11c).

Syringes are commonly used in the laboratory to transfer solutions or liquid reagents to reactions being performed under anhydrous conditions. The apparatus is usually fitted with a rubber septum, and the syringe needle is used to pierce the septum to deliver the liquid. You must be careful when inserting the needle into the septum to avoid bending it. The danger of bending can be minimized by supporting the sides of the needle with two fingers as you insert it into the septum.

A syringe is a sharp-pointed instrument, so you should be careful handling it. *Never* point an assembled syringe at yourself or anyone else, and do not leave needles pointed upward. You might get more than a simple needle prick as a result. You should clean a syringe immediately after use. This is done by pulling acetone or some other low-boiling solvent into the barrel and expelling it into an appropriate container. The rinsing should be repeated several times, whereupon the syringe should be dried by pumping the plunger to pull air through it or by attaching the barrel of the syringe to a water aspirator.

Dispensing Pumps and Automatic Pipets

Dispensing pumps and **automatic pipets** are devices used for accurately and quickly dispensing liquids. Because they are expensive, these devices are normally used to transfer liquids from containers, such as reagent bottles, that are being used by everyone in a laboratory.

Dispensing pumps (Fig. 2.12) are commonly used when quantities greater than about 0.5 mL must be dispensed. The plunger is made of Teflon, which allows you to use the unit with corrosive liquids and organic solvents, such as bromine in dichloromethane. The pump is loaded by pulling the plunger up as far as it will go, and the liquid that has been drawn into the glass reservoir is then expelled into the desired container by slowly depressing the plunger. Liquids of relatively low viscosity are usually expelled using only the weight of the plunger. More viscous liquids may require gently pushing the plunger down. If any liquid remains on the tip of the dispenser at the end of the transfer, the tip should be touched to the wall of the container to complete the operation.

Figure 2.12
Dispensing pump.

The length the plunger travels, and thus the amount of liquid ultimately delivered, is adjustable. In cases in which this liquid is the limiting reagent in a reaction, it should be transferred into a tared container, which is then reweighed to determine the precise amount of the reagent. To ensure maximum accuracy in measuring the volume of liquid delivered by the pump, be certain that an air bubble has not developed as the liquid is drawn into the reservoir and that the spout has no air voids. If either of these conditions is not met, the proper volume of liquid will not be dispensed. You can correct these problems by raising and depressing the plunger until no air bubbles are visible; any liquid that is delivered by this operation should be collected and disposed of in the appropriate waste container.

Automatic pipets are used for accurately delivering 10- to 1000-μL (0.01- to 1-mL) quantities of liquid and are particularly valuable in microscale procedures. Although several styles of these pipets are available, their operation is basically the same. The pipet comprises a barrel, which contains a spring-loaded plunger whose length of travel may be variable, and a disposable plastic tip into which the liquid is drawn (Fig. 2.13a). There are two "detent" or stop positions on the plunger; these control the filling and dispensing stages of the liquid transfer. You *must* attach the plastic tip securely, and you should *not* totally immerse the tip in the liquid being transferred. These precautions are necessary because liquid should *never* come into contact with the barrel and plunger because this may damage the pipet. If liquid does accidentally touch these surfaces, notify your instructor immediately.

The operation of the pipet is straightforward. First, hold the assembled pipet in the vertical position and depress the plunger to the *first* detent *before* inserting the tip into the liquid (Fig. 2.13a). Then insert the tip to a depth of 5–10 mm in the liquid and *slowly* release the plunger (Fig. 2.13b). If you allow the plunger to snap back to its original position, liquid may splash into the barrel of the pipet and damage the mechanism. Finally, dispense the liquid by placing the plastic tip, still held in a vertical position, against the inner wall of the tilted receiving vessel and slowly

Figure 2.13
Automatic pipet: (a) depressing plunger to first detent;
(b) releasing plunger to fill tip;
(c) redepressing plunger to first detent to deliver liquid;
(d) depressing plunger to second detent to complete delivery.

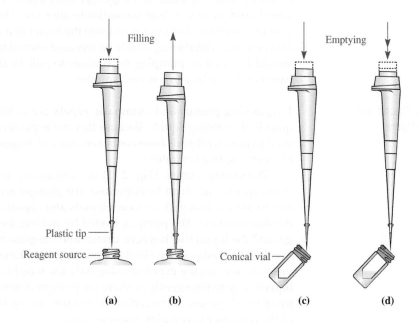

push the plunger down to the *first* detent (Fig. 2.13c); a second or two later, further depress the plunger to the *second* detent to expel the final drop of liquid from the plastic tip (Fig. 2.13d). Remove the pipet from the receiver and follow the directions of your instructor with respect to any further steps to be taken with the pipet and its tip.

2.6 WEIGHING METHODS

Figure 2.14
A top-loading balance with draft shield.

It will be necessary to measure quantities of reagents and reactants for the reactions you will perform. The techniques for measuring *volumes* of liquids are discussed in Section 2.5, and the *weight* of a liquid whose density is known is often most conveniently measured by transferring a known volume of the liquid to the reaction flask, using a graduated pipet or a syringe. This is especially true when the amount of a liquid to be weighed is less than 200–300 mg. However, for some liquids and all solids, weights are usually determined using a suitable **balance**. For quantities greater than 0.5 g (500 mg), a top-loading balance that reads accurately to the nearest 0.01 g (10 mg) is usually adequate. When performing reactions on the microscale, it is necessary to use a top-loading balance that has a draft shield and reads to the nearest 0.001 g (1 mg) (Fig. 2.14) or an analytical balance (Fig. 2.15) that reads to the nearest 0.0001 g (0.1 mg).

With modern electronic balances equipped with a taring device, it is possible to subtract the weight of a piece of paper or other container automatically from the combined weight to give the weight of the sample itself directly. For example, to weigh a liquid, tare a vial and remove it from the balance. Transfer the liquid from the reagent bottle to the vial with a pipet or a graduated pipet or syringe, and reweigh the vial. Be careful to avoid getting liquid on the outside of the vial or the balance pan. *Clean any spills promptly.*

To weigh a solid, place a piece of glazed weighing paper on the balance pan and press the tare device; the digital readout on the balance will indicate that the paper has "zero" weight. You should *never* weigh solids directly onto the balance pan. Using a microspatula, transfer the solid to be measured from its original container to the weighing paper until the reading on the balance indicates the desired weight. Do not pour the solid directly from the original bottle, because spills are more likely.

Figure 2.15
An analytical balance with draft shield.

Carefully transfer the solid to your reaction flask or other container. You may also weigh the solid directly into the reaction vessel by first taring the flask or conical vial. *Clean any spills promptly.*

2.7 MELTING-POINT METHODS AND APPARATUS

The laboratory practices and apparatus used to determine **melting points** of solids are discussed in this section, and the theory and use of melting points are described in Section 3.3. The task of determining the melting point of a compound simply involves heating a small amount of a solid and determining the temperature at which it melts. Many different types of heating devices can be used, but most utilize a capillary tube to contain the sample so that only a small amount of the sample is required.

Capillary Tubes and Sample Preparation

The first step in determining a melting point is transferring the sample into a melting-point capillary tube. Such tubes have one sealed end and are commercially available. The proper method for loading the sample into the capillary tube is as follows. Place a small amount of the solid on a clean watchglass and press the open end of the tube into the solid to force a small amount of solid (about 2–3 mm in height) into the tube (Fig. 2.16a); this operation should *not* be performed on filter paper, because fibers of paper as well as the solid may be forced into the tube. Then take a piece of 6- to 8-mm tubing about 1 m long, hold it vertically on a hard surface such as the bench top or floor, and drop the capillary tube down the larger tubing several times with the sealed end *down* (Fig. 2.16b). This packs the solid sample at the closed end of the capillary tube.

Melting-Point Determination

The melting point of the crystalline solid is determined by heating the packed capillary tube until the solid melts. Some representative devices for measuring melting points are presented in Figures 2.17–2.19. The most reproducible and accurate results are obtained by heating the sample at the rate of about 1–2 °C/min to ensure that heat is transferred to the sample at the same rate as the temperature increases and that the mercury in the thermometer and the sample in the capillary tube are in thermal equilibrium.

Many organic compounds undergo a change in crystalline structure just before melting, perhaps as a consequence of release of the solvent of crystallization. The solid takes on a softer, "wet" appearance, which may also be accompanied by shrinkage of the sample in the capillary tube. These changes in the sample should *not* be interpreted as the beginning of the melting process. Wait for the first tiny drop of liquid to appear.

Melting usually occurs over a range of a degree, perhaps slightly more. Accordingly, a **melting-point range** of a compound is typically reported with the lower temperature being that at which the first tiny drop of liquid appears and the higher temperature is that at which the solid has completely melted.

Melting-Point Apparatus

A simple type of melting-point apparatus is the **Thiele tube**, shown in Figure 2.17a. This tube is shaped such that the heat applied to a heating liquid in the sidearm by a burner is distributed evenly to all parts of the vessel by convection currents, so stirring is not required. Temperature control is accomplished by adjusting the flame

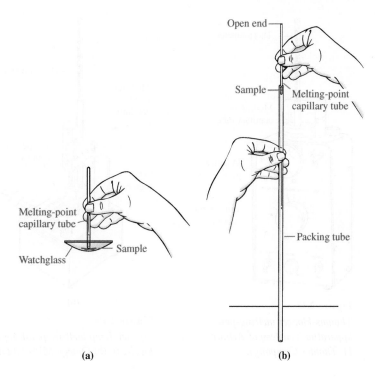

Figure 2.16

(a) Filling a melting-point capillary tube. (b) Packing the sample at the bottom of the tube.

(a)

(b)

Figure 2.17

(a) Thiele melting-point apparatus. (b) Arrangement of sample and thermometer for determining melting point.

(a)

(b)

produced by the microburner; this may seem difficult at first but can be mastered with practice.

Proper use of the Thiele tube is required to obtain reliable melting points. Secure the capillary tube to the thermometer at the position indicated in Figure 2.17b using either a rubber band or a small segment of rubber tubing. Be sure that the band

Figure 2.18
Thomas-Hoover melting-point apparatus (courtesy of Arthur H. Thomas Company).

(a)   **(b)**

Figure 2.19

(a) Mel-Temp melting-point apparatus (courtesy of Laboratory Devices). (b) Omega MPS 10 Digital melting-point apparatus.

holding the capillary tube on the thermometer is as close to the top of the tube as possible. Now support the thermometer and the attached capillary tube containing the sample in the apparatus either with a rubber stopper cork, as shown in Figure 2.17a, or by carefully clamping the thermometer so that it is immersed in the oil. The thermometer and capillary tube must *not* contact the glass of the Thiele tube. Since the oil will expand on heating, make sure that the height of the heating fluid is approximately at the level indicated in Figure 2.17a and that the rubber band is in the position indicated. Otherwise, the hot oil will come in contact with the rubber, causing the band to expand and loosen; the sample tube may then fall into the oil. Heat the Thiele tube at the rate of 1–2 °C/min to determine the melting point. The maximum temperature to which the apparatus can be heated is dictated by the nature of the heating fluid, a topic that is discussed in Section 2.9.

The Thiele tube has been replaced in modern laboratories by various **electric melting-point devices**, which are much more convenient to use. One common type of electric melting-point apparatus is the Thomas-Hoover unit shown in Figure 2.18. This particular device has a built-in vibrator to pack the sample in the capillary tube, and it also allows for the determination of the melting points of up to five samples simultaneously. The oil bath in this unit is heated and stirred. An electrical resistance heater is immersed in a container of silicone oil. The voltage across the heating element is varied by turning the large knob in front of the apparatus so that the oil is heated at a slow, controlled rate. A motor drives a stirrer in the oil bath to ensure even heating; the rate of stirring is controlled by a knob at the bottom of the unit. Some models are equipped with a movable magnifying lens system that gives the user a better view of the thermometer and the sample in the capillary tube.

The capillary tube containing the sample is inserted into the apparatus as illustrated in Figure 2.18.

The devices shown in Figure 2.19 are also electrically heated units but use a heated metal block rather than a liquid for transferring the heat to the capillary tube. A thermometer is inserted into a hole bored into the block, and the thermometer gives the temperature of the block and the capillary tube. Heating is accomplished by controlling the voltage applied to the heating element contained within the block.

2.8 BOILING-POINT METHODS AND APPARATUS

There are several techniques that may be used to determine the **boiling point** of a liquid, depending upon the amount of material available. When multigram quantities are available, the boiling point is typically determined by reading the thermometer during a simple distillation, which is described in Section 2.13. However, for smaller amounts of liquid there is sometimes not enough sample to distill, so other techniques have been developed. Two of these are described here.

Miniscale Method

An accurate boiling point may be determined with as little as 0.5–1.0 mL of liquid using the method illustrated in Figure 2.20. Working at a hood, place the liquid in a long, narrow Pyrex test tube, and add a small, black carborundum boiling stone; do not use a white marble chip, as bumping is more likely. Clamp the test tube and position the thermometer about 2 cm above the level of the liquid using a second clamp. Bring the liquid rapidly to a vigorous boil using a suitable heating device suggested by your instructor. You should see a reflux ring move up the test tube, and drops of the liquid should condense on the walls of the test tube. Control the amount of heating so the liquid does not boil out of the test tube. Be sure the bulb of the thermometer is fully immersed in the vapor of the boiling liquid long enough to allow the equilibration required for a good temperature reading to be obtained.

Figure 2.20
Miniscale technique to determine the boiling point of a liquid.

Refluxing vapor

Boiling liquid

Boiling stone

Figure 2.21
*Micro boiling-point apparatus:
(a) fusing a capillary tube;
(b) joining capillary tubes and
cutting off one end; (c) assembling
micro boiling-point apparatus
with correct placement of
ebullition and sample tubes,
and thermometer.*

Microscale Method

The boiling point of a liquid may be determined on amounts less than 0.5 mL by constructing a simple **micro boiling-point** apparatus as follows. First, prepare a capillary ebullition tube by taking a standard melting-point capillary tube, which is already sealed at one end, and making a seal in it about 1 cm from the open end, using a hot flame (Fig. 2.21a). Alternatively, join two melting-point capillary tubes by heating the closed ends in a hot flame; then make a clean cut about 1 cm below the point where the tubes have been joined (Fig. 2.21b). Second, seal a piece of 4- to 6-mm glass tubing at one end using a hot flame and cut it to a length about 1 cm longer than the capillary ebullition tube. These tubes are prepared most easily by using a glassblowing torch, and they may be provided by your instructor.

Attach the 4- to 6-mm tube to a thermometer with a rubber ring near the top of the tube. The bottom of the tube should be even with the mercury bulb of the thermometer. Put the capillary ebullition tube into the larger glass tube, and with a Pasteur pipet add the liquid whose boiling point you wish to determine until the level of the liquid is about 2 mm above the seal of the capillary tube (Fig. 2.21c).

Immerse the thermometer and the attached tubes in a heating bath. A Thiele tube (Fig. 2.17) is convenient for this purpose, but other heating baths can be used. *Be sure that the rubber ring is well above the level of the oil in the heating bath.* Heat the oil bath at the rate of about 5 °C/min until a *rapid* and *continuous* stream of bubbles comes out of the capillary ebullition tube. Before this occurs, you may see some bubbles form in an erratic fashion. This is not caused by boiling; rather, these bubbles are due to the expansion of air trapped in the capillary tube. You should see a marked change from the slow evolution of air bubbles to the rapid evolution of bubbles resulting from the boiling action of the liquid as the boiling point is reached. *However, this is **not** the boiling point!* Remove the heating source and allow the bath to cool slowly. As the liquid starts to rise into the capillary tube, note the temperature measured by the thermometer; *this is the boiling point of the liquid.* If the liquid rises sufficiently slowly into the capillary tube, note the temperatures at which the liquid starts to rise and at which the capillary tube is full. This will be the

boiling-point range of the liquid. Remove the capillary ebullition tube and expel the liquid from the small end by gently shaking the tube. Replace it in the sample tube and repeat the determination of the boiling point by heating the oil bath at the rate of 1–2 °C/min when you are within 10–15 °C of the approximate boiling point as determined in the previous experiment. Observed boiling points may be reproduced to within 1° C or 2 °C.

The physical basis of this technique is interesting. Before the liquid is heated, the capillary tube is filled with air. As the bath is heated, the air in the capillary tube is driven out and replaced with the vapor of the liquid. When the apparatus is heated until vigorous boiling of the liquid is observed, the actual boiling point of the liquid has been exceeded, and the air in the capillary tube has been completely replaced by the vapor of the liquid. On cooling, the vapor pressure of the liquid becomes equal to the external pressure, thus allowing the liquid to rise into the capillary tube. The temperature at which this occurs is, by definition, the boiling temperature of the liquid (Sec. 4.1).

2.9 HEATING METHODS

Heating is an important laboratory technique that is used in a variety of situations. For example, many chemical reactions require heating to proceed at a reasonable rate. Heating is also used to purify liquids by distillation, to remove volatile solvents during the work-up of a reaction, and to dissolve solids when purifying solid products by recrystallization.

Two general rules regarding heating are noted here. (1) Whatever device is being used to heat a liquid or a solid, you must arrange the apparatus so that the heating source can be *rapidly* removed to prevent accidents that may occur by over-heating. This normally means that the heating source is mounted on a ring clamp or a lab jack, either of which allows for quick removal of the device if necessary. (2) As a rule, the *safest* way to heat organic solvents is with a *flameless* heat source in a hood. This practice not only minimizes the chance of fire, but it also avoids filling the room with solvent vapors.

Heating with electrical devices is generally the method of choice in the organic laboratory, since it is much safer than using open flames. These devices are usually composed of two essential components: (1) a resistance element that converts electrical energy into thermal energy, and (2) a variable transformer for controlling the voltage across the element. They differ in the medium used to transfer heat from the element to the experimental apparatus, however. For example, air and oil are the heat transfer agents for a heat gun and an oil bath, respectively.

Although we describe below how burners are properly used for heating purposes, the experimental procedures in this textbook focus on *flameless* heating devices. Burners can be substituted for such devices in most cases, but not all. *Consult with your instructor if you have any questions regarding the suitability of substituting flames for other heating devices.*

Burners

Most chemistry laboratories are supplied with natural gas to fuel various types of burners. A burner provides the convenience of a rapid and reasonably inexpensive source of heat. However, many organic substances, especially solvents such as ether and hexane, are highly flammable, and you should always exercise

Figure 2.22
Laboratory burners: (a) Bunsen burner; (b) microburner.

good judgment when considering the use of a burner for heating volatile organic compounds. *Before using a burner to heat anything in the laboratory, consult with your instructor for the proper precautions and directions.* If an alternative mode of heating is available, choose it in preference to a burner. *Never use a burner to heat flammable materials in open containers such as beakers or Erlenmeyer flasks, because a fire may result.*

There are a number of situations in the laboratory where burners are appropriate heating sources. They may be used to heat Thiele tubes in the determination of melting or boiling points of organic substances and to heat a water bath to obtain and maintain temperatures from ambient to about 80 °C. When volatile solvents are not being used in the laboratory, burners may be safely substituted for heat guns to dry apparatus so a reaction may be conducted under anhydrous conditions (Sec. 2.26). On occasion, you will need a burner to bend glass tubing or to fashion a piece of glass apparatus. Burners can also be used to heat aqueous solutions that do not contain flammable substances or to heat higher-boiling liquids that are completely contained in round-bottom flasks either fitted with a reflux condenser (Sec. 2.22) or equipped for distillation as discussed in Sections 2.13–2.16. In these instances, it is important to lubricate the joints of the apparatus with a hydrocarbon or silicone grease to minimize the danger of leaking vapors and reduce the likelihood of the joints freezing.

You must be aware of what others are doing in the laboratory. Although you might be using a burner to perform a completely safe operation, someone nearby may be working with a very volatile, flammable solvent, some of which can creep along the bench top for several feet! These vapors or others in the room may be ignited explosively by an open flame.

Two common types of laboratory burners are pictured in Figure 2.22. The classic **Bunsen burner**, named after its inventor, is shown in Figure 2.22a. The needle valve at the bottom of the burner serves as a fine adjustment of the gas flow, and turning the barrel of the burner regulates the air flow; adjustment of gas and air flow provides control of the flame. In the **microburner** in Figure 2.22b, the air flow is adjusted at the baffle at the bottom of the burner, and the gas flow is adjusted at the gas valve on the laboratory bench.

Heating a flask with a burner may produce "hot spots" if most of the heat is applied to a small area on the bottom of the flask. Hot spots can lead to severe bumping, since the heat must be dispersed throughout the liquid by convection or by means of the turbulence caused by boiling. Hot spots can easily be avoided by holding the burner and slowly moving the flame over the bottom of the flask. Alternatively, a piece of wire gauze, which diffuses the heat reaching the flask, may be placed between the flame and the flask; the gauze is supported with an iron ring.

Heating Mantle

A common device for heating round-bottom flasks is a **heating mantle**. The mantle is typically fashioned either from a blanket of spun fiberglass partially covered with a flexible or rigid cover (Fig. 2.23a) or from a ceramic core contained in an aluminum housing (Fig. 2.23b). In either case, heat is provided by an electrical resistance coil embedded in the fiberglass or ceramic core, so when using these devices it is important not to spill liquid on them. The mantle may be equipped with a thermocouple so that its internal temperature can be monitored, but this is seldom done in undergraduate laboratories.

Figure 2.23
(a) Woven-glass heating mantle. (b) Heating mantle with ceramic core.

Oil Bath

Figure 2.24
Electrically heated oil bath.

Because heating mantles are constructed of nonferrous materials, they can be used in conjunction with magnetic stirring (Sec. 2.11), so simultaneous heating and stirring of a reaction mixture is possible. A variable transformer connected to an electrical outlet provides the power for heating the mantle. The cord from the mantle itself must *always* be plugged into a transformer, *never* directly into a wall outlet!

Most heating mantles have a hemispherical cavity into which the round-bottom flask is fitted. With those mantles having a spun fiberglass interior, a different size is required for each size of flask. However, if the hemisphere is a rigid ceramic material, it is possible to put a granular heat-transfer agent such as sand in the cavity so that flasks of sizes smaller than the diameter of the cavity can be accommodated (see "Sand Bath").

There are some drawbacks to using heating mantles. They heat up rather slowly, and it is difficult to obtain a given temperature or maintain a constant temperature. Heating mantles have a high heat capacity, so if it becomes necessary to discontinue heating suddenly (e.g., if a reaction begins to get out of control), *the heating mantle must be removed immediately from below the flask* to allow the flask to cool, either on its own or by means of a cooling bath; it is not sufficient simply to lower the voltage or turn off the electricity. After the mantle is removed, the electricity should be turned off at the transformer.

The amount of heat supplied by a heating mantle is moderated by the boiling liquid contained in the flask, because the hot vapors of the liquid transfer heat away from the mantle. If the flask becomes dry or nearly so during a distillation, the mantle can become sufficiently hot to melt the resistance wire inside, thus causing the mantle to "burn out." *Do not heat an empty flask with a heating mantle.* Moreover, most mantles are marked with a maximum voltage to be supplied, and this should not be exceeded.

Electrically heated **oil baths**, which typically contain either mineral oil or silicone oil, are commonly employed in the laboratory. These baths are heated either by placing the bath container on a hot plate or by inserting a coil of Nichrome resistance wire in the bath. In the latter case, the resistance wire serves as the heating element and is attached to a transformer with an electrical cord and plug (Fig. 2.24). Thermal equilibration within the bath is maintained by placing a paper clip or stir-bar in the bath and using a magnetic stirrer (Sec. 2.11).

Heating baths offer several important advantages over heating mantles. The temperature of the bath can be easily determined by inserting a thermometer in the liquid, and a given bath temperature may be obtained and accurately maintained by careful adjustment of the variable transformer. Although heat is transferred uniformly to the surface of the flask in the bath and there are no hot spots, there typically will be a temperature gradient of about 10 °C between the bath and the contents of the flask.

Some inconveniences are also encountered using oil baths. If the volume of heating liquid is fairly large, it may take a while to reach the desired bath temperature. The maximum temperature that may be safely attained in an oil bath is limited by the type of heating liquid being used. Silicone oils are more expensive but are generally preferable to mineral oils because they can be heated to 200–275 °C without reaching the **flash point**, the temperature at which a liquid can burst into flame, and without thickening through decomposition. Mineral oil should *not* be heated above about 200 °C because it will begin to smoke, and there is the potential danger of flash ignition of the vapors. Water must *not* be present in mineral and silicone oils, since at temperatures of about 100 °C, the water will boil, spattering hot oil. If water drops are present in the oil, change the heating fluid and clean and dry the container before refilling it.

A minor nuisance associated with oil baths is removing the film of mineral or silicone oils, both of which are water-insoluble, from the outer surface of the flask. This is best done by wiping the flask using a *small* amount of hexane or dichloromethane on a paper towel prior to washing the flask with soap and water.

Hot Plate

When flat-bottom containers such as beakers or Erlenmeyer flasks must be heated, convenient heat sources are **hot plates** (Fig. 2.25a) and **stirring hot plates** (Fig. 2.25b), which are hot plates with built-in magnetic stirrers; round-bottom flasks cannot be heated effectively with hot plates. The flat upper surface of the hot plate is heated by electrical resistance coils to a temperature that is controlled by a built-in voltage regulator, which is varied by turning a knob on the front of the unit. A hot plate generally should be limited to heating liquids such as water, mineral or silicone oil, and nonflammable organic solvents such as chloroform and dichloromethane. *Under no circumstances should a hot plate be used to boil highly flammable organic solvents*, because the vapors may ignite as they billow onto the surface of the hot plate or the electrical resistance coils inside the hot plate. Furthermore, many hot plates use a relay that turns the electricity on and off to maintain the desired temperature; because these relays are often not explosion-proof, they may produce sparks that can ignite fires.

Aluminum Block

An **aluminum block** is a convenient accessory used in conjunction with hot plates for reactions being performed in a microscale apparatus. The block has indentations of various sizes and shapes to accommodate conical vials and flasks (Fig. 2.26).

Hot plate control

Power light

(a)

Hot plate control
Stirrer control
Power light

(b)

Thermometer ——

Aluminum block

Figure 2.25
(a) Hot plate. (b) Stirring hot plate.

Figure 2.26
Aluminum block with conical vial in place.

There may also be a receptacle for inserting the bulb of a glass thermometer or the base of a metal dial thermometer, so the temperature of the block can be monitored. Although a glass thermometer may remain vertical without any further support, it should be held in place by a clamp or a wire to prevent breakage. It is not necessary to clamp a metal dial thermometer.

The temperature of the block is controlled by varying the power to the hot plate, and it is helpful to have a correlation between the two. You may calibrate the block by selecting four or five evenly spaced settings on the control dial of the hot plate, determining the temperature of the aluminum block at each setting, and recording both the setting and the corresponding temperature in your notebook for future reference.

Heat Gun

A useful device for heating apparatus, but not reaction mixtures, is a **heat gun** (Fig. 2.27), which is essentially a high-powered hair dryer. It is particularly useful for drying assembled apparatus, as this eliminates the need to use a flame. The stream of hot air from the heat gun is directed at the apparatus to drive atmospheric moisture out through an open port, which, after the entire apparatus is heated, is then fitted with a drying tube (Sec. 2.27) so that air is dried as it enters the cooling apparatus. A slow stream of dry nitrogen gas may be introduced during the heating process to help expel moisture from the apparatus. A drying tube or some other device such as a bubbler containing mineral oil must then be attached to the apparatus to prevent intrusion of atmospheric moisture as the apparatus cools and during the course of the reaction itself. For best results, you should start heating of the apparatus at a point most distant from the open port and work toward the opening.

The temperature of the air stream emerging from a heat gun is commonly from 300 °C to 500 °C, far higher than that of the typical handheld hair dryer. Consequently, you must use the heat gun prudently when drying apparatus held by plastic-containing devices such as Keck clips or clamps with rubber- or vinyl-coated jaws (Fig. 2.5), because the plastic can melt.

You should never point a heat gun in the direction of another person. Moreover, when using this device, you should be certain that there are no light-weight objects such as your starting material or reaction product in the vicinity that might become airborne on the powerful jet of air produced.

Figure 2.27
Heat gun.

Sand Bath

Sand baths are convenient devices for heating small volumes of material in small flasks. These baths are easily prepared by putting about 1–3 cm of sand in a Pyrex crystallizing or petri dish and then placing the dish on a hot plate (Fig. 2.28). The temperature in the sand bath is controlled by varying the heat setting on the hot plate. Alternatively, the sand may be placed in the ceramic well of a heating mantle, such as that shown in Figure 2.23b. The temperature of the sand bath in this case is controlled by the variable transformer attached to the mantle. The temperature may be monitored with a thermometer that is inserted into the bath to the same depth as the flask or conical vial being heated. Because sand is not a good conductor of heat, there is a temperature gradient in the bath, with the highest temperature being closest to the hot plate or heating mantle. This gradient can be exploited: The flask or vial can be deeply immersed in the sand bath for rapid heating; once the mixture has begun to boil, the flask or vial can be raised to slow the rate of reflux or boiling.

Although sand is cleaner than mineral and silicone oil, there are some limitations to using sand baths. For example, sand baths in glass containers are normally

Thermometer

Clamp

Pyrex dish
or beaker

Sand (1–3 cm depth)

Hot plate

Laboratory jack

Figure 2.28
Sand bath heated by hot plate.

not used to heat flasks to temperatures higher than about 200 °C because the glass might break. Furthermore, temperature control is difficult because sand is a poor heat conductor, and it is not possible to stir the bath. Finally, different devices may heat the sand bath to a different temperature gradient, so you must calibrate each heating device/sand bath combination for reproducible results. A procedure for doing this is discussed under "**Aluminum Block.**"

Water Bath

A **water bath** may be used when temperatures no higher than about 80–90 °C are desired. The water is contained in a wide-mouth vessel such as a beaker or crystallizing dish. The bath is preferably heated with an electrical hot plate (Fig. 2.29a), but a Bunsen burner can also be used. The bath should be supported on a ring clamp bearing a wire mesh to diffuse the flame (Fig. 2.29b). A useful "trick" to minimize the loss of water through evaporation is to cover the open portions of the bath with aluminum foil.

Steam Bath

Steam provides a safe source of heat when temperatures up to 100 °C are required. The steam outlet is connected to either a **steam bath** or a **steam cone** (Fig. 2.30), both of which have an outlet at the bottom to drain the condensed water. The tops of steam baths and cones are typically fitted with a series of overlapping concentric rings that may be removed in succession to provide openings of variable size. For example, if a rapid rate of heat transfer to a flask is desired, the rings are removed until up to one-half of the surface of the flask is immersed in the steam (Fig. 2.30a); cloth towels may be wrapped around the upper portion of the flask to insulate it. If a slower rate of heating is desired, the opening may be adjusted, so that less of the flask is in direct contact with the steam. To heat beakers and Erlenmeyer flasks, the opening should be small enough so that the container sits directly on top of the steam bath with only its lower surface exposed to steam.

Figure 2.29
Water bath: (a) heated by hot plate; (b) heated by Bunsen burner.

Figure 2.30
(a) Steam bath being used to heat a reaction mixture. (b) Steam cone. (c) Heating an Erlenmeyer flask on a steam cone, with a funnel attached to vacuum source for removing vapors.

When the steam valve is first opened, several minutes are usually required for the condensed water to drain out of the steam lines. Once the steam is issuing smoothly, the steam valve should be adjusted to provide a *slow*, steady flow of steam. There is no benefit to having a fast flow of steam, other than to fog your safety glasses and give everyone in the lab a sauna bath! Regardless of how fast the steam is flowing, the temperature will never exceed 100 °C. Another disadvantage of a fast flow of steam is that water will condense on your equipment and may even find its way into the reaction flask.

Steam baths are especially well suited as heat sources for the safe evaporation of volatile solvents used for a reaction or extraction (Sec. 2.29) or when purifying a solid product by recrystallization (Sec. 3.2). As an alternative to working at a hood, an inverted funnel attached to a vacuum source may be placed over the top of the container (Fig. 2.30c) to remove vapors and keep them from entering the room.

Microwave Ovens

Microwave ovens are attaining wider application as a heating source for organic reactions. Although these ovens rely upon the same technology as the microwave oven you may use at home to heat water or food, the ones used in the laboratory have been specially modified so they can be operated safely with organic solvents. Such ovens are explosion-proof and allow precise temperature control and uniformity of heating throughout the cavity of the oven. A number of commercial models that have a wide range of capacities and capabilities are available, and two different types are depicted in Figure 2.31. Some models feature temperature and pressure-feedback control, whereas others have modular accessories that allow for automated synthesis. Simple ovens are relatively inexpensive and are found in a number of academic laboratories, but more sophisticated models can be rather expensive and are normally found only in industrial laboratories.

The transfer of microwave energy to a molecule is extremely fast and results from the direct interaction of a molecule with the high-frequency, oscillating electric component of the microwave field. Because the absorption of microwave energy depends on the polarity of the molecule, the reactant molecules in

Courtesy of CEM Corporation

Courtesy of CEM Corporation

(b)

Figure 2.31   **(a)**

Commercial microwave reactors: (a) CEM Discover SP microwave synthesis instrument; (b) CEM MARS 6 Synthesis microwave system.

(a)   **(b)**

Courtesy of CEM Corporation

Figure 2.32
Reaction vessels: (a) assorted shapes of standard-taper ground glassware and sealable glass tubes; (b) array of special reaction tubes for multiple, simultaneous reactions.

an organic reaction often absorb microwave energy better than the solvent. This preferential transfer of microwave energy to the reacting molecules results in an *instantaneous temperature increase*, which cannot be measured because of its short lifetime and molecular nature, and a rapid reaction then ensues. Reactions that may require hours of heating in an oil bath may be completed in a matter of minutes in a microwave oven.

The use of microwave ovens as heat sources not only drastically reduces the heating times of organic reactions, but the reactions often proceed more efficiently and selectively than when conduction heating methods are used. This is because microwave energy is transferred uniformly and almost simultaneously to the entire sample, thus eliminating any hot spots that may result in side reactions. If sealed reaction vessels are used, superheating can be easily achieved, so reactions that might be otherwise difficult to induce can be performed with relative ease. Another benefit to using microwave heating in synthesis is that less solvent is required than for traditional heating methods.

Reactions heated in a microwave oven do not necessarily require special apparatus. For example, they can be conducted in vessels that range from simple laboratory glassware such as an Erlenmeyer flask covered with a watchglass to flasks of various shapes that are fitted with standard-taper glass joints (Fig. 2.32a). Specially designed pressure vessels can also be used. A number of tubes can also be mounted on a rotor so that multiple reactions can be conducted simultaneously (Fig. 2.32b).

2.10 COOLING TECHNIQUES

Sometimes it is necessary to cool a reaction mixture either to avoid undesired side reactions or to moderate the temperature of exothermic reactions that could become uncontrollable. Cooling is also often used to maximize the recovery of solid products during purification by recrystallization (Sec. 3.2) or to lower the temperature of hot reaction mixtures so that work-up procedures can be performed.

The most common cooling medium in the undergraduate laboratory is an ice-water bath. Liquid water is a more efficient heat transfer medium than ice, because it covers the entire surface area of the portion of the vessel that is immersed in the bath. Consequently, when preparing this type of bath, do not use ice alone.

An ice-water bath has an equilibrium temperature of 0 °C. For lower temperatures, an ice-salt bath can be prepared by mixing ice and sodium chloride in a proportion of about 3:1 to generate a temperature of approximately −20 °C. As the ice melts, the excess water should be removed and more ice and salt added to maintain this temperature. Still lower temperatures are possible using combinations of organic liquids and either dry ice (solid carbon dioxide) or liquid nitrogen; directions for preparing these types of baths can be found in Reference 4 of Chapter 1.

Immersing a vessel containing a warm reaction mixture in a container of tap water is useful for rapid cooling to room temperature so long as the glassware you are using is Pyrex. Consult with your instructor to ensure that this is the type of glass you have.

2.11   STIRRING METHODS

Heterogeneous reaction mixtures must be stirred to distribute the reactants uniformly and facilitate chemical reactions. Stirring also ensures thermal equilibration whenever the contents of a flask are being heated or cooled. If a mixture is boiling, the associated turbulence is usually sufficient to provide reasonable mixing; however, stirring a boiling mixture is an alternative to using boiling stones to maintain smooth boiling action and avoid bumping. Stirring is most effectively achieved using magnetic or mechanical stirring devices, but swirling is often sufficient.

Swirling

The simplest means of mixing the contents of a flask is **swirling**, which is accomplished by manually rocking the flask with a circular motion. If a reaction mixture must be swirled, carefully loosen the clamp(s) that support the flask and attached apparatus, and swirl the contents periodically during the course of the reaction. If the entire apparatus is supported by clamps attached to a single ring stand, the clamp(s) attached to the flask do not have to be loosened. Make sure all the clamps are tight, pick up the ring stand, and gently move the entire assembly in a circular motion to swirl the contents of the flask.

Magnetic Stirring

Egg-shape   Polygon

Spinvane

Figure 2.33
Magnetic stirbars and a spinvane.

Magnetic stirring is the most common technique for mixing the contents of a flask in the undergraduate laboratory. The equipment consists of a magnetic stirrer, which houses a large bar magnet that is spun by a variable-speed motor, and a stirbar (Fig. 2.33) that is contained in a round-bottom flask or conical vial. The metallic core of the stirbar is usually coated with a chemically inert substance such as Teflon, although glass is sometimes used for stirbars.

The stirbar is normally placed in the flask or vial *before* any other materials, such as solvents or reagents. In the case of introducing stirbars into a flask, you should *not* simply drop the stirbar into the flask, because you might crack or break it. Rather, tilt the flask and let the stirbar gently slide down the side.

A flat-bottom container such as a beaker or Erlenmeyer flask may be placed directly on top of the stirrer (Fig. 2.34a), whereas a round-bottom flask or a conical

Figure 2.34
(a) Magnetic stirring of the contents of Erlenmeyer flask. (b) Using a heating source with magnetic stirring of reaction mixture.

vial must be clamped above the stirrer (Fig. 2.34b). A flask containing a stirbar should be *centered* on the magnetic stirrer so that the stirbar rotates smoothly and does not wobble. As the stirrer motor turns, the stirbar rotates in phase with the motor-driven magnet. The stirring rate may be adjusted using the control dial on the stirring motor, but excessive speed should be avoided because it often causes the stirbar to wobble or "jump" rather than to rotate smoothly. Because the shapes of a conical vial and a stirbar are matched, wobbling is not a problem. Nonetheless, the stirring rate must still be adjusted to minimize splashing.

The use of magnetic stirring in conjunction with heating is illustrated in Figure 2.34b. This figure depicts apparatus in which the contents of a reaction vessel are simultaneously being heated and magnetically stirred, a common operation in the organic laboratory. A large stirbar or paper clip is used to stir the bath itself; this maintains a homogeneous temperature throughout the heating fluid. Other heating devices such as a sand bath, aluminum block, or heating mantle can be used as well, although monitoring of the heating source is more difficult with a mantle.

Mechanical Stirring

Thick mixtures and large volumes of fluids are most efficiently mixed using a **mechanical stirrer;** a typical setup is depicted in Figure 2.35. A variable-speed, explosion-proof, electric motor drives a stirring shaft and paddle that extend into the flask containing the mixture to be stirred. The motor should have high torque, so that it has sufficient power to turn the shaft and stir highly viscous mixtures. The stirrer shaft is usually constructed of glass, and the paddle, which agitates the contents of the flask, is constructed of an inert material such as stainless steel, Teflon, or glass. A glass paddle must be used to stir reaction mixtures containing active metals such as sodium or potassium. The paddle is easily removed from the

Stirring motor with
variable transformer

Stirrer shaft

Clamp

Addition or
separatory
funnel

Clamp

Three-neck
round-bottom
flask

Water out

Clamp

Condenser

Water in

Heating source

Stirring paddle

Laboratory jack

Figure 2.35
*Flask equipped with
mechanical stirring.*

shaft to facilitate cleaning, and different-sized paddles can be used according to
the size of the flask. The glass shaft and the inner bore of the standard-taper bearing are ground to fit each other precisely. A cup at the top of the bearing is used to
hold a few drops of silicone or mineral oil, which lubricates the shaft and provides
an effective seal.

The stirrer shaft is connected to the motor with a short length of heavy-walled
rubber tubing that is secured with twisted copper wire or a hose clamp. The motor
and shaft *must* be carefully aligned to avoid wear on the glass surfaces of the shaft
and bearing and to minimize vibration of the apparatus that could result in breakage. The bearing is held in place in the flask with either a rubber band or a clamp
so that it does not work loose while the motor is running. The rate of stirring is controlled by varying the speed of the motor with either a built-in or separate variable
transformer.

Various operations can be performed while using mechanical stirring. For
example, the flask in Figure 2.35 is a three-neck, standard-taper, round-bottom flask
that is equipped with an addition funnel and a condenser. This apparatus could be
used in cases where dropwise addition of a reagent to a stirred and heated reaction
mixture is required.

2.12 CENTRIFUGATION

Centrifugation is a useful technique to facilitate the separation of two immiscible
phases. The procedure simply involves spinning a sample in one or more tubes
at high speed in a bench-top centrifuge (Fig. 2.36). It is important to balance the

Rotor · Centrifuge tube

Speed control

Power switch

Figure 2.36
Centrifuge with centrifuge tube.

centrifuge before spinning to avoid excessive vibration of the centrifuge during spinning. For example, if the sample to be centrifuged is placed in a single centrifuge tube, you should then fill a second tube with an equal volume of solvent and place this tube opposite the sample tube in the centrifuge. After the sample has been spun, the phases may be separated by decantation (Sec. 2.19) or by removing a liquid phase with a Pasteur or filter-tip pipet (Sec. 2.5).

Centrifugation is sometimes more effective than filtration for removing suspended solid impurities in a liquid, especially when the particles are so fine they would pass through a filter paper. It may also be used for Craig tube filtration during a microscale recrystallization (Sec. 2.17). Centrifugation will also aid in separating the organic and aqueous layers when performing an extraction (Sec. 2.21), especially when emulsions are formed.

2.13 SIMPLE DISTILLATION

Simple distillation is a useful method for isolating a pure liquid from other substances that are not volatile. The experimental aspects of this technique are described in this section, whereas the theory and application of the method are discussed in detail in Section 4.3.

Miniscale Apparatus

Typical examples of laboratory apparatus for performing simple miniscale distillations are shown in Figure 2.37. The operation entails heating the liquid contained in the **stillpot**, also called the **distillation flask**, to its **boiling point**, which is defined as the temperature at which the total vapor pressure of the liquid is equal to the external pressure. The vapors then pass from the pot into a water- or air-cooled **condenser** where they condense to form the liquid phase that is collected in the **receiver**. The thermometer measures the temperature of the vapors. However, to obtain an accurate temperature reading, the thermometer must be carefully positioned so that the top of the mercury bulb is approximately even with the bottom of the sidearm outlet, as indicated in the inset in Figure 2.37a. To ensure an accurate temperature reading, a drop of condensate *must* adhere to the bottom of the mercury bulb. If a pure liquid is being distilled, the temperature read on the thermometer, which is termed the **head temperature**, will be identical to the temperature of the boiling liquid. Provided the pure liquid in the stillpot is not being superheated, the **pot temperature** will also be identical to the temperature of the boiling liquid, although the bath temperature will be higher. The head temperature corresponds to the boiling point of the liquid and will remain constant throughout the distillation. If a mixture of volatile compounds is being distilled, the head and pot temperatures will be different.

To avoid possible breakage and spillage, certain techniques must be followed when assembling the apparatus shown in Figure 2.37. The apparatus in Figure 2.37a is typically used for volumes of distillate in excess of about 10 mL. If a condenser is not necessary, the modified apparatus shown in Figure 2.37b, which comprises a distillation head and a bent vacuum adapter, is well suited for distilling quantities in the range of 1–10 mL; the shorter path length in this setup minimizes material losses. *You should thoroughly familiarize yourself with the following general guidelines before attempting to set up your apparatus.*

1. Assemble the equipment so that the distillation flask is elevated 15 cm or so above the bench to allow placement of a suitable heat source. Read Section 2.4 for a general discussion.

Figure 2.37
(a) Typical apparatus for miniscale simple distillation at atmospheric pressure or under vacuum. Inset shows correct placement of thermometer in stillhead. (b) Typical apparatus for miniscale shortpath distillation with ice bath for cooling receiver.

2. Review the relevant heating methods discussed in Section 2.9. Heating mantles may be used to heat the stillpot, but an oil bath is generally the best way to heat the flask, since the temperature of the bath, and hence the stillpot, is easier to control. The level of the liquid in the distillation flask should be below the level of the oil in the heating bath to minimize the risk of bumping. *Do not use a flame unless directed to do so by your instructor.* The flask should *never* be heated *directly* with a flame, because this may produce hot spots and cause bumping; the flask may only be heated with a flame *if* wire gauze is placed under the flask to diffuse the heat.

3. When distilling materials that have boiling points in excess of 125 °C, wrap the stillhead with glasswool and then aluminum foil to prevent excessive heat loss; failure to do so usually means that it will be necessary to heat the distillation flask to a higher temperature than would otherwise be necessary to drive over the last portion of product.

4. Place a stirbar in the distillation flask along with the liquid to be distilled *before* attaching the rest of the glassware; do not fill the flask more than *half*-full. *Start assembling the equipment by clamping the stillpot in position,* and attach the rest of the glassware by next putting the stillhead in place, then the condenser, and finally the vacuum adapter at the end of the condenser. Place a thin film of lubricant (Sec. 2.2) on each of the standard-taper glass joints before mating them. Joints must fit snugly so that flammable vapors do not leak from the apparatus into the room.

5. Note the location of the rigid clamps in Figure 2.37; *do not "overclamp" the apparatus.* An apparatus that is too rigidly clamped may become stressed to the point that it breaks during its assembly or the distillation. Align the jaws of the clamp *parallel* to the piece of glass being clamped so that the clamp may be tightened without twisting or torquing the glass and either breaking

it or pulling a joint loose. Tighten the clamps only enough to hold each piece securely in place. *Do not tighten the clamp unless the piece of glassware is correctly positioned and the clamp is properly aligned.* Always practice the rule: *Do not apply undue pressure to the glassware.*

6. Use Keck clips to hold the condenser, vacuum adapter, and receiver in place. Provide additional support for flasks of 100-mL capacity or larger; smaller flasks may also require additional support if they become more than half-full during a distillation. Clamp the receiver. An iron ring holding a piece of wire gauze or a cork ring may be used to provide additional support underneath the receiving flask.

7. Note the location of the "water in" and "water out" hoses on the condenser. The tube carrying the incoming water is *always* attached to the lower point to ensure that the condenser is filled with water at all times.

8. Adjust the water flow through the condenser to a *modest* flow rate. There is no benefit to a fast flow, and the increased pressure in the apparatus may cause a piece of rubber tubing to pop off, spraying water everywhere. Showers in the laboratory should be restricted to the emergency shower! It is good practice to *wire* the hoses to the condenser and to the water faucet to minimize the danger that they will break loose.

Microscale Apparatus

The volume of distillate in microscale distillations is typically less than 500 μL (0.5 mL), so it is important to minimize the surface area of glass to avoid unnecessary loss of material. This is accomplished by combining the stillhead and collector in a single piece called a **Hickman stillhead**, which has a reservoir for containing distillate and may have a port to facilitate removal of distillate (Fig. 2.38).

Figure 2.38
Microscale apparatus for distillation with external temperature monitoring.

Because the volume of distillate is small, it is somewhat difficult to measure the distillation temperature (boiling point) accurately at the microscale level. An *approximate* method is to determine the temperature of the heat source at the point when distillation occurs. This temperature is about 20 °C or so above the distillation temperature, so it is necessary to make the corresponding correction in the boiling point recorded for the distillate. Alternatively, a thermometer can be inserted through the condenser attached to the stillhead so the thermometer bulb is in the lower neck of the stillhead. A more accurate measurement of the boiling point of the distillate is possible using the technique of microscale boiling-point determination (Sec. 2.8).

A Pasteur or filter-tip pipet (Sec. 2.5) is used to remove distillate from the Hickman stillhead. If the head has a port, the pipet can be inserted at this point. If not, the condenser must first be removed and the pipet inserted through the top of the stillhead. Depending on the geometry of the stillhead, it may be necessary to use a Pasteur pipet having a curved tip to allow the tip to reach the bottom of the reservoir. A Bunsen or microburner (Sec. 2.9) may be used to produce the necessary bend.

2.14 FRACTIONAL DISTILLATION

The technique of **fractional distillation** is a useful method for isolating the individual pure liquid components from a mixture containing two or more volatile substances. This technique is described in this section, whereas the theory and application of this method are discussed in detail in Section 4.4. An apparatus for fractional distillation at atmospheric pressure or vacuum is shown in Figure 2.39. The principal difference between an apparatus for fractional and simple distillation is the presence of a **fractional distillation column** (Fig. 2.40) between the stillpot and the stillhead. This column is similar to a condenser, the major difference being that a fractionating column has a large outside jacket and some indentations at the male end to hold the packing in place. Unpacked distillation columns such as the Hempel column shown in Figure 2.40a can also be used as condensers, but condensers lack the indentations and cannot serve as packed distillation columns.

Before assembling the apparatus, clean and dry the inner tube of the distillation column to avoid contaminating your product. It is not necessary to dry the water jacket, because traces of water in it will not affect the distillation. Pack the column by adding the desired column packing, a small quantity at a time, through the top of the column while holding it vertical. The column packing is an inert material such as glass, ceramic, or metal pieces in a variety of shapes (helices, saddles, woven mesh, etc.). Common packings include glass tubing sections, glass beads, glass helices, or a stainless steel sponge. Some column packings are sufficiently large that they will stop at the indentations. If the packing falls through the column, put a small piece of glasswool, wire sponge, or wire screen into the column just above the indentations by pushing it down the column with a wooden dowel or a piece of glass tubing. The column packing should extend to the top of the water jacket but should not be packed too tightly.

Assemble the apparatus using the general guidelines for simple distillation apparatus (Secs. 2.4 and 2.13). *Read those instructions carefully before proceeding with setting up the apparatus in* Figure 2.39. Start by putting the liquid to be distilled and a stirbar in the distillation flask, and then clamp the flask in place. Do not fill the flask more than *half*-full. Attach the distillation column and make sure that it is *vertical*. Lubricate and tighten all the joints after the stillhead and condenser are in place. Clamp only the distillation flask and condenser. *Do not run water through the jacket of the distillation column.*

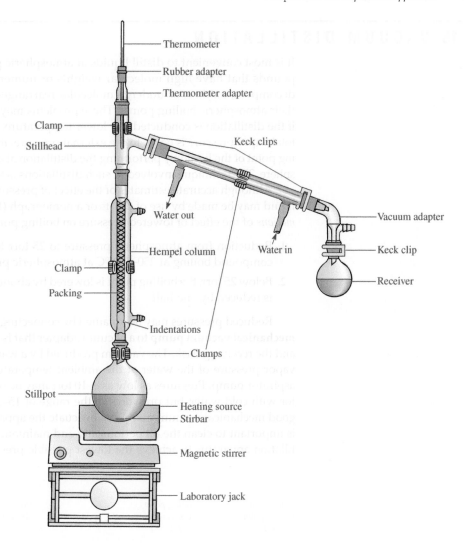

Figure 2.39
Miniscale apparatus for fractional distillation.

The following labels appear in the figure:

- Thermometer
- Rubber adapter
- Thermometer adapter
- Clamp
- Stillhead
- Keck clips
- Vacuum adapter
- Water out
- Water in
- Keck clip
- Hempel column
- Receiver
- Clamp
- Packing
- Indentations
- Clamps
- Stillpot
- Heating source
- Stirbar
- Magnetic stirrer
- Laboratory jack

Figure 2.40
Miniscale fractional distillation columns: (a) Hempel column filled with packing material; (b) Vigreux column.

- Raschig rings
- Indentations
- Indentations
- Indentations

(a)   (b)

2.15 VACUUM DISTILLATION

It is most convenient to distill liquids at atmospheric pressure (760 torr), but compounds that have high molecular weights or numerous functional groups may decompose, oxidize, or undergo molecular rearrangement at temperatures below their atmospheric boiling points. These problems may frequently be circumvented if the distillation is conducted at a lower temperature. Since a liquid boils when its total vapor pressure is equal to the external pressure, it is possible to lower the boiling point of the liquid by performing the distillation at a pressure *less* than one atmosphere. The technique involved in such distillations is termed **vacuum distillation**.

Although accurate estimates of the effect of pressure upon the boiling point of a liquid may be made by use of charts or a nomograph (Fig. 2.41), two useful *approximations* of the effect of lowered pressure on boiling points are:

1. Reduction from atmospheric pressure to 25 torr lowers the boiling point of a compound boiling at 250–300 °C at atmospheric pressure by about 100–125 °C.

2. Below 25 torr, the boiling point is lowered by about 10 °C each time the pressure is reduced by one-half.

Reduced pressures may be obtained by connecting a **water aspirator pump** or a **mechanical vacuum pump** to a vacuum adapter that is fitted between the condenser and the receiving flask. The vacuum produced by a water aspirator is limited by the vapor pressure of the water at the ambient temperature and the condition of the aspirator pump. Pressures as low as 8–10 torr may be obtained from a water aspirator with cold water, but pressures in the range of 15–25 torr are more common. A good mechanical vacuum pump can evacuate the apparatus to less than 0.01 torr; it is important to clean the oil periodically and maintain tight connections in the distillation apparatus to achieve the lowest possible pressures for a particular pump.

Figure 2.41

*Pressure–temperature alignment nomograph (dashed lines added for illustrative purposes). How to use the nomograph: Assume a reported boiling point of 120 °C at 2 torr. To determine the boiling point at 20 torr, connect 120 °C (column **a**) to 2 torr (column **c**) with a transparent plastic rule and observe where this line intersects column **b** (about 295 °c). This value corresponds to the normal boiling point. Next, connect 295 °C (column **b**) with 20 torr (column **c**) and observe where this intersects column **a** (160 °C). The approximate boiling point will thus be 160 °C at 20 torr.*

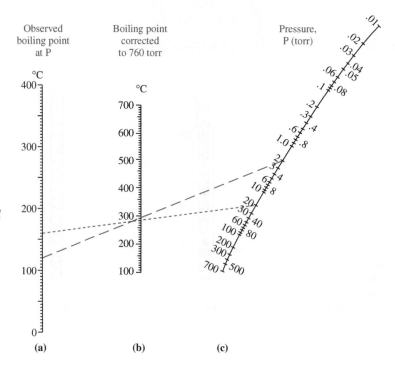

Observed boiling point at P

Boiling point corrected to 760 torr

Pressure, P (torr)

(a)   (b)   (c)

Some laboratories are equipped with "house" vacuum lines that are connected to a large central vacuum pump, but such vacuum systems seldom reduce the pressure to less than 50 torr. The pressure is measured by a device called a **manometer**.

The apparatus for vacuum distillations is a modification of that used for simple distillations (Fig. 2.37), and two options are shown in Figure 2.42. The setup depicted in Figure 2.42a is typically used for volumes of liquid in excess of about 10 mL. If a condenser is not required, the modified apparatus of Figure 2.42b can be

(a)

Figure 2.42

(a) Miniscale apparatus for vacuum distillation. (b) Modified apparatus for miniscale shortpath vacuum distillation.

(b)

used for distilling volumes of liquid in the range of 1–10 mL; the shorter pathlength of this apparatus minimizes loss of material.

The primary difference between the apparatus for simple and vacuum distillations is the attachment of the vacuum adapter to a vacuum source. Rather than being directly connected to the vacuum source, the adapter is normally linked to a heavy-walled sidearm flask. This flask has several functions. For example, if a water aspirator is being used to produce a vacuum, the flask serves as a **trap** to prevent water from backing up into the apparatus, which may occur if there is a sudden decrease in the water pressure in the aspirator. If it is equipped with a stopcock release valve, the flask also provides a means for connecting the apparatus and the manometer to the source of the vacuum and releasing the vacuum. Further discussion of traps is provided under "**Vacuum Filtration**" in Section 2.17.

The apparatus shown in Figure 2.42 is assembled (Sec. 2.4) according to the same general guidelines that are used for setting up the apparatus for simple distillation (Sec. 2.13). *Read those instructions carefully before proceeding with assembling the apparatus in Figure 2.42.* Because magnetic stirring (Sec. 2.11) is the most convenient means of obtaining an even rate of vaporization during the distillation, that is the method described here. However, if this option is not available, consult your instructor regarding alternative setups for vacuum distillation.

Start assembling the apparatus by tilting the stillpot to one side and sliding a stirbar gently down the side. Then add the liquid to the stillpot, making certain that the flask is no more than *half*-full. After lubricating all the ground-glass joints (Sec. 2.2), attach the stillhead and equip it with a thermometer adapter and thermometer. Continue the setup as described in Section 2.13, clamping the condenser, if one is used, and the receiver. Make sure all of the joints are tightly mated. Finally, connect the apparatus to the sidearm flask, which in turn must be connected to the vacuum source and the manometer.

The volume and density of the vapor formed by volatilization of a given amount of liquid are pressure-dependent. For example, the volume of vapor formed from vaporization of a drop of liquid will be about 20 times as great at 38 torr as it would be at 760 torr. During a vacuum distillation, large quantities of vapor are produced in the distillation flask, and these vapors enter the condenser with high velocity prior to condensing. Since the density of the vapor is much lower at reduced pressure than at atmospheric pressure, it may be difficult to control the rate of vaporization and thereby minimize the difference in pressure between the distillation flask and the manometer where the pressure must be accurately measured.

The manometer measures the pressure at which the distillation is being conducted; this value is important and is reported with the boiling point. For example, benzaldehyde boils at 180 °C at atmospheric pressure and at 87 °C at 35 torr, and these two boiling points are reported in the format bp 180 °C (760 torr) and 87 °C (35 torr). To obtain an accurate measure of the pressure, however, you must carefully *control the rate of distillation.* The problem arises because a drop of condensate forms from a larger volume of vapor as the pressure is lowered, thereby causing higher vapor velocities to enter the condenser. This creates a backpressure so that the pressure in the distillation flask and stillhead is higher than that measured on the manometer, which is located beyond the condenser and receiver and is therefore less sensitive to backpressure. The difference between the actual and measured pressure can be minimized by distilling the liquid at a slow but steady rate. Superheating the vapor can be avoided by maintaining the oil bath at a temperature no more than about 25–35 °C higher than the head temperature.

The following paragraphs provide a general procedure for performing a vacuum distillation using the apparatus shown in Figure 2.42a. *Read these instructions carefully prior to executing a vacuum distillation.*

1. *Never use glassware with cracks, thin walls, or flat bottoms, such as Erlenmeyer flasks, in vacuum distillations.* Hundreds of pounds of pressure may be exerted on the *total* exterior surfaces of systems under reduced pressure, even if only a water aspirator pump is used. Weak or cracked glassware may *implode,* and the air rushing into the apparatus will shatter the glassware violently in a manner little different from that of an explosion. *Examine the glassware carefully.*

2. Lubricate and seal *all* glass joints carefully during assembly of the apparatus; this will help avoid air leaks and provide lower pressure. The rubber fittings holding the thermometer in place must be tight. If necessary, the neoprene fittings normally used with the thermometer adapters may be replaced with short pieces of heavy-walled tubing. Do not use rubber stoppers elsewhere in the apparatus, since direct contact between the rubber and the hot vapors during distillation may extract impurities from the rubber and contaminate your product. The three-holed rubber stopper in the safety trap should fit snugly to the flask and the pieces of glass tubing. The safety trap must be made of heavy-walled glass and wrapped with electrical tape as protection in case it implodes. Heavy-walled vacuum tubing must be used for all vacuum connections. *Check the completely assembled apparatus to make sure that all joints and connections are tight.*

3. Now begin stirring the liquid and turn on the vacuum; if you are using a water aspirator pump, you should *fully open the water faucet.* The release valve on the safety flask should be *open. Slowly* close the release valve, but be prepared to reopen it if necessary. The liquid often contains small quantities of low-boiling solvents, so foaming and bumping are likely to occur. If this occurs, adjust the release valve until the foaming abates. This may have to be done several times until the solvent has been removed completely. When the surface of the liquid in the flask is relatively quiet, fully evacuate the system to the desired pressure. The release valve may have to be opened slightly until the desired pressure is obtained; a needle valve from the base of a Bunsen burner may be used for fine control of the pressure. Check the manometer for constancy of pressure. Note and record the pressure at which the distillation is being performed. This pressure should be monitored throughout the course of the distillation, even when fractions are not being collected. Begin heating the flask, and maintain the temperature of the bath so that the distillate is produced at the rate of 3–4 drops per minute. *Do not begin heating the flask until the system is fully evacuated and the vacuum is stable.* The best way to heat the stillpot is with an oil bath, since it is easier to control the temperature in the bath and distillation flask; however, other techniques (Sec. 2.9) may be suggested by your instructor.

4. If it is necessary to use multiple receivers to collect fractions of different boiling ranges, the distillation must be interrupted to change flasks. Remove the heating source with caution and allow the stillpot to cool somewhat. *Slowly* open the vacuum release valve to readmit air to the system. When atmospheric pressure is reached, change receivers, close the release valve, re-evacuate the system to the same pressure as before, reapply heat, and continue distilling. The operation may result in a different pressure in the fully evacuated system.

Figure 2.43
Multiflask adapter for miniscale vacuum distillation.

Periodically monitor and record the head temperature and the pressure, particularly just before and after changing receivers.

5. After the distillation is complete, discontinue heating, allow the pot to cool somewhat, slowly release the vacuum, and turn off the source of the vacuum.

One of the most inconvenient aspects of a vacuum distillation is the disruption of the distillation to change receivers. To eliminate this problem, you may wish to use a "cow" receiver that accommodates three or four round-bottom flasks (Fig. 2.43). These flasks are successively used as receivers by rotating them into the receiving position.

2.16 STEAM DISTILLATION

Steam distillation is a mild method for separating and purifying volatile liquid or solid organic compounds that are immiscible or insoluble in water. This technique is not applicable to substances that react with water, decompose on prolonged contact with steam or hot water, or have a vapor pressure of less than about 5 torr at 100 °C. The practical features and the apparatus of this technique are described in this section; its theory and applications are discussed in detail in Sections 4.5 and 4.6.

The two basic techniques commonly used for a steam distillation in the laboratory are differentiated on the basis of whether the steam is introduced from an external source or generated internally. For larger-scale reactions, the most common and most efficient method for conducting a steam distillation involves placing the organic compound(s) to be distilled in a round-bottom flask equipped with a Claisen adapter, a stillhead, and a water-cooled condenser, as depicted in Figure 2.44. The combination of introducing steam from an external source into the distillation flask via the inlet tube and the turbulence associated with the boiling action tends to cause occasional violent splashing, and the Claisen adapter is necessary to prevent the mixture from splattering into the condenser. During the course of the distillation, water may condense in the distillation flask and fill it to undesirable levels. This problem can be avoided by gently heating the stillpot with an appropriate heat source.

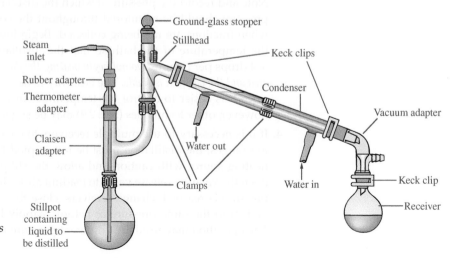

Figure 2.44
Apparatus for miniscale steam distillation. The steam tube is replaced with a stopper if steam is generated by direct heating.

Figure 2.45
Steam generator.

Safety tube
(35–50 cm long,
i.d. ≥ 8 mm)

Steam to
apparatus

Cork or
rubber
stopper

Round-
bottom
flask

Clamp

Water

Boiling
stones

Wire
gauze

Ring
support

Bunsen
burner

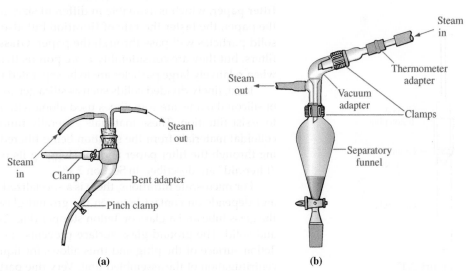

Figure 2.46
Water traps for use in steam distillations.

Steam
out

Steam
in

Clamp

Bent adapter

Pinch clamp

(a)

Steam
in

Thermometer
adapter

Steam
out

Vacuum
adapter

Clamps

Separatory
funnel

(b)

Steam may also be produced externally in a generator, as shown in Figure 2.45. The round-bottom flask is initially half-filled with water, and boiling stones are added *before* heating. The safety tube relieves internal pressure if steam is generated at too high a rate or the outlet tube to the apparatus becomes clogged. Steam is most conveniently obtained from a laboratory steam line. A trap (Fig. 2.46) is then placed between the steam line and the distillation flask to permit removal of any water and/or impurities present in the steam.

If only a small amount of steam is necessary to separate a mixture completely, a simplified method involving *internal steam generation* may be employed. Water is added directly to the distillation flask together with the organic compounds to be separated. The flask is equipped for steam distillation by setting up the apparatus as shown in Figure 2.44, *except* that the steam inlet tube is replaced with a stopper. The flask is then heated with a heating mantle or a Bunsen burner, and the distillate is collected. This technique is generally not applicable for distillations that require large amounts of steam, since it would be necessary to use an inappropriately large flask or to replenish the water in the flask frequently by using an addition funnel. However, this procedure is satisfactory and convenient for most miniscale reactions.

2.17 FILTRATION APPARATUS AND TECHNIQUES

Filtration is the primary technique used to separate solids from liquids. It is important to perform filtrations properly to avoid loss or contamination of your product, regardless of whether it is a solid or liquid.

Conceptually, the process of filtration is simple and involves passing a liquid containing a solid material through a barrier that is permeable only to the liquid. The liquid is collected in an appropriate receiver, and the solid remains on the barrier. However, there are many nuances to how filtrations are performed, and some of them are discussed below.

Filtration Media

— Glass or Teflon plug

— Ground-glass shoulder

— Glass tube

Figure 2.47
Craig tube assembly.

Various types of materials are used in conjunction with funnels to produce the barrier that is the key to separating solids from liquids. The most common of these is **filter paper**, which is available in different sizes and porosities. The more porous the paper, the faster the rate of filtration but also the greater the possibility that solid particles will pass through the paper. **Glasswool** and **cotton** may serve as filters, but they are considerably more porous than paper and thus are used only when relatively large particles are to be separated from a liquid.

Inert, finely divided solids such as silica gel and Celite, both of which are forms of silicon dioxide, are sometimes used along with filter paper, glasswool, or cotton to assist filtration. These materials generally function by *adsorption* of solids and colloidal materials from the solution being filtered, thereby preventing their passing through the filter paper and/or clogging its pores. Applications of Celite as a "filter-aid" are described in Section 2.18.

For microscale filtrations, there is a specialized technique that uses a Craig tube and depends on contact between the ground-glass strip on the inner shoulder of the glass tube and a glass or Teflon stopper (Fig. 2.47) to effect separation of liquid and solid. The ground-glass surface prevents complete contact with the glass or Teflon surface of the plug and thus allows for liquid to flow out of the tube upon centrifugation of the assembled unit. Very fine particles *are not* effectively removed by this technique.

Gravity Filtration

— Filter paper
— Ring support
— Stemmed funnel

— Erlenmeyer flask

Figure 2.48
Apparatus for miniscale gravity filtration.

Gravity filtration is the filtering technique most commonly used to remove solids such as impurities, decolorizing carbon (Sec. 2.18), or drying agents (Sec. 2.24) from liquids prior to crystallization, evaporation, or distillation. In addition to filter paper of the appropriate size, only a stemmed funnel, ring clamp, and a receiver (Fig. 2.48) are needed for this operation. The filter paper must be folded to fit into the funnel, and there are two ways of doing this. The simplest involves folding the circular piece of paper in half, then in half again so that corners 1 and 2 of Figure 2.49a are coincident, inserting the paper in the funnel, and opening the folded paper so that the solution can be poured through it. Before starting the filtration, it is helpful to first wet the filter paper with a small amount of the solvent being filtered. This serves to form a seal between the paper and the funnel and to keep the paper from collapsing on itself and closing the opening you have made in it.

The rate and ease of filtration are increased if a piece of **fluted filter paper** (Fig. 2.49g) is used, since fluting increases the surface area of paper contacted by

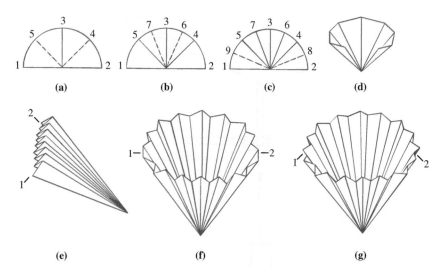

Figure 2.49
Folding filter paper.

(a) (b) (c) (d)

(e) (f) (g)

the liquid being filtered. Although such paper is commercially available, it may also be easily prepared from a flat piece of round filter paper. One method of fluting is shown in Figure 2.49 and involves the following sequence. Fold the paper in half, and then into quarters. Fold edge 2 onto 3 to form edge 4, and then 1 onto 3 to form 5 (Fig. 2.49a). Now fold edge 2 onto 5 to form 6, and 1 onto 4 to form 7 (Fig. 2.49b). Continue by folding edge 2 onto 4 to form 8, and 1 onto 5 to form 9 (Fig. 2.49c). The paper now appears as shown in Figure 2.49d. Do not crease the folds tightly at the center, because this might weaken the paper and cause it to tear during filtration. All folds thus far have been in the same direction. Now make folds in the *opposite* direction between edges 1 and 9, 9 and 5, 5 and 7, and so on, to produce the fanlike appearance shown in Figure 2.49e. Open the paper (Fig. 2.49f) and fold each of the sections 1 and 2 in half with reverse folds to form paper that is ready to use (Fig. 2.49g). Now that you have learned to fold filter paper, origami will seem simple!

To perform a filtration, the folded or fluted filter paper is first placed in a funnel and *wetted* with the appropriate solvent. The solution containing the solid is then poured onto the filter paper or fluted filter paper with the aid of a glass stirring rod to minimize dripping down the outside of the flask. As the solution passes through the paper, the solid remains on the filter paper. A typical experimental setup is shown in Figure 2.50. If you do not use a ring stand to support the funnel, you should insert a paper clip or a bent piece of wire between the funnel and the lip of the flask to avoid forming a solvent seal between the flask and the funnel. Such a seal will interfere with the filtration, since there will be no vent for the air displaced by the filtrate.

When volumes less than about 5 mL must be filtered, there is a danger of losing material on the filter paper, because the paper absorbs a significant volume of liquid. A Pasteur filtering pipet (Sec. 2.5) should be used in such cases. The liquid mixture to be filtered is transferred by pipet into the filtering pipet as shown in Figure 2.51. The plug should be rinsed with about 0.5 mL of solvent to maximize the recovery of the solute.

Glass rod

Ring support

Short-stemmed funnel

Erlenmeyer flask

Figure 2.50
Executing a miniscale gravity filtration.

Hot Filtration

Sometimes it is necessary to remove solid impurities from an organic compound that is only sparingly soluble in the solvent at room temperature. In such cases, the mixture must be heated, and **hot filtration** must be performed to remove the insoluble solids without leaving the desired material in the funnel, too. Such a situation commonly arises during purification of an organic solid by recrystallization

Figure 2.51
Gravity filtration with a Pasteur filtering pipet.

Figure 2.52
Miniscale hot filtration: (a) using steam cone for heating; (b) using hot plate for heating.

when decolorizing carbon (Sec. 2.18) or other insoluble residues must be removed; a filter-aid may be required to facilitate removal of finely divided solids.

Miniscale Apparatus

The apparatus for performing a miniscale hot filtration is shown in Figure 2.52. A small volume of the *pure* solvent is first placed in an Erlenmeyer flask containing a boiling stick or several boiling stones, and a short-stemmed or stemless filter funnel containing a piece of fluted filter paper is placed in the mouth of the flask.

A small stick or piece of thick paper is inserted between the flask and the funnel to allow the vapors of solvent to rise around the outside of the funnel. The flask is heated with either a steam bath or a hot plate until the solvent boils and the vapors rise up the sides of the flask and around the funnel; the solvent vapors heat the filter funnel to minimize problems of crystallization of the desired material in the funnel. A steam bath is preferred for flammable solvents that boil below 90 °C, whereas a hot plate can be used for all nonflammable solvents and, with proper care and in a hood, for flammable solvents boiling above 90 °C.

When the flask and funnel are hot, the hot solution is poured onto the filter paper. Following this, several milliliters of hot solvent are added to the flask that contained the original recrystallization solute and solvent. The additional hot solvent is poured onto the filter paper to ensure the complete transfer of material to the flask containing the filtered solution.

Microscale Apparatus

A microscale hot filtration is performed using a combination of a Pasteur filter-tip pipet (Fig. 2.9a) and a filtering pipet (Fig. 2.9b). The goal is to filter the hot solution containing solid contaminants as rapidly as possible so that none of the solute crystallizes before the filtrate reaches the receiver. This requires preheating both pipets by pulling hot solvent up into the barrel several times and then *quickly* transferring the hot solution into the top of the filtering pipet, as shown in Figure 2.53. As with miniscale gravity filtration, the filter should be rinsed with the solvent to maximize the recovery of the solute, since some of it may have crystallized on the cotton plug. However, no more than 0.2–0.5 mL of additional solvent should be used.

Vacuum Filtration

Vacuum filtration is a technique for collecting crystalline solids from solvents after recrystallization or precipitation. A typical apparatus is shown in Figure 2.54. Either a **Büchner funnel** or a smaller **Hirsch funnel** is used, with the latter being better suited for isolating quantities of solid ranging from 100 mg to 500 mg. The funnel is fitted to a vacuum filter flask using a neoprene adapter or a rubber stopper. The sidearm of the filter flask is connected to an aspirator or house vacuum line through a **trap** using heavy-walled rubber tubing. The trap prevents water from the aspirator from backing up into the filter flask if there is a loss of water pressure. The trap should be a *heavy-walled* bottle or a second vacuum filter flask

Figure 2.53
Microscale hot filtration.

Labels: Pasteur filter-tip pipet, Pasteur filtering pipet, Clamp, Cotton or glasswool plug, Beaker, Conical vial

Figure 2.54
Apparatus for miniscale vacuum filtration; inset shows Hirsch funnel, used for small quantities of solids.

Labels: Büchner funnel, Neoprene adapter, Clamp, Hirsch funnel, Heavy-walled filter flask, Heavy-walled tubing, Pinch clamp to release vacuum, Heavy-walled tubing, To aspirator, Filter trap (may be a second filter flask)

wrapped with electrical or duct tape. If a filter flask is used for the trap, it should be equipped with a two-holed stopper with its sidearm attached to the first filter flask with heavy-walled tubing. The glass tube that extends to within approximately 2–3 cm of the bottom of the filter trap is connected with heavy-walled tubing to the water aspirator pump so that any water that collects slowly in the trap is evacuated back through the aspirator pump and into the drain. If a sudden backup occurs, the vacuum should be released immediately so that the trap does not completely fill up with water. Any filtrate that may overflow the filter flask will also be collected in the trap. *The filter flask should be cleaned before doing the filtration,* since it may be necessary to save the filtrate.

A piece of filter paper is placed in the funnel so it lies flat on the funnel plate and covers all the small holes in the funnel. It should *not* extend up the sides of the funnel. A vacuum is applied to the system, and the filter paper is "wetted" with a small amount of pure solvent to form a seal with the funnel so that crystals do not pass around the edges of the filter paper and through the holes in the filter. The flask containing the crystals is swirled to suspend the crystals in the solvent, and the solution containing the crystals is poured slowly into the funnel. A stirring rod or spatula may be used to aid the transfer. The last of the crystals may be transferred to the funnel by washing them from the flask with some of the filtrate, which is called the **mother liquor**. When all the solution has passed through the filter, the vacuum is released slowly by opening the screw clamp or stopcock on the trap. The crystals are washed to remove the mother liquor, which contains impurities, by adding a small amount of *cold, pure solvent* to the funnel to just cover the crystals. Do not allow the filter paper to float off the surface of the funnel, as some of the crystals may be lost through the holes in the funnel. Vacuum is then reapplied to remove the wash solvent, and the crystals are pressed as dry as possible with a clean spatula or a cork while continuing to pull a vacuum on the funnel.

Most of the solvent may be evaporated from the crystals by allowing the vacuum to pull air through the crystals on the funnel for a few minutes. A clean spatula is used to scrape the crystals gently from the filter paper and then to transfer them to a clean watchglass. Be careful and do not scrape the filter paper too vigorously, because torn bits of filter paper or paper fuzz might contaminate your product. The crystals may be dried completely by allowing them to air-dry for a few hours or by leaving them loosely spread on a filter paper in a locker until the next class. The drying process may be accelerated by placing the crystals in an oven, but the temperature of the oven should be at least 20–30 °C below the melting point of the crystals. A vacuum desiccator can also be used to hasten the drying process. The vacuum desiccator should either be wrapped with electrical or duct tape or placed in a metal cage as protection should the desiccator implode. Specially designed desiccators permit heating samples under vacuum, but the temperature must again be kept about 20–30 °C below the melting point of the solid. Heat or vacuum desiccators should not be used to dry crystals of compounds that sublime readily (Sec. 2.20).

Vacuum filtration can also be used to remove undesired solids from a solution. If the solid is finely divided or colloidal, a pad of a filter-aid such as Celite is used to ensure complete removal of the solid. A pad of filter-aid can be formed on the filter funnel by first making a slurry of 0.5–1 g of filter-aid in a few milliliters of the solvent being used. The slurry is poured onto a filter paper in a Büchner or Hirsch funnel that is attached to a *clean* filter flask; vacuum is then slowly applied to draw the solvent through the filter paper, leaving a thin, even pad of the filter-aid; the pad should be about 2–3 mm thick. The solution containing the solid is then filtered as described above. Obviously this technique is useful for removing solid impurities *only* when the *solution* contains the substance of interest; a solid product is not collected in this manner, as it would be necessary to separate the desired material from the filter-aid. This would be a challenge even for Louis Pasteur (see the Historical Highlight *Discovery of Sterioisomers*, which is available online).

Craig Tube Filtration

Craig tube filtration is the microscale analog of vacuum filtration for isolating solid products. However, rather than using a vacuum to pull solvent through a filter, this technique depends on centrifugal forces to force the solvent through the tiny openings between a ground-glass surface and a plug seated against this surface.

The Craig tube itself consists of two components: a test tube-like glass vessel, the inner shoulder of which is ground-glass, and a plug made from glass or Teflon (Fig. 2.55a). In addition to the Craig tube, a centrifuge, centrifuge tubes, and a piece of wire are required.

The technique involves transferring a solution of the desired solid into the bottom portion of the Craig tube. After crystallization is complete, the top part of the Craig tube is attached, and a wire holder is affixed to the unit (Fig. 2.55b). The assembled unit is placed in an inverted centrifuge tube so that the top of the Teflon plug is at the bottom of the tube; the wire is looped over the lip of the tube (Fig. 2.55c). The entire assembly is then inverted carefully so that the two components of the Craig tube stay in close contact, thereby preventing solid from escaping past the plug into the centrifuge tube. After centrifugation, the solvent will be in the bottom of the centrifuge tube, whereas the desired solid remains in the Craig tube (Fig. 2.55d).

Figure 2.55
Filtration using the Craig tube.

    (a)   (b)   (c)   (d)

2.18 DECOLORIZING CARBON

A colored solution may result when a solid is dissolved in a suitable solvent for recrystallization. The color can arise from the compound itself or from the presence of colored impurities in the solution. Similarly, an organic liquid may be colored because of impurities or because of the nature of the compound itself. If you are in doubt about the cause of the color, you may try to remove it by adding activated **decolorizing carbon** to the solution or liquid and then removing the carbon by filtration; sometimes heating is required. Decolorizing carbon, which is commercially available under trade names such as Norite or Darco, is finely divided, activated charcoal with an extremely large surface area. Polar, colored impurities bind preferentially to the surface of decolorizing carbon; however, you should be aware that some of the desired compound may also be adsorbed, so excessive quantities of decolorizing carbon should *not* be used. After the decolorizing carbon is removed by filtration, the liquid or solution should be free of colored impurities; any remaining color is probably due to the compound itself.

Miniscale Technique

In a typical miniscale procedure, the solution, contained in an Erlenmeyer flask, is first brought to a temperature 10–20 °C *below* its boiling point. Decolorizing carbon is then added, and the hot solution is gently swirled. There are no firm rules concerning the amount of decolorizing carbon that should be used, but a good first approximation is to add about 0.5–1 g per 100 mL of solution. The decolorizing carbon is then removed by gravity filtration or hot gravity filtration, not vacuum filtration, using fluted filter paper; a filter-aid (Sec. 2.17) will facilitate complete removal of all of the carbon. Hot gravity filtration must be used whenever an organic solid is dissolved in a hot solvent, such as during a recrystallization. The solution or liquid must be refiltered if it contains any small black specks of carbon or any other solid material. Normally this extra step can be avoided if care is exercised when first filtering the mixture and a filter-aid is used. For example, the filter paper should not contain any tears or holes, and the mixture should be carefully poured onto the filter paper so that none of it runs down the side of the funnel. If the compound is known to be colorless and if the decoloration process does not remove the color, the procedure should be repeated.

Microscale Technique

The microscale procedure for decolorizing solutions is analogous to that used at the miniscale level. A small amount of powdered decolorizing carbon (10–20 mg) is added to the solution, which is contained in a conical vial or a test tube and is at a temperature 10–20 °C *below* its boiling point. After the mixture is gently swirled or stirred for several minutes, the decolorizing carbon is removed by hot filtration (Sec. 2.17). The particles of carbon are very fine, however, and some of them may pass through the cotton plug of the Pasteur filtering pipet, in which case the filtrations should be repeated, preferably with a clean filtering pipet. You may also make a small bed of a filter-aid such as Celite on top of the cotton plug to facilitate removal of the decolorizing carbon.

Pelletized carbon may be available in your laboratory, and you may be instructed to use it instead of the powdered form. Because of its larger size (pelletized carbon is about 10 mesh, or 2 mm in diameter, whereas powdered carbon is typically 100 mesh, or 0.14 mm in diameter), it is easier to separate from the solution. However, pelletized carbon is less efficient at removing colored contaminants because of its lower surface area.

2.19 DECANTING SOLUTIONS

Solid particles

Figure 2.56
Decantation.

Small amounts of dense solids are sometimes removed by simple **decantation.** This technique is a viable alternative to filtration for removing solid drying agents (Sec. 2.24), but it cannot be used to remove finely dispersed solids such as decolorizing carbon. To decant a liquid from a solid, the solid should first be allowed to settle to the bottom of the container. The container is then *carefully* tilted, and the liquid is *slowly* poured into a clean container, possibly with the aid of a glass stirring rod to minimize dripping down the outside of the flask (Fig. 2.56); the solid should remain in the original container. Decantation is preferable to gravity filtration when working with very volatile organic liquids, since filtration is likely to result in considerable evaporation and loss of material. When decanting from an Erlenmeyer flask, a *loosely packed* ball of glasswool can be put in the neck of the flask to help keep the solid in the flask.

One major disadvantage of decantation is that some liquid will remain in the flask; this problem may be minimized by using a Pasteur pipet to transfer the last few milliliters and rinsing the residue with several milliliters of pure solvent that are also transferred with a pipet.

2.20 SUBLIMATION

Like the vapor pressure of a liquid, the vapor pressure of a solid increases with temperature. If the vapor pressure of a solid is greater than the ambient pressure at its melting point, then the solid undergoes a direct-phase transition to the gas phase without first passing through the liquid state. This process is called **sublimation.**

The ease with which a molecule may escape from the solid to the vapor phase is determined by the strength of the intermolecular attractive forces between the molecules of the solid. Symmetrical structures have relatively uniform distributions of electron density, and they have small dipole moments compared to less symmetrical molecules. Since electrostatic interactions are the strongest intermolecular forces in the crystal lattice, molecules with smaller dipole moments will have higher vapor pressures. Sublimation thus is generally a property of relatively nonpolar substances having fairly symmetrical structures. Van der Waals forces, which are weaker than electrostatic ones, are also important and increase in magnitude with increasing molecular weight; hence, large molecules, even if symmetrical, tend not to sublime.

To be purified by sublimation, a compound must have a relatively high vapor pressure, and the impurities must have vapor pressures significantly lower than that of the compound being purified. If the impurities in a compound have similar vapor pressures, recrystallization (Sec. 3.2) or column chromatography (Sec. 6.3) may be used to purify the compound. Since few organic solids exhibit vapor pressures high enough to sublime at atmospheric pressure, most sublimations are performed at reduced pressure.

Two common types of sublimation apparatus are shown in Figure 2.57. In each case there is a chamber that may be evacuated using a water aspirator, house vacuum, or vacuum pump and a cold-finger in the center of the vacuum chamber to provide a surface upon which the sublimed crystals may form. The cold-finger is cooled by water (Fig. 2.57a) or another cooling medium such as ice/water or dry ice/acetone (Fig. 2.57b). A simple sublimation apparatus may be assembled

Figure 2.57
Miniscale sublimation apparatus:
(a) using water as the coolant;
(b) using low-temperature
coolants.

Figure 2.58
Simple sublimation apparatus:
(a) test tube sublimator;
(b) filter-flask sublimator.

inexpensively from two test tubes, one having a sidearm and rubber stoppers (Fig. 2.58a). In place of the test tube with a sidearm, a small vacuum-filtration flask may also be used (Fig. 2.58b). The cold-finger may be cooled either with running water or with ice chips. If the cold-finger is cooled with running water, the water hoses must be securely attached to the inlet and outlet of the condenser by means of a piece of twisted copper wire or a hose clamp.

Commercial sublimation adapters such as those shown in Figure 2.59 are available for microscale work. A Winston adapter is normally used with a sublimation chamber such as a 3- or 5-mL conical vial (Fig. 2.59a), whereas adjustable adapters may be used with 5- or 10-mL round-bottom flasks (Fig. 2.59b). In both cases a cold-finger that may be connected to a vacuum source is fitted to the chamber containing the substance to be sublimed, and the connection is secured with a standard-taper joint that is often accompanied by an O-ring and threaded collar (Fig. 2.59). The cold-finger can be filled with ice or dry ice. The advantage of using these adapters

Figure 2.59
Commercial microscale sublimation apparatus: (a) conical vial with Winston adapter; (b) round-bottom flask with adjustable adapter.

is that air- and water-tight seals are easier to obtain because of the standard-taper connections, and the one-piece construction makes it easier to remove the crystals from the cold-finger. However, they suffer the disadvantage of being more expensive than the simple apparatus previously described.

To purify an impure substance by sublimation, it is first placed at the bottom of the sublimation chamber. The sample is then heated under reduced pressure, using an oil or sand bath or a small flame, to a temperature *below the melting point* of the solid. The solid will be vaporized and transferred via the vapor phase to the surface of the cold-finger, where it condenses to form a pure solid. After the sublimation is complete, the pressure must be released carefully to avoid dislodging the crystals from the cold-finger with a surge of air. For similar reasons, care must be exercised when removing the cold-finger from the sublimation apparatus. The pure crystals are scraped from the cold-finger with a spatula.

2.21 EXTRACTION

A common technique for isolating and purifying the product of a chemical reaction involves liquid-liquid extraction, or simply **extraction**; the theory and applications are discussed in Chapter 5. This process involves transferring a solute from one solvent into another, because of its greater solubility in the second. The two solvents must be immiscible and form two distinct layers, and in general one layer is aqueous and the other is an organic solvent such as diethyl ether, hexane, or dichloromethane. Depending upon the amounts of material, the physical separation of the two immiscible phases will be performed in **separatory funnels** or **conical vials**.

Extractions Using Separatory Funnels: Miniscale Technique

Separatory funnels are available in many different shapes, ranging from pear-shaped to nearly conical (Fig. 2.60); the more elongated the funnel, the longer the time required for the two liquid phases to separate after the funnel is shaken. Although separatory funnels may be equipped with either a glass or Teflon stopcock, the latter does not require lubrication and thus is preferred because the solutions being separated do not get contaminated with stopcock grease. Separatory funnels are most commonly used during work-up procedures after completion of a chemical reaction. For example, they are used for extracting the desired product

Figure 2.60
Separatory funnels: (a) conical; (b) pear-shaped.

(a)

(b)

Point away from
yourself and others

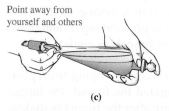

(c)

Figure 2.61
(a) Proper method for holding separatory funnel. (b) Shaking a separatory funnel. (c) Venting a separatory funnel.

from one immiscible liquid phase into another (Secs. 5.2 and 5.3) and for "washing" organic layers to remove undesired substances such as acids or bases from the desired organic compound.

There are a number of general guidelines for using separatory funnels that merit discussion:

1. **Filling Separatory Funnels.** The stopcock should be closed and a clean beaker placed under the funnel before any liquids are added to the funnel in case the stopcock leaks or is not completely closed. A separatory funnel should never be more than three-quarters full. The upper opening of the funnel is then stoppered with a ground-glass, plastic, or rubber stopper; most separatory funnels are now fitted with a ground-glass or plastic stopper.

2. **Holding and Using Separatory Funnels.** If the contents of the funnel are to be shaken, it is held in a specific manner. If the user is right-handed, the stopper should be placed against the base of the index finger of the left hand and the funnel grasped with the first two fingers and the thumb. The thumb and the first two fingers of the right hand can then be curled around the stopcock (Fig. 2.61a). Holding the funnel in this manner permits the stopper and the stopcock to be held tightly in place during shaking. A left-handed person might find it easier to use the opposite hand for each position. If you are lucky enough to be ambidextrous, take your choice!

3. **Shaking Separatory Funnels.** A separatory funnel and its contents should be shaken to mix the immiscible liquids as intimately as possible (Fig. 2.61b). The shaking process increases the surface area of contact between the immiscible liquids so that the equilibrium distribution of the solute between the two layers will be attained quickly; however, overly vigorous or lengthy shaking may produce **emulsions** (discussed below).

The funnel must be vented every few seconds to avoid the buildup of pressure within the funnel. Venting is accomplished by *inverting the funnel with the stopcock pointing upward and away from you and your neighbors* and slowly opening it to release any pressure (Fig. 2.61c). If the funnel is not carefully vented, liquid may be violently expelled, covering you and your laboratory partners with the contents. Venting is particularly important when using volatile, low-boiling solvents such as diethyl ether or methylene chloride; it is also necessary whenever an acid is neutralized with either sodium carbonate or sodium bicarbonate, because CO_2 is produced. If the funnel is not vented frequently, the stopper may be blown out accidentally; under extreme circumstances the funnel might blow up. Any sudden release of pressure is likely to result in the contents being lost and spattered on you and your coworkers—not a good way to make friends. The funnel may be vented simply by holding it as described previously and opening the stopcock by twisting the fingers curled around it without readjusting your grip on the funnel.

At the end of the period of shaking (1–2 min are usually sufficient if the shaking is vigorous), the funnel is vented a final time. It is then supported on an iron ring (Fig. 2.62), and the layers are allowed to separate. To prevent the iron ring from cracking or breaking the separatory funnel, the ring should be covered with a length of rubber tubing. This may be accomplished by slicing the tubing along its side and slipping it over the ring. Copper wire may be used to fix the tubing permanently in place. The lower layer in the separatory funnel is then carefully dispensed into a flask through the stopcock while the

Figure 2.62
Separatory funnel positioned on iron ring with beaker located to catch liquid if funnel leaks.

Ring support

Beaker

interface between the two layers is watched. If small quantities of insoluble material collect at the boundary between the layers and make it difficult to see this interface, it is best to remove these solids with the *undesired* liquid layer; a small amount of the desired layer is inevitably lost by this procedure. An alternative procedure is to remove the solids by gravity or vacuum filtration before separating the layers.

4. **Layer Identification.** It is important to ascertain which of the two layers in a separatory funnel is the aqueous layer and which is the organic. Because the layers should separate so that the denser solvent is on the bottom, knowledge of the densities of the liquids being separated provides an important clue as to the identity of each layer. This generalization is not foolproof, however, because a high concentration of a solute in one layer may reverse the relative densities of the two liquids. You must not confuse the identity of the two layers in the funnel and then discard the layer containing your product. *Both layers should always be saved until there is no doubt about the identity of each and the desired product has been isolated.*

Because one of the layers is usually aqueous and the other is organic, there is a simple and foolproof method to identify the two layers. Withdraw a few drops of the upper layer with a pipet and add these drops to about 0.5 mL of water in a test tube. If the upper layer is aqueous, these drops will be miscible with the water in the test tube and will dissolve, but if the upper layer is organic, the droplets will not dissolve.

5. **Emulsions.** Occasionally the two immiscible liquids will not separate cleanly into two distinct layers after shaking, because an **emulsion** may form that results from a colloidal mixture of the two layers. If prior experience leads you to believe that an emulsion might form, do not shake the funnel too vigorously; instead, swirl it gently to mix the layers. Encountering an emulsion during an experiment can be extraordinarily frustrating because there are no infallible, convenient procedures for breaking up emulsions. An emulsion left unattended for an extended period of time sometimes separates. However, it is usually more expedient to attempt one or more of the following remedies:

a. Add a few milliliters of a *saturated* solution of aqueous sodium chloride, commonly called brine, to the funnel and gently reshake the contents. This increases the ionic strength of the water layer, which helps force the organic material into the organic layer. This process can be repeated, but if it does not work the second time, other measures must be taken.

b. Filter the heterogeneous mixture by vacuum filtration through a thin pad of a filter-aid (Sec. 2.17), and return the filtrate to the separatory funnel. If a filter-aid is not used, the pores of the filter paper may become clogged, and the filtration will be slow. Sometimes emulsions are caused by small amounts of gummy organic materials whose removal will often remedy the problem.

c. Add a *small* quantity of water-soluble detergent to the mixture and reshake the mixture. This method is not as desirable as the first two techniques, particularly if the desired compound is in the water layer, because the detergent adds an impurity that must be removed later.

d. Intractable emulsions that appear to be stabilized by small trapped air bubbles are sometimes encountered during the work-up of phase-transfer reactions. If the separatory funnel is thick-walled, apply a gentle vacuum with a water aspirator to speed the separation of the phases.

e. If all these procedures fail, it may be necessary to select a different extraction solvent or seek the advice of your instructor.

Extractions Using Conical Vials and Centrifuge Tubes: Microscale Technique

Figure 2.63
Apparatus for microscale extraction: (a) conical vial; (b) screw-cap centrifuge tube.

The volumes of solvent are too small to use separatory funnels for extractions when reactions are conducted on a microscale level. In these instances, conical vials may be employed for volumes up to about 4 mL, and screw-cap centrifuge tubes may be used for volumes up to approximately 10 mL (Fig. 2.63). To avoid accidental spills, place the conical vial or centrifuge tube in a small beaker when you are not handling it.

Before using either a conical vial or a centrifuge tube for an extraction, ensure that it does not leak when shaken. This may be done easily by placing about 1 mL of water in the vial or tube and screwing the cap on securely. Shake the vial or tube vigorously and check for leaks.

Most extractions involve an aqueous and an immiscible organic phase, and it is important to mix the two layers thoroughly when performing an extraction. This may be achieved in several ways, the simplest of which is to shake the vial or tube vigorously. Carefully vent the vial or tube by loosening the cap to release any pressure. A vortex mixer works well if one is available in your laboratory. Adequate mixing can sometimes be achieved by stirring the mixture with a magnetic stirbar or a microspatula for 5–10 min. Another technique involves drawing the mixture into a Pasteur pipet and rapidly squirting it back into the vial or tube; be careful not to squirt so hard that the mixture spills out of the vial and onto the bench. Repeat this operation several times to ensure adequate extraction. Multiple extractions with fresh solvent should be performed to obtain maximum recovery of your product.

Depending upon the solvent used, the organic phase may be either the upper or lower layer, and a slightly different procedure is used for separating each. The more common situation is one in which the organic solvent is less dense than water, and it is this layer that is removed. For example, if you extract an aqueous solution with diethyl ether (ether), the organic layer is the upper one because ether is less dense than water. Alternatively, when a solvent that is heavier than water, such as dichloromethane, is used for the extraction, the lower layer is the organic layer, and it is the one removed.

Microscale Technique for Removing the Upper Organic Layer

The procedure for removing the upper organic layer using conical vials is outlined in Figure 2.64; centrifuge tubes may be substituted for the conical vials for larger volumes.

1. Place the aqueous phase containing the dissolved product in a 5-mL conical vial (Fig. 2.64a).

2. Add about 1 mL of diethyl ether, cap the vial securely, and mix the contents by vigorous shaking or by one of the other techniques outlined previously. Unscrew the cap slightly to vent the pressure in the vial. Allow the phases to separate completely so you can see the two layers. If you are using centrifuge tubes, centrifugation may facilitate this separation. The diethyl ether phase is the upper layer (Fig. 2.64b).

Figure 2.64

Extracting an aqueous solution using an organic solvent such as diethyl ether that is less dense than water: (a) the desired product is in the aqueous solution; (b) the aqueous phase is extracted with diethyl ether; (c) the lower (aqueous) phase is withdrawn; (d) the aqueous phase is transferred to a test tube; the ether layer remains in the conical vial; (e) the ether phase is transferred to a test tube, and the aqueous phase is returned to the extraction vial.

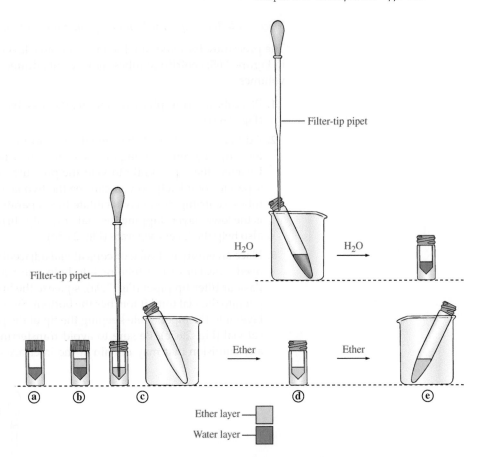

3. Place an empty test tube or conical vial adjacent to the extraction vial. You will need a beaker or test tube rack to hold a test tube. Attach a rubber bulb to a Pasteur filter-tip pipet (Fig. 2.9a), squeeze the bulb, and insert the tip of the pipet into the vial until it touches the bottom. Slowly withdraw the lower aqueous layer into the pipet while keeping the tip of the pipet on the bottom of the conical vial (Fig. 2.64c). Be sure to avoid transferring any of the organic layer or any emulsion that may remain at the interface of the two phases. Although a regular Pasteur pipet may also be used for this transfer, a filter-tip pipet is sometimes easier to control when removing small volumes of aqueous phase from the vial.

4. Transfer the aqueous phase you have drawn into the pipet into the empty test tube or conical vial, leaving the ether layer behind in the extraction vial (Fig. 2.64d). It is best to position the second container as close as possible to the extraction vial to avoid messy transfers and loss of your product.

5. Use a clean filter-tip pipet to transfer the ether layer from the extraction vial to a test tube for storage. Although a regular Pasteur pipet may also be used for this transfer, a filter-tip pipet is easier to control when removing small volumes of volatile organic solvents such as diethyl ether from the vial. Return the aqueous layer to the extraction vial using a Pasteur pipet (Fig. 2.64e).

6. Repeat steps 2–5 one or two times to achieve complete extraction of the desired material from the aqueous layer. Dry the organic layer with a suitable drying agent such as anhydrous sodium sulfate or magnesium sulfate (Sec. 2.25).

Microscale Technique for Removing the Lower Organic Layer

The procedure for removing the lower organic layer using conical vials is outlined in Figure 2.65; centrifuge tubes may be substituted for the conical vials for larger volumes.

1. Place the aqueous phase containing the dissolved product in a 5-mL conical vial (Fig. 2.65a).

2. Add about 1 mL of dichloromethane, cap the vial securely, and mix the contents by vigorous shaking or one of the other techniques outlined previously. Unscrew the cap slightly to vent the pressure in the vial. Allow the phases to separate completely so you can see the two layers. If you are using centrifuge tubes, centrifugation may facilitate this separation. The dichloromethane layer is the lower layer. Tapping the vial or gently stirring it with a microspatula may also help the layers separate (Fig. 2.65b).

3. Place an empty test tube or conical vial adjacent to the extraction vial. You will need a beaker or test tube rack to hold a test tube. Attach a rubber bulb to a Pasteur filter-tip pipet (Fig. 2.9a), squeeze the bulb, and insert the tip of the pipet into the vial until it touches the bottom. Slowly withdraw the lower organic layer into the pipet while keeping the tip of the pipet on the bottom of the conical vial (Fig. 2.65c). Be sure to avoid transferring any of the aqueous layer or any emulsion that may remain at the interface of the two phases. Although a

Figure 2.65

Extracting an aqueous solution using an organic solvent such as dichloromethane (CH_2Cl_2) that is denser than water: (a) the desired product is in the aqueous solution; (b) the aqueous phase is extracted with dichloromethane; (c) the lower (organic) phase is withdrawn; (d) the organic phase is transferred to a test tube; the aqueous phase remains in the conical vial.

Filter-tip pipet

Filter-tip pipet

CH_2Cl_2

Water

(a)   (b)   (c)   (d)

H_2O layer

CH_2Cl_2 layer

regular Pasteur pipet may also be used for this transfer, a filter-tip pipet is easier to control when removing small volumes of volatile organic solvents such as dichloromethane from the vial.

4. Transfer the organic phase you have drawn into the pipet into the *dry* empty test tube or *dry* conical vial, leaving the aqueous layer behind in the extraction vial (Fig. 2.65d). It is best to position the second container as close as possible to the extraction vial to avoid messy transfers and loss of your product.

5. Repeat steps 2–4 one or two times to achieve complete extraction of the desired material from the aqueous layer. Dry the organic layer with a suitable drying agent such as anhydrous sodium sulfate or magnesium sulfate (Sec. 2.25).

Layer Identification

It is important to ascertain which of the two layers in a conical vial is the aqueous layer and which is the organic. Because the layers will usually separate so that the denser solvent is on the bottom, knowledge of the densities of the liquids being separated provides an important clue as to the identity of each layer. This generalization is not foolproof, however, because a high concentration of a solute in one layer may reverse the relative densities of the two liquids. You must not confuse the identity of the two layers in the conical vial or test tube and then discard the layer containing your product. *Both layers should always be saved until there is no doubt about the identity of each and the desired product has been isolated.*

Since one of the layers is usually aqueous and the other is organic, there is a simple and foolproof method to identify the two layers: Add a few drops of water to the layer you believe to be the aqueous one. Watch closely as you add the water drops to see whether they dissolve and increase the volume of the layer. If the drops form a new layer on either the top or bottom of the layer you are testing, then the test layer is organic. Similarly, you can test to see if a layer is organic by adding a few drops of the organic solvent you are using, for example diethyl ether or dichloromethane. If the volume of the layer increases, it is organic, but if a new layer forms, the original layer is aqueous.

2.22 HEATING UNDER REFLUX

The term **heating under reflux** means that a reaction mixture is heated at its boiling point in a flask equipped with a **reflux condenser** to allow continuous return of the volatile materials to the flask; no solvent or reactant is removed or lost. Since the reaction can be conducted at higher temperatures using this technique, less time is required for its completion.

Miniscale Apparatus

A typical apparatus used for heating a miniscale reaction under reflux is shown in Figure 2.66a (see Sec. 2.4). The solvent, if any, and reactants are placed in a boiling flask, together with a stirbar; this flask should be set up about 15 cm or more from the bench top to allow for fast removal of the heating source if necessary. The flask is then fitted with a reflux condenser with water running slowly through it; the water hoses should be secured to the inlet and outlet of the condenser with copper wire or a hose clamp. *Do not stopper the top of the condenser, because a closed system should never be heated.* The heating source is then placed under the flask, and the

Figure 2.66
(a) Miniscale apparatus for heating a reaction mixture under reflux. (b) Miniscale apparatus for a reaction being heated under reflux and equipped for adding a liquid reagent or solution.

contents are slowly heated to the boiling point of the mixture; slow heating makes it possible to control any sudden exothermicity more readily. The volatile components of the mixture vaporize and reliquefy in the condenser, and the condensate returns to the boiling flask, where the process continues. A **reflux ring** appears in the condenser at the boundary of the two zones, and the condenser appears dry above this ring. The vapors should rise no more than about 3–6 cm above the bottom of the condenser. If they rise to the top of the condenser or are "seen" being emitted into the room, then you are probably overheating the flask. However, the flow of water through the condenser may also be insufficient, or the condenser may be too small to permit adequate cooling of the vapors. The problem may usually be corrected by lowering the temperature of the heat source or adjusting the flow of water through the condenser. Upon completion of the reaction, the heat is removed, and the flask is allowed to cool to room temperature. This ensures that all of the material in the condenser returns to the flask.

Many chemical reactions are executed by adding all the reagents, including solvent and catalyst, if any, to a reaction flask and heating the resulting mixture under reflux; however, sometimes one of the reactants must be added to the

Glasswool

Drying tube (optional)

Clamps

Water out

Condenser

Water in

Thermometer

Conical reaction vial

Spinvane

Heating source

Magnetic stirrer

Laboratory jack

Figure 2.67
Microscale apparatus for heating a reaction mixture under reflux.

reaction during reflux. For example, if a reaction is highly exothermic, its rate may be controlled by slowly adding one of the reagents. Other reactions may require that one of the reagents be present in high dilution to minimize the formation of side-products. Figure 2.66b shows a typical apparatus that allows a liquid reagent or solution to be added to the reaction flask using an addition funnel. The standard-taper separatory funnel available in many glassware kits can serve as an addition funnel, but there must be some provision for air to be admitted above the surface of the liquid to equalize the pressure inside; this can be accomplished by *not* placing the stopper on top of the funnel or by inserting a small piece of filter paper or copper wire between the stopper and the funnel. The flask can be heated to reflux during the addition, or the temperature in the flask can be maintained at room temperature or below with an ice-water bath and heated under reflux following completion of the addition.

Microscale Technique

The components of the apparatus for heating a reaction under reflux on the microscale are the same as for a preparative scale reaction. A common setup is shown in Figure 2.67. It is generally best to use a water-jacketed condenser to prevent any vapors from escaping; a short-air condenser is frequently inadequate for the more volatile organic solvents such as diethyl ether and dichloromethane.

2.23 GAS TRAPS

Some organic reactions release noxious gases that *should not* escape into the laboratory atmosphere, as this would create an unpleasant and perhaps unsafe environment. Such reactions should be performed in the hood, if space is available. Alternatively, a **gas trap** can be used.

Figure 2.68

Gas traps for miniscale apparatus: (a) trapping water-soluble gas in water; (b) removing gases through water aspirator.

Figure 2.69

Gas traps for microscale apparatus: (a) trapping water-soluble gas in water; (b) trapping water-soluble gas in moistened cotton or glasswool; (c) simplified modification for trapping water-soluble gas in moistened cotton or glasswool.

Miniscale

If the gas being generated is water-soluble, the trap shown in Figure 2.68a may be suitable. In this example, the vapors are carried into a stemmed funnel whose rim is just above or resting on the surface of the water. A second modification evacuates the undesired water-soluble gas through a house vacuum or water aspirator protected by a trap (Fig. 2.68b). This style of gas trap uses a vacuum adapter attached at the top of a reflux condenser. If the reaction being performed is sensitive to atmospheric moisture, the adapter can be filled with anhydrous calcium chloride, which dries the air passing through it. Only a slight vacuum is needed to make this type of trap effective, so do *not* turn the house vacuum or water aspirator on full-force.

Microscale

The style of trap portrayed in Figure 2.68a can also be used with a microscale apparatus. A typical apparatus of this sort is shown in Figure 2.69a. A simplified analog is shown in Figure 2.69b, wherein the water-soluble gas is trapped in the water covering the surface of a cotton or glasswool packing. In some cases, simply putting moistened cotton or glasswool in a drying tube attached to the apparatus (Fig. 2.69c) will suffice to trap the gas being evolved.

2.24 DRYING AGENTS

Drying a reagent, solvent, or product is a task that must be performed at some stage of nearly every reaction conducted in the organic chemistry laboratory. The techniques of drying solids and liquids are described in this and the following sections.

Drying Agents and Desiccants

Most organic liquids are distilled at the end of the purification process, and any residual moisture that is present may react with the compound during the distillation; water may also co- or steam-distill with the liquid and contaminate the distillate. To remove these small traces of moisture before distillation, **drying agents**, sometimes called **desiccants**, are used. There are two general requirements for a drying agent: (1) neither it nor its hydrolysis product may react chemically with the organic liquid being dried, and (2) it must be *completely* and *easily* removed from the dry liquid. A drying agent should also be efficient so that the water is removed in a reasonably short period of time.

Some commonly used drying agents and their properties are listed in Table 2.1. These desiccants function in one of two ways: (1) the drying agent interacts *reversibly* with water by the process of adsorption or absorption (Eq. 2.1), or (2) it reacts irreversibly with water by serving as an acid or a base.

With drying agents that function by reversible hydration, a certain amount of water will remain in the organic liquid in equilibrium with the hydrated drying agent. The lesser the amount of water left at equilibrium, the greater the efficiency of the desiccant. A drying agent that forms a hydrate (Eq. 2.1) must be *completely* removed by gravity filtration (Sec. 2.17) or by decantation (Sec. 2.19) *before* the dried liquid is distilled, since many hydrates decompose with loss of water at temperatures above 30–40 °C.

$$\text{Drying agent (solid)} + x\,H_2O\,(\text{liquid}) \rightleftharpoons \text{Drying agent} \cdot x\,H_2O\,(\text{solid}) \quad (2.1)$$
$$\text{Anhydrous} \qquad\qquad\qquad\qquad\qquad \text{Hydrate}$$

Drying agents that remove water by an irreversible chemical reaction are very efficient, but they are generally more expensive than other types of drying agents. Such drying agents are sometimes more difficult to handle and are normally used to remove *small* quantities of water from reagents or solvents prior to a chemical reaction. For example, phosphorus pentoxide, P_2O_5, removes water by reacting vigorously with it to form phosphoric acid (Eq. 2.2). Desiccants such as calcium hydride (CaH_2) and sodium (Na) metal also react vigorously with water. When CaH_2 or Na metal is used as a drying agent, hydrogen gas is evolved, and appropriate precautions must be taken to vent the hydrogen and prevent buildup of this highly flammable gas.

$$P_2O_5\,(\text{solid}) + 3\,H_2O\,(\text{liquid}) \longrightarrow 2\,H_3PO_4 \quad (2.2)$$

Of the drying agents listed in Table 2.1, *anhydrous* calcium chloride, sodium sulfate, and magnesium sulfate will generally suffice for the needs of this introductory laboratory course. Both sodium sulfate and magnesium sulfate have high capacities and absorb a large amount of water, but magnesium sulfate dries a solution

Table 2.1 *Table of Common Drying Agents, Their Properties, and Their Uses*

Drying Agent	Acid-Base Properties	Comments*
$CaCl_2$	Neutral	High capacity and fast action with reasonable efficiency; good preliminary drying agent; readily separated from dried solution because $CaCl_2$ is available as large granules; cannot be used to dry either alcohols and amines (because of compound formation) or phenols, esters, and acids (because drying agent contains some $Ca(OH)_2$).
Na_2SO_4	Neutral	Inexpensive, high capacity; relatively slow action and low efficiency; good general preliminary drying agent; preferred physical form is that of small granules, which may be easily separated from the dry solution by decantation or filtration.
$MgSO_4$	Weakly acidic	Inexpensive, high capacity, rapid drying agent with moderate efficiency; excellent preliminary drying agent; requires filtration to remove drying agent from solution.
H_2SO_4	Acidic	Good for alkyl halides and aliphatic hydrocarbons; cannot be used with even such weak bases as alkenes and ethers; high efficiency.
P_2O_5	Acidic	See comments under H_2SO_4; also good for ethers, aryl halides, and aromatic hydrocarbons; generally high efficiency; preliminary drying of solution recommended; dried solution can be distilled from drying agent.
CaH_2	Basic	High efficiency with both polar and nonpolar solvents, although inexplicably it fails with acetonitrile; somewhat slow action; good for basic, neutral, or *weakly* acidic compounds; cannot be used for base-sensitive substances; preliminary drying of solution is recommended; dried solution can be distilled from drying agent. *Caution:* Hydrogen gas is evolved with this drying agent.
Na or K	Basic	High efficiency; cannot be used on compounds sensitive to alkali metals or to base; special precautions and great care must be exercised in destroying excess drying agent to avoid fires; preliminary drying *required;* dried solution can be distilled from drying agent. *Caution:* Hydrogen gas is evolved with this drying agent.
BaO or CaO	Basic	Slow action but high efficiency; good for alcohols and amines; cannot be used with compounds sensitive to base; dried solution can be distilled from drying agent.
Molecular Sieve 3A or 4A[†]	Neutral	Rapid and highly efficient; preliminary drying recommended; dried solution can be distilled from drying agent if desired. *Molecular sieves* are aluminosilicates, whose crystal structure contains a network of pores of uniform diameter; the pore sizes of sieves 3A and 4A are such that only water and other small molecules such as ammonia can pass into the sieve; water is strongly adsorbed as water of hydration; hydrated sieves can be reactivated by heating at 300–320 °C under vacuum or at atmospheric pressure.

* Capacity, as used in this table, refers to the amount of water that can be removed by a given weight of drying agent; efficiency refers to the amount of water, if any, in equilibrium with the hydrated desiccant.
[†] The numbers refer to the nominal pore size, in Ångström units, of the sieve.

more completely. Calcium chloride has a low capacity, but it is a more efficient drying agent than magnesium sulfate. Do *not* use an unnecessarily large quantity of drying agent when drying a liquid, since the desiccant may adsorb or absorb the desired organic product along with the water. Mechanical losses on filtration or decantation of the dried solution may also become significant. The amount of drying agent required depends upon the quantity of water present, the capacity of the drying agent, and the amount of liquid to be dried.

2.25 DRYING ORGANIC SOLUTIONS

When an organic solvent is shaken with water or an aqueous solution during an extraction procedure, it will contain some dissolved water. The amount of water varies with the organic solvent. Of the solvents commonly used for extraction, diethyl ether dissolves the most water. If this water is not removed prior to evaporation of the organic solvent, significant quantities of water will remain in the product and may complicate further purification. It is therefore necessary to dry the solution using a suitable drying agent, usually *anhydrous* sodium sulfate or magnesium sulfate.

Miniscale Technique

Place the organic liquid to be dried in an Erlenmeyer flask of suitable size so that it will be no more than half-filled. Start by adding a small spatula-tip full of drying agent such as *anhydrous* sodium sulfate or magnesium sulfate and swirl the flask gently. Swirling increases the surface area for contact between the solid and liquid phases and generally facilitates drying. If the drying agent "clumps" or if liquid still appears cloudy after the solid has settled to the bottom of the flask, add more drying agent and swirl again. Repeat this process until the liquid appears clear and some of the drying agent flows freely upon swirling the mixture. An amount of drying agent that covers the bottom of the flask should normally be sufficient. After drying is complete, remove the drying agent either by gravity filtration (Sec. 2.17) or by decantation (Sec. 2.19). Rinse the drying agent once or twice with a small volume of the organic solvent and transfer the rinse solvent to the organic solution.

Microscale Technique

Place the organic liquid to be dried in a conical vial or a test tube of suitable size so that it will be no more than half-filled. Start by adding a microspatula-tip full of drying agent such as *anhydrous* sodium sulfate or magnesium sulfate; granular anhydrous sodium sulfate is the most convenient for microscale manipulations. Stir the mixture with a microspatula. Stirring increases the surface area for contact between the solid and liquid phases and generally facilitates drying. If the drying agent "clumps" or if the liquid still appears cloudy after the solid has settled to the bottom of the flask, add more drying agent and stir again. Repeat this process until the liquid appears clear and some of the drying agent flows freely upon stirring the mixture. After drying is complete, use a *dry* Pasteur filter-tip pipet (Fig. 2.9a) or a *dry* Pasteur pipet to transfer the organic solution to a *dry* conical vial or test tube. Rinse the drying agent once or twice with a small volume of the organic solvent and transfer the rinse solvent to the organic solution.

2.26 DRYING SOLIDS

Solid organic compounds must be dried because the presence of water or organic solvents will affect their weight, melting point, quantitative elemental analysis, and spectra. Since sources of protons must be excluded from some reactions, it is also necessary to remove all traces of moisture or protic solvents from a solid prior to performing such a reaction.

A solid that has been recrystallized from a volatile organic solvent can usually be dried satisfactorily by allowing it to air-dry at room temperature, provided it is not **hygroscopic** and thus absorbs moisture from the air. After the solid is collected on a Büchner or Hirsch funnel by vacuum filtration (Sec. 2.17), it is first pressed as dry as possible with a clean spatula or cork while air is pulled through the funnel and solid. The solid is then spread on a piece of filter paper, which absorbs the excess solvent, or on a clean watchglass and allowed to stand overnight or longer at room temperature.

Water is sometimes removed from a solid by dissolving the solid in a suitable organic solvent such as chloroform or toluene, removing any water by azeotropic distillation, and then recovering the solid by removal of the solvent. However, water is more commonly removed from organic solids using desiccators containing desiccants such as silica gel, phosphorus pentoxide, calcium chloride, or calcium sulfate. The desiccator may be used at atmospheric pressure or under a vacuum; however, if a vacuum is applied, the desiccator must be enclosed in a metal safety cage or wrapped with electrical or duct tape. Desiccators or tightly stoppered bottles containing one of these desiccants may also be used to store dry solids contained in small vials.

If the sample is hygroscopic or if it has been recrystallized from water or a high-boiling solvent, it must be dried in an oven operating at a temperature 20–30 °C *below* the melting or decomposition point of the sample. The oven-drying process can be performed at atmospheric pressure or under vacuum. Air-sensitive solids must be dried either in an inert atmosphere, such as in nitrogen or helium, or under vacuum. Samples to be submitted for quantitative elemental analysis are normally dried to constant weight by heating them under vacuum.

2.27 DRYING TUBES

Figure 2.70
Drying tubes: (a) straight; (b) bent.

Frequently it is necessary to protect a reaction mixture from moisture. There are advanced techniques for doing this that involve performing the reaction under an inert, dry atmosphere of nitrogen or argon gas, but a simple and moderately effective procedure simply utilizes a drying tube containing a drying agent. Two types of drying tubes are shown in Figure 2.70. The straight tube (Fig. 2.70a) and the bent tube (Fig. 2.70b) are commonly found in glassware kits for miniscale and microscale experiments, respectively. A drying tube is prepared by placing a loose plug of glasswool at the bottom of the tube and filling the tube with a *granular* desiccant such as anhydrous calcium chloride or calcium sulfate. The drying agent often contains a blue indicator that turns pink when it is no longer effective. Although it is not necessary to cap the straight tube with glasswool, a loose plug of glasswool should be placed in the other end of the bent drying tube to prevent the drying agent from falling out. Neither the desiccant nor the glasswool should be packed too tightly.

2.28 DRYING APPARATUS

Reactions involving the use of water-sensitive reagents must be performed under anhydrous conditions. Thus, prior to conducting such a reaction and prior to introducing any reagents, as much moisture as possible must be removed from the apparatus itself.

The glass parts of an apparatus, such as round-bottom flasks, conical vials, and glass syringes, may be conveniently dried by placing them for 15–20 min in a drying oven heated at about 150 °C. Remove the glass components from the oven using a pair of tongs, *not* your bare hands, and if possible place them in a desiccator to keep them dry as they cool to room temperature. After the glassware is cool, remove it from the desiccator and assemble the apparatus as quickly as possible to avoid reintroduction of moisture on the glass surface. If a desiccator is not available, then allow the glassware to cool only to the point where you can handle it, and assemble the apparatus as quickly as possible to avoid reintroduction of moisture on the glass surface.

An alternative method that may be used is to "flame-dry" the apparatus. If you are instructed to use this technique, be sure that no one is working with flammable solvents in the room. The apparatus, which must have one opening to the atmosphere, is gently warmed with a Bunsen burner or microburner (Sec. 2.9), a process called "flaming." A heat gun (Sec. 2.9) may also be used to heat the apparatus. The apparatus is heated first at the point most remote from the opening to the atmosphere, and the flame is gradually worked toward the opening. The moisture in the apparatus will be driven through the opening. If a condenser is part of the apparatus, no water should be in or flowing through it. While the apparatus is still warm, a filled drying tube (Sec. 2.27) is inserted into the opening, and the system is allowed to cool to room temperature. As cooling occurs, air from the atmosphere is drawn into the apparatus through the drying agent, which removes the moisture, and the dried apparatus is ready for use.

2.29 EVAPORATING SOLVENTS

For many experiments, it is necessary to remove the excess solvents to recover the product. Although an Erlenmeyer flask or beaker containing the solution could simply be left unstoppered in the hood until the solvent has evaporated, this is rather slow and impractical unless you like spending excessive amounts of time in the laboratory. A variety of techniques may be used to speed the process. Several other methods for removing solvents in miniscale experiments are shown in Figure 2.71, and the related microscale techniques are illustrated in Figure 2.72.

Miniscale Methods

The solvents in solutions contained in Erlenmeyer flasks can be evaporated by adding several boiling stones or a wooden boiling stick to the flask and heating the flask with a suitable heat source such as a steam bath, a hot plate, or a sand or oil bath so that the solvent boils gently. Magnetic stirring may also be used to prevent bumping. The evaporation rate can be increased by directing a stream of air or nitrogen into the flask (Fig. 2.71a), or by applying a vacuum over the flask with an inverted funnel to help draw away solvent vapors (Fig. 2.71b). The solution may also be placed in a filter flask with a wooden boiling stick. After the flask is subjected to a gentle vacuum, it is swirled over the heat source to facilitate smooth evaporation and reduce the possibility of bumping (Fig. 2.71c).

Another common method for removing solvent is by simple distillation (Sec. 2.13). Discontinue heating when only a small amount of solvent remains, and *do not overheat the stillpot*, as your product may decompose if heated too strongly. The stillpot should be cooled to room temperature, and the receiving flask containing the distillate should be replaced with an empty round-bottom flask. The apparatus should then be set up for vacuum distillation (Fig. 2.42b) and attached to a vacuum source, such as a water aspirator, for a few minutes to remove the last traces of solvent.

Figure 2.71
Miniscale techniques for evaporating solvents.

Microscale Methods

The solvents in solutions contained in a conical vial or test tube can be evaporated by heating the conical vial or test tube with a suitable heat source such as a sand bath or aluminum block. To prevent bumping, the solution should be agitated with magnetic stirring or with a boiling stone. The heat should also be adjusted so the solvent does not boil vigorously. The evaporation rate can be increased by directing a stream of air or nitrogen into the conical vial or test tube (Fig. 2.72a), or by applying a vacuum over the conical vial or test tube with an inverted funnel to help remove solvent vapors (Fig. 2.72b). The solution may also be placed in a test tube with a sidearm and a wooden boiling stick. After the test tube is subjected to a gentle vacuum, it is swirled over the heat source to facilitate smooth evaporation and reduce the possibility of bumping (Fig. 2.72c).

Another common method for removing solvent is by simple microscale distillation (Sec. 2.13). Discontinue heating when only a small amount of solvent remains, and *do not overheat the stillpot*, as your product may decompose if heated too strongly. The stillpot should be cooled to room temperature. Remove the Hickman stillhead and water condenser and insert an air condenser into the vial. Attach a thermometer adapter fitted with a glass tube to the top of the condenser and attach the tube to a vacuum source (Fig. 2.73). If you use a water aspirator, be sure to place

Figure 2.72
Microscale techniques for evaporating solvents.

Figure 2.73
Microscale apparatus for removing solvent under reduced pressure.

Figure 2.74
Rotary evaporator.

a trap between the condenser and the aspirator pump (Fig. 2.42b). While stirring the concentrated liquid, apply a gentle vacuum to the apparatus so the remainder of the organic solvent evaporates slowly under reduced pressure.

Rotary Evaporators

In a research laboratory and in many undergraduate laboratories, solvents are removed under reduced pressure, using an apparatus called a **rotary evaporator** (Fig. 2.74). This device is especially designed for the rapid evaporation of solvents without bumping. A variable-speed motor is used to rotate the flask containing the solvent being evaporated. While the flask is being rotated, a vacuum is applied and the flask may be heated. The rotation of the flask spreads a thin film of the solution on the inner surface of the flask to accelerate evaporation, and it also agitates the contents of the flask to reduce the problem of bumping. The rate of evaporation can be controlled by adjusting the vacuum, the temperature of the water bath, and the rate of rotation of the flask.

Solids

Recrystallization and Melting Points

One of the greatest visual pleasures for the practicing organic chemist is to observe the formation of a crystalline solid from solution. Obtaining beautifully formed crystals in myriad forms—prisms, cubes, platelets, rhomboids, and so on—gives a sense of accomplishment that is hard to match. This chapter gives the background to and experimental procedures for experiencing this thrill. Enjoy the experience!

3.1 INTRODUCTION

The organic chemist usually works with substances that are in the liquid or solid state. The purpose of this and the succeeding two chapters is to present an introduction to the theory and practice of the most important methods for separation and purification of mixtures of organic compounds in these two physical states. As you read these chapters, you should also refer to the corresponding sections of Chapter 2 that describe the apparatus used for the various techniques. References to the appropriate sections have been provided for your convenience.

A **pure compound** is a homogeneous sample consisting only of molecules having the same structure. However, substances believed to be pure on the basis of certain criteria may actually contain small amounts of contaminants. Indeed, the presence of impurities in what were believed to be pure compounds has sometimes led to incorrect structural assignments and scientific conclusions. The possibility of making such errors was of particular concern prior to the advent of the powerful chromatographic (Chap. 6) and spectroscopic (Chap. 8) techniques that have been developed since the 1950s. It is now relatively easy for the chemist to purify solids and liquids and to demonstrate their purity.

A compound formed in a chemical reaction or extracted from some natural source is rarely pure when initially isolated. For example, a chemical transformation intended to produce a single product almost invariably yields a reaction mixture containing a number of contaminants. These may include the products of side reactions, unchanged starting materials, inorganic materials, and solvents. Unfortunately, even chemicals that have been purchased are not always pure,

91

owing to the expense of the needed purification process or to decomposition that may occur during storage and shipment.

Organic chemists devote considerable effort to isolating pure products, the ultimate goal being to obtain a substance that cannot be further purified by any known experimental techniques. This chapter focuses on the purification of solids by recrystallization and their characterization by the physical property of melting points.

3.2 RECRYSTALLIZATION

Recrystallization of solids is a valuable technique to master, because it is one of the most common methods used to purify solids. Other techniques for purifying solids include **sublimation** (Sec. 2.20), **extraction** (Chap. 5), and **chromatography** (Chap. 6). Nevertheless, even when one of these alternative methods of purification has been used, the solid material thus isolated may still be recrystallized to achieve a higher state of purity.

The process of recrystallization involves dissolving the solid in an appropriate solvent at an elevated temperature and allowing the crystals to re-form on cooling, so that any impurities remain in solution. This technique, which is technically called **solution recrystallization** but is more commonly referred to simply as *recrystallization*, is discussed in this chapter. An alternative approach involves melting the solid in the absence of solvent and then allowing the crystals to re-form so that impurities are left in the melt. This method is not often used in the organic laboratory because the crystals often form from a viscous oil that contains the impurities and from which it is difficult to separate the desired pure solid. It is interesting to note, however, that this is the technique used to prepare the high-purity single crystals of silicon used in computer chips.

Almost all solids are *more* soluble in a *hot* than in a *cold* solvent, and solution crystallization takes advantage of this fact. Thus, if you first dissolve a solid in an amount of hot solvent insufficient to dissolve it when cold, crystals should form when the hot solution is allowed to cool. The extent to which the solid precipitates depends on the difference in its solubility in the particular solvent at temperatures between the extremes used. The upper extreme is determined by the boiling point of the solvent, whereas the lower limit is usually dictated by experimental convenience. For example, an ice-water bath is often used to cool the solution to 0 °C, whereas ice-salt and dry ice-acetone baths are commonly used to cool solutions to −20 °C and −78 °C, respectively (Sec. 2.10). The solid should be recovered with greater efficiency at these temperatures, provided the solvent itself does not freeze.

If the impurities present in the original solid mixture have dissolved and *remain* dissolved after the solution is cooled, isolation of the crystals that have formed should *ideally* provide pure material. Alternatively, the impurities may not dissolve at all in the hot solution and may be removed by filtration *before* the solution is cooled. The crystals that subsequently form should be purer than the original solid mixture. Solution recrystallization is seldom quite so simple in practice, but these two idealized generalizations do outline the basic principles of the technique.

Even after a solid has been recrystallized, it may still not be pure. Thus, it is important to determine the purity of the sample, and one of the easiest methods to do this is by determining the melting point of the solid. This technique is described in Section 3.3.

Solution recrystallization involves the following steps:

1. **Selection** of an appropriate solvent.
2. **Dissolution** of the solid to be purified in the solvent near or at its boiling point.

3. **Decoloration** with an activated form of carbon, if necessary, to remove colored impurities and **filtration** of the hot solution to remove insoluble impurities and the decolorizing carbon.

4. **Formation** of crystalline solid from the solution as it cools.

5. **Isolation** of the purified solid by filtration.

6. **Drying** the crystals.

Each step of the sequence is discussed in the following subsections, and representative experimental procedures are presented at the end of the discussion.

Selection of Solvent

The choice of solvent is perhaps the most critical step in the process of recrystallization because the correct solvent must be selected to form a product of high purity and in good recovery or yield. Consequently, a solvent should satisfy certain criteria for use in recrystallization: (**a**) The desired compound should be reasonably soluble in the *hot* solvent, about 1 g/20 mL (1 mg/20μL) being satisfactory, and **insoluble** or *nearly* insoluble in the *cold* solvent. Often the reference temperature for determination of the solubility in "cold" solvent is taken to be room temperature. This combination of solute and solvent allows dissolution to occur in an amount of solvent that is not unduly large and also permits recovery of the purified product in high yield. A solvent having these solubility properties as a function of temperature would be said to have a favorable **temperature coefficient** for the desired solute. (**b**) Conversely, the impurities should either be insoluble in the solvent at all temperatures *or* must remain at least moderately soluble in the cold solvent. In other words, if the impurities are soluble, the temperature coefficient for them must be *unfavorable;* otherwise the desired product *and* the impurities would both crystallize simultaneously from solution. (**c**) The boiling point of the solvent should be low enough so that it can be readily removed from the crystals. (**d**) The boiling point of the solvent should generally be lower than the melting point of the solid being purified. (**e**) The solvent should not react chemically with the substance being purified.

The chemical literature is a valuable source of information about solvents that may be used for recrystallizing known compounds. Nonetheless, even if a suitable solvent has been reported, you may still encounter problems because your compound forms an oil or fails to precipitate from solution at all. In such cases, using a "seed" crystal (see below), or concentrating the solution, or cooling the solution in an ice-water bath may lead to a successful recrystallization.

If the compound has *not* been prepared before, it is necessary to resort to trial-and-error techniques to find an appropriate solvent for recrystallization. The process of selection can be aided by consideration of some generalizations about solubility characteristics for classes of solutes. Polar compounds are normally soluble in polar solvents and insoluble in nonpolar solvents, for example, whereas nonpolar compounds are more soluble in nonpolar solvents. Such characteristics are summarized by the adage *"like dissolves like."* Of course, although a highly polar compound is unlikely to be soluble in a hot, nonpolar solvent, it *may* be very soluble in a cold, very polar solvent. In this case, a solvent of intermediate polarity *may* be suitable. The bottom line is that even experts who routinely perform recrystallizations often face challenges when selecting the appropriate solvent, so do not be frustrated if you too have problems.

The solvents commonly used in recrystallizations range widely in polarity, a property measured by the **dielectric constants** (ϵ), listed in Table 3.1. Those solvents with dielectric constants in the range of 2–3 are considered **nonpolar** and those with constants above 10 as **polar.** Solvents in the range of 3–10 are of intermediate polarity. Of the various solvents listed, petroleum ether deserves special mention because of its confusing common name. This solvent does *not* contain the ether functional

Table 3.1. *Common Solvents for Recrystallization**

Solvent	Boiling Point, °C (760 torr)	Freezing Point, °C[†]	Water Soluble	Dielectric Constant (ϵ)	Flammable
Petroleum ether[‡]	Variable		No	1.9	Yes
Diethyl ether	35		Slightly	4.3	Yes
Dichloromethane[§]	41		No	9.1	No
Acetone[‡]	56		Yes	20.7	Yes
Methanol[‡]	65		Yes	32.6	Yes
Tetrahydrofuran	65		Yes	7.6	Yes
Ethyl acetate[‡]	77		Yes	6.0	Yes
Ethanol (95%)[‡]	78		Yes	24.6	Yes
Cyclohexane	81	6	No	1.9	Yes
Water[‡]	100	0	N/A	78.5	No
1,4-Dioxane	101	11	Yes	2.2	Yes
Toluene[‡]	111		No	2.4	Yes

*Benzene has been omitted from this list, owing to its toxicity. Cyclohexane or toluene can usually be substituted for it.
[†]Freezing points not listed are below 0 °C.
[‡]Denotes solvents used most often.
[§]As a general rule, avoid the use of chlorocarbon solvents *if another equally good solvent is available.* Their toxicity is greater and their disposal is more difficult than for other types of solvents.

group at all. Rather, it is a mixture of volatile aliphatic hydrocarbons obtained from refining petroleum; the composition and boiling point of the mixture depends on the particular distillation "cut" obtained. Thus, the boiling range of this type of solvent is usually given, as in the description "petroleum ether, bp 60 to 80 °C (760 torr)."

Occasionally a *mixture* of solvents is required for satisfactory recrystallization of a solute. This mixture comprises only two solvents; one of these dissolves the solute even when cold and the other one does not. The logic of this will become clear when you read about this technique under "**Dissolution.**"

Chemical Structures of Common Recrystallization Solvents Found in Table 3.1

CH₂CH₃OH

Ethanol
(ethyl alcohol)

CH₃OH

Methanol
(methyl alcohol)

Cyclohexane

Toluene

CH₃CH₂OCH₂CH₃

Diethyl ether
(ether)

Tetrahydrofuran

1,4-Dioxane

CH₂Cl₂
Dichloromethane
(methylene chloride)

CH₃COCH₂CH₃
Ethyl acetate

CH₃CCH₃
Acetone

Dissolution

This step may involve the handling of relatively large volumes of volatile solvents. Although most solvents used in the organic laboratory are of relatively low toxicity, it is prudent to avoid inhaling their vapors. For this reason, the following operations are best performed in a hood. If this is not possible, clamping an inverted funnel over the recrystallization flask and connecting the stem of the funnel to a source of a vacuum (Fig. 2.71b) helps to minimize exposure. Another precaution is to choose the source of heat for the recrystallization carefully. As noted in Table 3.1, many solvents are highly flammable and should *never* be heated with an open flame; rather, water, steam, and oil baths or hot plates (Sec. 2.9) should be used.

Single Solvents

The procedures for performing dissolutions in *single* solvents at the mini- and microscales differ slightly, as described below. Discussion of the strategies for using *mixed* solvents for recrystallizations is presented as well.

Miniscale Technique

The solid to be purified is weighed and placed in an Erlenmeyer flask of appropriate size. A beaker is *not* a suitable vessel for recrystallization, because it has a relatively large surface area to which recrystallized product may stick, lowering the efficiency of recovery. Some crystals of the impure solid should always be retained, for they may be needed as **"seeds"** to induce crystallization. The flask should be equipped for magnetic stirring (Sec. 2.11) or contain either boiling stones or a wooden stick to prevent bumping of the solution while boiling.

A few milliliters of solvent are added to the flask, and the mixture is then heated (Sec. 2.9) to the boiling point. More solvent is added to the mixture in *small* portions using a Pasteur pipet (Sec. 2.5) until the solid just dissolves. It is important to let boiling resume after each addition so that a minimum amount of solvent is used to effect dissolution; *using excessive amounts of solvent decreases the recovery of the solute.* If it is necessary to perform a hot filtration (Sec. 2.17), it is advisable to add an additional 2–5% of solvent to prevent premature crystallization during this operation.

If adding solvent fails to dissolve any more solid, it is likely that insoluble impurities are present. These can be removed by hot filtration. Thus, to avoid using too much solvent and risking poor recovery of the purified solute, you should observe the dissolution process *carefully.* This is particularly important when only a relatively small quantity of solid remains, as this may be the material that is insoluble.

Microscale Technique

The protocol for dissolution at the microscale level mimics that at larger scales. The solid is weighed and placed in a test tube or Craig tube, with a few crystals of impure material being retained as "seeds" to induce crystallization. A boiling stone should be added to prevent bumping of the solution upon boiling. Alternatively, smooth boiling can be promoted by twirling a microspatula in the mixture.

A few drops of solvent, normally totaling no more than a milliliter, are added, and the mixture is heated (Sec. 2.9) to the boiling point. By means of a Pasteur pipet or filter-tip pipet (Sec. 2.5), additional solvent is added dropwise to the boiling mixture until the solid just dissolves. It is important to let boiling resume after each addition so that a minimum amount of solvent is used to effect dissolution; *using excessive amounts of solvent decreases the recovery of the solute.* If it is necessary to perform a hot filtration (Sec. 2.17), it is prudent to add an additional 2–5% of solvent to prevent premature crystallization during this operation.

If adding solvent fails to dissolve any more solid, it is likely that insoluble impurities are present. These can be removed by hot filtration. Thus, to avoid using too much solvent and risking poor recovery of the purified solute, you should observe the dissolution process *carefully.* This is particularly important when only a relatively small quantity of solid remains, as this may be the material that is insoluble.

Mixed Solvents

The same general approach used for single-solvent dissolution is followed at the mini- and microscale levels when mixed solvents are employed. However, there are two options for effecting dissolution once the solvents have been selected. In one, the solid to be purified is first dissolved in a *minimum* volume of the hot solvent in which it is soluble; the second solvent is then added to the *boiling* solution until the mixture turns cloudy. The cloudiness signals initial formation of crystals, caused by the fact that addition of the second solvent results in a solvent mixture in which the solute is less soluble. Finally, more of the first solvent is added *dropwise* until the solution clears.

Two further aspects of this option should be noted. First, the solution must be cooled slightly below the lower boiling point before the second solvent is added *if* this solvent has a boiling point *lower* than the first; otherwise the addition of this solvent could cause sudden and vigorous boiling of the mixture, and hot solvent might spew from the apparatus. Second, hot filtration should be performed if needed *before* addition of the second solvent; this will prevent crystallization during the filtration step. A potential disadvantage of this method for mixed-solvent recrystallization is that unduly large volumes of the second solvent may be required if excessive amounts of the initial solvent have been used.

In the second option, the solute is added to the solvent in which it is insoluble, and the mixture is heated near the boiling point of the solvent; the second solvent is then added in small portions until the solid just dissolves. As with recrystallization from a single solvent, it is generally wise to add 2–5% of additional solvent to prevent premature crystallization of the solute during hot filtration, if this step is necessary. The use of this approach to mixed-solvent recrystallization also has the disadvantage that using too much of the first solvent will require the addition of undesirably large volumes of the second one.

Decoloration and Hot Filtration

After dissolution of the solid mixture, the solution may be colored. This signals the presence of impurities if the desired compound is known to be colorless. If the compound is colored, contaminants may alter the color of the solution. For example, impurities should be suspected if the substance is yellow but the solution is green. The decoloration step is unnecessary if the solution is colorless, of course.

Colored impurities may often be removed by adding a small amount of decolorizing carbon (Sec. 2.18) to the hot, *but not boiling*, solution. If it were added to a boiling solution, the liquid would likely froth over the top of the flask, resulting in loss of product and possible injury. After the decolorizing carbon is added, the solution is heated to boiling for a few minutes while being continuously stirred or swirled to prevent bumping.

Completely removing decolorizing carbon by hot filtration (Sec. 2.17) may be difficult if the powdered rather than the pelletized form is used, because the finely divided solid particles may pass through filter paper or the cotton of a Pasteur filtering pipet (Sec. 2.5). If this occurs, the dark particles should be visible in the filtrate.

The following steps are recommended to assist in removing the decolorizing carbon during filtration. The hot decolorized solution is allowed to cool slightly below its boiling point, a small amount of a filter-aid such as Celite (Sec. 2.17) is added to *adsorb* the carbon, the mixture is briefly reheated to boiling, and then it is subjected to hot filtration. It may be necessary to repeat this procedure if some of the decolorizing carbon still passes through the filter paper and remains in the filtrate.

This technique for decolorizing solutions works because colored impurities as well as the compound being purified are adsorbed on the surface of the carbon particles. For electronic reasons, the colored substances adsorb more strongly to the surface. You should avoid using too much decolorizing carbon because the desired product itself may be adsorbed by it, so less product will be recovered.

Insoluble impurities, including dust and decolorizing carbon, are removed by gravity filtration of the hot solution (Sec. 2.17); this step is not necessary if the hot solution is clear and homogeneous. Gravity filtration is normally preferred to vacuum filtration because the latter technique may cause cooling and concentration of the solution, owing to evaporation of the solvent, and this may result in premature crystallization. A short-stemmed or stemless glass funnel should be used to minimize crystallization in the funnel, and using fluted filter paper (Sec. 2.17) will minimize crystallization on the filter. To keep liquid from flowing over the top of the funnel, the top of the paper should not extend above the funnel by more than 1–2 mm.

Crystallization

The hot solution of solute is allowed to cool slowly to room temperature, and crystallization should occur. During cooling and crystallization, the solution should be protected from airborne contaminants by covering the opening with a piece of filter paper, an inverted beaker, or by loosely plugging it with a clean cork. Rapid cooling by immersing the flask in water or an ice-water bath tends to lead to the formation of very small crystals that may adsorb impurities from solution. Generally the solution should not be disturbed as it cools, because this also leads to production of small crystals. The formation of crystals larger than about 2 mm should also be avoided because some of the solution may become occluded or trapped *within* the crystals. The drying of such crystals is more difficult, and impurities may be left in them. Should overly large crystals begin to form, brief, gentle agitation of the solution normally induces production of smaller crystals.

Failure of crystallization to occur after the solution has cooled somewhat usually means that either too much solvent has been used or the solution is **supersaturated.** A supersaturated solution can usually be made to produce crystals by *seeding.* A crystal of the original solid is added to the solution to induce crystallization, which may then be quite rapid. If no solid is available and a volatile solvent is being used, it is sometimes possible to produce a seed crystal by immersing the tip of a glass stirring rod or metal spatula in the solution, withdrawing it, and allowing the solvent to evaporate. The crystals that form on the end of the rod or spatula are then reinserted into the solution to initiate crystallization. Alternatively, crystallization can often be induced by using a glass rod to rub the inside surface of the crystallization vessel *at or just above* the air-solution interface. This should be done carefully, as excessive force may scratch or break the vessel or result in a broken rod.

Occasionally the solute will separate from solution as an oil rather than a solid. This type of separation, which is sometimes called **oiling out,** is undesirable for purification of solutes because the oils usually contain significant amounts of impurities. Two general approaches are helpful in solving this problem: (1) Oils may persist on cooling with no evidence of crystallization. These may often be induced to crystallize by scratching the oil against the side of the flask with a glass stirring rod at the interface of the oil and the solution. If this fails, several small seed crystals of the original solid may be added to the oil and the mixture allowed to stand for a period of time. Failure of these alternatives may necessitate separation of the oil from the solution and crystallizing it from another solvent. (2) Oils may form from the hot solution and then solidify to an amorphous mass at lower temperatures; in the meantime, crystals of the solute may precipitate from the mother liquor. Because the oil is not a pure liquid, the solid mass produced will be impure, as noted earlier. In a case such as this, the usual remedy is to reheat the entire mixture to effect dissolution, add a few milliliters of additional pure solvent, and allow the resulting solution to cool.

Filtration and Solvent Removal

The crystalline product is isolated by filtration. The technique for doing this varies depending on the scale on which the crystallization was performed.

Miniscale Technique

The solid product is isolated by vacuum filtration using a Büchner or Hirsch funnel and a clean, dry filter flask (Sec. 2.17). The crystals normally are washed with a small amount of *pure, cold* solvent, with the vacuum off; the vacuum is then reapplied to remove as much solvent as possible from the filter cake. Care must be taken in this step to ensure that the filter paper is not lifted off the bed of the filter while the vacuum is off, as this could result in loss of product when the vacuum is reapplied to remove the washes.

Microscale Technique

The crystalline product, contained in a Craig tube fitted with a plug, is isolated by centrifugation (Sec. 2.17). The solid is scraped off the plug back into the Craig tube and washed by stirring it briefly with a small amount of *pure, cold* solvent. The plug is then reinserted, and the solvent is once again removed by centrifugation.

Further cooling of the filtrate, sometimes called the **mother liquor,** in an ice-water or ice-salt bath (Sec. 2.10) may allow isolation of a second crop of crystals. The filtrate can also be concentrated by evaporating part of the solvent and cooling the residual solution. The crystals isolated as a second or even a third crop are likely to be less pure than those in the first. Consequently, the various crops should *not* be combined until their purity has been assessed by comparison of their melting points as described in Section 3.3.

Drying the Crystals

At the miniscale level, the final traces of solvent are removed by transferring the crystals from the filter paper of the Büchner or Hirsch funnel to a watchglass or vial. Alternatively, solids may also be transferred to fresh pieces of filter or weighing paper for drying. This is a less desirable option, however, because fibers of paper may contaminate the product when it is ultimately transferred to a container for submission to your instructor. For microscale operations, the solid may be left in the Craig tube used for the crystallization. It is good practice to protect the crystals from airborne contaminants by covering them with a piece of filter paper or using loosely inserted cotton or a cork to plug the opening of the vessel containing the solid.

Removing the last traces of solvent from the crystalline product may be accomplished by air- or oven-drying (Sec. 2.26). With the latter option, the temperature of the oven *must* be 20–30 °C below the melting point of the product; otherwise your crystals will become a puddle!

EXPERIMENTAL PROCEDURES

Recrystallization

Purpose To explore the techniques for recrystallizing solids.

SAFETY ALERT

1. **Wear safety glasses or goggles while performing the experiments.**

2. **Organic compounds are much more rapidly absorbed through the skin when they are in solution, particularly in water-soluble solvents such as acetone and ethanol. For this reason, do not rinse organic materials off your skin with solvents such as acetone; wash your hands thoroughly with soap and warm water instead.**

3. **Do not use a burner in these procedures unless instructed to do so. Most solvents used for recrystallization are flammable (Table 3.1).**

4. **When using a hot plate, do not set it at its highest value. A moderate setting will prevent overheating and the resultant bumping and splashing of materials from the flask. Do not employ hot plates for heating volatile or flammable solvents; rather, use a steam bath.**

5. **Do not inhale solvent vapors. If a hood is not available to you, clamp an inverted funnel just above the Erlenmeyer flask in which you will be heating solvents. Attach this funnel to a source of vacuum by means of rubber tubing (Fig. 2.71b).**

6. ***Never* add decolorizing carbon to a boiling solution; doing so may cause the solution to boil out of the flask. Add the carbon only when the temperature of the solvent is below the boiling temperature. This same precaution applies when using a filter-aid to assist in the removal of the carbon during the hot filtration step.**

A ■ *Solvent Selection*

Procedure

Preparation Refer to the online resources to answer Pre-Lab Exercises, access videos, and read the MSDSs for the chemicals used or produced in this procedure. Read or review Sections 2.5, 2.9, and 3.2, and review the solvent properties listed in Table 3.1. Although different criteria are used for defining solubility, plan to use the following definitions in this experiment: (a) soluble—20 mg of solute will dissolve in 0.5 mL of solvent; (b) slightly soluble—some but not all of the 20 mg of solute will dissolve in 0.5 mL of solvent; (c) insoluble—none of the solute appears to dissolve in 0.5 mL of solvent. *Be certain to record all your observations regarding solubilities in your notebook!*

Apparatus Test tubes (10-mm × 75-mm), hot-water (80–100 °C) or steam bath.

Protocol For *known* compounds, place about 20 mg (a microspatula-tip full) of the finely crushed solid in a test tube and add about 0.5 mL of water using a calibrated Pasteur pipet. Stir the mixture with a glass rod or microspatula to determine whether the solid is soluble in water at room temperature. If the solid is not completely soluble at room temperature, warm the test tube in the hot-water or steam bath, and stir or swirl its contents to determine whether the solid is soluble in hot water.

If any of your solutes are soluble in the hot solvent but only slightly soluble or insoluble at room temperature, allow the hot solution to cool slowly to room temperature and compare the quantity, size, color, and form of the resulting crystals with the original solid material.

Repeat the solubility test for the solutes using 95% ethanol and then petroleum ether (bp 60–80 °C, 760 torr). After completing these additional tests, record which of the three solvents you consider best suited for recrystallization of each of the solutes.

Compounds **1–4** contain a variety of functional groups that impart differing solubility properties to the molecules and are possible substrates on which to practice the technique of determining solubilities. Other known compounds may be assigned by your instructor.

1	2	3	4
Benzoic acid	Acetanilide	Naphthalene	Resorcinol

For *unknown* compounds, a *systematic* approach is important for determining their solubility. First, select the solvents from Table 3.1 to be used in the tests. It should not be necessary to test all the solvents, but you should consider trying those that are denoted with the symbol ‡ in the table. Your instructor may also suggest solvents to evaluate.

After selecting the solvents, obtain enough clean, dry test tubes so that there is one for each solvent to be tested. Place about 20 mg (a microspatula-tip full) of the finely crushed unknown in each test tube and add about 0.5 mL of a solvent to a tube containing the solid. Stir each mixture and determine the solubility of the unknown in each solvent at room temperature. Use the definitions of *soluble, slightly soluble, or insoluble* given earlier.

If the unknown is insoluble in a particular solvent, warm the test tube in the hot-water or steam bath. Stir or swirl the contents of the tube and note whether the unknown is soluble in hot solvent. If the solid is soluble in the hot solvent but only slightly soluble or insoluble at room temperature, allow the hot solution to cool to room temperature slowly. If crystals form in the cool solution, compare their quantity, size, color, and form with the original solid material and with those obtained from other solvents.

It is a good idea to test the solubility of a solute in a variety of solvents. Even though nice crystals may form in the first solvent you try, another one might prove better if it provides either better recovery or higher-quality crystals. To assist in determining the best solvent to use in recrystallizing an unknown, you should construct a table containing the solubility data you gather by the systematic approach described above.

If these solubility tests produce no clear choice for the solvent, mixed solvents might be considered. Review the discussion presented earlier in this section for the procedure for using a mixture of two solvents. Before trying any combinations of solvent pairs, take about 0.2 mL of each *pure* solvent being considered and mix them to ensure that they are miscible in one another. If they are not, that particular combination *cannot* be used.

B ▪ *Recrystallizing Impure Solids*

MINISCALE PROCEDURES

Preparation Refer to the online resources to answer Pre-Lab Exercises, access videos, and read the MSDSs for the chemicals used or produced in this procedure. Read or review Sections 2.6, 2.7, 2.9, 2.11, 2.17, and 2.18.

1. Benzoic Acid

Apparatus Two 50-mL Erlenmeyer flasks, graduated cylinder, ice-water bath, apparatus for magnetic stirring, vacuum filtration, and *flameless* heating.

Dissolution Place 1.0 g of impure benzoic acid in an Erlenmeyer flask equipped for magnetic stirring or with boiling stones. Measure 25 mL of water into the graduated cylinder and add a 10-mL portion of it to the flask. Heat the mixture to a gentle boil and continue adding water in 0.5-mL portions until no more solid appears to dissolve in the boiling solution. Record the total volume of water used; no more than 10 mL should be required. *Caution:* Because the sample may be contaminated with insoluble material, pay close attention to whether additional solid is dissolving as you add more solvent; if it is not, *stop adding solvent.*

Decoloration Pure benzoic acid is colorless, so a colored solution indicates that treatment with decolorizing carbon (Sec. 2.18) is necessary. *Caution:* Do *not* add decolorizing carbon to a *boiling* solution! Allow the solution to cool slightly, add a microspatula-tip full of carbon, and reheat to boiling for a few minutes. To aid in removing the finely divided carbon by filtration, allow the solution to cool slightly, add about 0.2 g of a filter-aid (Sec. 2.17), and reheat.

Hot Filtration and Crystallization If there are insoluble impurities or decolorizing carbon in the solution, perform a hot filtration (Sec. 2.17) using a 50-mL Erlenmeyer flask to receive the filtrate (Fig. 2.52). Rinse the empty flask with about 1 mL of *hot* water and filter this rinse into the original filtrate. If the filtrate remains colored, repeat the treatment with decolorizing carbon. Cover the opening of the flask with a piece of filter paper, an inverted beaker, or loose-fitting cork to exclude airborne impurities from the solution, and allow the filtrate to stand undisturbed until it has cooled to room temperature and no more crystals form.* To complete the crystallization, place the flask in an ice-water bath for at least 15 min.

Isolation and Drying Collect the crystals on a Büchner or Hirsch funnel by vacuum filtration (Fig. 2.54) and wash the filter cake with two small portions of *cold* water. Press the crystals as dry as possible on the funnel with a clean cork or spatula. Spread the benzoic acid on a watchglass, protecting it from airborne contaminants with a piece of filter paper, and air-dry it at room temperature or in an oven. Be certain that the temperature of the oven is 20–30 °C below the melting point of the product!

Analysis Determine the melting points of the crude and recrystallized benzoic acid, the weight of the latter material, and calculate your percent recovery using Equation 3.1.

$$\text{Percent recovery} = \frac{\text{Weight of pure crystals recovered}}{\text{Weight of original sample}} \times 100 \qquad (3.1)$$

2. Acetanilide

Apparatus Two 50-mL Erlenmeyer flasks, graduated cylinder, ice-water bath, apparatus for magnetic stirring, vacuum filtration, and *flameless* heating.

Dissolution Place 1.0 g of impure acetanilide in an Erlenmeyer flask equipped for magnetic stirring or with boiling stones. Measure 20 mL of water into the graduated cylinder and add a 10-mL portion of it to the flask. Heat the mixture to a gentle boil.

A layer of oil should form when the stated amount of water is added. (If you have not done so already, review the discussion of *Crystallization* in this section, focusing on how to crystallize compounds that form oils.) This layer consists of a solution of water in acetanilide. More water must be added to effect complete solution of the acetanilide in water. However, even if a homogeneous solution is produced at the boiling point of the mixture, an oil may separate from it as cooling begins. The formation of this second liquid phase is known to occur only under specific conditions: The acetanilide-water mixture must have a composition that is between 5.2% and 87% in acetanilide and be at a temperature above 80 °C. Because the solubility of acetanilide in water at temperatures near 100 °C exceeds 5.2%, a homogeneous solution formed by using the minimum quantity of water meets these criteria. Such a solution will yield an oil on cooling to about 83 °C; solid begins to form below this temperature.

Continue adding 3–5 mL of water in 0.5-mL portions to the boiling solution until the oil has completely dissolved. *Caution:* Because the sample may be contaminated with insoluble material, pay close attention to whether additional solid is dissolving as you add more solvent; if it is not, *stop adding solvent.* Once the acetanilide has just dissolved, add an additional 1 mL of water to prevent formation of oil during the crystallization step. If oil forms at this time, reheat the solution and add a little more water. Record the total volume of water used.

Continue the procedure by following the directions for *Decoloration, Hot Filtration and Crystallization, and Isolation and Drying* given for benzoic acid in Part 1.

Analysis Determine the melting points of the crude and recrystallized acetanilide, the weight of the latter material, and calculate your percent recovery using Equation 3.1.

3. Naphthalene

Apparatus Two 50-mL Erlenmeyer flasks, graduated cylinder, ice-water bath, apparatus for magnetic stirring, vacuum filtration, and *flameless* heating.

Dissolution Naphthalene may be conveniently recrystallized from methanol, 95% ethanol, or 2-propanol. Because these solvents are somewhat toxic and/or flammable, proper precautions should be taken. The sequence of steps up through the hot filtration should be performed in a hood if possible. Alternatively, if instructed to do so, position an inverted funnel connected to a vacuum source above the mouth of the flask being used for recrystallization (Fig. 2.71b).

Place 1.0 g of impure naphthalene in an Erlenmeyer flask equipped for magnetic stirring or with boiling stones and dissolve it in the minimum amount of boiling alcohol. *Caution:* Because the sample may be contaminated with insoluble material, pay close attention to whether additional solid is dissolving as you add more solvent; if it is not, *stop adding solvent.* Then add 0.5 mL of additional solvent to ensure that premature crystallization will not occur during subsequent transfers. Record the total volume of solvent used.

Continue the procedure by following the directions for *Decoloration, Hot Filtration and Crystallization, and Isolation and Drying* given for benzoic acid in Part 1; however, use the solvent in which you dissolved the naphthalene rather than water.

Analysis Determine the melting points of the crude and recrystallized naphthalene, the weight of the latter material, and calculate your percent recovery using Equation 3.1.

4. Unknown Compound

Apparatus Two 50-mL Erlenmeyer flasks, graduated cylinder, ice-water bath, apparatus for magnetic stirring, vacuum filtration, and *flameless* heating.

Dissolution *Accurately* weigh about 1 g of the unknown compound and transfer it to an Erlenmeyer flask equipped for magnetic stirring or with boiling stones. Measure about 15 mL of the solvent you have selected on the basis of solubility tests into a graduated cylinder and add 10 mL of it to the flask. Bring the mixture to a gentle boil using *flameless* heating unless water is the solvent, add a 1-mL portion of the solvent, and again boil the solution. Continue adding 3-mL portions of solvent, one portion at a time, until the solid has completely dissolved. Bring the solution to boiling after adding each portion of solvent. *Caution:* Because the sample may be contaminated with insoluble material, pay close attention to whether additional solid is dissolving as you add more solvent; if it is not, *stop adding solvent.* Record the total volume of solvent that is added.

Continue the procedure by following the directions for *Decoloration, Hot Filtration and Crystallization, and Isolation and Drying* given for benzoic acid in Part 1.

Analysis Determine the melting points of the crude and recrystallized unknown, the weight of the latter material, and calculate your percent recovery using Equation 3.1.

5. Mixed-Solvent Crystallization

Apparatus Two 25-mL Erlenmeyer flasks, graduated cylinder or calibrated Pasteur pipet, ice-water bath, apparatus for magnetic stirring, vacuum filtration, and *flameless* heating.

Dissolution Place 1.0 g of impure benzoic acid or acetanilide in an Erlenmeyer flask equipped for magnetic stirring or with boiling stones. Add 2 mL of 95% ethanol and heat the mixture to a gentle boil. If necessary, continue adding the solvent in 0.5-mL portions until no more solid appears to dissolve in the boiling solution. *Caution:* Because the sample may be contaminated with insoluble material, pay close attention to whether additional solid is dissolving as you add more solvent; if it is not, *stop adding solvent.* Record the total volume of 95% ethanol used.

Decoloration If the solution is colored, treat it with decolorizing carbon (Sec. 2.18) and a filter-aid according to the procedure given for benzoic acid in Part 1.

Hot Filtration and Crystallization If there are insoluble impurities or decolorizing carbon in the solution, perform a hot filtration (Sec. 2.17) using a 25-mL Erlenmeyer flask to receive the filtrate (Fig. 2.52). Rinse the empty flask with about 0.5 mL of *hot* 95% ethanol and filter this rinse into the original filtrate. If the filtrate remains colored, repeat the treatment with decolorizing carbon.

   Reheat the decolorized solution to boiling and add water dropwise from a Pasteur pipet until the boiling solution remains cloudy or precipitate appears; this may require adding several milliliters of water. Then add a few drops of 95% ethanol to produce a clear solution at the boiling point. Remove the flask from the heating source, and follow the same directions as given for benzoic acid in Part 1 to complete both this stage of the procedure and *Isolation and Drying.*

Analysis Determine the melting points of the crude and recrystallized product, the weight of the latter material, and calculate your percent recovery using Equation 3.1.

MICROSCALE PROCEDURES

Preparation Refer to the online resources to answer Pre-Lab Exercises, access videos, and read the MSDSs for the chemicals used or produced in this procedure. Read or review Sections 2.5, 2.6, 2.7, 2.9, 2.17, and 3.3. Plan to determine the melting points of both the purified and the crude solids as a way to show the benefits of recrystallization.

1. Benzoic Acid

Apparatus A 10-mm × 75-mm test tube, a calibrated Pasteur pipet and Pasteur filtering and filter-tip pipets, ice-water bath, apparatus for Craig tube filtration and *flameless* heating.

Dissolution Place 100 mg of benzoic acid in the test tube. Add 1 mL of water to the tube and heat the mixture to a gentle boil. To aid in dissolution, stir the mixture vigorously with a microspatula while heating; stirring also prevents bumping and possible loss of material from the test tube. If needed, add more solvent in 0.1- to 0.2-mL increments to dissolve any remaining solid. Bring the mixture to boiling and continue stirring after each addition. Once all the solid has dissolved, add an additional 0.1- to 0.2-mL portion of solvent to ensure that the solute remains in solution during transfer of the hot solution to a Craig tube. *Caution:* Because the sample may be contaminated with insoluble material, pay close attention to whether additional solid is dissolving as you add more solvent. If it is not, *stop adding solvent.* Record the approximate total volume of solvent used.

Decoloration If the solution is colored and the recrystallized product is known or suspected to be colorless, treatment with decolorizing carbon (Sec. 2.18) is necessary. *Caution:* Do *not* add decolorizing carbon to a *boiling* solution! Cool the solution slightly, add half of a microspatula-tip full of powdered carbon or, preferably, a pellet or two of decolorizing carbon, and reheat to boiling for a few minutes.

Hot Filtration and Crystallization Preheat a Pasteur filtering pipet by pulling hot solvent into the pipet. Then transfer the hot solution into this pipet with a Pasteur pipet or filter-tip pipet that has also been preheated with solvent, using the tared Craig tube as the receiver for the filtrate (Fig. 2.53). If decolorizing carbon or other insoluble matter appears in the Craig tube, pass the solution through the filtering pipet a second and, if necessary, a third time. Concentrate the clear solution in the Craig tube to the point of saturation by heating it to boiling. Rather than using a boiling stone to prevent superheating and possible bumping, continually stir the solution with a microspatula while heating. The saturation point will be signaled by the appearance of cloudiness in the solution and/or the formation of crystals on the microspatula at the air-liquid interface. When this occurs, add solvent dropwise at the boiling point until the cloudiness is discharged. Then remove the tube from the heating source, cap the tube with a loose plug of cotton to exclude airborne impurities, and allow the solution to cool to room temperature.★ If necessary, induce crystallization by gently scratching the surface of the tube at the air-liquid interface or by adding seed crystals, if these are available. To complete crystallization, cool the tube in an ice-water bath for at least 15 min.

Isolation and Drying Affix a wire holder to the Craig tube and, using the wire as a hanger, invert the apparatus in a centrifuge tube (Fig. 2.55). Remove the solvent by centrifugation, carefully disassemble the Craig tube, and scrape any crystalline product clinging to its upper section into the lower part. Protect the product from airborne contaminants by plugging the opening of the tube with cotton or by covering it with a piece of filter paper held in place with a rubber band. Air-dry the crystals to constant weight either at room temperature or in an oven. Be certain that the temperature of the oven is 20–30 °C below the melting point of the product!

Analysis Determine the melting points of the crude and recrystallized benzoic acid, the weight of the latter material, and calculate your percent recovery using Equation 3.1.

2. Acetanilide

Apparatus A 10-mm × 75-mm test tube, a calibrated Pasteur pipet and Pasteur filtering and filter-tip pipets, ice-water bath, apparatus for Craig tube filtration and *flameless* heating.

Dissolution Place 100 mg of acetanilide in the test tube. Add 1 mL of water to the tube and heat the mixture to a gentle boil. To aid in dissolution, stir the mixture vigorously with a microspatula while heating; stirring also prevents bumping and possible loss of material from the test tube. A layer of oil should form when the stated amount of water is added. An explanation for this phenomenon is found under the

Dissolution stage of Part 2 of the Miniscale Procedures. Continue adding water in 0.1-mL portions to the boiling solution until the oil has completely dissolved. *Caution:* Because the sample may be contaminated with insoluble material, pay close attention to whether additional solid is dissolving as you add more solvent; if it is not, *stop adding solvent.*

Once the acetanilide has just dissolved, add an additional 0.1 mL of water to prevent formation of oil during the crystallization step. If oil forms at this time, reheat the solution and add a little more water. Record the approximate total volume of water used.

Continue the procedure by following the directions for *Decoloration, Hot Filtration and Crystallization,* and *Isolation and Drying* given for benzoic acid in Part 1.

Analysis Determine the melting points of the crude and recrystallized acetanilide, the weight of the latter material, and calculate your percent recovery using Equation 3.1.

3. Naphthalene

Apparatus A 10-mm × 75-mm test tube, a calibrated Pasteur pipet and Pasteur filtering and filter-tip pipets, ice-water bath, apparatus for Craig tube filtration and *flameless* heating.

Dissolution Naphthalene may be conveniently recrystallized from methanol, 95% ethanol, or 2-propanol. Because these solvents are somewhat toxic and/or flammable, proper precautions should be taken. If instructed to do so, position an inverted funnel connected to a vacuum source above the mouth of the test tube being used for recrystallization (Fig. 2.71b) to minimize release of vapors into the laboratory.

Place 100 mg of naphthalene in the test tube and dissolve it in a minimum amount of boiling solvent. *Caution:* Because the sample may be contaminated with insoluble material, pay close attention to whether additional solid is dissolving as you add more solvent; if it is not, *stop adding solvent.* Then add 0.1 mL of additional solvent to ensure that premature crystallization will not occur during subsequent transfers. Record the total volume of solvent used.

Continue the procedure by following the directions for *Decoloration, Hot Filtration and Crystallization,* and *Isolation and Drying* given for benzoic acid in Part 1; however, rather than water, use the solvent in which you dissolved the naphthalene.

Analysis Determine the melting points of the crude and recrystallized naphthalene, the weight of the latter material, and calculate your percent recovery using Equation 3.1.

4. Unknown Compound

Apparatus A 10-mm × 75-mm test tube, a calibrated Pasteur pipet and Pasteur filtering and filter-tip pipets, ice-water bath, apparatus for Craig tube filtration and *flameless* heating.

Dissolution If instructed to do so, position an inverted funnel connected to a vacuum source above the mouth of the test tube being used for recrystallization (Fig. 2.71b) to minimize release of vapors into the laboratory. Place 100 mg of the unknown in the test tube and dissolve it in a minimum amount of boiling solvent you selected on the basis of solubility tests. *Caution:* Because the sample may be contaminated with insoluble material, pay close attention to whether additional solid is dissolving as you add more solvent; if it is not, *stop adding solvent.* Then add 0.1 mL of additional solvent to ensure that premature crystallization will not occur during subsequent transfers. Record the total volume of solvent used.

Continue the procedure by following the directions for *Decoloration, Hot Filtration and Crystallization,* and *Isolation and Drying* given for benzoic acid in Part 1; however, if you are not using water, use the solvent in which you dissolved the unknown.

Analysis Determine the melting points of the crude and recrystallized unknown, the weight of the latter material, and calculate your percent recovery using Equation 3.1.

5. Mixed-Solvent Crystallization

Apparatus A 10-mm × 75-mm test tube, a calibrated Pasteur pipet and Pasteur filtering and filter-tip pipets, ice-water bath, apparatus for Craig tube filtration and *flameless* heating.

Dissolution Place 100 mg of benzoic acid or acetanilide in the test tube. Add 0.2 mL of 95% ethanol and heat the mixture to a gentle boil. If necessary, continue to add 95% ethanol dropwise by Pasteur pipet until a homogeneous solution is obtained. *Caution:* Because the sample may be contaminated with insoluble material, pay close attention to whether additional solid is dissolving as you add more solvent; if it is not, *stop adding solvent.* Once all the solid has dissolved, add an additional 0.1- to 0.2-mL portion of solvent to ensure that the solute remains in solution during transfer of the hot solution to a Craig tube and, if necessary, decolorize the solution according to the directions in Part 1 for benzoic acid.

Hot Filtration and Crystallization Preheat a Pasteur filtering pipet by pulling hot solvent into the pipet. Then transfer the hot solution into this pipet with a Pasteur pipet or filter-tip pipet that has also been preheated with solvent, using the tared Craig tube as the receiver for the filtrate (Fig. 2.53). If decolorizing carbon or other insoluble matter appears in the Craig tube, pass the solution through the filtering pipet a second and, if necessary, a third time.

Reheat the decolorized solution to boiling and add water dropwise from a Pasteur pipet until the boiling solution remains cloudy or precipitate forms. Rather than using a boiling stone to prevent superheating and possible bumping, continually stir the solution with a microspatula while heating. Then add a drop or two of 95% ethanol to restore homogeneity. Remove the tube from the heating source, cap the tube with a loose plug of cotton to exclude airborne impurities, and allow the solution to cool to room temperature.[*] Follow the same directions as given for benzoic acid in Part 1 to complete both this stage of the procedure and *Isolation and Drying.*

Analysis Determine the melting points of the crude and recrystallized product, the weight of the latter material, and calculate your percent recovery using Equation 3.1.

Discovery Experiment *Formation of Polymorphs*

Develop an experimental procedure for forming polymorphs (see the Historical Highlight *Polymorphism,* which is available online) of *trans*-cinnamic acid or derivatives thereof by referring to the following publications: Bernstein, H. I.; Quimby, W. C., *J. Am. Chem. Soc.* **1943**, 65, 1845–1848; Cohen, M. D.; Schmidt, G. M. J.; Sonntag, F. I., *J. Chem. Soc.* **1964**, 2000–2013. Consult with your instructor before undertaking your proposed procedure.

WRAPPING IT UP

Flush any *aqueous filtrates* or *solutions* down the drain. With the advice of your instructor, do the same with the *filtrates* derived from use of *alcohols, acetone,* or other *water-soluble solvents.* Use the appropriate containers for the *filtrates* containing *halogenated solvents* or *hydrocarbon solvents.* Put *filter papers* in the container for nontoxic waste, unless instructed to do otherwise.

EXERCISES

1. Define or describe each of the following terms as applied to recrystallization:

   a. solution recrystallization

   b. temperature coefficient of a solvent

   c. the relationship between dielectric constant and polarity of a solvent

   d. petroleum ether   j. vacuum filtration

   e. mixed solvents   k. mother liquor

   f. solvent selection   l. filtrate

   g. filter-aid   m. solute

   h. hot filtration   n. solvent

   i. seeding   o. occlusion

2. Provide the IUPAC name for ethyl acetate and acetone.

3. For which of the following situations is it appropriate to perform a recrystallization? More than one answer may be circled.

   a. To purify an impure liquid.

   b. To purify an impure solid.

   c. The melting point of a compound is depressed and melts over a wide range.

   d. The melting point of a compound is sharp and agrees with the literature value.

   e. An impure solid is soluble in all possible recrystallization solvents that are available.

4. List the steps in the systematic procedure for miniscale recrystallization, briefly explaining the purpose of each step.

5. In performing a hot filtration at the miniscale level, what might happen if the filter funnel is not preheated before the solution is poured through it?

6. Describe the use of the following pieces of equipment during recrystallizations at the miniscale level.

   a. filter flask

   b. filter trap

   c. Büchner funnel

7. Why should the filter flask not be connected directly to a water aspirator pump when performing a vacuum filtration?

8. List the steps in the systematic procedure for microscale recrystallization, briefly explaining the purpose of each step.

9. Why is a Pasteur filter-tip pipet preferred to a regular Pasteur pipet for transferring hot volatile solutions?

10. In performing a hot filtration at the microscale level, what might happen if the Pasteur filtering and filter-tip pipets are not preheated before filtration?

11. Why is a Craig tube filtration ineffective for removing finely divided particles from a solution?

12. Briefly explain how a colored solution may be decolorized.

13. Briefly explain how insoluble particles can be removed from a hot solution.

14. When might Celite, a filter-aid, be used in recrystallization, and why is it used?

15. List five criteria that should be used in selecting a solvent for a recrystallization.

16. How does the principle of *like dissolves like* explain the differing solubilities of solutes in various solvents?

17. The following solvent selection data were collected for two different impure solids:

Solid A

Solvent	Solubility at Room Temperature	Solubility When Heated	Crystals Formed When Cooled
Methanol	Insoluble	Insoluble	—
Ethyl acetate	Insoluble	Soluble	Very few
Cyclohexane	Insoluble	Soluble	Many
Toluene	Insoluble	Soluble	Very few

Solid B

Solvent	Solubility at Room Temperature	Solubility When Heated	Crystals Formed When Cooled
Water	Soluble	—	—
Ethanol	Soluble	—	—
Dichloromethane	Insoluble	Insoluble	—
Petroleum ether	Insoluble	Insoluble	—
Toluene	Insoluble	Insoluble	—

Based on these results, what solvents or mixture of solvents might you consider using to recrystallize solids A and B? Explain.

18. Briefly describe how a mixture of sand and benzoic acid, which is soluble in hot water, might be separated to provide pure benzoic acid.

19. Look up the solubility of benzoic acid in hot water. According to the published solubility, what is the minimum amount of water in which 1 g of benzoic acid can be dissolved?

20. The solubility of benzoic acid at 0 °C is 0.02 g per 100 mL of water, and that of acetanilide is 0.53 g per 100 mL of water. If you performed either of these recrystallizations, calculate, with reference to the total volume of water you used in preparing the hot solution, the amount of material in your experiment that was unrecoverable by virtue of its solubility at 0 °C.

21. Assuming that either solvent is otherwise acceptable in a given instance, what advantages does ethanol have over 1-octanol as a crystallization solvent? hexane over pentane? water over methanol?

22. In the course of a synthesis of an important antibiotic, an impure solid was obtained as one of the intermediates. The solubility of this material in various solvents is shown below.

	Water	*Ethanol*	*Toluene*	*Petroleum Ether*	*2-Butanone*	*Acetic Acid*
Cold	Insoluble	Soluble	Insoluble	Insoluble	Soluble	Slightly soluble
Hot	Slightly soluble	Soluble	Soluble	Insoluble	Soluble	Soluble

   a. Which of the solvents above would be the most suitable for recrystallization of the impure solid? Explain your reasoning.

   b. Provide a reason why each of the following solvents would *not* be a suitable solvent for recrystallization of this material: ethanol, petroleum ether.

   c. Write the chemical structure for each of the solvents in the table above.

   d. Which solvent is the most polar? least polar?

23. The goal of the recrystallization procedure is to obtain purified material with a maximized recovery. For each of the items listed, explain why this goal would be adversely affected.

   a. In the solution step, an unnecessarily large volume of solvent is used.

   b. The crystals obtained after filtration are not washed with fresh cold solvent before drying.

   c. The crystals referred to in (**b**) are washed with fresh hot solvent.

   d. A large quantity of decolorizing carbon is used.

   e. Crystals are obtained by breaking up the solidified mass of an oil that originally separated from the hot solution.

   f. Crystallization is accelerated by immediately placing the flask of hot solution in an ice-water bath.

24. A second crop of crystals may be obtained by concentrating the vacuum filtrate and cooling. Why is this crop of crystals probably less pure than the first crop?

25. Explain why the rate of dissolution of a crystalline substance may depend on the size of its crystals.

26. In the technique of recrystallization, one step involves heating the solvent containing the solute to its boiling point and, after the solution is cooled to room temperature, cooling it further in an ice-water bath. Why is it important to operate at these temperature extremes during a recrystallization?

27. In the process of a recrystallization, if crystals do not form upon cooling the solution, it is often recommended that the inside of the flask be scratched at the air-liquid interface with a glass stirring rod. What purpose does this serve, and how does it work? What else might be done to induce crystallization?

28. Should some loss of sample mass be expected even after the most carefully executed recrystallization? Explain.

29. In general, what solvent should be used to rinse the filter cake during the vacuum filtration step of a recrystallization? Should this solvent be cooled prior to use?

30. Why do you seldom see high-boiling solvents used as recrystallization solvents?

31. At the end of a recrystallization, where should the *impurities* be located?

32. A student has been asked to recrystallize 1.0 g of impure stilbene from ethanol. Provide a set of standard step-by-step instructions for recrystallization of this sample so as to maximize the purity and yield obtained.

33. An important product from a multistep synthesis must be recrystallized to remove a small amount of an impurity. However, all the available solvents each individually fail to be suitable recrystallization solvents. Offer a solution to this problem using only the available solvents. (*Hint:* Consider binary solvents.)

34. A suspension of decolorizing carbon (charcoal) is often administered to poison victims.

    a. Speculate on the purpose decolorizing carbon serves in this particular application. (*Hint:* It is similar to the way in which decolorizing carbon is used in a recrystallization.)

    b. How is the charcoal ultimately removed from the victim?

3.3 PHYSICAL CONSTANTS: MELTING POINTS

Physical Constants

Physical constants of compounds are numerical values associated with measurable properties of these substances. These properties are *invariant* and are useful in the identification and characterization of substances encountered in the laboratory so long as accurate measurements are made under specified conditions such as temperature and pressure. Physical constants are useful only in the identification of *previously known* compounds, however, because it is not possible to predict the values of such properties accurately. Among the more frequently measured physical properties of organic compounds are **melting point (mp), boiling point (bp), index of refraction (n), density (d), specific rotation ([α])**, and **solubility.** Melting points, discussed below, boiling points, described in Section 4.2, and solubilities, outlined in Section 3.2, are the properties most commonly encountered. Specific rotation is discussed in Chapters 7 and 23 but applies only to molecules that are **optically active.** Whether the substance is known or unknown, such values, along with other properties like color, odor, and crystal form, should be recorded in the laboratory notebook.

The values of one or two of the common physical properties *may* be identical for more than one compound, but it is most unlikely that values of several such properties will be the same for two different compounds. Consequently, a list of physical constants is a highly useful way to characterize a substance. Extensive compilations of the physical constants are available. One of the most convenient is the *CRC Handbook of Chemistry and Physics,* which contains a tabulation of the physical constants and properties of a large number of inorganic and organic compounds. *The Handbook of Tables for Organic Compounds* is especially useful for organic compounds. Neither of these books is comprehensive; rather, they contain entries for only the more common organic and inorganic substances. So many compounds are known that multivolume sets of books are required to list their physical properties.

Melting Point of a Pure Substance

The melting point of a substance is defined as the temperature at which the liquid and solid phases exist in equilibrium with one another without change of temperature. Ideally, addition of heat to a mixture of the solid and liquid phases of a pure substance at the melting point will cause no rise in temperature until all the solid has melted. Conversely, removal of heat from the equilibrium mixture will produce no decrease in temperature until all the liquid solidifies. This means that the melting and freezing points of a pure substance are identical.

The melting point is usually expressed as the temperature *range* over which the solid starts to melt and then is completely converted to liquid. Consequently, rather than a melting *point,* what is actually measured is a **melting range,** although the two terms are often used interchangeably. If a crystalline substance is pure, it should melt over a narrow or sharp range, which will normally be about 1 °C if the melting point is determined carefully. The melting ranges reported for many "pure" compounds may be greater than 1 °C because the particular compound was not quite pure or the melting point was not measured properly. The process of melting may actually begin by "softening," as evidenced by an apparent shrinking of the solid, but such softening is difficult to observe. Thus, for our purposes, the *start* of melting is defined as the temperature at which the first tiny droplet of liquid can be detected. The *end* of melting is defined as the temperature at which no solid remains. If you observe that a substance melts over a range of 117–119 °C, for example, you should report this range rather than a single temperature such as 118 °C.

Effect of Impurities on Melting Points

Many solid substances prepared in the organic laboratory are initially impure, so the effect of impurities on melting-point ranges deserves further discussion. Although this topic is discussed in freshman chemistry textbooks, a brief review of its basic principles is given here.

The presence of an impurity generally *decreases* the melting point of a pure solid. This is shown graphically by the melting point–composition diagram of Figure 3.1, in which points *a* and *b* represent the melting points of pure *A* and *B*, respectively. Point *E* is called the **eutectic point** and is determined by the equilibrium composition at which *A* and *B* melt in constant ratio. In Figure 3.1, this ratio is 60 mol % *A* and 40 mol % *B*; an impure solid composed of *A* and *B* in this ratio would be called a **eutectic mixture.** The temperature at the eutectic point is designated by *e.*

Now consider the result of heating a solid mixture composed of 80 mol % *A* and 20 mol % *B*, a sample that might be considered as "impure *A*." As heat is applied to the solid, its temperature will rise. When the temperature reaches *e*, *A* and *B* will both begin to melt in the constant ratio defined by the composition at the eutectic point. Once all of the "impurity" *B* has melted, only solid *A* will be left in equilibrium with the melt. The remaining solid *A* will continue to melt as additional heat is

Figure 3.1
Melting point–composition diagram for two hypothetical solids, A and B.

supplied, and the percentage of *A* in the melt will increase, changing the composition of the melt from that of the eutectic mixture. This increases the vapor pressure of *A* in the solution according to Raoult's law (Eq. 4.2) and raises the temperature at which solid *A* is in equilibrium with the molten solution. The relationship between the equilibrium temperature and the composition of the molten solution is then represented by curve *EF* in Figure 3.1. When the temperature reaches *f*, no solid *A* will remain and melting of the sample will be complete. The impure sample *A* exhibits a melting "point" that extends over the relatively broad temperature range *e–f*. Because melting both begins and ends below the melting point of pure *A*, the melting point of *A* is said to be *depressed*.

The foregoing analysis is easily extended to the case in which substance *B* contains *A* as an impurity. In Figure 3.1, this simply means that the composition of the solid mixture is to the right of point *E*. The temperature during the melting process would follow curve *ED* or *EG*, and the melting range would now be *e–d* or *e–g*.

A sample whose composition is exactly that of the eutectic mixture (point *E*, Fig. 3.1) will exhibit a sharp melting point at the eutectic temperature. This means a eutectic mixture can be mistaken for a pure compound, because both have a sharp melting point.

From a practical standpoint, it may be very difficult to observe the initial melting point of solid mixtures, particularly with the capillary-tube melting-point technique used in the Experimental Procedure that follows. This is because the presence of only a minor amount of impurity means that only a tiny amount of liquid is formed in the stage of melting that occurs at the eutectic temperature. In contrast, the temperature at which the last of the solid melts (points *d* and *g*, Fig. 3.1) can be determined accurately. Consequently, a mixture containing smaller amounts of impurities will generally have both a higher final melting point and a narrower observed melting-point range than one that is less pure.

The broadening of the melting-point range that results from introducing an impurity into a pure compound may be used to advantage for identifying a pure substance. The technique is commonly known as a **mixed melting point** and is illustrated by the following example. Assume that an unknown compound *X* melts at 134–135 °C, and you suspect it is either urea, H_2NCONH_2, or *trans*-cinnamic acid, $C_6H_5CH{=}CHCO_2H$, both of which melt in this range. If *X* is mixed intimately with urea and the melting point of this mixture is found to be lower than that of the pure compound and pure urea, then urea is acting as an impurity, and the compound cannot be urea. If the mixture melting point is identical to that of the pure compound and of urea, the compound is identified as urea. Obviously, this procedure is useful in identifying compounds only when authentic samples of the likely possibilities are available.

A convenient and rapid method for ascertaining the purity of a solid is measuring its melting point. A narrow melting-point range ordinarily signals that the sample is *pure*, although there is a *low* probability that the solid is a eutectic mixture. If recrystallizing a sample changes an originally broad melting range to a narrow one, the reasonable conclusion is that the recrystallization was successful in purifying the solid. Should the melting-point range remain broad after recrystallization, the sample may be contaminated with solvent and additional drying is required. It is also possible that the recrystallization was not completely successful in removing impurities, in which case the solid should be recrystallized using the same solvent. If this fails to narrow the melting range satisfactorily, recrystallization should be performed with a different solvent.

Micro Melting-Point Methods

Determining accurate melting points of organic compounds can be time-consuming. Fortunately, micro methods are available that are convenient, require negligible amounts of sample, and give melting-point data that are satisfactory for most purposes. The technique using the capillary-tube melting-point procedure is the one used most commonly in the organic laboratory.

There are practical considerations in determining melting points, and some of them are briefly noted here. First, the observed melting-point range depends on several factors, including the quantity of sample, its state of subdivision, the rate of heating during the determination, and the purity and chemical characteristics of the sample. The first three factors can cause the observed melting-point range to differ from the true value because of the time lag for transfer of heat from the heating medium to the sample and for conduction of heat within the sample. For example, if the sample is too large, the distribution of heat may not be uniform, and inaccurate melting ranges will result. A similar problem of nonuniform heat distribution is associated with using large crystals. It will be difficult to pack the sample tightly in the capillary melting tube, and the airspace that results causes poor conduction of heat. If the rate of heating is too fast, the thermometer reading will lag behind the actual temperature of the heating medium and produce measurements that are low. The chemical characteristics of the sample may be important if the compound tends to decompose on melting. When this occurs, discoloration of the sample is usually evident, and it may be accompanied by gas evolution. The decomposition products constitute impurities in the sample, and the true melting point is lowered as a result. The reporting of melting points for compounds that melt with decomposition should reflect this, as in "mp 195 °C (dec)."

Proper technique for obtaining accurate melting points requires that the temperature of the melting-point device is increasing at a rate of only 2–3 °C/min. Consequently, in determining the melting point of a compound, valuable time can be wasted waiting for melting to occur if the proper slow rate of heating is being used on a sample whose melting point is unknown. It is considerably more efficient to prepare two capillary tubes containing the compound being studied and to determine a crude melting point, which will be lower than the true figure, by rapidly heating one of them. By then allowing the heating source to cool 10–15 °C below this *approximate* melting point, which will be lower than the true melting point, an accurate melting point can then be obtained with the second sample.

The accuracy of any type of temperature measurement ultimately depends on the quality and calibration of the thermometer. A particular thermometer may provide accurate readings in some temperature ranges but may be off by a degree or two in others. Melting points that have been determined using a calibrated thermometer may be reported in the form "mp 101–102 °C (corr.)," where "corr." is the abbreviation for "corrected"; the corresponding abbreviation for values obtained with an uncalibrated thermometer is "uncorr." for "uncorrected."

Calibration involves the use of standard substances for the measurement of the temperature at a series of known points within the range of the thermometer and the comparison of the observed readings with the true temperatures. The difference between the observed and the true temperature measurement provides a correction that must be applied to the observed reading. Calibration over a range of temperatures is necessary because the error is likely to vary at different temperatures.

EXPERIMENTAL PROCEDURES

Melting Points

Purpose To determine melting points using the capillary-tube method.

SAFETY ALERT

1. **Wear safety glasses or goggles and suitable protective gloves while performing the experiments.**

2. **If a burner is used in this experiment, be sure that no flammable solvents are nearby. Keep the rubber tubing leading to the burner away from the flame. Turn off the burner when it is not being used.**

3. **Some kinds of melting-point apparatus, such as the Thiele tube, use mineral or silicone oils as the heat transfer medium. These oils may not be heated safely if they are contaminated with even a few drops of water. Heating these oils above 100 °C may produce splattering of hot oil as the water turns to steam. Fire can also result if spattered oil comes in contact with open flames. Examine your Thiele tube for evidence of water droplets in the oil. If there are any, either change the oil or exchange tubes. Give the contaminated tube to your instructor.**

4. **Mineral oil is a mixture of high-boiling hydrocarbons and should not be heated above 200 °C because of the possibility of spontaneous ignition, particularly when a burner is used for heating. Some silicone oils may be heated to about 300 °C without danger (Sec. 2.9).**

5. **Be careful to avoid contact of chemicals with your skin. Clean up any spilled chemicals immediately with a brush or paper towel.**

6. **If you use a Thiele tube, handle it carefully when you are finished, because the tube cools slowly. To avoid burns, take care when removing it from its support.**

A ■ *Calibration of Thermometer*

Procedure

Preparation Refer to the online resources to answer Pre-Lab Exercises, access videos, and read the MSDSs for the chemicals used or produced in this procedure. Read or review Sections 2.7 and 2.9.

Apparatus Capillary melting-point tubes, packing tube, melting-point apparatus.

Protocol Carefully determine the capillary melting points of a series of standard substances. A list of suitable standards is provided in Table 3.2. The temperatures given in Table 3.2 correspond to the upper limit of the melting-point range for pure samples of these standards.

Plot the corrections in your notebook as deviations from zero versus the temperature over the range encompassed by the thermometer. This allows you to tell, for example, that at about 130 °C the thermometer gives readings that are 2 °C too low, or that at 190 °C the readings are about 1.5 °C too high. These values should then be applied to correct all temperature measurements taken. The corrections you obtain are valid only for the thermometer used in the calibration. If you break it, not only are you likely to be charged for a replacement, but you must repeat the calibration.

B ■ *Determining Capillary-Tube Melting Points*

Procedure

Preparation Refer to the online resources to answer Pre-Lab Exercises, access videos, and read the MSDSs for the chemicals used or produced in this procedure. Read or review Sections 2.7 and 2.9. Plan to prepare two melting-point tubes if you are determining the melting points of unknown compounds.

Apparatus Capillary melting-point tubes, packing tube, melting-point apparatus.

1. Pure Compound

Protocol Select one or more compounds from a list of available compounds of known melting point and determine the melting-point ranges. Repeat as necessary until you obtain accurate results and are confident with the technique.

2. Unknown Compound

Protocol Accurately determine the melting range of an unknown pure compound supplied by your instructor.

Table 3.2 *Standards for Thermometer Calibration*

Compound	Melting Point (°C)
Ice water	0
3-Phenylpropanoic acid	48.6
Acetamide	82.3
Acetanilide	114
Benzamide	133
Salicylic acid	159
4-Chloroacetanilide	179
3,5-Dinitrobenzoic acid	205

3. Mixed Melting Points

Sample Preparation Prepare a sample for mixed melting-point determination by introducing 5–10% of a second substance as an impurity into a solid whose melting point was determined in Part 1. Intimately mix the two components by grinding them together with a small mortar and pestle. Alternatively, use a small, clean watch glass and glass stirring rod or metal spatula to mix the components. Be careful not to apply too much pressure to the glass rod, however, because it is more fragile than a pestle and may break. Do not perform this operation on a piece of filter paper, because fibers from the paper may contaminate the sample.

Protocol Accurately determine the melting range of the sample and compare it to that of the major component of the mixture to study the effect of impurities on the melting range of a previously pure compound.

Discovery Experiment *Melting-Point Depression*

Obtain a pair of solids whose melting points are within 1 °C of one another and determine whether they indeed depress one another's melting points. Some possible examples are 3-furoic acid and benzoic acid, 1-naphthylacetic acid and 2,5-dimethylbenzoic acid, *trans*-cinnamic acid and 2-furoic acid, 2-methoxybenzoic acid and malic acid, or dimethyl fumarate, 9-fluorenone, and methyl 4-bromobenzoate. Consult with your instructor regarding these and other possibilities.

WRAPPING IT UP

Return any *unused samples* to your instructor or dispose of them in the appropriate container for nonhazardous organic solids. Discard the used *capillary tubes* in a container for broken glass; do *not* leave them in the area of the melting-point apparatus or throw them in wastepaper baskets.

C ▪ Who Else Has My Compound?

Discovery Experiment
Science is a collaborative enterprise, and teams of scientists often work together to make discoveries. It is also common for groups of researchers from different parts of the world to collaborate by sharing experimental data and exchanging samples. The groups must then establish that the samples with which they are working are identical.

How do you know if two compounds are the same? Among organic chemicals, there are many colorless (white) and colored solids and liquids, so a visual inspection is usually insufficient to determine identity. Rather, we must rely upon comparison of measured physical properties. In this experiment, you will be working with your classmates to discover which of you have the same compound and what that compound is. Through solving this puzzle, you will learn some basic techniques for analyzing and identifying organic compounds.

The first part of the puzzle involves measuring physical properties of your compound and comparing them with data obtained by other students. The second part is proving that all the samples with the same or similar properties are indeed identical. Under the *same* conditions, compounds having comparable properties *may* be the same, but compounds having different properties are clearly not. Completing the puzzle requires that the compound be identified. In this regard, *single* measurements by themselves are insufficient for identifying a compound. For example, just knowing that a compound is a clear liquid at room temperature and has a melting point of 0 °C rules out many possibilities, but a variety of compounds melt at 0 °C. Knowing that the compound melts at 0 °C, boils at 100 °C, has a density of 1 g/mL, and is soluble in methanol but not in hexane gives us greater confidence that the substance is water. Accordingly, as you collect more data, your confidence in matching compounds that may be the same will increase as will your confidence in establishing the identity of the compound.

You will be given a numbered sample of a colorless pure solid as your piece of the puzzle. The first goal in solving the puzzle is finding who else in the laboratory-has the same compound as you. To accomplish this task, you will determine some properties of your sample, such as its melting point (Sec. 3.3), its solubility in two solvents (water and aqueous sodium hydroxide), and its R_f on a thin-layer chromatography (TLC) plate (Sec. 6.2). You will then share your results by recording the data in a central location to find the other students who have a compound whose properties are the same as or similar to yours. Strategies for making unambiguous comparisons that will ultimately prove the identity of different samples are provided below.

Your laboratory notebook should contain all the data you obtained, as well as a brief compilation of the data reported by the other students whose data are comparable to yours. These labmates and you will have a second goal of determining exactly which compound your group has. To accomplish this objective, the group will be given a list of possible comparison substances to use for making a final identification of the compound. These substances will also be available for actually making the comparisons. As before, suggestions will be provided that allow you to prove the identity of your sample and the comparison compound.

EXPERIMENTAL PROCEDURES

Purpose To demonstrate the equivalence of different samples and prove their structure.

SAFETY ALERT

1. **Wear safety glasses or goggles and suitable protective gloves while performing the experiments.**

2. **The solvents used for TLC analysis are *highly* flammable, so do not handle these liquids near open flames.**

3. **The 2.5 *M* solution of sodium hydroxide is caustic. Do not allow it to come into contact with your skin. If this should happen, thoroughly rinse the affected area with water.**

Preparation Refer to the online resources to answer Pre-Lab Exercises, access videos, and read the MSDSs for the chemicals used or produced in this procedure. Review Sections 3.2, 3.3, and 6.2.

A ■ *Determining Melting Points*

Apparatus Apparatus for determining melting points.

Analysis Accurately determine the melting point of your unknown solid.★

B ■ *Testing Solubility*

Apparatus Two 10-mm × 75-mm test tubes, calibrated Pasteur pipet, microspatula.

Dissolution Place approximately 1 mL of water in the first test tube. Add about a microspatula-tip full of the unknown solid to the test tube and gently stir the mixture with the spatula. Place 1 mL of 2.5 *M* aqueous sodium hydroxide in the second test tube. Add about a microspatula-tip full of the unknown solid to this test tube and gently stir the mixture with the spatula. For both tests, record your observations in your laboratory notebook.★

C ■ *Thin-Layer Chromatography*

Apparatus Wide-mouth bottles with screw-top caps, 3-cm × 10-cm silica gel TLC plates, capillary pipet, 10-mm × 75-mm test tube, microspatula.

Setting Up Dissolve a microspatula-tip full of your compound in approximately 1 mL of acetone and stir the mixture to effect dissolution. Using a capillary pipet, spot the solution on each of three TLC plates.

Analysis Develop one plate in 100% ethyl acetate, one plate in hexane, and one plate in a 20:80 mixture of ethyl acetate:hexane (by volume). Visualize your plates under UV light, calculate the R_f-values, and record the results in your laboratory notebook and where other students can see them.★

D ■ *Comparing Data and Determining the Structure of the Unknown*

Apparatus Apparatus for determining melting points, several 10-mm × 75-mm test tubes, calibrated Pasteur pipets, microspatula, wide-mouth bottle with screw-top cap, 3-cm × 10-cm silica gel TLC plates, capillary pipet.

Group Analysis Working with members of your group, obtain mixed melting points for each combination of samples in the group to prove that they are the same compound.★
Using the solvent system you determined to be the best for your compound, perform a TLC analysis by spotting the TLC plate with two different samples and a third "co-spot" (Sec. 6.2) that has both samples spotted at the same location. This technique minimizes the possibility that differing R_f-values for the two single spots result from variations in where the spots were applied to the plate or to the concentration of the sample being spotted, as discussed in Section 6.2. The result of the

TLC analysis should reinforce the conclusion you reach on the identity of samples based on the results of measuring mixed melting point.

Using the melting point and TLC data your group has obtained and comparing them to the known compounds provided by your instructor, decide which compound you are likely to have and obtain an authentic sample of it. Demonstrate the identity of your sample and the known compound by performing appropriate mixed melting points, solubility tests, and TLC analysis.

WRAPPING IT UP

Put the organic solvents from TLC analysis in the appropriate container for nonhalogenated organic liquids. Dispose of the aqueous solutions from the solubility tests in the appropriate aqueous waste container. Discard TLC plates in the appropriate solid waste container and the capillary pipets in the broken glass container.

EXERCISES

Questions for Part B

1. Describe errors in procedure that may cause an observed capillary melting point of a pure compound
   a. to be *lower* than the correct melting point.
   b. to be *higher* than the correct melting point.
   c. to be *broad* in range (over several degrees).

2. Briefly define the following terms:
   a. vapor pressure as applied to melting
   b. melting point or melting-point range
   c. mixture or mixed melting point
   d. eutectic point
   e. eutectic mixture

3. Describe on a molecular level the difference between the two physical changes "melting" and "dissolving."

4. Answer the following questions about melting points.
   a. Why is the melting "point" of a substance actually a melting "range" and therefore should never be recorded as a single temperature?
   b. In theory, does a melting "point" exist? Explain your answer.

5. Explain how a eutectic mixture could be mistaken for a pure substance, and comment on whether encountering a eutectic mixture would be a frequent or infrequent occurrence.

6. Compound A and compound B have approximately the same melting point. State two ways in which a mixed melting point of these two compounds would be different from the melting point of either pure A or pure B.

7. Filter paper is usually a poor material on which to powder a solid sample before introducing it into a capillary melting-point tube because small particles of paper may end up in the tube along with the sample. Why is this undesirable, and how might the presence of paper in the sample make the melting-point determination difficult?

8. Some solids sublime before they melt, making a determination of a melting point impossible using a standard melting-point capillary tube. How could you modify your capillary tube to obtain a melting point for such a compound?

9. Some solids, in particular many amino acids, decompose upon melting. These compounds are often reported in the literature with the term "dec" following their melting-point range.

   a. How might the melting process appear different for this type of compound?

   b. Look up and record the melting point and structure for a compound that decomposes upon melting. Use a chemical handbook or a chemical catalog as the source of this information.

10. Criticize the following statements by indicating whether each is *true* or *false*, and if false, explain why.

    a. An impurity always lowers the melting point of an organic compound.

    b. A sharp melting point for a crystalline organic substance always indicates a pure single compound.

    c. If the addition of a sample of compound *A* to compound *X* does not lower the melting point of *X*, *X* must be identical to *A*.

    d. If the addition of a sample of compound *A* lowers the melting point of compound *X*, *X* and *A* cannot be identical.

11. The melting points of pure benzoic acid and pure 2-naphthol are 122.5 °C and 123 °C, respectively. Given a pure sample that is known to be either pure benzoic acid or 2-naphthol, describe a procedure you might use to determine the identity of the sample.

12. A student used the Thiele micro melting-point technique to determine the melting point of an unknown and reported it to be 182 °C. Is this value believable? Explain why or why not.

13. The melting point–composition diagram for two substances, *Q* and *R*, is provided in Figure 3.2, which should be used to answer the following questions.

    a. What are the melting points of pure *Q* and *R*?

    b. What are the melting point and the composition of the eutectic mixture?

    c. Would a mixture of 20 mol % *Q* and 80 mol % *R* melt if heated to 120 °C? to 160 °C? to 75 °C?

    d. A mixture of *Q* and *R* was observed to melt at 105–110 °C. What can be said about the composition of this mixture? Explain briefly.

Figure 3.2

Melting point–composition diagram for Exercise 13.

14. For the following melting points, indicate what might be concluded regarding the purity of the sample:

    a. 120–122 °C

    b. 46–60 °C

    c. 147 °C (dec)

    d. 162.5–163.5 °C

15. Suppose that the observed melting-point range of a solid was originally 150–160 °C, but this became 145–145.5 °C after recrystallization. How might you account for this observation?

16. An unknown solid was found to have a melting point of 160–161.5 °C. Given other information that was available for the unknown, it was thought to be one of the following compounds:

Salicyclic acid	158–160 °C
Benzanilide	161–163 °C
Triphenylmethanol	159.5–160.5 °C

    If you had access to pure samples of the above compounds and could only use a melting-point apparatus, devise an experiment to determine which of these three compounds corresponds to the identity of the unknown.

Questions for Part C

17. Two unknown compounds have the same R_f, and the co-spot produces one spot. Is this sufficient information to conclude that the compounds are the same? Why or why not?

18. Explain why R_f-values for the same compounds determined using the same solvent system can be different when reported.

19. Explain why melting points determined for the same compound can be different when reported by different members of the class.

20. Assume you observed a melting point of 157–159 °C for an unknown. Of the known standards, 1-naphthoic acid has a melting point of 157–160 °C and triphenylmethanol has a melting point of 159–160.5 °C.

    a. Explain how you might use TLC to decide which of the two compounds to test against your unknown.

    b. Explain how you might use solubility tests to decide which of the two compounds to test against your unknown.

21. The R_f of a compound is 0.35 when the TLC was developed in 20:80 ethyl acetate:hexane. Will the R_f be higher or lower when the TLC is developed using 50:50 ethyl acetate:hexane? Explain why.

22. Two students are measuring the melting point of the same compound.

    a. Student A observes a melting point of 157–158 °C when using a melting-point tube containing a sample 2 mm in height. However, student B obtains a melting point of 157–161 °C using a sample 5 mm in height. Explain why there is this difference in melting point.

    b. Student A observes a melting point of 130–135 °C, whereas student B reports a melting point of 157–158 °C. What is the source of the error in the measurement made by student A?

23. Explain why benzoic acid is insoluble in water at room temperature but is soluble in 2.5 M aqueous sodium hydroxide.

24. Carboxylic acids often streak on silica gel TLC plates. Adding a drop or two of acetic acid to the developing solvent mixture will reduce streaking.

    a. Explain why carboxylic acids streak on silica gel TLC plates.

    b. Explain why adding acetic acid reduces streaking.

    c. Amines also often streak on silica gel TLC. What might you add to reduce the streaking of an amine?

25. Two compounds are spotted on a TLC plate, and the plate is developed with hexane. Both compounds stay at the starting point ($R_f = 0$).

    a. Are the compounds necessarily identical?

    b. What could you do to get a nonzero value of R_f for these compounds?

26. Benzoic acid is soluble in both 2.5 M aqueous sodium hydroxide and saturated sodium bicarbonate solution but not in water. On the other hand, 4-hydroxyacetanilide is soluble in 2.5 M aqueous sodium hydroxide but not in saturated sodium bicarbonate solution or water. Explain the difference in solubilities between these two compounds.

CHAPTER 4

Liquids
Distillation and Boiling Points

The odors of both natural and synthetic organic chemicals surround us in our everyday life. Some, like the smells of fruits and many spices, are pleasant, whereas others, like those of decaying meat and rancid butter, can be rather unpleasant. We detect these odors because many organic chemicals are volatile, and in this chapter we explore the theory and practice of distillation, a technique that depends on our ability to coax these chemicals into the gas phase.

4.1 INTRODUCTION

The purification of solids by recrystallization and the use of melting points as a criterion of their purity are discussed in Chapter 3. The techniques used for the purification of liquids involve **simple, fractional, steam,** and **vacuum distillation,** and these are topics of this chapter. Boiling points are also discussed as a physical property that can be used as one means of determining the purity and identity of liquids.

Two of the laws, Dalton's and Raoult's, that you have encountered in previous courses in chemistry and/or physics are critical to understanding the phenomenon of distillation. We'll review them in this chapter and show how they are applied to all forms of distillation that you will find in this textbook. As you will see, the apparatus for accomplishing distillations is considerably more complex than that needed for recrystallizations (Chap. 3), but with a little experience you will be able to complete a distillation in as little time as it takes to perform a recrystallization.

4.2 BOILING POINTS OF PURE LIQUIDS

The molecules of a liquid are constantly in motion, and those at the surface are able to escape into the vapor phase. The consequences of vaporization of a liquid contained in a closed, *evacuated* system are considered first, and then the situation in which the system is open to the atmosphere is discussed.

In a closed evacuated system, the number of molecules in the gas phase will initially increase until the rate at which they reenter the liquid becomes equal to the rate at which they escape from it. At this point, no further *net* change is observed,

125

and the system is said to be in a state of **dynamic equilibrium.** The molecules in the gas phase are in rapid motion and continually collide with the walls of the vessel, which results in the exertion of pressure against the walls. The magnitude of this vapor pressure at a given temperature is called the **equilibrium vapor pressure** of the liquid at that temperature. Vapor pressure is temperature-dependent, as shown in Figure 4.1. As the temperature of the system rises, the average kinetic energy of the liquid molecules increases, thus facilitating their escape into the gas phase. The rate of reentry of gaseous molecules into the liquid phase also increases, and equilibrium is reestablished at the higher temperature. Because there are now more molecules in the gas phase than there were at the lower temperature, the vapor pressure of the system is greater.

There is a very important safety rule based on the fact that the pressure in a closed system *increases* as the temperature of the system rises. Such a system should *not* be heated unless an apparatus designed to withstand the pressure is used; otherwise, an *explosion* will result. For purposes of the first laboratory course in organic chemistry, the rule is quite simply *"Never heat a closed system!"*

Now suppose that a liquid sample at a particular temperature is placed in an *open* container so that the molecules of the vapor over the liquid are mixed with air. The **total pressure** above the liquid is defined by **Dalton's law of partial pressures** (Eq. 4.1) and is equal to the sum of the partial pressures of the sample and of air. The partial pressure of the sample is equal to its **equilibrium vapor pressure** at the given temperature. When the temperature of the liquid is raised, the equilibrium vapor pressure of the sample will rise, and the number of gas molecules that have escaped from the liquid by the process of evaporation will increase in the space above the liquid. This will have the net effect of displacing some of the air. At the higher temperature, the partial pressure of the sample will be a *larger* percentage of the *total* pressure. This trend will continue as the temperature of the liquid is further increased, until the equilibrium vapor pressure of the sample equals the total pressure. At this point, all of the air will have been displaced from the vessel containing the liquid. Entry of additional molecules from the liquid into the gas phase will only have the effect of further displacing those already in that phase; the partial pressure of the molecules of the sample will no longer increase.

$$P_{total} = P_{sample} + P_{air} \tag{4.1}$$

When the temperature of the liquid is such that the equilibrium vapor pressure of the sample equals the total pressure, the rate of evaporation increases dramatically, and bubbles form in the liquid. This is referred to as the boiling process, and the temperature associated with it is the **boiling point** of the liquid. Since the boiling

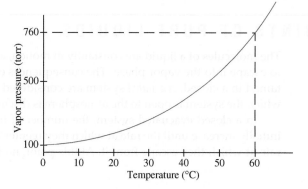

Figure 4.1
Graph of the dependence of vapor pressure on temperature for a liquid.

point is dependent upon the total pressure, that pressure must be specified when boiling points are reported, for example, "bp 132 °C (748 torr)." The boiling point of water is 100 °C *only* when the external pressure is 760 torr, which is 1 atm. The **normal** boiling point of a liquid is measured at 760 torr and is shown by the dashed lines in Figure 4.1 to be 60 °C. This figure also allows determination of the boiling point of the liquid at various total pressures. For example, at 500 torr the boiling point will be about 50 °C; the decrease in temperature results from the decrease in the total pressure in the system.

The dependence of vapor pressure upon total pressure, as reflected by the boiling point of a liquid, can be used to advantage in the following way. Suppose that a liquid has a normal boiling point such that it decomposes appreciably when heated to this temperature. Reduction in the total pressure of the system reduces the boiling point of the sample to a temperature at which it no longer decomposes. This technique, called **vacuum distillation,** is discussed in Section 2.15.

A **pure** liquid generally boils at a constant temperature or over a narrow temperature range, provided the total pressure in the system remains constant. However, the boiling point of a liquid is affected by the presence of both nonvolatile and volatile impurities, and most **mixtures** of liquids boil over a fairly wide range. These effects are discussed in Section 4.3.

EXPERIMENTAL PROCEDURES

Boiling Points of Pure Liquids

Purpose To determine the boiling points of pure liquids at atmospheric pressure.

SAFETY ALERT

1. **Wear safety glasses or goggles and suitable protective gloves while performing the experiments.**
2. **Volatile organic liquids are flammable, so use open flames, if needed, with *great* care. Consult with your instructor *before* lighting a burner.**
3. **Use paper towels to clean up spilled liquids; discard the towels as directed by your instructor. Avoid contact of organic liquids with your skin; if this happens, wash the affected area thoroughly with soap and water.**

MINISCALE PROCEDURE

Preparation Refer to the online resources to answer Pre-Lab Exercises, access videos, and read the MSDSs for the chemicals used or produced in this procedure. Read or review Sections 2.8 and 2.9.

Apparatus Thermometer, 10-mm × 75-mm test tube, two clamps, apparatus for *flameless* heating.

Setting Up Obtain a liquid from your instructor and determine its boiling point following the technique described in Section 2.8, using the apparatus shown in Figure 2.20 and *flameless* heating.

MICROSCALE PROCEDURE

Preparation Refer to the online resources to answer Pre-Lab Exercises, access videos, and read the MSDSs for the chemicals used or produced in this procedure. Read or review Sections 2.8 and 2.9.

Apparatus Thiele melting-point apparatus, micro boiling-point apparatus, a Bunsen burner or microburner.

Setting Up Determine the boiling point of the liquid(s) assigned by your instructor. Follow the technique presented in Section 2.8 for using the micro boiling-point apparatus (Fig. 2.21). Use a Bunsen burner or a microburner for heating. In the event you do not know the boiling point of the liquid, *first* determine an approximate boiling-point range by heating the Thiele tube (Fig. 2.17a) *fairly rapidly*. Repeat the measurement by heating the tube until the temperature is 20–30 °C below the approximate boiling point, and then heat the sample at a rate of 4–5 °C/min to obtain an accurate value. It may be desirable to repeat this procedure to obtain a more reliable boiling point.

WRAPPING IT UP

Unless instructed otherwise, return the organic liquids to the appropriate bottle for either nonhalogenated liquids or halogenated liquids.

EXERCISES

1. Refer to Figure 4.1 and answer the following:
   a. What total pressure would be required in the system in order for the liquid to boil at 45 °C?
   b. At about what temperature would the liquid boil when the total pressure in the system is 300 torr?

2. Describe the relationship between escaping tendency of liquid molecules and vapor pressure.

3. Define the following terms:
   a. boiling point
   b. normal boiling point
   c. Dalton's law of partial pressures
   d. equilibrium vapor pressure

4. Using Dalton's law, explain why a fresh cup of tea made with boiling water is not as hot at higher altitudes as it is at sea level.

5. At a given temperature, liquid A has a higher vapor pressure than liquid B. Which liquid has the higher boiling point?

6. Explain why the boiling point at 760 torr of a solution of water, bp 100 °C (760 torr), and ethylene glycol (HOCH$_2$CH$_2$OH), bp 196–198 °C (760 torr), exceeds 100 °C. For purposes of your answer, consider ethylene glycol as a *nonvolatile* liquid.

7. Why should there be no droplets of water in the oil of a heating bath?

8. Why is the micro boiling-point technique not applicable for boiling points in excess of 200 °C if *mineral* oil rather than *silicone* oil is the heating fluid in the Thiele tube?

9. A *rotary evaporator* (Sec. 2.29) is a device frequently used in the laboratory to remove solvent quickly under vacuum.

    a. Why is it possible to effect the removal of solvent at temperatures below their normal boiling points using this device?

    b. Other than being faster than simple distillation for removing a given volume of solvent, why might this type of distillation be preferred for isolating a desired product?

4.3 SIMPLE DISTILLATION

Simple distillation allows separation of distillates from less volatile substances that remain as pot residue at the completion of the distillation. In the ideal case, only a single component of the mixture will be volatile, so the distillate will be a pure compound. Real life is rarely ideal, however, and it is more common that several volatile components comprise the mixture. Simple distillation allows isolation of the various components of the mixture in acceptable purity if the *difference* between the boiling points of each pure substance is greater than 40–50 °C. For example, a mixture of diethyl ether, bp 35 °C (760 torr), and toluene, bp 111 °C (760 torr), could easily be separated by simple distillation, with the ether distilling first. Organic chemists frequently use this technique to separate a desired reaction product from the solvents used for the reaction or its work-up. The solvents are usually more volatile than the product and are readily removed from it by simple distillation.

To understand the principles of distillation, a review of the effect of impurities on the vapor pressure of a pure liquid is necessary. The discussion starts with consideration of the consequences of having *nonvolatile* impurities present and then turns to the more common case of contamination of the liquid with other *volatile* substances.

Consider a homogeneous solution composed of a nonvolatile impurity and a pure liquid; for the present purpose, these are taken as sugar and water, respectively.

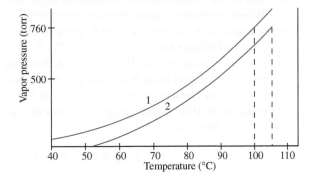

Figure 4.2
Diagram of dependence of vapor pressure on temperature. Curve 1 is for pure water, and Curve 2 is for a solution of water and sugar.

As a nonvolatile impurity, the sugar reduces the vapor pressure of the water because it lowers the concentration of the volatile constituent in the liquid phase. The consequence of this is shown graphically in Figure 4.2. In this figure, Curve 1 corresponds to the dependence of the temperature upon the vapor pressure of *pure* water and intersects the 760-torr line at 100 °C. Curve 2 is for a *solution* having a particular concentration of sugar in water. Note that the presence of the nonvolatile impurity reduces the vapor pressure at any temperature by a constant amount, in accord with Raoult's law as discussed later. The temperature at which this curve intersects the 760-torr line is higher because of the lower vapor pressure, and consequently the temperature of the boiling solution, 105 °C, is higher.

Despite the presence of sugar in the solution, the **head temperature** (Sec. 2.13) in the distillation will be the same as for pure water, namely, 100 °C (760 torr), since the water condensing on the thermometer bulb is now uncontaminated by the nonvolatile impurity. The **pot temperature** will be *elevated*, however, owing to the decreased vapor pressure of the solution (Fig. 4.2). As water distills, the pot temperature will progressively rise because the concentration of the sugar in the stillpot increases, further lowering the vapor pressure of the water. Nevertheless, the head temperature will remain constant, just as though pure water were being distilled.

The quantitative relationship between vapor pressure and composition of homogeneous liquid mixtures is known as **Raoult's law** (Eq. 4.2). The factor P_X represents the partial pressure of component X, and it is equal to the vapor pressure, P_X°, of pure X at a given temperature times the mole fraction N_X of X in the mixture. The **mole fraction** of X is defined as the fraction of *all* molecules present in the liquid mixture that are molecules of X. It is obtained by dividing the number of moles of X in a mixture by the sum of the number of moles of all components (Eq. 4.3). Raoult's law is strictly applicable only to **ideal solutions,** which are defined as those in which the interactions between *like* molecules are the same as those between *unlike* molecules. Fortunately, many organic solutions approximate the behavior of ideal solutions, so the following mathematical treatment applies to them as well.

$$P_X = P_X^\circ N_X \tag{4.2}$$

$$N_X = \frac{nX}{nX + nY + nZ + \cdots} \tag{4.3}$$

Note that the partial vapor pressure of X above an ideal solution depends *only* on its mole fraction in solution, and it is completely independent of the vapor pressures of the other volatile components of the solution. If all components other than X are nonvolatile, the total vapor pressure of the mixture will be equal to the partial pressure of X, since the vapor pressure of nonvolatile compounds may be taken as zero. Accordingly, the distillate from such a mixture will always be pure X. This is the case for the distillation of a solution of sugar and water, as discussed earlier.

When a mixture contains two or more volatile components, the *total* vapor pressure is equal to the *sum* of the partial vapor pressures of each such component. This is known as **Dalton's law** (Eq. 4.4), where P_X, P_Y, and P_Z refer to the vapor pressures of the volatile components. The process of distilling such a liquid mixture may be significantly different from that of simple distillation, because the vapors above the liquid phase will now contain some of each of the volatile components. Separation of the liquids in this case may require the use of fractional distillation, which is discussed in Section 4.4.

$$P_{\text{total}} = P_X + P_Y + P_Z + \cdots \tag{4.4}$$

EXPERIMENTAL PROCEDURES

Simple Distillation

Purpose To demonstrate the technique for purification of a volatile organic liquid containing a nonvolatile impurity.

SAFETY ALERT

1. **Wear safety glasses or goggles and suitable protective gloves while performing the experiments.**

2. **Cyclohexane is highly flammable, so be sure that burners are not being used in the laboratory. Use *flameless* heating (Sec. 2.9).**

3. **Examine your glassware for cracks and other weaknesses before assembling the distillation apparatus. Look for "star" cracks in round-bottom flasks, because these can cause a flask to break upon heating.**

4. **Proper assembly of glassware is important to avoid possible breakage and spillage or the release of distillate vapors into the room. Be certain that all connections in the apparatus are tight before beginning the distillation. Have your instructor examine your setup after it is assembled.**

5. **The apparatus used in these experiments *must* be open to the atmosphere at the receiving end of the condenser. *Never heat a closed system,* because the pressure buildup may cause the apparatus to explode!**

6. **Be certain that the water hoses are *securely* fastened to your condensers so that they will not pop off and cause a flood. If heating mantles or oil baths are used for heating in this experiment, water hoses that come loose may cause water to spray onto electrical connections or into the heating sources, either of which is potentially dangerous.**

7. **Avoid excessive inhalation of organic vapors at all times.**

MINISCALE PROCEDURE

Preparation Refer to the online resources to access videos, Pre-Lab Exercises, and read the MSDSs for the chemicals used or produced in this procedure. Read or review Sections 2.2, 2.4, 2.9, 2.11, and 2.13.

Apparatus A 25-mL round-bottom flask, ice-water bath, apparatus for simple distillation, magnetic stirring, and *flameless* heating.

Setting Up Place 10 mL of cyclohexane containing a nonvolatile dye in the round-bottom flask. Add a stirbar to the flask to ensure smooth boiling, and assemble the simple distillation apparatus shown in Figure 2.37a. Be sure to position the thermometer in the stillhead so the *top* of the mercury thermometer bulb is level with the *bottom* of the sidearm of the distillation head. Have your instructor check your apparatus *before* you start heating the stillpot.

Distillation Start the magnetic stirrer and begin heating the stillpot. As soon as the liquid begins to boil *and the condensing vapors have reached the thermometer bulb,* regulate the heat supply so that distillation continues steadily at a rate of 2–4 *drops per second;* if a drop of liquid cannot be seen suspended from the end of the thermometer, the rate of distillation is too *fast.* As soon as the distillation rate is adjusted and the head temperature is constant, note and record the temperature. Continue the distillation and periodically record the head temperature. Discontinue heating when only 2–3 mL of impure cyclohexane remains in the distillation flask. Record the volume of distilled cyclohexane that you obtain.

Optional Procedure

You may be required to perform this distillation using the shortpath apparatus discussed in Section 2.13 and illustrated in Figure 2.37b. After assembling the equipment, make certain that the *top* of the mercury in the thermometer bulb is level with the *bottom* of the sidearm of the distillation head. The preferred way to collect the distillate in this distillation is to attach a dry round-bottom flask to the vacuum adapter and put a drying tube containing calcium chloride on the sidearm of the adapter to protect the distillate from moisture. The receiver should be cooled in an ice-water bath to prevent loss of product by evaporation and to ensure complete condensing of the distillate.

MICROSCALE PROCEDURE

Preparation Refer to the online resources to answer Pre-Lab Exercises, access videos, and read the MSDSs for the chemicals used or produced in this procedure. Read or review Sections 2.3, 2.4, 2.9, 2.11, and 2.13.

Apparatus A Pasteur pipet, apparatus for simple distillation, magnetic stirring, and *flameless* heating.

Setting Up Transfer 2 mL of impure cyclohexane containing a nonvolatile dye to a 5-mL conical vial. Equip the vial with a spinvane, the Hickman stillhead, and a condenser as shown in Figure 2.38. Place the apparatus in the heating source and have your instructor check your apparatus before you start heating the vial.

Distillation Start the magnetic stirrer and begin heating the stillpot. Increase the temperature of the heating source until vapors of distillate begin to rise into the Hickman stillhead and condense into the flared portion of the head. Control the rate of heating so that the vapor line rises no more than halfway up the upper portion of the head; otherwise, distillate may be lost to the atmosphere. Be certain to record the bath temperature at which the cyclohexane distills. You may also measure the approximate temperature of the vapors by inserting a thermometer through the condenser and stillhead to a point in the top third of the conical vial just above the boiling liquid. This will give you an approximate boiling point of the cyclohexane. Terminate heating when about 0.3 mL of the original solution remains. Disconnect the vial from the Hickman stillhead and, using a Pasteur pipet, transfer the distillate to a properly labeled screw-cap vial.

WRAPPING IT UP

Unless directed otherwise, return the *distilled* and *undistilled cyclohexane* to a bottle marked "Recovered Cyclohexane."

EXERCISES

1. Define the following terms:
   a. simple distillation
   b. head temperature
   c. pot temperature
   d. Raoult's law
   e. ideal solution
   f. Dalton's law

2. Sketch and completely label the apparatus required for a simple distillation.

3. Why should you never heat a closed system, and how does this rule apply to a distillation?

4. Explain the role of the stirbar that is normally added to a liquid that is to be heated to boiling.

5. In a miniscale distillation, the top of the mercury bulb of the thermometer placed at the head of a distillation apparatus should be adjacent to the exit opening to the condenser. Explain the effect on the observed temperature reading if the bulb is placed (a) below the opening to the condenser or (b) above the opening.

6. Distillation is frequently used to isolate the nonvolatile organic solute from a solution containing an organic solvent. Explain how this would be accomplished using a simple distillation.

7. Using Raoult's and Dalton's laws, explain why an aqueous NaCl solution will have a higher boiling point than pure water.

8. At 100 °C, the vapor pressures for water, methanol, and ethanol are 760, 2625, and 1694 torr, respectively. Which compound has the highest normal boiling point and which the lowest?

4.4 FRACTIONAL DISTILLATION

It is easy to separate a volatile compound from a nonvolatile one by simple distillation (Sec. 4.3). The same technique may also be used to separate volatile compounds from one another if their boiling points differ by at least 40–50 °C. If this is not the case, the technique of fractional distillation is normally used to obtain each volatile component of a mixture in pure form. The theoretical basis of this technique is the subject of the following discussion, and a potential practical application of fractional distillation that could lessen the air pollution produced by internal combustion engines is described in the Historical Highlight *Reducing Automobile Emissions*, which is available online.

Theory

For simplicity, we'll only consider the theory for separating *ideal solutions* (Sec. 4.3) consisting of two volatile components, designated *X* and *Y*. Solutions containing more than two such components are often encountered, and their behavior on distillation may be understood by extension of the principles developed here for a binary system.

The vapor pressure of a compound is a measure of the ease with which its molecules escape the surface of a liquid. When the liquid is composed of two volatile components, in this case *X* and *Y*, the number of molecules of *X* and of *Y* in a given volume of the vapor above the mixture will be proportional to their respective partial vapor pressures. This relationship is expressed mathematically by Equation 4.5, where N_X'/N_Y' is the ratio of the mole fractions of *X* and *Y* in the *vapor* phase. The mole fraction of each component may be calculated from the equations $N_X' = P_X/(P_X + P_Y)$ and $N_Y' = P_Y/(P_X + P_Y)$. The partial vapor pressures, P_X and P_Y, are determined by the composition of the liquid solution according to Raoult's law (Eq. 4.2). Since the solution boils when the sum of the partial vapor pressures of X and Y is equal to the external pressure, as expressed by Dalton's law (Eq. 4.4), the boiling temperature of the solution is determined by its composition.

$$\frac{N'_X}{N'_Y} = \frac{P'_X}{P'_X} = \frac{P^\circ_X N_x}{P^\circ_Y N_Y} \tag{4.5}$$

The relationship between temperature and the composition of the liquid and vapor phases of ideal binary solutions is illustrated in Figure 4.3 for mixtures of benzene, bp 80 °C (760 torr), and toluene, bp 111 °C (760 torr). The lower curve, the **liquid line,** gives the boiling points of all mixtures of these two compounds. The upper curve, the **vapor line,** is calculated using Raoult's law and defines the composition of the vapor phase in equilibrium with the boiling liquid phase at the same temperature. For example, a mixture whose composition is 58 mol % benzene and 42 mol % toluene will boil at 90 °C (760 torr), as shown by point A in Figure 4.3. The composition of the vapor in equilibrium with the solution when it *first* starts to boil can be determined by drawing a horizontal line from the *liquid line* to the *vapor line;* in this case, the vapor has the composition 78 mol % benzene and 22 mol % toluene, as shown by point B in Figure 4.3. This is a key point, for it means that at any given temperature the *vapor phase is richer in the more volatile component than is the boiling liquid with which the vapor is in equilibrium.* This phenomenon provides the basis of **fractional distillation.**

When the liquid mixture containing 58 mol % benzene and 42 mol % toluene is heated to 90 °C (760 torr), its boiling point, the vapor formed initially contains 78 mol % benzene and 22 mol % toluene. If this first vapor is condensed, the condensate would also have this composition and thus would be much richer in benzene than the original liquid mixture from which it was distilled. After this vapor is removed from the original mixture, the liquid remaining in the stillpot will contain a smaller mol % of benzene and a greater mol % of toluene because more benzene

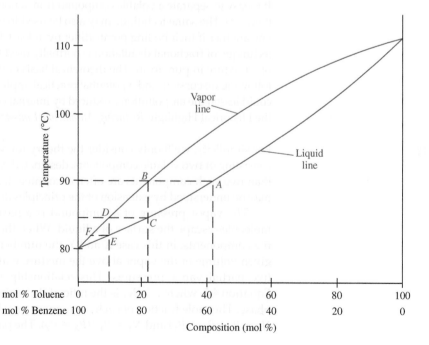

Figure 4.3

Temperature–composition diagram for binary mixture of benzene and toluene.

than toluene was removed by vaporization. The boiling point of the liquid remaining in the distilling flask will rise as a result. As the distillation continues, the boiling point of the mixture will steadily increase until it approaches or reaches the boiling point of pure toluene. The composition of the distillate will change as well and will ultimately consist of "pure" toluene.

Now let's return to the first few drops of distillate that are obtained by condensing the vapor initially formed from the original mixture. This condensate, as noted earlier, has a composition identical to that of the vapor from which it is produced. Were this liquid to be collected and then redistilled, its boiling point would be the temperature at point C, namely 85 °C; this boiling temperature is easily determined by drawing a vertical line from the vapor line at point B to the liquid line at point C, which corresponds to the composition of the distillate initially produced. The first distillate obtained at this temperature would have the composition D, 90 mol % benzene and 10 mol % toluene; this composition is determined from the intersection with the vapor line of the horizontal line from point C on the liquid line.

In theory, this process could be repeated again and again to give a very small amount of pure benzene. Similarly, collecting the *last* small fraction of each distillation and redistilling it in the same stepwise manner would yield a very small amount of pure toluene. If larger amounts of the initial and final distillates were collected, reasonable quantities of materials could be obtained, but a large number of individual simple distillations would be required. This process would be extremely tedious and time-consuming. Fortunately, the repeated distillation can be accomplished almost automatically in a single operation by using a **fractional distillation column,** the theory and use of which are described later in this section.

Most homogeneous solutions of volatile organic compounds behave as ideal solutions, but some of them exhibit *nonideal* behavior. This occurs because unlike molecules are affected by the presence of one another, thereby causing deviations from Raoult's law for *ideal* solutions (Eq. 4.2). When nonideal solutions have vapor pressures *higher* than those predicted by Raoult's law, the solutions are said to exhibit *positive* deviations from it; solutions having vapor pressures lower than predicted are thus considered to represent *negative* deviations from the law. In the present discussion, we'll consider only positive deviations associated with binary solutions, as such deviations are generally most important to organic chemists.

To produce positive deviations in a solution containing two volatile liquids, the forces of attraction between the molecules of the two components are *weaker* than those between the molecules of each individual component. The combined vapor pressure of the solution is thus *greater* than the vapor pressure of the pure, more volatile component for a particular range of compositions of the two liquids. This situation is illustrated in Figure 4.4, in which it may be seen that mixtures in the composition range between X and Y have boiling temperatures *lower* than the boiling temperature of either pure component. The *minimum-boiling* mixture, composition Z in Figure 4.4, may be considered as though it is a third component of the binary mixture. It has a *constant* boiling point because the vapor in equilibrium with the liquid has a composition *identical* to that of the liquid itself. The mixture is called a **minimum-boiling azeotrope.** Fractional distillation of such mixtures will *not* yield both of the components in pure form; rather, only the azeotropic mixture *and* the component present in *excess* of the azeotropic composition will be produced from the fractionation. For example, pure ethanol cannot be obtained by fractional distillation of aqueous solutions containing less than 95.57% ethanol, the azeotropic composition, even though the boiling point of this azeotrope is only 0.15 °C below that of pure ethanol. Since optimal fractional distillations of aqueous solutions containing

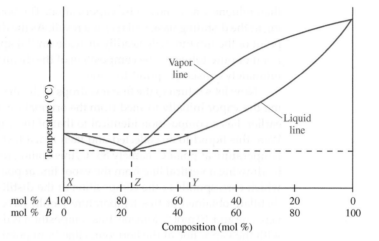

Figure 4.4
Temperature–composition diagram for minimum-boiling azeotrope.

less than 95.57% ethanol yield this azeotropic mixture, "95% ethyl alcohol" is readily available. Pure or "absolute" ethanol is more difficult to obtain from aqueous solutions. However, it can be prepared by removing the water chemically, through the use of a drying agent such as molecular sieves (Sec. 2.24), or by distillation of a ternary mixture of ethanol-water-benzene.

Azeotropic distillation is a useful technique for removing water from organic solutions. For example, toluene and water form an azeotrope having a composition of 86.5 wt % toluene and 13.5 wt % water, and so distillation of a mixture of these two effectively removes water from a mixture. This technique is used in the Experimental Procedure of Section 18.4 for driving an equilibrium in which water is being formed to completion. Azeotropic distillation may also be used to dry an organic liquid that is to be used with reagents that are sensitive to the presence of water. This application is found in the Experimental Procedure of Section 15.2, in which *anhydrous p*-xylene is required for a Friedel-Crafts alkylation reaction.

Fractional Distillation Columns and Their Operation

There are many types of fractional distillation columns, but all can be discussed in terms of a few fundamental characteristics. The column provides a vertical path through which the vapor must pass from the stillpot to the condenser before being collected in the receiver (Fig. 2.39). This path is significantly longer than in a simple distillation apparatus. As the vapor from the stillpot rises up the column, some of it condenses *in the column* and returns to the stillpot. *If the lower part of the distilling column is maintained at a higher temperature than the upper part of the column*, the condensate will be partially revaporized as it flows down the column. The uncondensed vapor, together with that produced by revaporization of the condensate in the column, rises higher and higher in the column and undergoes a repeated series of condensations and revaporizations. This repetitive process is equivalent to performing a number of simple distillations *within* the column, with the vapor phase produced in each step becoming increasingly richer in the *more* volatile component; the condensate that flows down the column correspondingly becomes richer in the less volatile component.

Each step along the path *A-B-C-D-E-F* of Figure 4.3 represents a *single* ideal distillation. One type of fractional distillation column, the bubble-plate column, was designed to effect one such step for each **plate** it contained. This led to the description of the efficiency of any fractional distillation column in terms of its equivalency to such a column in **theoretical plates.** Another index of the separating efficiency of a fractional distillation column is the **HETP**, which stands for *height equivalent* to a *theoretical plate* and is the vertical length of a column that is necessary to obtain a separation efficiency of one theoretical plate. For example, a column 60 cm long with an efficiency of 30 plates has an HETP value of 2 cm. Such a column would usually be better for research purposes than a 60-plate column that is 300 cm long (HETP = 5 cm) because of the small liquid capacity and **hold-up** of the shorter column. "Hold-up" refers to the condensate that remains in a column during and after distillation. When small amounts of material are to be distilled, a column must be chosen that has an efficiency, HETP, adequate for the desired separation and also a low to moderate hold-up.

As stated earlier, equilibrium between liquid and vapor phases must be established in a fractional distillation column so that the more volatile component is selectively carried to the top of the column and into the condenser, where the vapor condenses into the distillate. After all of the more volatile component is distilled, the less volatile one remains in the column and the stillpot; the heat supplied to the stillpot is then further increased to distill the second component. The most important requirements for performing a successful fractional distillation are (a) intimate and extensive contact between the liquid and the vapor phases in the column, (b) maintenance of the proper temperature gradient along the column, (c) sufficient length of the column, and (d) sufficient difference in the boiling points of the components of the liquid mixture. Each of these factors is considered here.

a. The desired contact between the liquid and vapor phases can be achieved by filling the column with an inert material having a large surface area. Examples of suitable packing materials include glass, ceramic, or metal pieces. Figure 2.40a shows a Hempel column packed with Raschig rings, which are pieces of glass tubing approximately 6 mm long. This type of column will have from two to four theoretical plates per 30 cm of length, if the distillation is carried out sufficiently slowly to maintain equilibrium conditions. Another type of fractional distillation column is the Vigreux column (Fig. 2.40b), which is useful for small-scale distillations of liquid where low hold-up is of paramount importance. A 30-cm Vigreux column will only have 1–2 theoretical plates and consequently will be less efficient than the corresponding Hempel column. The Vigreux column has the advantage of a hold-up of less than 1 mL as compared with 2–3mL for a Hempel column filled with Raschig rings.

b. **Temperature gradient** refers to the difference in temperature between the top and bottom of the column. The maintenance of the proper temperature gradient within the column is particularly important for an effective fractional distillation. Ideally, the temperature at the bottom of the column should be approximately equal to the boiling temperature of the solution in the stillpot, and it should decrease continually in the column until it reaches the boiling point of the more volatile component at the head of the column. The significance of the temperature gradient is seen in Figure 4.3, where the boiling temperature of the distillate decreases with each succeeding step, for example, *A* (90 °C) to *C* (85 °C) to *E* (82 °C).

The necessary temperature gradient from stillpot to stillhead in most distillations will be established *automatically* by the condensing vapors *if* the rate of distillation is properly adjusted. Frequently, this gradient can be maintained only by insulating the column with a material such as glasswool around the outside of the column. Insulation helps reduce heat losses from the column to the atmosphere. Even when the column is insulated, an insufficient amount of vapor may be produced to heat the column if the stillpot is heated too slowly, so that little or no condensate reaches the head. This rate must then be increased, but it must be kept below the point where the column is flooded. A **flooded column** is characterized by a column or "plug" of liquid that may be observed within the distillation column, often at the joint between it and the stillpot.

Factors directly affecting the temperature gradient in the column are the rate of heating of the stillpot and the rate at which vapor is removed at the stillhead. If the heating is too vigorous or the vapor is removed too rapidly, the entire column will heat up almost uniformly, and there will be no fractionation and thus no separation of the volatile components. On the other hand, if the stillpot is not heated strongly enough and if the vapor is removed too slowly at the top, the column will flood with returning condensate. Proper operation of a fractional distillation column thus requires *careful* control of the heat supplied to the stillpot and of the rate at which the distillate is removed at the stillhead. This rate should be *no more than one drop* every 2–3 sec.

The ratio of the amount of condensate returning to the stillpot and the amount of vapor removed as distillate per unit time is defined as the **reflux ratio.** A ratio of 10:1, for example, means that 10 drops of condensate return to the stillpot for each drop of distillate that is obtained. In general, the higher the reflux ratio, the more efficient the fractional distillation.

c. Correct column length is difficult to determine in advance of performing a fractional distillation. The trial-and-error technique must normally be used, and if a particular column does not efficiently separate a certain mixture, a longer column or a different type of column or column packing must be selected.

d. The difference in boiling points between the two pure components of a mixture should be no less than 20–30 °C in order for a fractional distillation to be successful when a Hempel column packed with Raschig rings or a similar type of packing is used. As mentioned previously, modifications in column length and type may result in the successful separation of mixtures having smaller boiling point differences.

In summary, the most important variables that can be controlled experimentally in a fractional distillation are proper selection of the column and column packing, adequate insulation of the column, and careful control of the rate of heating so as to provide the proper reflux ratio and a favorable temperature gradient within the column. Under such conditions, two different temperature **plateaus** will be observed in the fractional distillation of a typical binary mixture (Fig. 4.5). The head temperature should first rise to the normal boiling point of the more volatile component and remain there until that component is mostly removed (Fig. 4.5, Fraction 1). The head temperature may then drop somewhat, indicating that the more volatile component has largely been removed. As additional heat is provided to the stillpot, the less volatile component will begin to distill, and the head temperature will rise to the boiling point of the second component (Fig. 4.5, Fraction 2). If the separation is efficient, the volume of this fraction, which contains a mixture of the two components, will be small. The head temperature should then remain constant at the normal boiling point of the less volatile component until most of it has distilled (Fig. 4.5, Fraction 3).

Figure 4.5
Progress curve for typical fractional distillation.

EXPERIMENTAL PROCEDURES

Fractional Distillation of a Binary Mixture

Purpose To demonstrate the technique for the separation of two volatile organic liquids.

> ### SAFETY ALERT
>
> 1. **Wear safety glasses or goggles and suitable protective gloves while performing the experiments.**
>
> 2. **Cyclohexane and toluene are highly flammable, so be sure that burners are not being used in the laboratory. Use *flameless* heating (Sec. 2.9).**
>
> 3. **Examine your glassware for cracks and other weaknesses before assembling the distillation apparatus. Look with particular care for "star" cracks in round-bottom flasks, because these can cause a flask to break upon heating.**
>
> 4. **Proper assembly of glassware is important to avoid possible breakage and spillage and to avoid the release of distillate vapors into the room. Be certain that all connections in the apparatus are tight *before* beginning the distillation. Have your instructor examine your setup after it is assembled.**
>
> 5. **The apparatus used in these experiments *must* be open to the atmosphere at the receiving end of the condenser. *Never heat a closed system,* because the pressure buildup may cause the apparatus to explode!**
>
> 6. **Be certain that the water hoses are securely fastened to your condensers so that they will not pop off and cause a flood. If heating mantles or oil baths are to be used, water hoses that come loose may cause water to spray onto electrical connections or into the heating sources, either of which is potentially dangerous to you and to those who work around you.**
>
> 7. **Avoid excessive inhalation of organic vapors at all times.**

MINISCALE PROCEDURE

Preparation Refer to the online resources to answer Pre-Lab Exercises, access videos, and read the MSDSs for the chemicals used or produced in this procedure. Read or review Sections 2.2, 2.4, 2.9, 2.11, and 2.14.

Apparatus A 50-mL round-bottom flask, apparatus for fractional distillation, magnetic stirring, and *flameless* heating.

Setting Up Place 10 mL of cyclohexane and 20 mL of toluene in the round-bottom flask, and add a stirbar to ensure smooth boiling. Equip this flask for fractional distillation as shown in Figure 2.39. Pack a Hempel or similar distillation column, using the type of packing specified by your instructor. When packing the column, *be careful not to break off the glass indentations at the base of the column.* Do not pack the column too tightly, because heating a fractional distillation apparatus equipped with a column that is too tightly packed is analogous to heating a closed system. Insulate the fractionating column by wrapping it with glasswool. The position of the thermometer in the stillhead is particularly important; the *top* of the mercury thermometer bulb should be level with the *bottom* of the sidearm of the distillation head. Clean and dry three 25-mL containers, which may be bottles or Erlenmeyer flasks, for use as receiving flasks, and label them *A, B,* and *C.* Place receiver *A* so that the tip of the vacuum adapter extends inside the neck of the container to minimize evaporation of the distillate. Have your instructor check your assembled apparatus before heating the stillpot.

Distillation Start the magnetic stirrer and begin heating the stillpot using the heating method specified by your instructor. As the mixture is heated, the head temperature will rise to 81 °C (760 torr), which is the normal boiling point of cyclohexane, and distillation will begin. Regulate the heat so that distillation continues steadily at a rate *no faster* than *one drop of distillate every 1–2 sec;* if a drop of liquid cannot be seen suspended from the end of the thermometer, the rate of distillation is *too fast.*

The head temperature will remain at 81 °C for a period of time, but eventually it will either rise or drop slightly. Receiver *A* should be left in place until this increase or decrease in temperature is observed. As soon as the temperature deviates from 81 °C by more than ±3 °C, change to receiver *B* and increase the amount of heat supplied to the stillpot. The temperature will again start to rise, and more liquid will distill. Leave receiver *B* in place until the temperature reaches 110 °C (760 torr), which is the normal boiling point of toluene, and change to receiver *C.* Continue the distillation until 1–2 mL of liquid remains in the stillpot, and then discontinue heating.

Analysis Record the volumes of the distillate collected in each receiver by means of a graduated cylinder. Allow the liquid in the column to drain into the stillpot, and then record the volume of this pot residue. If instructed to do so, submit or save 0.2-mL samples of each fraction *A, B,* and *C* for GLC or GC-MS analysis (Sec. 6.4).

Discovery Experiment   *Comparative Fractional Distillations*

Compare the results of a fractional distillation using a packed column, as described in the above experimental procedure, with those obtained using a simple distillation apparatus, an unpacked column, and one packed with a material different from that used originally.

Discovery Experiment   *Fractionation of Alternative Binary Mixtures*

Subject other binary mixtures to fractional distillation using a variety of apparatus. These might include 50:50 mixtures of acetone and 1,4-dioxane, hexane and heptane, hexane and toluene, heptane and ethyl benzene, acetone and toluene, and tetrahydrofuran and 1-butanol. You might also wish to explore the efficacy of fractional distillation for separating a binary mixture having components whose boiling points are separated by less than 30 °C. Examples might be tetrahydrofuran and acetone, ethyl acetate and 1,4-dioxane, and ethanol and methanol.

Discovery Experiment   *Fractional Distillation of Unknowns*

Obtain a 50:50 mixture of two unknown solvents from your instructor. These solvents will differ in boiling point by more than 20 °C. Possible solvents include hexane, cyclohexane, heptane, octane, toluene, ethyl benzene, acetone, methanol, 1-butanol, tetrahydrofuran, 1,4-dioxane, ethyl acetate, and others listed by your instructor. Look up the boiling points for each of these solvents. Perform a fractional distillation using a packed column as described in the above experimental procedure. Prepare a distillation curve with the boiling point on the vertical axis and the volume on the horizontal axis. Based upon your experimentally determined boiling points for the two liquids, identify the components of your mixture.

WRAPPING IT UP

Unless directed to do otherwise, pour the *pot residue* into the container for non-halogenated organic liquids and return the *distillation fractions* to a bottle marked "Recovered Cyclohexane and Toluene."

EXERCISES

1. Define the following terms:

   a. fractional distillation   **f.** mole fraction

   b. head temperature   **g.** height equivalent to a theoretical plate (HETP)

   c. pot temperature   **h.** temperature gradient

   d. Raoult's law   **i.** Dalton's law

   e. ideal solution   **j.** reflux ratio

2. Specify whether a simple distillation or a fractional distillation would be more suitable for each of the following purifications, and briefly justify your choice.

   a. Preparing drinking water from sea water.

   b. Separating benzene, bp 80 °C (760 torr), from toluene, bp 111 °C (760 torr).

   c. Obtaining gasoline from crude oil.

   d. Removing diethyl ether, bp 35 °C (760 torr), from *p*-dichlorobenzene (s), mp 174–175 °C.

3. Sketch and completely label the apparatus required for fractional distillation.

4. Explain why a packed fractional distillation column is more efficient than an unpacked column for separating two closely boiling liquids.

5. If heat is supplied to the stillpot too rapidly, the ability to separate two liquids by fractional distillation may be drastically reduced. In terms of the theory of distillation presented in the discussion, explain why this is so.

6. Explain why the column of a fractional distillation apparatus should be aligned as near to the vertical as possible.

7. Explain the role of the stirbar normally added to a liquid that is to be heated to boiling.

8. The top of the mercury bulb of the thermometer placed at the head of a distillation apparatus should be adjacent to the exit opening to the condenser. Explain the effect on the observed temperature reading if the bulb is placed (a) below the opening to the condenser or (b) above the opening.

9. Calculate the mole fraction of each compound in a mixture containing 15.0 g of cyclohexane and 5.0 g of toluene.

10. In the fractional distillation of your mixture of cyclohexane and toluene, what can be learned about the efficiency of the separation on the basis of the relative volumes of fractions *A, B,* and *C*?

11. **a.** A mixture of 60 mol % *n*-propylcyclohexane and 40 mol % *n*-propylbenzene is distilled through a simple distillation apparatus; assume that no fractionation occurs during the distillation. The boiling temperature is found to be 157 °C (760 torr) as the first small amount of distillate is collected. The standard vapor pressures of *n*-propylcyclohexane and *n*-propylbenzene are known to be 769 torr and 725 torr, respectively, at 157.3 °C. Calculate the percentage of each of the two components in the first few drops of distillate.

    b. A mixture of 50 mol % benzene and 50 mol % toluene is distilled under exactly the same conditions as in Part **a.** Using Figure 4.3, determine the distillation temperature and the percentage composition of the first few drops of distillate.

    c. The normal boiling points of *n*-propylcyclohexane and *n*-propylbenzene are 156 °C and 159 °C, respectively. Compare the distillation results in Parts **a** and **b.** Which of the two mixtures would require the more efficient fractional distillation column for separation of the components? Why?

12. Examine the boiling point–composition diagram for mixtures of toluene and benzene given in Figure 4.3.

    a. Assume you are given a mixture of these two liquids of composition 70 mol % toluene and 30 mol % benzene and that it is necessary to effect a fractional distillation that will afford at least some benzene of greater than 99% purity. What would be the *minimum* number of theoretical plates required in the fractional distillation column chosen to accomplish this separation?

    b. Assume that you are provided with a 20-cm Vigreux column having an HETP of 10 cm to distill a mixture of 48 mol % benzene and 52 mol % toluene. What would be the composition of the first small amount of distillate that you obtained?

13. At 50 °C, the vapor pressures for methanol and ethanol are 406 torr and 222 torr, respectively. Given a mixture at 50 °C that contains 0.2 mol of methanol and 0.1 mol of ethanol, compute the partial pressures of each liquid and the total pressure.

14. Figure 4.6 shows a temperature (°C) versus composition diagram for a mixture of acetone and ethyl acetate at 760 torr. You can calculate the mole fraction of ethyl acetate at any given point of the plot by subtracting the mole fraction of acetone from 1 because of the relationship mole fraction of acetone + mole fraction of ethyl acetate = 1. Answer the following questions using Figure 4.6.

    a. Specify the normal boiling point of acetone and of ethyl acetate.

    b. A mixture of acetone and ethyl acetate of unknown ratio begins to boil at 65 °C (760 torr). What is the composition of this binary mixture in terms of the mole fraction of acetone and that of ethyl acetate?

    c. For a mixture comprised of 50 mol % acetone and 50 mol % ethyl acetate:
       i. At what temperature will the mixture begin to boil?
       ii. When the mixture begins to boil, what is the mol % acetone in the vapor?

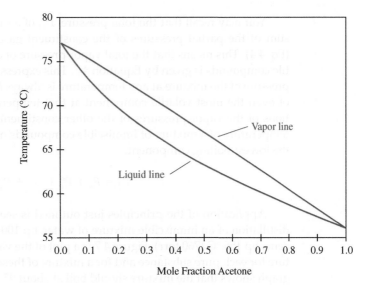

Figure 4.6
Temperature–composition diagram for acetone and ethyl acetate (Exercise 14).

  iii. What is the boiling point of the liquid formed by condensation of the vapor from **(ii)** above?
  iv. How many theoretical plates (distillation stages) will be necessary to isolate acetone of at least 90% purity?
 d. A mixture of acetone and ethyl acetate of unknown molar ratio is heated to boiling. Immediately after boiling commences, a sample of the vapor is found to contain 0.8 mol fraction of ethyl acetate. What is the composition in mol % of the original binary mixture and what is its boiling point?
 e. Would you expect a solution comprised of 30 mol % acetone and 70 mol% ethyl acetate to boil above or below *ca.* 67 °C in Denver, CO, which prides itself for being at an elevation of one mile above sea level? Explain your answer.

4.5 STEAM DISTILLATION

The separation and purification of *volatile* organic compounds that are **immiscible** or nearly immiscible with water can often be accomplished by steam distillation. The technique normally involves the codistillation of a mixture of organic liquids and water, although some organic solids can also be separated and purified by this means. Of the various distillation methods, steam distillation is utilized least frequently, owing to the rather stringent limitations on the types of substances for which it can be used. These limitations as well as the virtues of this technique are revealed by considering the principles underlying steam distillation.

Theory and Discussion

The partial pressure P_i of each component i of a mixture of *immiscible*, volatile substances at a given temperature is equal to the vapor pressure P_i° of the pure compound at the same temperature (Eq. 4.6) and does not depend on the mole fraction of the compound in the mixture. In other words, each component of the mixture vaporizes independently of the others. This behavior contrasts sharply with that exhibited by solutions of miscible liquids, for which the partial pressure of each constituent of the mixture depends on its mole fraction in the solution (Raoult's law, Eq. 4.2).

$$P_i = P_i^\circ \tag{4.6}$$

You may recall that the total pressure, P_T, of a mixture of gases is equal to the sum of the partial pressures of the constituent gases, according to Dalton's law (Eq. 4.4). This means that the total vapor pressure of a mixture of immiscible, volatile compounds is given by Equation 4.7. This expression shows that the total vapor pressure of the mixture at any temperature is always *higher* than the vapor pressure of even the most volatile component at that temperature, owing to the contributions of the vapor pressures of the other constituents in the mixture. The boiling temperature of a mixture of immiscible compounds must then be *lower* than that of the lowest-boiling component.

$$P_T = P_a^\circ + P_b^\circ + \cdots + P_i^\circ \qquad (4.7)$$

Application of the principles just outlined is seen in an analysis of the steam distillation of an immiscible mixture of water, bp 100 °C (760 torr), and bromobenzene, bp 156 °C (760 torr). Figure 4.7 is a plot of the vapor pressure versus temperature for each pure substance and for a mixture of these compounds. Analysis of this graph shows that the mixture should boil at about 95 °C (760 torr), the temperature at which the total vapor pressure equals standard atmospheric pressure. As theory predicts, this temperature is below the boiling point of water, which is the lowest-boiling component in this example.

The ability to distill a compound at the relatively low temperature of 100 °C or less by means of a steam distillation is often of tremendous value, particularly in the purification of substances that are heat-sensitive and would therefore decompose at higher temperatures. It is useful also in the separation of compounds from reaction mixtures that contain large amounts of nonvolatile residues such as inorganic salts.

The composition of the condensate from a steam distillation depends upon the molar masses of the compounds being distilled and upon their respective vapor pressures at the temperature at which the mixture steam-distils. To illustrate this, consider a mixture of two immiscible components, *A* and *B*. If the vapors of *A* and *B* behave as ideal gases, the ideal gas law can be applied and Equations 4.8a and 4.8b are obtained. In these two expressions, $P°$ is the vapor pressure of the pure liquid, V is

Figure 4.7

Vapor pressure–temperature graph for pure bromobenzene, pure water, and a mixture of bromobenzene and water.

the volume in which the gas is contained, g is the weight in grams of the component in the gas phase, M is its molar mass, R is the gas constant, and T is the absolute temperature in kelvins (K). Dividing Equation 4.8a by Equation 4.8b gives Equation 4.9.

$$P_A^{\circ}V_A = (g_A/M_A)(RT) \tag{4.8a}$$

$$P_B^{\circ}V_B = (g_B/M_B)(RT) \tag{4.8b}$$

$$\frac{P_A^{\circ}V_A}{P_B^{\circ}V_B} = \frac{g_A/M_A(RT)}{g_B/M_B(RT)} \tag{4.9}$$

Because the RT factors in the numerator and denominator are identical and because the volume in which the gases are contained is the same for both ($V_A = V_B$), these terms in Equation 4.9 cancel to yield Equation 4.10.

$$\frac{\text{Grams of } A}{\text{Grams of } B} = \frac{(P_A^{\circ})(\text{molar mass of } A)}{(P_B^{\circ})(\text{molar mass of } B)} \tag{4.10}$$

Now let the immiscible mixture of A and B consist of bromobenzene and water, whose molar masses are 157 g/mol and 18 g/mol, respectively, and whose vapor pressures at 95 °C, as determined from Figure 4.6, are 120 torr and 640 torr, respectively. The composition of the distillate at this temperature can be calculated from Equation 4.10 as shown in Equation 4.11. This calculation indicates that on the basis of weight, *more* bromobenzene than water is contained in the steam distillate, even though the vapor pressure of the bromobenzene is *much* lower at the temperature of the distillation.

$$\frac{g_{\text{bromobenzene}}}{g_{\text{water}}} = \frac{(120)(157)}{(640)(18)} = \frac{1.64}{1} \tag{4.11}$$

Organic compounds generally have molar masses much higher than that of water, so it is possible to steam-distill compounds having vapor pressures of only about 5 torr at 100 °C with a fair efficiency on a weight-to-weight basis. Thus, *solids* that have vapor pressures of at least this magnitude can be purified by steam distillation. Examples are camphor, used in perfumes, and naphthalene, present in some brands of mothballs. The rather high vapor pressures of these solids is evidenced by the fact that their odors are detectable at room temperature.

In summary, steam distillation provides a method for separating and purifying moderately volatile liquid and solid organic compounds that are insoluble or nearly insoluble in water from nonvolatile compounds. Although relatively mild conditions are used in steam distillation, it cannot be used for substances that decompose on prolonged contact with steam or hot water, that react with water, or that have vapor pressures of 5 torr or less at 100 °C, all of which are significant limitations to the method.

4.6 STEAM DISTILLATION: ISOLATION OF CITRAL FROM LEMON GRASS OIL

Citral is a naturally occurring oil mainly comprised of two isomeric unsaturated aldehydes, geranial (**1**) and neral (**2**), that are extremely difficult to separate. These isomers differ only in the spatial orientation of the substituents about the carbon-carbon double bond that bears the aldehyde moiety (–CHO). This qualifies **1** and **2** as **stereoisomers** and more specifically as **diastereomers,** the definitions of which are found in Section 7.1. In addition, they are sometimes also called **geometric isomers,** an older term that is used to describe stereoisomers that differ because the three-dimensional distribution of their substituents is the result of a double bond.

Citral possesses a lemonlike odor and taste. Although this odor and taste is pleasant to humans, it is less attractive to other species. For example, ants secrete citral toward off potential predators, so citral is functioning as a **defense pheromone.** There are several commercial applications of citral. For example, it

1	**2**
Geranial	Neral

3
Vitamin A

may be added to perfumes whenever a lemonlike essence is desired, and it is used as an intermediate for the synthesis of vitamin A (**3**).

The commercial importance of citral has stimulated an extensive search for its presence in nature. One source is the oil from the skins of lemons and oranges, although it is only a minor component of this oil. However, citral is the major constituent of the oil obtained from lemon grass, and in fact 75–85% of the crude oil derived from pressing lemon grass is this natural product.

Citral is a chemically labile substance, and its isolation therefore presents a challenge. This task is simplified by the fact that citral, bp 229 °C (760 torr), is relatively volatile and has low solubility in water. These properties make it a suitable candidate for steam distillation, a technique that allows distillation of citral from crude lemon grass oil at a temperature below 100 °C, which is far below its normal boiling point. Neutral conditions are maintained in steam distillation, as is the partial exclusion of atmospheric oxygen, so the possibility of oxidation and/or polymerization of citral is minimized.

Steam distillation can be performed by two different methods, as described in Section 2.16. The first involves using an external steam source, which may be either a laboratory steam line or a steam generator, and passing the steam through a mixture of lemon grass oil and water. This method has some advantages, but it is experimentally more difficult than an alternative method, which involves heating a mixture of lemon grass oil and water and collecting the distillate. Although this latter procedure is simpler from the experimental standpoint, there are some limitations to its application. For example, in the steam distillation of only slightly volatile substances, a large initial volume of water will be required, or water must be added as the distillation proceeds, perhaps by means of an addition funnel. Nonetheless, the technique for internal generation of steam is well suited for this experiment.

EXPERIMENTAL PROCEDURES

Steam Distillation of Citral from Lemon Grass Oil

Purpose To isolate the natural product citral using the technique of steam distillation.

SAFETY ALERT

1. **Wear safety glasses or goggles and suitable protective gloves while performing the experiment.**

2. **Steam distillation involves the use of glassware that becomes very hot. Exercise care when handling hot glassware.**

3. **Be certain that the steam distillate is cooled below 30 °C before extracting it with diethyl ether; otherwise excessive pressure may develop in the separatory funnel and may blow out the stopper.**

4. **Diethyl ether is *extremely* flammable. Be certain that there are no flames in the vicinity during its use and its removal from the citral.**

MINISCALE PROCEDURE

Preparation Refer to the online resources to answer Pre-Lab Exercises, access videos, and read the MSDSs for the chemicals used or produced in this procedure. Read or review Sections 2.2, 2.4, 2.9, 2.13, 2.16, 2.19, 2.21, 2.29, and 4.4.

Apparatus A 125-mL separatory funnel, a 125-mL round-bottom flask, apparatus for steam distillation using an internal steam source, simple distillation, and *flameless* heating.

Setting Up Determine the weight of 2.5 mL of lemon grass oil, and then place it and 60 mL of water in the round-bottom flask. Equip this flask for steam distillation as shown in Figure 2.44, except replace the steam inlet tube with a stopper; use the heating source suggested by your instructor. Have your instructor check your assembled apparatus before continuing the experiment.

Distillation Heat the liquid to boiling and adjust the heat source so that the distillation proceeds as rapidly as possible; avoid applying excess heat to the stillpot, especially when the volume of water in the flask has been reduced below about 25 mL. Continue the distillation until oil droplets no longer appear in the distillate, which should occur after about 30 mL of distillate has been collected. Allow the distillate to cool to room temperature or below, using an ice-water bath if necessary.★

Isolation Transfer the cooled distillate to a separatory funnel and extract it sequentially with two 15-mL portions of diethyl ether. The funnel should be vented *frequently* to avoid the buildup of pressure. Transfer the organic layer from the separatory funnel to a 50-mL Erlenmeyer flask, and add about 0.5–1.0 g of *anhydrous* calcium chloride. Allow the ethereal solution to remain in contact with the drying agent until the organic layer is dry, as evidenced by its being completely clear. If the experiment is stopped at this point, loosely stopper the flask and store it in a hood; *never leave flasks containing diethyl ether in your locker drawer.*★

Decant the dried organic solution into a *tared* 125-mL round-bottom flask. Equip the flask for simple distillation and evaporate the solvent using simple distillation. Alternatively, one of the techniques in Section 2.29 as directed by your instructor may be used to concentrate the solution. After the ether is completely removed, the pot residue is crude citral.

Analysis Determine the weight of citral isolated and calculate the percentage recovery of citral based on the weight of the original sample of lemon grass oil. If instructed to do so, save or submit a sample of citral for GLC or GC-MS analysis (Sec. 6.4). You may be asked to perform one or more of the following tests on citral.

1. Test for a carbon-carbon double bond. Perform at least one of the qualitative tests for unsaturation described in Section 4.7A.

2. Test for an aldehyde. Perform the chromic acid qualitative test provided in Section 4.7B.

3. Analyze the citral using gas chromatography to assess the purity of the product.

WRAPPING IT UP

Flush the *aqueous solution* remaining in the distillation flask down the drain. Do the same with the *aqueous steam distillate* once you have completed the extraction.

EXERCISES

1. Why was a steam distillation rather than a simple distillation performed in the isolation of citral from lemon grass oil?

2. What type of product is expected from the reaction of citral with Br_2/CH_2Cl_2? With chromic acid?

3. Provide structures for the semicarbazone and the 2,4-dinitrophenylhydrazone of citral.

4. Why does the citral float on the surface of the aqueous distillate rather than sinking to the bottom?

5. Explain why the substitution of 1-propanol, bp 97 °C (760 torr), for water in a steam distillation would not work.

6. Suppose that you are to steam-distill a sample of a natural product whose vapor pressure at 100 °C is known to be half that of citral. What consequence would this have on the amount of distillate required per mole of the natural product present?

7. Both geranial (**1**) and vitamin A (**3**) are members of the class of natural products called **terpenes**. This group of compounds has the common characteristic of being biosynthesized by linkage of the appropriate number of five-carbon units having the skeletal structure shown below. Determine the number of such units present in each of these terpenes and indicate the bonds linking the various individual units.

$$
\begin{array}{ccc}
C & & C \\
\diagdown & & \diagup \\
& C-C & \\
\diagup & & \\
C & &
\end{array}
$$

8. In the reduction of nitrobenzene to aniline (Eq. 4.12), the product is readily isolated from the reaction mixture by steam distillation.

a. From this information, what do you know about the *miscibility* of aniline in water?

b. At approximately what temperature will aniline and water codistil?

c. What is the normal boiling point of aniline?

$$\langle\!\!\!\bigcirc\!\!\!\rangle\text{---NO}_2 \xrightarrow[\text{(reduction)}]{[H]} \langle\!\!\!\bigcirc\!\!\!\rangle\text{---NH}_2 \tag{4.12}$$

9. Using Raoult's and Dalton's laws, explain how the boiling point of a binary mixture depends on the miscibility of the two liquids.

4.7 QUALITATIVE ANALYSIS

There are a number of simple chemical tests that are easy to perform and provide valuable information regarding the presence of particular types of functional groups in an organic compound. Two types of procedures are described below. One of them allows you to test for the presence of unsaturation in the form of a carbon-carbon double or triple bond, functionalities that characterize alkenes and alkynes, respectively, and the other provides a way to detect the presence of an aldehyde group.

A. Tests for Unsaturation

The two common qualitative tests for unsaturation involve the reaction of the compounds with *bromine in dichloromethane* and with *potassium permanganate*, respectively. In both cases, a positive test is denoted by decoloration of the reagent.

1. Bromine in Dichloromethane

Bromine adds to the carbon-carbon double bond of alkenes to produce dibromoalkanes (Eq. 4.13) and reacts with alkynes to produce tetrabromoalkanes (Eq. 4.14).

When this reaction occurs, molecular bromine is rapidly consumed, and its characteristic dark red-orange color disappears so long as bromine is not added in excess. The nearly *instantaneous* disappearance of the bromine color is a positive test for unsaturation. The test is ambiguous, however, because some alkenes do not react with bromine, and some react very slowly. In the case of a negative test, the Baeyer procedure using potassium permanganate (Sec. 4.7A2) should be performed to confirm the absence of a carbon-carbon double bond.

$$\text{R}_2\text{C}\!\!=\!\!\text{CR}_2 + \text{Br}_2 \xrightarrow{\text{CH}_2\text{Cl}_2} \text{R}_2\overset{\overset{\displaystyle \text{Br}}{|}}{\text{C}}\!\!-\!\!\overset{\underset{\displaystyle \text{Br}}{|}}{\text{CR}}_2 \tag{4.13}$$

An alkene   Bromine   A dibromide
             (red-orange)   *(colorless)*

$$\text{RC}\!\!\equiv\!\!\text{CR} + \text{Br}_2 \xrightarrow{\text{CH}_2\text{Cl}_2} \overset{\overset{\displaystyle \text{Br}\ \ \text{Br}}{|\ \ |}}{\text{RC}}\!\!-\!\!\overset{\underset{\displaystyle \text{Br}\ \ \text{Br}}{|\ \ |}}{\text{CR}} \tag{4.14}$$

An alkyne   A tetrabromide
                  (colorless)

More details about the reaction of bromine in dichloromethane with alkenes are found in Section 10.4.

EXPERIMENTAL PROCEDURES

Bromine Test for Unsaturation

SAFETY ALERT

1. **Wear safety glasses or goggles and suitable protective gloves while performing the experiments.**

2. **Bromine is a hazardous chemical that may cause serious chemical burns. Do not breathe its vapors or allow it to come into contact with your skin. Perform all operations involving the transfer of solutions of bromine at a hood. If you get bromine on your skin, immediately flood the affected area with warm water and soak it in 0.6 M sodium thiosulfate solution for up to 3 h if the burn is particularly serious.**

3. **Do not use acetone to rinse glassware containing residual amounts of bromine, because acetone reacts readily with bromine to form α-bromoacetone, a severe lachrymator. Rather, follow the procedure described in Wrapping It Up.**

Preparation Refer to the online resources to read the MSDSs for the chemicals used or produced in this procedure.

Dissolve 50 mg or 1–2 drops of the compound in dichloromethane, and add to this solution 0.1 *M* bromine in dichloromethane dropwise. Rapid disappearance of the red-orange color of bromine to give a colorless solution is a positive test for unsaturation in the form of a carbon-carbon double or triple bond.

WRAPPING IT UP

Decolorize any *solutions* in which the color of *bromine* is visible by dropwise addition of cyclohexene; then discard the resulting mixtures in the container for halogenated organic liquids.

2. Potassium Permanganate

A second qualitative test for unsaturation, the Baeyer test, depends on the ability of potassium permanganate to oxidize a carbon-carbon double bond to give an alkanediol (4.15) or a carbon-carbon triple bond to yield carboxylic acids (Eq. 4.16).

The permanganate ion is destroyed in the reaction, and a brown precipitate of MnO_2 is produced. The disappearance of the characteristic purple color of the permanganate ion is a positive test for unsaturation. However, care is needed in interpreting this test: Compounds containing certain other functional groups, such as alcohols, aldehydes, and amines, also decolorize permanganate ion because they undergo oxidation.

$$R_2C{=}CR_2 + MnO_4^- \xrightarrow{H_2O} R_2\overset{\overset{\displaystyle OH}{|}}{C}{-}\overset{\overset{\displaystyle OH}{|}}{C}R_2 + \underline{MnO_2} \qquad (4.15)$$

An alkene	Permanganate	A glycol	Manganese
	(purple)	*(colorless)*	dioxide
			(brown)

$$R^1C\equiv CR^2 + MnO_4^- \xrightarrow{H_2O} R^1CO_2H + R^2CO_2H + \underline{MnO_2} \qquad (4.16)$$

An alkyne   Permanganate   Carboxylic acids   Manganese
         (purple)   *(colorless)*   dioxide
                                              (brown)

EXPERIMENTAL PROCEDURES

Baeyer Test for Unsaturation

SAFETY ALERT

Wear suitable protective gloves to prevent solutions of potassium permanganate from contacting your skin. These solutions cause unsightly stains on your hands for several days. If this oxidant comes in contact with the skin, immediately flood the affected area with warm water.

Preparation Refer to the online resources to read the MSDSs for the chemicals used or produced in this procedure.

Dissolve 1 or 2 drops of the compound in 2 mL of 95% ethanol, and then add 0.1 M $KMnO_4$ solution dropwise; note the results. Count the number of drops added before the purple color of the permanganate ion persists. Ethanol does not react significantly with potassium permanganate within five minutes at room temperature. For a **blank determination**, count the number of drops of aqueous permanganate that can be added to 2 mL of 95% ethanol before the color persists. A significant difference in the number of drops required in the two cases is a positive test for unsaturation.

WRAPPING IT UP

Filter all *solutions* to remove manganese dioxide. Place the *filter paper* and the *filter cake* containing manganese salts in the container for heavy metals. Flush all *filtrates* down the drain.

B. Test for Aldehyde Function

Chromic Acid

One of several qualitative tests for the presence of an aldehyde function depends on its susceptibility to oxidation to a carboxylic acid group by chromic acid, H_2CrO_4, which is orange-red in color. Because chromic acid is unstable when stored for extended periods of time, it is prepared as needed by dissolving chromic anhydride in aqueous sulfuric acid (Eq. 4.17). The sulfuric acid serves as a proton source for the oxidation reaction in which the aldehyde is converted to a carboxylic acid (Eq. 4.18). As with the tests described for unsaturation, the oxidation results in a color change. In this case, the orange-red color of the solution owing to the presence of Cr^{6+} changes to green upon the formation of Cr^{3+} as the oxidation proceeds. Formation of a green-colored precipitate is also observed in some cases.

$$CrO_3 + H_2O \xrightarrow[\text{(orange-red)}]{} H_2CrO_4 \qquad (4.17)$$

$$RCHO \xrightarrow[\text{(orange-red)}]{H_2CrO_4/H_2SO_4} RCO_2H + Cr_2(SO_4)_3 + \textit{Green} \text{ solution or precipitate} \qquad (4.18)$$

Chromic acid also oxidizes primary and secondary alcohols (Eqs. 4.19 and 4.20, respectively); it gives no visible reaction with ketones and tertiary alcohols (Eqs. 4.20 and 4.21, respectively) under the conditions of the test.

$$RCH_2OH \xrightarrow[\text{(orange-red)}]{H_2CrO_4} [RCHO] \xrightarrow[\text{(orange-red)}]{H_2CrO_4} RCO_2H + \textit{Green} \text{ solution or precipitate}$$

A 1° alcohol

$$(4.19)$$

$$R_2CHOH \xrightarrow[\text{(orange-red)}]{H_2CrO_4} R_2CO + \textit{Green} \text{ solution} \xrightarrow[\text{\textbackslash\textbackslash}]{H_2CrO_4} \text{No visible reaction}$$
or precipitate

A 2° alcohol

$$(4.20)$$

$$R_3COH \xrightarrow[\text{\textbackslash\textbackslash}]{H_2CrO_4} \text{No visible reaction}$$

$$(4.21)$$

The stoichiometry of oxidation of an aldehyde to a carboxylic acid is depicted in Equation 4.22 and that for conversion of a 1° alcohol ($R^1 =$ alkyl or aryl, $R^2 = H$) or 2° alcohol ($R^1 = R^2 =$ alkyl or aryl) is shown in Equation 4.23, where you can see that three moles of sulfuric acid are consumed in the process.

$$3 \text{ RCHO} + 2 \text{ H}_2CrO_4 + 3 \text{ H}_2SO_4 \longrightarrow 3 \text{ RCO}_2H + Cr_2(SO_4)_3 + 5 \text{ H}_2O \qquad (4.22)$$

$$3 \overset{\overset{\displaystyle OH}{|}}{R^2-CH-R^1} + 2 \text{ H}_2CrO_4 + 3 \text{ H}_2SO_4 \longrightarrow 3 \overset{\overset{\displaystyle O}{\|}}{R^2-C-R^1} + Cr_2(SO_4)_3 + 8 \text{ H}_2O$$

$$(4.23)$$

Thus, the chromic acid reagent gives a clear-cut distinction between primary and secondary alcohols and aldehydes on the one hand and tertiary alcohols and ketones on the other. Aldehydes may be distinguished from primary and secondary alcohols by means of other tests that will not be discussed here.

EXPERIMENTAL PROCEDURES

Chromic Acid Test for Aldehydes and 1° and 2° Alcohols

SAFETY ALERT

1. **Wear safety glasses or goggles and suitable protective gloves while performing the experiments.**

2. **When preparing and handling solutions of chromic acid, be careful to keep the acids from contacting your skin. Chromic acid causes unsightly stains on your hands for several days and may cause severe chemical burns. If the oxidant comes in contact with your skin, immediately flood the affected area with warm water and rinse it with 5% sodium bicarbonate solution.**

3. **The preparation of chromic acid requires diluting a paste of chromic anhydride and *concentrated* sulfuric acid with water. Be certain to add the water slowly to the acid and swirl the mixture to ensure continuous mixing. Swirling keeps the denser sulfuric acid from layering at the bottom of the flask and avoids possible splattering that might occur because of the heat generated when the two layers are suddenly mixed.**

Preparation Refer to the online resources to read the MSDSs for the chemicals used or produced in this procedure.

If the chromic acid reagent is not available, prepare it as follows. Add 1 g of chromic anhydride (CrO_3) to 1 mL of *concentrated* sulfuric acid and stir the mixture until a smooth paste is obtained. Dilute the paste cautiously with 3 mL of distilled water, and stir this mixture until a clear orange solution is obtained.

Dissolve 1 drop of a liquid or about 10 mg of a solid alcohol or carbonyl compound in 1 mL of reagent-grade acetone. Add 1 drop of the acidic chromic anhydride reagent to the acetone solution and shake the tube to mix the contents. A positive reaction is indicated by disappearance of the orange color of the reagent and the formation of a green or blue-green precipitate or emulsion.

Aliphatic aldehydes and primary and secondary alcohols give a positive test within 5 sec; aromatic aldehydes require 30–45 sec. Color changes occurring after about 1 min should not be interpreted as positive tests; other functional groups such as ethers and esters may slowly hydrolyze under the conditions of the test, releasing alcohols that in turn provide "false-positive" tests. Tertiary alcohols and ketones produce no visible change in several minutes. Phenols and aromatic amines give dark precipitates, as do aromatic aldehydes and benzylic alcohols having hydroxyl or amino groups on the aromatic ring.

WRAPPING IT UP

Add sodium sulfite to the *aqueous solution* of chromium salts to destroy excess Cr^{6+}. Make the solution slightly basic, to form chromium hydroxide, and isolate this salt by vacuum filtration through a bed of a filter-aid. Place the *filter paper* and the *filter cake* in the container for heavy metals; flush the *filtrate* down the drain.

Extraction

Extraction is a technique commonly used in organic chemistry to separate a material you want from those you do not. While the term extraction may be unfamiliar to you, the process is actually something you commonly perform. For example, many of you probably start the day with an extraction when you brew a pot of coffee or tea, especially after a long night of studying. By heating coffee grounds or tea leaves with hot water, you extract the caffeine, together with other water-soluble compounds such as dark-colored tannins, from the solid material. You can then drink the liquid, which is certainly more enjoyable than eating coffee grounds or tea leaves, to ingest the caffeine and benefit from its stimulating effect. Similarly, when you make a soup, the largely aqueous liquid portion contains numerous organic and inorganic compounds that have been extracted from spices, vegetables, fish, or meat, and these give your culinary creation its distinctive flavor. In the procedures found in this chapter, you will have an opportunity to develop your existing experimental skills further by isolating organic compounds using different types of extractions.

5.1 INTRODUCTION

The desired product from a reaction is frequently part of a mixture, and its isolation in pure form can be a significant experimental challenge. Two of the more common methods for separating and purifying organic liquids and solids are **recrystallization** and **distillation**; these procedures are discussed in Chapters 3 and 4, respectively. Two other important techniques available for these purposes are **extraction** and **chromatography**. As you will see here and in Chapter 6, both of these methods involve partitioning of compounds between two *immiscible* phases. This process is termed **phase distribution** and can result in separation of compounds if they distribute differently between the two phases.

Distribution of solutes between phases is the result of **partitioning** or **adsorption** phenomena. Partitioning involves the difference in solubilities of a substance in two immiscible solvents—in other words, *selective dissolution*. Adsorption, on the other hand, is based on the *selective attraction* of a substance in a liquid or gaseous mixture to the surface of a solid phase. The various chromatographic techniques depend on both of these processes, whereas the process of extraction relies only on partitioning.

Extraction involves *selectively* removing one or more components of a solid, liquid, or gaseous mixture into a separate phase. The substance being extracted will partition between the two immiscible phases that are in contact, and the ratio of its distribution between the phases will depend on the relative solubility of the solute in each phase.

155

5.2 THEORY OF EXTRACTION

Liquid-liquid extraction is one of the most common methods for removing an organic compound from a mixture. This process is used by chemists not only in the isolation of natural products but also in the isolation and purification of products from most chemical reactions.

The technique involves distributing a solute, A, between two immiscible liquids, S_x, the **extracting phase**, and S_0, the **original phase**. The immiscible liquids normally encountered in the organic laboratory are water and some organic solvent, such as diethyl ether, $(C_2H_5)_2O$, or dichloromethane, CH_2Cl_2. At a given temperature, the amount of A, in g/mL, in each phase is expressed *quantitatively* in terms of a constant, K, commonly called the **partition coefficient** (Eq. 5.1). Strictly speaking, the volumes of *solution* should be used in the definition of [A], but if the solutions are dilute, only slight errors result if volumes of *solvent* are used. Furthermore, a close *approximation* of the partition coefficient K may be obtained by simply dividing the solubility of A in the extracting solvent S_x by the solubility of A in the original solvent S_0.

$$K = \frac{[A] \text{ in } S_x}{[A] \text{ in } S_0} \tag{5.1}$$

The process of liquid-liquid extraction can be considered a competition between two immiscible liquids for solute A, with solute A distributing or partitioning between these two liquids when it is in contact with both of them. The mathematical expression of Equation 5.1 shows that *at equilibrium*, the ratio of concentrations of A in the two phases will always be constant.

Equation 5.1 leads to several important predictions. For example, if $K > 1$, solute A will be mainly in the *extracting* solvent, S_x, so long as the *volume*, V_x, of this solvent is at least equal to the volume, V_0, of the original solvent S_0; the amount of solute remaining in S_0 will depend on the value of K. By recasting Equation 5.1 into Equation 5.2, it is easily seen that if the volumes of S_x and S_0 are equal, the value of K is simply the ratio of the grams of A in each solvent. Because the product of the ratios A_x/A_0 and V_0/V_x must be constant for a given solvent pair, increasing the volume of extracting solvent S_x will result in a net *increase* in the amount of solute A in S_x.

$$K = \frac{\text{grams of } A \text{ in } S_x}{\text{grams of } A \text{ in } S_0} \times \frac{\text{mL of } S_0}{\text{mL of } S_x} = \frac{A_x}{A_0} \times \frac{V_0}{V_x} \tag{5.2}$$

There are practical, economic, and environmental reasons, however, that limit the quantities of organic solvents that can realistically be used in extractions. The question thus arises as to whether it is better to perform a single extraction with all of the solvent or to perform several extractions with smaller volumes. For example, if an organic compound is dissolved in 10 mL of water, and only 30 mL of diethyl ether is used, will *three* extractions with 10-mL portions of ether provide better recovery of solute than a *single* one with 30 mL?

Applying Equation 5.2 would provide the answer, but the process is tedious for multiple extractions. This equation can be generalized, however, to accommodate multiple extractions in terms of the fraction, F_A, of solute A remaining in the original solvent S_0 after n extractions, using a constant volume of an immiscible solvent S_x for each extraction. Thus, the amount of solute remaining in S_0 after one extraction with S_x is obtained by rearranging Equation 5.3 to Equation 5.4. The fraction, F_A, of A still in the original solvent is obtained through Equation 5.5, in which C_i and C_f are the *initial* and *final* concentrations, respectively. For a second extraction,

Equations 5.6–5.8 result. Equation 5.8 can be generalized to Equation 5.9 when n extractions are performed.

$$K = \left(\frac{A_0 - A_1}{V_x}\right)\left(\frac{V_0}{A_1}\right) \tag{5.3}$$

where A_0 = amount (grams) of solute in S_0 before extraction

$A_1 = (A_0 - A_x)$ = amount (grams) of solute in S_0 after extraction

V_0 and V_x = volume (mL) of original and extracting solvents, respectively

$$A_1 = \frac{(A_0 V_0)}{KV_x + V_0} \tag{5.4}$$

$$F_A = \frac{C_f}{C_i} = \frac{V_0}{KV_x + V_0} \tag{5.5}$$

where $C_i = A_0/S_0$ and $C_f = A_1/S_0$, respectively

$$K = \left(\frac{A_1 - A_2}{V_x}\right)\left(\frac{V_0}{A_2}\right) \tag{5.6}$$

$$K = A_0\left(\frac{V_0}{KV_x + V_0}\right)^2 \tag{5.7}$$

$$F_{A'} = \frac{C_f}{C_i} = \left(\frac{V_0}{KV_x + V_0}\right)^2 \tag{5.8}$$

$$F_A = \frac{C_f}{C_i} = \left(\frac{V_0}{KV_x + V_0}\right)^n \tag{5.9}$$

Performing calculations according to Equation 5.9 for the case of $K = 5$, we obtain $F_A = 1/16$ for a single extraction with 30 mL of diethyl ether and $1/216$ for three extractions with 10-mL portions. This means that 6.3% of A remains in the aqueous phase when *one* extraction is performed, whereas the value drops to only 0.5% in the case of *three* successive extractions with the *same* total volume of solvent. The amount of A that could be isolated is increased by some 6% with multiple extractions. Setting K at 2 gives corresponding values for F_A of $1/7$ and $1/27$, respectively, which translates to a 10% increase in the quantity of A that is removed from the aqueous layer with multiple extractions.

From these calculations, it is clear that multiple extractions become increasingly important as the value of K decreases. The improved recovery of solute from multiple extractions, though, must be balanced with the practical consideration that the relatively small increase in recovery may not justify the additional time and solvent required to perform multiple extractions unless the product is of great value. Importantly, the partition coefficients K are generally large for the extractions required in the procedures provided in this textbook, so only one or two extractions are usually required.

Selection of the appropriate extracting solvent is obviously a key for successfully using this technique to isolate and purify compounds, and important guidelines for making the correct choice are summarized here.

1. The extracting solvent *must not react* in a chemically irreversible way with the components of the mixture.

2. The extracting solvent *must be immiscible,* or nearly so, with the original solution.

3. The extracting solvent *must selectively remove* the desired component from the solution being extracted. That is, the partition coefficient K of the component being removed must be *high,* while the partition coefficients of all other components should be *low.*

4. The extracting solvent *should be readily* separable from the solute. Use of a volatile solvent facilitates its removal by simple distillation or one of the other techniques outlined in Section 2.29.

5.3 BASE AND ACID EXTRACTIONS

The discussion of Section 5.2 focuses on partitioning *one* substance between two *immiscible* solvents. Now consider what happens if a mixture of *two or more* compounds is present in a given volume of solvent S_0 and an extraction using a solvent S_x is performed. If the partition coefficient K of one component, A, is *significantly* greater than 1.0 and if those of other components are *significantly* less than 1.0, the majority of A will be in S_x, whereas most of the other compounds will remain in S_0. Physical separation of the two solvents will give a partial separation, and thus purification, of the solute A from the other components of the mixture.

Solutes differing significantly in polarity should have very different coefficients K for partitioning between nonpolar and polar solvents. For example, consider the distribution of two organic compounds, the first neutral and nonpolar, and the second ionic and polar, between a nonpolar solvent and a polar solvent. If a solution of these compounds in the nonpolar solvent is shaken with the polar solvent, the neutral compound will preferentially partition into the *nonpolar* phase, whereas the polar constituent will preferentially partition into the *polar* phase. Separating the two phases effects a separation of the two solutes.

Carboxylic acids and phenols (**1** and **3**, Eqs. 5.10 and 5.11, respectively) are two classes of organic compounds containing **functional groups** that are **polar** and **hydrophilic** (water-loving). Unless they contain fewer than about six carbon atoms, such compounds are generally either *insoluble* or only *slightly soluble* in water because of the **hydrophobic** (water-avoiding) properties of the carbon-containing portion, R or Ar, of the molecule. They are *soluble* in common organic solvents like dichloromethane or diethyl ether that have at least modest polarity (Table 3.1). Consequently, carboxylic acids or phenols dissolved in diethyl ether, for example, will largely remain in that phase when the solution is extracted with water.

$$R-C \underset{O}{\overset{O-H}{\big|}} \quad + \quad B^- Na^+ \text{ (aq)} \rightleftharpoons R-C \underset{O}{\overset{O^-}{\big|}} \quad Na^+ + B-H \qquad (5.10)$$

1		**2**
A carboxylic acid	A base	A carboxylate
Water *insoluble*	(as sodium salt)	(as sodium salt)
$K_{\text{water/org}} < 1$		Water *soluble*
		$K_{\text{water/org}} > 1$

$$\text{Ar—O} \overset{\displaystyle}{\underset{\displaystyle H}{}} \quad \text{B}^- \text{ Na}^+ \text{ (aq)} \rightleftharpoons \text{Ar—O}^- \text{ Na}^+ + \text{ B—H} \qquad (5.11)$$

3   **4**

A phenol (Ar = aryl)   A base   A phenoxide
Water *insoluble*   (as sodium salt)   (as sodium salt)
$K_{\text{water/org}} < 1$ Water *soluble*
                                                  $K_{\text{water/org}} > 1$

Now consider what happens if the organic solution is extracted with a *basic* aqueous solution. If the base, B^-, is strong enough, the organic acid **1** or **3** will be converted into the corresponding **conjugate base 2** or **4** (Eqs. 5.10 and 5.11). Because it is a salt, the conjugate base is highly polar, and $K_{\text{water/org}} > 1$. Thus the conjugate bases of organic acids may be selectively extracted from an organic phase into an aqueous phase. If the basic extract is then neutralized by an acid such as hydrochloric acid, the conjugate base **2** or **4** will be protonated to regenerate the organic acid **1** or **3** (Eqs. 5.12 and 5.13). Because the acid is water insoluble, it will appear as either a precipitate or a second layer if it is a liquid. The desired organic acid may then be recovered by filtration or separation of the layers.

$$\text{R—C} \overset{\displaystyle O^-}{\underset{\displaystyle O}{}} \text{Na}^+ + \text{H—Cl (aq)} \rightleftharpoons \text{R—C} \overset{\displaystyle O—H}{\underset{\displaystyle O}{}} + \text{Na}^+ \text{Cl}^- \qquad (5.12)$$

2   **1**
Water *soluble*   Water *insoluble*

$$\text{Ar—O}^- \text{ Na}^+ + \text{H—Cl} \rightleftharpoons \text{Ar—O} \underset{\displaystyle H}{} + \text{Na}^+ \text{Cl}^- \qquad (5.13)$$

4   **3**
Water *soluble*   Water *insoluble*

Thus, two water-insoluble organic compounds, **HA**, which is acidic, and **N**, which is neutral, that are dissolved in an organic solvent may be separated by selectively extracting the acidic compound into a basic aqueous phase. After the aqueous and organic phases are separated, **HA** is recovered from the aqueous phase upon neutralization, and **N** is obtained by removing the organic solvent (Fig. 5.1).

The choice of the base is determined by the acidity of the organic acid **HA**. This is defined in aqueous solution by K_a in Equation 5.14 and is often expressed as **pK_a** (Eq. 5.15).

$$K_{a(\text{HA})} = \frac{[\text{A}^-][\text{H}_3\text{O}^+]}{[\text{HA}]} \qquad (5.14)$$

$$\text{p}K_a = -\log_{10} K_a \qquad (5.15)$$

The next step in determining what base is needed is to consider the equilibrium shown in Equation 5.16, in which **HA** is the organic acid, B^- is the base being used, and **HB** is the **conjugate acid** of this base. The **equilibrium constant**, K_{eq}, for this process is given by Equation 5.17. Given the expressions for K_a for **HA** and **HB** (Eqs. 5.14 and 5.18, respectively), Equation 5.17 transforms to Equation 5.19. From this equation we see that $\log_{10}$ of the equilibrium constant, K_{eq}, equals p$K_{a(\text{HB})}$ − p$K_{a(\text{HA})}$. We can predict how effective a particular base B^- will be in converting an

Figure 5.1
Separating an acidic compound and a neutral compound.

organic acid **HA** to its anion simply by knowing the relative acidities of **HA** and **HB**. The equilibrium for Equation 5.14 will lie to the right; namely, $\log_{10} K_{eq}$ will be >1 whenever the acidity of **HA** is *greater* than that of **HB**.

$$HA + B^- \underset{}{\overset{K_{eq}}{\rightleftharpoons}} A^- + HB \tag{5.16}$$

$$K_{eq} = \frac{[A^-][HB]}{[HA][B^-]} \tag{5.17}$$

$$K_{a(HB)} = \frac{[B^-][H_3O^+]}{[HB]} \tag{5.18}$$

$$K_{eq} = \frac{K_{a(HA)}}{K_{a(HB)}} \text{ and } \log_{10} K_{eq} = \log_{10} K_{a(HA)} - \log_{10} K_{a(HB)} \tag{5.19}$$

Let's now consider two specific examples of mixtures of acidic and neutral compounds. Assume that you have a solution of benzoic acid (**5**) and naphthalene (**7**), a neutral compound, in diethyl ether and that you wish to separate the two. Both compounds and the solvent are insoluble in water at room temperature. The strategy is to convert **5** to its water-soluble conjugate base **6** by using an aqueous base to extract the organic solution. The pK_a of benzoic acid is 4.2, so any base whose *conjugate acid* has a pK_a *greater* than this will yield a value of $\log_{10} K_{eq} > 1$ (Eq. 5.19). Aqueous hydroxide, whose conjugate acid is water (pK_a 15.7), would give $\log_{10} K_{eq} = 10.5$, which means that this base will be extremely effective for the extraction of **5** into the aqueous phase. Indeed, aqueous hydroxide is commonly used for extracting organic acids into an aqueous phase.

Other aqueous bases could also be used. For example, carbonate (**8**) and bicarbonate (**9**) would also work because their conjugate acids, **9** and **10** (Eq. 5.20), have pK_as of 10.3 and 6.4, respectively. These bases are not as strong as hydroxide, so on the purely *thermodynamic* basis described by Equation 5.19, they would provide a lower value of $\log_{10} K_{eq}$. However, both **8** and **9** may be converted to carbonic acid (**10**) upon di- and monoprotonation, respectively (Eq. 5.20). Carbonic acid is *kinetically* unstable and decomposes to carbon dioxide and water (Eq. 5.21), and this disrupts the equilibrium of Equation 5.16, driving it to the *right* as **HB**, in the form of **10**, undergoes decomposition. The net result would be efficient deprotonation of **5** by these bases.

$$\tag{5.20}$$

8	**9**	**10**
Carbonate	Bicarbonate	Carbonic acid

$$H_2CO_3 \longrightarrow CO_2 + H_2O \tag{5.21}$$

It is important to recognize that the conversion of **10** into carbon dioxide and water presents a practical problem when **8** and **9** are used to deprotonate acids for purposes of extraction. If a separatory funnel is being used for the extraction, the gas pressure that results from the formation of carbon dioxide can blow out the stopcock and stopper, spraying the contents on you, your neighbors, and the benchtop. Alternatively, if a screw-cap vessel is being used, the contents may be ejected when the cap is loosened to relieve the gas pressure. Accordingly, it is preferable to use hydroxide for basic extractions as a general rule.

Turning to a second example, assume you have a solution of 2-naphthol (**11**), which is a phenol, and naphthalene (**7**) in diethyl ether. The pK_a of **11** is 9.5, so in this case either aqueous hydroxide or carbonate would give $K_{eq} > 1$; bicarbonate (**8**) is too weak a base to effect the deprotonation. Because aqueous hydroxide would provide $K_{eq} \gg 1$, it is the base of choice for this extraction.

11	**12**
2-Naphthol	2-Naphthoxide

The examples discussed so far have involved separation of compounds having different acidities from neutral compounds. However, organic bases, which are usually amines, must sometimes be separated from mixtures containing organic acids and neutral compounds. Before considering how to perform such separations, it is useful to regard amines as being derivatives of ammonia where the hydrogen atoms on the nitrogen atom are replaced with substituted carbon atoms that are designated as R or Ar. Like the functional groups of carboxylic acids and phenols, the amino functional group is polar and hydrophilic, and amines having more than six carbon atoms are either insoluble or only slightly soluble in water because of the hydrophobic properties of the R or Ar groups.

Amines are often soluble in aqueous acid at a pH < 4 because they are converted into their respective ammonium salts, the conjugate acids of the amines. This process is illustrated in Equation 5.22, wherein aqueous HCl serves as the acid. The enhanced ionic character of the ammonium salt, as we saw before with the salts of carboxylic acids and phenols, makes it water soluble ($K_{water/org} > 1$), whereas its parent base is not ($K_{water/org} < 1$). When the acidic solution is neutralized by adding aqueous base, the ammonium ion is deprotonated and converted into the original water-insoluble organic base (Eq. 5.23) that will now either precipitate from solution or form a separate layer. The amine may then be recovered by filtration or by separation of the layers.

$$R-\ddot{N}H_2 + H-Cl \text{ (aq)} \longrightarrow R\overset{+}{N}H_2-H \ Cl^- \tag{5.22}$$

(Water *insoluble*)	(Water *soluble*)
$K_{water/org} < 1$	$K_{water/org} > 1$

$$\underset{\substack{(\text{Water } soluble) \\ K_{\text{water/org}} > 1}}{\overset{+}{R}\overset{}{N}H_2\!-\!H\ Cl^- + HO^-\ Na^+\,(aq)} \longrightarrow \underset{\substack{(\text{Water } insoluble) \\ K_{\text{water/org}} < 1}}{R\!-\!\ddot{N}H_2 + H_2O + NaCl} \tag{5.23}$$

The question now is: How can the two water-insoluble organic compounds, **B**, which is basic, and **N**, which is neutral, that are dissolved in an organic solvent be separated? This problem is analogous to the earlier one of separating two water-insoluble organic compounds, **HA**, which is acidic, and **N**, which is neutral (see Fig. 5.1). Namely, the basic compound can be selectively extracted into an acidic aqueous phase as the soluble salt HB$^+$. After the aqueous and organic phases are separated, **B** is recovered from the aqueous phase upon neutralization, and **N** is obtained by removing the organic solvent (Fig. 5.2).

One example of an amine is 4-nitroaniline (**13**), which is converted into the corresponding anilinium ion **14** by aqueous hydrochloric acid. Although 4-nitroaniline is a weak base with a pK_b of 13.0, its conjugate acid **14** has a pK_a of 1.0 and is a relatively strong organic acid. Hence, it is necessary to use a strong aqueous acid such as 6 M HCl, in which the acid is actually the hydronium ion, H$_3$O$^+$, having a pK_a of about -1.7, to convert **13** efficiently into its salt **14**.

13	**14**
4-Nitroaniline	4-Nitroanilinium hydrochloride

In the experimental procedures that follow, three types of separations of neutral, acidic, and basic compounds are described. The first involves separating a mixture of benzoic acid (**5**) and naphthalene (**7**), using aqueous hydroxide for the extraction. The flowchart for this separation corresponds to that of Figure 5.1. The second procedure involves separating a mixture of **5, 7,** and 2-naphthol (**11**). In this case, **5** is first *selectively* removed from the solution by extraction with aqueous bicarbonate, which *does not* deprotonate **11**. A second basic extraction using aqueous hydroxide removes **11** from the organic solution. The flowchart for this sequence is depicted in Figure 5.3.

The last procedure involves separating a three-component mixture containing benzoic acid (**5**), naphthalene (**7**), and 4-nitroaniline (**13**). First **13** is *selectively* removed from the organic solution by extraction with aqueous hydrochloric acid, and then **5** is removed by extraction with aqueous base, leaving **7** in the organic solution. The flowchart for this sequence of operations is shown in Figure 5.4. All procedures demonstrate the power of using the *pH*-dependence of water solubility of organic compounds as a basis for separation. Such procedures for the preparative separation of organic compounds are much easier than the chromatographic separations you will learn about in Chapter 6.

Figure 5.2
Separating a basic compound and a neutral compound.

Figure 5.3
Separating a neutral compound and two compounds having differing acidities.

Figure 5.4
Separating an acidic, a basic, and a neutral compound.

EXPERIMENTAL PROCEDURES

Base and Acid Extractions

Purpose To separate multicomponent mixtures as a function of pH of solution.

SAFETY ALERT

1. Wear safety glasses or goggles and suitable protective gloves while performing the experiments.
2. Do not allow any of the chemicals used in this experiment to come in contact with your skin. If they do, *immediately* wash the affected areas with soap and water.
3. Diethyl ether is *extremely* volatile and flammable. Be certain there are *no flames* in your vicinity when using it for extraction and when removing it from the "Neutral Compound."
4. Use caution in handling wet glassware, because it can easily slip out of your grasp.
5. When extracting a solution, vent the separatory funnel or vial *often* to avoid a buildup of pressure.

MINISCALE PROCEDURE

Preparation Refer to the online resources to answer Pre-Lab Exercises, access videos, and read the MSDSs for the chemicals used or produced in this procedure. Read or review Sections 2.7, 2.10, 2.11, 2.13, 2.17, 2.21, 2.24, 2.25, 2.29, 3.2, and 3.3.

A. One-Base Extraction

Apparatus Separatory funnel, ice-water bath, apparatus for vacuum filtration, simple distillation, and *flameless* heating.

Dissolution Obtain 2 g of a mixture of benzoic acid and naphthalene. Dissolve the mixture by swirling it with 30 mL of diethyl ether in an Erlenmeyer flask. If any solids remain, add more diethyl ether to effect complete dissolution. Transfer the solution to the separatory funnel.

Extraction Extract the solution with a 15-mL portion of 2.5 M (10%) aqueous sodium hydroxide. *Be sure to hold both the stopcock and the stopper of the funnel tightly and frequently vent the funnel by opening its stopcock* (Fig. 2.61). Identify the aqueous layer by the method described in Section 2.21, transfer it to an Erlenmeyer flask labeled "Hydroxide Extract," and transfer the organic solution into a clean Erlenmeyer flask containing two spatula-tips full of *anhydrous* sodium sulfate and labeled "Neutral Compound." Let this solution stand for about 15 min, occasionally swirling it to hasten the drying process. If the solution remains cloudy, add additional portions of sodium sulfate to complete the drying process.★

Precipitating and Drying Cool the "Hydroxide Extract" in an ice-water bath. Carefully acidify this solution with 3 M hydrochloric acid, so that the solution is distinctly acidic to pHydrion paper.★ Upon acidification, a precipitate should form; cool the mixture for 10–15 min to complete the crystallization.

Collect the precipitate by vacuum filtration (Fig. 2.54) using a Büchner or Hirsch funnel. Wash the solid on the filter paper with a small portion of *cold* water. Transfer the solid to a labeled watchglass, cover it with a piece of filter or weighing paper, and allow the product to air-dry until the next laboratory period. Alternatively, dry the solid in an oven having a temperature of 90−100 °C for about 1 h. After drying, transfer the benzoic acid to a dry, *tared* vial.

Separate the "Neutral Compound" from the drying agent by decantation (Fig. 2.56) or gravity filtration through a cotton plug (Fig. 5.5) into a 100-mL round-bottom flask. Remove the solvent by simple distillation (Fig. 2.37). Alternatively, use rotary evaporation or one of the other techniques described in Section 2.29. Allow the residue to cool to room temperature to solidify, scrape the contents of the flask onto a piece of weighing paper to air-dry, and then transfer it to a dry, *tared* vial.

Analysis and Recrystallization Determine the weight and melting point of each crude product. If you know the relative amounts of benzoic acid (**5**) and naphthalene (**7**) in the original mixture, calculate the percent recovery of each compound. Recrystallize them according to the general procedures provided in Section 3.2 and determine the melting points of the purified materials.

B. Two-Base Extraction

Apparatus Separatory funnel, ice-water bath, apparatus for vacuum filtration, simple distillation, and *flameless* heating.

Dissolution Obtain 2 g of a mixture of benzoic acid, 2-naphthol, and naphthalene. Dissolve the mixture by swirling it with 30 mL of diethyl ether in a 125-mL Erlenmeyer flask. If any solids remain, add more diethyl ether to effect complete dissolution.

Ring support
Cotton plug
Cork ring

Figure 5.5
Miniscale gravity filtration through a cotton plug.

Extraction Extract the solution with a 20-mL portion of 1.25 *M* (10%) aqueous sodium bicarbonate (Fig. 2.61). *Caution:* Gaseous carbon dioxide is generated! To prevent accidental loss of material, do the following: After adding the aqueous bicarbonate to the funnel, swirl the *unstoppered* funnel until all foaming has subsided. Then stopper the funnel and, *holding both the stopcock and the stopper of the separatory funnel tightly,* invert the funnel and *immediately* vent it by opening the *stopcock.* Finally, shake the funnel with *frequent* venting until gas is no longer evolved. Separate the layers, transferring the aqueous layer to a 125-mL Erlenmeyer flask labeled "Bicarbonate Extract."★ To ensure properly identifying the aqueous layer, consult Section 2.21. Return the organic solution to the separatory funnel.

Now extract the organic solution with a 20-mL portion of *cold* 2.5 *M* (10%) aqueous sodium hydroxide, venting the funnel frequently during the process. Transfer the aqueous layer to a 125-mL Erlenmeyer flask labeled "Hydroxide Extract." Transfer the organic solution to a 50-mL Erlenmeyer flask labeled "Neutral Compound" and containing two spatula-tips full of *anhydrous* sodium sulfate. Loosely cork the flask containing the organic solution, and let it stand for at least 15 min with occasional swirling to promote drying. If the solution remains cloudy, add additional portions of sodium sulfate to complete the drying process.★

Precipitating and Drying For the "Bicarbonate" and "Hydroxide Extracts," follow the directions given in Part A for the "Hydroxide Extract." Also apply the protocol provided in Part A for the "Neutral Compound" to the corresponding solution obtained in this procedure.

Analysis and Recrystallization Determine the weight and melting point of each crude product. If you know the relative amounts of benzoic acid (**5**), naphthalene (**7**), and 2-naphthol (**11**) in the original mixture, calculate the percent recovery of each compound. Recrystallize them according to the general procedures provided in Section 3.2 and determine the melting points of the purified materials.

C. Acid-Base Extraction

Apparatus Separatory funnel, ice-water bath, apparatus for vacuum filtration, simple distillation, and *flameless* heating.

Dissolution Obtain 1.5 g of a mixture of benzoic acid, 4-nitroaniline, and naphthalene. Dissolve the mixture by swirling it with about 40 mL of dichloromethane in an Erlenmeyer flask. Transfer the solution to the separatory funnel.

Extraction Extract the organic solution three times using 15-mL portions of 3 *M* hydrochloric acid. *Be sure to hold both the stopcock and the stopper of the funnel tightly and frequently vent the funnel by opening its stopcock* (Fig. 2.61). Identify the aqueous layer by the method described in Section 2.21. Combine the three acidic aqueous layers from the extractions in an Erlenmeyer flask labeled "HCl Extract." Return the organic layer to the separatory funnel and extract it two times using 15-mL portions of 3 *M* sodium hydroxide solution. Combine these two aqueous layers in a second flask labeled "Hydroxide Extract." Transfer the organic solution into a clean Erlenmeyer flask containing two spatula-tips full of *anhydrous* sodium sulfate and labeled "Neutral Compound." Let this solution stand for about 15 min, occasionally swirling it to hasten the drying process. If the solution remains cloudy, add additional portions of sodium sulfate to complete the drying process.★

Precipitating and Drying Cool the "HCl Extract" and the "Hydroxide Extract" in an ice-water bath. Neutralize the "HCl Extract" with 6 *M* sodium hydroxide and add a little excess base to make the solution distinctly basic to pHydrion paper. Neutralize the "Hydroxide Extract" with 6 *M* hydrochloric acid and add a little excess acid to make the solution distinctly acidic to pHydrion paper.★ Upon neutralization, a precipitate should form in each flask.

Collect the precipitate in the flasks labeled "HCl Extract" and "Hydroxide Extract" by vacuum filtration (Fig. 2.54) using a Büchner or Hirsch funnel. Wash each solid on the filter paper with a small portion of *cold* water. Transfer each solid to a labeled watchglass, cover it with a piece of filter or weighing paper, and allow the product to air-dry until the next laboratory period. Alternatively, dry the solid in an oven having a temperature of 90−100 °C for about 1 h. After drying, transfer the 4-nitroaniline and the benzoic acid each to a different dry, *tared* vial.

Separate the "Neutral Compound" from the drying agent by decantation (Fig. 2.56) or gravity filtration through a cotton plug (Fig. 5.5) into a 100-mL round-bottom flask. Remove the solvent by simple distillation (Fig. 2.37). Alternatively, use rotary evaporation or one of the other techniques described in Section 2.29. Allow the residue to cool to room temperature to solidify, scrape the contents of the flask onto a piece of weighing paper to air-dry, and then transfer it to a dry, *tared* vial.

Analysis and Recrystallization Determine the weight and melting point of each crude product. If you know the relative amounts of benzoic acid (**5**), naphthalene (**7**), and 4-nitroaniline (**13**) in the original mixture, calculate the percent recovery of each compound. Recrystallize them according to the general procedures provided in Section 3.2 and determine the melting points of the purified materials.

MICROSCALE PROCEDURE

Preparation Refer to the online resources to answer Pre-Lab Exercises, access videos, and read the MSDSs for the chemicals used or produced in this procedure. Read or review Sections 2.7, 2.9, 2.10, 2.17, 2.21, 2.25, 2.29, 3.2, and 3.3.

A. One-Base Extraction

Apparatus A 5-mL conical vial, Pasteur filter-tip pipets, ice-water bath, 3-mL Craig tube, apparatus for Craig tube filtration and *flameless* heating.

Dissolution Obtain 0.2 g of a mixture of benzoic acid and naphthalene. Dissolve the mixture by swirling it with 2 mL of diethyl ether in the conical vial. If any solids remain, use a filter-tip pipet to add enough diethyl ether to effect complete dissolution.

Extraction Add 0.5 mL of *cold* 2.5 *M* (10%) aqueous sodium hydroxide to the vial and cap it. Shake the vial vigorously for a minute or two, occasionally unscrewing the cap slowly to relieve any pressure. Allow the layers to separate and, using a Pasteur pipet, transfer the *aqueous* layer to a Craig tube labeled "Hydroxide Extract." To ensure properly identifying the aqueous layer, consult Section 2.21.

Precipitating and Drying Add a microspatula-tip full of sodium sulfate to the organic solution in the conical vial and dry the solution according to the directions provided in Section 2.25. Once the organic solution is dry, use a filter-tip pipet to transfer it to a test tube. Evaporate the diethyl ether using *flameless* heating; stir the solution with a microspatula or use a boiling stone to prevent bumping during evaporation. Perform this step at a hood or use a funnel attached to a vacuum source for removing vapors (Fig. 2.72b). Scrape the isolated naphthalene onto a piece of weighing paper to air-dry and then transfer it to a dry, *tared* vial.

Cool the "Hydroxide Extract" in an ice-water bath and, with stirring, add *concentrated* hydrochloric acid dropwise with a Pasteur pipet until the solution is distinctly acidic to pHydrion paper.★ Allow the acidified solution to remain in the bath to complete the crystallization. Then isolate the benzoic acid by Craig tube filtration (Fig. 2.55). Rinse the solid back into the Craig tube, with 0.5 mL of cold water, stir the mixture briefly, and repeat the filtration. Transfer the benzoic acid to a watchglass, cover it with a piece of filter paper, and allow the product to air-dry until the next laboratory period. After drying it, put the isolated solid in a dry, *tared* vial.

Analysis and Recrystallization Determine the weight and melting point of each crude product. If you know the relative amounts of benzoic acid (**5**) and naphthalene (**7**) in the original mixture, calculate the percent recovery of each compound. Recrystallize them according to the general microscale procedures provided in Section 3.2 and determine the melting points of the purified materials.

B. Two-Base Extraction

Apparatus A 5-mL conical vial, Pasteur filter-tip pipets, ice-water bath, 3- and 5-mL Craig tubes, apparatus for Craig tube filtration and *flameless* heating.

Dissolution Obtain 0.3 g of a mixture of benzoic acid, 2-naphthol, and naphthalene. Dissolve the mixture by swirling it with 2 mL of diethyl ether in the conical vial. If any solids remain, use a filter-tip pipet to add enough diethyl ether to effect complete dissolution.

Extraction Extract the solution with a 1-mL portion of 1.25 *M* (10%) aqueous sodium bicarbonate. *Caution:* Gaseous carbon dioxide is generated! To prevent accidental loss of material, do the following: After adding the first portion of aqueous bicarbonate to the vial, stir the two-phase mixture vigorously with a microspatula or stirring rod until all foaming has subsided. Then cap the vial, shake it gently for 2–3 sec, and loosen the cap *slowly* to relieve any pressure. Retighten the cap and shake the vial vigorously for about 1 min, occasionally venting the vial by loosening its cap. Allow the layers to separate and, using a Pasteur pipet, transfer the *aqueous* layer to a 5-mL Craig tube labeled "Bicarbonate Extract." To ensure properly identifying the aqueous layer, consult Section 2.21.★

Now extract the organic solution with a 0.5-mL portion of *cold* 2.5 *M* (10%) aqueous sodium hydroxide, following the same protocol as that given in part **A,** placing the aqueous layer in a 3-mL Craig tube labeled "Hydroxide Extract."

Precipitating and Drying Use the same procedure as that given in part **A** for *separately* acidifying the "Bicarbonate" and "Hydroxide" extracts and isolating and drying the two solids that result.

Analysis and Recrystallization Determine the weight and melting point of each crude product. If you know the relative amounts of benzoic acid (**5**), naphthalene (**7**), and

2-naphthol (**11**) in the original mixture, calculate the percent recovery of each compound. Recrystallize the products according to the general microscale procedures provided in Section 3.2 and determine the melting points of the purified materials.

C. Acid-Base Extraction

Apparatus A 5-mL conical vial, Pasteur and filter-tip pipets, ice-water bath, four 13-mm × 10-mm test tubes, apparatus for vacuum filtration and *flameless* heating.

Dissolution Obtain 0.25 g of a mixture of benzoic acid, 4-nitroaniline, and naphthalene. Dissolve the mixture by swirling it with about 3 mL of dichloromethane in the conical vial.

Extraction Add 1.0 mL of 3 *M* hydrochloric acid to the vial and cap it. Shake the vial vigorously for a minute or two, occasionally unscrewing the cap to relieve any pressure. Allow the layers to separate and, using a Pasteur pipet, transfer the *aqueous* layer to a test tube labeled "HCl Extract." To ensure properly identifying the aqueous layer, consult Section 2.21. Repeat this extraction with 1.0 mL of 3 *M* hydrochloric acid, and again transfer the aqueous layer to the test tube labeled "HCl Extract." Now extract the organic solution once with 1.0 mL of 3 *M* sodium hydroxide solution, and transfer the *aqueous* layer to a *second* test tube labeled "Hydroxide Extract." Transfer the organic solution into a clean test tube labeled "Neutral Compound."★

Precipitating and Drying Add two microspatula-tips full of *anhydrous* sodium sulfate to the organic solution in the test tube "Neutral Compound" and dry the solution according to the directions provided in Section 2.25. Once this organic solution is dry, use a Pasteur or a filter-tip pipet to transfer it to another test tube. Evaporate the dichloromethane using *flameless* heating; stir the solution with a microspatula or use a boiling stone to prevent bumping during evaporation. Perform this step at a hood or use a funnel attached to a vacuum source for removing vapors (Fig. 2.72b). Scrape the isolated naphthalene onto a piece of weighing paper to air-dry and then transfer it to a dry, *tared* vial.

   Cool the test tubes containing the "HCl Extract" and the "Hydroxide Extract" in an ice-water bath. Neutralize the "HCl Extract" with 6 *M* sodium hydroxide and add a little excess base to make the solution distinctly basic to pHydrion paper. Neutralize the "Hydroxide Extract" with 6 *M* hydrochloric acid and add a little excess acid to make the solution distinctly acidic to pHydrion paper. Upon neutralization, a precipitate should form in each test tube; cool each tube for 10–15 min to complete the crystallization.★ Isolate the 4-nitroaniline and the benzoic acid by vacuum filtration using a Hirsch funnel (Fig. 2.54). Rinse each solid back into its *respective* test tube, with 0.5 mL of *cold* water, stir the mixture briefly, and repeat the filtration. Transfer the 4-nitroaniline and the benzoic acid each to a separate labeled watchglass, cover both with a piece of filter paper, and allow the products to air-dry until the next laboratory period. Alternatively, dry the solid in an oven having a temperature of 90–100 °C for about 1 h. After drying is complete, transfer the 4-nitroaniline and the benzoic acid each to a different dry, *tared* vial.

Analysis and Recrystallization Determine the weight and melting point of each crude product. If you know the relative amounts of benzoic acid (**5**), naphthalene (**7**), and 4-nitroaniline (**13**) in the original mixture, calculate the percent recovery of each compound. Recrystallize them according to the general procedures provided in Section 3.2 and determine the melting points of the purified materials.

Discovery Experiment *Separation of Unknown Mixture by Extraction*

Obtain an unknown mixture comprising up to four different compounds to be separated using one-base extraction, two-base extraction, acid-base extraction, or a combination of these separation techniques. The mixture might contain, for example, one compound from each of the following four groups: (1) *N*-(4-nitrophenyl) benzamide, 9-fluorenone, fluorene, anthracene, phenanthrene, or naphthalene; (2) *p*-cresol, 2-naphthol, or 4-methoxyphenol; (3) benzoic acid, 2-methylbenzoic acid, 2-chlorobenzoic acid, 4-chlorobenzoic acid, or salicyclic acid; (4) 3-nitroaniline, 4-nitroaniline, or 2-methyl-4-nitroaniline. Devise a scheme that may be used for separating this unknown mixture into its individual components; the procedure you devise should work irrespective of the number of components in the mixture. After recrystallizing each of the components, determine their melting points so that you can identify each compound.

Discovery Experiment *Isolation of Ibuprofen*

Devise and perform an extraction procedure that may be used to separate pure 2-(4′-isobutylphenyl) propionic acid (ibuprofen) from the over-the-counter medications Advi or Nuprin. Determine the melting point of the material you isolate to confirm its identity.

WRAPPING IT UP

Place the used *filter papers* in a container for nontoxic solid waste. Flush the *acidic* and *basic* filtrates down the drain. Pour any *diethyl ether* that has been isolated into a container for nonhalogenated organic liquids. Pour any *dichloromethane* that has been isolated into a container for halogenated organic liquids.

EXERCISES

General Questions

1. Using Equation 5.9, show that three extractions with 5-mL portions of a solvent give better recovery than a single extraction with 15 mL of solvent when $K = 0.5$.

2. Show mathematically how Equation 5.17 can be converted to Equation 5.19.

3. Define the following terms:

   a. immiscible liquid phases
   b. distribution or partition coefficient
   c. adsorption
   d. absorption
   e. conjugate base of an acid, HA
   f. conjugate acid of a base, B⁻
   g. liquid-liquid extraction
   h. hydrophilic

4. Explain why swirling or shaking a solution and its drying agent hastens the drying process.

5. Assume that the partition coefficient, K, for partitioning of compound *A* between diethyl ether and water is 3; that is, *A* preferentially partitions into ether.

a. Given 400 mL of an aqueous solution containing 12 g of compound A, how many grams of A could be removed from the solution by a *single* extraction with 200 mL of diethyl ether?

b. How many total grams of A can be *removed* from the aqueous solution with three successive extractions of 67 mL each?

6. Based on the principle of "like dissolves like," indicate by placing an "x" in the space those compounds listed below that are likely to be soluble in an organic solvent like diethyl ether or dichloromethane and those that are soluble in water.

Compound	Organic Solvent	Water
a. $C_6H_5CO_2Na$	_____	_____
b. Naphthalene (7)	_____	_____
c. Anthracene (15)	_____	_____
d. Phenol (16)	_____	_____
e. C_6H_5ONa	_____	_____
f. Aniline (17)	_____	_____
g. NaCl	_____	_____
h. CH_3CO_2H	_____	_____

15
Anthracene

16
Phenol

17
Aniline

7. Benzoic acid (5) is soluble in diethyl ether but not water; however, benzoic acid is extracted from diethyl ether with aqueous sodium hydroxide.

a. Complete the acid-base reaction below by writing the products of the reaction.

$+ \ NaOH \ (aq) \ \longrightarrow \ \rule{2cm}{0.4pt} \ + \ \rule{2cm}{0.4pt}$

5

b. In the reaction of Part **a,** label the acid, the base, conjugate acid, and conjugate base.

c. Indicate the solubility of benzoic acid and its conjugate base in diethyl ether and in water.

8. Aniline (**17**), an amine, is soluble in diethyl ether but not water; however, aniline is extracted from diethyl ether with aqueous hydrochloric acid.

a. Complete the acid-base reaction below by writing the products of the reaction.

b. In the reaction of Part **a,** label the acid, the base, conjugate acid, and conjugate base.

c. Indicate the solubility of aniline and its conjugate acid in diethyl ether and in water.

9. Naphthalene (**7**) is soluble in diethyl ether, but it is insoluble in water regardless of the solution pH. Explain why this compound cannot be readily ionized in aqueous solution.

10. There are three common functional groups in organic chemistry that are readily ionized by adjusting the pH of the aqueous solution during an extraction. Name and write the chemical structure of these three functional groups, and show each of them in both their neutral and ionized forms.

11. **a.** The pK_a of benzoic acid (**5**) is 4.2. Show mathematically that this acid is 50% ionized at pH 4.2.

b. Use the result of Part **a** to explain why precipitation of **5** is incomplete if the pH of an aqueous solution of benzoate ion (**6**) is lowered only to pH 7 by adding acid.

12. Consider the base monosodium phosphate, Na^+ $(HO)_2P(=O)O^-$. (a) Write the structure of the conjugate acid of this base. (b) Given that the pK_a of this conjugate acid is 2.1, explain why an aqueous solution of monosodium phosphate would be ineffective for extracting benzoic acid (**5**) from a diethyl ether solution.

13. Provide a flowchart analogous to those in Figures 5.1–5.4 for separating a diethyl ether solution containing anthracene (**15**), benzoic acid (**5**), 2-naphthol (**11**), and 4-nitroaniline (**13**). All compounds are solids when pure.

14. The equilibrium for phenol (**16**), sodium phenoxide (**18**), sodium bicarbonate, and carbonic acid is shown below:

| **16** | | | **18** | |
| Phenol | Sodium bicarbonate | | Sodium phenoxide | Carbonic acid |

a. The pK_as for phenol and carbonic acid are 10.0 and 6.4, respectively. Determine the K_{eq} for this reaction.

b. Based upon your answer to Part **a,** would sodium bicarbonate be a suitable base for separating phenol from a neutral organic compound via an aqueous extraction?

One-Base Extraction

15. What practical consideration makes aqueous hydroxide rather than aqueous bicarbonate (**9**) the preferred base for extracting benzoic acid (**5**) from diethyl ether?

16. Naphthalene (**7**) has a relatively high vapor pressure for a solid (because of its volatility, naphthalene is the active ingredient in some brands of mothballs). In view of this, what might happen if you placed naphthalene under vacuum or in an oven for several hours to dry it?

17. When benzoic acid (**5**) is partitioned between diethyl ether and aqueous sodium hydroxide solution and the aqueous layer is separated, acidification of the aqueous solution yields a precipitate.

a. Using arrows to symbolize the flow of electrons, show the reaction of benzoic acid (**5**) with hydroxide and draw the structure of the product of this reaction.

b. Using arrows to symbolize the flow of electrons, show the reaction of the product of the reaction in Part **a** with aqueous acid and draw the structure of the product of this reaction.

c. Why does the organic product of the reaction in Part **b** precipitate from aqueous solution?

Two-Base Extraction

18. Why does the sequence for extracting the diethyl ether solution of benzoic acid (**5**), naphthalene (**7**), and 2-naphthol (**11**) start with aqueous bicarbonate and follow with aqueous hydroxide rather than the reverse order?

19. What would be the consequence of acidifying the basic extract containing **12** only to pH 10?

20. Why is caution advised when acidifying the bicarbonate solution of sodium benzoate (**6**)?

Acid-Base Extraction

21. Why is *anhydrous* sodium sulfate added to the organic solution remaining after the extractions with 6 *M* HCl and 6 *M* NaOH?

22. The pK_as of benzoic acid (**5**) and 4-nitroanilinium hydrochloride (**14**) are 4.2 and 1.0, respectively.

    a. Determine the K_{eq} for the reaction of benzoic acid (**5**) with 4-nitroaniline (**13**).

    b. Based upon your answer to Part **a**, do you think a significant amount of salt would form from mixing equimolar amounts of benzoic acid (**5**) and 4-nitroaniline (**13**)? Explain your reasoning.

5.4 EXTRACTION OF A NATURAL PRODUCT: TRIMYRISTIN

Solvent extraction is a valuable method for obtaining a desired substance from its natural source. A familiar example is the hot-water extraction of caffeine (**19**) and the various oils that constitute the flavors of freshly brewed coffee and tea from coffee beans and tea leaves. As contrasted to the **liquid-liquid** extractions described in Section 5.2, this process is an example of **solid-liquid** extraction. The theory underlying it is the same, however. Because most organic compounds we wish to isolate are *insoluble* in water, organic solvents such as diethyl ether, dichloromethane, ethanol, and acetone are used for extracting **natural products**, that is, compounds found in nature.

19

Caffeine

Extraction of natural materials often produces complex mixtures of products, as described in the Historical Highlight *Natural Products*, which is available online, so additional operations are required to separate and purify individual compounds. These procedures may involve techniques in which the compounds are separated on the basis of their acidic or basic properties (Sec. 5.3) or by chromatographic methods (Chap. 6). Thus, the isolation of pure natural products normally involves a series of complex and time-consuming operations. Fortunately there are some exceptions to this general rule. The isolation of trimyristin (**20**) from nutmeg is one of them.

Trimyristin, which occurs in many vegetable **fats** and **oils**, is a **triester** of glycerol (**21**) and the **fatty acid** myristic acid (**22**). Most naturally occurring fats and oils are esters of **21** and straight-chain carboxylic acids, the most common of which contain 14–20 carbon atoms. In fact, the only difference between a fat and an oil is whether it is a solid or liquid at room temperature. The lack of any double bonds in the carbon chain of myristic acid makes **20** a member of the family of **saturated fats**, which allegedly increase the risk of heart disease if present in the diet in excessive amounts.

Nutmeg is a hard, aromatic seed of an East Indian tree (*Myristica fragrans*) that is the source of trimyristin in this experiment. It has been a valued spice ever since its discovery in the Spice Islands of Indonesia by Portuguese sea captains over four centuries ago. The chemical makeup of nutmeg is somewhat unusual in that extraction of the ground seeds with diethyl ether yields **20** in high purity without contamination by other structurally related esters of glycerol and fatty acids. Although the simple procedure described here is *not* typical of that usually required for isolating natural products, it nicely demonstrates the general technique of extracting biological materials from natural sources.

$$\text{CH}_2\text{—O—}\overset{\displaystyle\overset{O}{\|}}{\text{C}}\text{—(CH}_2)_{12}\text{CH}_3$$
$$\text{CH—O—}\overset{\displaystyle\overset{O}{\|}}{\text{C}}\text{—(CH}_2)_{12}\text{CH}_3$$
$$\text{CH}_2\text{—O—}\overset{\displaystyle\overset{O}{\|}}{\text{C}}\text{—(CH}_2)_{12}\text{CH}_3$$

20	**21**	**22**
Trimyristin	Glycerol	Myristic acid

$$\text{CH}_2\text{OH}$$
$$\text{CHOH}$$
$$\text{CH}_2\text{OH}$$

$$\text{HO—}\overset{\displaystyle\overset{O}{\|}}{\text{C}}\text{—(CH}_2)_{12}\text{CH}_3$$

EXPERIMENTAL PROCEDURES

Isolation of Trimyristin from Nutmeg

Purpose To apply the technique of extraction for isolating a natural product.

SAFETY ALERT

1. **Wear safety glasses or goggles and suitable protective gloves while performing the experiments.**

2. **Diethyl ether and acetone are both highly volatile and flammable solvents. Use no flames or electrical dryers in the laboratory during this experiment.**

3. **Monitor the heating of diethyl ether under reflux so that the ring of condensate *remains in the lower third of the condenser*. This will ensure that no vapors escape from the condenser into the room. Very little heat will be required regardless of the heating source used to maintain reflux.**

MINISCALE PROCEDURE

Preparation Refer to the online resources to answer Pre-Lab Exercises, access videos, and read the MSDSs for the chemicals used or produced in this procedure. Read or review Sections 2.7, 2.9, 2.10, 2.11, 2.13, 2.17, 2.21, 2.22, 2.29, 3.2, and 3.3.

Apparatus A 25- and a 50-mL round-bottom flask, ice-water bath, apparatus for heating under reflux, magnetic stirring, simple distillation, vacuum filtration, and *flameless* heating.

Setting Up Place about 4 g of ground nutmeg in the 50-mL round-bottom flask containing a spinvane and add 10 mL of diethyl ether. If only seeds of nutmeg are available, grind the seeds in a mortar and pestle or a blender. Attach a water-cooled condenser to the flask, and set up the apparatus for heating under reflux (Fig. 2.66a).

Extraction and Isolation Heat the mixture under gentle reflux for approximately 0.5 h. Allow the contents of the flask to cool to room temperature.★ Filter the mixture by gravity filtration through a fluted filter paper into the 25-mL round-bottom flask (Fig. 2.50), rinsing any residue remaining in the 50-mL flask onto the filter paper with an additional 2–4 mL of diethyl ether.

Equip the flask containing the filtrate for simple distillation (Fig. 2.37a), and remove the diethyl ether by distillation. Alternatively, use rotary evaporation or one of the other techniques described in Section 2.29. Dissolve the resulting yellow oil in 3–4 mL of acetone while warming the mixture (*no flames!*). Immediately pour the hot solution into a 25-mL Erlenmeyer flask. Allow the solution to stand at room temperature for about 0.5 h. If crystallization has not occurred during this time, gently scratch the flask at the air-liquid interface.★ Complete crystallization by cooling the flask in an ice-water bath for an additional 15 min. Collect the trimyristin on a Büchner or Hirsch funnel by vacuum filtration (Fig. 2.54). Transfer it to a watchglass or piece of clean filter or weighing paper for air-drying.

Analysis Weigh the trimyristin and determine its melting point. The reported melting point is 55–56 °C. Based on the original weight of nutmeg used, calculate the percent recovery.

WRAPPING IT UP

Once any residual diethyl ether has evaporated from the filter cake, place the used filter paper and its contents in a container for nonhazardous solids. Any diethyl ether that has been isolated should be poured into a container for nonhalogenated organic liquids. Either flush the acetone-containing filtrate down the drain or pour it into a container for nonhalogenated organic liquids.

MICROSCALE PROCEDURE

Preparation Refer to the online resources to answer Pre-Lab Exercises, access videos, and read the MSDSs for the chemicals used or produced in this procedure. Read or review Sections 2.5, 2.7, 2.9, 2.10, 2.13, 2.17, 2.21, 2.22, 2.29, 3.2, and 3.3.

Apparatus A 5-mL conical vial or 10-mL round-bottom flask, ice-water bath, Pasteur filter-tip and filtering pipets, tared 5-mL Craig tube, apparatus for heating under reflux, magnetic stirring, and *flameless* heating.

Setting Up Place 1 g of ground nutmeg in the vial or flask equipped for magnetic stirring. Add 2–3 mL of diethyl ether, attach the reflux condenser, and set up the apparatus for heating under reflux (Fig. 2.67).

Extraction and Isolation Heat the heterogeneous mixture for 0.5 h under *gentle* reflux with stirring, being certain not to lose diethyl ether through the top of the condenser. Allow the mixture to cool to room temperature★ and then gravity-filter the ethereal solution into a Craig tube, using Pasteur filter-tip and filtering pipets (Fig. 2.51). Use a filter-tip pipet to rinse the reaction vessel with 0.5–1 mL of techni-cal diethyl ether, and use the pipet to transfer the rinsing to the Craig tube.★

Add a small boiling stone or a boiling stick to the Craig tube and, working at the hood, gently evaporate the diethyl ether using *flameless* heating. If hood space is not available, perform the evaporation under an inverted funnel attached to a water aspirator (Fig. 2.69b). After the diethyl ether has evaporated, add 1 mL of acetone to the residual yellow oil and warm the mixture briefly (*no flames!*) to effect solution. Allow the solution to cool to room temperature and remain at that temperature for about 0.5 h.★ If crystallization has not occurred during this time, gently scratch the tube at the air-liquid interface.

Complete crystallization by cooling the Craig tube in an ice-water bath for an additional 15 min, and then isolate the product by Craig tube filtration (Fig. 2.55). Carefully disassemble the Craig tube, scrape any crystalline material clinging to the glass or Teflon plug into the tube, and air-dry the product.

Analysis Weigh the trimyristin and determine its melting point. The reported melt-ing point is 55–56 °C. Based on the original weight of nutmeg used, calculate the percent recovery.

WRAPPING IT UP

Either flush the *acetone-containing filtrate* down the drain or pour it into a container for nonhalogenated organic liquids.

EXERCISES

1. Why is diethyl ether rather than acetone chosen as the extraction solvent?

2. Why is finely divided rather than whole nutmeg specified in this extraction?

3. What are the structural features, functional groups, and physical properties of trimyristin that place it in the class of natural products known as saturated fats?

4. Trimyristin is only one component of nutmeg. How are the other components of nutmeg removed in this experiment?

5. Considering the solubility properties of trimyristin (**20**), would water be a good choice as an extracting solvent for obtaining **20** from nutmeg?

6. Judging from your examination of trimyristin (**20**), is this substance responsible for the odor of nutmeg?

7. Why is ether added to the nutmeg and then removed at the end of the experiment?

8. What is the purpose of adding acetone to the crude product?

9. Indicate which of the following are reasons for performing a simple distillation in this procedure (more than one answer may be correct):

   a. To obtain pure diethyl ether.

   b. To isolate the nonvolatile solute.

   c. To isolate a volatile impurity.

    d.   To make extra work for the student.

    e.   To remove the solvent from the solute.

10. A certain plant material is known to contain mainly trimyristin and tripalmitin in approximately equal amounts. Tripalmitin, the structure of which you should look up in a reference source, has mp 66–67 °C. Extraction of these two compounds from the plant material with diethyl ether gave an oil after removal of the solvent, and this oil was difficult to crystallize. Explain this result. (*Hint:* See Sec. 3.3.)

11. In the Miniscale Procedure,

    a.   why is gravity rather than vacuum filtration used to separate the ethereal extract from the residual nutmeg?

    b.   why is fluted rather than plain filter paper used for the filtration procedure?

12. In the Microscale Procedure,

    a.   why is it important that a gentle rather than vigorous reflux rate be used?

    b.   why is it preferred to transfer the ethereal extract with a Pasteur filter-tip pipet rather than a regular Pasteur pipet?

Chromatography

Oh, no, you forgot to put the cap on your ballpoint pen, so you now have a big inkspot on the pocket of your best shirt as a "reward" for your negligence. You dab it with a wet cloth, only to see the spot grow even larger. New colors seem to be appearing as the spot widens. Congratulations! You've just performed a form of chromatography in which you have crudely separated some of the dyes that comprise the ink. In this chapter, we'll be learning some of the basic chromatographic techniques that are used to separate mixtures by taking advantage of the differential distribution of the individual compounds between two immiscible phases—in the case of your shirt, these were its fibers and the water you used in an effort to remove the stain. These techniques are now so powerful and sophisticated that even enantiomers (Sec. 7.1) may be separated efficiently by such means.

6.1 INTRODUCTION

We described the common laboratory techniques of recrystallization, distillation, and extraction for purifying organic compounds in Chapters 3, 4, and 5. In many cases, however, the mixtures of products obtained from chemical reactions do not lend themselves to ready separation by any of these techniques because the physical properties of the individual components are too similar. Fortunately, there are a number of chromatographic procedures available that we can use to effect the desired purification, and some of them are described in this chapter.

The word **chromatography** was first used to describe the colored bands observed when a solution containing plant pigments is passed through a glass column containing an adsorbent packing material. From that origin, the term now encompasses a variety of separation techniques that are widely used for analytical and preparative purposes.

All methods of chromatography operate on the principle that the components of a mixture will distribute unequally between two immiscible phases, which is also the basis for separations by extraction (Chap. 5). The **mobile phase** is generally a liquid or a gas that flows continuously over the fixed **stationary phase,** which may be a solid or a liquid. The individual components of the mixture have different affinities for the mobile and stationary phases, so a dynamic equilibrium is established in which each component is selectively, but temporarily, removed from the mobile phase by binding to the stationary phase. When the equilibrium concentration of that substance in the mobile phase decreases, it is released from the stationary phase and the process continues. Since each component partitions between the two

179

Figure 6.1
*Separation of mixture
by chromatography.*

phases with a different equilibrium constant or **partition coefficient,** the components divide into separate regions termed **migratory bands** (Fig. 6.1). *The component that interacts with or binds more strongly to the stationary phase moves more slowly in the direction of the flow of the mobile phase.* The attractive forces that are involved in this selective adsorption are the same forces that cause attractive interactions between any two molecules: electrostatic and dipole-dipole interactions, hydrogen bonding, complexation, and van der Waals forces.

The chromatographic methods used by modern chemists to identify and/or purify components of a mixture are characterized by the nature of the mobile and stationary phases. For example, the techniques of **thin-layer** (TLC), **column,** and **high-performance liquid chromatography** (HPLC) each involve *liquid-solid* phase interactions. **Gas-liquid partition chromatography** (GLC), also known as **gas chromatography** (GC), involves distributions between a mobile *gas* phase and a stationary *liquid* phase coated on a solid support. These important techniques can be used as tools to analyze and identify the components in a mixture as well as to separate the mixture into its pure components for preparative purposes. Although there are other chromatographic techniques, such as ion exchange and paper chromatography, a review of those methods is beyond the scope of this discussion.

6.2 THIN-LAYER CHROMATOGRAPHY

Thin-layer chromatography (TLC) is a form of **solid-liquid adsorption** chromatography and is an important technique in organic chemistry for rapid analysis of small quantities of samples, sometimes as little as 10^{-9} g. Thus, TLC is frequently used to monitor the progress of reactions and of preparative column chromatographic separations as well as to determine the optimal combinations of solvent and adsorbent for such separations (Sec. 6.3). An important limitation to TLC, and to column chromatography as well, is that it cannot be used on volatile compounds having boiling points below about 150 °C (760 torr).

To execute a TLC analysis, a *small* amount of the sample to be analyzed, or a solution of it, is first applied to a solid **adsorbent** bound to a rectangular glass or plastic plate (Fig. 6.2a). The adsorbent serves as the stationary phase. Next, the plate, with its spotted end down, is placed in a closed jar, called a **developing chamber** (Fig. 6.3). The chamber contains a *saturated atmosphere* of a suitable **eluant** or **eluting solvent,** which is the mobile phase and may be comprised of either a single solvent or mixture of two or more. A folded filter paper is often used to help maintain solvent equilibration in the chamber. It is important that the level of solvent in the chamber be *below* the location of the spot on the TLC plate. Otherwise, the sample

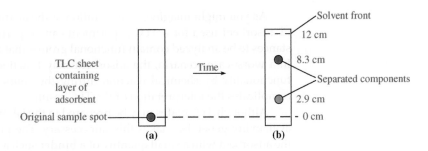

Figure 6.2
Thin-layer chromatography.
(a) Original plate loaded
with sample. (b) Developed
chromatogram.

Figure 6.3
TLC chamber. (a) Folded filter
paper to be placed in developing
chamber for solvent equilibration.
(b) Developing chamber with filter
paper and TLC plate.

would be removed from the plate by dissolution in the bulk solvent, thereby ruining the analysis.

When the plate is placed in the chamber, the solvent begins to ascend the plate. The individual components of the sample, which initially were at the bottom, or "origin," are carried up the plate along with the eluting solvent at a rate that is dependent on their relative affinities for the solid adsorbent: more weakly adsorbed compounds move up the plate faster than do those that are more strongly adsorbed. Ideally, when the solvent front has nearly reached the top of the TLC plate, which normally takes only a few minutes, each component of the original sample will appear as a separate spot on the plate (Fig. 6.2b). This completes the development of the chromatogram, and the plate is then removed from the developing chamber for analysis.

Adsorbents

The solid adsorbent in TLC is usually alumina (Al_2O_3) or silica gel (silicic acid, $SiO_2 \times H_2O$), both of which are polar. Alumina is the more polar of the two and is commercially available in three forms: neutral, acidic, and basic. Acidic and basic alumina are sometimes used to separate basic and acidic compounds, respectively, but neutral alumina is the most common form of this adsorbent for TLC. Silica gel, which is slightly acidic, is the adsorbent used in the experimental procedures described in this section.

As you might imagine, consideration of the acidic or basic character of the solid adsorbent used for a TLC experiment can be particularly important if the substances to be analyzed contain functional groups that are sensitive to acids or bases. In a worst-case scenario, the adsorbent may function as a catalyst to destroy the functionality by chemical reaction during the course of the analysis; this greatly complicates the interpretation of the TLC results.

Although TLC plates may be prepared in the laboratory, their ready commercial availability generally makes this unnecessary. The plates are produced by mixing the adsorbent with a small quantity of a **binder** such as starch or calcium sulfate and spreading the mixture as a layer approximately 250 μ thick on the supporting plate. The binder is needed for proper adhesion of the thin layer of adsorbent to the plate. TLC plates should be dried in an oven for an hour or more at 110 °C prior to use to remove any atmospheric moisture adsorbed on them. This is necessary because the activity of the adsorbent and its effectiveness in separating the components of a mixture are decreased because water occupies binding sites on the surface of the solid.

The strength of adsorption of organic compounds to the stationary phase of the TLC plate depends on the polarity and nature of the adsorbent and the type of functional groups present in the compounds. Substances containing carboxyl groups and other polar functional groups are more strongly adsorbed than are those containing less polar moieties, such as those present in alkenes and alkyl halides, as the **elutropic series** shown in Figure 6.4 indicates. When a sample contains compounds having highly polar functionalities, more polar eluants may be required to effect a TLC analysis, as discussed next.

Eluant

Selecting the eluant or the mobile liquid phase is an important decision that must be made when planning a TLC analysis. The best solvent or combination of solvents is determined by trial and error, so several TLC plates may need to be prepared and developed using different eluants to determine the optimal conditions.

Certain criteria guide the selection of the eluant. For example, an effective eluting solvent must readily dissolve the solute but not compete with it for binding sites on the stationary phase. If the mixture to be separated is not soluble in the solvent, the individual components may remain adsorbed at the origin. Another criterion is that a solvent should not be too polar because it may bind strongly to the adsorbent and force the solute to remain in the mobile phase. In such circumstances, the components will move rapidly up the TLC plate, offering little opportunity to establish the solid-liquid equilibria required for separation. Consequently, the eluting solvent must be significantly *less* polar than various components of the mixture to obtain an effective separation. As a rule, the relative ability of different solvents to move a given substance up a TLC plate is termed **eluting power** and is generally found to follow the order shown in Figure 6.5.

Sometimes a TLC analysis produces a streak on the plate rather than one or more spots. This may result from the presence of so many compounds in the sample

Increasing adsorption on polar stationary phases

Figure 6.4
Elutropic series for polar stationary phases.

$$RCO_2H > ROH > RNH_2 > R^1R^2C{=}O > R^1CO_2R_2 > OCH_3 > RR^1R^2C{=}CR^3R^4 > RHal$$

Increasing eluting power with *polar* stationary phases

Water
Methanol
Ethanol
1-Propanol
Acetone
Ethyl acetate
Diethyl ether
Chloroform
Dichloromethane
Toluene
Hexane
Petroleum ether

Increasing eluting power with *nonpolar* stationary phases

Figure 6.5
Eluting power of solvents as function of polarity of stationary phases.

that the individual spots are not separated, and they appear as a streak instead of a series of spots. Using a different eluant may eliminate the streaking and produce individual spots. Another cause of streaking may be using a solution of the sample to spot the plate that is too concentrated. In this case, diluting the solution may minimize streaking. Finally, compounds such as carboxylic acids, RCO_2H, and amines, RHN_2, that are strongly adsorbed to the alumina or silica gel coating of the TLC plate tend to produce streaks rather than spots. The streaks observed on a TLC plate by such compounds occur because of the equilibrium between neutral and ionized forms of the respective functional groups. One easy way to address this problem is to "deactivate" the adsorbent by adding some acetic acid to the eluant when analyzing carboxylic acids and some ammonium hydroxide in the case of amines; only a drop or two of the acid or base is normally required.

Experimental Technique

The general protocol for preparing and developing a TLC plate was described earlier. We have not yet discussed the means for detecting or "visualizing" the separated components of a mixture in a TLC analysis. There are several ways of doing this. The easiest situation is that in which the compounds being separated are colored, so visual detection is easy, a "no-brainer," so to speak. Many organic compounds are colorless, however, and a variety of methods have been developed to detect their presence on the plate:

1. Compounds that **fluoresce** may be located by placing the plate under an ultraviolet light. Since the spots disappear when the light is removed, it is necessary to circle the spots with a pencil to have a permanent record of the chromatogram. There are also commercially available plates that contain a fluorescent material as part of their coating; compounds that do not fluoresce but do absorb ultraviolet light then appear as dark spots under ultraviolet light.

2. The chromatographic plate may be sprayed with a variety of reagents such as sulfuric acid, potassium permanganate, phosphomolybdic acid, and ninhydrin; these reagents react with the individual components to produce colored or dark spots.

3. The chromatographic plate may be exposed to iodine vapor by placing it in a closed chamber containing several crystals of iodine. As the iodine forms complexes with the various organic compounds, the brown spots appear on the plate. Since the process is reversible and the spots fade, it is necessary to circle the spots with a pencil to have a permanent record of the chromatogram.

Once the separation of the components of the mixture is complete and the individual spots have been detected, the **retention factor** (R_f) of each compound may be calculated as shown below for the chromatogram pictured in Figure 6.2b:

$$R_f = \frac{\text{distance traveled by substance}}{\text{distance traveled by solvent}}$$

$$R_f \text{ (compound 1)} = \frac{2.9 \text{ cm}}{12 \text{ cm}} = 0.24$$

$$R_f \text{(compound 2)} = \frac{8.3 \text{ cm}}{12 \text{ cm}} = 0.69$$

The R_f-value for a compound is a physical constant for a given set of chromatographic conditions, so the adsorbent and the eluting solvent should be recorded along with the experimentally determined R_f-values.

There are many important applications of TLC in contemporary organic chemistry. For example, TLC is commonly used to identify components of an unknown mixture by running chromatograms of the unknown sample side by side with known standards. Another application is to demonstrate potential identity of different samples by analyzing two or more samples on the same TLC plate. If they are identical, the samples should have the same R_f-value. Because there could be minor variations in these values due to slightly different locations of the initial spots on the plate or due to differences in the concentration of the samples, *co-spotting*, a technique in which two different samples are spotted on top of one another, may be used to eliminate such variations. Appearance of two spots after the plate is developed proves that the samples are different; observing a single spot is indicative that the samples are identical but in itself does not prove this definitively. The co-spotting technique is applied in the Discovery Experiment in Part C of Section 3.3.

Multiple aliquots of samples collected from chromatographic columns (Sec. 6.3) may be analyzed by TLC to follow the course of a chromatographic separation. Alternatively, it is possible to follow the progress of a reaction using TLC to monitor the disappearance of starting material or the appearance of product; samples are simply withdrawn from a reaction mixture and subjected to TLC analysis.

Applications

Two experiments are presented here to demonstrate the TLC technique. The first involves the separation of the pigments present in spinach leaves. A variety of other sources, including crushed tomato pulp or carrot scrapings, as well as leaves from grasses, shrubs, and trees, may be substituted; however, *waxy* leaves are not acceptable.

In the second experiment, the **diastereomers** (geometric isomers) (Sec. 7.1) *syn*-azobenzene (**1**) and *anti*-azobenzene (**2**) are separated. Commercially available azobenzene consists predominantly of the more stable *anti-* form, but this isomer may be isomerized to the less stable *syn-* isomer by irradiation with ultraviolet light or sunlight. Since the colors of the two isomers differ, they may be detected visually. The course of the reaction and the effectiveness of the irradiation is followed by placing spots of irradiated and nonirradiated samples side by side on the TLC strip.

| **1** | **2** |
| *syn*-Azobenzene | *anti*-Azobenzene |

Separation of Spinach Pigments by TLC

Purpose To identify solvent mixtures that will separate the colored components in spinach leaves using thin-layer chromatography.

SAFETY ALERT

1. **Wear safety glasses or goggles and suitable protective gloves while performing the experiment.**

2. **Petroleum ether, ethanol, and acetone are highly volatile and flammable solvents. Be certain there are *no flames* in the vicinity during this experiment.**

Procedure

Preparation Refer to the online resources to answer Pre-Lab Exercises, access videos, and read the MSDSs for the chemicals used or produced in this procedure. Read or review Section 2.21.

Apparatus A wide-mouth bottle with a screw-top cap, small mortar and pestle, 2-cm × 10-cm silica gel TLC plates, capillary pipets.

Setting Up Prepare a developing chamber by placing a folded filter paper lengthwise in a wide-mouth bottle (Fig. 6.3). As directed by your instructor or by working in teams, prepare several mixtures of eluants that contain different ratios of varying pairs of the following solvents: petroleum ether, bp 60–80 °C (760 torr), chloroform, acetone, and ethanol. For example, prepare 10 mL of a 70:30 mixture of petroleum ether and acetone to use as one eluant. Add an amount of the eluant to the developing chamber so that it forms a 1-cm layer on the bottom of the container. Screw the cap onto the bottle *tightly,* and *shake* the container *well* to saturate the atmosphere of the chamber with vapors of the solvent.

Preparing the Sample Using a small mortar and pestle, thoroughly grind a spinach leaf in a mixture of 4 mL of petroleum ether and 2 mL of ethanol. Transfer the liquid extract to a test tube using a Pasteur pipet and *swirl* the extract gently with an equal volume of water. Do not shake the test tube, because emulsions (Sec. 2.21) are easily formed, and these can be a source of great frustration to you. Remove and discard the aqueous layer; if you do not know which layer is the aqueous one, perform the necessary test (Sec. 2.21). Wash the organic layer with water two more times to remove the ethanol and any water-soluble materials that are present in the leaves. Transfer the petroleum ether extract to the Erlenmeyer flask and add several spatula-tips full of *anhydrous* sodium sulfate. After 5–10 min, decant the solution from the drying agent. If the solution is not deeply colored, concentrate it using a gentle stream of air or nitrogen to remove some of the solvent.

Preparing and Developing a Plate Obtain a 2-cm × 10-cm strip of silica gel chromatogram sheet *without* a fluorescent indicator. Handle the strip *only* by the sides to avoid contaminating the plate with oils from your hands. Place a pencil dot in the middle of the sheet about 1 cm from one end. Using a capillary pipet, apply a spot of pigment solution over the pencil dot by *lightly* and *briefly* applying the tip of the pipet to the surface of the plate; you may blow gently on the plate as the sample is applied. Do not allow the spot to diffuse to a diameter of more than 1−2 mm during application of the sample. Perform the spotting process an additional four or five times, allowing the solvent of each drop to evaporate before adding the next. When the spot has thoroughly dried, place the strip in the developing chamber containing the first eluant; be careful not to splash solvent onto the plate. The spot *must be above* the solvent level. Allow the solvent front to move to within 2−3 mm of the top of the strip and then remove the strip. Mark the position of the solvent front with a pencil, and allow the plate to air-dry. Repeat this process with other eluants as directed by your instructor.

Analysis A good separation will reveal as many as eight colored spots. These are the carotenes (orange), chlorophyll *a* (blue-green), the xanthophylls (yellow), and chlorophyll *b* (green). Calculate the R_f-values of all spots on the plate(s) you developed. Compile these data, together with those of others if you worked as a team, and determine which solvent mixture provided the best and which the worst separations. Draw a picture to scale of the plates you developed that showed the best and worst separations and include them in your notebook as a permanent record.

Discovery Experiment *Effect of Solvent Polarity on Efficiency of Separation*

Explore the possible consequences of using only petroleum ether, a nonpolar solvent, for extracting the pigments from spinach leaves. For comparative purposes, use the same solvent mixture for developing the TLC plate as you did in the original procedure.

Discovery Experiment *Analysis of Plant Pigments from Various Sources*

Investigate the distribution of pigments obtained from other sources of green leaves, such as beets, chard, trees, and shrubs. Avoid the use of waxy leaves, as their coating makes extraction and isolation of the pigments more difficult.

WRAPPING IT UP

    Put the unused eluants containing only *petroleum ether, acetone,* and *ethanol* in the container for nonhalogenated organic liquids and any unused eluants containing *chloroform* in the container for halogenated organic liquids. Discard the dry chromatographic plates in the nonhazardous solid waste container.

EXPERIMENTAL PROCEDURE

Separation of Syn- and Anti-Azobenzenes by TLC

Purpose To identify solvent mixtures that will separate *syn-* and *anti*-azobenzenes using thin-layer chromatography.

SAFETY ALERT

1. **Wear safety glasses or goggles and suitable protective gloves while performing the experiment.**

2. **Petroleum ether, ethanol, and acetone are highly volatile and flammable solvents. Be certain there are *no flames* in the vicinity during this experiment.**

3. **Since azobenzene is a suspected carcinogen, avoid contacting it with your skin or ingesting it.**

Procedure

Preparation Refer to the online resources to answer Pre-Lab Exercises, access videos, and read the MSDSs for the chemicals used or produced in this procedure.

Apparatus A wide-mouth bottle with a screw-top cap, 3-cm × 10-cm silica gel TLC plates, capillary pipet.

Setting Up Prepare a developing chamber by placing a folded filter paper lengthwise in a wide-mouth bottle (Fig. 6.3). As directed by your instructor, prepare three or four mixtures of eluants that contain different ratios of varying pairs of the following solvents: petroleum ether, bp 60−80 °C (760 torr), chloroform, acetone, and ethanol. For example, prepare 10 mL of a 90:10 mixture of petroleum ether and chloroform to use as one eluant. Add an amount of the eluant to the developing chamber so that it forms a 1-cm layer on the bottom of the container. Screw the cap onto the bottle *tightly*, and *shake* the container *well* to saturate the atmosphere of the chamber with vapors of the solvent.

Preparing and Developing a Plate Obtain three or four 3-cm × 10-cm strips of silica gel chromatogram sheets *without* a fluorescent indicator. Handle the strip *only* by the sides to avoid contaminating the plate with oils from your hands. Place one pencil dot about 1 cm from the left side and about 1 cm from one end of one sheet and another about 1 cm from the right side the same distance from the bottom as the first. Using a capillary pipet, carefully apply a *small* spot of a 10% solution of commercial azobenzene in toluene, which you should obtain from your instructor, over one of the pencil dots. Do not allow the spot to diffuse to a diameter of more than 1−2 mm during application of the sample. Repeat this process for each strip. Allow the spots to dry and then expose the plates to sunlight for one to two hours (or a sunlamp for about 20 min).

   When the irradiation is complete, apply another spot of the *original* solution on the plate over the second pencil dot in the same manner as just described and allow each strip to dry. Place a strip in the developing chamber, being careful not to splash solvent onto the plate. Both spots *must be above* the solvent level. Allow the solvent to move to within approximately 2−3 mm of the top of the strip and then remove the strip. Repeat this process for each additional strip using a different eluting solvent as directed by your instructor. Mark the position of the solvent front with a pencil, and allow the plate to air-dry.

Analysis Note the number of spots arising from each of the two original spots. Pay particular attention to the relative intensities of the two spots nearest the starting point in each of the samples; these are *syn*-azobenzenes. Calculate the R_f-values of each of the spots on your developed plate. In your notebook, include a picture of the developed plate drawn to scale as a permanent record. Identify the solvent mixture that gave the best separation of *syn*- and *anti*-azobenzene.

Discovery Experiment   *Analysis of Analgesics by TLC*

Design and execute an experimental procedure for testing over-the-counter analgesics such as Excedrin, Tylenol, and Advil for the presence of caffeine (**3**) and/or acetaminophen (**4**). A 50:50 (v:v) mixture of ethanol and dichloromethane can be used to extract the active ingredients.

3
Caffeine

4
Acetaminophen

WRAPPING IT UP

Put the *unused eluants* containing mixtures of only *petroleum ether, acetone,* and *ethanol* in the container for nonhalogenated organic liquids and any unused mixtures containing *chloroform* in the container for halogenated organic liquids. Put the dry chromatographic plates in the hazardous solid waste container, since they contain small amounts of azobenzene.

EXERCISES

1. Explain why TLC is not suitable for use with compounds that have boiling points below about 150 °C (760 torr).

2. What may occur if a mixture containing a component that is very sensitive to acidic conditions is subjected to a TLC analysis in which silica gel serves as the stationary phase?

3. In a TLC experiment, why should the spot not be immersed in the solvent in the developing chamber?

4. Explain why the solvent must not be allowed to evaporate from the plate during development.

5. Explain why the diameter of the spot should be as small as possible.

6. Which of the two diastereomers of azobenzene would you expect to be more thermodynamically stable? Why?

7. From the results of the TLC experiment with the azobenzenes, describe the role of sunlight.

8. A student obtained the silica gel TLC plate shown in Figure 6.6 by spotting samples of Midol, caffeine, and acetaminophen on the plate and eluting with petroleum ether:chloroform (9:1 v:v).

    a.   What are the R_f-values of acetaminophen and of caffeine, respectively?

    b.   Based on this TLC analysis, what are the ingredients in a tablet of Midol?

    c.   What are the mobile and stationary phases, respectively, in this TLC experiment?

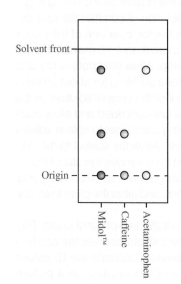

Figure 6.6
TLC analysis of mixture.

d. No spots were observed visually when the TLC plate was removed from the developing chamber. How might the student effect visualization of the spots?

e. Another student accidentally used Midol PM in her experiment and observed only one spot. Speculate as to which spot was absent and offer a possible explanation for the difference in this student's result.

9. Carboxylic acids often produce a streak rather than a spot when analyzed on silica gel TLC plates. Adding a drop or two of acetic acid to the eluant reduces streaking.

a. Explain why carboxylic acids may streak on silica gel TLC plates and why adding acetic acid to the eluent reduces streaking.

b. Amines also often streak on silica gel TLC. What might you add to the eluant to reduce such streaking?

6.3 COLUMN CHROMATOGRAPHY

Figure 6.7
Chromatography column.

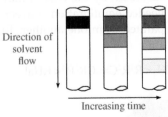

Figure 6.8
Separation of mixture by column chromatography.

Column chromatography is another form of **solid-liquid adsorption** chromatography and depends on the same fundamental principles as does thin-layer chromatography (TLC, Sec. 6.2), as you will see from the discussion that follows. It has an advantage over TLC in that multigram amounts of mixtures can be separated but has the disadvantage that this technique requires considerably more time to perform. The larger scale of separations using this technique makes it valuable for purifying components of a reaction mixture so that one of them may be purified and used in a subsequent chemical reaction.

In column chromatography, a multicomponent mixture is typically dissolved in a small amount of an appropriate solvent and applied to the top of a packed column containing a finely divided, active solid **adsorbent** that serves as the stationary phase (Fig. 6.7). Next an **eluant** or **eluting solvent,** which is the mobile phase, is passed down the column. The individual components of the mixture, which were initially adsorbed on the stationary phase at the top of the column, begin to move downward with the eluting solvent (Fig. 6.8). These components travel at different rates depending on their relative affinities for the packing material; a more weakly adsorbed compound is eluted faster from the column than is a more strongly adsorbed compound. The individual components are collected in separate containers as they exit from the bottom of the column in bands. The solvent is then removed from each fraction by evaporation to provide the pure components, which are characterized and identified by determining their physical constants (Chaps. 3 and 4) and spectral properties (Chap. 8).

As with TLC (Sec. 6.2), the chromatographically separated bands are easily observed when all of the components of a mixture are colored. Many organic compounds are colorless, however, and other methods are required for detecting the banks as they elute from the column. For those organic compounds that absorb ultraviolet or visible light (Sec. 8.4), **electronic detectors** that measure differences in the absorption of light as the solvent exits the column are used to locate the bands of the individual components. Detectors that measure differences in the **refractive index** of the **eluate** are also used to identify the different bands; such detectors do not rely on absorption of light by the organic components.

If a detector is not available, the progress of the chromatographic separation can be conveniently followed using TLC to analyze the eluate at regular intervals.

Direction of solvent flow

Increasing time

Another method, albeit more laborious, involves collecting small, equal fractions of the eluate from the column in a series of flasks. The solvent in each flask is then evaporated, and the presence or absence of a solute in the residue provides a means of locating the bands of each component. If there is adequate separation of the different bands, a given flask will often contain only one constituent of the original mixture. That component will normally appear in a number of consecutive fractions, however.

Adsorbents

Alumina and silica gel are the most commonly used adsorbents for column chromatography, just as they are for TLC. The quality of these adsorbents is high in that they have *uniform particle size* and *high specific area*. The higher the specific area, the faster the equilibrium of the solute between the mobile and solid phases is established and the narrower the bands. High specific areas on the order of several hundred m^2/g are common for good grades of alumina and silica gel.

As noted in the discussion of TLC, the strength of the adsorption of an organic compound to the solid support depends on the polarity and nature of the adsorbent as well as on the nature of the functional groups present in the molecule. When **normal-phase column chromatography** is performed, a polar stationary phase such as alumina or silica gel is used in combination with organic solvents as the mobile phase or eluant. Under these conditions, the **elutropic series** described for TLC in Section 6.2 applies.

In **reverse-phase column chromatography,** the packing material for the stationary phase consists of glass beads coated with a nonpolar hydrocarbon film, and mixtures of water and organic solvents are generally used as the eluting solvents. Under these conditions, nonpolar organic molecules are more strongly attracted to the nonpolar stationary phase, whereas polar solutes are more strongly attracted to the mobile phase. The order of elution is then the *reverse* of that shown earlier, with the more polar components of a mixture eluting more rapidly than the less polar ones. The elutropic series then becomes the reverse (Fig. 6.9). Reverse-phase chromatography may sometimes be used to separate mixtures that are inseparable by normal-phase chromatography.

Eluant

The most efficient method for determining the optimal solvent system(s) for a specific column chromatographic separation often is to perform a series of trial separations using TLC. These trials can be run quickly, and the amount of material needed is small.

The same criteria for selecting an eluant for TLC apply to normal-phase column chromatography. For example, if the mixture to be separated is not soluble or only slightly soluble in the eluant being used, the components will remain permanently adsorbed to the stationary phase near the top of the column. Conversely, if the components are too readily displaced from the adsorbent by the eluant, they will move down the column too rapidly to allow the needed equilibration between the stationary and mobile phases and will not be separated from one another. To

$$RCO_2H < ROH < RNH_2 < R^1R^2C{=}O < R^1CO_2R^2 < ROCH_3 < R^1R^2C{=}CR^3R^4 < RHal$$

Figure 6.9
Elutropic series for nonpolar stationary phases.

Increasing adsorption on nonpolar stationary phases

Water
Methanol
Ethanol
1-Propanol
Acetone
Ethyl acetate
Diethyl ether
Chloroform
Dichloromethane
Toluene
Hexane
Petroleum ether

Increasing eluting power with *nonpolar* stationary phases

Increasing eluting power with *polar* stationary phases

Figure 6.10
Eluting power of solvents as function of polarity of stationary phases.

allow equilibration in normal-phase column chromatography, the eluting solvent must be *less* polar than the components of the mixture. Although the relative eluting power of solvents for normal-phase column chromatography is the same as that for TLC when alumina or silica gel is used as an adsorbent, that for reverse-phase chromatography is the reverse (Fig. 6.10).

Experimental Technique

With this general discussion as background, we can now discuss some of the experimental aspects of performing column chromatography. The optimal adsorbent and eluting solvent(s) typically are first determined using TLC, and then the column is packed with the adsorbent. The amount of adsorbent used to prepare the column varies according to the differences in partition coefficients and polarities of the individual components in the chromatographic system. For simple separations, it is possible to use as little as 10 g of adsorbent for each 1 g of the mixture, but when the components of the mixture have similar polarities, this ratio must be increased to as much as 100–200:1; a ratio of about 25:1 is a convenient starting point. As a general rule of thumb, the ratio of height-to-diameter for the packed column should be about 8:1.

The column is made of glass and is fitted with a stopcock or a segment of polyethylene tubing with a screw clamp to control the flow of solvent through the column. The column is prepared by first inserting a plug of cotton or glasswool into the small end of the column using a long glass rod or a piece of wire; this prevents the adsorbent from washing out of the bottom. A layer of white sand approximately 1 cm deep is then added to provide an even bed for the adsorbent (Fig. 6.7).

Proper packing of the column is vital to the success of column chromatography because this step ultimately determines the efficiency of separation. Two general protocols are followed for this operation. The first of these, the **dry-pack** method, involves pouring the dry adsorbent slowly into a vertical glass column half-filled with the solvent that will serve as the eluant. The other technique is the **wet-pack** method, in which a slurry of the adsorbent in the eluting solvent is added to the column; this is the preferred procedure when silica gel is the adsorbent. With both methods, the column must be constantly tapped as the solid settles through the liquid to ensure even and firm packing of the adsorbent and to remove any entrapped air bubbles. Some solvent may be drained from the column during this operation, but the liquid level in the column should *never* be allowed to fall below the top of the adsorbent. If this occurs, the air bubbles that form in the column will allow *channeling,* which results in poor separations because the components are eluted as ragged rather than sharp bands. *Uniform packing of the adsorbent is essential* so the solvent will move down the column with a horizontal front.

To complete packing the column, a layer of sand is normally placed on top of adsorbent, as shown in Figure 6.7. The purpose of the sand is twofold: (1) It allows the sample to flow evenly onto the surface of the adsorbent, and (2) it prevents disruption of the packing material as eluting solvent is added.

Mixtures of solids are typically dissolved in a *minimal* volume of a solvent before being transferred to the top of the column; a liquid mixture may be transferred directly to the column. It is important to distribute the sample evenly on the surface of the adsorbent and to use as little solvent as possible in loading the column. This procedure ensures that the bands that form during development of the chromatogram will be narrow, thereby providing the best possible separation. If too much solvent is used to dissolve the sample, the initial band will be broad, and poor resolution of the mixture may result.

Once the sample has been loaded onto the column, there are several different techniques that may be used to elute its components. In a **simple elution** experiment, a *single* solvent is passed through the column during the entire course of the separation. This procedure works well for the separation of mixtures containing only two or three compounds having similar polarities. However, the more common chromatographic procedure is **stepwise** or **fractional elution.** In this technique, a series of increasingly more polar solvents is used to elute the mixture from the column. A nonpolar solvent such as petroleum ether or hexane is first used to move the least polar component of the mixture down the column, while the others remain at or near the top of the column. After elution of the first band, the polarity of the eluant is increased using either pure solvents or combinations of mixed solvents so that the bands are individually eluted from the column. Systematic and gradual increases in solvent polarity are sometimes essential so that individual bands remaining on the column separate and do not co-elute. As the polarity of the solvent system is increased, those components of the mixture that are more tightly adsorbed on the column will begin to move. As a rule of thumb, a volume of solvent approximately equal to three times the column volume is passed through the column prior to switching to a solvent of higher polarity.

The separation is monitored using one of a variety of methods. Unfortunately, many organic compounds are not highly colored, as we noted above, and sophisticated devices for their detection are rarely available in the undergraduate laboratory. The most effective technique for following the separation is to collect fractions of equal volume in tared flasks, to concentrate the solvent, and to reweigh the flasks. The fractions containing the different bands may then be easily identified by the relative amounts of solute in each flask. One may also use TLC to monitor the separation.

In the experiment that follows, column chromatography will be used to separate fluorene (**5**) from an oxidation product, 9-fluorenone (**6**). One of these compounds is white, and the other is yellow. Consequently, the progress of the chromatography may be followed by evaporation of the solvent at periodic intervals as well as by observing the slower-moving yellow band.

5
Fluorene
White

6
9-Fluorenone
Yellow

Column Chromatography

Purpose To separate fluorene and 9-fluorenone by column chromatography.

SAFETY ALERT

1. **Wear safety glasses or goggles and suitable protective gloves while performing the experiment.**

2. **Petroleum ether is a highly volatile and flammable mixture of low-boiling hydrocarbons. During the preparation and development of the chromatographic column, be certain that there are *no flames* in the vicinity.**

Procedure

Preparation Refer to the online resources to answer Pre-Lab Exercises, access videos, and read the MSDSs for the chemicals used or produced in this procedure. Read or review Sections 2.13 and 2.29.

Apparatus A *dry* 50-mL glass buret or chromatography column about 1 cm in diameter and 25 cm long and fitted with either a Teflon stopcock or a short piece of polyethylene tubing and a screw clamp, three 25- or 50-mL Erlenmeyer flasks, watchglass, Pasteur pipet, apparatus for simple distillation. Consult with your instructor regarding whether the glassware, sand, and alumina require oven-drying prior to use.

Setting Up Clamp the glass buret or chromatography column in a vertical position with its *ungreased* stopcock, preferably Teflon, *closed.* Using a piece of glass tubing, insert a small plug of cotton or glasswool loosely into the bottom of the column. Cover the cotton or glasswool plug with enough clean, *dry* sand to form a layer about 1 cm thick, and add approximately 25 mL of petroleum ether, bp 30−60 °C (760 torr). Place a funnel on top of the column, and *slowly* add 5 g of *dry* neutral alumina to the column while constantly tapping the buret. A rubber "tapping mallet" may be made by placing a pencil in a one-hole rubber stopper. When this process has been completed, wash the inner walls of the column with additional petroleum ether to remove any alumina that may adhere to the sides. Cover the alumina with a 1-cm layer of clean sand, and open the stopcock to allow the solvent to drain into an Erlenmeyer flask until the solvent level reaches just to the *top of the alumina.* The column is now ready for the addition of the sample mixture (Fig. 6.11).

Figure 6.11
Setup for column chromatography.

Separation Obtain a sample of an approximately 1:1 mixture of fluorene (**5**) and 9-fluorenone (**6**) and accurately determine its melting-point range. In a small test tube, suspend about 0.1 g of this mixture in 0.5 mL of petroleum ether and slowly add just enough dichloromethane with a Pasteur pipet to effect solution. Using a Pasteur or filter-tip pipet, carefully transfer this solution directly to the top of the column. Open the stopcock until the liquid level is at the top of the alumina. *Do not allow the solvent to drain below the level of the alumina,* as air bubbles and channels might develop in the solid support. Add approximately 1−2 mL of fresh petroleum ether to the top of the column, and again allow the liquid to drain to the top of the alumina.

Fill the buret with approximately 20 mL of fresh petroleum ether, open the stop-cock, and collect the eluant in an Erlenmeyer flask. Follow the progress of the chromatography by collecting a drop or two of eluant on a watchglass with every 5 mL that elutes from the column. When the solvent evaporates, any white solid present will be visible on the watchglass. You can determine when all of the white solid has been eluted using this visualization technique. Your instructor might also direct you to follow the chromatography by TLC using 15% dichloromethane in petroleum ether as the developing solvent (Sec. 6.2). Most of the white solid should elute in a volume of 15–20 mL of petroleum ether, and slow movement of a yellow band down the column should occur. Wash any of the white solid from the tip of the column into your collection flask with fresh petroleum ether.

When all of the white solid has eluted from the column, change the collection flask to another clean Erlenmeyer flask. Elute the column with about 5 mL of petroleum ether, and then change the eluant to dichloromethane, a more polar solvent. Watch the progress of the yellow band as it now proceeds rapidly down the column. When this yellow band just reaches the bottom of the column, change to a third clean Erlenmeyer flask. The intermediate fraction should not contain significant amounts of solid; verify this by evaporating a few drops on a watchglass. Continue eluting the column with dichloromethane until the eluant is colorless; approximately 10 mL will be required.

Isolation The first and third fractions should contain pure samples of fluorene and 9-fluorenone. Most of the solvent in these fractions may be removed by simple distillation (Sec. 2.13) or by one of the techniques outlined in Section 2.29. Attach the flask to a water aspirator or house vacuum to remove the last traces of solvent under reduced pressure.

Analysis When the crystals of each of the purified compounds are completely dry, determine their melting points and weights. Use your experimentally determined melting points to identify the order of elution of fluorene and fluorenone from the column and record your observations and conclusions in your notebook. Based upon the weight of the mixture used, calculate the percent recovery of fluorene and 9-fluorenone.

Discovery Experiment

Column Chromatographic Separation of Benzyl Alcohol and Methyl Benzoate

Design and execute an experimental procedure for separating benzyl alcohol (**7**) from methyl benzoate (**8**) by normal-phase column chromatography.

7
Benzyl alcohol

8
Methyl benzoate

WRAPPING IT UP

Place recovered *petroleum ether* in the container for nonhalogenated organic liquids, but pour the recovered *dichloromethane* into a container for halogenated organic liquids. Spread out the *alumina adsorbent,* which is wet with organic solvent, in a hood to dry and then place it in the nonhazardous solid waste container.

EXERCISES

1. What are two important characteristics of the silica gel and alumina that are used for column chromatography?

2. Explain why it is unwise to use column chromatography on compounds having boiling points below about 150 °C (760 torr).

3. Define the following terms:
   a. eluate
   b. eluant
   c. adsorbent
   d. reverse-phase chromatography

4. In normal-phase column chromatography, which solvent has more eluting power: petroleum ether or dichloromethane? In what way is the eluting power of a solvent related to its polarity?

5. In reverse-phase column chromatography, which phase is the more polar: the stationary phase or the mobile phase?

6. How does the principle of "like dissolves like" explain the affinity of a compound for the mobile phase relative to the stationary phase?

7. When separating a mixture by normal-phase column chromatography, why is it better to change from a less polar solvent to a more polar solvent rather than the opposite?

8. If the polarity of the eluant is to be increased during column chromatography, the increase is normally made gradually. However, in this procedure, *pure* dichloromethane rather than intermediate mixtures of petroleum ether and dichloromethane was added to the column. Why is this variation from the usual technique acceptable in this case?

9. State two major differences between TLC and column chromatography.

10. If you had 5.0 g of material that needed to be purified, would you opt for using TLC or column chromatography to purify your material? Explain your answer.

11. Why is it preferable to use a Teflon or an ungreased stopcock rather than a greased stopcock on a column used for column chromatography?

12. Why should care be exercised in the preparation of the column to prevent air bubbles from being trapped in the adsorbent?

13. Why is a layer of sand placed above the cotton plug prior to the addition of the column packing material?

14. Does fluorene or 9-fluorenone move faster down the column when petroleum ether is used as the eluant? Why?

15. Consider the structures of fluorene (**5**) and 9-fluorenone (**6**).
    a. Other than the aromatic rings, specify what functional group, if any, is present in **6**.
    b. Predict which compound is more polar and explain why.
    c. Which compound would you expect to elute from a normal-phase column first and why?

16. In the separation of fluorene and fluorenone by column chromatography, what is the stationary phase? The mobile phase?

17. The observed melting point of the 1:1 mixture of fluorene and 9-fluorenone should be relatively sharp, although lower than the melting point of either of the pure compounds. On the other hand, a 3:1 mixture has a broad melting-point range of about 60–90 °C. Explain these observations. (*Hint:* See Sec. 3.3.)

18. A mixture containing compounds **9–11** was separated by normal-phase column chromatography, using neutral alumina as the stationary phase and petroleum ether as the eluant. Predict the order in which **9–11** will elute from the column.

19. Most peptides (polymers of amides constructed from amino acids) are very polar molecules; therefore, they are most successfully purified by reverse-phase chromatography.

    a. In this type of column chromatography, is the mobile or stationary phase more polar?

    b. Provide an explanation for why reverse-phase chromatography is more appropriate in this application than normal-phase chromatography.

20. A typical eluting solvent for reverse-phase column chromatography is acetonitrile (CH_3CN), water, and a buffer. In what way is this mobile phase different from the one you used to separate fluorene (**5**) from 9-fluorenone (**6**)?

6.4 GAS-LIQUID CHROMATOGRAPHY

Gas-liquid chromatography (GLC), which is also called **gas chromatography** (GC), is a technique that may be used to separate mixtures of volatile compounds whose boiling points may differ by as little as 0.5 °C. It can also be applied as an analytical tool to identify the components of a mixture or in preparative applications when quantities of the pure components are desired.

Basic Principles of Gas-Liquid Chromatography

Gas-liquid chromatography operates on the principle of partitioning the components of a mixture between a mobile gaseous phase and a stationary liquid phase. In practice, a sample is injected into a heated chamber where it is immediately vaporized and carried through a column by a flowing inert gas such as helium or nitrogen, which is called the **carrier gas.** This gaseous mixture is the **mobile phase.** The column is packed with a finely divided solid support that has been coated with a viscous, high-boiling liquid, which serves as the **stationary phase.** As the mobile phase moves through the column, its components are continuously partitioned between the two phases. Those components that show a higher affinity for the mobile phase move through the column more quickly, whereas those with a stronger attraction to the stationary phase migrate more slowly, and separation occurs. As with a fractional distillation column (Sec. 4.4), a GLC column may be characterized by the number of theoretical plates it provides. GLC columns typically have many more theoretical plates, however, so they can effect separations that would be impossible using fractional distillation.

Table 6.1 *Common Stationary Phases for Gas-Liquid Chromatography*

Liquid Phase	Type	Property	Maximum Temperature Limit, °C
Carbowax 20M	Hydrocarbon wax	Polar	250
OV-17	Methylphenyl silicone	Intermediate polarity	300
QF-1	Fluorosilicone	Intermediate polarity	250
SE-30	Silicone gum rubber	Nonpolar	375

The **retention time** of a component is the elapsed time required for the compound to pass from the point of injection to the detector, and it may be used for purposes of identification. The retention time of a component is *independent* of the presence or absence of other components in the sample mixture. There are four experimental factors that influence retention time of a compound: (1) the *nature* of the stationary phase, (2) the *length* of the column, (3) the *temperature* of the column, and (4) the *flowrate* of the inert carrier gas. Thus, for a particular column, temperature, and flowrate, the retention time will be the same for a specific compound.

Although a large number of **stationary liquid phases** are available, only a few are widely used (Table 6.1). Each liquid phase has a maximum temperature limit above which it cannot be used. This temperature depends upon the stability and volatility of the liquid phase; at higher temperatures, the liquid phase will vaporize and "bleed" from the column with the mobile phase.

The differences in the *partition coefficients* (Secs. 5.2 and 5.3) of the individual components of a mixture in GLC depend primarily upon the differences in solubility of each of the components in the liquid phase. Two important factors determining the solubility of a gas in a liquid are its vapor pressure at the **ambient temperature,** which is the temperature of the column, and the magnitude of its interactions with the liquid phase. Regarding the first factor, the solubility of a gas in a liquid decreases as its vapor pressure increases. This means the more volatile components of a gaseous mixture tend to pass through the column more rapidly and elute before the less volatile ones. The impact of the second factor is understandable from the principle that *like dissolves like* (Sec. 3.2). Polar samples are most effectively separated by using a polar liquid phase, whereas nonpolar compounds are best separated using a nonpolar liquid phase.

The stationary liquid phase is normally coated as a thin film on an inert solid support. The support is composed of small, uniformly meshed granules, so that a large surface area of the liquid phase is available for contact with the vapor phase to ensure efficient separation. Some common types of solid supports include Chromosorb P and Chromosorb W, which are composed of diatomaceous earth. The surface areas of these supports vary from 1 m^2/g to 6 m^2/g. Columns are now commercially available with a wide variety of liquid phases on different solid supports. An alternative method of supporting the liquid phase is used in capillary columns. In these columns, the liquid is coated directly onto the inner walls of the tubing. These columns are highly efficient but relatively expensive.

In general, the efficiency, or **resolution,** of a column increases with increasing length and decreasing diameter. Increasing the pathlength increases the difference in retention times between bands, whereas decreasing the diameter of the column gives rise to narrower bands. With a small band separation, as measured from the band centers, wide bands are more likely to overlap (Fig. 6.12a) than narrow bands (Fig. 6.12b).

Figure 6.12
Effect of bandwidth on resolution.
(a) Wide bands. (b) Narrow bands.

The two other experimental factors that may be varied to alter the degree of separation of the bands are the *temperature* at which the column is maintained and the *flowrate* of the carrier gas. Increasing the temperature results in shorter retention times, because the solubility of gases in liquids decreases with increasing temperature. The partition coefficients are affected, and the bands move through the column at a faster rate. Higher flowrates also cause retention times to decrease. In spite of the decreased resolution and band separation obtained at higher temperatures and flowrates, these conditions are sometimes necessary for substances that would otherwise have very long retention times.

Instrumentation

All commercially available gas-liquid chromatographs (GLCs) have a number of basic features in common. These are illustrated schematically in Figure 6.13. Parts **1–5** are associated with supplying the dry **carrier gas,** usually helium or nitrogen, and allowing an operator to control its flow. The mixture to be separated is injected using a gas-tight syringe through a septum into the **injection block (6),** an individually heated chamber in which the sample is immediately vaporized. The sample then enters the flowing stream of carrier gas and is swept into and through the **column (7),** which is located in an **oven (8)** and consists of coiled aluminum, stainless steel, or glass tubing containing an appropriate stationary phase. In the column, the individual components separate into bands that ultimately pass through a **detector (9),** producing a signal whose voltage is proportional to the amount of material other than carrier gas present in the mobile phase. One type of detector that is commonly used is the **thermal conductivity detector** (TCD), which operates on the basis of differences in the thermal conductivity of the mobile phase as a function of its composition. A **flame ionization detector** (FID) is much more sensitive and functions by detecting the number of ions produced by passing the mobile phase through a hydrogen flame. The **recorder (10)** plots the changes in voltage measured by the detector as a function of time to give the **gas chromatogram.** The vapors then pass from the detector into either the atmosphere or a collection device at the **exit port (11).**

The GLCs available in the introductory organic laboratory often can only be operated at a constant oven temperature, the so-called **isothermal** mode of operation. Some GLCs, however, have a temperature-programming option, which allows the temperature of the oven, and consequently that of the column, to be varied over a range of temperatures. This is particularly useful when the sample being analyzed contains components having widely varying boiling points. Thus, the oven

Figure 6.13
Schematic diagram of apparatus for GLC.

Figure 6.14

Example of temperature versus time program for a GLC analysis.

temperature may be held at a constant temperature for a specified period of time at the beginning of an analysis but then may be increased to higher temperatures as a function of time. For example, the temperature program illustrated in Figure 6.14 involved an initial temperature of 100 °C. After 5 minutes, the temperature was ramped up to 125 °C at a rate of 5 °C/min, held at that temperature for 5 additional minutes, and then further ramped up at a rate of 10 °C/min to a final temperature of 175 °C for completing the analysis. The programming option allows the higher-boiling components of the mixture to elute in a reasonable period of time because rates of elution increase with increasing temperatures. If the analysis had been performed at the higher temperature, the lower-boiling components might have eluted too quickly to be separated from each other.

Another modification of a GLC involves its direct connection to a mass spectrometer (MS) to produce a hybrid instrument commonly called a **GC-MS**. This combination provides a powerful analytical technique because the GC-MS combines the separating power of GLC with the ability of mass spectrometry (Sec. 8.5) to determine molar masses using very small amounts of material (see the Historical Highlight *Who's Taking What? Analysis of Biological Fluids for Illegal Substances*, which is available online). Thus, a GC-MS allows the mass spectrum of each component of a mixture of volatile compounds to be obtained as each individual component emerges from the GLC. If the components of a mixture are known, as they would be if you were analyzing a fraction obtained by normal-phase column chromatography of a mixture of benzyl alcohol (**7**) and methyl benzoate (**8**) (Optional Discovery Experiment, Sec. 6.3), you could use the information provided by the mass spectrum, as discussed below, to confirm the identity and order of elution from the chromatographic column of each of these colorless compounds. If the components are unknown, on the other hand, knowledge of the molar mass of each compound, coupled with information about the origin of the mixture being analyzed, may allow assignment of structures to the unknown substances. Other spectral data (Chap. 8) are generally required before structural assignments can be made, however.

A partial schematic of a GC-MS is shown in Figure 6.15; the elements of the GLC that precede the column have been omitted but are the same as those shown in Figure 6.13. By comparing Figures 6.13 and 6.15, you see that a GC-MS has the outlet of the column (**1**) leading to a **stream splitter** (**2**), which sends part of the eluant to a **detector** (**3**), usually an FID, and part of it to a **mass spectrometer** (**4**); **recorders** (**5, 6**) provide the necessary records of when components are eluting from the GLC and what their mass spectra are. Both the GLC and the MS units are vented to the atmosphere (**7**).

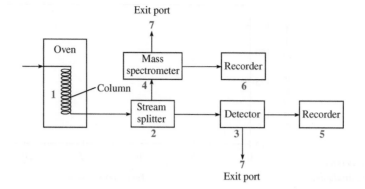

Figure 6.15
*Partial schematic diagram of a
GC-MS.*

Experimental Techniques

Qualitative Analysis. The *retention time* of a pure compound is constant under a specified set of experimental conditions, including the column, temperature, and flowrate. Consequently, this property may be used as a *first* step to identify an unknown compound or the individual components in a mixture. In a typical experiment, an unknown compound or mixture is injected into the injection port of a GLC, and the retention time(s) of the component(s) is (are) measured. A series of known samples are then injected under the same conditions. Comparison of the retention times of the standard samples with those of the unknown allows a preliminary identification of the component(s) of the unknown. A convenient way of confirming that the retention times of a standard and the unknown are the same involves injecting a sample prepared by combining equal amounts of the two. If a *single* peak is observed in the chromatogram, the *retention times* of the standard and the unknown are identical. However, observation of the same retention time for a known and an unknown substance is a *necessary but not sufficient* condition to establish identity, because it *is* possible for two different compounds to have the same retention time. Independent confirmation of the identity of the unknown by spectral (Chap. 8) or other means is imperative.

An example of the use of GLC as a qualitative, analytical tool is illustrated in Figure 6.16. These sets of peaks represent a gas chromatographic separation of the distillation fractions of a mixture of cyclohexane and toluene similar to those obtained in the distillation experiment described in Section 4.3. The notations A, B, and C refer to the three fractions taken in that experiment. The individual peaks in the mixture may be identified by comparing their retention times with those of pure cyclohexane and pure toluene; the peak with the shorter retention time in the mixture is cyclohexane, whereas the peak with the longer retention time is toluene.

The GC-MS analysis of a mixture of benzyl alcohol (**7**, bp 215.3 °C, 760 torr) and methyl benzoate (**8**, bp 199.6 °C, 760 torr) is provided in Figure 6.17. Figure 6.17a is the GLC trace of the mixture. Because the GLC column being used separates mixtures on the basis of the relative boiling points of its components, Peak A, which emerges first, should be **8**, whereas Peak B should be **7**. This could be confirmed with a GLC if the retention times of **7** and **8** were determined separately. Such additional analyses would not be necessary with GC-MS, however, because the masses of the two compounds are different; that of **7** is 108.14 atomic mass units (amu), and that of **8** is 136.15 amu. Figures 6.17b and 6.17c are the mass spectra corresponding to Peaks A and B, respectively. In Figure 6.17b we see that the molecular ion is at *m/z* 108, whereas it is at *m/z* 136 in Figure 6.17c. This proves that Peaks A and B are benzyl alcohol and methyl benzoate, respectively; our expectation of the reverse

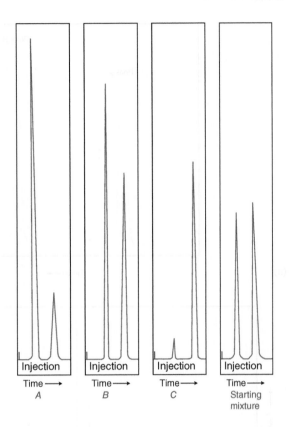

Figure 6.16
GLC analysis of the distillation fractions from the distillation experiment in Section 4.3.

order of elution, based on considering boiling points, has been proven wrong. The stationary phase of the column must interact more strongly than expected with **8.** In any case, the GC-MS analysis has provided unambiguous identification of the two peaks through a *single* analysis of the mixture. Moreover, integration of the peaks in the GLC tract would provide a quantitative measure of their relative proportions in the mixture being analyzed.

A feature that may complicate the interpretation of the mass spectrum from a GC-MS analysis may be seen in Figure 6.17c. This feature is a small peak at *m/z* 146, a value that is not associated with either of the components contained in our known mixture of **7** and **8.** Although the substance that is the source of the peak could be an impurity in our sample, it is not. Rather, it is a peak produced from the "stationary" phase of the GLC column being used; the temperature at which the GLC was operating caused a small amount of the high-boiling material comprising this phase to "bleed" from the column into the MS, which results in minor peaks appearing in the spectrum. Such peaks can be confusing if you are not prepared for their possible presence.

A second example illustrating the power of an analysis by GC-MS rather than GLC alone is that of a mixture of methyl benzoate (**8**) and diethyl malonate (**12**), as seen in Figure 6.18. The boiling point of **12** (199.3 °C, 760 torr) is essentially identical to that of **8,** and considering the differing polarities expected for these two esters, predicting their order of elution from a GLC column (Fig. 6.18a) is problematic. However, the mass spectra associated with the two peaks eluting at 150 and 164 seconds in the GLC trace show that Peak A must be diethyl malonate (Fig. 6.18b), whereas Peak B is methyl benzoate (Fig. 6.18c).

(a)

Time ⟶

(b)

Figure 6.17
GC-MS analysis of a mixture of benzyl alcohol and methyl benzoate. (a) GLC trace. Column and conditions: 0.5-mm × 15-m, 0.25-μ film of BPX5 (DB-5); initial column temperature: 50 °C (2.0 min); ramp rate: 15 °C/min; final column temperature: 260 °C (30 min); flowrate: 1.5 mL/min. (b) MS spectrum of Peak A. (c) MS spectrum of Peak B.

(c)

(a)

(b)

Figure 6.18

GC-MS analysis of a mixture of methyl benzoate and diethyl malonate. (a) GLC trace. Column and conditions: 0.5-mm × 15-m, 0.25-μ film of BPX5 (DB-5); initial column temperature: 50 °C (2.0 min); ramp rate: 15° C/min; final column temperature: 260 °C (30 min); flowrate: 1.5 mL/min. (b) MS spectrum of Peak A. (c) MS spectrum of Peak B.

(c)

You may be perplexed by the fact that the molecular ion from **12** appears to have m/z 161 rather than the expected value of 160.17. This could be the result of poor calibration of the instrument, so that the m/z values are inaccurate; proper calibration of the instrument is critical for proper interpretation of GC-MS data. The actual explanation in this case, though, is that diethyl malonate apparently has a propensity to be protonated under the conditions of the mass spectrometric analysis, so it is this ion rather than the molecular ion that is being detected. This is yet another phenomenon that can make interpretation of GC-MS data difficult for a beginner. In any event, a single GC-MS analysis of the mixture of **8** and **12** allows unambiguous determination of the order of elution from the GLC column of these esters, just as we saw in the analysis of the mixture of **7** and **8**.

$$CH_3CH_2O \overset{\displaystyle O}{\overset{\|}{C}} CH_2 \overset{\displaystyle O}{\overset{\|}{C}} OCH_2CH_3$$

12

Diethyl malonate

Quantitative Analysis. The voltage output of the detector is related to the mole fraction of the material being detected in the vapor, so there is a correlation between the *relative areas* under the peaks in the chromatogram and the *relative amounts* of each of the components in the mixture. The quantitative evaluation of the chromatogram thus requires reliable methods for determining these peak areas.

An **electronic integrator,** which measures the intensity of detector output as a function of time, is the most accurate method for determining peak areas. However, these devices are expensive, so they are usually found only in research laboratories. Since the thickness and density of chart paper are reasonably uniform, another means of determining the relative areas involves carefully cutting the peaks out with scissors; the peak areas are then assumed to be proportional to their weight, as measured on an analytical balance. The original chromatogram should be saved as a permanent record, so the peaks should be cut from a photocopy of the chromatogram.

If the peaks are symmetrical, as are those shown in Figure 6.16, the areas may be approximated by assuming them to be equilateral triangles. The area of a symmetrical peak is then determined by multiplying the width of the peak at its half-height times its height. The percentage of each component in the mixture may be computed as the area of the peak corresponding to that component, expressed as a percentage of the sum of the areas of all peaks in the chromatogram. A sample calculation of this type is shown in Figure 6.19.

Although the peak areas are related to the mole fraction of the component in the mobile phase, they are not *quantitatively* related because the response of the detector varies with the class of the compound. Not all compounds have the same thermal conductivity (TCD), nor do they ionize in a hydrogen flame to form the same types or number of ions (FID). Thus, it is necessary to *correct* the measured areas in the chromatogram using the appropriate **response factor** to obtain an accurate quantitative analysis of the mixture. Although response factors for different compounds may be determined experimentally, approximate values are published in monographs on gas chromatography. The response factors for thermal conductivity and flame ionization detectors for compounds that you may encounter in this experiment are given in Table 6.2. Notice that the correction factors vary more

Areas: $A = 17 \times 8 = 136 \text{ mm}^2$
           $B = 34 \times 17 = 578$
           $C = 25 \times 12 = 300$
           ———————————
           Total area = 1014 mm^2

$\%A = \dfrac{136}{1014} \times 100 = 13.4\%$

$\%B = \dfrac{578}{1014} \times 100 = 57.0\%$

$\%C = \dfrac{300}{1014} \times 100 = 29.6\%$
           ———————————
                                100%

Figure 6.19
Determination of percentage composition of a mixture by GLC.

Table 6.2 *Weight (Wf) and Mol (Mf) Correction Factors for Some Representative Substances**

	Thermal Conductivity		Flame Ionization	
Substance	Wf	Mf	Wf	Mf
Benzene	1.00	1.00	1.00	1.00
Toluene	1.02	0.86	1.01	0.86
Ethylbenzene	1.05	0.77	1.02	0.75
Isopropylbenzene	1.09	0.71	1.03	0.67
Ethyl acetate	1.01	0.89	1.69	1.50
n-Butyl acetate	1.10	0.74	1.48	0.99
Heptane	0.90	0.70	1.10	0.86
o-Xylene	1.08	0.79	1.02	0.75
m-Xylene	1.04	0.76	1.02	0.75
p-Xylene	1.04	0.76	1.02	0.75
Ethanol	0.82	1.39	1.77	3.00
Water	0.71	3.08	—	—

*McNair, H.M.; Bonelli, E.J. *Basic Gas Chromatography*, 5th ed., Varian Aerograph, Walnut Creek, CA, 1969.

widely for flame ionization detectors than for thermal conductivity detectors. In the experimental section, you have the opportunity to calculate the response factor for an unknown compound.

To analyze a mixture of substances quantitatively when **weight factors** are known, the peak area for each component is simply *multiplied* by the weight factor for that particular compound. The resulting *corrected* areas are used to calculate the percentage composition of the mixture according to the procedure outlined in Figure 6.19. Note that using these factors provides the composition on a *weight percentage* basis.

Calculation of the composition of a mixture on a mole percentage basis requires the use of **mole factors**, M_f. These are obtained by dividing the weight factors by the molar masses of each component of the standard solution and normalizing the resulting numbers. A sample calculation utilizing mole correction factors is provided below in the analysis of a mixture of ethanol, heptane, benzene, and ethyl acetate with a GLC equipped with a thermal conductivity detector. The last column shows the percentage composition calculated directly from the measured peak areas, without correction for detector response. The dramatic differences in the calculated composition with and without this correction, as noted in the last two columns, underscore the importance of this correction for quantitative analysis.

Compound	Area (A) (mm^2)	Uncorrected % (A/207.1 × 100)	M_f	A × M_f	Mol % (A × M_f/194 × 100)
Ethanol	44.0	21.2	1.39	61.2	31.5
Heptane	78.0	37.7	0.70	54.6	28.1
Benzene	23.2	11.2	1.00	23.2	11.9
Ethyl acetate	61.9	29.9	0.89	55.4	28.5
Total	207.1	100		194.4	100

Use of Syringes. Liquid samples are injected into the heated injection port of a gas chromatograph using a gas-tight syringe with a capacity of 1–10 μL. The sample is either injected neat or dissolved in a volatile liquid such as diethyl ether or pentane. The sample should not contain nonvolatile substances that may eventually clog the injection port or contaminate the stationary phase of the column.

Gas-tight syringes are precision-made and expensive. You should handle them carefully and adhere to the following procedure when using them. To fill the syringe, draw slightly more than the desired volume of the sample into the barrel by withdrawing the plunger as needed, point the syringe needle up, push out the excess liquid, and then wipe the tip of the needle with a tissue. To inject the sample, insert the needle *straight* into the septum as far as it will go, *push the plunger all the way in with one quick motion,* and remove the syringe from the septum while holding the plunger in place. If the sample is injected slowly, it will produce a wider band on the column and the peaks will be broadened. Be careful to *avoid bending the needle or the plunger* during the injection. It is important to *clean the syringe immediately after use.* Rinse it with a volatile solvent such as acetone and dry it by drawing a stream of air through it.

EXPERIMENTAL PROCEDURES

A ■ *Qualitative and Quantitative Analyses of a Mixture of Compounds by GLC*

Purpose To identify individual components of a mixture and to quantify the relative amounts of each by gas-liquid chromatography.

SAFETY ALERT

1. **Wear safety glasses or goggles and suitable protective gloves while performing the experiments.**

2. **The solvents and other liquids used in sample mixtures for GLC analysis are flammable and volatile. There should be *no open flames* used in the vicinity of these liquids.**

Procedure

Preparation Refer to the online resources to answer Pre-Lab Exercises, access videos, and read the MSDSs for the chemicals used or produced in this procedure. Read or review Section 2.5.

Apparatus Gas chromatograph, gas-tight syringe.

Setting Up Since there is considerable variation in the operating procedures of different commercial gas chromatographs, consult your instructor for specific directions for using the instrument. Obtain a mixture of liquid compounds from your instructor. A suitable mixture would be one containing varying amounts of some or all of the following compounds: ethanol, ethyl acetate, toluene, *n*-butyl acetate, ethylbenzene, and isopropylbenzene.

Injection Following the directions of your instructor, use the syringe to inject a 1- to 5-μL sample of your mixture into the gas-liquid chromatograph to obtain a gas chromatogram. In the same way, obtain a chromatogram of pure samples of ethanol, ethyl acetate, toluene, *n*-butyl acetate, ethylbenzene, and isopropylbenzene under the same instrumental conditions used for the mixture.

Qualitative Analysis Identify each of the components in your mixture as follows: (1) Compare the retention time of each of the components of the mixture with the retention times of the authentic samples of pure compounds. (2) Verify your assignments for one or more components of the mixture by preparing a series of new sample mixtures, each of which contains one volume of the pure known compound with two volumes of the original mixture. Prepare a gas chromatogram of this new mixture. The peak of the mixture that has been amplified in the new chromatogram corresponds to the pure compound that has been added.

Quantitative Analysis Use the quantitative method recommended by your instructor to determine the relative areas under the peaks in your chromatogram of the mixture. Correct these values by multiplying the measured area by the appropriate weight or molar correction factors for the type of detector in the GLC. These factors may be found in Table 6.2 or will be provided by your instructor. Determine the relative amounts, in weight and mole percentages, of each of the components of the mixture.

Discovery Experiment   *Analysis of Factors Affecting Retention Times*

Design and execute experiments that explore how factors such as flowrate and temperature influence the GLC retention times of a standard mixture of compounds.

Discovery Experiment   *Effect of Stationary Phase on Separation of a Mixture*

Design and execute experiments that explore how the nature of the stationary phase of a GLC column influences the separation of a mixture of volatile compounds having a range of polarities. Consult with your instructor regarding the variety of GLC columns that would be available for your investigation.

WRAPPING IT UP

Pour any unused *volatile organic compounds* into the appropriate container for non-halogenated liquids or halogenated hydrocarbons; put any mixture that contains a halogenated hydrocarbon in the container for halogenated hydrocarbons.

B ■ *Determining GLC Response Factors*

Discovery Experiment   **Purpose** To determine the weight and mole correction factors for compounds using flame ionization or thermal conductivity detectors.

SAFETY ALERT

1. **Wear safety glasses or goggles and suitable protective gloves while performing the experiments.**

2. **The solvents and other liquids that are used in sample mixtures for GLC analysis are flammable and volatile. *No flames* should be used in the vicinity of these liquids.**

Procedure

Preparation Refer to the online resources to answer Pre-Lab Exercises, access videos, and read the MSDSs for the chemicals used or produced in this procedure. Read or review Section 2.5.

Apparatus Gas chromatograph, 1-mL syringe, gas-tight syringe.

Setting Up Since there is considerable variation in the operating procedures of different commercial gas chromatographs, consult your instructor for specific directions to use the instrument in your laboratory. Look up the densities of the compound whose weight and mole response factors are to be determined and the standard substance whose response factors are known. For example, the standard sample may be any of the compounds in Table 6.2. Prepare a test solution by combining 0.5 mL of the compound whose response factors are to be determined and 0.5 mL of the standard in a small vial or test tube, and swirl the solution to mix the two components *completely.* Assuming the volumes of the two liquids are additive, calculate the weight fraction and mole fraction composition of the mixture.

Injection Following the directions of your instructor, use a gas-tight syringe to inject a 1- to 5-μL sample of the test mixture into the GLC to obtain a chromatogram. Repeat this process for each of the compounds for which you are to obtain weight and mole response factors.

Analysis Use the quantitative method recommended by your instructor to determine the areas under the two peaks in your gas chromatogram of the test mixture. Divide the weight fraction and mole fraction of each of the two components by the area under the peak for that component. Normalize the value for the weight and mole factors for the unknown by the corresponding weight and mole factors for the standard, and record these values in your notebook.

Discovery Experiment   *Molar Response Factors of Isomers*

Using the protocol of this experimental procedure, determine whether the molar response factors, M_f, for either a pair of acyclic or a pair of cyclic isomers are identical.

Discovery Experiment   *Molar Response Factors of Non-Isomeric Compounds*

Using the protocol of this experimental procedure, determine whether the molar response factors, M_f, are identical for a pair of non-isomeric compounds, one of which is a hydrocarbon and the other of which contains a heteroatom, that have the same molecular weight. Possible examples might be (a) cyclohexane, (b) hexane and 2- or 3-pentanone, and (c) ethylbenzene and benzaldehyde.

WRAPPING IT UP

Pour any unused *volatile organic compounds* into the appropriate container for non-halogenated liquids or halogenated hydrocarbons; put any mixture that contains a halogenated hydrocarbon in the container for halogenated hydrocarbons.

EXERCISES

1. Define the following terms:
   a. stationary phase
   b. mobile phase
   c. carrier gas
   d. retention time
   e. solid support
   f. thermal conductivity

2. Refer to the GLC traces given in Figure 6.16. These are analyses of the various fractions collected during the fractional distillation of the mixture of cyclohexane and toluene. The weight and mole correction factors (flame ionization detector) for cyclohexane are 0.84 and 0.78, respectively, and those for toluene are given in Table 6.2. Accurately determine both the weight percent and the mole percent compositions of the distillation fractions *A, B,* and *C.*

3. Benzene (1 g, 12.5 mmol) is allowed to react with 1-chloropropane (1 g, 12.5 mmol) and $AlCl_3$. The product (1.2 g) is subjected to analysis on a GLC equipped with a thermal conductivity detector. The chromatogram shows two product peaks identified as *n*-propylbenzene (area = 65 mm^2; W_f = 1.06) and isopropylbenzene (area = 113 mm^2; W_f = 1.09). Calculate the percent yield of each of the two isomeric products obtained in this reaction. Note that since each of the products has the same molar mass of 120, the use of weight factors gives both weight and mole percent composition.

Benzene   1-Chloropropane

n-Propylbenzene   Isopropylbenzene

4. A gas chromatogram of the organic components of a sample of beer using a column that separates compounds on the basis of their relative boiling points provides a GLC trace with several peaks. Two of the smaller peaks, with retention times of 9.56 and 16.23 minutes, are believed to be ethyl acetate and ethyl butyrate, respectively.

   a. From the above information, which component of the sample, ethyl acetate or ethyl butyrate, elutes faster? What are the reported boiling points of these two substances?

   b. What GLC experiment(s) could you perform to confirm the identity of the peaks at 9.56 and 16.23 minutes?

   c. The major component of the sample elutes first. What organic compound present in beer do you suspect is responsible for this peak? Is your speculation consistent with the reported boiling point of the compound you are proposing? Explain.

   d. Suggest two ways in which you could adjust the conditions of the experiment so as to reduce the retention time for all components in the sample.

5. In the Olympics and other major athletic competitions, a GC-MS is used to screen athletes for their use of banned substances such as steroids (see the Historical Highlight *Who's Taking What? Analysis of Biological Fluids for Illegal Substances*, which is available online). If an analytical method has been developed and the retention time for a particular anabolic steroid has been determined, how might you use this method to determine the possibility of the use of anabolic steroids by an athlete?

CHAPTER 7

Stereoisomers

Most organic molecules are three-dimensional, and the physical and chemical properties of such molecules are determined by the relative orientations of the various atoms and functional groups that comprise the molecule. Information about the shapes of organic molecules is important because it may help us understand how two molecules interact with each other; this knowledge may have practical applications. For example, an enzyme binds its substrate because of favorable bimolecular interactions that result from shape complementarity of the two three-dimensional molecules. In such cases, knowledge of the shape of the enzyme and the substrate can lead to the design and development of novel molecules that bind to the active site of an enzyme, thereby blocking its mode of action. These small organic molecules may be useful drugs for treating diseases. Similarly, our senses of smell and taste are dependent upon specific interactions between small organic molecules and three-dimensional receptors in our nose and on our tongue. Hence, our ability to enjoy a healthy life and to appreciate the odors and flavors we experience every day is a consequence of the shapes of molecules. So while you are studying stereoisomers, don't forget that it is the complementarity of molecular shapes that enables you to smell the roses and to enjoy a delicious meal!

7.1 INTRODUCTION

Understanding the three-dimensional properties of organic molecules is an essential part of organic chemistry. The experiments in this chapter are designed to provide an introduction to **stereoisomers**. Such isomers have molecular skeletons that are identical, but they differ in the three-dimensional orientation of their atoms in space. The two broad subclasses of stereoisomers that are of importance in organic chemistry are **conformational** isomers and **configurational** isomers.

Conformational isomers, as illustrated by the two **Newman projections 1** and **2** for 1,2-dibromoethane, are stereoisomers that are interconverted by rotation about

211

single bonds. The interconversion of such isomers is usually a low-energy process and occurs rapidly at room temperature and below. Consequently, isolation of conformational isomers is *rarely* possible.

Configurational isomers, in contrast, are stereoisomers whose interconversion normally requires the breaking and remaking of a chemical bond, usually a high-energy process. An example of this type of stereoisomerism is seen with carvone, a natural product for which the two representations **3** and **4** can be written. These are related as **nonsuperimposable mirror images**, so **3** and **4** are defined as **enantiomers**. Molecules having a nonsuperimposable mirror image are said to be **chiral**, a term that is derived from the Greek word *cheir*, meaning "hand." Because hands are perhaps the best-known example of nonsuperimposable mirror images, we use the word "chiral" to characterize molecules that have a "handedness." A companion term, **achiral**, describes all molecules having **superimposable mirror images**.

3   4
Carvone   Carvone

Enantiomers have identical physical and chemical properties in an achiral environment. Their differential effect on plane-polarized light, however, is an important exception to this general rule: Enantiomers rotate the plane of such light an equal number of degrees but in *opposite* directions. For this reason they are sometimes called **optical isomers** and are said to possess **optical activity**. Under most circumstances, the chiral compounds that you will prepare or use in the laboratory will be 50:50 mixtures of the two enantiomers, a composition referred to as the **racemate** or the **racemic mixture**. An equimolar mixture of **3** and **4** would thus be called a racemate and would produce *no* net rotation of plane-polarized light.

Frequently the source of chirality in an organic molecule is the presence of a **center of chirality**, also called a **stereocenter**, which is usually a carbon atom bearing four different substituents. This is the case for the atom marked with an asterisk in **3** and **4**. The substituents are a hydrogen atom, an isopropenyl group, and two methylene groups. The methylene groups are distinct from each other because one of them is bound to a carbon atom of a carbon-carbon double bond, and the other to a carbon atom doubly bound to oxygen.

A chiral molecule containing two or more stereocenters may exist as **diastereomers**, which are configurational isomers that are *not* enantiomers. Consider 1,2-cyclohexanediol, **5**, for example, which contains the stereocenters marked by asterisks. Each center consists of a carbon atom bound to a hydroxyl group, a hydrogen atom, a methylene group, and the carbon atom of the other center of chirality. Three configurational isomers, **6–8**, can then be written for **5**. The *cis* isomer **6** is clearly different from the *trans* isomers **7** and **8**, which are enantiomers of one another and diastereomers of **6**. Moreover, even though it contains two chirality centers, **6** has a *superimposable mirror image*. We could therefore variously describe this isomer as being achiral, optically inactive, or **meso**. The last term describes those configurational isomers that have two or more chirality centers *and* a superimposable mirror image. It is possible to separate **6** from the enantiomers **7** and **8** because *diastereomers have different, although sometimes very similar, physical and chemical properties.*

5	**6**	**7**	**8**
1,2- Cyclohexanediol	*cis*-1,2- Cyclohexanediol	*trans*-1,2- Cyclohexanediol	*trans*-1,2- Cyclohexanediol

Even isomeric pairs of *achiral* molecules may exist as separable stereoisomers. Examples include *cis*- and *trans*-, or (Z)- and (E)-, alkenes such as **9** and **10**, and cyclic compounds such as **11** and **12**. The molecules are nonsuperimposable, yet they are *not* mirror images; that is, they are *not* enantiomers. Rather, they are diastereomers and have different chemical and physical properties.

9	**10**
Dimethyl maleate	Dimethyl fumarate

11	**12**
cis-1,4-Dimethylcyclohexane	*trans*-1,4-Dimethylcyclohexane

In contrast to the case with diastereomers, separation or **resolution** of enantiomers is generally more difficult because of the identity of their chemical properties; their differing effect on plane-polarized light is not helpful for separation. Nonetheless, various experimental techniques have been developed to accomplish resolutions, and one of them is presented in Section 7.6.

The procedures in this chapter are chosen to study (a) separating diastereomers by chromatography (Sec. 7.2), (b) converting one diastereomer into another (Sec. 7.3), (c) evaluating some chemical and physical properties of enantiomers (Sec. 7.4), and (d) resolving a racemate (Sec. 7.6). A description of the technique of polarimetry is given in Section 7.5 to support this last study.

7.2 SEPARATION OF DIASTEREOMERIC 1,2-CYCLOHEXANEDIOLS

Cyclic 1,2-diols can be produced stereoselectively from cycloalkenes. For example, a racemic mixture of the enantiomeric *trans*-1,2-cyclohexanediols **7** and **8** is produced by reaction of a peracid with cyclohexene and hydrolysis of the intermediate epoxide **13** (Eq. 7.1). Oxidation of this alkene by permanganate, on the other hand, gives *cis*-1,2-cyclohexanediol (**6**), as shown in Equation 7.2. The intermediate in this process is presumably the cyclic manganese-containing species **14**, which is not isolated.

$$\text{(7.1)}$$

13

Cyclohexene   Cyclohexene oxide

$$\text{(7.2)}$$

14

Permanganate ion

In this experiment, a commercial mixture containing the *meso*-isomer **6** and the *racemate* **7/8** will be separated into its components. This mixture is produced by the catalytic hydrogenation of catechol (**15**) (Eq. 7.3). Because the enantiomeric *trans*-diols **7** and **8** have identical physical properties, except for their effect on plane-polarized light, they will not be resolved from one another in the procedure of this section. The *meso*-isomer **6**, on the other hand, being diastereomeric to **7** and **8**, has physical and chemical properties that differ sufficiently to allow its separation from the other two stereoisomers.

$$\text{(7.3)}$$

15

Catechol

The separation of diastereomers from one another may often be done solely on the basis of differing solubilities, but this is not the case with the 1,2-cyclohexanediols because they have similar solubilities. The desired separation is possible by chromatographic means, however, and the technique of TLC (Sec. 6.2) is used here, although column (Sec. 6.3) or gas chromatography (Sec. 6.4) would also work.

A challenge to the experimentalist is determining which spot on the TLC plate corresponds to the racemic mixture of *trans* isomers and which to the *cis* diastereomer. In other words, it is one thing to separate different compounds and quite another to assign specific structures to them. This common problem can be addressed in a variety of ways. For example, you might extract the portions of the TLC plate containing each of the isomers and determine the melting point or IR or NMR spectral characteristics (Chap. 8) of the solid remaining after removal of solvent. Spectral characterization is difficult to do in this case because only extremely small amounts of material would be obtained from the TLC plate. In addition, the melting points of the diastereomers are similar, 100–101 °C for **6** and 103–104 °C for **7** and **8**; thus, traces of impurities in the *trans* isomer could easily lower its melting range (Sec. 3.3) to that of the *cis*, making unambiguous assignment of structure impossible.

The method used to make the needed structural assignment in this experiment is to compare the "unknown" compounds with an authentic specimen of "known" material. Specifically, by spotting the TLC plate with authentic samples of either *cis*- or *trans*-1,2-cyclohexanediol as well as the commercial mixture, you will be able to make a direct comparison of R_f-values and determine which component of the mixture is which.

EXPERIMENTAL PROCEDURE

Separation of Diastereomeric 1,2-Cyclohexanediols

Purpose To demonstrate that diastereomers may be readily separated by chromatography.

SAFETY ALERT

1. **Wear safety glasses or goggles and suitable protective gloves while performing the experiments.**

2. **Acetone and petroleum ether are *highly* flammable; *do not have any flames in the vicinity* when they are being used.**

3. **Do not allow iodine to come in contact with your skin and do not inhale its vapor; as it is corrosive and *toxic*.**

Procedure

Preparation Refer to the online resources to answer Pre-Lab Exercises, access videos, and read the MSDSs for the chemicals used or produced in this procedure. Read or review Section 6.2.

Apparatus Capillary pipets, 2-cm × 10-cm silica gel TLC plate, TLC chamber.

Setting Up Obtain or prepare approximately 5% solutions in acetone of a commercial mixture of *cis*- and *trans*-1,2-cyclohexanediol and of pure *trans*-1,2-cyclohexanediol. Place small spots of each solution side by side on the TLC plate.

Elution and Analysis Develop the chromatogram (Fig. 6.3b) using a mixture of 75% petroleum ether, bp 60–80 °C (760 torr), and 25% 2-propanol by volume. Remove the plate from the solvent mixture when the solvent front approaches the top of the plate, mark the position reached by the solvent front, air-dry the plate for a few minutes, and place it in a closed container with a few iodine crystals to make the spots visible. Remove the plate and circle the location of any spots. Record the R_f-value of each spot.

Discovery Experiment   *Solvent Effects on R_f-Values*

Investigate the effect of changing the ratios of the two solvents used for the separation of the 1,2-cyclohexanediols on the R_f-values for the diastereomers. One way to report the relative effectiveness of various solvent mixtures in effecting the separation would be to report values of ΔR_f for each mixture used. Consult with your instructor before carrying out any experiments.

WRAPPING IT UP

Transfer any excess of the *eluting solvents* to a container for nonhalogenated organic liquids. Dispose of the used *TLC plates* in a container for nonhazardous materials.

EXERCISES

1. The maximum number of stereoisomers that are possible for a given molecule can be determined from the formula 2^n, where n is the number of stereocenters in the molecule.

   a. Using this formula, what is the maximum number of stereoisomers that could exist for 1,2-cyclohexanediol (**5**)?

   b. Why are there only three stereoisomers of 1,2-cyclohexanediol (**5**) rather than the number you calculated in part **a**?

2. A *meso* compound is achiral by virtue of the fact that it contains an internal mirror plane of symmetry, so the molecule is superimposable on its mirror image and hence does not have an enantiomer. Use a dashed line to show the internal plane of symmetry present in the structure of *cis*-1,2-cyclohexanediol (**6**).

3. Indicate the stereochemical relationship (i.e., enantiomers or diastereomers) between each of the following pairs of stereoisomers of 1,2-dibromocyclohexane, and indicate whether each pair will have identical or different physical properties.

   a.   and

   b.   and

   c.   and

4. Which, if any, of the molecules in Exercise 3 are *meso* compounds?

5. By comparing the R_f-values of the separated spots from the mixture of the *cis* and *trans* diols with the R_f-value of the *trans* isomer, decide which isomer is adsorbed more strongly by the silica gel. Give a reason for the difference in adsorptivity of the two isomers.

6. It may happen that the separation of the isomers by TLC is not complete—one of the spots may be found to be smeared out to some extent. Considering this, which one of the isomers could more easily be obtained pure by column chromatography?

7. What would be the consequence of using pure petroleum ether as the eluting solvent?

8. What would be the consequence of using pure 2-propanol as the eluting solvent?

9. In Section 7.1, it was stated that interconversion of configurational isomers requires the cleavage and reformation of a chemical bond. Indicate any such bonds in *cis*-1,2-cyclohexanediol (**6**) that upon being broken and then reformed would afford the enantiomers **7** and **8**. Be sure to describe what must occur stereochemically during the process of cleavage and remaking of a bond to effect the interconversion of **6** to the other two diastereomers.

10. **a.** Consider the bond dissociation energies of the various bonds in **6**, and then predict which bond would be easiest to break for the conversion of **6** to **7** and **8**.

   b. Estimate the $\Delta H^{\ddagger}$ for this process.

7.3 ISOMERIZATION OF DIMETHYL MALEATE TO DIMETHYL FUMARATE

Dimethyl maleate (**9**) and dimethyl fumarate (**10**) are examples of *cis* and *trans* isomers, called **geometric isomers**. Their names do not reflect this isomeric relationship, however, because the parent diacids from which they can be prepared were discovered and named before this stereochemical fact was known. Being diastereomers, these diesters would be expected to have different physical and chemical properties, and indeed they do. The difference in their physical states at room temperature is the most dramatic: The maleate **9** is a *liquid* with a freezing point of about $-19\,°C$, whereas the fumarate **10** is a *solid* with a melting point of 103–104 °C.

9	10
Dimethyl maleate	Dimethyl fumarate

As discussed in Section 7.1, the interconversion of configurational isomers involves the cleavage and subsequent remaking of a chemical bond. The diastereomers **9** and **10** could be equilibrated by breaking the π component of the carbon-carbon double bond, rotating about the single bond that remains, and remaking the π-bond. The required bond cleavage can be promoted (a) thermally, (b) photochemically, or (c) chemically. The latter possibility involves the reversible addition of a reagent to one of the carbon atoms involved in the carbon-carbon double bond, resulting in breaking of the π-bond.

The goal of this experiment is to formulate a mechanism by which dimethyl maleate (**9**) is isomerized to dimethyl fumarate (**10**) by the action of heat, light, or molecular bromine. Three different experiments will be performed to study the reaction. This involves exposing two samples of **9**, one of which contains a catalytic amount of bromine, to light and treating a third sample with bromine but keeping the mixture in the dark at room temperature. By thoughtful analysis of your results, you should be able to discover which of the three modes of breaking the π-bond, which were designated (a)–(c) in the previous paragraph, is responsible for the isomerization. The analysis will also allow you to determine whether polar intermediates or radical species are involved in the isomerization.

There will be residual bromine at the end of the procedure, which must be destroyed prior to isolating the crystalline **10**. A convenient way to do this is by adding cyclohexene, which reacts with bromine to produce a dibromide (Eq. 7.4), to the reaction mixture.

$$\text{Cyclohexene} \quad + \quad Br_2 \longrightarrow \quad trans\text{-}1,2\text{-Dibromocyclohexane} \tag{7.4}$$

EXPERIMENTAL PROCEDURES

Isomerization of Dimethyl Maleate to Dimethyl Fumarate

Purpose To study the mechanism of a geometric isomerization.

SAFETY ALERT

1. **Wear safety glasses or goggles and suitable protective gloves while performing the experiments.**

2. **Bromine is a hazardous chemical that may cause serious chemical burns. Do not breathe its vapors or allow it to come into contact with the skin. Perform all operations involving the transfer of solutions of bromine at a hood. If you get bromine on your skin, *immediately* wash the affected area with soap and warm water and soak the skin in 0.6 *M* sodium thiosulfate solution, for up to two hours if the burn is particularly serious.**

3. **Bromine reacts with acetone to form α-bromoacetone, $BrCH_2COCH_3$, a powerful *lachrymator*. Do not rinse glassware that might contain residual bromine with acetone! Rather, follow the procedure described in "Wrapping It Up."**

MINISCALE PROCEDURE

Preparation Refer to the online resources to answer Pre-Lab Exercises, access videos, and read the MSDSs for the chemicals used or produced in this procedure. Read or review Sections 2.9, 2.17, and 3.2.

Apparatus Three 13-mm × 100-mm test tubes, incandescent light source, apparatus for vacuum filtration and *flameless* heating.

Setting Up Measure 0.5 mL of dimethyl maleate into each of the test tubes. Working at the hood and wearing suitable protective gloves, add enough of a 0.6 *M* solution of bromine in dichloromethane dropwise with the filter-tip pipet to *two* of the tubes to give an orange solution. Add an equal volume of dichloromethane to the third test tube. Loosely cork all the test tubes.

Isomerization and Isolation Place one of the tubes containing bromine in the dark, and expose the other two tubes to strong light. Place the tubes about 20 cm from the light source to prevent boiling the solvent and popping the cork off the tubes. If decoloration of a solution occurs, add an additional portion of the bromine solution. After 30 min, record in which test tube(s) crystals have appeared and add 1–2 drops of cyclohexene to the tubes containing bromine to destroy it. Add 2–3 mL of 95% ethanol to any tubes containing solid, and heat the mixture to effect dissolution. Then allow the solution to cool slowly to room temperature. Isolate the precipitate by vacuum filtration using a Hirsch funnel (Fig. 2.54). Press the precipitate as dry as possible on the filter disk with the aid of a clean cork or spatula, transfer it to a watchglass or piece of weighing paper, and air-dry it.

Analysis Weigh the dimethyl fumarate (**10**) and calculate its yield, based on the amount of dimethyl maleate used in the solution(s) that produced **10**. Determine its melting point and compare the result with the reported melting point. If the observed melting point is low, recrystallize the product from ethanol.

MICROSCALE PROCEDURE

Preparation Refer to the online resources to answer Pre-Lab Exercises, access videos, and read the MSDSs for the chemicals used or produced in this procedure. Read or review Sections 2.9, 2.17, and 3.2.

Apparatus Two Craig tubes, 10-mm × 75-mm test tube, incandescent light source, apparatus for Craig tube filtration and *flameless* heating.

Setting Up Using a Pasteur pipet, dispense two drops of dimethyl maleate into each of the Craig tubes and the test tube. Working at the hood, add enough of a 0.1 *M* solution of bromine in dichloromethane dropwise to the *Craig tubes* to produce an orange-colored solution. Add sufficient dichloromethane to the *test tube* so that the volume of solution in the tube approximately equals that in the Craig tubes. Loosely cork each of the tubes.

Isomerization and Isolation Place one of the Craig tubes in a dark location, and expose the other two solutions to strong light. Place the tubes about 20 cm from the light source to prevent boiling the solvent and popping the cork off the tubes. Should decoloration of a solution occur, add an additional portion of the bromine solution. After 20 min, record in which test tube(s) crystals have appeared and add 1 drop of cyclohexene to the tubes containing bromine to destroy it. Add 0.5–1 mL of 95% ethanol to any tubes containing solid, and heat the mixture to effect dissolution. Then allow the solution to cool slowly to room temperature.

Filter any solutions containing solids by Craig tube filtration (Fig. 2.55). Transfer the product to a watchglass or piece of weighing paper, and air-dry it.

Analysis Weigh the dimethyl fumarate (**10**). Determine its melting point and compare the result with the reported melting point. If the observed melting point is low, recrystallize the product from ethanol using a Craig tube.

Discovery Experiment   *Iodine as a Catalyst for Isomerization*

Devise a protocol for determining whether iodine will promote the isomerization of dimethyl maleate to dimethyl fumarate, and use the same three conditions as used with bromine. You may also wish to explore the relative rates at which isomerization occurs with the two different halogens. Consult with your instructor before performing any experiments.

Discovery Experiment   *Assessing Purities of Dimethyl Maleate and Fumarate*

Devise analyses that will allow you to assess the purity of the dimethyl maleate used and of the dimethyl fumarate produced in the isomerization experiment.

WRAPPING IT UP

Decolorize any *solutions or containers in which the color of bromine* is visible by dropwise addition of cyclohexene, and then discard the resulting solutions in a container for halogenated organic liquids. Flush the *ethanolic filtrate* from the recrystallization process down the drain with water.

EXERCISES

1. What observation allows exclusion of thermal cleavage of the π-bond as the initiation step for isomerization of **9** to **10**?

2. What observation allows exclusion of direct photoexcitation of **9** as the means of initiating its conversion to **10**?

3. Depending on the type of chemical reaction, molecular bromine (Br_2) can be a source of Br^+, Br^-, or $Br^\cdot$.

   a. Provide a Lewis dot structure for each of the three species of mono atomic bromine.

   b. Using curved arrows to symbolize the flow of electrons, show how Br^+, Br^-, and $Br^\cdot$ may be formed from Br_2.

   c. Based upon the results of this experiment, indicate whether Br^+, Br^-, or $Br^\cdot$ initiates the isomerization of dimethyl maleate to dimethyl fumarate. Explain your reasoning.

4. Using curved arrows to symbolize the flow of electrons, write a stepwise mechanism for the isomerization of **9** to **10**. Be certain to account specifically for the roles, if any, that light and bromine play in promoting this process.

5. Explain why decoloration of a solution of bromine and **9** is slow.

6. State which one of the two diesters **9** and **10** is predicted to be more stable thermodynamically, and support your prediction.

7. What type of configurational isomers are **9** and **10**, enantiomers or diastereomers? How is your answer consistent with the fact that these two compounds have different physical properties?

8. The double bonds of **9** and **10** contain both a σ- and a π-component. Explain why rupture of the π-bond rather than the σ-bond is the more probable event in the isomerization process.

7.4 PROPERTIES OF THE ENANTIOMERS OF CARVONE

Enantiomers have *identical* physical properties except with respect to their effect on plane-polarized light. Their chemical properties are also identical, with one *important* exception: One enantiomer undergoes a chemical change at a rate *different* from that of its mirror image *if* the reaction occurs in a stereochemically asymmetric environment, such as the active site of an enzyme or a biological receptor. This is because the active sites of enzymes and receptors are themselves chiral and, as a consequence, complex preferentially with one member of a pair of enantiomers. The complexed enantiomer will then undergo the enzyme-promoted reaction faster than will its mirror image.

A dramatic example of such differential complexation is the olfactory response to the two enantiomers of carvone. (*R*)-(−)-Carvone (**3**) smells like spearmint and is the principal component of spearmint oil, which also contains minor amounts of limonene (**16**) and *α*- and *β*-phellandrene (**17** and **18**, respectively). On the other hand, (*S*)-(+)-carvone (**4**), along with limonene, is found in caraway seed and dill seed oils and has been shown to be the compound largely responsible for the characteristic odor of these oils. This remarkable difference in how we sense these two enantiomers is because the odor receptors in our noses are chiral environments that are linked to the nervous system. The type of receptor site that complexes with a particular enantiomer will determine what odor is detected by the brain.

| **3** | **4** | **16** | **17** | **18** |
| (*R*)-(−)-Carvone | (*S*)-(+)-Carvone | Limonene | α-Phellandrene | β-Phellandrene |

It is unusual for both enantiomers of a compound to occur naturally, and it is extremely rare for each of them to be available from readily accessible sources such as spearmint and caraway oils. This fortunate situation enables the separate isolation of **3** and **4** from their respective sources by vacuum distillation (Sec. 2.15). Because of their ready isolation, both of these enantiomers are commercially available.

The investigative experiments that you perform on the enantiomeric carvones will depend upon the time and facilities that are available. If samples of both enantiomers are available, you can verify their different odors. Bear in mind, however, that 8–10% of the population cannot detect the difference, and you may be in this group. Should a gas-liquid chromatograph (Sec. 6.4) be available, analyses may be performed on samples of spearmint, caraway oils, and dill seed as well as solutions containing one or both of the enantiomers **3** and **4**. Typical chromatograms are given in Figures 7.1–7.3.

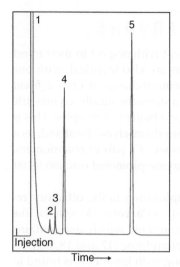

Figure 7.1
GLC analysis of spearmint oil (numbers in parentheses are retention times in minutes and peak areas, respectively). Peak 1: solvent; peak 2: unknown; peak 3: phellandrenes (2.16, 17819); peak 4: limonene (2.46, 84889); peak 5: (R)-(−)-carvone (5.13, 248316).

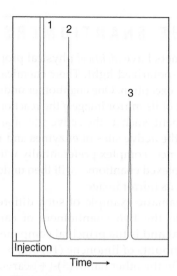

Figure 7.2
GLC analysis of caraway seed oil (numbers in parentheses are retention times in minutes and peak areas, respectively). Peak 1: solvent; peak 2: limonene (2.46, 125719); peak 3: (S)-(+)-carvone (5.13, 385950).

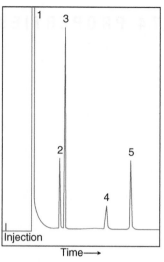

Figure 7.3
GLC analysis of dill seed oil (numbers in parentheses are retention times in minutes and peak areas, respectively). Peak 1: solvent; peak 2: unknown (2.27, 49418); peak 3: limonene (2.44, 139972); peak 4: unknown (4.15, 17091); peak 5: (S)-(+)-carvone (5.07, 78223).

Note that under the conditions of these analyses, limonene (**16**) and the phellandrenes **17** and **18** elute *before* **3** and **4** because their boiling points, 170–177 °C (760 torr), are lower than that of the enantiomers of carvone (bp 232–233 °C, 760 torr).

$$2\ Br_2/CH_2Cl_2 \quad \text{(Red solution)} \qquad (7.5a)$$

Colorless

$$KMnO_4/H_2O \quad \text{(Purple solution)} \qquad + MnO_2\downarrow \quad (7.5b)$$

Carvone

Colorless   Brown precipitate

2,4-Dinitrophenylhydrazine   Carvone 2,4-dinitrophenylhydrazone

The presence of a carbon-carbon double bond in carvone can be shown by performing a test for unsaturation using bromine in dichloromethane (Eq. 7.5a) and aqueous potassium permanganate (Eq. 7.5b). Confirmation of the carbonyl function in **3** and/or **4** and of the structural assignment may be obtained by making the 2,4-dinitrophenylhydrazone (2,4-DNP) of carvone (Eq. 7.6) and comparing its melting point to that reported (193–194 °C) for authentic material. By performing a mixture melting point (Sec. 3.3) of the 2,4-DNPs prepared from both **3** and **4**, you can determine whether one enantiomeric derivative depresses the melting point of the other.

EXPERIMENTAL PROCEDURES

A. Properties of the Enantiomeric Carvones

Discovery Experiment   **Purpose** To explore the physical and chemical properties of enantiomers.

SAFETY ALERT

1. **Wear safety glasses or goggles and suitable protective gloves while performing the experiments.**

2. **Bromine is a hazardous chemical that may cause serious chemical burns. Do not breathe its vapors or allow it to come into contact with the skin. Perform that all operations involving the transfer of solutions of bromine at a hood. If you get bromine on your skin, *immediately* wash the affected area with soap and warm water and soak the skin in 0.6 *M* sodium thiosulfate solution, for up to 3 h if the burn is particularly serious.**

3. **Bromine reacts with acetone to form α-bromoacetone, BrCH₂COCH₃, a powerful *lachrymator*. Do not rinse glassware that might contain residual bromine with acetone! Rather, follow the procedure described in Wrapping It Up.**

4. **Do not allow solutions of potassium permanganate to contact your skin. These solutions cause unsightly stains on your skin and last several days. If contact occurs, wash the affected area thoroughly with warm water.**

Procedure

Preparation Refer to the online resources to answer Pre-Lab Exercises, access videos, and read the MSDSs for the chemicals used or produced in this procedure. Read or review Sections 6.3, 4.7A, and 4.7B.

1. **Odor.** Compare the odor of the carvone obtained from spearmint oil with that of the carvone from caraway seed oil.

2. **Tests for Unsaturation.** Obtain samples of **3** and/or **4**, and test for the presence of a carbon-carbon double bond using the procedures in Section 4.7A and/or Section 4.7B. Record your observations.

3. **Gas-Liquid Chromatography.** Obtain gas chromatograms of spearmint and caraway oils and of the enantiomeric carvones. Co-inject an authentic sample of **3** or **4** with one of the natural oils to determine which peak in the oil corresponds to carvone. Use the data to determine the retention time of the carvones and the relative percentage of carvone **3** or **4** compared to the total amount of the three compounds **16–18** that may be present in the essential oils. A capillary column with dimensions of 0.25 mm × 25 m with 100% dimethylpolysiloxane as the stationary phase using a flowrate of 1.2 mL/min at an oven temperature of 110°C is suitable for the analysis.

WRAPPING IT UP

Transfer the *pot residues* from the distillation of spearmint and/or caraway oil to the container for nonchlorinated organic wastes. After performing the prescribed tests on your distillate, put the isolated *natural products* into appropriately labeled containers. Place the *dichloromethane solution* from the bromine test for unsaturation in a container for halogenated organic liquids; put the *manganese dioxide* from the Baeyer test for unsaturation in a container for heavy metals.

B. Formation of Carvone 2,4-Dinitrophenylhydrazone

Arylhydrazines ($ArNHNH_2$), especially 2,4-dinitrophenylhydrazine, are commonly used to make crystalline derivatives of carbonyl compounds. The formation of carvone 2,4-dinitrophenylhydrazone is shown in Equation 7.6, and the mechanism of this transformation is typical of that followed by the reactions of aldehydes and ketones with a number of compounds, RNH_2, that may be considered derivatives of ammonia. Thus, the overall reaction involves initial acid-catalyzed addition of the elements of N–H across the carbonyl π-bond to afford a tetrahedral intermediate that subsequently undergoes dehydration to form a hydrazone (Eq. 7.7). Arylhydrazines are valuable reagents for both classifying and forming derivatives of aldehydes and ketones because the solid products obtained from these test reactions may be used as derivatives to determine the identity of a carbonyl compound; 2,4-dinitrophenylhydrazine is particularly useful in this regard. Arylhydrazines will give a positive test, which is indicated by formation of an insoluble product, for either an aldehyde or a ketone.

$$R_2C{=}O \quad + \quad H_2NNH{-}Ar \quad \underset{\xrightarrow{\hspace{1cm}}}{\overset{H_3O^+}{\rightleftharpoons}} \quad R_2C\overset{\displaystyle OH}{\underset{\underset{NHAr}{\overset{|}{N}H}}{\overset{|}{\diagdown}}} \quad \rightleftharpoons \quad R_2C{=}NNHAr \quad + \quad H_2O \quad (7.7)$$

An aldehyde   2,4-Dinitrophenyl-   A 2,4-dinitrophenylhydrazone
or ketone   hydrazine

$$Ar =$$

EXPERIMENTAL PROCEDURES

Preparation of Carvone 2,4-Dinitrophenylhydrazone

SAFETY ALERT ⚠

1. **Wear safety glasses or goggles and suitable protective gloves while performing the experiment.**

2. *Concentrated* **sulfuric acid may produce severe chemical burns, and 2,4-dinitrophenylhydrazine is a possible mutagen and may cause skin sensitization. If this acid or the ethanolic solution of 2,4-dinitrophenylhydrazine comes in contact with your skin, immediately flood the affected area with water and rinse it with 5% sodium bicarbonate solution.**

Preparation Refer to the online resources to read the MSDSs for the chemicals used or produced in this procedure.

Apparatus Two 50-mL Erlenmeyer flasks, graduated cylinder, apparatus for magnetic stirring and *flameless* heating.

Setting Up If a solution of the arylhydrazine in acid is not supplied, prepare it by dissolving 0.6 g of 2,4-dinitrophenylhydrazine in 3 mL of *concentrated* sulfuric acid. Slowly add this solution, with stirring, to a mixture of 4.5 mL of water and 15 mL of 95% ethanol. Stir the solution vigorously and then filter it to remove any undissolved solids. Dissolve a 400- to 500-mg sample of **3** or **4** in 20 mL of 95% ethanol, and add this solution to 15 mL of the 2,4-dinitrophenylhydrazine reagent.

Reaction Shake or stir the mixture vigorously; if a precipitate does not form immediately, let the solution stand for 15 min. If recrystallization is necessary, use a mixture of ethanol and ethyl acetate or aqueous ethanol as the solvent.

Determine the melting point of the 2,4-dinitrophenylhydrazone. Obtain a sample of the 2,4-DNP prepared from the enantiomer opposite from yours. Make a 1:1 mixture of the enantiomeric 2,4-DNPs by intimately mixing the two, and determine the melting point of the mixture. Compare the result with the melting point you originally obtained.

WRAPPING IT UP

Neutralize and then filter any excess 2,4-*dinitrophenylhydrazine solution*. Put the filter cake in the container for nonhazardous solids. Flush the *filtrate*, as well as *filtrates* obtained from recrystallization, down the drain.

EXERCISES

Questions for Part A

1. Why would a positive test for unsaturation on spearmint, caraway seed, or dill seed oil *not* prove that the enantiomeric carvones contain unsaturation?

2. Bromine is known to add to carbon-carbon double bonds in an *anti* fashion; namely, one bromine atom adds from one face of the carbon-carbon double bond and the other bromine adds from the opposite face. Draw all of the possible stereoisomeric forms of the tetrabromo products obtained upon the addition of bromine to (*R*)-(−)-carvone (3) (see Eq. 7.5a).

3. What is it about the properties of enantiomers that makes it easy, at least for most people, to distinguish the enantiomers of carvone by smell but difficult to distinguish them experimentally in the laboratory? What common laboratory technique could be used to distinguish the two enantiomers of carvone?

4. If you developed a new type of GLC column that *could* distinguish between *R*- and *S*-carvone, in what fundamental way would this column be different from the one you used for the GLC analysis in this experiment?

5. There is no *meso* form for carvone. What symmetry element is absent in carvone that would allow for the existence of a *meso* form?

6. Based on the results of the GLC analysis of the enantiomeric carvones, would you expect the *boiling point* of a *mixture* of the enantiomers to be the same as that of a *single* enantiomer? Explain.

7. Assume that the peak areas given in Figures 7.1 and 7.2 for limonene (**16**), the phellandrenes **17** and **18**, and the carvones **3** and **4** reflect the relative *molar* amounts of these compounds. Compute the percentage of the carvone present in spearmint and caraway seed oils.

8. Figure 7.3 is a gas chromatogram of dill seed oil.

    a. Using the same assumption as that in Exercise 7, compute the percentage of the carvone present in this oil.

    b. Explain why the retention time of the carvone in this oil does *not* allow assignment of which enantiomer is present.

Questions for Part B

9. Identify the center of chirality in limonene (**16**) and draw the (*R*)-enantiomer of this molecule.

10. Show that the phellandrenes **17** and **18** are chiral.

11. In Section 7.1 it was stated that interconversion of configurational isomers involves the cleavage and remaking of a chemical bond. Indicate the four different bonds in **3** that could be broken and then remade to produce **4**. For each bond, be sure to describe what must occur stereochemically during the process of bond cleavage/remaking that effects the interconversion of one enantiomer to the other.

12. When a student attempted to prepare the 2,4-DNP of (*S*)-(+)-carvone directly from dill seed oil, a red-orange solid was obtained that melted at 110–115 °C even after recrystallization. This result suggests that a mixture of DNPs has been produced and that they are not easily separated by recrystallization. How does the GLC analysis of this oil (Fig. 7.3) support the proposition that there could be a significant amount of a carbonyl-containing contaminant of carvone in dill seed oil? What might you do experimentally to support your proposal?

13. What melting point should you expect from a mixture of the 2,4-DNPs of (*R*)-(−)-carvone (**3**) and (*S*)-(+)-carvone (**4**) if both derivatives are pure?

7.5 POLARIMETRY

A solution of chiral molecules exhibits optical activity (Sec. 7.1) because there is a net rotation of the plane of polarized light as the light passes through the sample. An explanation of the physical basis for this phenomenon is beyond the scope of this discussion, but it is important to understand that achiral molecules do *not* exhibit this same property.

Of primary interest to organic chemists is the fact that each member of a pair of enantiomers rotates plane-polarized light by exactly the same amount but in *opposite* directions. This means that a solution containing equal amounts of two enantiomers will cause *no* net rotation of the light, because the rotation in one direction by molecules of one enantiomer will be exactly offset with an opposite rotation by those of the other. Thus, a solution containing enantiomers can produce a rotation only if the amount of one enantiomer exceeds that of its mirror image.

The **observed rotation** expressed by the value α depends on a number of factors: (a) the nature of the chiral compound; (b) its *concentration*, if in solution, or *density*, if a neat liquid; (c) the *pathlength* of the cell containing the sample; (d) the *temperature* at which the measurement is taken; (e) the nature of the *solvent*, if one is used; and (f) the *wavelength* of plane-polarized light used in the measurement. Factors (b) and (c) determine the average number of chiral molecules in the path of the light and, in combination with the wavelength, have the greatest effect on the magnitude of the rotation of a sample.

The sign of a is defined by convention. When you view through the sample *toward* the source of the plane-polarized light, a clockwise rotation of the plane of the light is taken as positive or **dextrorotatory**, whereas a counterclockwise rotation is negative or **levorotatory**.

Optical rotation represents a **physical constant** of a chiral compound *if* the variables listed above are considered. By specifying the temperature and wavelength at which the measurement is taken, and dividing the observed rotation by the factors that define the average number of molecules in the light path, a constant called the **specific rotation**, [α], is obtained. This is expressed mathematically by Equation 7.8.

$$[\alpha]_\lambda^T = \frac{\alpha}{l \times c} \text{ or } \frac{\alpha}{l \times d} \tag{7.8}$$

where

[α] = specific rotation (degrees)
λ = wavelength (nanometers, nm)
T = temperature (°C)
α = observed rotation (degrees)
l = pathlength of cell (decimeters, dm)
c = concentration (g/100 mL of solution)
d = density (g/mL, neat sample)

A specific rotation might be reported as $[\alpha]_{490}^{25} - 19.6°$ ($c = 1.8$, $CHCl_3$). Alternatively, the specific rotation might be written as $[\alpha]_D^{23} + 39.2°$ ($c = 12.1$, CH_3OH). The "D" refers to the fact that a sodium lamp emitting radiation at 589 nm, the D line of sodium, was used as the light source. The chemical formulas given with c specify the solvent used for the measurement.

The basic components of all **polarimeters**, the instruments used for measuring optical rotations, are shown in Figure 7.4. Ordinary light first passes through a polarizer, such as a Nicol prism or a sheet of Polaroid film, and the plane-polarized light that emerges passes through the sample tube. If the tube contains an optically active substance, the plane of the light will be rotated either clockwise or counterclockwise. The degree of rotation can be determined by turning the analyzer, another Nicol prism or piece of Polaroid film, until the intensity of the light reaching the detector matches that observed in the absence of any of the optically active compounds in the sample tube. The human eye or a photoelectric cell constitutes the detector. Finally, using Equation 7.8 allows the conversion of the observed rotation a to the specific rotation [α].

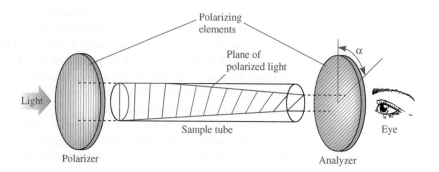

Figure 7.4
Schematic illustration of optical rotation (α = ~−45°).

7.6 RESOLUTION OF RACEMIC 1-PHENYLETHANAMINE

Enantiomers cannot normally be physically separated (see the Historical Highlight *Discovery of Stereoisomers*, which is available online) because they have identical properties, except toward plane-polarized light (Sec. 7.1). However, *diastereomers* do have different physical properties, such as solubility, melting and boiling points, and adsorptive characteristics in chromatography. Recognition of this characteristic of diastereomers has led to a general strategy for separating enantiomers that involves three steps: (1) conversion of the enantiomers to diastereomers by way of a chemical reaction; (2) separation of these diastereomers by any of a number of standard techniques, such as recrystallization and chromatography; and (3) regeneration of one or both of the pure enantiomers from the corresponding diastereomer. This is illustrated schematically in Equation 7.9, where R and S represent the enantiomers to be separated and R' stands for a *single* enantiomer of a compound, termed a **resolving agent**, with which both R and S can react. To convince yourself that RR' and SR' are actually *diastereomers*, you need only show that the *enantiomer* of RR', for instance, would be written as SS'.

$$R + S \xrightarrow{R'} RR' + SR' \xrightarrow[\text{technique}]{\text{Separation}} \begin{array}{c} RR' \xrightarrow{\text{Regeneration}} R + R' \\ SR' \xrightarrow{\text{Regeneration}} S + R' \end{array} \qquad (7.9)$$

| Racemic mixture | Diastereomeric mixture | Separated diastereomers | Pure enantiomers |

Nature is the source of a number of different resolving agents that are used to prepare separable diastereomeric pairs from enantiomers. In this experiment, we use (+)-tartaric acid (**19**), which is produced from grapes during the production of wine, to resolve racemic 1-phenylethanamine (**20**), as shown in the resolution scheme outlined in Figure 7.5. Conversion of the enantiomers of **20** into separable diastereomers involves an acid-base reaction with **19**. The diastereomeric salts that result have *different* solubilities in methanol and are separable by fractional crystallization. The individual enantiomers of **20** are then regenerated by neutralization of the corresponding salts.

In principle, it would be possible to obtain both enantiomers of a chiral compound in high optical purity by this technique, but this generally would require an undesirably large number of crystallizations. Fortunately, one or two crystallizations usually suffice to allow isolation, in reasonable optical purity, of the enantiomer derived from the less soluble salt.

Measuring the optical rotation (Sec. 7.5) of the 1-phenylethanamine recovered in this experiment allows you to determine whether it is the (+)- or (−)-enantiomer and the extent to which optical purification is achieved. If you are to do this part of the experiment, consult with your instructor for special experimental directions.

$$(+)C_6H_5-\underset{\underset{CH_3}{|}}{CH}-NH_2 \; + \; (-)C_6H_5-\underset{\underset{CH_3}{|}}{CH}-NH_2$$

20

1-Phenylethanamine
Racemic mixture of enantiomers

$$(+)HO_2C-\underset{\underset{OH}{|}}{CH}-\underset{\underset{OH}{|}}{CH}-CO_2H$$

19

$$\left[\begin{array}{cc} (+)HO_2CCH-\underset{\underset{OH}{|}}{CH}-\underset{\underset{OH}{|}}{CO_2^-} & (+)C_6H_5-\underset{\underset{CH_3}{|}}{CH}-\overset{+}{N}H_3 \\[2em] + \\[1em] (+)HO_2CCH-\underset{\underset{OH}{|}}{CH}-\underset{\underset{OH}{|}}{CO_2^-} & (-)C_6H_5-\underset{\underset{CH_3}{|}}{CH}-\overset{+}{N}H_3 \end{array}\right] \Biggr\} \text{Diastereomers}$$

Separate by fractional crystallization from methanol

$$\left[(+)HO_2CCH-\underset{\underset{OH}{|}}{CH}-\underset{\underset{OH}{|}}{CO_2^-} \quad (+)C_6H_5-\underset{\underset{CH_3}{|}}{CH}-\overset{+}{N}H_3\right]$$

$$\left[(+)HO_2CCH-\underset{\underset{OH}{|}}{CH}-\underset{\underset{OH}{|}}{CO_2^-} \quad (-)C_6H_5-\underset{\underset{CH_3}{|}}{CH}-\overset{+}{N}H_3\right]$$

↓ NaOH   ↓ NaOH

$$(+)NaO_2CCH-\underset{\underset{OH}{|}}{CH}-CO_2Na \; + \; (+)C_6H_5-\underset{\underset{CH_3}{|}}{CH}-NH_2$$

$$(+)NaO_2CCH-\underset{\underset{OH}{|}}{CH}-CO_2Na \; + \; (-)C_6H_5-\underset{\underset{CH_3}{|}}{CH}-NH_2$$

(1) Extract with diethyl ether   (1) Extract with diethyl ether
(2) Distill   (2) Distill

$$(+)C_6H_5-\underset{\underset{CH_3}{|}}{CH}-NH_2$$

$$(-)C_6H_5-\underset{\underset{CH_3}{|}}{CH}-NH_2$$

Figure 7.5
Resolution of (±)-1-phenylethanamine.

EXPERIMENTAL PROCEDURE

Resolution of Racemic 1-Phenylethanamine

Purpose To demonstrate a method for resolving enantiomers.

SAFETY ALERT

1. **Wear safety glasses or goggles and suitable protective gloves while performing the experiments.**

2. **Methanol is flammable; when heating it, use *flameless* heating.**

3. **Avoid contact of 14 *M* NaOH with skin, as this solution is corrosive. Wear suitable protective gloves when handling this solution. If skin contact occurs, wash the area with copious amounts of water.**

4. **Diethyl ether is extremely volatile and flammable. When using it for extractions or distilling it, be certain that there are *no flames* in the vicinity.**

MINISCALE PROCEDURE

Preparation Refer to the online resources to answer Pre-Lab Exercises, access videos, and read the MSDSs for the chemicals used or produced in this procedure. Read or review Sections 2.6, 2.9, 2.11, 2.13, 2.15, 2.17, 2.25, and 2.29.

Apparatus A 500-mL Erlenmeyer flask, apparatus for magnetic stirring, vacuum filtration, and *flameless* heating.

Setting Up Dissolve an *accurately* weighed sample of approximately 5 g of racemic 1-phenylethanamine in 35 mL of methanol and determine its specific rotation [α] using a polarimeter and Equation 7.8. The amounts of 1-phenylethanamine and methanol specified here give satisfactory results when a polarimeter sample tube with a light path of 2 dm or more and a volume of 35 mL or less is used. If a tube of shorter light path and/or larger volume is used, either increase the amount of racemic 1-phenylethanamine or combine your resolved product with that of other students to give a solution having sufficient concentration to allow accurate measurement of the optical rotation. Consult with your instructor about the type of polarimeter sample tube that is available.

Diastereomer Formation and Isolation Place 15.6 g of (+)-tartaric acid and 210 mL of methanol in the Erlenmeyer flask and heat the mixture to boiling using *flame-less* heating. To the stirred hot solution, *cautiously* add the solution recovered from the polarimeter and 7.5 g of racemic 1-phenylethanamine. Allow the solution to cool slowly to room temperature and to stand *undisturbed* for 24 h or until the next laboratory period.★ The amine hydrogen tartrate should separate in the form of white prismatic crystals. If the salt separates in the form of needlelike crystals, the mixture should be reheated *until all* the crystals have dissolved and then allowed to cool slowly. If any prismatic crystals of the salt are available, use them to seed the solution.

Collect the crystals of the amine hydrogen tartrate by vacuum filtration using a Büchner funnel (Fig. 2.54); wash them with a small volume of *cold* methanol. Do *not* discard the filtrate (see Wrapping It Up). Dissolve the crystals in four times their weight of water, and add 8 mL of 14 *M* sodium hydroxide solution. Extract the resulting mixture with four 35-mL portions of diethyl ether. Do *not* discard the aqueous solution (see Wrapping It Up). Wash the combined ethereal extracts with 25 mL of saturated aqueous sodium chloride, and then dry the ethereal solution over several spatula-tips full of *anhydrous* sodium sulfate for 10–15 min; swirl the flask; if the solution remains cloudy after this period of time, add additional portions of sodium sulfate. Decant or filter the dried solution from the desiccant.★

Remove the diethyl ether from the solution by simple distillation. (Alternatively, use one of the techniques discussed in Section 2.29.) Then distill the residual 1-phenylethanamine under vacuum with a shortpath apparatus (Fig. 2.37b). It will be necessary to use electrical heating for the distillation because the amine has a boiling point of 94–95 °C (28 torr).

Analysis Determine the yield of the resolved 1-phenylethanamine (**20**). Combine your product with that of another student to give a total of 3 g of **20**. Weigh the combined product accurately, and then transfer it *quantitatively* into about 30 mL of methanol. Measure the volume of the methanolic *solution* accurately. It is the *weight,* in grams, of the 1-phenylethanamine divided by the *volume* of the *solution,* in mL, that provides the concentration *c* for Equation 7.8. Transfer the solution to the polarimeter sample tube, measure the observed rotation, and determine the specific rotation of the sample by using Equation 7.8. Optically pure (*R*)-(+)-1-phenylethanamine has $[\alpha]_D^{20}$ + 30° (*c* = 10, CH₃OH).

WRAPPING IT UP

Transfer the recovered *diethyl ether* to a container for nonhalogenated organic liquids, and put the *methanolic filtrate* from isolation of the less soluble amine tartrate either in this same container or in a special container from which the more-soluble amine tartrate could be recovered; consult your instructor for specific directions. Place the *methanolic solution of 1-phenylethanamine* either in a container so labeled or in the one labeled for nonhalogenated organic liquids. Transfer the *sodium hydroxide solution* into a container so labeled; the tartaric acid salt it contains can be recovered.

EXERCISES

1. The agent used in this experiment to resolve racemic 1-phenylethylamine is (+)-tartaric acid (**19**).

19
(+)-Tartaric acid

   a. How many stereocenters does **19** contain?

   b. Draw *all* possible stereoisomers for **19** and indicate whether each will be optically active or not. Indicate whether the possible pairs of stereoisomers of tartaric acid are enantiomers, diastereomers, or *meso*.

   c. Which, if any, of the other stereoisomers of **19** might be used to resolve racemic 1-phenylethylamine?

   d. Would there be any possible advantage to using a different stereoisomer of **19** to resolve racemic 1-phenylethylamine? Explain your answer.

   e. Is the optical purity of the resolving agent important? Explain your answer.

2. Consider the reaction of racemic 1-phenylethylamine with racemic tartaric acid.

   a. How many stereochemically distinct salts will be formed?

   b. What is the stereochemical relationship (i.e., enantiomeric, diastereomeric, or identical) between the different possible pairs of products formed in this reaction?

   c. What pairs of salts should be separable by crystallization and what pairs would be inseparable by crystallization?

3. When a label is prepared for the product that has been isolated, it should include the magnitude of the specific rotation as well as its sign. Suppose the observed rotation was found to be 180°. How could you determine whether the rotation was (+) or (−)?

4. Given the value for the specific rotation of optically pure 1-phenylethanamine, calculate the optical purity of the sample of resolved amine for which you obtained a rotation.

5. How might the enantiomeric purity of the product be increased?

6. Describe clearly the point in the experimental procedure at which the major part of the other enantiomer was removed from your product.

7. How could the (+)-tartaric acid be recovered so that it can be reused to resolve additional racemic amine?

8. The absolute configuration of one of the stereocenters in (+)-tartaric acid is *R*. Explain why the absolute configuration at the other stereocenter must also be *R* and provide a perspective drawing of the (*R, R*)-enantiomer of tartaric acid.

9. The absolute configuration of (+)-1-phenylethanamine is *R*. Make a perspective drawing of this configuration.

10. Suppose you had prepared a racemic organic *acid* and it was necessary to resolve it. Outline a procedure to illustrate how this could be done.

CHAPTER 8

Spectral Methods

We live in a world of electromagnetic radiation, which have energies ranging from those carried by γ- and X-rays to those associated with radio waves and microwaves. We're worried about damage to ourselves and our environment that might be caused by undesired ultraviolet radiation, which now penetrates the atmosphere because of the "ozone hole" and ozone depletion caused by the presence of man-made halocarbons in the stratosphere. And what about the concerns that the radiation from cell phones, which seem to be glued to the ears of many of our friends, may cause cancer? Yes, radiation may be harmful, but it obviously has myriad benefits as well. Where would we be without solar radiation, for example? You are surely aware of many practical applications of radiation of various wavelengths. For instance, *infrared* and *microwave* radiation are used for heating food, and radio waves are the key to *magnetic resonance imaging* (MRI), a noninvasive technique for assessing soft tissues in the body as described in the Historical Highlight *Medical Diagnostics via Nuclear Magnetic Resonance Spectroscopy,* which is available online. A variety of spectroscopic techniques that depend on radiation of specific wavelengths are described in this chapter, and you will soon see that they comprise an important component in the arsenal of methods that we have for analyzing organic compounds.

8.1 INTRODUCTION

As a beginning organic chemist, you will first master the skills and techniques for executing organic reactions and isolating the products in pure form. The structures of the starting materials and products that you encounter will usually be known, so you will not be faced with the necessity of determining the structures of your products. However, you may be required to determine the structure of a compound whose structure is known by your instructor but not by you. Indeed, solving the structures of compounds is a challenge commonly faced by organic chemists, who must identify compounds that are isolated from new chemical reactions or from natural sources.

Historically, the method for determining the exact structures of organic compounds was a difficult and time-consuming undertaking. A variety of solubility and classification tests were first performed to identify the functional groups present (Chap. 25, which is available online only). Typically, the structure of the unknown substance was then elucidated by performing a series of chemical reactions to convert it into another compound whose structure was already known. Fortunately, powerful spectroscopic methods are now available that have greatly simplified the task of structural elucidation. Perhaps the techniques of greatest importance to the practicing organic chemist are **infrared** (IR) and **nuclear magnetic resonance** (NMR)

235

spectroscopy, although **ultraviolet-visible** (UV-Vis) spectroscopy, **mass spectrometry** (MS), and **X-ray diffractometry** are also widely used for structure determination. With the notable exception of X-ray diffractometry, unequivocal definition of a structure is seldom possible by performing only a single type of spectroscopic analysis; rather, a combination of different analyses is generally required. We emphasize IR and NMR spectroscopy (Secs. 8.2 and 8.3, respectively) in this textbook because the necessary instrumentation is commonly available to students in the introductory organic laboratory course. However, in addition to these spectroscopic techniques, UV-Vis spectroscopy (Sec. 8.4) and MS (Sec. 8.5) will be discussed because of their importance to the practice of organic chemistry. You can find a number of invaluable resources and **spectral databases** on the Web. The URL of one of the most extensive of these (i.e., Spectral Database for Organic Compounds, SDBS) is found at http://sdbs.db.aist.go.jp/sdbs/cgi-bin/cre_index.cgi.

Most of the spectral methods described in this chapter depend on the absorption of radiation to produce an excitation that transforms the original molecule to a state of higher energy. The particular type of excitation that occurs is determined by the amount of energy associated with the radiation. The relationship between the energy of radiation (in kcal/mol), E, and its frequency, v, or wavelength, λ, is expressed in Equation 8.1, where N is Avogadro's number (6.023×10^{23}), h is Planck's constant (1.5825×10^{-37} kcal/sec), c is the velocity of light (2.998×10^{14} μm/sec), and v and λ are in Hertz (1.0 Hz = 1.0 cycle/sec) and micrometers (μm = 10^{-6} m), respectively. Because of the reciprocal relationship between v and λ, the *higher* the frequency or the *lower* the wavelength, the *higher* is the energy associated with the radiation.

$$E = Nhv = Nhc/\lambda \qquad (8.1)$$

In IR spectroscopy, the absorption of radiation corresponds to transitions among different molecular **vibrational-rotational levels** within the same electronic state, normally the ground state, of the molecule; these excitations require energies of 1–36 kcal/mol. The data from IR spectroscopy are most useful for determining the presence or absence of **functional groups** in a compound. For instance, examining the appropriate regions of an IR spectrum will show whether or not carbon-carbon multiple bonds, aromatic rings, carbonyl groups, or hydroxyl groups are present. This technique does not give quantitative information regarding the elemental composition of a compound, nor does it allow assignment of an exact structure to an unknown compound unless the IR spectrum of the unknown is shown to be *identical* to that of a known compound.

The phenomenon associated with NMR spectroscopy involves transitions among **nuclear spin states** in a magnetic field; these excitations are of very low energy, less than 1 cal/mol. NMR spectroscopy provides information regarding the number and arrangement of various atoms in a molecule, although not all types of nuclei found in organic molecules are detectable by this method. Application of this technique to the analysis of hydrogen atoms in molecules, which is called ^{1}H NMR spectroscopy, allows determination of the number of **nearest neighbors** to a hydrogen atom as well as the presence of certain functional groups, such as carbon-carbon multiple bonds and carbonyl groups. A companion method involving carbon-13 atoms, and abbreviated as ^{13}C NMR spectroscopy, permits analysis of the different types of carbon atoms present in a molecule and may be useful for detecting certain functional groups whose presence is not indicated in the ^{1}H NMR spectrum.

To give you valuable experience in interpreting IR and NMR spectroscopic data, reproductions of these spectra are provided for most of the reagents and products encountered in the preparative experiments found in this book, as are exercises associated with these spectra. Moreover, the spectra are available online (refer to the online resources for this book), and tools are available there for analyzing them in more detail than is possible for the reproductions. After you understand the basis for these two techniques, as discussed in the next two sections, you should use the spectra on the website to help you become an expert in interpreting IR and NMR data.

Ultraviolet-visible (UV-Vis) spectroscopy is based on electronic transitions, processes that require higher energies than either IR or NMR spectroscopy. For example, visible spectroscopy involves electronic excitations having energies of 38–72 kcal/mol, whereas those associated with ultraviolet spectroscopy are still greater, being 72–143 kcal/mol. UV-Vis spectroscopy is most useful for detecting the presence of systems of π-bonds that are **conjugated**, that is, systems in which two or more multiple bonds are directly linked by single bonds, as in 1,3-dienes, aromatic rings, and 1,3-enones. Because of this structural limitation, this technique is of less value to modern organic chemists than either IR or NMR spectroscopy. It finds use in the study of the kinetics of chemical reactions, however, because of the rapidity with which the spectra can be obtained.

In Figure 8.1 we summarize the relationships between the energies associated with IR, NMR, and UV-Vis spectroscopies, the corresponding wavelengths of light providing these energies according to Equation 8.1, and the molecular effect associated with each technique. The microwave region is included for the sake of completeness, although organic chemists do not typically characterize compounds by microwave spectroscopy. However, they do use the energy associated with radiation in the microwave region to heat reaction mixtures, using commercially available microwave ovens to do so (Sec. 2.9).

Mass spectrometry is a useful technique for determining the molar masses and elemental compositions of organic molecules. This technique involves ionizing the sample in a high vacuum and then detecting the gaseous ions after they have passed through a magnetic field, which "sorts" the ions according to their relative masses and charges. Such information is of great value in identifying the component being analyzed. Because MS requires that the sample be ionized, the energies required

Increasing energy

Spectral region	Ultraviolet	Visible	Infrared	Microwave	Radio
Wavelength (µm)	0.20–0.40 (200–400 nm)	0.40–0.75 (400–750 nm)	0.75–50	50–5×10^4	5×10^4 – 2×10^9
Energy (kcal/mol)	143–71.5	71.5–38.1	38.1–0.57	0.57–5.7×10^{-4}	5.7×10^{-4} – 1.4×10^{-8}
Molecular effect.	Promotion of electron from filled to unfilled orbital, e.g., $\pi \rightarrow \pi^*$.	Same as UV except that absorption of light is visible (colored) to the human eye.	Stretching and bending of interatomic bonds.	Molecular rotations.	Realignment of nuclear spins in a magnetic field.

Figure 8.1

A portion of the electromagnetic spectrum.

are much higher than those used in IR, NMR, and UV-Vis, none of which involves ionization. For example, producing ionization by bombarding the sample with an electron beam (Eq. 8.2) requires electrons having energies of some 70 electron volts (1600 kcal/mol).

$$P + e^- \rightarrow P^{\ddagger} + 2e^- \tag{8.2}$$

Molecule M+ (Parent ion)

8.2 INFRARED (IR) SPECTROSCOPY

Introduction

The span of the infrared portion of the electromagnetic spectrum is from about 0.8 μm (3.8 $\times$ 10^{14} Hz) to 50 μm (6 $\times$ 10^{12} Hz). Within this range, the region from 2.5 μm to 20 μm is of most interest to organic chemists because most of the common functional groups absorb IR radiation in this range. Although wavelength, in μm, was once used to express the location of an IR absorption, current practice is to use another unit, **wavenumber** or **reciprocal centimeter,** written as $\tilde{\nu}$ or cm^{-1}. The relationship between wavelength and wavenumber is expressed by Equation 8.3. Thus, the normal IR range of 4000–500 cm^{-1} corresponds to 2.5–20 μm. As with frequency, ν, the energy associated with radiation *increases* as $\tilde{\nu}$ increases. It is imprecise to call wavenumbers "frequencies," despite the mathematical relationship between the two, although we do so for convenience.

$$\text{Wavenumber (cm}^{-1}) = 10{,}000/\lambda \ (\mu\text{m}) \tag{8.3}$$

Principles

The effect of IR radiation on an organic molecule may be qualitatively understood by imagining that the covalent bonds between atoms are analogous to groups of molecular springs that are constantly undergoing **stretching, twisting**, and **bending** (Fig. 8.2). The characteristic frequencies, and thus the energies, of these vibrations depend on the masses of the atoms involved and the type of chemical bond joining the atoms. Indeed, a good approximation of the *stretching* frequency of bonds can be obtained by applying **Hooke's law** for simple **harmonic oscillators** (Eq. 8.4). The value of the **force constant** k, which essentially reflects the strength of the bond between the atoms A and B, is about 5 $\times$ 10^5 dyne cm^{-1} for single bonds and twice and three times this figure for double and triple bonds, respectively.

$$\tilde{\nu} = \frac{1}{2\pi c}\sqrt{\frac{k}{m^*}} \tag{8.4}$$

where $\tilde{\nu}$ = frequency of absorption in cm^{-1}
c = speed of light
k = force constant of bond
m^* = reduced mass of atoms joined by bond = $\dfrac{m_A m_B}{m_A + m_B}$

Figure 8.2
Some vibrational modes of covalently bonded atoms.

| Symmetric stretch | Asymmetric stretch | Symmetric in-plane bend | Asymmetric in-plane bend |

The validity of Equation 8.4 for predicting stretching frequencies is illustrated by its use to give $\tilde{\nu}_{C-H}$ as 3040 cm^{-1} (see Exercise 2); the experimental value for this vibration is in the range of 2960–2850 cm^{-1}. Other consequences of Equation 8.4 are noteworthy. For instance, experimentally the wavenumber for stretching of a C–D bond is about 2150 cm^{-1}, a lower value than that for a C–H bond; this difference reflects the effect of atomic mass on the absorption frequency as accounted for by m^* in Equation 8.3 (see Exercise 3). Similarly, the appearance of the stretching vibration of a carbon-carbon double bond at about 1620 cm^{-1} and that of a carbon-carbon triple bond at approximately 2200 cm^{-1} is because k (Eq. 8.4) approximately doubles in going from a double to a triple bond; this increases the energy required to stretch the bond between the atoms.

Because different energies are required to cause the molecular vibrations associated with various types of bonds, each functional group has a characteristic vibrational frequency. When the energy of the radiation corresponds to that associated with a particular stretching or bending vibration, energy from the incident light will be absorbed. The efficiency of this absorption depends on several factors that are beyond the scope of the present discussion. Suffice it to say that the greater the change in the **dipole moment** associated with a particular vibration, the more efficient the transfer of energy to the molecule and the stronger the observed absorption.

Most modern IR instruments are designed to measure the amount of light transmitted *through* the sample, the **transmittance**, rather than the **absorbance** of light *by* the sample. Consequently, the **infrared spectra** portrayed in this textbook are plots of transmittance, expressed in *percent* as defined in Equation 8.5, versus the wavenumber of the **incident radiation** (I_0), as seen in Figure 8.3. The lower the %T, the greater the amount of incident radiation being absorbed by the sample. For example, in the IR spectrum shown in Figure 8.3, the strongest absorption occurs at about 1675 cm^{-1}. Although you might consider this absorption to be a minimum, it is referred to as a "peak" in the parlance of IR spectroscopy. The energy required to produce the observed molecular excitations *decreases* in going from left to right in the spectrum.

$$\%T = I/I_0 \times 100 \tag{8.5}$$

I = intensity of radiation transmitted through sample at a particular wavelength

I_0 = intensity of radiation incident on sample at same wavelength

Figure 8.3

IR spectrum of 3-methyl-2-cyclohexenone (neat)

What happens to a molecule once it has absorbed infrared radiation? The absorption causes an increase in the total energy of the molecule, thereby increasing the **amplitude** of the particular molecular vibration that was responsible for absorption of the radiation. However, no irreversible change occurs in the molecule, as the excess energy is quickly dissipated in the form of heat. Obtaining an IR spectrum thus does *not* permanently change the molecular structure.

Practical Considerations

Transmittance

As a general rule, the IR spectrum of a liquid or solid sample should show detectable peaks for relatively weak absorptions, but there should be no absorptions so strong that they go off-scale. Experimentally, a good operating principle is to modify the sample being analyzed so the *most* intense peak in its spectrum is close to 0%*T*; this maximizes the possibility that even weak peaks will be observable. This practice is exemplified in Figure 8.3, where the strongest absorption has a %*T* of nearly zero. The two experimental variables determining the absolute intensities of the peaks in an IR spectrum are the concentration of the sample and the length of the cell containing it. These factors define the number of molecules that are in the path of the radiation as it passes through the sample; the more molecules there are, the greater the intensity of an absorption. With modern IR spectrometers, the intensities of the peaks on the printed spectrum may also be amplified electronically.

Cells

There are two general types of sample cells used for IR spectroscopy: cells in which the pathlength or thickness of the sample is fixed, and those in which the pathlength is variable. The latter type of cell is expensive, but a simple version of it is two transparent salt plates, or windows, between which the sample is sandwiched (Figs. 8.4 and 8.5). The cell is easily prepared by putting a drop of sample on one plate and placing the second plate on top of the film that results. The assembled plates are then put into a holder that fits into the spectrometer. If a sample is too thick, some of it may be removed or the plates may be squeezed together—*carefully* to avoid breaking them—to produce a thinner film. However, if the sample is too thin, the plates can be separated and more sample can be added. The plates are then rejoined so as to leave the film of the sample as thick as possible.

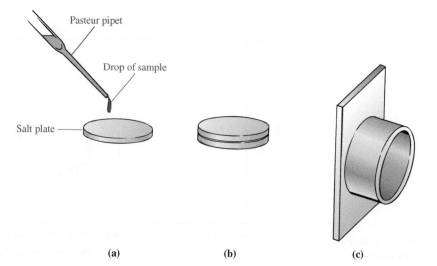

Pasteur pipet

Drop of sample

Salt plate

(a)   (b)   (c)

Figure 8.4
Simple demountable IR cell.
(a) Salt plate for sample.
(b) Assembled pair of salt plates.
(c) Cell holder.

Figure 8.5

A demountable IR cell. (a) Detail showing correct assembly of cell. (b) Completely assembled cell.

Front plate

Neoprene gasket

Window

Sample

Window

Neoprene gasket

Back plate

(a)

(b)

A **fixed-thickness cell** is constructed of two transparent salt plates separated by a plastic or metal gasket, which defines the thickness of the sample contained in the cell. One such cell is shown in Figure 8.6. A syringe is normally used to load the cell, which should be laid at an angle on top of a pen or pencil to keep an air bubble from forming within the cell as the sample is introduced. If a bubble is observed, more sample must be added until the bubble disappears.

Plugs

Pencil

Opening for filling cell cavity with sample

Teflon or metal spacer

Opening to cell cavity

Window

Metal plates held together with screws

Salt plates

(a)

Metal plates held together with screws

Plugged opening to cell cavity

Window

Teflon or metal spacer

Salt plates

Plugged opening for filling cell cavity with sample

(b)

Figure 8.6

A fixed-thickness IR cell. (a) Cell with plugs removed for cleaning or filling. (b) Completely assembled cell.

Proper care and handling of the windows used in IR cells are *very* important. The plates are generally a clear fused salt such as sodium chloride or potassium bromide. If these plates are exposed to moisture, the windows will become cloudy. Cloudy plates decrease the amount of light transmitted through the cell, so you should avoid breathing on or touching the faces of the windows when handling assembled IR cells or the salt plates themselves. Moreover, any substances, including samples, that come in contact with the cells must be *dry*. This means that any materials used to clean an IR cell or salt plates must also be *dry*. Suitable solvents for cleaning cells include *absolute* ethanol, dichloromethane, and chloroform. It goes without saying that woe be to the student who cleans an IR cell with water; this is a very expensive way to make saline solution!

Permanent fixed-thickness cells are dried by passing a slow stream of *dry* nitrogen or air through them, whereas demountable cells and salt plates may also be dried by *gently* wiping or blotting them with soft laboratory tissue to avoid scratching the faces of the cells or plates. The clean, dry cells or plates are stored in a desiccator for protection from atmospheric moisture.

Liquid Samples

The IR spectrum of a liquid sample is most conveniently obtained on the pure liquid, called **a neat sample**. Since concentration is not a variable in such samples, peak intensities may be modified only by changing the pathlength. This is done in one of two ways. One approach is to use a fixed-thickness cell with a pathlength of 0.020–0.030 mm. This short pathlength generally provides a spectrum in which strong absorptions, like those associated with carbonyl groups, give peaks that remain on the chart paper, rather than going to $0\%T$. If the spectrum is not suitable, however, different fixed thicknesses may be used. The other option is to use salt plates and to adjust the thickness of the film between them by gently pressing the plates together. Salt plates are *not* appropriate for use with volatile samples because some or all of the sample will evaporate as the spectrum is run, and inaccurate relative peak intensities will result. Alternatively, a liquid sample may be analyzed as a solution or absorbed into a microporous film, as described in the next section.

Solid Samples

There are several options for preparing a solid sample for IR spectroscopy. These include the preparation of **solutions, castings, KBr pellets**, and **microporous polyethylene films**, or the use of an accessory for obtaining **attenuated total reflectance** IR spectra. Each of these techniques is described in turn.

1. **Solution**. A common technique for obtaining an IR spectrum of a solid is to analyze it in solution. The preferred solvents are chloroform and carbon disulfide, in that order. The solubility of the sample in the selected solvent must be in the range of 5–10 wt % because solutions of such concentration generally provide suitable spectra in a fixed-thickness cell having a pathlength of 0.1 mm.

    Two fixed-thickness cells having identical pathlengths, so-called **matched cells**, are used if a *double-beam* IR instrument is available. One cell contains the solution and is referred to as the **sample cell**, whereas the other contains pure solvent and is the **reference cell**. The purpose of the reference cell is to allow cancellation of the absorption bands due to the solvent itself so that they do not appear in the spectrum.

If a **Fourier transform (FT) IR** instrument is used, a reference cell is not required because the IR spectrum of the pure solvent is previously stored in the memory of the spectrometer. The software that controls the instrument then *subtracts* those absorptions that are due to the solvent from the spectrum. However, as is evident from the IR spectrum in Figure 8.7a, this subtraction may be imperfect; this is most notable at about 750 and 1220 cm^{-1}, where the solvent chloroform absorbs strongly.

2. **Casting.** A solution of a solid may be "cast" onto a sodium chloride disc to provide a thin film of solid upon evaporation of the solvent. Appropriate solvents include diethyl ether, dichloromethane, and chloroform.

   To prepare a casting, about two drops of a solution are applied evenly to the disc, and the solvent is allowed to evaporate. The disc is then placed in a holder in the spectrometer, and the spectrum is obtained (Fig. 8.7b). If the film is too thin to provide a useful spectrum, additional solution may be added to the disc or a more concentrated solution of the solid may be used. On the other hand, if the sample is so thick that the absorption bands are too intense, the disc may need to be cleaned with a solvent and the casting procedure repeated with a more dilute solution. It is important to remember that sodium chloride is **hygroscopic**, so water or 95% ethanol should *not* be used for preparing solutions or cleaning the discs.

3. **Potassium Bromide (KBr) Pellet.** Potassium bromide may be fused to produce a nearly transparent salt plate. This is done by subjecting the KBr to high pressures, in the range of 5–10 tons/in^2, to generate the heat required for the salt to become plastic and flow; releasing the pressure then provides a clear plate. A KBr pellet containing the compound may thus be prepared by intimately mixing about 1 mg of sample with 100 mg of *dry,* powdered KBr and subjecting the resulting mixture to high pressure in a die. Mixing is done with a mortar and pestle made of agate rather than ceramic materials, because ceramics may be abraded and contaminate the pellet with nontransparent particles. Once a pellet having satisfactory transparency has been produced, the IR spectrum is obtained. If the intensity of the bands in the spectrum is unsatisfactory, another pellet must be prepared in which the concentration of the solid sample has been modified accordingly.

   Potassium bromide is *extremely* hygroscopic, so it must be stored in a tightly capped bottle in a desiccator except when needed for preparation of a pellet. Even with the greatest precautions, it is difficult to prepare a pellet that is completely free of water. Possible contamination of a KBr pellet with water may cause problems in interpreting IR absorptions in the range 3600–3200 cm^{-1} because the O–H stretching vibration of water occurs in this region. Consequently, moisture in the pellet may partially or completely obscure the absorptions due to the O–H and N–H stretching vibrations of alcohols and amines, which occur in this same region (Fig. 8.7c). Further, the appearance of an absorption in the aforementioned range may lead to an erroneous conclusion about the possible presence of O–H and N–H groups in the sample.

4. **Microporous Polyethylene Films.** Disposable IR cards are available to which samples can be directly applied for infrared analysis (Fig. 8.8). These convenient IR cards have two 19-mm circular apertures containing a thin microporous film of chemically resistant polyethylene. Cards having microporous polytetrafluoroethylene are also available for special applications. The cards with polyethylene films may be used for infrared analysis from 4000 cm^{-1} to 400 cm^{-1} except

(a) In chloroform.

(b) As casting on NaCl disc.

(c) As KBr pellet.

(d) On microporous polyethylene film (IR card).

Figure 8.7
IR spectra of acetanilide.

(e) Using ATR device.

for the region of aliphatic C–H stretching that occurs between 3000 cm^{-1} and 2800 cm^{-1}. In this region of an FT-IR spectrum, there are several sharp spikes at about 2840 cm^{-1} and 2910 cm^{-1} that result from the imperfect subtraction of the stored spectrum of the polyethylene film, as shown in Figure 8.7d.

To prepare a sample for IR analysis, two drops of a solution of the compound in a volatile organic solvent are simply applied to the film using a pipet or syringe. Because of the unique nature of the microporous films, the solution is absorbed into the film. The solvent evaporates within minutes at room temperature, leaving the solute evenly dispersed in the polyethylene matrix. Nonvolatile liquid samples may also be applied to the film. The cards are then mounted directly into the slot on the spectrometer that is normally used for holding reusable cells. Because the cards are expensive, you should use them twice before either taping them on your laboratory notebook or discarding them.

5. **Reflectance Spectroscopy.** The technique of **reflectance spectroscopy** is another means for obtaining IR spectra of solid and liquid samples, and we'll discuss one form of it, **attenuated total reflectance** (ATR). This method is based on

Figure 8.8
Polyethylene IR sample cards for two samples.

monitoring radiation as it is *reflected from* a sample. The sample is placed on or between plates that focus the incident light on it and allow passage of the reflected light to the detector. The reflected light, called the *evanescent wave*, behaves as if it has penetrated the sample to depths varying from a fraction of a wavelength to several wavelengths. If the sample absorbs some of the evanescent wave, the intensity of the wave is decreased (attenuated) at the specific wavelengths that are characteristic of the absorption bands of the sample, much the same as when transmission of light *passing through* a sample is being measured (Figs. 8.7a–d), as discussed previously.

Specialized adapters (Fig. 8.9) for this technique fit into the sample compartment of the FT-IR instrument, and the resulting spectra are called ATR spectra. As with transmission spectra, they are plotted as %T versus cm^{-1}. The appearance of ATR spectra is similar but not identical to transmission spectra, as seen by comparing the ATR spectrum of acetanilide (Fig. 8.7e) with those of Figures 8.7a–d. Although the position and shapes of the peaks are the same, the intensities of the peaks are not, and the overall range of %T may be less with ATR spectra. The peak intensities are different in the two techniques, because ATR spectra depend on such parameters as index of refraction, the angle of incident light, and the depth of wave penetration as a function of wavelength, none of which is important in transmission spectra. The range of %T for an ATR spectrum is also less because of the limited depth of penetration of the evanescent wave into the sample and on other instrumental parameters. However, the shapes and positions of the peaks are unchanged, so the protocol for interpreting an ATR spectrum is the same as that for a transmission spectrum.

To obtain an ATR spectrum using an adapter such as that shown in Figure 8.9, the sample is placed on top of the sample stage and the press is lowered to produce intimate contact between the sample and the crystal plate. The apparatus is now set up to obtain the desired spectrum.

Spectral Calibration

If you are not using an FT-IR spectrometer, it is good laboratory practice to calibrate the experimentally obtained IR spectra against a standard that has absorption bands at known positions. Calibration is necessary because problems such

Figure 8.9
Device for obtaining ATR spectra.

Press assembly

Sample stage

© 2014 PIKE Technologies

as misalignment of the chart paper in the instrument or mechanical slippage may result in absorptions appearing at wavenumbers that are incorrect. Polystyrene (Fig. 8.10) is a commonly used standard and is typically available as a thin, transparent film in a cardboard holder that fits into the spectrometer. After the spectrum of the sample is obtained, the standard is inserted in place of the sample cell; the reference cell, if used, is removed; and at least one band of the standard is recorded on the spectrum of the sample. In the case of polystyrene, the absorption at 1601 cm^{-1} is typically recorded, although other bands may be used.

Once the reference peak of the standard is measured, the reported positions of the absorptions of the sample are adjusted accordingly. For example, if the 1601 cm^{-1} peak of polystyrene appears at 1595 cm^{-1} on the spectrum, the wavenumbers for the bands in the sample are corrected by *adding* 6 cm^{-1} to the values recorded on the chart paper. Because the source of the error may vary over the range, it is not strictly correct to use the same correction factor throughout the spectrum, but for most purposes this is a satisfactory protocol. For more precise work, calibration peaks should be recorded at several different locations in the IR spectrum.

Figure 8.10
IR spectrum of polystyrene film.

Use and Interpretation of IR Spectra

An IR spectrum of a compound may be used to *identify* the *functional groups* present in a molecule, *evaluate* the *purity* of a known compound, and *determine* the *structure* of an unknown substance. These valuable applications are discussed in the following paragraphs.

Analysis of the IR spectrum of an unknown compound provides useful information regarding the presence or absence of different functional groups. As a general rule, absorptions in the approximate range of 4000–1250 cm^{-1}, the so-called **functional group region**, involve vibrational excitations that are associated with particular functional groups. Thus, the stretching mode of the carbonyl group of a ketone appears in the region of 1760–1675 cm^{-1}, and that of a carbon-carbon double bond occurs in the vicinity of 1680–1610 cm^{-1}. A compilation of various functional groups and the region of the IR spectrum in which they absorb is provided in Table 8.1. Table 8.2 is a more detailed version of Table 8.1. Note that the *absence* of an absorption may have as much significance as the *presence* of one. For example, if there is no strong band in the carbonyl region of the spectrum, functional groups such as aldehyde, ketone, amide, and ester are probably absent.

Assessing purity involves comparing the spectrum of the sample with that of the pure material; the observation of extra bands in the sample is a sign that contaminants are present. On the whole, the technique is not sensitive to low levels

Table 8.1 *Abbreviated Table of Infrared Absorption Ranges of Functional Groups*

Bond	Type of Compound	Frequency Range, cm^{-1}	Intensity
C–H	Alkanes	2850–2970	Strong
C–H	Alkenes	1340–1470	Strong
		3010–3095	Medium strong
		675–995	
C–H	Alkynes (–C≡C–H)	3300	Strong
C–H	Aromatic rings	3010–3100	Medium
		690–900	Strong
O–H	Monomeric alcohols, phenols	3590–3650	Variable
	Hydrogen-bonded alcohols, phenols	3200–3600	Variable, sometimes broad
	Monomeric carboxylic acids	3500–3650	Medium
	Hydrogen-bonded carboxylic acids	2500–3200	Broad
N–H	Amines, amides	3300–3500	Medium
C=C	Alkenes	1610–1680	Variable
C=C	Aromatic rings	1500–1600	Variable
C≡C	Alkynes	2100–2260	Variable
C–N	Amines, amides	1180–1360	Strong
C=N	Nitriles	2210–2280	Strong
C–O	Alcohols, ethers, carboxylic acids, esters	1050–1300	Strong
C=O	Aldehydes, ketones, carboxylic acids, esters	1675–1760	Strong
NO$_2$	Nitro compounds	1500–1570	Strong
		1300–1370	Strong

of impurities, however, because levels of contamination in the range of 1–5% may easily go undetected.

Determining the structure of an unknown depends on the generally accepted premise that *if the IR spectra of two samples are superimposable, then the samples are identical.* The criterion of **superimposability** is a *very* demanding requirement because it means that the intensity, shape, and location of *every* absorption in the two spectra *must be identical.* This is difficult to achieve unless the *same* cells and instrument are used to obtain the IR spectra of *both* samples. Nevertheless, finding that the IR spectrum of an unknown material is very similar to the spectrum of a known material, even if the spectra are not obtained with the same instrument and cells, makes it probable that the structures of two samples are similar, if not identical. Alternative methods, such as NMR spectroscopy (Sec. 8.3), preparing derivatives (Secs. 25.6–18, which are available online only), and mixed melting points (Sec. 3.3), are then used to determine whether or not the samples are indeed identical.

Absorptions that occur in the region from 1250 cm^{-1} to 500 cm^{-1} usually arise from a complex *vibrational-rotational* excitation of the *entire* molecule. This portion of the spectrum is typically *unique* for a particular compound and is aptly described as the **fingerprint region**. Although the IR spectra of similar molecules may be comparable in the functional group region from 4000 cm^{-1} to 1250 cm^{-1}, it is highly unlikely that this similarity will extend throughout the fingerprint region, as is illustrated in Figure 8.11.

Figure 8.11

IR spectra of (a) 2,3-dimethylbutane and (b) 2-methylbutane (neat).

Table 8.2 *Detailed Table of Characteristic Infrared Absorption Frequencies*

The hydrogen stretch region (3600–2500 cm^{-1}). Absorption in this region is associated with the stretching vibration of hydrogen atoms bonded to carbon, oxygen, and nitrogen. Care should be exercised in interpreting very weak bands because these may be overtones of strong bands occurring at frequencies from 1800 cm^{-1} to 1250 cm^{-1}, which is one-half the value of the weak absorption. Overtones of bands near 1650 cm^{-1} are particularly common.

$\tilde{\nu}$ (cm^{-1})	Functional Group	Comments
(1) 3600–3400	O–H stretching Intensity: variable	3600 cm^{-1} (sharp) unassociated O–H, 3400 cm^{-1} (broad) associated O–H; both bands frequently present in alcohol spectra; with strongly associated O–H (CO$_2$H or enolized β-dicarbonyl compound), band is very broad (about 500 cm^{-1} with its center at 2900–3000 cm^{-1}).
(2) 3400–3200	N–H stretching Intensity: medium	3400 cm^{-1} (sharp) unassociated N–H, 3200 cm^{-1} (broad) associated N–H; an NH$_2$ group usually appears as a doublet (separation about 50 cm^{-1}); the N–H of a secondary amine is often very weak.
(3) 3300	C–H stretching of an alkyne Intensity: strong	The *complete* absence of absorption in the region from 3300 cm^{-1} to 3000 cm^{-1} indicates the absence of hydrogen atoms bonded to C=C or C≡C and *usually* signals the lack of unsaturation in the molecule. Because this absorption may be very weak in large molecules, care should be exercised in this interpretation. In addition to the absorption at about 3050 cm^{-1}, aromatic compounds frequently show *sharp* bands of medium intensity at about 1500 *and* 1600 cm^{-1}.
(4) 3080–3010	C–H stretching of an alkene Intensity: strong to medium	
(5) 3050	C–H stretching of an aromatic compound Intensity: variable; usually medium to weak	
(6) 3000–2600	OH strongly hydrogen-bonded Intensity: medium	A very broad band in this region superimposed on the C–H stretching frequencies is characteristic of carboxylic acids.
(7) 2980–2900	C–H stretching of an aliphatic compound Intensity: strong	As in previous C–H entries (3–5), *complete* absence of absorption in this region indicates absence of hydrogen atoms bound to tetravalent carbon atoms. The tertiary C–H absorption is weak.
(8) 2850–2760	C–H stretching of an aldehyde Intensity: weak	Either one or two bands *may* be found in this region for a single aldehyde function in the molecule.

The triple-bond region (2300–2000 cm^{-1}). Absorption in this region is associated with the stretching vibration of triple bonds.

$\tilde{\nu}$ (cm^{-1})	Functional Group	Comments
(1) 2260–2215	C≡N Intensity: strong	Nitriles conjugated with double bonds absorb at *lower* end of frequency range; nonconjugated nitriles appear at *upper* end of range.
(2) 2150–2100	C≡C Intensity: strong in *terminal* alkynes, variable in others	This band is absent if the alkyne is symmetrical and will be very weak or absent if the alkyne is nearly symmetrical.

Table 8.2 *Detailed Table of Characteristic Infrared Absorption Frequencies (Continued)*

The double-bond region (1900–1550 cm^{-1}). Absorption in this region is *usually* associated with the stretching vibration of carbon-carbon, carbon-oxygen, and carbon-nitrogen double bonds.

$\tilde{\nu}\,(cm^{-1})$	Functional Group	Comments
(1) 1815–1770	C=O stretching of an acid chloride Intensity: strong	Carbonyls conjugated with double bonds absorb at *lower* end of range; nonconjugated carbonyls appear at *upper* end of range.
(2) 1870–1800 and 1790–1740	C=O stretching of an acid anhydride Intensity: strong	*Both bands* are present; *each band* is altered by ring size and conjugation to approximately the same extent noted for ketones.
(3) 1750–1735	C=O stretching of an ester or lactone Intensity: very strong	This band is subject to all the structural effects discussed for ketones; thus, a conjugated ester absorbs at about 1710 cm^{-1} and a γ-lactone absorbs at about 1780 cm^{-1}.
(4) 1725–1705	C=O stretching of an aldehyde or ketone Intensity: very strong	This value refers to carbonyl absorption frequency of acyclic, nonconjugated aldehyde or ketone having no electronegative groups, for example, halogens, near the carbonyl group; because this frequency is altered in a predictable way by structural alterations, the following generalizations are valid: (a) *Effect of conjugation:* Conjugation of carbonyl group with an aryl ring or carbon-carbon double or triple bond *lowers* the frequency by about 30 cm^{-1}. If the carbonyl group is part of cross-conjugated system (unsaturation on each side of the carbonyl group), the frequency is lowered by about 50 cm^{-1}. (b) *Effect of ring size:* Carbonyl groups in six-membered and larger rings exhibit approximately the same absorption as acyclic ketones; carbonyl groups in rings smaller than six absorb at *higher* frequencies: For example, a cyclopentanone absorbs at about 1745 cm^{-1} and a cyclobutanone at about 1780 cm^{-1}. The effects of conjugation and ring size are additive: A 2-cyclopentenone absorbs at about 1710 cm^{-1}, for example. (c) *Effect of electronegative atoms:* An electronegative atom, especially oxygen or halogen, bound to the α-carbon atom of an aldehyde or ketone usually raises the position of the carbonyl absorption frequency by about 20 cm^{-1}.
(5) 1700	C=O stretching of an acid Intensity: strong	Conjugation *lowers* this absorption frequency by about 20 cm^{-1}.
(6) 1690–1650	C=O stretching of an amide or lactam Intensity: strong	Conjugation *lowers* the frequency of this band by about 20 cm^{-1}. The frequency of the band is *raised* about 35 cm^{-1} in γ-lactams and 70 cm^{-1} in β-lactams.
(7) 1660–1600	C=C stretching of an alkene Intensity: variable	Conjugated alkenes appear at *lower* end of range, and absorptions are medium to strong; nonconjugated alkenes appear at *upper* end of range, and absorptions are usually weak. The absorption frequencies of these bands are raised by ring strain but to a lesser extent than noted with carbonyl functions.
(8) 1680–1640	C=N stretching Intensity: variable	This band is usually weak and difficult to assign.

Table 8.2 *Detailed Table of Characteristic Infrared Absorption Frequencies (Continued)*

The hydrogen bending region (1600–1250 cm⁻¹). Absorption in this region is commonly due to bending vibrations of hydrogen atoms attached to carbon and to nitrogen. These bands generally do not provide much useful structural information. In the listing, the bands that are most useful for structural assignment are marked with an **asterisk**.

$\tilde{\nu}$ (cm⁻¹)	Functional Group	Comments
(1) 1600	NH$_2$ bending Intensity: strong to medium	In conjunction with bands in the 3300 cm⁻¹ region, this band is often used to characterize primary amines and amides.
(2) 1540	NH bending Intensity: generally weak	In conjunction with bands in the 3300 cm⁻¹ region, this band is used to characterize secondary amines and amides. This band, like the N–H stretching band in the 3300 cm⁻¹ region, may be weak in secondary amines.
(3)* 1520 and 1350	NH$_2$ coupled stretching bands Intensity: strong	This pair of bands is usually intense.
(4) 1465	CH$_2$ bending Intensity: variable	Intensity of this band varies according to the number of methylene groups present; the more such groups, the more intense the absorption.
(5) 1410	CH$_2$ bending of carbonyl containing component Intensity: variable	This absorption is characteristic of methylene groups adjacent to carbonyl functions; its intensity depends on the number of such groups present in the molecule.
(6)* 1450 and 1375	CH$_3$ bending Intensity: strong	The band of lower frequency (1375 cm⁻¹) is usually used to characterize a methyl group. If two methyl groups are bound to one carbon atom, a characteristic doublet (1385 cm⁻¹ and 1365 cm⁻¹) is observed.
(7) 1325	CH bending Intensity: weak	This band is weak and often unreliable.

The fingerprint region (1250–600 cm⁻¹). The fingerprint region of the spectrum is generally rich in detail and contains many bands. This region is particularly diagnostic for determining whether an unknown substance is identical to a known substance, the IR spectrum of which is available. It is not practical to make assignments to all these bands because many of them represent combinations of vibrational modes and therefore are very sensitive to the overall molecular structure; moreover, many single-bond stretching vibrations and a variety of bending vibrations also appear in this region. Suggested structural assignments in this region must be regarded as tentative and are generally taken as corroborative evidence in conjunction with assignments of bands at higher frequencies.

$\tilde{\nu}$ (cm⁻¹)	Functional Group	Comments
(1) 1200	—O Intensity: strong	It is not certain whether these strong bands arise from C–O bending or C–O stretching vibrations. One or more strong bands are found in this region in the spectra of alcohols, ethers, and esters. The relationship indicated between Intensity: strong structure and band location is approximate, and any structural assignment based on this relationship must be regarded as tentative. Esters often exhibit one or two strong bands between 1170 cm⁻¹ and 1270 cm⁻¹.
(2) 1150	C–O Intensity: strong	
(3) 1100	CH–O Intensity: strong	

Table 8.2 *Detailed Table of Characteristic Infrared Absorption Frequencies (Continued)*

$\tilde{\nu}\ (cm^{-1})$	Functional Group	Comments
(4) 1050	CH_2-O Intensity: strong	
(5) 985 and 910	 C–H bending Intensity: strong	This pair of strong bands characterizes a terminal vinyl group.
(6) 965	 C–H bending Intensity: strong	This strong band is present in the spectra of *trans*-1,2-disubstituted ethylenes.
(7) 890	$C=CH_2$ C–H bending Intensity: strong	This strong band characterizes a 1,1-disubstituted ethylene group and its frequency may be *increased* by 20–80 cm^{-1} if the methylene group is bound to an electronegative group or atom.
(8) 810–840	 Intensity: strong	Very unreliable; this band is not always present and frequently seems to be outside this range because substituents are varied.
(9) 700	 Intensity: strong	This band, attributable to a *cis*-1,2-disubstituted ethylene, is unreliable because it is frequently obscured by solvent absorption or other bands.
(10) 750 and 690	 C–H bending Intensity: strong	These bands are of limited value because they are frequently obscured by solvent absorption or other bands. They are most useful when independent evidence leads to a structural assignment that is complete except for positioning of aromatic substituents.
(11) 750	 C–H bending Intensity: strong	
(12) 780 and 700	 and 1,2,3- Intensity: very strong	

Table 8.2 *Detailed Table of Characteristic Infrared Absorption Frequencies (Continued)*

$\tilde{\nu}\,(cm^{-1})$	Functional Group	Comments
(13) 825	and 1, 2, 4- Intensity: strong	
(14) 1400–1000	C–F Intensity: strong	The position of these bands is sensitive to structure. As a result, they are not particularly useful because the presence of halogen is more easily detected by chemical methods. The bands are usually strong.
(15) 800–600	C–Cl Intensity: strong	
(16) 700–500	C–Br Intensity: strong	
(17) 600–400	C–I Intensity: strong	

It is important to learn to identify functional groups on the basis of the IR spectra of organic compounds. A useful way to do this is to examine the spectra of known compounds and correlate the key absorption bands with the functional groups responsible for them with the aid of Tables 8.1 and 8.2. Analysis of the IR spectra of the starting materials and products associated with the various experimental procedures in this textbook is an excellent way to sharpen your skills at interpreting such spectra.

EXERCISES

1. Derive the mathematical relationship between v and $\tilde{\nu}$ by using Equations 8.1 and 8.3 and point out why it is imprecise to use the terms "frequency" and "wavenumber" interchangeably.

2. Compute the reduced mass, m^*, for the C–H bond and use Equation 8.4 to confirm that $\tilde{\nu}_{C-H}$ is predicted to be 3040 cm^{-1}.

3. Compute the reduced mass, m^*, of a C–D bond and then use Equation 8.4 to determine the wavenumber at which the C–D stretching vibration occurs, assuming the corresponding vibration for the C–H group is 3000 cm^{-1}.

4. Assume that the force constant k for a carbon-carbon double bond is twice that of a carbon-carbon single bond. Use Equation 8.4 to determine the wavenumber at which the C–C stretching vibration occurs, assuming the corresponding vibration for the C=C group is 1640 cm^{-1}. Repeat this calculation for the case of a carbon-carbon triple bond, for which k is the appropriate multiple of that for the carbon-carbon double bond.

5. Water has its O–H stretching mode at about 3600 cm^{-1}. Use Equation 8.5 to calculate at approximately what wavenumber "heavy" water, D$_2$O, would have the corresponding O–D stretch. Show your work.

6. **a.** If the strengths of a carbon-carbon triple bond and a carbon-nitrogen triple bond are assumed to be the same, as reflected in the force constant k

(Eq. 8.4), predict which of the two functionalities would appear at the higher wavenumber in the IR spectrum and support your prediction.

b. Is this in agreement with experimental fact? (See Table 8.2.)

7. Define the following terms:

   a. percent transmittance   **e.** functional group
   b. neat liquid   **f.** stretching vibrational mode
   c. hygroscopic   **g.** bending vibrational mode
   d. fixed-thickness cell   **h.** superimposability (as applied to IR spectra)

8. Explain why the windows of IR cells and KBr pellets should not be exposed to moisture.

9. Explain why the windows of IR cells should not be subjected to excessive pressure.

10. Why and when is an IR spectrum calibrated?

11. Name two solvents that are appropriate and two that are inappropriate for cleaning the faces of IR cells.

12. What are the two main drawbacks to obtaining IR spectra of solids as KBr pellets?

13. When obtaining the IR spectra of liquids or solids using IR cards with microporous polyethylene films, there are two sharp spikes at 2840 cm^{-1} and 2910 cm^{-1}. What causes these spikes?

14. What portions of an IR spectrum are referred to as the functional group region and the fingerprint region?

15. What difficulties would be associated with obtaining a solution IR spectrum on a double-beam spectrometer without using a reference cell containing the pure solvent?

16. What problems might be encountered in attempting to make a KBr pellet of an organic solid that has crystallized as a hydrate?

17. From one of the sources of IR spectra listed in the references at the end of this section, obtain a spectrum of acetanilide contained in a Nujol mull. (A Nujol mull is a slurry prepared by intimately mixing a solid sample with mineral oil.) Compare this spectrum to the ATR spectrum of Figure 8.7e and explain the source(s) of any differences.

18. Figures 8.12–8.18 are the IR spectra of the compounds shown on the individual spectra. Assign as many of the major absorptions as possible in the spectrum of each compound, using Tables 8.1 and 8.2.

Refer to the online resources and use the spectra viewer and Tables 8.1–8.8 as needed to answer the blue-numbered questions on spectroscopy.

Figure 8.12

IR spectra of phenylethyne (neat).

Figure 8.13
IR spectrum of n-*butyl acetate.*

Figure 8.14
IR spectrum of N,N-*dimethylacetamide (neat).*

Figure 8.15
IR spectrum of benzoyl chloride (neat).

Figure 8.16
IR spectrum of diethyl ether (neat).

Figure 8.17
IR spectrum of benzonitrile (neat).

Figure 8.18
IR spectrum of N-methylaniline (neat).

REFERENCES

1. Colthup, N. B.; Daly, L. H.; Wiberley, S. E. *Introduction to Infrared and Raman Spectroscopy*, 3rd ed., Academic Press, New York, 1990.

2. Roelges, N. P. G. *A Guide to the Complete Interpretation of Infrared Spectra of Organic Compounds*, John Wiley & Sons, Chichester and New York, 1994.

3. Silverstein, R. M.; Webster, F. X.; Kiemle, D. *Spectrometric Identification of Organic Compounds*, 7th ed., John Wiley & Sons, New York, 2005.

4. Crews, P.; Rodriquez, J.; Jaspers, M. *Organic Structure Analysis*, Oxford University Press, New York, 1998.

5. Keller, R. J., ed. *Sigma Library of FT–IR Spectra*, Sigma Chemical Company, St. Louis, MO, 1986.

6. Pouchert, C. J., ed. *Aldrich Library of FTIR Infrared Spectra*, 2nd ed., vol. I–III, Aldrich Chemical Co., Milwaukee, WI, 1997. Compilation of over 18,000 spectra.

7. *Sadtler Standard Infrared Prism Spectra*, vol. 1–123, Sadtler Research Laboratories, Philadelphia, PA. Compilation of 91,000 spectra.

8. *Sadtler Standard Infrared Grating Spectra*, vol. 1–123, Sadtler Research Laboratories, Philadelphia, PA. Compilation of 91,000 spectra.

8.3 NUCLEAR MAGNETIC RESONANCE (NMR) SPECTROSCOPY

Introduction

Nuclear magnetic resonance (NMR) spectroscopy is probably the single most powerful spectroscopic method available to the modern organic chemist for analyzing compounds. This technique depends on the property of **nuclear spin** that is exhibited by certain nuclei when they are placed in a magnetic field. Some such nuclei commonly found in organic compounds are ^1H, ^2H, ^{19}F, ^{13}C, ^{15}N, and ^{31}P. Nuclei such as ^{12}C, ^{16}O, and ^{32}S do *not* have nuclear spin and thus cannot be studied by NMR spectroscopy.

Principles

The theory underlying the technique of NMR spectroscopy is presented in any comprehensive lecture textbook of organic chemistry and is thus not discussed in detail here. It is important to recall, however, that the observation of absorptions in the NMR spectrum of a particular nucleus is due to *realignment of nuclear spins* in an **applied magnetic field**, H_0, as shown in Parts (b) and (c) of Figure 8.19. The energy associated with the realignment of spin or spin flip, as it is often called, is expressed in Equation 8.6 and depends on the strength of the applied field H_0 and the **magnetogyric ratio** γ characteristic of the particular nucleus being examined. In a modern NMR spectrometer operating at 90–500 MHz, the value of H_0 is in the range of 21,000–117,000 gauss (2.1–11.7 Tesla); this means that the energy required for the transition is less than 0.1 cal/mol for a nucleus such as ^1H.

$$\Delta E = h\gamma H_0/2\pi \qquad (8.6)$$

Figure 8.19

Spin properties of the hydrogen nucleus. (a) Rotation of hydrogen nucleus and its magnetic moment. (b) Magnetic moment of nucleus aligned with applied external magnetic field. (c) Magnetic moment of nucleus aligned against applied magnetic field.

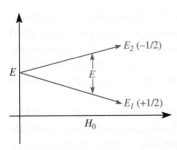

Figure 8.20

Energy difference between spin states as a function of strength of magnetic field, H_0.

Another way of illustrating the relationship between the strength of the magnetic field H_0 and the energy difference between spin states is given in Figure 8.20. Here it is seen that for a nucleus having two spin states, labeled $\pm 1/2$, where the plus and minus signs refer to nuclear spins aligned with and against the applied magnetic field, H_0, respectively; ΔE increases as H_0 increases. Energies of this sort are associated with radiation in the **radio-frequency (rf) range** of the electromagnetic spectrum; consequently, a radio-frequency oscillator is incorporated into the NMR spectrometer to provide the energy for the nuclear spin excitation. When the relationship between the magnetic field and the frequency of the oscillator is that defined by Equation 8.7, the **resonance condition** is achieved, and the transition from one nuclear spin state to the other can occur.

$$h\upsilon = h\gamma H_0/2\pi \quad \text{or} \quad \upsilon = \gamma H_0/2\pi \quad\quad (8.7)$$

where ΔE = energy difference between two spin states
$\quad h$ = Planck's constant
$\quad \gamma$ = magnetogyric ratio (a constant characteristic of a particular type
$\quad\quad$ of nucleus)
$\quad H_0$ = strength of applied external magnetic field
$\quad \upsilon$ = frequency of oscillator

To meet the resonance condition of Equation 8.7, H_0 could be held constant and υ varied, or υ could be kept constant and H_0 changed. In **continuous-wave** (CW) spectrometers, the earliest form of commercial NMR instruments, the latter option is more common, and the spectrum is obtained by slowly sweeping through the range of field strengths required to produce resonance at a particular oscillator frequency. Modern instruments, called **Fourier transform** (FT) spectrometers, operate with a pulse technique in which all resonance frequencies are produced simultaneously while H_0 is held constant; this technique allows collection of spectral data in much less time than with a CW instrument.

Figure 8.21
^{1}H NMR spectrum of 1-nitro propane (300 MHz, CDCl$_3$).

Proton Magnetic Resonance (^{1}H NMR) Spectroscopy

The hydrogen nucleus, ^{1}H, has a nuclear spin, I_z, of 1/2 and the NMR spectral technique involving it is called **proton magnetic resonance** spectroscopy, abbreviated as ^{1}H NMR. Before discussing this type of NMR spectroscopy further, it is important to note that in the terminology of NMR spectroscopy, a covalently bound hydrogen atom is often imprecisely called a proton. Rather than being purists, we will frequently use this term for the sake of convenience.

Consideration of Equations 8.6 and 8.7 reveals that the *same* energy would be required to produce the resonance condition for *all* hydrogen nuclei in a molecule *if* all of these nuclei were *magnetically identical*. In other words, if the magnetic environment as defined by H_0 were identical for all of the hydrogen atoms, the same frequency v would be needed for the spin flip. The consequence of this would be the appearance of a *single* absorption band in the ^{1}H NMR spectrum, a most uninformative result. Fortunately, the nuclei are *not* all magnetically equivalent because the three-dimensional electronic structure of a molecule produces variations in the magnetic environments of the hydrogen atoms in it (see "**Chemical Shift**" later in this section). This means that achieving the resonance condition requires *varying* amounts of energy and results in a spectrum containing valuable information for determining molecular structure.

Let's now analyze the ^{1}H NMR spectrum of 1-nitropropane (**1**) shown in Figure 8.21. The significant features illustrated in this spectrum are **chemical shift**, **spin-spin splitting**, and **peak integration**, each of which is discussed in detail. Before doing so, however, let's first take a brief overview of the general format for presenting NMR data.

$$\underset{\textbf{1}}{\overset{\displaystyle H_a \quad H_b \quad H_c}{\underset{\displaystyle H_a \quad H_b \quad H_c}{H_a{-}C{-}C{-}C{-}NO_2}}}$$

In Figure 8.21, the strength of the applied magnetic field H_0 is plotted along the horizontal axis and *increases* from left to right in the spectrum. This direction is called **upfield**; the opposite one is termed **downfield**. As an example, the group of peaks labeled B could be described as being upfield of the resonances labeled C. The *energy* needed to achieve resonance also *increases* in going from left to right. The intensity of the absorption band is plotted on the vertical axis and *increases* from the bottom, or baseline, of the plot.

A scale is printed horizontally across the bottom of the spectrum to define the location of peaks in the spectrum relative to those of a reference compound. The **delta-** or **δ-scale**, which is in units of **parts per million**, ppm, is used for the ^{1}H NMR spectra in this textbook. The reference compound most commonly used is **tetramethylsilane**, $(CH_3)_4Si$, which is abbreviated as TMS. TMS is an inert volatile liquid that is added directly to solutions of the sample and thus serves as an **internal standard**. The essential aspects of ^{1}H NMR spectra are presented under the headings "Chemical Shift," "Spin-Spin Splitting," and "Integration," respectively. The ^{1}H NMR spectra of most of the organic compounds you will be encountering in this and subsequent chapters are provided at the website associated with this textbook. These spectra may be integrated and expanded to facilitate their analysis.

Chemical Shift

Three groups of peaks, centered at about 1.0, 2.0, and 4.4 on the δ-scale, are seen in the spectrum of Figure 8.21. These values are the chemical shifts, in ppm, of the three chemically distinct types of hydrogen nuclei in 1-nitropropane (**1**). More precise values for chemical shifts may be determined by analyzing the spectrum using the software available online (refer to the online resources for this book). Predicting whether the various protons in a molecule are chemically distinct, and thus will *probably* appear at different chemical shifts, is an important part of understanding ^{1}H NMR spectra. Fortunately, determining how many chemically different protons are present in a molecule is straightforward.

The first step in the analysis is to group the protons according to their **connective equivalency**, placing in the same set all protons having identical molecular connectivities. The next step is to assess whether *connectively* equivalent nuclei are **chemically equivalent** and thus likely to have identical chemical shifts. This procedure is illustrated by further consideration of **1**. Examining **1** reveals that it contains three connectively unique sets of protons, labeled H_a, H_b, and H_c, respectively. *Hydrogen nuclei that are connectively distinct are chemically nonequivalent* and are defined as being **heterotopic**. As a general rule, such nuclei have *different* chemical shifts from one another, although some of the peaks may overlap.

The second step in the analysis involves evaluating connectively *identical* nuclei for their possible chemical nonequivalency. This is done by a **substitution test** in which it is supposed that a hydrogen nucleus is replaced with a **probe nucleus** that can be distinguished from other nuclei in the same set. The stereoisomeric relationship between such substituted species is then assessed.

The substitution procedure is presented in Figure 8.22 for protons of types H_a and H_c in **1**. As shown in Figure 8.22a, the hypothetical replacement of H_{a1}, H_{a2}, and H_{a3} with deuterium (**D**) as the probe nucleus provides three new representations that are *identical* to one another. The identity is established by rotating about the bond joining the methyl group adjacent to the methylene group and seeing that the resulting molecules are *superimposable* upon one another. Such hydrogens are defined as being **homotopic** and are *chemically equivalent*. Consequently, they will all have the *same* chemical shift in the ^{1}H NMR spectrum.

(a)

(b)

Figure 8.22
Topicity analysis of 1-nitro propane by substitution.

Applying the same substitution process to H_{c1} and H_{c2} gives *enantiomeric* representations, as seen in Figure 8.22b, and the two hydrogen atoms are described as being **enantiotopic**. Enantiotopic nuclei are also *chemically equivalent* and will have the same chemical shift, provided they are in an *achiral* environment, such as that provided by the solvents commonly used in NMR studies. It is left as an exercise to demonstrate that the hydrogen atoms of type H_b are also enantiotopic.

Another type of topicity is possible when a molecule contains a **center of chirality** (Sec. 7.1), or is subject to **restricted rotation**, a phenomenon usually associated with the presence of a π-bond or a ring. The first case is illustrated by considering 2-chlorobutane (**2**) as seen in Figure 8.23; although only a single enantiomer is shown, applying the same analysis to its mirror image would give the same result. The molecule has a chiral center at C2, the effect of which makes the connectively equivalent hydrogen atoms at C3 chemically *nonequivalent*. The nonequivalency is revealed by performing the substitution test on these protons, as illustrated in Figure 8.23. This generates **diastereomers**, thereby defining the two protons at C3 as **diastereotopic**. Because diastereotopic nuclei *may* have chemical shifts that are different from one another, there are *five* rather than four sets of chemically nonequivalent protons in **2**. These are the methine proton at C2, the two *diastereotopic* protons at C3, and the two different methyl groups, each of which has three *homotopic* protons.

Diasterotopic hydrogens

Figure 8.23
Topicity analysis of 2-chlorobutane by substitution.

Figure 8.24
Topicity analysis of 4-methyl-1-pentene by substitution.

Chemical nonequivalency resulting from restricted rotation is seen in 4-methyl-1-pentene (**3**) (Fig. 8.24). Applying the substitution test shows the two vinylic hydrogen atoms at C1 are *diastereotopic* and thus chemically nonequivalent. This means that **3** contains *three* distinct sets of vinylic protons, each of which appears at a different chemical shift, as seen in Figure 8.25. Applying the substitution test in this instance is simply a formal way of proving that the geminal vinylic protons of an *unsymmetrically* substituted alkene, $H_2C=CR^1R^2$, are *not* equivalent to one another. For example, relative to R^1, one of these hydrogen atoms will be *cis* and the other *trans*; their nonequivalency is determined rapidly and simply from this stereochemical relationship.

$H_2C=CHCH_2CH(CH_3)_2$

Figure 8.25
^1H NMR spectrum of 4-methyl-1-pentene (300 MHz, $CDCl_3$).

The presence of diasterotopic hydrogens in a molecule may complicate the appearance of a ^{1}H NMR spectrum. However, the simple procedure of looking for chiral centers and for the features that may enforce restricted rotation in the molecule minimizes erroneous analyses caused by the increased number of resonances associated with diastereotopicity.

Let's now return to our analysis of 1-nitropropane (**1**), in which we previously identified three chemically nonequivalent sets of protons. As a result, we predict that three different resonances should appear in its ^{1}H NMR spectrum, just as is observed (Fig. 8.21). The chemical shifts of the three groups of resonances vary according to the magnetic environment of each hydrogen nucleus in the molecule. This primarily depends on two factors: (1) the externally applied magnetic field, and (2) circulation of the electrons within the molecule.

The strength of the field H_0 is *uniform* throughout the molecule, so it *cannot* be responsible for the magnetic nonequivalence of the different nuclei. On the other hand, the shielding effect produced by circulation of electrons is *not* uniform because the electron density varies within the molecule. This is a critical point because the **internal electric field** of the molecule induces an **internal magnetic field**, H_i, about the various nuclei in the molecule. The induced magnetic field *opposes* the applied magnetic field, so the **effective magnetic field**, H_e, at a given nucleus is *less* than H_0 (Eq. 8.8); this phenomenon is the source of **diamagnetic shielding** in the molecule.

$$H_e = H_0 - H_i \tag{8.8}$$

The strength of H_i varies in the same way as that of the electric field, thus making H_e different for the various types of protons in the sample. For example, the greater H_i is, the less H_e is, so a larger external field must be applied to achieve resonance. In other words, the more magnetic shielding experienced by a nucleus, the higher is the external field H_0 required to produce the resonance condition at a particular oscillator frequency.

The consequences of the magnitude of H_i on the location of resonances in ^{1}H NMR spectra are seen in Figure 8.21. The nitro group, NO_2, serves as an electron sink, causing the electric field to be lower for those nuclei near it. Correspondingly, the value of H_i is less for these nuclei than for others more distant from the nitro function. The protons labeled H_c in **1** experience the *least* diamagnetic shielding and thus resonate at the *lowest* value of H_0. The methyl protons H_a are the *most* **shielded** because they are furthest from the nitro group and appear at the highest field (Fig. 8.21). Put another way, the nearer a proton is to the electronegative nitro group, the more **deshielded** it is relative to other hydrogen atoms in the molecule, and the further **downfield** it appears. On this basis, the various groups of peaks in the spectrum are assigned as shown below.

$$\overset{\text{A}\qquad\text{B}\qquad\text{C}}{CH_3 - CH_2 - CH_2 - N\underset{O^-}{\overset{O}{\diagup\diagdown}}}$$

$$\delta\ \ 1.02\quad 2.05\quad 4.37$$

The locations of the three groups of peaks in 1-nitropropane are expressed in units of ppm on the *δ-scale*, as was noted earlier. This scale, calculated according to Equation 8.9, is defined such that measurement of chemical shifts is *independent* of the frequency of the rf oscillator of the NMR spectrometer. The numerator is the shift, in Hz, of a resonance of the sample relative to that of the reference compound, which is usually TMS for ^{1}H NMR spectra. The sign of the shift by convention is taken to be *positive* if the resonance appears *downfield* of TMS and *negative* if it is *upfield*. The shift is divided by the corresponding oscillator frequency, and for convenience in reporting chemical shifts, the result is multiplied by 10^6 to provide units of ppm. Alternatively, the relative positions of peaks in the spectrum could simply be expressed in Hz relative to TMS. This is unsatisfactory because the chemical shift would then depend on the frequency of the oscillator being used in the spectrometer. Consequently, reporting shifts in Hz would require specifying the frequency at which the data were obtained.

$$\delta = \frac{\text{chemical shift (in Hz)} \times 10^6}{\text{oscillator frequency (in Hz)}} \tag{8.9}$$

To illustrate the difference between measuring chemical shifts in terms of δ versus Hz, consider the case in which a resonance is observed *100 Hz* downfield of TMS when a spectrometer operating at 100 MHz is used. On the δ-scale, this corresponds to a shift of *1.0 ppm*. Had the spectrum been measured on a 300-MHz instrument, the resonance would appear at *300 Hz* rather than at *100 Hz* downfield of TMS; however, the chemical shift would remain at *1.0 ppm* on the δ-scale.

Almost all types of protons appear *downfield* of TMS and thus have chemical shifts that are *positive* values because of the sign convention noted above. However, some organic compounds contain hydrogen nuclei that are *more* shielded than those in TMS and therefore appear *upfield* of it. The chemical shifts are negative in such cases.

The consistent effect that particular functional groups have on chemical shifts has led to tabulations correlating these shifts with the structural features responsible for them. Table 8.3 is one such compilation, and Table 8.4 is an expanded listing of typical chemical shifts for a variety of specific types of hydrogen atoms. Because the shifts are given in units of ppm, they apply to all NMR spectrometers regardless of the frequency at which the instrument is operating.

Table 8.3 *Chemical Shifts of Hydrogen Atoms Attached to Various Functional Groups*

Functional Group; Hydrogen Type Shown as H	*δ, ppm*	*Functional Group; Hydrogen Type Shown as H*	*δ, ppm*
TMS, $(CH_3)_4Si$	0	Alcohols, ethers	
Cyclopropane	0–1.0	$HO{-}\overset{\textstyle\mid}{\underset{\textstyle\mid}{C}}{-}\mathbf{H}$	3.4–4
Alkanes		$RO{-}\overset{\textstyle\mid}{\underset{\textstyle\mid}{C}}{-}\mathbf{H}$	3.3–4
RCH_3	0.9		
R_2CH_2	1.3		
R_3CH	1.5		
Alkynes		Acetals	
$-\overset{\textstyle\mid}{C}{=}\overset{\textstyle\mid}{C}{-}\mathbf{H}$ (vinyl)	4.6–5.9	acetal structure	5.3
$-\overset{\textstyle\mid}{C}{=}\overset{\textstyle\mid}{C}{-}\mathbf{CH_3}$ (allyl)	1.7		
Alkynes		Esters	
$-C{\equiv}C{-}\mathbf{H}$	2–3		
$-C{\equiv}C{-}\mathbf{CH_3}$	1.8	$R{-}\overset{\textstyle O}{\overset{\|}{C}}{-}O{-}\overset{\textstyle\mid}{\underset{\textstyle\mid}{C}}{-}\mathbf{H}$	3.7–4.1

Table 8.3 *Chemical Shifts of Hydrogen Atoms Attached to Various Functional Groups (Continued)*

Functional Group; Hydrogen Type Shown as H	δ, ppm	Functional Group; Hydrogen Type Shown as H	δ, ppm
Aromatic			
Ar–**H**	6–8.5	RO–C(=O)–C–**H**	2–2.6
Ar–C–**H** (benzyl)	2.2–3		
Fluorides, F–C–**H**	4–4.5	Carboxylic acids	
		HO–C(=O)–C–**H**	2–2.6
Chlorides			
Cl–C–**H**	3–4	R–C(=O)–O–**H**	10.5–12
Cl–C(Cl)–**H**	5.8	Ketones	
		R–C(=O)–C–**H**	2–2.7
Bromides, Br–C–**H**	2.5–4	Aldehydes	
		R–C(=O)–**H**	9–10
Iodides, I–C–**H**	2–4	Amides	
		R–C(=O)–N–**H**	5–8
Nitroalkanes, O$_2$N–C–**H**	4.2–4.6	Alcohols, R–O–**H**	4.5–9
		Phenols, Ar–O–**H**	4–12
		Amines, R–N**H**$_2$	1–5

Table 8.4 *Compilation of ^{1}H NMR Absorptions for Various Molecules*

Listed below are the ^{1}H NMR chemical shifts observed for the protons of a number of organic compounds. The shifts are classified according to whether they are methyl, methylene, or methine types of hydrogen atoms. The atom shown in **bold** is responsible for the absorptions listed.

Methyl Absorptions

Compound	δ, ppm	Compound	δ, ppm
CH$_3$NO$_2$	4.3	CH$_3$CHO	2.2
CH$_3$F	4.3	CH$_3$I	2.2
(CH$_3$)$_2$SO$_4$	3.9	(CH$_3$)$_3$N	2.1
C$_6$H$_5$COOCH$_3$	3.9	CH$_3$CON(CH$_3$)$_2$	2.1
C$_6$H$_5$–O–CH$_3$	3.7	(CH$_3$)$_2$S	2.1
CH$_3$COOCH$_3$	3.6	CH$_2$=C(CN)CH$_3$	2.0

Table 8.4 *Compilation of ^1H NMR Absorptions for Various Molecules (Continued)*

Methylene Absorptions

Compound	δ, ppm	Compound	δ, ppm
CH_3OH	3.4	CH_3COOCH_3	2.0
$(CH_3)_2O$	3.2	CH_3CN	2.0
CH_3Cl	3.0	CH_3CH_2I	1.9
$C_6H_5N(CH_3)_2$	2.9	$CH_2=CH–C(CH_3)=CH_2$	1.8
$(CH_3)_2NCHO$	2.8	$(CH_3)_2C=CH_2$	1.7
CH_3Br	2.7	CH_3CH_2Br	1.7
CH_3COCl	2.7	$C_6H_5C(CH_3)_3$	1.3
CH_3SCN	2.6	$C_6H_5CH(CH_3)_2$	1.2
$C_6H_5COCH_3$	2.6	$(CH_3)_3COH$	1.2
$(CH_3)_2SO$	2.5	$C_6H_5CH_2CH_3$	1.2
$C_6H_5CH=CHCOCH_3$	2.3	CH_3CH_2OH	1.2
$C_6H_5CH_3$	2.3	$(CH_3CH_2)_2O$	1.2
$(CH_3CO)_2O$	2.2	$CH_3(CH_2)_3 X (X=Cl, Br, I)$	1.0
$C_6H_5OCOCH_3$	2.2	$CH_3(CH_2)_4CH_3$	0.9
$C_6H_5CH_2N(CH_3)_2$	2.2	$(CH_3)_3CH$	0.9

Methylene Absorptions

Compound	δ, ppm	Compound	δ, ppm
$EtOCOC(CH_3)=CH_2$	5.5	$C_6H_5CH_2N(CH_3)_2$	10.0
CH_2Cl_2	5.3	$CH_3CH_2SO_2F$	9.9
CH_2Br_2	4.9	CH_3CH_2I	9.8
$(CH_3)_2C=CH_2$	4.6	$C_6H_5CH_2CH_3$	3.3
$CH_3COO(CH_3)C=CH_2$	4.6	CH_3CH_2SH	3.3
$C_6H_5CH_2Cl$	4.5	$(CH_3CH_2)_3N$	3.1
$(CH_3O)_2CH_2$	4.5	$(CH_3CH_2)_2CO$	2.6
$C_6H_5CH_2OH$	4.4	$BrCH_2CH_2CH_2Br$	2.4
$CF_3COCH_2C_3H_7$	4.3	Cyclopentanone (α-CH_2)	2.4
$Et_2C(COOCH_2CH_3)_2$	4.1	Cyclohexene (α-CH_2)	2.4
$HC≡C–CH_2Cl$	4.1	Cyclopentane	2.4
$CH_3COOCH_2CH_3$	4.0	Cyclohexane	2.0
$CH_2=CHCH_2Br$	3.8	$CH_3(CH_2)_4CH_3$	2.0
$HC≡CCH_2Br$	3.8	Cyclopropane	1.5
$BrCH_2COOCH_3$	3.7	C_6H_5CHO	1.4
CH_3CH_2NCS	3.6	$4-ClC_6H_4CHO$	1.4
CH_3CH_2OH	3.6	$4-CH_3OC_6H_4CHO$	0.2
$CH_3CH_2CH_2Cl$	3.5	CH_3CHO	9.7
$(CH_3CH_2)_4N^+I^-$	3.4	Pyridine (α-H)	8.5

Table 8.4 *Compilation of ^1H NMR Absorptions for Various Molecules (Continued)*

Methylene Absorptions

Compound	δ, ppm	Compound	δ, ppm
CH_3CH_2Br	3.4	$1,4-C_6H_4(NO_2)_2$	8.4
$C_6H_5CH=CHCOCH_3$	7.9	*p*-Benzoquinone	6.8
C_6H_5CHO	7.6	$C_6H_5NH_2$	6.6
Furan (α-**H**)	7.4	Furan (β-**H**)	6.3
Naphthalene (β-**H**)	7.4	$CH_3CH=CHCOCH_3$	5.8
$1,4-C_6H_4I_2$	7.4	Cyclohexene (vinylic **H**)	5.6
$1,4-C_6H_4Br_2$	7.3	$(CH_3)_2C=CHCH_3$	5.2
$1,4-C_6H_4Cl_2$	7.2	$(CH_3)_2CHNO_2$	4.4
C_6H_6	7.3	Cyclopentyl bromide (**H** at C1)	4.4
C_6H_5Br	7.3	$(CH_3)_2CHBr$	4.2
C_6H_5Cl	7.2	$(CH_3)_2CHCl$	4.1
$CHCl_3$	7.2	C_6H_5CCH	2.9
$CHBr_3$	6.8	$(CH_3)_3CH$	1.6

The *magnitude* of chemical shifts is primarily related to the *electronegativity* of any functional groups that are near the proton being observed and the nature of *induced magnetic fields*, H_i, in molecules having π-bonds. Both effects are associated with circulation of electrons in the molecule and the interaction of the resulting induced magnetic field with the applied field H_0 (Eq. 8.8).

You can see the influence of electronegativity by comparing the chemical shifts of the methyl halides, CH_3–X, in which the δ-values are 4.3, 3.0, 2.7, and 2.2 for X = F, Cl, Br, and I, respectively (Table 8.4). That of π-electrons is seen in the relative chemical shifts of the aliphatic protons of alkanes versus the vinylic protons of alkenes, as in ethane (δ 0.9) and ethylene (δ 5.25). The hydrogen nuclei of ethylene are *deshielded* partly because an sp^2-hybridized carbon atom is more electronegative than one that is sp^3-hybridized. Moreover, the vinylic protons are further deshielded because they lie in a region of the induced field H_i where the lines of force *add* to H_0, as shown in Figure 8.26a. A similar analysis of the induced field accounts for the fact that an alkynic proton, as in acetylene (δ 1.80), resonates at *higher* field than does a vinylic hydrogen, even though an sp-hybridized carbon atom is the *most* electronegative of all carbon atoms found in organic compounds. In alkynes, the acetylenic proton is in a portion of the magnetic field induced by the circulating π-electrons where the lines of force *oppose* the applied field (Fig. 8.26b), thereby increasing the shielding of this type of nucleus. The protons of aromatic compounds, as in benzene (δ 7.27), resonate at lower fields than the vinylic protons of alkenes because the induced magnetic field responsible for deshielding is greater owing to the *cyclic* nature of the circulation of π-electrons (Fig. 8.26c); this phenomenon is called the **ring-current effect**.

Spin-Spin Splitting

Information beyond that regarding the magnetic environment of a nucleus is available from analysis of the NMR spectrum. For example, the **spin-spin splitting** pattern shown by a particular type of hydrogen atom provides information about the

(a)

(b)

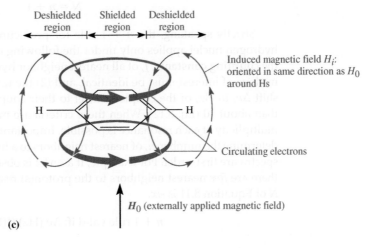

Figure 8.26

Effect on chemical shifts of induced field from circulation of π-electrons. (a) Deshielding of vinylic protons. (b) Shielding of acetylenic protons. (c) Deshielding of aromatic protons.

(c)

number of its **nearest neighbors**. Nearest neighbors are elements that have nuclear spin and, in most cases, are *no more* than *three* bonds, or *two* atoms, away from the atom of interest. For our purposes, hydrogen is the primary element of interest for analyzing splitting patterns because compounds containing other magnetically active nuclei are only rarely encountered in the experiments in this textbook. The presence of ^{13}C in organic compounds is *not* a factor in the analysis because this isotope, which also has a nuclear spin of 1/2, is present at the level of only about 1%. Thus, peaks due to spin-spin splitting with carbon atoms are generally too weak to be observed.

Observation of coupling between atoms having nuclear spins requires that the nuclei must *not* be magnetically equivalent. By definition, **magnetic equivalency** means that *every* nucleus in a particular set of chemical shift-equivalent nuclei is coupled equally to *every* other nucleus of the spin system. Chemical shift-nonequivalent nuclei are always magnetically nonequivalent and couple with one another. As discussed earlier, this type of nonequivalence is generally seen for protons that are chemically distinct by virtue of being *heterotopic* or *diastereotopic*. That protons may be chemically equivalent but magnetically nonequivalent is illustrated by a discussion later in this section.

Analysis of 1-nitropropane (**1**) in this context reveals that the type *a* hydrogen atoms have *two* nearest neighbors in the form of the two protons of type *b* that are magnetically active and chemically nonequivalent. Nuclei of type *c* also have two such nearest neighbors, whereas those of type *b* nuclei have *five*.

The **splitting** or **coupling pattern** of each chemically distinct type of nucleus may be predicted in the following way. In the general case, an atom *A* having nuclear spin I_z, where $I_z = 1/2, 1, 3/2$, and so on, and coupling with another atom *B* will split the resonance of *B* into the number of peaks given by Equation 8.10. When *A* is ^1H, ^{13}C, ^{19}F, or ^{31}P, this expression simplifies to the $n + 1$ rule expressed in Equation 8.11 because $I_z = 1/2$ for all these nuclei. The expected splitting patterns for 1-nitropropane (**1**), according to Equation 8.11, are *three* peaks each for H_a and H_c and six peaks for H_b. This is precisely the splitting pattern seen in Figure 8.21.

$$N = 2nI_z + 1 \tag{8.10}$$

where N = number of peaks observed for absorbing atom *B*
n = number of magnetically equivalent nearest neighbor atoms *A*
I_z = nuclear spin of *A*

$$N = n + 1 \tag{8.11}$$

Strictly speaking, the $n + 1$ rule for predicting splitting patterns of coupled hydrogen nuclei applies only under the following circumstances: (1) The values of the **coupling constants**, *J*, of all nearest neighbor hydrogen nuclei coupling with the nucleus of interest must be identical, and (2) the ratio of the difference in chemical shift $\Delta\nu$, in Hz, of the coupled nuclei to their coupling constant *J* must be greater than about 10 (Eq. 8.12). When these criteria are met, a **first-order analysis** of the multiplicity of each resonance is possible. Importantly, Equation 8.11 may be used to determine the number, *n*, of nearest neighbors to a hydrogen nucleus when ^1H NMR spectra are first-order. For example, if a *sextet* is observed in the ^1H NMR spectrum, there are *five* nearest neighbors to the proton(s) responsible for that sextet because *N* of Equation 8.11 is *six*.

$$n + 1 \text{ rule valid if: } \Delta\nu \text{ (Hz)}/J\text{(Hz)} \geq \sim 10 \tag{8.12}$$

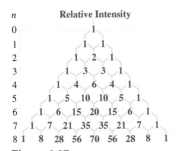

n	Relative Intensity
0	1
1	1 1
2	1 2 1
3	1 3 3 1
4	1 4 6 4 1
5	1 5 10 10 5 1
6	1 6 15 20 15 6 1
7	1 7 21 35 35 21 7 1
8	1 8 28 56 70 56 28 8 1

Figure 8.27

Pascal's triangle showing relative intensities of first-order multiplets; coupling constants of all nearest neighbors, n, *are assumed to be equal.*

The *relative intensities* of the peaks in a *first-order* multiplet that obeys Equation 8.11 may be approximated by applying Pascal's triangle (Fig. 8.27). For example, if a hydrogen atom has two nearest neighbors, then the triplet that results is predicted to have intensities in the ratio of about 1:2:1. If there are five nearest neighbors, the corresponding ratio for the sextet is 1:6:15:20:15:6:1. For reasons we'll not discuss, the prediction of relative peak intensities by this method becomes more accurate as the ratio $\Delta v/J$ increases.

Because most ^{1}H NMR spectra in this book were obtained at 300 MHz, many may be interpreted as first-order spectra, and the peak intensities of the multiplets are predictable by applying Pascal's triangle. For example, reconsidering the spectrum of Figure 8.21, we find that $J_{ab} = J_{bc} \cong 7$ Hz. The difference in chemical shifts for the sets of coupled nuclei is $\Delta v_{AB} = 315$ Hz and $\Delta v_{BC} = 700$ Hz, respectively. Application of Equation 8.12 gives values of $315/7 = 45$ and $700/7 = 100$, respectively, for the ratios of interest. The fact that both ratios exceed the criterion of Equation 8.12 means that the splitting patterns are predicted by the $n + 1$ rule, as evidenced in Figure 8.21. Moreover, the relative intensities of each of the peaks in the multiplets in Figure 8.21 have the approximate values predicted from Pascal's triangle.

As the term $\Delta v/J$ decreases, the observed spectra become **second-order** and are more complicated. The splitting patterns may no longer be predictable by the $n + 1$ rule, and the intensities of the peaks in the multiplets cannot be approximated by using Pascal's triangle. This trend may be illustrated by examining the ^{1}H NMR spectra for 1-butanol (**4**) in Figures 8.28 and 8.29 and pentane (**5**) in Figure 8.30.

4

1-Butanol

5

Pentane

Figure 8.28

^{1}H NMR spectrum of 1-butanol in CDCl$_3$ (300 MHz).

Figure 8.29
¹H NMR spectrum of 1-butanol in DMSO-d₆ (300 MHz).

Figure 8.30
¹H NMR spectrum of pentane (300 MHz, CDCl₃).

In 1-butanol (**4**), nuclei of types H_e and H_d produce multiplets centered at δ 0.94 ppm and at about δ 1.4 ppm, respectively; J_{de} is about 7 Hz (Fig. 8.28). Using a 300-MHz spectrometer, the ratio $\Delta\nu/J_{de}$ is greater than 10, so the criterion of Equation 8.12 is satisfied. The nuclei of type H_e appear as a triplet, and the relative intensities of the three peaks approximate that expected from Pascal's triangle, namely, 1:2:1. In contrast, although the numbers of peaks in the multiplets for protons of types H_d and H_c correspond to those predicted by the $n + 1$ rule, the relative intensities of the peaks are no longer approximated by using Pascal's triangle. This is because the criterion of Equation 8.12 is *not* satisfied. The difference in chemical shifts of the two types of protons is only 48 Hz ($300 \times (1.56 - 1.40)$) and their coupling constant, J_{cd}, is about 7, so the ratio $\Delta\nu/J_{cd}$ is about 7 (48/7), and the ratios of the peaks in the two multiplets cannot be predicted by applying simple first-order rules.

Assuming that $J_{bc} = J_{ab}$, applying the $n + 1$ rule to nuclei of type H_b and H_a leads to a predicted splitting pattern of a quartet for H_b and a triplet for H_a. However, what we actually observe are a somewhat broadened and distorted quartet and a triplet centered at δ 3.65 and 1.63 ppm, as seen in Figure 8.28. The broadening of these peaks is not due to any second-order effects. Rather, the spectrum was obtained under conditions in which the hydroxylic proton H_a is undergoing rapid exchange with other molecules of the alcohol or with adventitious water present in the sample. As a result of this rapid exchange, the spin information carried by the hydroxylic proton is partially lost, and the peaks arising from protons that are spin-coupled with the hydroxyl proton are broadened. Spectra obtained under conditions that suppress this exchange, such as using CD_3SOCD_3 (DMSO-d_6) rather than $CDCl_3$ as the solvent, allow us to observe the coupling, and the quartet at δ 3.38 for H_a and the triplet at δ 4.33 for H_b are sharp, as illustrated in Figure 8.28. Comparison of the spectra of **4** in Figures 8.28 and 8.29 also reveals that the chemical shifts for protons in a molecule *can* vary with the solvent, the most significant change being for H_a, so it is always important to indicate the solvent used to prepare the sample when reporting NMR spectral data.

In the spectrum of pentane (**5**) (Fig. 8.30), we see that nuclei of type H_a resonate at about δ 0.90 ppm, whereas those of types H_b and H_c appear at about δ 1.30 ppm. Because the ratio $\Delta v/J_{ab}$ is greater than 10, the condition of Equation 8.12 is satisfied, and the peak for the H_a is a triplet according to the $n + 1$ rule. However, the ratio $\Delta v/J_{bc}$ is much less than 10, and the splitting pattern for the nuclei of types H_b and H_c is complex and cannot be interpreted by first-order analysis and the $n + 1$ rule.

Thus far, we have predicted splitting patterns by grouping hydrogens according to their chemical shift equivalence and then determining the number of nearest neighbors n. There are cases, however, in which nuclei may have equivalent chemical shifts but are *not* magnetically equivalent; this may complicate the appearance and interpretation of the spectrum. This phenomenon is often seen in the ^{1}H NMR spectra of aromatic compounds.

For example, analyzing 4-bromonitrobenzene (**6**) by the substitution test leads to the conclusion that there are two sets of homotopic hydrogen atoms, H_aH_a' and H_bH_b'. The two protons in each set have equivalent chemical shifts. However, the coupling constants between all members of the spin set are *not* equivalent. Namely, $J_{H_aH_b} \equiv J_{H_{a'}H_{b'}}$ and $J_{H_{ab'}} \equiv J_{H_{a'}H_{b'}}$, but $J_{H_{a'}H_{b'}} \neq J_{H_{a'}H_b}$ and $J_{H_aH_b} \neq J_{H_aH_{b'}}$. This makes H_a and $H_{a'}$ **magnetically nonequivalent**, as are H_b and $H_{b'}$. We note that the pairs of nuclei $H_aH_{b'}$ and $H_{a'}H_b$ are not nearest neighbors, yet coupling may occur between them. Nuclear spin interactions that occur over more than three bonds are termed **long-range coupling**, a phenomenon that is most frequently observed in systems having **conjugated π-bonds**.

6

4-Bromonitrobenzene

The effect of magnetic nonequivalency is evident in the ^{1}H NMR spectrum of **6** (Fig. 8.31). Carefully examining the resonances for each of the doublets reveals additional splitting that is the result of magnetic nonequivalency within the set.

Figure 8.31

¹H NMR spectrum of 4-bromonitrobenzene (300 MHz, CDCl₃).

Figure 8.32

The geminal Karplus correlation. J_{HH} for CH_2 groups as a function of H–C–H angle, θ.

Nevertheless, the aromatic multiplets still have the overall appearance of the doublets expected from the simple first-order analysis using the $n + 1$ rule.

The *magnitude* of coupling constants, J_{HH}, is *independent* of the oscillator frequency, so units of Hz rather than δ are used. The measured values depend on two main factors: the number of bonds intervening between the coupled nuclei, and the bond angle or dihedral angle between the nuclei. The angular dependence of **geminal coupling**, which is coupling between magnetically different protons on the *same* atom, is shown graphically in Figure 8.32, in which θ is the bond angle between the coupled nuclei. The situation for **vicinal coupling**, in which the coupled nuclei are on *adjacent* atoms and have a **dihedral angle**, ϕ, is presented in Figure 8.33. Some typical ranges for coupling constants are provided in Figure 8.34.

Coupling constants provide valuable information about the structure of organic compounds. If two different protons are on adjacent carbon atoms, the coupling

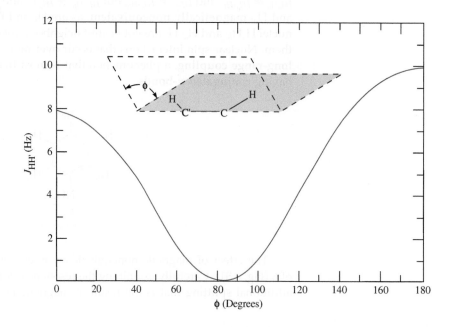

Figure 8.33

The vicinal Karplus correlation. Relationship between dihedral angle, ϕ, and coupling constant for vicinal protons.

Figure 8.34

Examples of the values of coupling constants J_{HH} for selected functional groups.

constant between the two has only one value, that is, $J_{ab} \equiv J_{ba}$, because the protons are mutually coupled with each other. Thus, if the coupling constants of two multiplets suspected to be the result of two particular nuclei coupling with one another are *not* the same, the two nuclei are *not* coupled. This fact helps in making structural assignments.

It is difficult to measure coupling constants using the 300-MHz spectra reproduced in this textbook because the spectra have been reduced, making measurement of the distance between peaks subject to large error. However, using modern high-field NMR spectrometers, the values of coupling constants for simple splitting patterns are easily determined from digital printouts of chemical shifts of the various peaks. As noted previously, to enable you to gain experience in interpreting ^{1}H NMR data, the ^{1}H NMR spectra of all the organic reactants and products encountered in subsequent chapters are found online (refer to the online resources for this book). By expanding the spectra, you will be able to measure the coupling constants accurately.

Data provided in the form of $\Delta\delta$ must first be converted to Hz before computation of the coupling constant, J. For example, if the peaks of a quartet are found to differ by 0.04 ppm in a spectrum obtained on a 300-MHz instrument, the coupling constant would be 12 Hz by application of Equation 8.13, where $\Delta\delta$ is 0.04 and the *instrument frequency* is 300×10^6 Hz.

$$J(\text{Hz}) = (\Delta\delta \times 10^{-6}) \times \text{instrument frequency} \qquad (8.13)$$

Integration

The areas of the peaks in a ^{1}H NMR spectrum are important measures of the *relative* numbers of the different types of hydrogen nuclei in a molecule because these peak areas are a linear function of the number of nuclei producing the peaks. The areas are determined by an electronic integration that is typically plotted as a stepped trace that appears on the spectrum, as seen in Figure 8.21.

In practice, you would first record the ^{1}H NMR spectrum and then plot the integration on the same sheet of paper. The vertical distance that the integration rises over a peak or group of peaks is a direct measure of the area under the peak(s) and thus of the relative number of protons producing the resonances. You would usually reset the integrator to zero after each set of resonances has been measured, as in Figure 8.21, but this need not be done, as seen in Figure 8.28. Resetting the integrator allows the height of each step in the integration curve to be higher, so the experimental error in measuring the step is less. In this textbook, the numerical values of the integrations rather than the plotted integrals will normally be provided below sets of peaks.

Once you determine the heights of each step, you can compute the *relative* number of hydrogen atoms producing the steps by dividing the integration heights by the smallest height measured to give a ratio of the various types of protons. For example, in the spectrum of 1-nitropropane (**1**), the heights of the steps in Figure 8.21 are represented by the distances *ab*, *ac*, and *ad*. The relative heights are measured to obtain the ratio of 1.55:1.0:1.0 for groups *A*, *B*, and *C* in the spectrum. Because the number of hydrogen atoms present in a molecule must be integral, this ratio is converted to whole numbers by multiplying *each* value by an *integer* that gives the **relative ratio** of the types of protons in whole numbers. *If* you know the molecular formula of the molecule, the integer is chosen to give a *sum* of the values in the ratio corresponding to the *absolute* number of hydrogen nuclei present. There are seven hydrogen atoms in 1-nitropropane, so *two* is used as the multiplier to provide the **absolute ratio** 3.10:2.0:2.0, the sum of which is 7.1. Because the accuracy of electronic integration is generally no better than 5–10%, this is within experimental error of the total number of protons in the molecule.

Analysis of ^{1}H NMR Spectra

General Procedure

The proper analysis of the ^{1}H NMR spectrum of a compound may provide a wealth of information regarding its structure. Determining the *chemical shift* of each peak or group of peaks allows speculation about the *type* of functional group to which the hydrogen nucleus producing the resonance is attached. The *spin-spin splitting pattern* and the $n + 1$ rule give information concerning the *number of nearest neighbor hydrogen atoms* for the proton(s) producing a particular absorption. *Integration* allows evaluation of the peak areas of each set of peaks and defines the *relative numbers of each type* of proton present in the molecule; if the molecular formula of the compound is known, the *absolute* number of each type may also be computed.

Appropriate steps for complete analysis of a ^{1}H NMR spectrum are as follows:

1. Determine the relative numbers of the different types of hydrogen nuclei by measuring the integrated step height for each peak or group of peaks. Convert the resulting *relative* ratio to an *absolute* ratio if the molecular formula is known. The *absolute* values of the integrations are provided for most of the spectra in this textbook by the numbers in parentheses located below the respective sets of peaks in the spectra.

2. Measure the *chemical shifts* of the different sets of resonances and use the data to try to deduce the functional groups that might be present in the molecule. This process is greatly aided if the molecular formula and/or other spectral data, such as IR and ^{13}C NMR spectra, are available.

3. Analyze the spin-spin splitting patterns of each group of resonances to determine the number of nearest neighbors associated with each type of hydrogen nucleus.

Applying these steps *may* allow assignment of the structure of an unknown compound. However, without additional information about a substance, such as its molecular formula or the chemical reaction(s) by which it was formed, only a partial structural determination may be possible. Other chemical or spectral data may be required to make a complete assignment. In any case, structural assignments based *exclusively* on interpretation of spectra *must* be confirmed by comparing the spectra of the unknown with those of an authentic specimen, if the compound is known, or by chemically converting the unknown to a known compound.

Analysis of an Unknown

An example of the use of this procedure for making a structural assignment follows. Other examples may be found in the exercises at the end of this section.

Problem

Provide the structure of the compound having the molecular formula C_4H_9Br and whose ^1H NMR spectrum is given in Figure 8.35.

Solution

For purposes of this analysis, the groups of peaks are referred to as representing nuclei of types *A–C* in going from left to right, or upfield, in the spectrum.

Integration. The *relative* ratio of the integrals for H_a, H_b, and H_c is 2.1:1.0:5.7, respectively. Given the molecular formula, the *absolute* ratio must be 2:1:6, for a total of nine hydrogen atoms. Thus, there are *two* hydrogen atoms of type *A*, *one* of type *B*, and *six* of type *C*.

Chemical Shifts. The δ-values at the center of the individual multiplets are approximately 3.3 (H_a), 2.0 (H_b), and 1.0 (H_c). The low-field doublet for H_a presumably reflects the nearby presence of the electronegative bromine atom. Since there are *two*

Figure 8.35
^1H NMR spectrum for unknown compound (300 MHz, $CDCl_3$).

such nuclei, the partial structure, $-CH_2Br$, may be written. The appearance of H_c at about δ 1.0 is consistent with these nuclei being associated with a methyl group. Because there are six such hydrogen atoms, *two* methyl groups must be present. Another partial structure, $(CH_3)_2C-$, may then be written. Because these two partial structures account for *all* of the carbon atoms required by the molecular formula and all but *one* of the hydrogen atoms, the missing atom must be a *methine* hydrogen atom located on the carbon atom bearing the two methyl groups and the bromomethyl substituent. Although the chemical shift of this methine hydrogen atom is outside the range for such nuclei as given in Table 8.3, reference to Table 8.4 shows that the methyl resonances are shifted *downfield* when there is a halogen atom β to the carbon atom bearing the hydrogen atom of interest; this is evidenced in the δ-value of 1.7 for the methyl group of ethyl bromide, CH_3CH_2Br. The same effect applies to methine hydrogen atoms. Enough information is now available to enable assignment of the structure of the unknown as 1-bromo-2-methylpropane (isobutyl bromide), $(CH_3)_2CHCH_2Br$.

In this case, the determination of the structure may be made simply on the basis of molecular formula, chemical shifts, and integration; analysis of the spin-spin splitting patterns, as in Part (c), is required only to confirm the conclusion regarding the structure of the unknown and not for initially generating the structure. More typically, the information contained in the splitting patterns is crucial to a first formulation of the structure of an unknown.

Spin-Spin Splitting Patterns. The fact that the nuclei of types *A* and *B* appear as a doublet means that they have *one* nearest neighbor, which is consistent with the fact that both H_a and H_c in 1-bromo-2-methylpropane should be coupled with the methine hydrogen atom H_b but *no* others. Conversely, H_b must be coupled with both the other types of protons, which total eight. Application of the $n + 1$ rule dictates nine peaks in the splitting pattern of the methine hydrogen atom. The two "missing" members of this multiplet are not apparent because they are the weak, outermost peaks, and the amplification at which the spectrum was plotted was insufficient for them to be seen. If the multiplet for H_b is viewed at significantly higher amplification, all nine peaks may be observed.

There is an important lesson to be learned from the fact that fewer peaks are observed than would be expected from the $n + 1$ rule: When multiplets of six or more are predicted, one or more of the outermost absorptions of the multiplet is frequently too weak (see Pascal's triangle, Figure 8.27) to appear under the normal operating conditions for the spectrometer. This fact must be taken into account when using spin-spin splitting patterns to interpret the spectrum of an unknown compound.

Determination of Structure. Although the present problem was solved by using only two of the three basic types of information available from an integrated ^1H NMR spectrum, the usual case would require careful interpretation of all three. The strategy is to generate partial structures that are consistent with the available data and then to attach these pieces in various ways that are consistent with the observed splitting patterns and the valences of the bonded atoms. In a majority of cases, particularly with molecules having a total of only 20 atoms or so and one or more functional groups, the actual structure of the compound may be correctly assigned. Once proposed, a possible structure must be checked carefully to ascertain that it is consistent with all of the available data. If not, the structure is incorrect and alternative possibilities must be explored and tested against the data until one is found that is consistent with *all* of the spectral information.

EXERCISES

1. Define, explain, or give an example of the following:

   a. tetramethylsilane (TMS) as an internal standard

   b. (δ) delta-scale

   c. upfield shift

   d. $n + 1$ rule

   e. chemical shift

   f. integration curve

   g. *relative* ratio of integrated peaks

   h. *absolute* ratio of integrated peaks

   i. homotopic hydrogen atoms

   j. enantiotopic hydrogen atoms

   k. diastereotopic hydrogen atoms

   l. coupling constant

   m. chemically equivalent protons

   n. diamagnetic shielding

   o. heterotopic hydrogen atoms

2. Why is it important to specify the solvent that is used when reporting chemical shifts?

3. Demonstrate that the H_b atoms in 1-nitropropane (**1**) are enantiotopic.

4. Label the sets of connectively equivalent sets of protons in the molecules (**a**)–(**j**) according to whether they are homotopic, enantiotopic, or diastereotopic. Using Tables 8.3 and 8.4, predict the approximate chemical shifts, in δ, and the splitting patterns expected for each chemically distinct type of hydrogen atom that you identify. Assume in this analysis that the coupling constants of all nearest neighbors are identical.

   a. CH_3CHCH_3
          |
          Cl

   b. $C_6H_5CHCH_3$
          |
          Br

   c. $(CH_3)_2CHCH(CH_3)_2$

   d. $(CH_3)_3CCCH_2CH_3$ (with C=O)

   e.

   f. $CLCH_2CH_2CH_2CL$

   g.

   h. $(CH_3)_2CHC$—OH (with C=O)

   i.

   j.

5. By applying the criterion of Equation 8.12, predict whether the multiplicities of the following sets of coupled nuclei, H_a and H_b, could be predicted by a first-order analysis. Note the units used to report the chemical shifts of the nuclei and the frequency of the *rf* oscillator for the instrument used to obtain the shifts.

   a. δ_{H_a} 1.3 ppm, δ_{H_b} 3.9 ppm, J_{ab} = 8.5 Hz, 200-MHz instrument

   b. δ_{H_a} 0.8 ppm, δ_{H_b} 1.0 ppm, J_{ab} = 2.5 Hz, 200-MHz instrument

    c. δ_{H_a} 0.8 ppm, δ_{H_b} 1.0 ppm, J_{ab} = 8.5 Hz, 200-MHz instrument

    d. δ_{H_a} 0.8 ppm, δ_{H_b} 1.0 ppm, J_{ab} = 8.5 Hz, 500-MHz instrument

    e. δ_{H_a} 0.8 ppm, δ_{H_b} 1.0 ppm, J_{ab} = 2.5 Hz, 200-MHz instrument

6. The ^{1}H NMR chemical shifts, splitting patterns, and *relative* numbers of hydrogen atoms for three compounds are provided in **a–c**. Deduce one or more structures consistent with these data. You may find it helpful to sketch these spectra on a sheet of paper.

    a. C_4H_9Br: δ 1.04 (6 H) doublet, δ 1.95 (1 H) multiplet, δ 3.33 (2 H) doublet

    b. $C_3H_6Cl_2$: δ 2.2 (2 H) quintet, δ 3.75 (1 H) triplet

    c. $C_5H_{11}Br$: δ 0.9 (3 H) doublet, δ 1.8 (1.5 H) complex multiplet, δ 3.4 (1 H) triplet

7. Compute the *relative* and *absolute* ratios for the different types of hydrogens in each pair of molecules shown below.

8. Figures 8.36–8.42 are of the compounds whose structures are shown on the spectra. Interpret these spectra as completely as possible in terms of the observed chemical shifts, integrations, and splitting patterns.

9. Provide structures of the compounds **7–9**, whose ^{1}H NMR spectra are given in Figures 8.43–8.45, respectively.

    a.   compound **7**, $C_{10}H_{14}$

    b.   compound **8**, C_3H_8O

    c.   compound **9**, C_8H_9Cl

Refer to the online resources and use the spectra viewer and Tables 8.1–8.8 as needed to answer the blue-numbered questions on spectroscopy.

Figure 8.36
^{1}H NMR spectrum of carvone (300 MHz, CDCl$_3$).

Figure 8.37
¹H NMR spectrum of ethyl phenyl ether (300 MHz, CDCl₃).

Figure 8.38
¹H NMR spectrum of 2-butanone (300 MHz, CDCl₃).

Figure 8.39
¹H NMR spectrum of butanoic acid (300 MHz, CDCl₃).

Figure 8.40
*¹H NMR spectrum of
N-ethylaniline (300 MHz,
CDCl₃).*

Figure 8.41
*¹H NMR spectrum of ethyl
acetate (300 MHz, CDCl₃).*

Figure 8.42
*¹H NMR spectrum of butanal
(300 MHz, CDCl₃).*

Figure 8.43
*¹H NMR spectrum (300 MHz, CDCl₃) for compound **7**.*

Figure 8.44
*¹H NMR spectrum (300 MHz, CDCl₃) for compound **8**.*

Figure 8.45
*¹H NMR spectrum (300 MHz, CDCl₃) for compound **9**.*

REFERENCES

1. Silverstein, R. M.; Webster, F. X.; Kiemle, D. *Spectrometric Identification of Organic Compounds*, 7th ed., John Wiley & Sons, New York, 2005.

2. Crews, P.; Rodriquez, J.; Jaspars, M. *Organic Structure Analysis*, Oxford University Press, New York, 1998.

3. King, R. W.; Williams, K. R. "The Fourier Transform in Chemistry-NMR," *J. Chem.* Ed., 1989, *66*, A213–A219, A243–A248.

4. King, R. W.; Williams, K. R. "The Fourier Transform in Chemistry-NMR," *J. Chem.* Ed., 1990, *67*, A93–A99, A100–A105, A125–A137.

5. Breitmaier, E. *Structure Elucidation by NMR in Organic Chemistry: A Practical Guide*, 3rd rev. ed., John Wiley & Sons, Chichester and New York, 2002.

6. Pouchert, C. J., ed. *Aldrich Library of NMR Spectra*, vol. I–II, Aldrich Chemical Co., Milwaukee, WI, 1983.

7. Pouchert, C. J.; Behnke, J., eds. *Aldrich Library of ^{13}C and ^1H NMR Spectra*, vol. I–III, Aldrich Chemical Co., Milwaukee, WI, 1993.

8. *Sadtler Nuclear Magnetic Resonance Spectra*, vol. 1–118, Sadtler Research Laboratories, Philadelphia, PA. Compilation of 64,000 ^1H NMR spectra.

9. *Sadtler 300 MHz Proton NMR Standards*, vol. 1–24, Sadtler Research Laboratories, Philadelphia, PA. Compilation of 12,000 300 MHz ^1H NMR spectra.

Carbon-13 Nuclear Magnetic Resonance (^{13}C NMR) Spectroscopy

Carbon-12, ^{12}C, has no **nuclear spin**, but its carbon-13 isotope, ^{13}C, has a nuclear spin of 1/2, as does hydrogen, ^1H. The **natural abundance** of ^{13}C is about 1.1%, so unless substances are prepared from precursors having artificially high levels of this isotope, only 1.1% of the carbon atoms in a compound will undergo the *spin flip* or *resonance* that is characteristic of the NMR experiment. In other words, only about one in a hundred carbon atoms of a sample having ^{13}C present at natural abundance will produce an absorption in the NMR spectrum.

This low level of the NMR-active isotope of carbon makes it more difficult to obtain a suitable **carbon nuclear magnetic resonance** (^{13}C NMR) spectrum. For example, whereas it is usually possible to measure a ^1H NMR spectrum in a few minutes, it may take tens of minutes or even hours to accumulate enough data to produce a ^{13}C NMR spectrum in which the signal-to-noise ratio is high enough to observe the resonances due to the carbon atoms. If possible, larger sample sizes are therefore used to expedite data acquisition for a ^{13}C NMR spectrum.

The principles of ^{13}C NMR spectroscopy are the same as those for ^1H NMR spectroscopy. When placed in an external magnetic field, ^{13}C nuclei adopt one of two spin states, whose energy difference is determined `by the strength of the field (Fig. 8.19). Exposing these nuclei to electromagnetic radiation that has the appropriate energy produces the resonance condition (Eq. 8.7). This means that the spin flip associated with transition from one energy level to another (Fig. 8.20) may occur.

Like hydrogen nuclei, the ^{13}C nuclei exist in different electronic environments in the molecule, and this gives rise to *chemical shifts* that are characteristic of the magnetic environments of the various carbon atoms. As in ^1H NMR spectroscopy, tetramethylsilane, $(CH_3)_4Si$, is commonly used as the reference compound for measuring chemical shifts in ^{13}C NMR spectra. Most ^{13}C resonances are *downfield* from that of TMS and are given *positive* values by convention. These chemical

Figure 8.46
Broadband proton-decoupled
^{13}C NMR spectrum of 2-butanone
(75 MHz, $CDCl_3$).

shifts are computed according to Equation 8.8 and are reported as parts per million (ppm) on a δ scale (Fig. 8.46), again in analogy to the procedure used with ^1H NMR shifts.

Modern instruments are capable of measuring both ^1H and ^{13}C NMR spectra. When used for the latter purpose, the spectrometer is usually operated in a mode that permits decoupling of all of the proton(s) that are *nearest neighbors* to the ^{13}C atom, a process termed **broadband proton decoupling**. Applying this technique provides ^{13}C NMR spectra that do *not* show the effects of the spin-spin splitting observed in ^1H NMR spectra. Because of the low natural abundance of this isotope in organic compounds, the probability of having two such atoms bound to one another is only *one* in *ten thousand*, so coupling between adjacent ^{13}C atoms is not observed in the typical spectrum. Proton-decoupled ^{13}C NMR spectra are thus extremely simple, as they consist of a single sharp resonance line for each magnetically distinct carbon atom.

The characteristics of a ^{13}C spectrum are exemplified in the spectrum of 2-butanone (Fig. 8.46) where the four different types of carbon atoms produce four separate resonances. Note that three peaks near δ 79.5 ppm in the spectrum are associated with the solvent, deuterochloroform, $CDCl_3$. The single carbon atom of this solvent produces more than one peak because of *deuterium*-carbon splitting, which is *not* eliminated by the broadband decoupling technique. Deuterium has a nuclear spin, I_z, of 1, so applying Equation 8.10 predicts a *triplet* for the carbon atom, as is observed experimentally.

A specialized technique associated with ^{13}C NMR spectroscopy bears mention here because it provides information regarding the number of hydrogen atoms attached to a given carbon atom. This technique is referred to as **distortionless enhancement of polarization**, commonly abbreviated as DEPT. Although discussion of the basis for this technique is beyond the scope of this textbook, DEPT spectra are easy to interpret. Specifically, if there is an *odd* number of hydrogen atoms attached to a carbon atom, as is the case for a methyl (CH_3) or methane (CH) carbon atom, the DEPT spectrum will have a *positive* peak at the same chemical shift as observed in a normal ^{13}C NMR spectrum. If an *even* number of hydrogen atoms is attached, as with a methylene (CH_2) carbon atom, a *negative* peak is seen in the DEPT spectrum, and *no* peak is observed when *no* hydrogen atom is attached to a carbon atom. The DEPT spectrum of 2-butanone is provided in

Figure 8.47
DEPT ^{13}C NMR spectrum of 2-butanone (75 MHz, CDCl$_3$).

Figure 8.46. The lower trace is the usual ^{13}C spectrum, whereas the upper one is the DEPT spectrum. As expected, there are two positive peaks representing the resonances for the two magnetically distinct methyl groups and one negative peak for the methylene group of the molecule. Note that the peak at about δ 210 in Figure 8.45 is absent in the DEPT spectrum, proof that this resonance is due to the carbonyl carbon atom.

In contrast to decoupled spectra, *coupled* ^{13}C NMR spectra may be extremely complex and difficult to interpret because the magnitudes of ^{1}H–^{13}C coupling constants are on the order of 120–230 Hz, so overlapping of peaks may become a serious problem. However, special instrumental techniques provide the type of information that would normally be derived by analysis of splitting patterns, namely the number of hydrogen atoms bound to the carbon atom of interest. For example, **off-resonance decoupling** simplifies the spectrum as is illustrated in Figure 8.47, where the multiplets for each carbon atom correspond to the number of peaks predicted by the same $n + 1$ rule (Eq. 8.11) that is used to predict spin-spin splitting patterns in ^{1}H NMR spectra. The difference is that n corresponds to the number of hydrogen nuclei that are *directly bound* to the carbon atom rather than the number of *nearest neighbors* as defined for predicting the splitting patterns for proton spectra. Even *off-resonance decoupled* spectra may be complicated, so only the *chemical shifts* derived from *broadband decoupled* spectra are given for the starting materials and products in this textbook.

The chemical shifts of carbons atoms are *much more sensitive* to their environments than are those of protons, so the range of chemical shifts for ^{13}C resonances is much greater than for ^{1}H. Specifically, whereas most types of hydrogen atoms resonate within 10 ppm downfield of TMS, the chemical shifts of carbon atoms occur over a range of some 220 ppm downfield of TMS. This means that it is unlikely that two different types of carbon atoms will resonate at exactly the same chemical shift. Consequently, the number of peaks in a proton-decoupled ^{13}C NMR spectrum may *tentatively* be interpreted as being equal to the number of different types of carbon atoms present in the molecule. This principle is illustrated both by the observation of four peaks in the ^{13}C NMR spectrum of 2-butanone (Fig. 8.47) and of six peaks in that of methyl benzoate (Fig. 8.48). In the latter case,

Figure 8.48
13C NMR spectrum of methyl benzoate (75 MHz, CDCl3).
13C NMR: δ 51.8, 128.5, 129.7, 130.5, 132.9, 166.8.

the carbon atoms *ortho* and *meta* to the ester function are magnetically equivalent, so there are only four resonances for the carbon atoms of the aromatic ring.

As with ^{1}H NMR spectroscopy, tables of chemical shifts have been developed for carbon atoms in different environments; one example is provided by Table 8.5. Examination of the data in the table shows that some of the structural features in a molecule that produce downfield chemical shifts in ^{1}H NMR spectra do the same in ^{13}C NMR spectra. For example, an electronegative substituent such as a carbonyl group or hetero-atom on a carbon atom causes a downfield shift in the resonance, relative to a saturated hydrocarbon, just as it would with hydrogen nuclei. This effect is caused by the *deshielding* effect of the electronegative moiety. A second factor in defining chemical shifts in ^{13}C NMR spectra is the hybridization of the carbon atom, as is seen by comparing the range of chemical shifts for the sp^3- (10–65 ppm), sp^2- (115–210 ppm), and sp-hybridized (65–85 ppm) carbon atoms of the general structures shown in Table 8.5.

Finally, the nature of a substituent G that is zero (α-effect), one (β-effect), or two (γ-effect) atoms away from the carbon atom of interest, as shown in **10,** may also affect the chemical shift. Such substituent effects are remarkably additive, so it is possible to predict the expected chemical shifts for carbon atoms in a molecule with reasonable accuracy. Table 8.6 is a compilation of some of these additivity effects for acyclic alkanes. These may be used in conjunction with the data in Table 8.7 to predict ^{13}C chemical shifts for a compound. Similar tables of additivity effects for other classes of organic compounds are published in the references at the end of this section.

$$\overset{\gamma}{\diagup}\overset{\alpha}{\diagup}\underset{\beta}{\diagdown}\diagup G$$

10

The following example illustrates the method for making these predictions. Consider 1-chlorobutane (**11**), $CH_3CH_2CH_2CH_2Cl$. The computed chemical shifts, δ,

Table 8.5 ^{13}C *Chemical Shifts in Carbon-13 Magnetic Resonance (^{13}C NMR) Spectroscopy*

Absorbing Carbon Atom (shown as **C** or **Ar**)	Approximate δ, ppm
R**CH₂**CH₃	13–16
RCH₂**CH₃**	16–25
R₃**CH**	25–38
CH₃C—R (C=O)	30–32
CH₃C—OR (C=O)	20–22
R**CH₂**–Cl	40–45
R**CH₂**–Br	28–35
R**CH₂**–NH₂	37–45
R**CH₂**–OH	50–65
R**C**≡CH	67–70
RC≡**CH**	74–85
R**CH**=CH₂	115–120
RCH=**CH₂**	125–140
R**C**≡N	118–125
ArH	125–150
R**C**—NR′R″ (C=O)	160–175
R**C**—OR′ (C=O)	170–175
R**C**—OH (C=O)	175–185
R**C**—H (C=O)	190–200
R**C**—CH₃ (C=O)	205–210

are determined from the observed chemical shifts (Table 8.7) of 13.4 ppm and 25.2 ppm for C1 (C4) and C2 (C3), respectively, for butane as follows:

C1: 13.4 + 31 = 44.4 ppm   C2: 25.2 + 11 = 36.2 ppm

C3: 25.2 − 4 = 21.2 ppm   C4: 13.4 + 0 = 13.4 ppm

The observed chemical shifts for **11** are provided below the structure. The agreement is remarkable and shows the impressive predictive power of this approach.

$$CH_3—CH_2—CH_2—CH_2—Cl$$
13.4   20.4   35.0   44.6
11
1-Chlorobutane

Table 8.6 *Incremental Substituent Effects (ppm) on Replacement of H by Y in Alkanes; Y is Terminal or Internal** *(+ downfield, − upfield)*

	α		β		
Y	Terminal	Internal	Terminal	Internal	γ
CH$_3$	+9	+6	+10	+8	−2
CH=CH$_2$	+20		+6		−0.5
C;CH	+4.5		+5.5		−3.5
COOH	+21	+16	+3	+2	−2
COOR	+20	+17	+3	+2	−2
COR	+30	+24	+1	+1	−2
CHO	+31		0		−2
Phenyl	+23	+17	+9	+7	−2
OH	+48	+41	+10	+8	−5
OR	+58	+51	+8	+5	−4
OCOR	+51	+45	+6	+5	−3
NH$_2$	+29	+24	+11	+10	−5
NHR	+37	+31	+8	+6	−4
NR$_2$	+42		+6		−3
CN	+4	+1	+3	+3	−3
F	+68	+63	+9	+6	−4
Cl	+31	+32	+11	+10	−4
Br	+20	+25	+11	+10	−3
I	−6	+4	+11	+12	−1

*Add these increments to the shift values of the appropriate carbon atom in Table 8.7.
Source: Reference 1.

Returning now to the broadband-decoupled spectrum of 2-butanone (Fig. 8.46), the resonance that is far downfield near δ 210 is assignable to the carbonyl carbon atom (Table 8.5). The remaining three resonances near δ 37, 30, and 8 ppm may be assigned either (a) by recognizing that the carbonyl group causes the carbon atoms to which it is directly attached to be shifted farther downfield than those atoms more distant from it (α-effect) or (b) by using the data of Tables 8.6 and 8.7 to compute the expected chemical shifts. These computations are provided below. Again, comparison with the experimentally observed shifts provided in Figure 8.46 shows the value of the method of substituent additivity in assigning resonances to the carbon atoms associated with them. When data from DEPT spectra are available, the task of making ^{13}C assignments is greatly simplified. In the case of 2-butanone, the DEPT spectrum (Fig. 8.47) allows immediate identification of the chemical shifts

Table 8.7 *^{13}C Shifts for Some Linear and Branched-Chain Alkanes (ppm from TMS)*

Compound	C1	C2	C3	C4	C5
Methane	−2.3				
Ethane	5.7				
Propane	15.8	16.3	15.8		
Butane	13.4	25.2	25.2		
Pentane	13.9	22.8	34.7	22.8	13.9
Hexane	14.1	23.1	32.2	32.2	23.1
Heptane	14.1	23.2	32.6	29.7	32.6
Octane	14.2	23.2	32.6	29.9	29.9
Nonane	14.2	23.3	32.6	30.0	30.3
Decane	14.2	23.2	32.6	31.1	30.5
2-Methylpropane	24.5	25.4			
2-Methylbutane	22.2	31.1	32.0	11.7	
3-Methylpentane	11.5	29.5	36.9	18.8 (3-methyl group)	

Source: Reference 1.

for the carbonyl and methylene carbon atoms; differentiation between the two methyl groups of the molecule, however, requires use of the strategies (a) and/or (b) described above.

C1: −2.3 + 30 = 27.7 ppm   C2: cannot be computed from data in Tables 8.6 and 8.7

C3: 5.7 + 31 = 36.7 ppm   C4: 5.7 + 1 = 6.7 ppm

The relative intensities of the various peaks are *not* identical in the ^{13}C NMR spectrum of 2-butanone, even though a single carbon atom is responsible for each resonance. The physical basis for this is beyond the scope of this discussion. Although there is *not* a 1:1 correlation between the number of carbon atoms producing a particular resonance and the intensity of that resonance, there is a *rough* correlation between intensities as reflected in peak heights and whether or not any hydrogen atoms are on the carbon atom resonating at a particular chemical shift. As a general rule, if there are *no* hydrogen atoms attached to the carbon atom, the intensity of the resonance associated with that atom will be relatively low. This relationship is seen in Figure 8.46, although you should note that the signal at δ 208.3, which is the resonance for the carbonyl carbon atom, has been electronically amplified by a factor of 2 relative to all other resonances in the spectrum. In contrast, note that the intensity of the methylene carbon atom is greater than that of either of the carbon atoms of the methyl groups in the molecule, a fact showing that there is *no general* relationship between the number of hydrogen nuclei on a carbon atom and the intensity of the absorption for that atom. In short, although integrated ^1H NMR spectra provide invaluable information about the number of the various types of hydrogen nuclei present, this type of information is not available for ^{13}C NMR spectra unless special techniques are used.

Consideration of the ^{13}C NMR spectrum (Fig. 8.48) of methyl benzoate (**12**) shows how assigning peaks may be done on the basis of information in Table 8.5 and use of intensities as a measure of whether or not a carbon atom bears any hydrogens. First of all, there are a total of *six* magnetically distinct carbon atoms in the molecule because of its symmetry. According to the data in Table 8.5, the resonance at δ 166.8 must be due to the carbonyl carbon atom, C1′, and, as expected, is of relatively low intensity. Similarly, the table supports assigning the absorption at δ 51.8 to the methoxy carbon atom, C2′.

The remaining four resonances are in the aromatic region of the ^{13}C NMR spectrum and may be assigned as follows. Based on its low intensity, the peak at δ 130.5 is probably due to C1 of the aromatic ring. The carbon atoms *ortho* and *para* to an ester group should be *downfield* of those for the *meta* carbon atoms because the ester moiety deshields the *ortho* and *para* positions by delocalizing π-electrons, as illustrated by resonance structures **12a–12c** and by inductively withdrawing electrons from the ring via σ-bonds. This means that the resonance at δ 128.5 is for the two meta carbon atoms. A *tentative* assignment of the remaining two resonances at δ 129.7 and 132.9 in the aromatic region is made on the basis that there are twice as many *ortho* as *para* carbon atoms. Thus the more intense resonance at δ 129.7 is for C2 and that at δ 132.9 for C4. These assignments are consistent with those calculated using the substituent additivity effects provided in Table 8.8.

Methyl benzoate

Table 8.8 *Incremental Shifts of the Aromatic Carbon Atoms of Monosubstituted Benzenes (ppm from Benzene at 128.5 ppm, + downfield, − upfield); Carbon Atom of Substituents (ppm from TMS)*

Substituent	C1 (Attachment)	C2	C3	C4	C of Substituent (ppm from TMS)
H	0.0	0.0	0.0	0.0	
CH_3	9.3	+0.7	−0.1	−2.9	21.3
CH_2CH_3	+15.6	−0.5	0.0	−2.6	29.2 (CH_2), 15.8 (CH_3)
$CH(CH_3)_2$	+20.1	−2.0	0.0	−2.5	34.4 (CH), 24.1 (CH_3)
$C(CH_3)_3$	+22.2	−3.4	−0.4	−3.1	34.5 (C), 31.4 (CH_3)
$CH=CH_2$	+9.1	−2.4	+0.2	−0.5	137.1 (CH), 113.3 (CH_2)
CH≡CH	−5.8	+6.9	+0.1	+0.4	84.0 (C), 77.8 (CH)
C_6H_5	+12.1	−1.8	−0.1	−1.6	
CH_2OH	+13.3	−0.8	−0.6	−0.4	64.5
OH	+26.6	−12.7	+1.6	−7.3	
OCH_3	+31.4	−14.4	+1.0	−7.7	54.1
OC_6H_5	+29.0	−9.4	+1.6	−5.3	

Table 8.8 *Incremental Shifts of the Aromatic Carbon Atoms of Monosubstituted Benzenes (ppm from Benzene at 128.5 ppm, + downfield, − upfield); Carbon Atom of Substituents (ppm from TMS) (Continued)*

Substituent	C1 (Attachment)	C2	C3	C4	C of Substituent (ppm from TMS)
O‖OCCH₃	+22.4	−7.1	−0.4	−3.2	23.9 (CH₃), 169.7 (C=O)
O‖CH	+8.2	+1.2	+0.6	+5.8	192.0
O‖CCH₃	+7.8	−0.4	−0.4	+2.8	24.6 (CH₃), 195.7 (C=O)
O‖CC₆H₅	+9.1	+1.5	−0.2	+3.8	196.4 (C≡O)
O‖COH	+2.9	+1.3	+0.4	+4.3	168.0
O‖COCH₃	+2.0	+1.2	−0.1	+4.8	51.0 (CH₃), 166.8 (C=O)
O‖CCl	+4.6	+2.9	+0.6	+7.0	168.5
C≡N	−16.0	+3.6	+0.6	+4.3	119.5
NH₂	+19.2	−12.4	+1.3	−9.5	
N(CH₃)₂	+22.4	−15.7	+0.8	−11.8	40.3
O‖NHCCH₃	+11.1	−9.9	+0.2	−5.6	
NO₂	+19.6	−5.3	+0.9	+6.0	
F	+35.1	−14.3	+0.9	−4.5	
Cl	+6.4	+0.2	+1.0	−2.0	
Br	−5.4	+3.4	+2.2	−1.0	
I	−32.2	+9.9	+2.6	+7.3	
SO₂NH₂	+15.3	−2.9	+0.4	+3.3	

Source: Reference 1.

EXERCISES

1. Based on the substituent effects shown in Table 8.8, calculate the chemical shifts expected for the various carbon atoms of the aromatic ring of methyl benzoate (**12**).

2. The ^{13}C NMR spectrum of 3-methyl-2-butanone, $CH_3C(=O)CH(CH_3)_2$, exhibits peaks at δ 18.2, 27.2, 41.6, and 211.2. Provide assignments of the various resonances to the carbon atoms responsible for them, using Tables 8.5–8.7.

3. Compute the ^{13}C NMR chemical shifts expected for 2-methyl-1-propanol, $(CH_3)_2CHCH_2OH$.

REFERENCES

1. Whitesell, J. K.; Minton, M. A. *Stereochemical Analysis of Alicyclic Compounds by C-13 NMR Spectroscopy*, Chapman & Hall, London, 1987.

2. Silverstein, R. M.; Webster, F. X.; Kiemle, D. *Spectrometric Identification of Organic Compounds*, 7th ed., John Wiley & Sons, New York, 2005.

3. Crews, P.; Rodriquez, J.; Jaspers, M. *Organic Structure Analysis*, Oxford University Press, New York, 1998.

4. Breitmaier, E. *Structure Elucidation by NMR in Organic Chemistry: A Practical Guide*, 3rd rev. ed., John Wiley & Sons, Chichester and New York, 2002.

5. Pouchert, C. J.; Behnke, J., eds. *Aldrich Library of ^{13}C and ^{1}H NMR Spectra*, vol. I–III, Aldrich Chemical Co., Milwaukee, WI, 1993.

6. ^{13}C *NMR Spectra*, vol. 1–210, Sadtler Research Laboratories, Philadelphia, PA. Compilation of 42,000 ^{13}C NMR spectra.

Preparing Samples for NMR Spectroscopy

Organic chemists normally obtain NMR spectra of liquid samples contained in special glass tubes of high-precision bore. Although spectra may be obtained of pure liquids having low viscosities, the substrate of interest, regardless of its normal physical state, is generally dissolved in a deuterated solvent. The spectra of viscous samples are unsatisfactory because broad absorptions may be observed, causing a loss of resolution between the peaks.

All modern NMR spectrometers operate in the pulsed Fourier transform (FT) mode; for technical reasons, this means that deuterated solvents must be used. Deuterochloroform, $CDCl_3$, is the most common such solvent, although a number of other, albeit more expensive, solvents such as acetone-d_6, $(CD_3)_2CO$, dimethyl sulfoxide-d_6, CD_3SOCD_3, and benzene-d_6, C_6D_6 are also available. Deuterium oxide, D_2O, may be used if the sample is water-soluble.

Proton-containing solvents are generally not appropriate because the intense resonance due to the hydrogen atoms of the solvent may obscure absorptions of the sample itself. Although deuterium, ^{2}H, also possesses the property of nuclear spin, its resonance appears in a region of the NMR spectrum different from that of ^{1}H. Replacing all of the hydrogen atoms of solvent with deuterium removes such resonances from the ^{1}H NMR spectrum, but for practical reasons, weak absorptions due to residual protium is normally observed. This is because common organic solvents containing 100 atom % D are too expensive for routine spectral work. For example, the approximate cost of $CDCl_3$ having 100 atom % D is $1.97/g, whereas that having 99.8 atom % D is $0.24/g.

Minor resonances due to solvent must therefore be taken into account when interpreting ^{1}H NMR spectra. Moreover, in solvents in which there is more than one deuterium atom on a carbon atom, the residual *protium* on that atom will have a splitting pattern that reflects the presence of the deuterium atoms, which have a nuclear spin, I_z, of 1. The resonance of the residual hydrogen present in deuteroacetone appears as a *quintet* centered at δ 2.17, for example. This pattern arises from the coupling of a single residual hydrogen with the two deuterium atoms that would be present on the same carbon atom if the acetone-d_6 has 99 atom % or greater deuterium content. You may confirm that the multiplet for the residual hydrogen should be a quintet by using Equation 8.10.

Of course, the *carbon* atom(s) of an NMR solvent appear in the ^{13}C NMR spectrum whether or not the solvent is deuterated. As with residual protium, splitting of the carbon resonance(s) of the solvent by attached deuterium atoms is observed.

The appearance of *three* peaks for the carbon atom of $CDCl_3$ illustrates this (Figs. 8.46–8.48). Care should therefore be exercised when interpreting ^{13}C NMR spectra to avoid mistaking peaks due to solvent for those of the sample itself. If overlap of sample and solvent peaks is suspected, it may be necessary to select a different solvent for a ^{13}C NMR spectrum.

The concentration of the solutions appropriate for obtaining NMR spectra is in the range of 5–15% by weight; about 1 mL of solution is needed to fill an NMR tube to the proper level, although as little as 0.6 mL can be used. The tube must be scrupulously clean and dry, and the solution must be free of undissolved solids arising from the sample itself or even dust. Furthermore, as a consequence of contact of the solute or the solvent with metals such as the iron in "tin" cans, trace amounts of ferromagnetic impurities may contaminate the solution; the result is a spectrum that has only broad, weak absorptions.

A solution that contains any solid material must be filtered *prior* to measuring its NMR spectrum. This is done most easily by inserting a small plug of glasswool into a Pasteur pipet and then filtering the solution through the plug into the NMR tube. Following their use, NMR tubes should be cleaned and thoroughly dried. They are best stored in a closed container or in an inverted position to minimize the possibility that particles of dust will contaminate the inside of the tube.

8.4 ULTRAVIOLET AND VISIBLE SPECTROSCOPY

Introduction

Absorption of light of the proper energy produces electronic transitions within molecules, and these transitions are the basis for UV-Vis spectroscopy. The absorptions occur in the ultraviolet and/or visible regions, which are adjacent to one another in the electromagnetic spectrum (Fig. 8.1). We can visually detect absorption of light by organic compounds in the visible region because the compounds appear colored as a result; in contrast, colorless compounds absorb energy in the UV region, and the phenomenon is not detectable by the human eye.

Principles

The phenomenon underlying UV-Vis spectroscopy is the excitation of an electron from a lower- to a higher-energy electronic state. The two kinds of electronic excitations of greatest value to organic chemists involve promotion of an electron

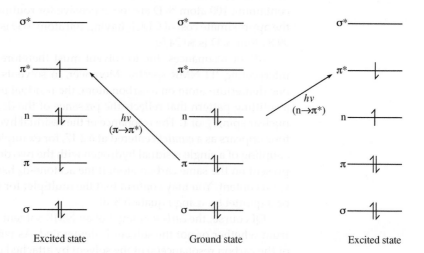

Figure 8.49
Energy diagram showing electronic transitions.

originally in either a nonbonding molecular orbital (an *n*-electron) or a *σ*-bonding molecular orbital (a *π*-electron) into an antibonding molecular orbital (Fig. 8.49). By referring to Figure 8.1, you can see that the energies needed for such excitations range from 38 kcal/mol to more than 140 kcal/mol, values corresponding to light in the wavelength range of 750–200 nm (Eq. 8.1). Shorter wavelengths of light are required for excitation of *σ*-type electrons, and the resulting spectra, although important for understanding some fundamental molecular properties, are not of much value for assigning structures to organic molecules. As seen in Figure 8.49, *π*-type antibonding molecular orbitals, designated *π**, are lower in energy than the corresponding *σ**-orbitals. Consequently, the transitions caused by absorption of light in the UV-Vis region involve populating the *π**-state by excitation of an electron from an *n*-orbital ($n \rightarrow \pi^*$) or a *π*-orbital ($\pi \rightarrow \pi^*$).

An excited state produced by promotion of an electron to an antibonding molecular orbital can return to the ground state in a process variously called *relaxation* or *decay*. Although the relaxation may involve radiationless loss of energy in the form of heat, it may also occur with emission of light. An interesting example of the latter is associated with the chemiluminescence of luminol (Sec. 20.4).

Practical Considerations

A UV-Vis spectrum is subject to a number of variables, among which are solvent, concentration of the solution being examined, and the pathlength of the cell through which the light passes. The amount of light absorbed by a particular solution is quantitatively defined by the Beer-Lambert law (Eq. 8.14).

$$A = \log \frac{I_0}{I} = \epsilon c l \qquad (8.14)$$

where A = absorbance or optical density

I_0 = intensity of radiation incident on sample at specified wavelength

I = intensity of radiation transmitted through sample at same wavelength

ϵ = molar extinction coefficient

c = concentration of sample in mol/liter

l = length of cell in cm

In a UV-Vis spectrum, absorbance, A, is plotted as a function of the wavelength, λ, of the incident radiation. The concentration (c) and cell length (l) are known, and the absorbance (A) can be determined from the spectrum at each wavelength, so the molar extinction coefficient (ϵ) may be calculated from Equation 8.14. These coefficients typically range from 10 to 100,000. Citral (**13**), a natural product that can be obtained by steam-distilling lemon grass oil (Sec. 4.5), has the UV spectrum provided in Figure 8.50. Because **13** is colorless, there is no absorption in the visible region of the spectrum. Two separate traces are shown on the spectrum because solutions having different concentrations were needed so that maxima for both strong and weak absorptions could be observed.

For reporting a UV-Vis spectrum, only the wavelength, λ_{max}, and intensity, ϵ or $\log \epsilon$, of any maxima are typically given, along with the solvent in which the measurement was made. The solvent should be stated because the values of both ϵ and λ_{max} are solvent-dependent. The critical information contained in Figure 8.50 could thus be expressed in the following way:

$$\lambda_{max}^{cyclohexane} \text{ 230 nm, } \log \epsilon \text{ 4.06; 320 nm, } \log \epsilon \text{ 1.74}$$

The absorption maxima in **13** are due to the *α,β*-saturated carbonyl **chromophore**, where chromophore is defined as the functional group(s) responsible for the

Figure 8.50
UV spectrum of citral.

absorption. The more intense absorption at 230 nm arises from a $\pi \to \pi^*$ electronic excitation, whereas that at 320 nm results from an $n \to \pi^*$ excitation.

Preparing Samples for Ultraviolet-Visible Spectroscopy

Almost all UV-Vis spectra are obtained on solutions, and a wide variety of organic solvents as well as water may be used for UV-Vis spectroscopy. These solvents share the common property of not absorbing significantly at wavelengths greater than about 200 nm, as indicated in Table 8.9. The wavelength provided for each solvent is the so-called cutoff point, below which the solvent has appreciable absorption, making it unusable for this type of spectroscopy. "Technical" and "reagent-grade" solvents often contain light-absorbing impurities and should be purified before use, but more expensive "spectral-grade" solvents need no purification.

The concentration of the solution should be such that the observed value of A is in the range of 0.3–1.5 to provide the greatest accuracy in the measurement. The approximate concentration should be estimated from the value of ϵ for any

Table 8.9 *Solvents for Ultraviolet-Visible Spectroscopy*

Solvent	Useful Spectra Range (Lower Limit in nm)
Acetonitrile	<200
Chloroform	245
Cyclohexane	205
95% Ethanol	205
Hexane	200
Methanol	205
Water	200

chromophores present, from which c can be determined using Equation 8.14. As a rough rule of thumb, 0.01–0.001 M solutions will give absorbances of appropriate magnitude for excitations of low intensity (log ϵ of about 1.0). This solution may then be diluted to permit more intense absorptions to be observed in the desired range for A.

The cells for containing the solutions must be quartz for ultraviolet spectra because other varieties of glass absorb light in the UV region of the spectrum; those for visible spectroscopy are made of less expensive borosilicate glass. Because this latter type of glass is opaque to light in the UV region, it is not suitable for UV cells.

The concentration of solute must be accurately known so that you may precisely measure A and hence of ϵ. This requires the solute to be accurately weighed, *quantitatively* transferred to a volumetric flask, and precisely diluted. Furthermore, introducing even a minute amount of an intensively absorbing impurity into the solution may have a dramatic effect on the observed UV spectrum, so it is critical to use clean glassware. To minimize contamination, the cells should be thoroughly rinsed both before and after use with the solvent being used for the measurement, and the outside optical surfaces of the cells must be clean and free of fingerprints. Moreover, you should not use acetone as a solvent because trace residues of acidic or basic catalysts on the quartz may promote an aldol condensation to form trace quantities of 4-methyl-3-penten-2-one (**14**). Because this enone has a high value of ϵ, its presence may interfere with obtaining a precise measurement of ϵ. The cells are best cleaned using the same pure solvent that is being used for the solute; they must also be rinsed thoroughly when changing from one concentration to another.

14
4-Methyl-3-penten-2-one

Uses of Ultraviolet-Visible Spectroscopy

As mentioned earlier, UV-Vis spectroscopy may be used to help determine the structure of an unknown compound provided the compound contains a chromophore that absorbs in the UV-Vis region. A more common use of this spectroscopic technique, however, is for making quantitative measurements of the rates of reactions. Equation 8.14 dictates that if the quantities λ, l, and A are known for a solution, it is possible to determine c, the concentration of the absorbing species in solution. You can determine the magnitude of ϵ experimentally by preparing solutions of the sample of known concentration and then determining A for each of these solutions in a cell of known pathlength. Because the extinction coefficient and the pathlength, l, are known and have constant values, any change in A for a particular solution over time must result from a variation in c. By monitoring A as a function of time, you may determine the time-dependence of the concentration of the absorbing species, thus enabling you to measure the rate at which it is formed or consumed.

Quantitative determination of the relative rates of electrophilic bromination of substituted benzenes (Eq. 8.15), as described in Section 15.5, relies on this application of UV-Vis spectroscopy. The rates of reaction are found by following the disappearance of molecular bromine, whose maximum is in the visible region at 400 nm.

$$G-\hexagon + Br_2 \longrightarrow G-\hexagon{Br} + HBr \qquad (8.15)$$

EXERCISES

1. Define the following terms:

   a. $\pi \rightarrow \pi^*$ transition   e. visible spectrum

   b. I_0 f. "cut-off" point of a solvent

   c. conjugated π-system   g. electronic transition

   d. chromophore   h. absorbance

2. Give three properties of a good solvent for UV-Vis applications.

3. Why should acetone not be used to rinse quartz UV cells?

4. Suppose you were not careful in preparing a sample for UV-Vis analysis, so that its actual concentration, c, in methanol was 0.00576 g/L rather than the 0.00600 g/L that you thought it was.

   a. What is the error, in mg/L, in the mass of the sample present in solution?

   b. Suppose that a maximum was observed at a wavelength of 278 nm for this solution and that the value of A at this maximum was 0.623 when a 1-cm cell was used.

      i. Using the value of c that you thought you had for the solution, calculate ϵ for the sample, assuming that its molar mass is 100.

      ii. Show how the information regarding this maximum could be expressed.

      iii. What is the % error in ϵ associated with the inaccuracy in your knowing the actual concentration of the solution?

5. For each of the maxima given in the UV spectra of Figures 8.51–8.56, determine the value of A and of log ϵ.

REFERENCES

1. Silverstein, R. M.; Webster, F. X.; Kiemle, D. *Spectrometric Identification of Organic Compounds*, 7th ed., John Wiley & Sons, New York, 2005.

2. Crews, P.; Rodriquez, J.; Jasper, M. *Organic Structure Analysis*, Oxford University Press, New York, 1998.

3. *Sadtler Standard Ultraviolet Spectra*, vol. 1–170, Sadtler Research Laboratories, Philadelphia, PA. Compilation of 148,140 UV Spectra.

4. Phillips, J. P.; Bates, D.; Feuer, H.; Thyagarajan, B. S., eds. *Organic Electronic Spectra Data*, vol. I–XXXI, John Wiley & Sons, New York, 1996.

Figure 8.51
UV spectrum of 4-methyl-3-penten-2-one.

Figure 8.52
UV spectrum of 2-furaldehyde.

Figure 8.53
UV spectra of cis- (solid line) and trans- (dashed line) stilbene. Concentration 0.0500 g/L; pathlength 1.0 cm.

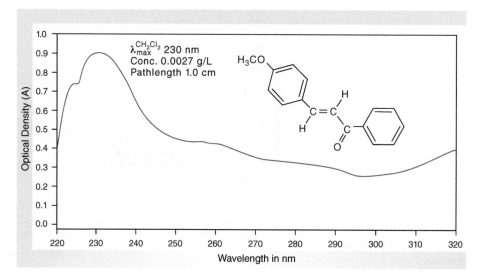

Figure 8.54
UV spectrum of trans- p-anisalacetophenone.

Figure 8.55
UV spectrum of 4,4-dimethyl-2-cyclohexen-1-one.

Figure 8.56
UV spectrum of styrene.

8.5 MASS SPECTROMETRY

Introduction

The characterization and identification of the structures of organic molecules are greatly aided by the application of various spectroscopic techniques such as IR (Sec. 8.2), NMR (Sec. 8.3), and UV-Vis (Sec. 8.4), all of which have the virtue of *not* destroying the sample being studied. Mass spectrometry (MS) is another important spectroscopic method used by organic chemists, but the small amount (~10^{-6} g) of sample needed for a measurement is destroyed in the process. Although this instrumental technique has a variety of applications, two of its most important are the determination of molar masses (molecular weights) and elemental compositions. It is *faster*, more *accurate*, and requires a much smaller amount of sample than the classic methods for obtaining these data. Mass spectrometers, because of their expense, are generally unavailable in the undergraduate organic laboratory. However, less expensive mass spectrometers that are combined with GLCs (GC-MS instruments, Sec. 6.5) are becoming more prevalent in teaching laboratories, so we'll give a brief introduction to mass spectrometry and the interpretation of mass spectra.

Principles

Instrumentation

Before considering the specific structural information that a mass spectrometer can provide, let's develop a general understanding of how a mass spectrometer operates. Most mass spectrometers consist of a few basic components, the design, operation, and sophistication of which distinguish one type of mass spectrometer from another. These components are represented in Figure 8.57. Other instruments that can separate mixtures, such as a gas or a liquid chromatograph, may also be used in tandem with a mass spectrometer. As seen in the figure, many of the components are enclosed in an evacuated chamber where pressures as low as 10^{-7} torr are maintained. This is necessary because the analysis requires generation of highly reactive ionic species in the gas phase, and it is critical that their concentration remain low to minimize bimolecular reactions.

The inlet system allows introduction of a very small amount (e.g., μg or μL) of a sample, which may be a solid, pure liquid, solution, or gas, into the ionization chamber. Ions are produced by one of several means, including bombardment with high-energy electrons, chemical methods, and irradiation with lasers. The ions are transported in the vapor phase into the mass analyzer. By means of a magnetic field, the analyzer sorts the ions in a manner so that they reach the detector according to

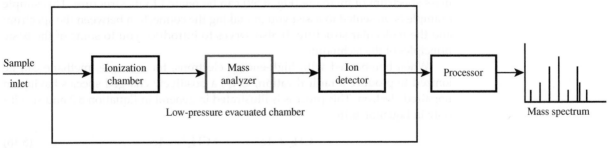

Figure 8.57
Block diagram of mass spectrometer.

☐ Ions/fragments other than +1.

◯ +1 ions, larger *m/z* represented by darker fill color.

Figure 8.58

Function of mass analyzer.

Figure 8.59

Mass spectrum of methane.

their *mass-to-charge ratio*, designated *m/z*; only ions of a single mass reach the detector at any given time. This principle is illustrated in Figure 8.58.

The detector of a mass spectrometer converts the intensity of the ion beam into an electronic signal that can be transformed by the processor into useful formats for storage or display. The intensity of the ion beam and thus of the signal is determined by the number of the ions of a given *m/z* that reach the detector. A common way to display the processed information is as a **mass spectrum**, in which the intensity of the signal is displayed as a function of the *m/z* of the ions arriving at the detector (Fig. 8.59). From such a display, molar mass and, in some cases, structural features can be determined.

Mass Spectra

To illustrate the value of mass spectrometry, let's consider how the peaks in the mass spectrum of methane (Fig. 8.59) can be related to its structure. This simple example is presented to assist you in making the connection between the spectrum and the molecular structure. It also serves to introduce you to some of the basic principles of the technique.

When bombarded with high-energy electrons, for example, methane can be ionized to produce a **radical cation**, CH_4^+, a positively charged species having an unpaired electron. This process is illustrated in general in Equation 8.2 and specifically in Equation 8.16.

$$CH_4 + 1e^- \longrightarrow CH_4^+ + 1e^-$$

Methane   *m/z* 16

(8.16)

The CH_4^+ ion has the same mass as the original molecule and is called the **molecular** or **parent ion**. This ion, with $z = +1$, reaches the detector and produces a signal at *m/z* 16, the molar mass of methane. The molecular ion is unstable, however, and can fragment in the ionization chamber into other cations, referred to as **daughter ions**, that are sorted according to their *m/z* before reaching the detector. In the case of methane, fragmentation occurs simply via loss of hydrogen atoms from the molecular ion as seen in Equation 8.17. The resulting mass spectrum (Fig. 8.59) shows the parent and daughter ions predicted in Equation 8.17. The relative intensities of the various peaks is related to the probability of a particular ion being formed and reaching the detector.

$$\underset{m/z\,16}{CH_4^+} -H^\cdot \longrightarrow \underset{m/z\,15}{CH_3^+} -H^\cdot \longrightarrow \underset{m/z\,14}{CH_2^+} -H^\cdot \longrightarrow \underset{m/z\,13}{CH^+} -H^\cdot \longrightarrow \underset{m/z\,12}{C^+} \quad (8.17)$$

As you might surmise, molecules more complex than methane produce more complicated fragmentation patterns, as illustrated by the mass spectrum of 2-methylbutane (Fig. 8.60). A generalized representation of the fragmentation of the parent ion and daughter ions into combinations of smaller charged and neutral species for such molecules is shown in Equations 8.18 and 8.19. In these equations and the subsequent discussion, the radical cations are represented in a simplified form in which we omit the symbol · for the unpaired electron.

$$P^+ : A^+ + B^\cdot \quad \text{or} \quad P^\cdot + B^+ \quad (8.18)$$

$$A^+ \longrightarrow C^+ + D^\cdot \quad \text{or} \quad C^\cdot + D^+ \quad (8.19)$$

To assist your understanding of mass spectra, you need to be aware of an important phenomenon in MS that arises from the fact that many elements are comprised of more than one stable or nonradioactive isotope. For example, ^{13}C NMR spectroscopy (Sec. 8.3) depends on the presence of this isotope, along with its more common

Figure 8.60
Mass spectrum of 2-methylbutane.

^{12}C isotope, in organic molecules. Because mass spectrometers are readily able to distinguish between ions having a difference of m/z of one atomic mass unit (amu), a molecular ion and the various fragments derived from it may appear as multiple peaks because of the presence of stable isotopes. Thus, you will observe peaks in a spectrum having values of m/z that are one, two, or more atomic mass units greater than that of the molecular ion, P^+; such peaks are labeled $P^+ + 1$, $P^+ + 2$, and so on. The mass spectrum of carbon monoxide, for example, has a peak at m/z 28 that corresponds to the molar mass of P^+ for ^{12}C^{16}O. However, the spectrum also has another peak at m/z 29; this is the $P^+ + 1$ peak and reflects the presence of ^{13}C and ^{17}O in some molecules of the sample. The intensity of this peak is small, approximately 1% of that for the molecular ion, and reflects the low **natural abundance** of these two isotopes.

Table 8.10 lists the relative abundances of the stable isotopes of carbon and hydrogen, as well as those of some of the heteroatoms commonly found in organic molecules. The percentages are derived by defining the abundance of the isotope of lowest atomic mass as 100% and relating the abundances of its heavier isotopes to this figure. Referring to this table will be valuable when you are interpreting mass spectra, particularly when the sample contains bromine and chlorine, both of which are comprised of high percentages of two isotopes.

The fragmentation of the molecular ion normally affords a large number of peaks in the mass spectrum, making its detailed interpretation difficult. Consequently, your use of the spectral data may be limited to determining the molar mass of the sample. You can extract more information from the spectrum, however, by applying principles that you may have encountered in learning about organic reaction mechanisms. For example, there is a preference for a molecular ion to undergo bond breaking so as to provide an allyl, benzyl, or tertiary carbocation, each of which is a relatively stable ion. There is also a tendency for thermodynamically stable neutral molecules, such as water, ammonia, dinitrogen, carbon monoxide, carbon dioxide, ethylene, and acetylene, to be formed by fragmentation. These are not observed, because only ions are detectable in the mass spectrum, but their loss from a molecular ion to give the corresponding fragments gives clues to the nature of functional groups that may be present in the sample. Consequently, determining the m/z values of fragments may provide valuable information about the structural features present in an unknown compound.

Table 8.10 *Natural Abundance of Various Isotopes*

Element	Relative Abundance (%) of Lowest-Mass Isotope	Relative Abundance (%) of Other Isotopes	
Carbon	100 ^{12}C	1.08 ^{13}C	
Hydrogen	100 ^{1}H	0.016 ^{2}H	
Nitrogen	100 ^{14}N	0.38 ^{15}N	
Oxygen	100 ^{16}O	0.04 ^{17}O	0.20 ^{18}O
Sulfur	100 ^{32}S	0.78 ^{33}S	4.40 ^{34}S
Chlorine	100 ^{35}Cl	32.5 ^{37}Cl	
Bromine	100 ^{79}Br	98.0 ^{81}Br	

Application of these principles is important for interpreting the mass spectrum of 2-methylbutane (**15**) (Fig. 8.60). Several peaks reveal information in support of its structure. For example, the molecular ion, P^+, is seen at m/z 72, which is the molar mass you would calculate for **15** by using atomic masses of 12.000 for carbon and 1.000 for hydrogen. The small $P^+ + 1$ peak at m/z 73 results from the presence of ^2H and ^{13}C in the compound. A $P^+ + 2$ peak is not observed because it is statistically unlikely that a molecule of 2-methylbutane would contain two atoms of either ^2H or ^{13}C or one of each of these isotopes. As you can see, the most intense peak in the spectrum is not the molecular ion, but rather is a fragmentation peak having m/z 43. This peak is defined as the **base peak**; its relative intensity is set as 100%, and the intensities of all other peaks in the spectrum are measured relative to it.

As seen in Scheme 8.1, the origin of several of the more intense daughter ions observed in Figure 8.60 may be explained by considering the possible ways in which the 2-methylbutane cation radical (**16**) may fragment. The ion **16** may lose a methyl group from C2 in two ways to form either 2-butyl cation radical (**17**) and methyl radical (**18**) or 2-butyl radical (**19**) and methyl cation radical (**20**). Based on the known relative stabilities of carbocations, we can reason that forming **17**, a secondary carbocation, is preferred to forming **20**, a primary carbocation. Indeed, the low-intensity peak at m/z 15 is due to **20**, with the much more intense peak at m/z 57 being **17**. The molecular ion **16** may also fragment through cleavage of the C2 and C3 bond to give either 2-propyl cation radical (**21**) and ethyl radical (**22**) or the 2-propyl radical (**23**) and the ethyl cation radical (**24**). The formation of **21**, m/z 43, is expected to be favored over that of **24**, m/z 29, based on the relative stabilities of these two carbocations. The other peaks in the spectrum result from phenomena that we will not discuss, but some of them are the $P^+ + 1$ and $P^+ + 2$ peaks of the various cation radicals.

Our discussion thus far has assumed that a mass spectrometer can distinguish between ions that differ in mass by one amu, where the m/z values are measured to

Scheme 1
Fragmentation of 2-methylbutane cation radical.

the nearest whole number. In the jargon of MS, these are called *low-resolution* mass spectrometers. Much more expensive instruments, called *high-resolution* mass spectrometers, can measure the m/z ratios to ± 0.001 or ± 0.0001 amu and provide very accurate measurements of molar masses. These data may then be used to determine elemental composition, which is extremely valuable information when you are trying to assign a structure to an unknown compound.

EXERCISES

1. Define the following terms:

   a. base peak

   b. daughter ion

   c. stable isotope

   d. molecular ion

   e. fragmentation, as applied to MS

   f. natural abundance, as applied to isotopes

   g. cation radical

   h. molar mass

2. In the mass spectrum of 2,2-dimethylpropane, the base peak occurs at m/z 57, peaks of moderate intensity are found at m/z 29 and 43, and a weak peak occurs at m/z 15.

   a. What is m/z for P^+?

   b. Using curved arrows to symbolize the flow of electrons, account for the specified fragmentation peaks.

3. The mass spectrum of 3,3-dimethylheptane consists of peaks at the following values of m/z; relative peak intensities are in parentheses: 43 (100%), 57 (100%), 71 (90%), 29 (40%), 99 (15%), and 113 (very low). The parent peak is not observed.

   a. Indicate what fragments are responsible for each value given of m/z.

   b. Explain why the peak at 113 is of very low intensity.

   c. Calculate the value of m/z for the parent peak and suggest why it is not observed.

4. A student performed a reaction in which bromobenzene was converted to the corresponding Grignard reagent, which then reacted with methyl benzoate (Eq. 8.20). Mass spectra (Figs. 8.61–8.63) were obtained of three components of the reaction mixture, which were expected to be triphenylmethanol (**25**), benzophenone (**26**), and biphenyl (**27**). Confirm that these products were indeed formed by assigning the appropriate mass spectrum to each compound. For each spectrum, (a) specify the parent and base peaks; (b) identify the compound represented by the mass spectrum; and (c) predict the structure of the ion for each of the peaks for which a value of m/z is provided.

| | 25 | 26 | 27 |

Figure 8.61

Mass spectrum of component of reaction mixture (Exercise 4).

Figure 8.62

Mass spectrum of component of reaction mixture (Exercise 4).

Figure 8.63

Mass spectrum of component of reaction mixture (Exercise 4).

REFERENCES

1. Silverstein, R. M.; Webster, F. X.; Kiemle, D. *Spectrometric Identification of Organic Compounds*, 7th ed., John Wiley & Sons, New York, 2005.

2. Crews, P.; Rodriquez, J.; Jasper, M. *Organic Structure Analysis*, Oxford University Press, New York, 1998.

3. Johnstone, R. A. W.; Rose, M. E. *Mass Spectrometry for Chemists and Biochemists*, 2nd ed., Cambridge University Press, Cambridge, UK, 1996.

4. McLafferty, F. W.; Stauffer, D. B. *The Wiley/NBS Registry of Mass Spectral Data*, vol. I–VII, John Wiley & Sons, New York, 1989.

CHAPTER 9

Alkanes

In this chapter you will be investigating the reactivity of the carbon-hydrogen bond in **acyclic (noncyclic)** and **cyclic alkanes**, which are saturated hydrocarbons. This class of compounds contains only carbon-carbon single bonds and carbon-hydrogen bonds. To the extent there is a **functional group** in such molecules, it would be the C–H bond. But in reality, this type of bond is rather inert chemically unless it is made more reactive, or "activated," by being part of a molecule that has a functional group such as a carbon-carbon multiple bond (Chap. 10), halogen atom (Chap. 14), hydroxyl function (Chap. 16), or carbonyl group (Chap. 18) present. This lack of reactivity, even toward strong acids and bases, severely limits the chemical transformations that can be accomplished. However, we'll see that there are ways by which we can effect certain types of reactions of unactivated C–H bonds and produce more chemically interesting products, that is, "functionalized" compounds, as a result.

9.1 INTRODUCTION

The fact that alkanes are chemically unreactive toward most reagents makes them of limited use as practical starting materials for most laboratory syntheses. However, chemical reactions that occur by **free-radical chain** mechanistic pathways do allow the introduction of certain functional groups into an alkane. For example, alkanes are converted into alkyl chlorides or bromides, R–X (X = Cl or Br, respectively), by a free-radical process in which a mixture of the alkane and halogen is heated at 200–400 °C or is irradiated with ultraviolet light (Eq. 9.1). Under these conditions, the σ-bond of molecular chlorine or bromine undergoes **homolytic cleavage** (Eq. 9.2) to generate chlorine and bromine atoms. These conditions provide sufficient energy to promote homolytic cleavage of the σ-bond of molecular chlorine or bromine to generate chlorine and bromine atoms, respectively, which are **free radicals** (Eq. 9.2); the amount of energy required to effect this reaction is called the **bond dissociation energy**. Generating chlorine and bromine atoms is essential to initiating the reaction between an alkane and molecular chlorine or bromine to form alkyl halides. These may then be transformed into a variety of other functional groups. An environmentally damaging reaction associated with the formation of chlorine atoms by photochemical degradation of chlorocarbons is discussed in the Historical Highlight *Keeping it Cool*, which is available online.

$$\text{R}\!-\!\text{H} + \text{X}_2 \xrightarrow{\text{heat or h}\nu} \text{R}\!-\!\text{X} + \text{H}\!-\!\text{X} \qquad (9.1)$$
$$\text{X = Cl or Br} \qquad \text{Alkyl halides}$$
$$\text{(mixture)}$$

$$\text{X}\!-\!\text{X} \xrightarrow{\text{heat or h}\nu} \text{X}^{\bullet} + \text{X}^{\bullet} \qquad (9.2)$$

309

The procedures described in Sections 9.2 and 9.3 illustrate methods for transforming alkanes to alkyl chlorides and bromides by **free-radical substitution** reactions. These experiments give you an opportunity to analyze the mixtures of products that are obtained and to gain insights about the relative reactivities of different types of hydrogen atoms toward chlorine and bromine radicals.

9.2 CHLORINATION USING SULFURYL CHLORIDE

Chlorine atoms may be generated from molecular chlorine under mild conditions using a *catalytic* amount of an **initiator**, In–In. Thus, homolysis of a molecule of initiator occurs upon irradiation or gentle heat to give free radicals, In• (Eq. 9.3). These free radicals may then react with molecular chlorine to produce In–Cl and a chlorine atom (Eq. 9.4) to initiate the **free-radical chain reaction**. For safety and convenience, sulfuryl chloride, SO_2Cl_2, rather than molecular chlorine is used in this experiment as the source of chlorine radicals.

$$\text{In—In} \xrightarrow{\text{heat or hv}} \text{In}^{\bullet} + \text{In}^{\bullet} \tag{9.3}$$

$$\text{In}^{\bullet} + \text{Cl—Cl} \longrightarrow \text{In—Cl} + \text{Cl}^{\bullet} \tag{9.4}$$

The first step in our procedure for initiating the free-radical chain reaction is the homolysis of 1,1'-azobis[cyclohexanenitrile] (**1**), abbreviated as ABCN, to form nitrogen and the free radical **2** (Eq. 9.5). The rate of this reaction is sufficiently fast at 80–100 °C to generate enough chlorine atoms to initiate the chain process. The radical **2** then attacks sulfuryl chloride to generate chlorine atoms and SO_2 according to Equations 9.6 and 9.7. The series of reactions depicted in Equations 9.5–9.7 comprise the **initiation** steps of the reaction.

Initiation

$$\tag{9.5}$$

1
ABCN

                2

$$\text{In}^{\bullet} + \text{Cl—S(=O)(=O)—Cl} \longrightarrow \text{In—Cl} + {}^{\bullet}\text{S(=O)(=O)—Cl} \tag{9.6}$$

Sulfuryl chloride

$${}^{\bullet}\text{S(=O)(=O)—Cl} \longrightarrow SO_2 + \text{Cl}^{\bullet} \tag{9.7}$$

Propagation steps are the next to occur in the overall process. These include abstraction of a hydrogen atom from the hydrocarbon by a chlorine atom to produce a new free radical, $R^{\bullet}$ (Eq. 9.8), which attacks sulfuryl chloride to yield the alkyl chloride and the radical $^{\bullet}SO_2Cl$ (Eq. 9.9). In the final propagation step, this radical fragments into SO_2 and another chlorine atom (Eq. 9.10), thereby producing a radical chain reaction.

Propagation

$$Cl^{\bullet} + H\!\!-\!\!R \longrightarrow R^{\bullet} + H\!\!-\!\!Cl \qquad (9.8)$$

$$R^{\bullet} + Cl\!\!-\!\!\overset{\displaystyle O}{\underset{\displaystyle O}{\overset{\|}{\underset{\|}{S}}}}\!\!-\!\!Cl \longrightarrow R\!\!-\!\!Cl + {}^{\bullet}\!\overset{\displaystyle O}{\underset{\displaystyle O}{\overset{\|}{\underset{\|}{S}}}}\!\!-\!\!Cl \qquad (9.9)$$

$$^{\bullet}\!\overset{\displaystyle O}{\underset{\displaystyle O}{\overset{\|}{\underset{\|}{S}}}}\!\!-\!\!Cl \longrightarrow SO_2 + Cl^{\bullet} \qquad (9.10)$$

You might assume that this reaction, once initiated, continues until either sulfuryl chloride or the alkane, whichever is the **limiting reagent** (Sec. 1.6), is entirely consumed. In practice, the chain reaction is interrupted by a series of side reactions known as **termination** reactions (Eqs. 9.11–9.13), and the initiation process must be continued throughout the course of the reaction.

Termination

$$Cl^{\bullet} + {}^{\bullet}Cl \longrightarrow Cl\!\!-\!\!Cl \qquad (9.11)$$

$$R^{\bullet} + {}^{\bullet}Cl \longrightarrow R\!\!-\!\!Cl \qquad (9.12)$$

$$R^{\bullet} + {}^{\bullet}R \longrightarrow R\!\!-\!\!R \qquad (9.13)$$

In summary, the distinct stages of a **free-radical chain mechanism** are:

1. **Initiation.** Radicals are formed in *low* concentration from neutral molecules, resulting in a *net increase* in the concentration of free radicals within the system.

2. **Propagation.** Radicals produced in the first step react with molecules to yield new molecules and new radicals; there is *no net change* in the concentration of radicals.

3. **Termination.** Various radicals combine to give molecules, resulting in a *net decrease* in radical concentration and a decrease in the rate of the reaction.

Free-radical halogenation of saturated compounds generally yields products of both monosubstitution and polysubstitution. For example, chlorination of

1-chlorobutane (**3**), the substrate you will use in the experimental procedure of this section, gives the *dichloro* compounds **4–7**, which are products of monosubstitution (Eq. 9.14), as well as *trichloro* compounds. Polysubstitution may be minimized by using an excess of the substrate undergoing halogenation as compared to the amount of halogenating agent present. Thus the molar ratio of substrate to chlorinating agent is more than two in our procedure. This strategy has a logical basis: statistically speaking, a chlorine atom is more likely to collide with a molecule of **3** than with a dichlorobutane.

$$CH_3CH_2CH_2CH_2Cl \xrightarrow[SO_2Cl_2/heat]{ABCN} ClCH_2(CH_2)_2CH_2Cl + CH_3\overset{\overset{\displaystyle Cl}{|}}{C}HCH_2CH_2Cl$$

$$\quad\quad\quad \textbf{3} \quad\quad\quad\quad\quad\quad\quad\quad\quad\quad\quad\quad \textbf{4} \quad\quad\quad\quad\quad\quad \textbf{5}$$

$$\quad \text{1-Chlorobutane} \quad\quad\quad\quad\quad \text{1,4-Dichlorobutane} \quad \text{1,3-Dichlorobutane}$$

(9.14)

$$+ CH_3CH_2\overset{\overset{\displaystyle Cl}{|}}{C}HCH_2Cl + CH_3(CH_2)_2CHCl_2$$

$$\quad\quad\quad\quad\quad \textbf{6} \quad\quad\quad\quad\quad\quad\quad \textbf{7}$$

$$\quad\quad\quad \text{1,2-Dichlorobutane} \quad \text{1,1-Dichlorobutane}$$

The mechanistic step determining which product will ultimately be formed in a free-radical halogenation reaction is the one in which a hydrogen atom is abstracted (Eq. 9.8). In the case of 1-chlorobutane, the ratio of the different isomers **4–7** produced is dictated by the relative energies of the **transition states** for the various C–H abstractions. These energies, in turn, are controlled by a **statistical factor** and an **energy factor**. The statistical factor is determined by the number of hydrogen atoms whose replacement will give a specific *constitutional isomer*, whereas the energy factor is related to the strength of the type of C–H bond being broken and is called the **relative reactivity**. In hydrocarbons such as butane, $CH_3(CH_2)_2CH_3$, the relative reactivity of **secondary** (2°) and **primary** (1°) hydrogen atoms (Fig. 9-5, Sec. 9.3) toward chlorine atoms is 3.3:1.0 and reflects the fact that a secondary C–H bond is *weaker* than a primary one. However, the presence of the chloro substituent means that these values do *not* apply to the corresponding types of hydrogens in **3**. The electron-withdrawing effect of this substituent is known to *increase* the strength of the C–H bonds near to it, thereby *lowering* their relative reactivity. By analyzing the data from this experiment, you should be able to determine if this is true and to derive corrected values for the relative reactivities if it is.

Using sulfuryl chloride for chlorination results in formation of equimolar amounts of sulfur dioxide and hydrochloric acid (Eq. 9.15), both of which are lost as gases from the reaction mixture. The resulting loss of mass provides a convenient means for monitoring the progress of the chlorination because the reaction is complete when the theoretical loss of mass has occurred. This can be calculated on the basis that sulfuryl chloride is the *limiting reagent* in the reaction.

$$R{-}H + SO_2Cl_2 \xrightarrow{\text{Initiator}} R{-}Cl + SO_2 + HCl$$

(9.15)

EXPERIMENTAL PROCEDURES

Free-Radical Chain Chlorination of 1-Chlorobutane

Purpose To perform a free-radical chain reaction and determine relative reactivities of C–H bonds toward chlorine atoms.

SAFETY ALERT

1. **Wear safety glasses or goggles and suitable protective gloves while performing the experiments.**

2. **Sulfuryl chloride is a potent lachrymator and reacts violently with water to generate sulfuric acid and hydrochloric acid. *Exercise great care* when handling this chemical: *Make sure that your apparatus is dry.* Do *not* allow it to contact your skin or inhale its vapors; measure out this chemical in a hood. Vessels containing sulfuryl chloride should be kept stoppered whenever possible!**

3. **Be certain that all connections in your apparatus are tight prior to heating the reaction mixture and that your gas trap is functioning to prevent venting of noxious gases into the laboratory.**

4. **Carbon dioxide is generated in the separatory funnel when the reaction mixture is washed with aqueous sodium carbonate. To relieve any gas pressure that develops, be sure to vent the funnel frequently when shaking it.**

MINISCALE PROCEDURE

Preparation Refer to the online resources to answer Pre-Lab Exercises, access videos, and read the MSDSs for the chemicals used or produced in this procedure. Review Sections 2.11, 2.14, 2.21, 2.22, 2.23, 2.25, 2.28, and 6.4.

Apparatus A *dry* 25-mL round-bottom flask, gas trap, separatory funnel, ice-water bath apparatus for magnetic stirring, heating under reflux, and *flameless* heating.

Setting Up Assemble the apparatus for heating under reflux, and attach a gas trap such as that in Figure 2.65 to the top of the condenser. Equip the round-bottom flask with a stirbar and add 0.1 g of 1,1'-azobis(cyclohexanenitrile) (ABCN). Working *at the hood,* sequentially measure 5 mL of 1-chlorobutane and 2 mL of sulfuryl chloride into the flask. Stopper and weigh the flask, and then attach it to the reflux condenser, working quickly to minimize introducing vapors of sulfuryl chloride into the atmosphere.

Reaction Heat the solution under gentle reflux for 20 min with stirring and then allow the reaction mixture to cool below the reflux temperature. Add a second 0.1-g portion of ABCN and continue heating the stirred mixture under reflux for another 10 min. Weigh the flask and consult with your instructor if the resulting mixture has not lost at least 90% of the expected mass after this second period of reflux.

Work-Up and Isolation Cool the reaction mixture in an ice-water bath and, working at the hood, *cautiously* pour it with stirring into 15 mL of ice-cold saturated aqueous sodium chloride (brine) contained in an Erlenmeyer flask. Transfer the resulting two-phase solution to a separatory funnel and separate the layers.★ Wash the organic layer in the funnel with 10 mL of 0.5 *M* sodium carbonate solution, venting the funnel *frequently*. Using pHydrion paper, determine whether the aqueous layer (Sec. 2.21) is basic. If not, repeat washing the organic layer with 10-mL portions of sodium carbonate solution until the aqueous washes are *basic*. Wash the organic layer with 15 mL of brine and transfer it to an Erlenmeyer flask containing several spatula-tips full of *anhydrous* sodium sulfate.★ Swirl the flask occasionally during 10–15 min to hasten drying, which is indicated by the liquid becoming clear. If the liquid remains cloudy, add additional portions of sodium sulfate with swirling. Carefully decant the organic layer into a dry, *tared* container.

Analysis Weigh the reaction mixture and use the result to calculate the percentage yield of material you recovered. This calculation requires that you first determine the theoretical mass of product dichlorides *and* unreacted 1-chlorobutane expected from the reaction.

Analyze the reaction mixture by gas chromatography or GC-MS or submit samples to your instructor for such analysis. The peak corresponding to 1-chlorobutane should be "off-scale" or nearly so, so that the peaks for the isomeric dichlorobutanes are sufficiently large for an accurate analysis of their areas. Assign the four product peaks to the isomers responsible for them and calculate the relative amounts of each isomer present in the mixture. A typical GLC trace of the reaction mixture is depicted in Figure 9.1.

Obtain IR and ¹H NMR spectra of your starting material and an NMR spectrum of your product mixture; compare them with those of authentic samples (Figs. 9.2–9.4).

MICROSCALE PROCEDURE

Preparation Refer to the online resources to answer Pre-Lab Exercises, access videos, and read the MSDSs for the chemicals used or produced in this procedure. Review Sections 2.4, 2.11, 2.14, 2.21, 2.22, 2.23, 2.25, 2.28, and 6.4.

Figure 9.1
GLC analysis of mixture from chlorination of 1-chlorobutane (numbers in parentheses are boiling points at 760 torr and peak areas, respectively). Peak 1: solvent; peak 2: 1-chlorobutane (77–78 °C, 2719); peak 3: 1,1-dichlorobutane (114–115 °C, 487); peak 4: 1,2-dichlorobutane (121–123 °C, 1771); peak 5: 1,3-dichlorobutane (131–133 °C, 3367); peak 6: 1,4-dichlorobutane (161–163 °C, 1137); column and conditions: 0.5-mm × 30-m, 0.25-μ film of SE-54; 80 °C, 40 mL/min.

Apparatus *Dry* 3-mL screw-cap conical vial, gas trap, 5-mL conical vial or screw-cap centrifuge tube, apparatus for magnetic stirring, heating under reflux, and *flameless* heating.

Assembly Assemble the apparatus for heating under reflux, and attach a gas trap, such as that of Figure 2-69, to the top of the condenser. Place a spinvane in the vial and add 10 mg (0.010 g) of 1,1'-azobis(cyclohexanecarbonitrile) (ABCN). Working *at the hood,* sequentially add 0.5 mL of 1-chlorobutane and 0.2 mL (200 μL) of sulfuryl chloride to the vial. Cap and weigh the vial, and then attach it to the reflux condenser, working quickly to minimize introducing vapors of sulfuryl chloride into the atmosphere.

Reaction Heat the solution under gentle reflux with stirring for about 20 min and then allow the reaction mixture to cool below the reflux temperature. Add a second 0.010-g portion of ABCN and continue heating the stirred mixture under reflux for another 10 min. Weigh the flask, and consult with your instructor if the resulting mixture has not lost at least 90% of the expected mass after this second period of reflux.

Work-Up and Isolation Remove the vial from the apparatus and cap it.★ Working at the hood, *cautiously* pour the reaction mixture with stirring into 1 mL of ice-cold saturated aqueous sodium chloride (brine) contained in the 5-mL conical vial or centrifuge tube. Allow the layers to separate, and stir or shake the mixture briefly. Remove the lower layer with a Pasteur pipet and determine whether it is the organic or aqueous layer (Sec. 2.21).★ *Do not* discard *either* layer until you know which one is the organic layer.

Carefully add 1 mL of aqueous 0.5 *M* sodium carbonate to the organic layer (*caution:* foaming), shake the mixture with venting as necessary, and remove the lower layer. Using pHydrion paper, determine whether the aqueous layer is basic. If not, repeat washing the organic layer with 1-mL portions of sodium carbonate solution until the aqueous washes are *basic.* Shake the organic layer with 1 mL of brine, and carefully transfer it to a dry vial containing one or two microspatula-tips full of *anhydrous* sodium sulfate. Stir the resulting mixture occasionally during 10–15 min to hasten drying, which is indicated by the liquid becoming clear. If the liquid remains cloudy, add additional portions of sodium sulfate with stirring. Carefully transfer the dried mixture to a dry, *tared* sample vial using a Pasteur filter-tip pipet.

Analysis Weigh the reaction mixture and use the result to calculate the percentage yield of material you recovered. This calculation requires that you first determine the theoretical weight of product dichlorides *and* unreacted 1-chlorobutane expected from the reaction.

Analyze the reaction mixture by gas chromatography or GC-MS or submit samples to your instructor for such analysis. The peak corresponding to 1-chlorobutane should be "off-scale" or nearly so, so that the peaks for the isomeric dichlorobutanes are sufficiently large for an accurate analysis of their areas. Assign the four product peaks to the isomers responsible for them and calculate the relative amounts of each isomer present in the mixture. A typical GLC trace of the reaction mixture is depicted in Figure 9.1.

Obtain IR and NMR spectra of your starting material and an NMR spectrum of your product mixture; compare them with those of authentic samples (Figs. 9.2–9.4).

Discovery Experiment   *Chlorination of Heptane*

Using the procedure described above as your model, design and execute an experimental protocol for monochlorinating heptane and determining the ratio of chloroheptanes produced. Also describe how you would analyze the mixture for the presence of polychloro isomers. Check with your instructor before undertaking any experimental work.

Discovery Experiment   *Chlorination of 2,3-Dimethylbutane*

Using the procedure described above as your model, design and execute an experimental protocol for monochlorinating 2,3-dimethylbutane and determining the ratio of chlorodimethylbutanes produced. Also describe how you would analyze the mixture for the presence of polychloro isomers. Check with your instructor before undertaking any experimental work.

WRAPPING IT UP

Combine *all of the aqueous solutions.* If the resulting solution is acidic, neutralize it carefully with solid sodium carbonate and flush it down the drain with a large excess of water. Rinse the *glass wool* from the gas trap with aqueous 0.5 *M* sodium carbonate and discard it in the trash. Put the *sodium sulfate* in a container for solids contaminated with alkyl halides.

EXERCISES

1. Define and provide a specific example of each of the following terms:
   a. propagation step
   b. initiation step
   c. bond homolysis
   d. limiting reagent
   e. secondary hydrogen atom
2. Why is 1-chlorobutane not classified as a hydrocarbon?
3. Which, if any, of the dichlorobutanes contain a center of chirality?
4. What is the fundamental difference between a reaction mechanism that uses single-headed arrows versus one that uses half-headed (fishhook) arrows?

         Single-headed   Fishhook

5. Using curved arrows to symbolize the flow of electrons, write the complete mechanism for the formation of 1,3-dichlorobutane from 1-chlorobutane using ABCN (**1**) and sulfuryl chloride. Clearly label each step as to whether it is part of the initiation, propagation, or termination steps.
6. Why is only a catalytic amount of ABCN (**1**) needed in this reaction?
7. Write at least one product that might be derived from ABCN (**1**) and explain why the potential contamination of the reaction mixture by such a product is of little concern.

8. Why is the amount of sulfuryl chloride used *less* than the amount theoretically required to convert all the starting material to monosubstituted products?

9. Why is it prudent to use caution when pouring the crude reaction mixture into brine? Why should the reaction mixture and brine be *ice-cold* for this step?

10. In the work-up procedure, the organic layer was washed with a solution of sodium carbonate.

    a. Why was this wash necessary?

    b. Write the reaction that occurs during this wash and specify the gas that is evolved.

    c. Why is it especially important to vent a separatory funnel when performing this wash?

11. How might a loss of mass greater than theoretically calculated occur in a chlorination reaction performed according to the procedure of this section?

12. What physical property of 1,4-dichlorobutane accounts for it being the isomer with the longest retention time in the GLC analysis?

13. Using a relative reactivity for 1° to 2° hydrogen atoms of 1:3.3,

    a. calculate the percentage of each chlorobutane expected from monochlorination of 1-chlorobutane, and

    b. determine the ratio of 1° to 2° dichlorobutanes to be expected.

14. a. Using your own data or those provided in Figure 9.1, calculate the ratio of 1° to 2° dichlorobutanes obtained from monochlorination of 1-chlorobutane.

    b. Why does this ratio *not* agree with that found in Exercise 13?

15. Based on your calculation of the percentage of the various dichlorobutanes formed in the reaction, determine the values for the relative reactivity for the various types of hydrogen atoms in 1-chlorobutane. Assign the relative reactivity of the methyl hydrogen atoms as 1.0.

16. Specify the factors determining the proportion of monochlorinated products derived from 1-chlorobutane.

17. Chlorine and ABCN (**1**) are good initiators for free-radical chain reactions.

    a. What characterizes a molecule that is a good initiator?

    b. Why is it unnecessary to use a stoichiometric amount of the initiator?

    c. Two other common radical initiators are shown below. Using curved arrows to symbolize the flow of electrons, show the bond homolysis that occurs when these initiators are heated or irradiated.

Benzoyl peroxide   2,2′-Azobis[2-methylpropanenitrile]

18. A linear geometry is portrayed in the structures of both **1** and, as shown in Exercise 17, 2,2′-azobis [2-methylpropanenitrile]. Using "R" to represent the alkyl substituents on the nitrogen atoms, provide a representation of the thermodynamically more favored geometry of these azo compounds. Justify your answer.

19. Indicate from which of the three compounds shown below it should be easiest to prepare a *single* pure monochlorinated product, and explain why.

20. Consider the monochlorination of heptane using ABCN (**1**) and sulfuryl chloride.

   a. Write the structures for and the names of the possible monochlorination products that could be produced.

   b. Circle the center(s) of chirality, if any, that are present in the isomers that you wrote in Part **a**.

   c. Assume that you have been provided with a pure sample of each monochloroheptane but that the samples were unlabeled. Discuss how you could use a combination of ^{1}H and ^{13}C NMR spectroscopic data to prove the structure of each sample.

21. Chlorofluorocarbons (CFCs) have been found to contribute to the problem of ozone depletion. In the stratosphere, CFCs initiate a radical chain reaction that can consume 100,000 ozone molecules per CFC molecule.

   a. Freon 12, CF_2Cl_2, is a typical CFC. Using curved arrows to symbolize the flow of electrons, show the radical initiation step that occurs when light interacts with Freon 12 in the stratosphere. (*Hint:* Identify the weakest bond in the molecule.)

   b. Write the products formed when the radical produced in the reaction in Part **a** reacts with a molecule of ozone, O_3, in the first propagation step. (*Hint:* Oxygen, O_2, and another radical are produced.)

22. Consider the NMR data for 1-chlorobutane (Fig. 9.3).

   a. In the ^{1}H NMR spectrum, assign the various resonances to the hydrogen nuclei responsible for them.

   b. For the ^{13}C NMR data, assign the various resonances to the carbon nuclei responsible for them.

23. Figure 9.4 is the ^{1}H NMR spectrum of a mixture of 1,*x*-dichlorobutanes obtained by free-radical chlorination of 1-chlorobutane.

   a. Assign the triplets centered near δ 1.0 and 1.1 and the doublet centered near δ 1.6 to the hydrogen nuclei of the three different isomers from which they arise. (*Hint:* Write the structures of the isomeric 1,*x*-dichlorobutanes and predict the multiplicity and approximate chemical shift of each group of hydrogens.)

   b. The integrated areas of the selected multiplets in the spectrum are as follows: δ 1.0, 2 units, δ 1.1, 9 units, δ 1.6, 32 units, δ 3.6, 26.8 units, δ 3.7, 26.8 units, δ 4.0, 3.3 units, δ 5.8 (a triplet *not* visible in the spectrum), 0.7 units. Based on your assignments in Part **a** and on the integration data, calculate the approximate percentage of each isomer present in the mixture. Explain your determination as completely as possible.

Refer to the online resources and use the spectra viewer and Tables 8.1–8.8 as needed to answer the blue-numbered questions on spectroscopy.

24. Compare the ^{1}H NMR spectrum of the mixture of 1,*x*-dichlorobutanes (Fig. 9.4) with that of 1-chlorobutane (Fig. 9.3). What features in Figure 9.4 are consistent with the incorporation of a second chlorine atom in this experiment and with the proposal that a *mixture* of dichlorobutanes, rather than a single isomer, is formed?

SPECTRA

Starting Materials and Products

Figure 9.2
IR spectrum of 1-chlorobutane (neat).

$CH_3(CH_2)_2CH_2Cl$

Figure 9.3
NMR data for 1-chlorobutane (CDCl$_3$).

$CH_3(CH_2)_2CH_2Cl$

(a) ^{1}H NMR spectrum (300 MHz).
(b) ^{13}C NMR data: δ 13.4, 20.3, 35.0, 44.6.

Figure 9.4
¹H NMR spectrum of mixture of 1,x-dichlorobutanes (CDCl₃, 300 MHz).

9.3 BROMINATION: SELECTIVITY OF HYDROGEN ATOM ABSTRACTION

Hydrocarbons undergo free-radical chain bromination (Eq. 9.16) through a mechanism analogous to that for free-radical chlorination (Sec. 9.2). The **initiation** step involves the thermal or photochemical **homolysis** of molecular bromine to form bromine atoms, Br• (Eq. 9.17); the energy required for this is referred to as the **bond dissociation energy**. **Propagation** steps then produce the **substitution** product, R–Br, and regenerate Br• to continue the chain reaction (Eqs. 9.18 and 9.19). The **termination** steps are analogous to those of Equations 9.11–9.13 (Sec. 9.2).

$$R\!-\!H + Br_2 \xrightarrow{\;h\nu\;} R\!-\!Br + H\!-\!Br \qquad (9.16)$$
$$\text{(colorless) (reddish} \qquad \text{(colorless) (colorless)}$$
$$\text{orange)}$$

$$Br\!-\!Br \xrightarrow{\;\text{heat or } h\nu\;} Br^\bullet + Br^\bullet \qquad (9.17)$$

$$Br^\bullet + H\!-\!R \longrightarrow R^\bullet + H\!-\!Br \qquad (9.18)$$

$$R^\bullet + Br\!-\!Br \longrightarrow R\!-\!Br + Br^\bullet \qquad (9.19)$$

As with chlorination, the mechanistic step involving abstraction of a hydrogen atom by Br (Eq. 9.18) determines which product will be formed. The likelihood that a particular type of hydrogen atom will be successfully abstracted is defined by the relative energies of the **transition states** for the various C–H abstractions. These energies, in turn, are controlled by a **statistical factor** and an **energy factor**. The statistical factor is determined by the number of hydrogen atoms whose replacement will give a specific *constitutional isomer*, whereas the energy factor is related to the

Figure 9.5
Classification of hydrogen atoms in hydrocarbons.

strength of the type of C–H bond being broken and is called the **relative reactivity**. In the experiment of this section, you will determine the relative rates of the free-radical chain reaction of molecular bromine with several hydrocarbons containing different types of hydrogen atoms. You should be able to deduce the order of relative reactivity of *seven* different types of hydrogens toward Br• by carefully considering the results of your observations.

As summarized by the structures in Figure 9.5, hydrogen atoms in hydrocarbons are classified according to the type of carbon atom to which they are attached. Specifically, hydrogens bound to sp^3-hybridized carbons are categorized as **aliphatic;** all hydrogen atoms in the hydrocarbon family of alkanes belong to this group. Aliphatic hydrogens may be further classified as being **primary (1°), secondary (2°),** or **tertiary (3°)** depending upon the number of other carbon atoms attached to the reference carbon atom; primary carbon atoms are bound to one other carbon atom, secondary to two, and tertiary to three. An aliphatic hydrogen atom attached to an sp^3-hybridized carbon atom that in turn is bound to a vinylic or aromatic carbon atom is called **allylic** or **benzylic,** respectively. Combining these classifications gives the terms primary, secondary, or tertiary aliphatic, allylic, or benzylic hydrogen atoms. Hydrogen atoms attached to sp^2-hybridized carbon atoms are classified as either **vinylic** or **aromatic,** depending upon whether the carbon atom is part of a normal double bond or an aromatic ring. Hydrogen atoms are called **acetylenic** when bound to an sp-hybridized carbon atom.

The hydrocarbons **8–13** contain various types of hydrogen atoms and are the substrates for our experiment. Measuring their relative reactivities toward Br• in free-radical chain bromination is straightforward for two reasons. First, molecular bromine is colored, so the progress of bromination is easily monitored by observing the disappearance of color as the bromine is consumed (Eq. 9.16). Second, because the relative rates of hydrogen abstraction by Br• are highly sensitive to the strength of the C–H bond being broken, the substrates have rates of free-radical chain bromination that differ enough to allow ready determination of the qualitative order of their reactivity.

CH_3	CH_2CH_3	$CH(CH_3)_2$	$C(CH_3)_3$
8	**9**	**10**	**11**
Toluene	Ethylbenzene	Isopropylbenzene	*tert*-Butylbenzene

	CH_3
12	**13**
Cyclohexane	Methylcyclohexane

Experimentally, you will allow an *excess* of a hydrocarbon to react with molecular bromine and will measure the time required for disappearance of the color of the bromine. The reactions are performed in dichloromethane solution using two different conditions: (a) room temperature with no illumination, and (b) room temperature with illumination. The period of time required for the decoloration depends upon the reaction conditions and the relative reactivities of the various hydrocarbons; some of the substrates may not react at all under the specified conditions. The ratios of the different reaction times thus represent the relative rates of the free-radical chain bromination reaction.

EXPERIMENTAL PROCEDURE

Relative Rates of Free-Radical Chain Bromination

Discovery Experiment   **Purpose** To determine the relative reactivities of different types of hydrogen atoms toward bromine atoms.

SAFETY ALERT

1. **Wear safety glasses or goggles and suitable protective gloves while performing the experiments.**

2. ***Bromine is a hazardous chemical that may cause serious chemical burns.*** **Do *not* breathe its vapors or allow it to come into contact with skin. Perform all operations involving the transfer of the pure liquid or its solutions at a hood. If you get bromine on your skin, wash the area immediately with soap and warm water and soak the affected area in 0.6 *M* sodium thiosulfate solution for up to 3 h if the burn is particularly serious.**

3. **Keep the 1 *M* bromine in dichloromethane solution in the hood. Dispense 0.5-mL portions of it, using a pipet pump, calibrated Pasteur filter-tip pipet, or a buret fitted with a Teflon stopcock.**

4. Bromine reacts with acetone to produce the powerful **lachrymator** α-bromoacetone, $BrCH_2COCH_3$. Do *not* rinse glassware containing residual bromine with acetone!

5. Avoid inhalation of the vapors of any the materials being used in this experiment.

Procedure

Preparation Refer to the online resources to answer Pre-Lab Exercises, access videos, and read the MSDSs for the chemicals used or produced in this procedure. Review Sections 2.5 and 2.10.

Construct a table in your notebook with the following four main headings: (a) "Hydrocarbon"; (b) "Types of Hydrogen Atoms," entries under which will include the terms 1° aliphatic, 2° benzylic, and so on; (c) "Conditions," with the subheadings RT, RT/hν; and (d) "Elapsed Time," entries under which will be the time required for reaction as measured by decoloration.

Carefully plan your execution of this experiment, because proper labeling of test tubes and managing of time once the reactions have been started are critical. You will perform two separate experimental trials, with one trial starting 5–10 min after the other.

Apparatus Fourteen 13-mm × 100-mm test tubes and corks, six 10-mm × 75-mm test tubes, six Pasteur pipets, filter-tip pipet, ice-water bath, light source.

Setting Up Organize *twelve* 13-mm × 100-mm test tubes into pairs and label each tube in the pair with the name or number of the hydrocarbon **8–13** that you will add to it. Measure 2.5 mL of dichloromethane into each of these tubes and then add 0.5 mL (10 drops) of the hydrocarbon for which it is labeled. Label the *two* remaining 13-mm × 100-mm test tubes "Control" and add 3 mL of dichloromethane to them. Place the *seven* pairs of labeled tubes in a beaker. Working at the hood, dispense 0.5 mL of a 1 *M* solution of bromine in dichloromethane into each of the *six* small test tubes and into one of the larger tubes labeled "Control."

Bromination In rapid succession, add the solution of bromine to each of the hydrocarbon-containing tubes. Do this with agitation to ensure good mixing and record the time of the additions. After all the additions are made, loosely cork each tube to prevent loss of solvent and bromine. Monitor all the tubes, including that labeled "Control," and record the elapsed time required for the discharge, if any, of the color of each solution.

After about 5 min, again dispense 0.5 mL of 1 *M* bromine solution into the six small test tubes used originally for this reagent *and* also into the remaining tube labeled "Control." Working as before, add the solution of bromine to each of the hydrocarbon-containing tubes. Place the six tubes containing the resulting solutions and the tube labeled "Control" in a beaker located 14–16 cm above an unfrosted 100- or 150-watt light bulb so that all the solutions are exposed to the same amount of light. Monitor these seven tubes and record the elapsed time required for the discharge, if any, of the color of each solution.

Terminate your observations of both sets of samples after about 1 h.

Analysis Evaluate the intensity of color remaining in any of the tubes relative to that of the "Control" samples and record the results in your notebook. Obtain IR and ^{1}H NMR spectra of your starting materials and compare them with those of authentic samples (Figs. 9.6–9.17).

WRAPPING IT UP

Decolorize any *solutions* in which the color of bromine is visible by the dropwise addition of cyclohexene; then discard the resulting solutions together with *all other solutions* in a container for halogenated organic liquids.

EXERCISES

1. Draw the structure of the major monobrominated product expected from each of the hydrocarbons used in this experiment.

2. Comment on the need for light in order for bromination to occur with some of the hydrocarbons.

3. Why is a "Control" sample needed in this experiment?

4. Perform the calculations necessary to demonstrate that bromine is indeed the limiting reagent (Sec. 1.6). The densities and molar mass needed to complete the calculations can be found in the abbreviated MSDSs on the website or various handbooks of chemistry.

5. How is polyhalogenation minimized in this experiment?

6. Using curved arrows to symbolize the flow of electrons, show the mechanism for the propagation steps that convert **10** into the corresponding monobromide, and provide two possible termination steps for the process.

7. Answer the questions below for substrates **8–13**.

    a. Which is statistically the most likely to form a monobromide? Briefly explain your answer.

    b. Which is the most likely to form a single isomer of a monobromide? Briefly explain your answer.

    c. Which is expected to produce the *most* stable radical? Briefly explain your answer.

    d. Which is expected to produce the *least* stable radical? Briefly explain your answer.

8. Arrange the six hydrocarbons **8–13** in *increasing* order of reactivity in free-radical chain bromination.

9. a. On the basis of the observed order of reactivity of the hydrocarbons, deduce the order of reactivity of the seven different types of hydrogens found in these compounds, that is, (1) primary aliphatic, (2) secondary aliphatic, (3) tertiary aliphatic, (4) primary benzylic, (5) secondary benzylic, (6) tertiary benzylic, and (7) aromatic.

    b. Outline the logic you used to arrive at your sequence in Part **a**.

10. Consider the free-radical halogenation of an alkane. You may wish to refer to your lecture textbook and to tables of bond dissociation energies when answering these questions.

    a. Why is the bromination much more regioselective than chlorination?

    b. Why are fluorinations extremely dangerous? (*Hint:* Consider heats of reaction of the propagation steps.)

    c. Why is it difficult to generate an alkyl iodide by free-radical chain halogenation?

11. How does the Hammond postulate explain

    a. the greater selectivity for bromination versus chlorination of an alkane?

    b. the fact that the relative reactivity of C–H bonds of an alkane for free-radical chain halogenation is 3° > 2° > 1°?

12. Why is a 1° benzylic radical more stable than a 1° aliphatic radical?

13. Consider the IR spectra of compounds **8–11** (Figs. 9.6, 9.8, 9.10, and 9.12), all of which contain a phenyl ring. What absorption(s) in the spectra denote the presence of this ring?

14. Consider the NMR data for compounds **8–13** (Figs. 9.7, 9.9, 9.11, 9.13, 9.15, and 9.17).

    a. In the ^{1}H NMR spectra, assign the various resonances to the hydrogen nuclei responsible for them.

    b. In the ^{13}C NMR spectra of these compounds, assign the various resonances to the carbon nuclei responsible for them.

Refer to the online resources and use the spectra viewer and Tables 8.1–8.8 as needed to answer the blue-numbered questions on spectroscopy.

SPECTRA

Starting Materials

Figure 9.6
IR spectrum of toluene (neat).

Figure 9.7
NMR data for toluene (CDCl₃).

(a) ¹H NMR spectrum (300 MHz).
(b) ¹³C NMR data: δ 21.4, 125.3, 128.2, 129.0, 137.8.

Figure 9.8
IR spectrum of ethylbenzene (neat).

Figure 9.9
NMR data for ethylbenzene (CDCl₃).

(a) ¹H NMR spectrum (300 MHz).
(b) ¹³C NMR data: δ 15.6, 29.1, 125.7, 127.9, 128.4, 144.2.

Figure 9.10
IR spectrum of isopropylbenzene (neat).

Figure 9.11
NMR data for isopropylbenzene (CDCl₃).

(a) ¹H NMR spectrum (300 MHz).
(b) ¹³C NMR data: δ 24.1, 34.2, 125.8, 126.4, 128.4, 148.8.

Figure 9.12
IR spectrum of tert-butylbenzene (neat).

Figure 9.13
NMR data for tert-butylbenzene
($CDCl_3$).

(a) ^{1}H NMR spectrum (300 MHz).
(b) ^{13}C NMR data: δ 31.3, 34.5, 125.1, 125.3, 128.0, 150.8.

Figure 9.14
IR spectrum of cyclohexane (neat).

Figure 9.15
NMR data for cyclohexane
($CDCl_3$).

(a) ^{1}H NMR spectrum (300 MHz).
(b) ^{13}C NMR datum: δ 27.3.

Figure 9.16
IR spectrum of methylcyclohexane (neat).

Figure 9.17
NMR data for methylcyclohexane (CDCl₃).

(a) ¹H NMR spectrum (300 MHz).
(b) ¹³C NMR data: δ 22.9, 26.6, 26.7, 33.0, 35.6.

Alkenes

As discussed in the preceding chapter, alkanes are saturated hydrocarbons. You encounter such compounds in your daily life, especially in the form of fuels to power automobiles, trains, planes, and many power plants that generate electricity. These fuels also contain compounds that are known as alkenes, which are unsaturated hydrocarbons. In this chapter, you will discover that alkenes are much more interesting organic compounds because they undergo a variety of different chemical reactions, not just the radical reactions that are typical of alkanes. Thus, although you may have considered the chemistry of alkanes "boring," you will find that the chemistry of alkenes is rich and characterized by a number of fascinating transformations of the carbon-carbon double bond, the functional group found in alkenes.

10.1 INTRODUCTION

A **functional group** is an atom or group of atoms that governs the chemical and physical properties of a family of compounds. The introduction and manipulation of these functional groups are major objectives in modern organic chemistry. In this chapter, we will explore the chemistry of **alkenes, 1**, which are organic compounds possessing a polarizable carbon-carbon double bond, a π-bond, as the functional group. Methods for introducing a carbon-carbon double bond into a molecule from alkyl halides and alcohols are presented first, and then some of the addition reactions that characterize this functional group are examined.

Elimination reactions are among the most common ways to produce a carbon-carbon π-bond. For example, the elements of hydrogen halide, H–X, may be eliminated from an **alkyl halide, 2**. The functional group of an alkyl halide is a carbon-halogen, single bond, **C–X**, and the process by which the carbon-halogen bond and an adjacent carbon-hydrogen bond are converted into a carbon-carbon π-bond via **dehydrohalogenation** is an example of a **functional group transformation**. A carbon-carbon π-bond may also be formed by removing the elements of water from an **alcohol, 3**, in which a **C–OH** single bond is the functional group; this reaction is called **dehydration**. Although other aspects of the chemistry of alkyl halides and alcohols will be presented in Sections 14.1–14.5 and 16.2, a brief introduction to these families is essential to understanding how they may be used as starting materials for the synthesis of alkenes.

1

An alkene

2

X = Cl, Br, I
An alkyl halide

3

An alcohol

10.2 DEHYDROHALOGENATION OF ALKYL HALIDES

The electronegative halogen atom of an alkyl halide polarizes the carbon-halogen bond so the carbon atom bears a partial positive charge, δ^+, and the halogen atom a partial negative charge, δ^-. This polarization may be transmitted through the σ-bond network, a phenomenon referred to as an **inductive effect**, to enhance the acidity of hydrogen atoms on the β-carbon atom. Removing a proton from this atom, and simultaneous departure of the halide ion, which is a **leaving group**, from the α-carbon atom forms the carbon-carbon π-bond of the alkene in a **concerted reaction** (Eq. 10.1). An important characteristic of a good leaving group is that it should be a *weak* base; in contrast, the base B:⁻ should be *strong*. The **transition state 4** for the reaction is shown in Equation 10.2, in which the curved arrows symbolize the flow of electrons. As shown in **4**, the preferred **dihedral angle** between the β-hydrogen being removed and the leaving group X is 180°, an angular relationship called **anti-periplanar**. This transformation of alkyl halides to form alkenes is called **dehydrohalogenation**.

An alkyl halide   An alkene   (10.1)

4

Transition state

The concerted reaction depicted in Equation 10.2 is classified as an **E2** process, where E stands for **elimination** and 2 refers to the **molecularity** of the rate-determining step of the reaction. For E2 processes, the rate of the reaction depends upon the concentrations of the organic substrate and the base (Eq. 10.3), so *both* reactants are involved in the transition state of the rate-determining step. This **bimolecularity** is illustrated in **4** (Eq. 10.2).

$$\text{Rate} = k_2[\text{alkyl halide}][\text{B:}^-] \qquad (10.3)$$

Because it bears a partial positive charge, the α-carbon atom of an alkyl halide is **electrophilic** and thus also subject to attack by **nucleophiles**, which, as **Lewis bases**, are electron-rich and frequently anionic species. This process produces **substitution** rather than elimination products (Eq. 10.4), so a possible competition between the two types of reactions must be considered. Examining transition states **4** and **5** for the two reactions (Eqs. 10.2 and 10.4) suggests that steric factors may play a critical role in the competition. In support of this hypothesis, it is observed experimentally that elimination is favored as the degree of substitution on the α-carbon atom increases. The resulting steric hindrance inhibits direct attack of the base, or nucleophile, on the α-carbon atom, and removal of the β-proton and formation of elimination products becomes favored. Thus, when the α-carbon atom is *tertiary*, elimination to form alkenes is generally the only observable reaction.

$$\text{B:}^- \;\; \underset{\text{H}}{\overset{\text{RCH}_2}{\text{C}}}\!\!\overset{\delta^+\;\delta^-}{\cdots}\text{X} \longrightarrow \left[\underset{\text{H H}}{\overset{\text{CH}_2\text{R}}{\overset{\delta^-}{\text{B}}\text{-} \text{-} \text{-} \overset{\delta^-}{\underset{}{\text{C}}}\text{-} \text{-} \text{X}}} \right]^{\ddagger} \longrightarrow \text{B}\underset{\text{H}}{\overset{\text{CH}_2\text{R}}{\text{-}\text{C}}}\text{H} \; + \; \text{X:}^- \qquad (10.4)$$

5

Transition state

Dehydrohalogenation may give a mixture of products if the halogen is unsymmetrically located on the carbon skeleton. For example, 2-bromo-2-methylbutane (**6**), the substrate you will use in this experiment, yields both 2-methyl-2-butene (**7**) and 2-methyl-1-butene (**8**) on reaction with strong base (Eq. 10.5). Because such elimination reactions are normally *irreversible* under these experimental conditions, the alkenes **7** and **8** do *not* undergo equilibration subsequent to their production. Consequently, the ratio of **7** and **8** obtained is defined by the relative rates of their formation. These rates, in turn, are determined by the relative free energies of the two transition states, **9** and **10**, respectively, rather than by the relative free energies of the alkenes **7** and **8** themselves.

$$\underset{\substack{|\\ \text{H}}}{\overset{\substack{\text{CH}_3\\|}}{\text{CH}_2\text{-}\underset{\substack{|\\\text{Br}}}{\text{C}}\text{-}\underset{\substack{|\\\text{H}}}{\text{CH}}\text{-}\text{CH}_3}} \xrightarrow{\text{base}} \underset{\substack{|\\\text{H}}}{\overset{\substack{\text{CH}_3\\|}}{\text{CH}_2\text{-}\text{C}}\text{=}\text{CH}\text{-}\text{CH}_3} \; + \; \underset{\substack{|\\\text{H}}}{\overset{\substack{\text{CH}_3\\|}}{\text{CH}_2\text{=}\text{C}\text{-}\text{CH}\text{-}\text{CH}_3}} \qquad (10.5)$$

6   **7**   **8**

2-Bromo-2-methylbutane   2-Methyl-2-butene   2-Methyl-1-butene

$$\left[\begin{array}{c} \overset{\delta^-}{\text{Br}} \\ \text{H}_3\text{C}\diagup\overset{\alpha}{\text{C}}\diagdown\text{CH}_3 \\ \text{H}_3\text{C}\diagdown\overset{\beta}{\text{C}}\diagup\text{H} \\ | \\ \text{H} \\ | \\ \text{B:}\,\delta^- \end{array} \right]^{\ddagger} \qquad\qquad \left[\begin{array}{c} \overset{\delta^-}{\text{Br}} \\ \text{H}_3\text{C}\diagup\overset{\alpha}{\text{C}}\diagdown\text{CH}_2\text{CH}_3 \\ \text{H}\diagdown\overset{\beta}{\text{C}}\diagup\text{H} \\ | \\ \text{H} \\ | \\ \text{B:}\,\delta^- \end{array} \right]^{\ddagger}$$

9   **10**

In the absence of complicating factors, the predominant product in an E2 elimination is the more highly substituted alkene, which is **7** in the present example; this observation is the source of Zaitsev's rule. The trend is observed because an increase in the number of alkyl substituents on the double bond almost always increases the stability—that is, lowers the free energy—of an alkene. Those factors that stabilize the product alkenes also play a role in stabilizing the respective transition states in which *partial* double-bond character is developing between the two carbon atoms. Thus, the enthalpy of activation, $[\Delta H^{\ddagger}]$, for forming the more stable alkene **7** is less than that for the less stable alkene **8**.

The relative free energies of **transition states** of competing elimination reactions may be influenced by **steric factors** that increase the energies of some transition states relative to others. For instance, the less stable alkene may become the major product if steric factors raise the energy of the transition state leading to the more highly substituted alkene more than that for forming the less highly substituted. This may occur when sterically demanding substituents on the carbon atom from which the proton is being removed offer hindrance to the approaching base. Removal of a proton from a less substituted carbon to give a less substituted alkene then becomes more favorable. As more sterically demanding, so-called bulky bases are used, abstraction of the more accessible proton will also be favored, thereby leading to the formation of even greater amounts of the less substituted alkene.

Transition states **9** and **10** serve as an illustrative example of these principles. In **9**, which is the transition state leading to the formation of **7**, there can be an unfavorable steric interaction between the methyl group on the β-carbon atom and the base. There is no comparable steric interaction in **10**, which produces the thermodynamically less stable **8**. Consequently, as the steric bulk of the base is increased, **8** will be formed in increased amounts.

In the two experiments that follow, you will explore these principles by performing base-promoted elimination reactions of 2-bromo-2-methylbutane (**6**). Using this tertiary alkyl bromide eliminates problems that might be associated with forming byproducts from substitution reactions and allows a study of effects on product ratios when bases having different steric requirements are used. The dehydrohalogenation of **6** may be effected by using various base/solvent combinations. For example, some members of your class may use potassium hydroxide as the base and 1-propanol as the solvent for the dehydrohalogenation of **6**, whereas others may use potassium *tert*-butoxide as the base and *tert*-butyl alcohol as the solvent. By comparing the ratios of the alkenes produced by the two different methods, you can explore how varying the reagents and solvents can influence the course of a chemical reaction.

EXPERIMENTAL PROCEDURES

Base-Promoted Elimination of an Alkyl Halide

Discovery Experiment   **Purpose** To demonstrate the formation of alkenes by dehydrohalogenation and assess product distributions arising from use of different bases.

SAFETY ALERT

1. **Wear safety glasses or goggles and suitable protective gloves while performing the experiments.**

2. **Most of the chemicals used in this experiment are highly flammable, so do not handle these liquids near open flames.**

3. **The solutions used in this experiment are *highly* caustic. Do not allow them to come in contact with your skin. If this should happen, flood the affected area with water and then thoroughly rinse the area with a solution of *dilute* acetic acid.**

A ■ *Elimination with Alcoholic Potassium Hydroxide*

MINISCALE PROCEDURE

Preparation Refer to the online resources to answer Pre-Lab Exercises, access videos, and read the MSDSs for the chemicals used or produced in this procedure. Review Sections 2.2, 2.4, 2.9, 2.10, 2.11, 2.14, 2.22, 2.27, and 6.4.

Apparatus A 10-mL and a 50-mL round-bottom flask, drying tube, ice-water bath, apparatus for fractional distillation, magnetic stirring, and *flameless* heating.

Setting Up Place 25 mL of a 4 *M* solution of potassium hydroxide in 1-propanol in the 50-mL flask containing a stirbar. Attach a drying tube to the flask. Add 2.5 mL of 2-bromo-2-methylbutane and then equip the flask for fractional distillation. Be sure to lubricate the ground-glass joint connecting the Hempel column to the flask with a thin layer of stopcock grease. Lubricating this joint is *particularly* important in this experiment because the strong base being used may cause the joint to freeze. To increase the cooling efficiency of the Hempel column, fill it with Raschig rings, coarsely broken glass tubing, coarse steel wool, or another packing material as directed by your instructor. Using a short piece of tubing, fit the vacuum adapter holding a 10-mL receiving flask with a drying tube and immerse the receiving flask in an ice-water bath. Attach water hoses to the *Hempel column* and circulate water through the jacket of the column during the period of reflux for this reaction.

Elimination and Isolation Heat the reaction mixture under gentle reflux with stirring for a period of 1–1.5 h.★ Allow the mixture to cool below its boiling point, and drain the water from the jacket of the Hempel column. Connect water hoses to the *condenser* so that the apparatus is now set for fractional distillation. Leave in the receiving flask any low-boiling distillate that has condensed and continue to cool this flask in an ice-water bath. Distill the product mixture, collecting all distillate boiling *below* 45 °C, and transfer the product to a tared sample bottle with a tight-fitting stopper or cap.

Analysis Weigh the distillate and calculate the yield of products. Then put the container in an ice-water bath until all tests on the distillate have been completed. Test the distillate for unsaturation using the bromine and Baeyer tests (Secs. 4.7A1 and 4.7A2, respectively). Analyze your distillate by GLC or submit a sample of it for analysis. After obtaining the results, calculate the relative percentages of the two isomeric alkenes formed; assume that the response factors are identical for the two alkenes. A typical GLC analysis of the products from this elimination is shown in Figure 10.1a. Obtain IR and NMR spectra of your starting material and product. Compare your spectra with those of authentic samples of 2-bromo-2-methylbutane (Figs. 10.2 and 10.3) and of the two possible alkenes (Figs. 10.4–10.7). The ^{1}H NMR spectrum of the product mixture obtained from a representative experiment is given in Figure 10.8.

MICROSCALE PROCEDURE

Preparation Refer to the online resources to answer Pre-Lab Exercises, access videos, and read the MSDSs for the chemicals used or produced in this procedure. Review Sections 2.3, 2.4, 2.9, 2.11, 2.13, 2.22, 2.27, and 6.4.

Figure 10.1

Typical GLC traces of the products of elimination of 2-bromo-2-methylbutane. Assignments and peak areas (in parentheses): peak 1: 2-methyl-1-butene; peak 2: 2-methyl-2-butene. (a) Elimination with KOH; peak 1 (82210), peak 2 (120706). (b) Elimination with KOC(CH$_3$)$_3$; peak 1 (94260), peak 2 (85140); column and conditions: 50-mm × 30-m, 0.25-μ film of SE-54; 80 °C, 40 mL/min.

(a)   (b)

Apparatus A 5-mL conical vial, ice-water bath, apparatus for simple distillation, magnetic stirring, and *flameless* heating.

Setting Up Place a spinvane in the conical vial. Using a calibrated pipet, add 2.5 mL of a 4 *M* solution of potassium hydroxide in 1-propanol. Avoid getting any of this solution on the ground-glass joint of the vial. With a clean syringe or calibrated pipet, measure 1 mL of 2-bromo-2-methylbutane into the vial. Thoroughly mix the contents by stirring for a few seconds. Equip the vial for microscale distillation and circulate cold water through the condenser. Be sure to lubricate the ground-glass joint connecting the Hickman distillation head to the vial with a thin layer of stopcock grease. Lubricating this joint is *particularly* important because otherwise the strong base being used may cause the joint to freeze.

Elimination and Isolation Heat the reaction vial with a bath that has been preheated to about 75 °C (bath temperature). The vial should be positioned so that the levels of the top of the bath and of the liquid in the reaction vial are equal. Stir the mixture at 75 °C for 1 h and then slowly raise the temperature of the bath to about 90 °C over a period of about 20 min, collecting all of the distillate until about 2.5 mL of liquid remains in the vial. It should not be necessary to heat the bath higher than 90–95 °C. Using a Pasteur or a filter-tip pipet, transfer the distillate to a dry sample vial or a test tube with a tight-fitting cap or stopper.

Analysis Weigh the distillate and calculate the yield of products. Then put the container in an ice-water bath until all tests on the distillate have been completed. Test the distillate for unsaturation using the bromine and Baeyer tests (Secs. 4.7A1 and

4.7A2, respectively). Analyze your distillate by GLC or submit a sample of it for analysis. After obtaining the results, calculate the relative percentages of the two isomeric alkenes formed; assume that the response factors are identical for the two alkenes. A typical GLC trace of the products from this elimination is shown in Figure 10.1a. Obtain IR and NMR spectra of your starting material and product. Compare your spectra with those of authentic samples of 2-bromo-2-methylbutane (Figs. 10.2 and 10.3) and of the two possible alkenes (Figs. 10.4–10.7). The ^{1}H NMR spectrum of the product mixture obtained from a representative experiment is given in Figure 10.8.

B ■ *Elimination with Potassium* **tert-Butoxide**

MINISCALE PROCEDURE

Preparation Refer to the online resources to answer Pre-Lab Exercises, access videos, and read the MSDSs for the chemicals used or produced in this procedure. Review Sections 2.2, 2.4, 2.9, 2.10, 2.11, 2.14, 2.22, 2.27, and 6.4.

Apparatus A *dry* 50-mL round-bottom flask, drying tube, ice-water bath, apparatus for fractional distillation, magnetic stirring, and *flameless* heating.

Setting Up Equip the flask with a stirbar. Working quickly, add 25 mL of a 1 *M* solution of potassium *tert*-butoxide in *tert*-butyl alcohol. This reagent is *very* moisture-sensitive, so exposure to the atmosphere *must* be minimized. Then add 2.5 mL of 2-bromo-2-methylbutane to the solution and equip the flask for fractional distillation. Complete setup according to the directions in the Miniscale Procedure of Part **A**.

Elimination and Isolation Follow the procedure given in the Miniscale Procedure of Part **A** for completing the reaction.

Analysis Weigh the distillate and calculate the yield of products. Then put the container in an ice-water bath until all tests on the distillate have been completed. Test the distillate for unsaturation using the bromine and Baeyer tests (Secs. 4.7A1 and 4.7A2, respectively). Analyze your distillate by GLC or submit a sample of it for analysis. After obtaining the results, calculate the relative percentages of the two isomeric alkenes formed; assume that the response factors are identical for the two alkenes. A typical GLC trace of the products from this elimination is shown in Figure 10.1b. Obtain IR and NMR spectra of your starting material and product. Compare your spectra with those of authentic samples of 2-bromo-2-methylbutane (Figs. 10.2 and 10.3) and of the two possible alkenes (Figs. 10.4–10.7). The ^{1}H NMR spectrum of the product mixture obtained from a representative experiment is given in Figure 10.9.

Discovery Experiment   *Elimination of Alternate Non-Terminal Alkyl Halides*

Obtain a sample of another non-terminal bromoalkane such as 2-bromoheptane from your instructor. Team up with another student, with one of you performing a base-promoted dehydrobromination using alcoholic potassium hydroxide according to the procedure in Part **A** and the other using potassium *tert*-butoxide according to the procedure in Part **B**. Calculate your yield and analyze the products by GLC to identify and to determine the ratio of the three possible alkenes. Compare the ratio of products obtained using the two different bases and explain any differences. Be sure to consult with your instructor before undertaking any experimental procedures.

WRAPPING IT UP

Neutralize the *residue remaining in the stillpot* with 10% aqueous hydrochloric acid; then flush it down the drain. Place the *dichloromethane solution* from the bromine test for unsaturation in a container for halogenated organic liquids; put the *manganese dioxide* from the Baeyer test for unsaturation in a container for heavy metals.

EXERCISES

1. Create a reaction profile (potential energy diagram) analogous to that in Figure 13.1, making it consistent with the information provided regarding the relative amounts of **7** and **8** (Eq. 10.5) and the relative energies of **9** and **10**.

2. Demonstrate that the amount of 4 M potassium hydroxide solution used for the elimination provides an excess of base relative to the alkyl bromide.

3. How might the ratio of alkenes obtained in the *tert*-butoxide-promoted elimination change if the *tert*-butyl alcohol contains water?

4. Why does the excess of base used in these eliminations favor the E2 over the E1 mechanism for elimination?

5. What is the solid material that precipitates as the eliminations proceed?

6. If all the elimination reactions in the experimental section had proceeded by the E1 mechanism, would the results have been different from those actually obtained? Why?

7. From the results of the experiments that were performed and/or from the GLC data in Figure 10.1, what conclusions can be drawn concerning the effect of relative base size upon product distribution?

8. In the miniscale procedures, what purpose is served by packing the Hempel column with Raschig rings or other packing materials?

9. Draw the Newman projections for the three staggered conformations about the C2–C3 bond of 2-bromoheptane and indicate which of these possesses an antiperiplanar relationship between the β-hydrogen atom and the bromine atom.

10. If the steric demands of the leaving group in the 2-methyl-2-butyl system were greater than those of a methyl group, why would 2-methyl-1-butene be expected to be formed in greater amounts than if the leaving group were smaller than methyl, regardless of which base is used? Use Newman projections to support your explanation.

11. It has been shown that there is an important stereoelectronic requirement for an anti-periplanar relationship between the leaving group and the β-hydrogen during an E2 elimination. Predict the structure of the main product of the dehydrobromination of *trans*-2-bromomethylcyclohexane using alcoholic potassium hydroxide. (*Hint:* The elimination may not occur via the lowest energy chair conformer.)

12. What is the main side reaction that competes with elimination when a primary alkyl halide is treated with alcoholic potassium hydroxide, and why does this reaction compete with elimination of a primary alkyl halide but not a tertiary alkyl halide?

Refer to the online resources and use the spectra viewer and Tables 8.1–8.8 as needed to answer the blue-numbered questions on spectroscopy.

13. Consider the NMR data for 2-bromo-2-methylbutane (Fig. 10.3).

    a. In the ^{1}H NMR spectrum, assign the various resonances to the hydrogen nuclei responsible for them.

    b. For the ^{13}C NMR data, assign the various resonances to the carbon nuclei responsible for them.

14. Consider the spectral data for 2-methyl-1-butene (Figs. 10.4 and 10.5).

    a. In the functional group region of the IR spectrum, specify the absorptions associated with the carbon-carbon double bond and the hydrogens attached to the C1 carbon atom.

    b. In the ^{1}H NMR spectrum, assign the various resonances to the hydrogen nuclei responsible for them.

    c. For the ^{13}C NMR data, assign the various resonances to the carbon nuclei responsible for them.

15. Consider the spectral data for 2-methyl-2-butene (Figs. 10.6 and 10.7).

    a. In the functional group region of the IR spectrum, specify the absorptions associated with the carbon-carbon double bond and the hydrogen attached to the C3 carbon atom.

    b. In the ^{1}H NMR spectrum, assign the various resonances to the hydrogen nuclei responsible for them.

    c. For the ^{13}C NMR data, assign the various resonances to the carbon nuclei responsible for them.

16. Discuss the differences observed in the IR and NMR spectra of 2-bromo-2-methylbutane and of 2-methyl-1-butene and 2-methyl-2-butene that are consistent with dehydrohalogenation occurring in this experiment.

17. Referring to the spectra on the website for Figures 10.5 and 10.7–10.9, calculate the percentage compositions of the mixtures of isomeric methylbutenes obtained from the reaction of 2-bromo-2-methylbutane with potassium hydroxide and potassium *tert*-butoxide. In the spectra of the mixtures, the integration of the resonances in the region of δ 5.0 ppm has been electronically amplified in the upper stepped line so that the relative areas of the two multiplets in that region can be more accurately measured.

SPECTRA

Starting Materials and Products

Figure 10.2

*IR spectrum of 2-bromo-
2-methylbutane (neat).*

Figure 10.3

*NMR data for 2-bromo-
2-methylbutane (CDCl₃).*

(a) ¹H NMR spectrum (300 MHz).
(b) ¹³C data: δ 10.7, 33.8, 40.3, 69.2.

Figure 10.4

*IR spectrum of 2-methyl-1-butene
(neat).*

Figure 10.5
NMR data for 2-methyl-1-butene (CDCl₃).

(a) ¹H NMR spectrum (300 MHz).
(b) ¹³C data: δ 12.5, 22.3, 31.0, 108.8, 147.5.

Figure 10.6
IR spectrum of 2-methyl-2-butene (neat).

Figure 10.7
NMR data for 2-methyl-2-butene (CDCl₃).

(a) ¹H NMR spectrum (300 MHz).
(b) ¹³C data: δ 13.4, 17.3, 25.6, 118.8, 132.0.

Figure 10.8
¹H NMR spectrum (CDCl₃,
300 MHz) of the product mixture
from the elimination of 2-bromo-
2-methylbutane with potassium
hydroxide.

Figure 10.9
¹H NMR spectrum (CDCl₃,
300 MHz) of the product mixture
from the elimination of 2-bromo-
2-methylbutane with potassium
tert-butoxide.

10.3 DEHYDRATION OF ALCOHOLS

The nonbonding lone pairs of electrons on the oxygen atom of the hydroxyl group, which is the functional group of alcohols, serve as **Lewis bases** that may complex with **Brønsted-Lowry** and **Lewis acids**, which are defined as proton-donating and electron pair-accepting acids, respectively. For example, the protonation of an alcohol, **11**, is rapid and reversible and produces an **oxonium ion, 12**, in which the positively charged oxygen atom further polarizes the carbon-oxygen single bond. In contrast to the hydroxide ion, a strong base and a poor leaving group, water is a weak base and an excellent leaving group. Hence, **12** may undergo facile elimination and substitution reactions upon reaction with a base or **nucleophile** Nu:⁻ (Eq. 10.6).

(10.6)

The loss of water that effects **dehydration** of the oxonium ion **12** to produce an alkene, **13**, may occur by mechanistic pathways like those shown in Equations 10.7 and 10.8. With *secondary* and *tertiary* alcohols, **12** undergoes **endothermic** heterolysis via the transition state **12‡** to produce the **carbocation 15** and water. Rapid transfer of a proton from an adjacent carbon atom to a weak base such as water then gives **13** (Eq. 10.7). The fragmentation of **12** to give **15** is kinetically a **first-order** process and is the **rate-determining step** (rds) of the overall reaction. Consequently, the reaction is classified as **E1**, which stands for *elimination unimolecular*. Because the formation of the carbocation **15** is an endothermic process, the transition state **12‡** is productlike and is characterized by extensive heterolysis of the carbon-oxygen bond with simultaneous development of partial positive charge on the carbon atom. Thus, the ease of dehydration of alcohols by the E1 mechanism is in the order $3° > 2° \gg 1°$, paralleling the relative stabilities of carbocations. In the case of *primary* alcohols, ionization of the oxonium ion **12** would produce a highly unstable $1°$ carbocation (**12**, $R^3 = R^4 = H$), so the alternate E2 (*elimination bimolecular*) mechanism of Equation 10.8 applies.

(10.7)

(10.8)

As illustrated in Equations 10.6–10.8, each of the steps along the reaction pathway is reversible, so an alkene may undergo acid-catalyzed hydration to form an alcohol. In practice, reversal of the dehydration may be avoided by removing the alkene, whose boiling point is always *lower* than the parent alcohol, from the reaction mixture by distillation. This technique shifts the equilibrium to the right and maximizes the yield of alkene. You may recall that shifting an equilibrium of a reversible reaction to the right by removing one of the products as it is formed follows from the **Le Châtelier principle**.

According to Equation 10.6, elimination and substitution reactions can be competing pathways (see also Sec. 14.3), but the relative importance of each may often be controlled through proper choice of reaction conditions and reagents. In the case of the acid-catalyzed dehydration of secondary and tertiary alcohols, these competing reactions involve combining the intermediate carbocation with the **conjugate base**, A$^-$, of the acid catalyst A–H or with some other nucleophile such as solvent. For example, if hydrochloric acid were chosen as the catalyst for dehydration, the chloride ion could trap the carbocation to give an alkyl chloride (Eq. 10.9). This side reaction thus not only produces an undesired by-product but also consumes the acid catalyst. Consequently, the desired dehydration is terminated because formation of the alkyl chloride is irreversible under the conditions normally used for dehydration.

$$\underset{\underset{R}{|}}{\overset{\overset{R}{|}}{R-C-OH}} \xrightarrow[-H_2O]{HCl} \underset{\underset{R}{|}}{\overset{\overset{R}{|}}{R-C^+}} \underset{Cl:^-}{} \longrightarrow \underset{\underset{R}{|}}{\overset{\overset{R}{|}}{R-C-Cl}} \qquad (10.9)$$

The side reaction depicted in Equation 10.9 is avoided when the formation of the substitution product is reversible. For this reason, sulfuric acid is often used to effect the dehydration of alcohols in the undergraduate laboratory because the intermediate alkyl bisulfate (Nu:$^-$ = HSO$_3$O$^-$) substitution product readily reionizes to the intermediate carbocation under the reaction conditions (Eq. 10.10). Subsequent loss of a proton then gives the desired alkene.

$$\underset{\underset{H}{|}}{\overset{\overset{R^2\ R^3}{|\ \ |}}{R^1-C-CH-OH}} \underset{-H_2O}{\overset{H_3O^+}{\rightleftharpoons}} \underset{\underset{H}{|}}{\overset{\overset{R^2\ R^3}{|\ \ |}}{R^1-C-CH}} \overset{-H^+}{\rightleftharpoons} \underset{R^1}{\overset{R^2}{>}}C=CHR^3 \qquad (10.10)$$

Elimination product

Nu:$^-$

$$\underset{\underset{H\ \ Nu}{|\ \ \ \ |}}{\overset{\overset{R^2\ \ R^3}{|\ \ \ \ \ |}}{R^1-C-CH}}$$

Substitution product

Two or more isomeric alkenes may be formed in E1 dehydration reactions. For example, consider the unsymmetrically substituted carbocation **17**, formed from alcohol **16** by protonation and subsequent loss of water (Eq. 10.11). Loss of H$_A$ produces **18**, whereas loss of H$_B$ yields **19**. The relative amounts of the two alkenes produced depends upon the relative free energies of the transition states for the loss of the H$_A$ and H$_B$. In these **transition states**, the C–H bond is partially broken, and the C–C double bond is partially formed, as depicted in **20** and **21**. Since the major difference between these two transition states is the substitution on the incipient carbon-carbon double bond, their relative energies parallel the relative stabilities of the product alkenes. Consequently, the lower-energy transition state is the one leading to the more highly substituted and more stable alkene, which is **19** in this generalized case. This is the mechanistic basis for Zaitsev's rule. Furthermore, as a general rule, the (*E*)-isomer of a pair of diastereomers (geometric isomers) predominates over the (*Z*)-isomer for steric reasons.

$$R^1-CH=CH-\underset{\underset{H_B}{|}}{\overset{\overset{R^2}{|}}{C}}-R^3$$

18

$$R^1-\underset{\underset{H_A}{|}}{CH}-CH-\underset{\underset{H_B}{|}}{\overset{\overset{R^2}{|}}{C}}-R^3 \quad\underset{\longleftarrow}{\overset{-H_2O}{\longrightarrow}}\quad R^1-\underset{\underset{H_A}{|}}{CH}-\overset{+}{CH}-\underset{\underset{H_B}{|}}{\overset{\overset{R^2}{|}}{C}}-R^3$$

16   **17**

$$\overset{-H_A^+}{\nearrow}$$
$$\overset{-H_B^+}{\searrow}$$

(10.11)

$$R^1-\underset{\underset{H_A}{|}}{CH}-CH=\overset{\overset{R^2}{|}}{C}-R^3$$

19

$$\left[R^1-\underset{\underset{\delta^+ H_A}{\vdots}}{CH}\overset{\delta^+}{=\!=}CH-\underset{\underset{H_B}{|}}{\overset{\overset{R^2}{|}}{C}}-R^3\right]^{\ddagger} \qquad \left[R^1-\underset{\underset{H_A}{|}}{CH}-CH\overset{\delta^+}{=\!=}\underset{\underset{\delta^+ H_B}{\vdots}}{\overset{\overset{R^2}{|}}{C}}-R^3\right]^{\ddagger}$$

20   **21**

The distribution of alkenes from the acid-catalyzed dehydration of alcohols is not always predictable from the orientational factors just described. Rather, the intermediate carbocation may rearrange through the migration of either a **hydride**, $H:^-$, or an **alkyl group**, $R:^-$, from a carbon atom adjacent to the cationic center. The rearrangement is more likely if the new carbocation is more stable than the original ion. The subsequent loss of a proton from the rearranged carbocation according to the orientation effects already discussed then leads to the additional alkenes.

$$(CH_3)_2CHCH_2\underset{\underset{OH}{|}}{CH}-CH_3 \quad\overset{H_3O^+}{\underset{\Delta}{\longrightarrow}}\quad (CH_3)_2CHCH_2CH=CH_2 \quad + \quad \underset{H}{\overset{(CH_3)_2CH}{>}}C=C\underset{CH_3}{\overset{H}{<}}$$

22   **23**   **24**

4-Methyl-2-pentanol   4-Methyl-1-pentene   *trans*-4-Methyl-2-pentene

(10.12)

$$\underset{H}{\overset{(CH_3)_2CH}{>}}C=C\underset{H}{\overset{CH_3}{<}} \quad + \quad \underset{H_3C}{\overset{H_3C}{>}}C=C\underset{H}{\overset{CH_2CH_3}{<}} \quad + \quad \underset{H}{\overset{H}{>}}C=C\underset{CH_2CH_2CH_3}{\overset{CH_3}{<}}$$

25   **26**   **27**

cis-4-Methyl-2-pentene   2-Methyl-2-pentene   2-Methyl-1-pentene

The acid-catalyzed dehydration of 4-methyl-2-pentanol (**22**), which is one of the experiments in this section, is shown in Equation 10.12 and nicely illustrates the preceding principles. Dehydration of **22** produces a complex mixture of isomeric alkenes, including 4-methyl-1-pentene (**23**), *trans*-4-methyl-2-pentene (**24**), *cis*-4-methyl-2-pentene (**25**), 2-methyl-2-pentene (**26**), and 2-methyl-1-pentene (**27**). A pathway to **23–25** involves deprotonating the intermediate carbocation **28**. However, forming **26** and **27** requires intervention of carbocations **29** and **30** (see Exercise 13).

$$(CH_3)_2CH-CH_2-\overset{+}{C}HCH_3 \qquad (CH_3)_2CH-\overset{+}{C}H-CH_2CH_3 \qquad (CH_3)_2\overset{+}{C}-CH_2-CH_2CH_3$$
$$\textbf{28} \qquad\qquad\qquad \textbf{29} \qquad\qquad\qquad \textbf{30}$$

In contrast to **22**, cyclohexanol (**31**) undergoes acid-catalyzed dehydration *without* rearrangement to yield a single product (Eq. 10.13). This reaction may be accompanied by acid-catalyzed polymerization of the desired cyclohexene, however.

$$\text{(10.13)}$$

31
Cyclohexanol   Cyclohexene

EXPERIMENTAL PROCEDURES

Dehydration of Alcohols

Discovery Experiment   **Purpose** To determine the product distribution in the acid-catalyzed dehydration of alcohols to alkenes.

> **SAFETY ALERT**
>
> 1. **Wear safety glasses or goggles and suitable protective gloves while performing the experiments.**
>
> 2. **The majority of materials, particularly the product alkenes, that will be handled during this experiment are highly flammable, so use *flameless* heating and be certain that there are no open flames in your vicinity.**
>
> 3. **Several experimental operations require pouring, transferring, and weighing chemicals and reagents that cause burns on contact with your skin. Should acidic solutions accidentally come in contact with your skin, immediately flood the affected area with water, then wash it with 5% sodium bicarbonate solution.**

A ▪ *Dehydration of 4-Methyl-2-Pentanol*

MINISCALE PROCEDURE

Preparation Refer to the online resources to answer Pre-Lab Exercises, access videos, and read the MSDSs for the chemicals used or produced in this procedure. Review Sections 2.4, 2.9, 2.10, 2.11, 2.13, 2.14, 2.22, 2.25, 2.27, and 6.4.

Apparatus A 25-mL and two 10-mL round-bottom flasks, drying tube, ice-water bath, apparatus for simple and fractional distillation, magnetic stirring, and *flameless* heating.

Setting Up Place a stirbar and 4.0 mL of 4-methyl-2-pentanol in the 25-mL round-bottom flask. Add 2.5 mL of 9 *M* sulfuric acid and thoroughly mix the liquids by gently swirling the flask. Equip the flask for fractional distillation and use an ice-water bath to cool the receiving flask.

Elimination Heat the reaction mixture and collect all distillates while maintaining the head temperature below 90 °C. If the reaction mixture is not heated too strongly, the head temperature will stay below 80 °C for most of the reaction. When about 2.5 mL of liquid remains in the reaction flask, discontinue heating. Transfer the distillate to an Erlenmeyer flask and add several spatula-tips full of *anhydrous* potassium carbonate to neutralize any acid and to dry the distillate.★ Occasionally swirl the mixture during a period of 10–15 min to hasten drying; add additional portions of *anhydrous* potassium carbonate if the liquid remains cloudy.

Isolation Transfer all of the dried organic mixture into a dry 10-mL round-bottom flask by decantation or by using a Pasteur pipet. Add a stirbar and equip the flask for simple distillation (Fig. 2.37a). Use a tared 10-mL flask as the receiver. Protect the receiver from atmospheric moisture by equipping the vacuum adapter with a drying tube and cool the receiver in an ice-water bath.

Isolate the alkenes by simple distillation. The expected products of the reaction are 4-methyl-1-pentene, *cis*-4-methyl-2-pentene, *trans*-4-methyl-2-pentene, 2-methyl-1-pentene, and 2-methyl-2-pentene. Collect the fraction boiling between 53 °C and 69 °C (760 torr) in a pre-weighed, dry 10-mL receiving flask.

Analysis Weigh the distillate and determine the yield of products. Test the distillate for unsaturation using the bromine and Baeyer tests (Secs. 4.7A1 and 4.7A2, respectively). Analyze your distillate by GLC or submit a sample of it for analysis. After obtaining the results, calculate the relative percentages of the isomeric alkenes formed; assume that the response factors are identical for the isomers. A typical GLC trace of the alkenes from this elimination is shown in Figure 10.10. Obtain IR and NMR spectra of your starting material and compare your spectra with those of an authentic sample (Figs. 10.11 and 10.12).

B ■ *Dehydration of Cyclohexanol*

MINISCALE PROCEDURE

Preparation Refer to the online resources to answer Pre-Lab Exercises, access videos, and read the MSDSs for the chemicals used or produced in this procedure. Review Sections 2.9, 2.10, 2.11, 2.13, 2.14, 2.22, 2.27, and 6.4.

Apparatus A 25-mL and two 10-mL round-bottom flasks, drying tube, ice-water bath, apparatus for simple and fractional distillation, magnetic stirring, and *flameless* heating.

Figure 10.10
GLC trace of the product mixture from the dehydration of 4-methyl-2-pentanol. Assignments and peak areas (in parentheses): peak 1: diethyl ether (solvent for analysis); peak 2: 4-methyl-1-pentene (3030); peak 3: cis- and trans-4-methyl-2-pentene (51693); peak 4: 2-methyl-1-pentene (2282); peak 5: 2-methyl-2-pentene (9733). Column and conditions: 0.25-mm × 30-m APT-Hep-Tex; 37 °C, 35 mL/min.

Setting Up Place a stirbar and 5.0 mL of cyclohexanol in the 25-mL round-bottom flask. Add 2.5 mL of 9 *M* sulfuric acid and thoroughly mix the liquids by gently swirling the flask. Equip the flask for fractional distillation and use an ice-water bath to cool the receiving flask.

Elimination and Isolation Follow the same protocol as provided in Part **A**, with the *single exception* that the product should be collected over the range of 80–85 °C (760 torr). Note that water and cyclohexene form a **minimum-boiling azeotrope**, which is a mixture of the two compounds that distills *below* the boiling point of either pure compound (see Figure 10.4). This makes it vital that the crude product be dried *thoroughly* prior to the simple distillation.

Analysis Weigh the distillate and, assuming that it is pure cyclohexene, determine the yield of product. Test the distillate for unsaturation using the bromine and Baeyer tests (Secs. 4.7A1 and 4.7A2, respectively). Analyze your distillate by GLC or submit a sample of it for analysis. Obtain IR and ¹H NMR spectra of your starting material and product and compare them with those of authentic samples (Figs. 10.23–10.26).

MICROSCALE PROCEDURE

Preparation Refer to the online resources to answer Pre-Lab Exercises, access videos, and read the MSDSs for the chemicals used or produced in this procedure. Review Sections 2.4, 2.9, 2.13, 2.22, 2.25, and 6.4.

Apparatus A 3-mL conical vial, apparatus for magnetic stirring, simple distillation, and *flameless* heating.

Setting Up Place a spinvane in the conical vial. Using a calibrated Pasteur pipet, add 1 mL of cyclohexanol and 0.5 mL of 9 *M* sulfuric acid to the vial. Thoroughly mix the contents by briefly stirring or swirling the liquid for a few seconds. Equip the vial for microscale distillation and circulate cold water through the condenser.

Chapter 10 ■ *Alkanes* 349

Elimination and Isolation Heat the reaction vial in a bath that has been preheated to about 120 °C (bath temperature). The vial should be positioned so that the levels of the top of the bath and of the liquid in the reaction vial are equal. Slowly raise the temperature of the bath as necessary to maintain distillation of the product and continue distilling until about 0.5–0.7 mL of liquid remains in the vial. It should not be necessary to heat the bath higher than about 130–135 °C. Dismantle the apparatus, and remove the distillate from the Hickman stillhead using a Pasteur or filter-tip pipet. Transfer the distillate to a dry sample vial or test tube, and add several microspatula-tips full of *anhydrous* potassium carbonate to neutralize any acid and to dry the distillate. Occasionally swirl the mixture for 5–10 min to hasten drying and add additional portions of *anhydrous* potassium carbonate if the liquid remains cloudy.

Analysis Test the distillate for unsaturation using the bromine and Baeyer tests (Secs. 4.7A1 and 4.7A2, respectively). Analyze your distillate by GLC or submit a sample of it for analysis. Obtain IR and ^{1}H NMR spectra of your starting material and product and compare them with those of authentic samples (Figs. 10.23–10.26).

Discovery Experiment *Elimination of Stereoisomeric Alcohols*

Obtain a sample of another alcohol such as 4-methylcyclohexanol or 2-methyl-cyclohexanol, both of which are typically available as a mixture of *cis-* and *trans-*isomers, from your instructor. Perform the acid-catalyzed dehydration of the alcohol you are given according to the procedure in Part **B**. Calculate your yield and analyze the product(s) by capillary GLC. In the case of 2-methylcyclohexanol, identify the alkenes formed and determine their ratio. Consult with your instructor before undertaking any experiments.

WRAPPING IT UP

Dilute the *residue remaining in the stillpot* with water, carefully neutralize it with sodium carbonate, and flush it down the drain with large quantities of water. Dry the *potassium carbonate* on a tray in the hood and flush it down the drain or place it in a container for nonhazardous solids. Pour the *dichloromethane solution* from the bromine test for unsaturation in a container for halogenated organic liquids, and put the *manganese dioxide* from the Baeyer test for unsaturation in a container for heavy metals.

EXERCISES
General Questions

1. Create a reaction profile (potential energy diagram) analogous to that in Figure 13.1, making it consistent with the information provided regarding the relative amounts of **18** and **19** (Eq. 10.11) and the relative energies of **20** and **21**.

2. Define the Le Châtelier principle.

3. Why is the boiling point of the parent alcohol higher than that of the product alkene?

4. Why are the distillates obtained in the initial step of the dehydration reaction dried over *anhydrous* potassium carbonate?

5. Why is the head temperature kept below 90 °C in the initial step of the dehydration reaction?

6. Why is it necessary to separate a dried organic solution from the drying agent prior to distilling the solution?

© 2016 Cengage Learning. All Rights Reserved. May not be scanned, copied or duplicated, or posted to a publicly accessible website, in whole or in part.

7. In principle, the equilibrium in the dehydration of an alcohol could be shifted to the right by removal of water. Why is this tactic not a good option for the dehydration of 4-methyl-2-pentanol and cyclohexanol?

8. The loss of a proton from suitably substituted carbocations can provide both the *trans-* and *cis-*isomers of the resulting alkene, but the *trans-*isomer normally predominates. For example, deprotonation at C3 of the 2-pentyl carbocation produces mainly *trans-*pentene. By analyzing the relative energies of the conformational isomers of the carbocation that lead to the two isomeric 2-pentenes, explain why the *trans-*isomer is formed preferentially. This analysis is aided by the use of Newman projections based on the partial structure shown.

9. Why is it proposed that the alcohol functional group is protonated by acid before dehydration can occur via either an E1 or an E2 mechanism?

10. List two of the major differences in the dehydration of an alcohol by an E1 and E2 reaction mechanism.

11. What is the driving force for the rearrangement of intermediate carbocations during the dehydration of alcohols via the E1 mechanism?

12. Give a detailed mechanism for the dehydration reaction shown. Use curved arrows to symbolize the flow of electrons.

$$\text{[structure]} \xrightarrow[\text{–}H_2O]{H_2SO_4\text{(cat.)}} \text{[structure]}$$

Questions for Part A

13. Give a detailed mechanism explaining the formation of **23–25** from the dehydration of 4-methyl-2-pentanol (**22**). Use curved arrows to symbolize the flow of electrons.

14. A mechanism for formation of **26** and **27** can be developed from rearrangement of the carbocation **28** by way of hydride shifts.

   a.   Provide a mechanism by which **26** may be formed from **28** through a hydride shift followed by deprotonation. Use curved arrows to symbolize the flow of electrons.

   b.   Show how your mechanism in Part **a** can be elaborated to provide for formation of **27**.

   c.   An alternative pathway for formation of the alkene **26** involves addition of a proton to **24** or **25** and deprotonation of the resulting carbocation. Similarly, **27** could be produced by the addition of a proton to **26** followed by deprotonation of the intermediate carbocation. Write out mechanisms for these transformations using curved arrows to symbolize the flow of electrons.

   d.   How might you experimentally distinguish the two types of mechanisms for the formation of **26** and **27**?

15. The dehydration of 4-methyl-2-pentanol (**22**) gives a mixture of the five isomeric alkenes **23–27**.

   a. Which of the alkenes **23–27** would you expect to be the major product? Explain the basis for your answer.

   b. Provide a rationale for why only small amounts of 4-methyl-1-pentene (**23**) and 2-methyl-1-pentene (**27**) are produced in this reaction.

16. Near the end of the dehydration of 4-methyl-2-pentanol, a white solid may precipitate from the reaction mixture. What is the solid likely to be?

17. Give the structure, including stereochemistry, of the product of adding bromine (refer to Sec. 10.6) to *trans*-4-methyl-2-pentene (**24**) and to *cis*-4-methyl-2-pentene (**25**). Relative to one another, are these products identical, enantiomeric, or diastereomeric (Sec. 7.1)? Give your reasoning.

Refer to the online resources and use the spectra viewer and Tables 8.1–8.8 as needed to answer the blue-numbered questions on spectroscopy.

18. Consider the spectral data for 4-methyl-2-pentanol (Figs. 10.11 and 10.12).

   a. In the functional group region of the IR spectrum, specify any absorptions associated with the alcohol function.

   b. In the ^{1}H NMR spectrum, assign the various resonances to the hydrogen nuclei responsible for them.

   c. For the ^{13}C NMR data, assign the various resonances to the carbon nuclei responsible for them.

19. Consider the spectral data for 4-methyl-1-pentene (Figs. 8.25, 10.13, and 10.14).

   a. In the IR spectrum, specify the absorptions associated with the carbon-carbon double bond and the vinylic hydrogen atoms.

   b. In the ^{1}H NMR spectrum, assign the various resonances to the hydrogen nuclei responsible for them.

   c. For the ^{13}C NMR data, assign the various resonances to the carbon nuclei responsible for them.

20. Consider the spectral data for *trans*-4-methyl-2-pentene (Figs. 10.15 and 10.16).

   a. In the IR spectrum, specify the absorptions, if any, associated with the carbon-carbon double bond and with the *trans* relationship of the vinylic hydrogen atoms.

   b. In the ^{1}H NMR spectrum, assign the various resonances to the hydrogen nuclei responsible for them.

   c. For the ^{13}C NMR data, assign the various resonances to the carbon nuclei responsible for them.

21. Consider the spectral data for *cis*-4-methyl-2-pentene (Figs. 10.17 and 10.18).

   a. In the IR spectrum, specify the absorptions, if any, associated with the carbon-carbon double bond and with the *cis* relationship of the vinylic hydrogen atoms.

   b. In the ^{1}H NMR spectrum, assign the various resonances to the hydrogen nuclei responsible for them.

   c. For the ^{13}C NMR data, assign the various resonances to the carbon nuclei responsible for them.

22. Consider the spectral data for 2-methyl-1-pentene (Figs. 10.19 and 10.20).

    a. In the IR spectrum, specify the absorptions associated with the carbon-carbon double bond and with the hydrogens attached to the C1 carbon atom.

    b. In the ^{1}H NMR spectrum, assign the various resonances to the hydrogen nuclei responsible for them.

    c. For the ^{13}C NMR data, assign the various resonances to the carbon nuclei responsible for them.

23. Consider the spectral data for 2-methyl-2-pentene (Figs. 10.21 and 10.22).

    a. In the IR spectrum, specify the absorptions associated with the carbon-carbon double bond and with the vinylic hydrogen atom.

    b. In the ^{1}H NMR spectrum, assign the various resonances to the hydrogen nuclei responsible for them.

    c. For the ^{13}C NMR data, assign the various resonances to the carbon nuclei responsible for them.

24. Discuss the differences observed in the IR and NMR spectra of 4-methyl-2-pentanol and the various methylpentenes that are consistent with dehydration occurring in this experiment.

Questions for Part B

25. Define the term *minimum-boiling azeotrope* and explain why it is impossible to separate an azeotropic mixture completely by distillation.

26. Why is it particularly important that the crude cyclohexene be dry prior to its distillation?

27. Give the structure, including stereochemistry, of the product of addition of bromine (refer to Sec. 10.6) to cyclohexene. Should it be possible, at least in principle, to resolve this dibromide into separate enantiomers? Explain your answer.

28. Provide a detailed mechanism for the acid-catalyzed dehydration of cyclohexanol. Use curved arrows to symbolize the flow of electrons.

29. Cyclohexyl carbocation undergoes attack by nucleophiles. With this in mind, propose a mechanism whereby cyclohexene might form the dimer **31** in the presence of H_3O^+. Note that this type of reaction leads to polymerization of cyclohexene.

31

Refer to the online resources and use the spectra viewer and Tables 8.1–8.8 as needed to answer the blue-numbered questions on spectroscopy.

30. Consider the spectral data for cyclohexanol (Figs. 10.23 and 10.24).

    a. In the functional group region of the IR spectrum, specify any absorptions associated with the alcohol function.

    b. In the ^{1}H NMR spectrum, assign the various resonances to the hydrogen nuclei responsible for them.

    c. For the ^{13}C NMR data, assign the various resonances to the carbon nuclei responsible for them.

31. Consider the spectral data for cyclohexene (Figs. 10.25 and 10.26).

    a. In the IR spectrum, specify the absorptions associated with the carbon-carbon double bond and with the vinylic hydrogen atoms.

    b. In the ¹H NMR spectrum, assign the various resonances to the hydrogen nuclei responsible for them.

    c. For the ¹³C NMR data, assign the various resonances to the carbon nuclei responsible for them.

32. Discuss the differences observed in the IR and NMR spectra of cyclohexanol and cyclohexene that are consistent with dehydration occurring in this experiment.

SPECTRA

Starting Materials and Products

The ¹H NMR spectrum of 4-methyl-1-pentene is provided in Figure 8.25.

Figure 10.11

IR spectrum of 4-methyl-2-pentanol (neat).

Figure 10.12

NMR data for 4-methyl-2-pentanol (CDCl₃).

(a) ¹H NMR spectrum (300 MHz).
(b) ¹³C data: δ 22.5, 23.2, 23.9, 24.9, 48.8, 65.8.

Figure 10.13
IR spectrum of 4-methyl-1-pentene (neat).

Figure 10.14
^{13}C *NMR data for 4-methyl-1-pentene (CDCl$_3$). Chemical shifts: δ 22.3, 28.1, 43.7, 115.5, 137.8.*

$(CH_3)_2\ CHCH_2CH{=}CH_2$

Figure 10.15
IR spectrum of trans-4-methyl-2-pentene *(neat).*

Figure 10.16
NMR data for trans-4-methyl-2-pentene *(CDCl$_3$).*

(a) ^1H NMR spectrum (300 MHz).
(b) ^{13}C data: δ 17.6, 22.7, 31.5, 121.6, 139.4.

Figure 10.17
IR spectrum of cis-4-methyl-2-pentene (neat).

Figure 10.18
NMR data for cis-4-methyl-2-pentene (CDCl₃).

(a) ¹H NMR spectrum (300 MHz).
(b) ¹³C data: δ 12.7, 23.1, 26.4, 121.4, 138.6.

Figure 10.19
IR spectrum of 2-methyl-1-pentene (neat).

Figure 10.20

NMR data for 2-methyl-1-pentene (CDCl₃).

(a) ¹H NMR spectrum (300 MHz).

(b) ¹³C data: δ 13.8, 20.8, 22.3, 40.0, 109.7, 146.0.

Figure 10.21

IR spectrum of 2-methyl-2-pentene (neat).

Figure 10.22

NMR data for 2-methyl-2-pentene (CDCl₃).

(a) ¹H NMR spectrum (300 MHz).

(b) ¹³C data: δ 14.5, 17.5, 21.6, 25.7, 126.9, 130.6.

Figure 10.23
IR spectrum of cyclohexanol (neat).

Figure 10.24
NMR data for cyclohexanol (CDCl₃).

(a) ¹H NMR spectrum (300 MHz).
(b) ¹³C data: δ 24.5, 25.9, 35.5, 70.1.

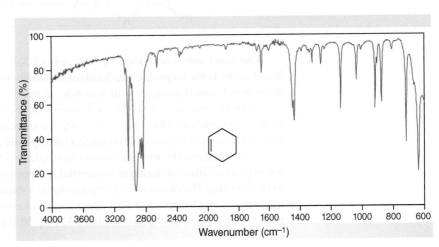

Figure 10.25
IR spectrum of cyclohexene (neat).

Figure 10.26
NMR data for cyclohexene (CDCl₃).

(a) ¹H NMR spectrum (300 MHz).
(b) ¹³C data: δ 22.9, 25.2, 127.3.

10.4 ADDITION REACTIONS OF ALKENES: OVERVIEW

Alkenes are useful starting materials for organic syntheses because they undergo a large variety of reactions involving their functional group, the carbon-carbon double or π-bond. One of the typical reactions of alkenes is the **addition** of a reagent X–Y across the π-bond (Eq. 10.14). Such additions occur because of two principal factors. First, the strength of the carbon-carbon π-bond is 60–65 kcal/mol, a value substantially lower than the 80–100 kcal/mol that is typical for the σ-bond strengths of other atoms bound to carbon. Adding a reagent across the double bond is therefore usually *exothermic*, because one π-bond is being replaced by two σ-bonds. Second, the π-electrons of the double bond are more loosely held than σ-electrons, so they are more polarizable. Consequently, the π-electron "cloud" is readily distorted through electrostatic interaction with an electron-deficient reagent, and this polarization enhances the reactivity of the alkene toward attack.

$$\begin{array}{c} R^1 \\ \diagdown \\ R^2 \end{array} C = C \begin{array}{c} R^3 \\ \diagup \\ R^4 \end{array} + X-Y \longrightarrow R^2 - \underset{\underset{X}{|}}{\overset{\overset{R^1}{|}}{C}} - \underset{\underset{R^4}{|}}{\overset{\overset{Y}{|}}{C}} - R^3 \qquad (10.14)$$

The most common mode of addition of X–Y across the π-bond of alkenes involves an **ionic stepwise mechanism** in which an **electrophile**, E^+, first attacks the π-bond, functioning as a nucleophile, to produce a carbocation. In such cases, X–Y may be rewritten as E–Nu, wherein E is the *less* electronegative of the two atoms. The **nucleophile**, Nu:⁻, then attacks the carbocation to produce the final product. The overall process is termed **electrophilic addition.**

To understand the mechanism of electrophilic addition, it is helpful to consider this type of addition in terms of sequential reactions of Lewis acids and Lewis bases. In the first step, the electrons of the polarizable π-bond, a Lewis base, attack the electrophile E^+, an electron-deficient species and a Lewis acid, to give either an acyclic planar carbocation **32** or a cyclic cation **33** (Eq. 10.15). If carbocation **32** is formed as an intermediate in the electrophilic addition to an *unsymmetrically substituted* alkene, the electrophile E^+ generally adds to the terminus of the double bond that produces the *more stable carbocation* **32** according to **Markovnikov's rule** (see the Historical Highlight *Additions Across Carbon-Carbon π-Bonds*, which is available online). Subsequent reaction of this carbocation, a Lewis acid, with the nucleophile $Nu:^-$, which is electron-rich and a Lewis base, then gives the addition product. When the positively charged carbon atom of **32** is substituted with three different groups, the nucleophile may attack from *both* faces of the planar ion by paths **a** and **b** to give a mixture of **34** and **35**, which are stereoisomers (Sec. 7.1). Acyclic carbocations like **32** may also rearrange via 1,2-hydride or 1,2-alkyl migration to form a more stable carbocation (Sec. 10.3); subsequent reaction of this isomeric carbocation with the nucleophilic partner produces adducts isomeric with **34** and **35**. Some electrophilic reagents add to a carbon-carbon double bond to form a cyclic cation **33**; such cations tend *not* to rearrange. The nucleophile $Nu:^-$ may then attack **33** by paths **c** and **d** to give mixtures of isomers **36** and **37**.

(10.15)

The general principles set forth in Equation 10.15 may be illustrated by considering several examples. The addition of the unsymmetrical reagent hydrogen bromide, H–Br, to an alkene is shown in Equation 10.16, and the acid-catalyzed addition of water, or **hydration**, of an alkene is depicted in Equation 10.17. In both reactions, the π-bond is protonated to form the more stable intermediate carbocation according to Markovnikov's rule. This carbocation may then rearrange to a more stable carbocation (Sec. 10.3), or, as shown in the present case, may undergo direct nucleophilic attack from *both* faces of the planar carbocation to give the observed alkyl bromide or alcohol. These two examples of electrophilic additions to alkenes are **regioselective** processes because the two unsymmetrical reactants combine *predominantly* in one orientational sense to give one **regioisomer** preferentially.

(10.16)

$$R^1-\underset{\underset{}{\overset{R^2}{C}}}{}=\overset{R^3}{CH} + H^+ \xrightarrow[H_2O]{} \left[R^1-\underset{\underset{H_2O:}{\overset{}{\underset{+}{C}}}}{\overset{R^2}{\underset{}{}}}-\overset{R^3}{\underset{H}{CH}} \right] \longrightarrow R^1-\underset{\underset{HO^+}{\overset{R^2}{C}}}{}-\underset{H}{\overset{R^3}{CH}} \longrightarrow R^1-\underset{\underset{HO}{\overset{R^2}{C}}}{}-\underset{H}{\overset{R^3}{CH}} + B-H \quad (10.17)$$

As mixture of
stereoisomers

Since the ionic addition of hydrogen bromide and the acid-catalyzed addition of water to an alkene proceed via planar carbocations, the formation of products derived from rearranged carbocations is relatively common. In contrast, the bromination of alkenes, which is illustrated by the addition of bromine to cyclopentene (Eq. 10.18), occurs via a **cyclic bromonium ion 38** related to **33**, so skeletal rearrangements of such cations are *not* generally observed. Moreover, the intervention of this cyclic ion dictates that the stereochemistry of the addition is completely in the *anti* sense, as shown. The reddish color of bromine is discharged upon addition to an alkene, making this reaction a useful qualitative test for unsaturation (Sec. 4.7A1).

$$(10.18)$$

Cyclopentene	Bromine	A cyclic bromonium ion	*trans*-1,2-
(colorless)	*(red-brown)*	**38**	Dibromocyclopentene
			(colorless)

The addition of reagents X–Y to carbon-carbon π-bonds may also proceed via a **concerted mechanism** in which each new σ-bond is formed simultaneously on the *same* face of the π-bond. The stereochemistry of such reactions is necessarily *syn*. For example, the reaction of potassium permanganate, which is purple, with an alkene such as cyclohexene proceeds via *syn*-addition of permanganate ion across the π-bond to give **39**, which is colorless. Subsequent decomposition of **39** gives a *cis*-1,2-diol and manganese dioxide, the brown precipitate that is observed as the other product of the reaction (Eq. 10.19). This decoloration of potassium permanganate by alkenes forms the basis of the Baeyer qualitative test for the presence of carbon-carbon π-bonds (Sec. 4.7A2).

$$(10.19)$$

Cyclohexene	Potassium		*cis*-Cyclohexane-	Manganese
(colorless)	permanganate	**39**	1,2-diol	dioxide
	(purple)		*(colorless)*	*(brown)*

10.5 ADDITION OF HYDROBROMIC ACID TO ALKENES

In this experiment, you will study the **ionic addition** of hydrobromic acid, H–Br, to 1-hexene to ascertain whether 2-bromohexane (**41**) is produced preferentially according to Markovnikov's rule (Eq. 10.20). This type of reaction is normally rather difficult to perform in the undergraduate laboratory for several reasons. First, common alkenes are immiscible with *concentrated* hydrobromic acid, and the reaction is sluggish if the layers are not mixed efficiently. Second, H–Br is a *strong* acid that protonates water extensively to give the hydronium ion, H_3O^+. Because a hydronium ion is a *weaker* acid than undissociated hydrobromic acid, it is unable to protonate the alkene rapidly under mild reaction conditions. Moreover, the presence of water in the reaction mixture introduces the possibility of competing acid-catalyzed addition of water to the alkene (Eq. 10.17, Sec. 10.4). These problems are reduced by using *anhydrous* hydrogen bromide, but the highly corrosive nature of this gas makes it difficult to handle and use.

$$CH_3(CH_2)_3CH\!=\!CH_2 + H\!-\!Br \longrightarrow CH_3(CH_2)_3\underset{Br}{CH}\!-\!\underset{H}{CH_2} \qquad (10.20)$$

40
1-Hexene

41
2-Bromohexane

A convenient solution to these experimental difficulties entails the addition of a catalytic amount of a quaternary ammonium salt such as methyltrioctylammonium chloride, $CH_3(n\text{-}C_8H_{17})_3N^+Cl^-$, also known as Aliquat 336, to the heterogeneous mixture of the aqueous acid and the alkene. The tetraalkyl ammonium salt partitions between the aqueous and organic phases because of the amphoteric nature of the catalyst: It is **lipophilic** or **nonpolar-loving** due to the alkyl groups and **hydrophilic** or **polar-loving** because of the ionic ammonium function, respectively. By forming a complex such as **42** with the H–Br, the quaternary ammonium salt extracts H–Br from the aqueous phase, transporting it into the organic phase and the presence of the alkene (Scheme 10.1). The quaternary ammonium salt repartitions into the aqueous phase to complete the catalytic cycle. The transfer of the H–Br into the organic phase essentially dehydrates the acid, making it more reactive toward the alkene so that the addition becomes possible.

Scheme 10.1

$$H\!-\!Br + R_4\overset{+}{N}X^- \rightleftharpoons R_4\overset{+}{N}X^-\text{----}H\!-\!Br$$
42

Aqueous phase

Phase interface

Organic phase

$$CH_3(CH_2)_3\underset{Br}{CH}\!-\!\underset{H}{CH_2} + R_4\overset{+}{N}X^- \xleftarrow{\text{1-hexene}} R_4\overset{+}{N}X^-\text{----}H\!-\!Br$$
42

Compounds that promote the transport of reagents between immiscible layers by means of **ion pairs** like **42** are called **phase-transfer catalysts**. Their presence can have dramatic effects on the rates of bimolecular reactions between reagents contained in immiscible phases: Rate accelerations of 10^4 to 10^9 are common. One factor that determines the overall rate of a reaction involving phase-transfer catalysis is the efficiency of the partitioning of the reagents and reactants between the two phases. This is a function of the total surface area at the interface of the phases. To increase this area, the reaction mixture must be *vigorously* agitated by stirring or shaking to promote emulsification and the formation of tiny droplets of the immiscible layers.

As depicted in Scheme 10.2, the mechanism of the reaction presumably involves protonation of 1-hexene to afford the secondary carbocation **43**. Attack of bromide ion on this ion then leads to 2-bromohexane (**41**). The carbocation may also rearrange by way of a hydride shift (Sec. 10.3) to provide a different secondary carbocation, **44**, which would provide 3-bromohexane (**45**) upon reaction with bromide ion. Alternatively, it may deprotonate to form 2-hexene (**46**), addition of H–Br to which could afford both **41** and **45**. In this experiment, you will determine the regiochemistry of the addition of H–Br to the unsymmetrical alkene 1-hexene (**40**) and thereby assess whether this ionic reaction proceeds according to Markovnikov's rule.

Scheme 10.2

$$CH_3(CH_2)_3\ CH{=}CH_2 + H{-}Br{-}{-}X^- \ ^+NR_4 \longrightarrow CH_3(CH_2)_2CH{-}\overset{+}{C}HCH_2{-}H \xrightarrow{Br^-} \mathbf{41}$$

with the structures **43**, **44**, **45**, **46** and associated intermediates shown below:

$$\underset{\mathbf{45}}{\underset{\text{3-Bromohexane}}{CH_3(CH_2)_2\overset{\overset{Br}{|}}{C}H{-}CH_2CH_3}} \xleftarrow{Br^-} \underset{\mathbf{44}}{CH_3CH_2\overset{+}{C}H{-}\underset{H}{C}H{-}CH_3} \xleftrightarrow{Br{-}H} \underset{\mathbf{46}}{\underset{\text{2-Hexene}}{CH_3(CH_2)_2CH{=}CHCH_3}}$$

with **43** labeled, and arrows ~H⁻ and –H⁺ connecting the intermediates.

EXPERIMENTAL PROCEDURES

A. Addition of Hydrobromic Acid to 1-Hexene

Discovery Experiment   **Purpose** To determine the regiochemistry of the polar addition of H–Br to an alkene.

SAFETY ALERT

1. **Wear safety glasses or goggles and suitable protective gloves while performing the experiments.**

2. *Concentrated* **hydrobromic acid (47–49%) is a** *highly corrosive* **and** *toxic* **material. Working at a hood, measure out the amount of acid required and avoid inhaling its vapors. If the acid comes in contact with your skin, flood the affected area immediately and thoroughly with water and rinse it with 5% sodium bicarbonate solution.**

3. Quaternary ammonium salts are *toxic* substances and can be absorbed through the skin. Should they accidentally come in contact with your skin, wash the affected area immediately with large amounts of water.

4. *If a flame must be used in the distillation step, be certain that all joints in the apparatus are well lubricated and tightly mated because 1-hexene and the solvents used, particularly petroleum ether, are highly flammable. Flameless heating is preferred.*

MINISCALE PROCEDURE

Preparation Refer to the online resources to answer Pre-Lab Exercises, access videos, and read the MSDSs for the chemicals used or produced in this procedure. Review Sections 2.4, 2.8, 2.9, 2.11, 2.13, 2.17, 2.19, 2.21, 2.22, and 6.4.

Apparatus A 25- and a 50-mL round-bottom flask, separatory funnel, apparatus for heating under reflux, simple distillation, magnetic stirring, and *flameless* heating.

Setting Up Combine 3.0 mL of 1-hexene, 14 mL of 48% aqueous hydrobromic acid, and 1.0 g of methyltrioctylammonium chloride in the 50-mL round-bottom flask. Equip the flask with a water-cooled condenser and set up the apparatus heating under reflux.

Addition and Work-Up With *rapid* stirring, which is critical to the success of the reaction, heat the heterogeneous reaction mixture under gentle reflux for 2 h and then allow the mixture to cool to room temperature.★

Carefully transfer the two-phase mixture to a separatory funnel, rinse the reaction flask with 15 mL of petroleum ether (bp 60–80 °C, 760 torr), and add the rinse to the separatory funnel. Shake the funnel thoroughly and allow the layers to separate. Verify that the lower one is the aqueous phase and remove it. Sequentially wash the organic phase with two 15-mL portions of 10% sodium bicarbonate solution. Vent the funnel *frequently* because gas is evolved and excessive pressure must *not* develop in the funnel. Transfer the organic layer to an Erlenmeyer flask and dry it with swirling over several spatula-tips full of *anhydrous* sodium sulfate for at least 0.5 h.★ Add additional portions of *anhydrous* sodium sulfate if the liquid remains cloudy.

Isolation Decant or gravity-filter the dried solution into a 25-mL round-bottom flask, equip the flask for simple distillation, and distill the product. Carefully control the rate of heating throughout the course of this distillation because severe foaming can occur. *Do not attempt to remove the solvent too rapidly*, as this may result in excessive loss of product due to foaming.

After the first fraction containing solvent and unreacted 1-hexene is removed over a range of room temperature to 80 °C (760 torr),★ change the receiving flask and collect the bromohexane as a single fraction in a tared container. Because the volume of product is likely to be only 1–2 mL, an accurate boiling point may be difficult to obtain, and all material that distills above 110–115 °C (760 torr) should be collected to obtain a reasonable yield.

Analysis Weigh the distillate and determine the yield of crude product. Determine the structure of the principal product of the reaction using IR and/or NMR spectroscopic methods or by GLC, assuming authentic samples of the isomers are available.

Figure 10.27
GLC traces of the bromohexanes.
(a) Mixture of petroleum ether
(peak 1), 2-bromohexane (peak 2),
and 1-bromohexane (peak 3).
(b) Distillate from addition of HBr
to 1-hexene containing petroleum
ether (peak 1). Column and con-
ditions: 0.25-mm × 25-m AT-1
Helifex 100% dimethylsiloxane;
110 °C, 40 mL/min.

An important difference in the ^{1}H NMR spectra for 2-bromohexane and 3-bromo-hexane is that the multiplets for the hydrogen atom on the same carbon atom as the bromine are centered at δ 4.13 and δ 4.00, respectively. A GLC trace of a mixture of 1- and 2-bromohexanes is given in Figure 10.27a, along with a chromatogram of the product of the reaction (Fig. 10.27b); note that 2- and 3-bromohexane are *not* separated under the GLC conditions used. Obtain IR and ^{1}H NMR spectra of your starting material and product and compare them with those of authentic samples (Figs. 10.28–10.33). Alternatively, determine whether or not Markovnikov addition has occurred by subjecting the product to the silver nitrate or sodium iodide/acetone tests for classifying alkyl halides (Secs. 10.5B1 and 10.5B2, respectively).

  Discovery Experiment   *Analysis of Bromohexanes*

Using authentic samples of 2- and 3-bromohexane and after consulting with your instructor, attempt to find GLC conditions that allow separation of these isomers. This may require the use of other types of capillary columns than that employed to obtain the traces of Figure 10.27. If your investigation is successful, analyze your distillate for the presence of 3-bromohexane.

MICROSCALE PROCEDURE

Preparation Refer to the online resources to answer Pre-Lab Exercises, access videos, and read the MSDSs for the chemicals used or produced in this procedure. Review Sections 2.4, 2.8, 2.9, 2.11, 2.13, 2.17, 2.19, 2.21, 2.22, and 6.4.

Apparatus A 5-mL conical vial, screw-cap centrifuge tube, apparatus for heating under reflux, simple distillation, magnetic stirring, and *flameless* heating.

Setting Up and Addition Equip the conical vial with a spinvane and add to it 500 μL (0.5 mL) of 1-hexene, 2 mL of 48% aqueous hydrobromic acid, and 150 mg of methyltrioctylammonium chloride. Equip the vial with a water-cooled condenser and assemble the apparatus for heating under reflux. With *rapid* stirring, which is critical to the success of the reaction, heat the heterogeneous mixture under reflux for about 2 h, and then allow the reaction mixture to cool to room temperature. ★

Work-Up Add 1 mL of petroleum ether (bp 60–80 °C, 760 torr) to the reaction mixture and shake the vial, venting as necessary to relieve pressure. Allow the layers to separate. Remove the aqueous layer with a Pasteur pipet. Do *not* discard this liquid until you are certain that it is indeed the aqueous layer!

Add 0.5 mL of 10% sodium bicarbonate solution, stir the mixture (*Caution:* gas evolution!), and remove the aqueous layer after allowing separation. Using pHydrion paper, check whether this layer is basic. If it is not, repeat washing the organic layer with aqueous sodium carbonate until the aqueous washes are basic. Finally, dry the reaction mixture over several microspatula-tips full of *anhydrous* sodium sulfate for about 15 min; swirl the vial occasionally to hasten drying. Add additional portions of *anhydrous* sodium sulfate if the liquid remains cloudy.

Distillation Using a filter-tip pipet, transfer the dried organic layer into a clean 5-mL conical vial equipped with a stirbar. Attach the vial to a Hickman stillhead equipped with a water-cooled condenser (Fig. 2.38) and perform a microscale distillation. The first fraction should contain solvent and unchanged 1-hexene and should distill at a bath temperature below 110 °C. Remove this fraction from the still, and collect a second fraction, using a bath temperature of 165 °C. This fraction should mainly consist of the desired product. Transfer this distillate to a tared screw-cap vial.

Analysis Weigh the distillate and determine the yield of crude product. Determine the structure of the principal product of the reaction using IR and/or NMR spectroscopic methods or by GLC, assuming authentic samples of the isomers are available. An important difference in the ^{1}H NMR spectra for 2-bromohexane and 3-bromohexane is that the multiplets for the hydrogen atom on the same carbon atom as the bromine are centered at δ 4.13 and δ 4.00, respectively. A GLC trace of a mixture of 1- and 2-bromohexanes is given in Figure 10.27a, along with a chromatogram of the product of the reaction (Fig. 10.27b); note that 2- and 3-bromohexane are *not* separated under the GLC conditions used. Obtain IR and ^{1}H NMR spectra of your starting material and product and compare them with those of authentic samples (Figs. 10.28–10.33). Alternatively, determine whether or not Markovnikov addition has occurred by subjecting the product to the silver nitrate or sodium iodide/acetone tests for classifying alkyl halides (Secs. 10.5B1 and 10.5B2, respectively).

Discovery Experiment   *Analysis of Bromohexanes*

Using authentic samples of 2- and 3-bromohexane and after consulting with your instructor, attempt to find GLC conditions that allow separation of these isomers. This may require the use of other types of capillary columns than that employed to obtain the traces of Figure 10.27. If your investigation is successful, analyze your distillate for the presence of 3-bromohexane.

WRAPPING IT UP

Carefully (foaming may occur!) combine the *aqueous layers and washes*, neutralize if necessary with sodium carbonate, and flush them down the drain with excess water. Since the *first distillation fraction* may contain some halogenated product, pour it in a container for halogenated organic solvents. Evaporate the petroleum ether from the *sodium sulfate* used as drying agent in the hood, and place the solid in a container for nonhazardous solids. Dilute the *silver nitrate test solution* with water and flush it down the drain with excess water unless instructed to do otherwise.

B. Qualitative Analysis of Alkyl Halides

As described in Section 4.7, there are a number of simple chemical tests that are easy to perform and that provide valuable information regarding the presence of particular types of functional groups. With respect to alkyl halides, such tests are useful in determining whether the compound in question is a primary, secondary, or tertiary halide. The following discussion is limited to two qualitative tests: (a) reaction with **alcoholic silver nitrate** solution and (b) reaction with **sodium iodide in acetone**.

1. Alcoholic Silver Nitrate

If a compound is known to contain bromine, chlorine, or iodine, information concerning the structure of the alkyl halide may be obtained from reacting the substance with alcoholic silver nitrate. The overall reaction is shown in Equation 10.21.

$$RX + AgNO_3 \xrightarrow{\text{ethanol}} AgX + RONO_2 \qquad (10.21)$$

An alkyl halide   Silver nitrate   A silver halide   An alkyl nitrate

Mechanistically, this reaction is of the S_N1 type. As noted in Section 14.2, tertiary halides are more reactive in an S_N1 reaction than are secondary halides, which are in turn more reactive than primary halides. Thus, the rate of precipitation of silver halide in this test is expected to correspond to the ease of ionization of the alkyl halide to a carbocation according to the order primary > secondary > tertiary. From a practical standpoint, these differences are best determined by testing authentic samples of primary, secondary, and tertiary halides with silver nitrate in separate test tubes and comparing the results.

Alkyl bromides and iodides react more rapidly than chlorides, and the latter may even require warming to produce a reaction in a reasonable time period. Aryl halides are unreactive toward the test reagent, as are vinyl or alkynyl halides in general. Allylic and benzylic halides, even when primary, show reactivities as great as or greater than tertiary halides because of resonance stabilization of the intermediate allyl or benzyl carbocations.

EXPERIMENTAL PROCEDURE

Silver Nitrate Test

Discovery Experiment   **Purpose** To assess the degree of an alkyl halide.

SAFETY ALERT

1. **Wear safety glasses or goggles and suitable protective gloves while performing the test.**

2. **Avoid spilling alcoholic silver nitrate on your skin because staining may result. If you do come into contact with the solution, immediately flood the affected area with water.**

Preparation Refer to the online resources to read the MSDSs for the chemicals used or produced in this procedure.

Apparatus Test tubes, calibrated Pasteur pipets, apparatus for *flameless* heating.

Setting Up Place about 2 mL of a 0.1 *M* solution of silver nitrate in 95% ethanol in a test tube.

Reaction Add 1 drop of the alkyl halide to the test tube. If no reaction is observed within 5 min at room temperature, warm the mixture in a beaker of boiling water and observe any change. Note the color of any precipitate: silver chloride is white but turns purple on exposure to light, silver bromide is pale yellow, and silver iodide is dark yellow. If there is any precipitate, add several drops of 1 *M* nitric acid solution to it and note any changes; silver halides are insoluble in acid.

To determine expected reactivities, test known primary, secondary, and tertiary halides in this manner. If possible, use alkyl iodides, bromides, and chlorides so that differences in halogen reactivity can also be observed.

WRAPPING IT UP

Neutralize and then filter all test mixtures and place the *filter cake* in a container for recovered silver halides. Flush the *filtrate* down the drain.

2. Sodium Iodide Test

Another method for distinguishing between primary, secondary, and tertiary halides is a test using sodium iodide in acetone. This test complements the alcoholic silver nitrate test, and when the two are used together, it is possible to determine the gross structure of the attached alkyl group with reasonable accuracy.

The basis of this test is that neither sodium chloride nor sodium bromide is very soluble in acetone, whereas sodium iodide is. Mechanistically, these reactions are S_N2 substitutions (Secs. 14.2 and 14.4) in which iodide ion is the nucleophile (Eqs. 10.22 and 10.23); the order of reactivity expected is primary > secondary > tertiary.

$$RCl + NaI \xrightarrow{\text{acetone}} RI + \underline{NaCl} \tag{10.22}$$

$$RBr + NaI \xrightarrow{\text{acetone}} RI + \underline{NaBr} \tag{10.23}$$

In this test, *primary bromides* give a precipitate of sodium bromide in about 3 min at room temperature, whereas the *primary* and *secondary chlorides* must be heated to about 50 °C before reaction occurs. *Secondary* and *tertiary bromides* react at 50 °C, but the *tertiary chlorides* fail to react in a reasonable time. This test is necessarily limited to bromides and chlorides.

EXPERIMENTAL PROCEDURE

Sodium Iodide Test

Discovery Experiment   **Purpose** To assess the degree of an alkyl halide.

SAFETY ALERT

1. Wear safety glasses or goggles and suitable protective gloves while performing the test.

Preparation Refer to the online resources to read the MSDSs for the chemicals used or produced in this procedure.

Apparatus Test tubes, calibrated Pasteur pipets, ice-water bath, 25- or 50-mL beaker, apparatus for *flameless* heating.

Setting Up Place 1 mL of the sodium iodide-acetone test solution in a test tube, and add 2 drops of the chloro or bromo compound. If the compound is a solid, dissolve about 50 mg of it in a minimum volume of acetone and add this solution to the reagent. Shake the test tube to effect mixing, and allow the solution to stand for 3 min at room temperature. Note whether a precipitate forms; if no change occurs after 3 min, warm the mixture in a beaker of water at 50 °C. After 6 min of heating, cool the solution to room temperature and note whether any precipitate forms. Occasionally a precipitate forms immediately after combination of the reagents; this represents a positive test only if the precipitate remains after the mixture is shaken and allowed to stand for 3 min.

Perform this reaction with a series of primary, secondary, and tertiary chlorides and bromides. Note in all cases the differences in reactivity as evidenced by the rate of formation of sodium bromide or sodium chloride.

WRAPPING IT UP

Place all *solutions* and *solids* involved in this test in the container for halogenated liquids, unless instructed to do otherwise.

EXERCISES

Questions for Part A

1. Define the following terms:

   a. lipophilic   b. hydrophobic   c. hydrophilic

2. What structural feature of an alkene makes it nucleophilic?

3. It was stated that aqueous H–Br exists predominantly as H_3O^+ Br^-. Confirm the validity of this statement by calculating the equilibrium constant for the process shown below. The values in parentheses are the pK_as of the indicated acids. (*Hint:* See Sec. 5.3.)

$$H\!-\!Br + H_2O \rightleftharpoons H_3O^+ + Br^-$$
$$\quad(-8) \qquad\qquad\qquad (-1.7)$$

4. The yield of 2-bromohexane is much higher when the reaction mixture is stirred than when it is not. Why is this?

5. Account mechanistically for the fact that polar addition of H–Br to 1-hexene preferentially forms 2- rather than 1-bromohexane.

6. Provide an explanation for the order of carbocation stability: 3°>2°>1°.

7. Specify those structural characteristics of a phase-transfer catalyst that allow it to be soluble in both organic and aqueous solvents.

8. Show how a phase-transfer catalyst can interact with H–Br via electrostatic, hydrogen bonding, dipole-dipole, and van der Waals interactions.

9. When methyltrioctylammonium chloride is used as the phase-transfer catalyst, formation of 2-*chloro*hexane as a by-product is possible. Why would contamination of the desired 2-bromohexane by the chloro compound be a minor concern in this experimental procedure?

10. Why is methyltrioctylammonium chloride only partially soluble in 1-hexene? In aqueous hydrobromic acid?

11. Outline a procedure that would allow monitoring of the course of this addition reaction as a function of time. In other words, how might the reaction mixture be analyzed periodically so that you could determine when the reaction is complete?

12. In the work-up of this procedure, why is it necessary to wash the organic phase with a solution of sodium bicarbonate? What gas is evolved?

13. Devise an experiment that would demonstrate that a phase-transfer catalyst accelerates the rate of reaction between 1-hexene and aqueous hydrobromic acid. Why would it be difficult to perform the ionic addition of hydrogen bromide to 1-pentene using the procedure outlined in this experiment?

14. Refer to Figure 10.27a and determine whether the stationary phase used for the GLC analysis shown separates the bromohexanes on the basis of their relative boiling points or of some other property. Explain your answer.

15. Write the structure of the major product expected from the ionic addition of H–Br to each of the alkenes shown.

a.
b.
c.

16. Using curved arrows to symbolize the flow of electrons, provide a mechanism for the reaction of hydrogen bromide with each of the alkenes in Exercise 15. Provide a rationale for any reactions expected to proceed with skeletal rearrangement.

Refer to the online resources and use the spectra viewer and Tables 8.1–8.8 as needed to answer the blue-numbered questions on spectroscopy.

17. Consider the spectral data for 1-hexene (Figs. 10.28 and 10.29).

   a. In the IR spectrum, specify the absorptions associated with the carbon-carbon double bond and the vinylic hydrogen atoms.

   b. In the ^1H NMR spectrum, assign the various resonances to the hydrogen nuclei responsible for them.

   c. For the ^{13}C NMR data, assign the various resonances to the carbon nuclei responsible for them.

18. Consider the NMR data for 1-bromohexane (Fig. 10.31).

    a. In the ^{1}H NMR spectrum, assign the various resonances to the hydrogen nuclei responsible for them.

    b. For the ^{13}C NMR data, assign the various resonances to the carbon nuclei responsible for them.

19. Consider the NMR data for 2-bromohexane (Fig. 10.33).

    a. In the ^{1}H NMR spectrum, assign the various resonances to the hydrogen nuclei responsible for them.

    b. For the ^{13}C NMR data, assign the various resonances to the carbon nuclei responsible for them.

20. Discuss the differences observed in the IR and NMR spectra of 1-hexene and 2-bromohexane that are consistent with addition of HBr occurring in this experiment.

21. Assume you obtained the ^{1}H NMR spectrum of a mixture of 1- and 2-bromohexane and found that the integrations for the multiplets centered at δ 3.4 and 4.1 were 20 units and 40 units, respectively. What is the relative ratio of these two compounds in the sample?

22. The ^{1}H NMR spectrum of the crude bromohexane isolated by distillation in one experiment reveals a multiplet centered at δ 4.00, which is slightly upfield of the multiplet centered at δ 4.13 for 2-bromohexane.

    a. Determine whether this multiplet may signal the presence of 3-bromohexane by looking up the ^{1}H NMR spectrum of this isomer in either the Aldrich or Sadtler catalogs of reference spectra.

    b. The integrated area of the multiplet centered at δ 4.00 is 45 units, whereas that of the multiplet centered at δ 4.13 is 200 units. Assuming that the upfield multiplet represents 3-bromohexane, what is its percentage in the mixture of bromohexanes?

Question for Part B

23. What is the mechanistic basis (a) of the silver nitrate and (b) of the sodium iodide in acetone tests for differentiating 1° and 2° alkyl bromides from each other?

SPECTRA

Starting Materials and Products

CH$_3$(CH$_2$)$_3$CH=CH$_2$

Figure 10.28

IR spectrum of 1-hexene (neat).

Figure 10.29
NMR data for 1-hexene (CDCl₃).

(a) ¹H NMR spectrum (300 MHz).
(b) ¹³C data: δ 14.0, 22.5, 31.6, 33.8, 114.3, 139.2.

Figure 10.30
IR spectrum of 1-bromohexane (neat).

Figure 10.31
NMR data for 1-bromohexane (CDCl₃).

(a) ¹H NMR spectrum (300 MHz).
(b) ¹³C data: δ 14.0, 22.7, 28.1, 31.2, 33.1, 33.4.

Figure 10.32
IR spectrum of 2-bromohexane (neat).

Figure 10.33
NMR data for 2-bromohexane (CDCl₃).

(a) ¹H NMR spectrum (300 MHz).
(b) ¹³C data: δ 13.9, 22.1, 26.5, 29.9, 41.0, 51.0.

10.6 BROMINATION OF ALKENES

The reaction of bromine with (*E*)-stilbene (**47**) to give *meso*-stilbene dibromide (**48**) as the major product (Eq. 10.24) is another example of an electrophilic addition reaction of alkenes. The addition of bromine to many alkenes is a **stereospecific** reaction that proceeds by *anti* addition to the double bond. However, the addition of bromine to **47** is *not* stereospecific because small amounts of *dl*-stilbene dibromide (**49**) are also formed in this reaction. The formation of *meso*-stilbene dibromide presumably occurs via the nucleophilic attack of bromide on the intermediate cyclic bromonium ion, **50**. The possible interconversion of **50** and the acyclic carbocation **51** (Eq. 10.25) is one possible way to account for the presence of *dl*-stilbene dibromide in the product.

The bromination of (*E*)-stilbene using a solution of bromine in dichloromethane involves a chlorinated solvent. This protocol violates a principle of "green chemistry" (see the Historical Highlight *Green Chemistry*, which is available online) because halogenated compounds are environmental hazards as a general rule. Thus, alternative methods have been developed for producing bromine in nonhalogenated solvents. One of these involves the *in situ* formation of bromine by oxidation of hydrobromic acid with hydrogen peroxide (Eq. 10.26). Because stilbene is basically insoluble in water, a product of the reaction and the solvent for the hydrogen peroxide, an organic solvent that is environmentally more benign than a halogenated liquid and dissolves both the alkene and water is needed. Ethanol or other alcohols serve this purpose in the green experiment described in this section.

The dibromides **48** and **49** are *diastereomers* and thus have different physical and chemical properties (Sec. 7.1). Indeed, their melting points differ by 100 °C. This makes it very easy to identify the major product of the bromination.

$$\text{(10.24)}$$

47	48	49
(*E*)-Stilbene	*meso*-Stilbene dibromide	*dl*-Stilbene dibromide
(*colorless*)	(*colorless*)	(*colorless*)

$$\text{(10.25)}$$

50   **51**

$$HBr + H_2O_2 \longrightarrow Br_2 + H_2O \qquad (10.26)$$

Another example of a stereochemical outcome from brominating an alkene is illustrated by the transformation of (*E*)-cinnamic acid (**52**) to 2,3-dibromo-3-phenyl-propanoic acid (Scheme 10.3). Rather than using molecular bromine for this reaction, as is the case for bromination of (*E*)-stilbene (Eq. 10.21), pyridinium tribromide (**53**), a solid that is easier to handle in the laboratory, is used as the source of active bromine.

Unlike the bromination of (*E*)-stilbene, whereby the *meso* isomer **48** and the enantiomeric pair **49** are the possible products (Eq. 10.21), two different pairs of enantiomers, **54** and **55**, of 2,3-dibromo-3-phenylpropionic acid are possible for the corresponding reaction with (*E*)-cinnamic acid. What distinguishes the pair of enantiomers **54a** and **54b** from the **55a** and **55b** pair is their relative stereochemistry. For example, the stereogenic carbon atoms of **54a** and **54b** have the *2R,3R* and *2S,3S* configurations, respectively, whereas those in **55a** and **55b** have the *2S,3R* and *2R,3S* configurations, respectively (Scheme 10.3). Thus, the pairs of enantiomers **54** and **55** are diastereomers of one another.

Although several systems are commonly used to name the relative stereochemistry of adjacent stereogenic centers, one of the least ambiguous systems is *"like"* (*l*) and *"unlike"* (*u*). Accordingly, the pair of enantiomers **54** in which both stereocenters have the same configuration, either both *R* or both *S*, are *l*-2,3-dibromo-3-phenylpropanoic acid and enantiomers **55** are *u*-2,3-dibromo-3-phenylpropanoic acid. The fact that the *like* and *unlike* diastereomers have very different melting points makes it relatively easy to determine the major product that is formed from the addition of bromine to **52**. You will be able to accomplish this determination following the experimental procedure provided in this section.

Scheme 10.3

52
(E)-Cinnamic acid

53
Pyridinium tribromide

54a
(2R, 3R)
like mp = 93.5−95°C

54b

55a
(2S, 3R)

54b
(2S, 3R)
unlike mp = 202−204°C

2,3-Dibromo-3-phenylpropanoic acid

EXPERIMENTAL PROCEDURES

A. Bromination of (E)-Stilbene

Discovery Experiment   **Purpose** To determine the stereochemistry of the electrophilic addition of bromine to an alkene.

SAFETY ALERT

1. **Wear safety glasses or goggles and suitable protective gloves while performing the experiments.**

2. *Bromine is a hazardous chemical that may cause serious chemical burns.* **Do not breathe its vapors or allow it to come into contact with the skin. Perform all operations involving the transfer of the pure liquid or its solutions at a hood. If you get bromine on your skin, wash the area immediately with soap and warm water and soak the affected area in 0.6 *M* sodium thiosulfate solution, for up to 3 h if the burn is particularly serious.**

3. **Dispense the 1 *M* bromine in dichloromethane solution from burets or similar devices fitted with Teflon stopcocks and located in hoods.**

4. Bromine reacts with acetone to produce the powerful **lachrymator** α-bromoacetone, $BrCH_2COCH_3$. Do *not* rinse glassware containing residual bromine with acetone!

5. All parts of this experiment should be conducted in a hood if possible.

MINISCALE PROCEDURE

Preparation Refer to the online resources to answer Pre-Lab Exercises, access videos, and read the MSDSs for the chemicals used or produced in this procedure. Review Sections 2.7, 2.11, and 2.17.

Apparatus A 25-mL round-bottom flask, apparatus for magnetic stirring and vacuum filtration.

Setting Up Equip the flask with a stirbar and add to it 0.9 g of (*E*)-stilbene and 10 mL of dichloromethane. Stir or swirl the mixture to effect dissolution.

Bromination and Isolation Measure 5 mL of freshly prepared 1 *M* bromine in dichloromethane directly from the dispenser into the round-bottom flask. Swirl the flask gently during the addition to mix the contents. After the addition of the bromine solution is complete, stopper the reaction flask loosely and stir the mixture for 15 min. Isolate the product by vacuum filtration. Wash the product with one or two 1-mL portions of *cold* dichloromethane until it is white. Transfer the product to a watchglass or a piece of filter or weighing paper and allow it to air-dry. Recrystallize the product from xylenes.

Analysis Weigh the product and determine the yield. Measure the melting point to determine the stereochemistry of the bromination. *Caution:* Do *not* use mineral oil as the heating fluid for this determination! Obtain IR and ^1H NMR spectra of your starting material and product and compare them with those of authentic samples (Figs. 10.34–10.37).

MICROSCALE PROCEDURE

Preparation Refer to the online resources to answer Pre-Lab Exercises, access videos, and read the MSDSs for the chemicals used or produced in this procedure. Review Sections 2.7, 2.11, and 2.17.

Apparatus A 5-mL conical vial, apparatus for magnetic stirring, vacuum filtration, and Craig tube filtration.

Setting Up Equip the vial with a spinvane and add to it 180 mg of (*E*)-stilbene and 2 mL of dichloromethane. Stir or swirl the mixture to effect dissolution.

Bromination and Isolation Measure 1 mL of freshly prepared 1 *M* bromine in dichloromethane directly from the dispenser into the vial. Swirl the vial gently during the addition to mix the contents. After the addition of the bromine solution to the vial is complete, loosely screw a cap onto the vial and stir the mixture for 15 min. Isolate the product by vacuum filtration using a Hirsch funnel. Wash the product with several drops of *cold* dichloromethane until it is white. Transfer the product to a watchglass or a piece of filter or weighing paper and allow it to air-dry. Using a Craig tube, recrystallize the product from xylenes.

Analysis Weigh the product and determine the yield. Measure the melting point to determine the stereochemistry of the bromination. *Caution:* Do *not* use mineral oil as the heating fluid for this determination! Obtain IR and ^{1}H NMR spectra of your starting material and product and compare them with those of authentic samples (Figs. 10.34–10.37).

Discovery Experiment

Green Experiment

Bromination of (E)-Stilbene: The Green Approach

Purpose To demonstrate the stereochemistry of the electrophilic addition of bromine, generated *in situ*, to an alkene using an environmentally more benign process.

SAFETY ALERT

1. **Wear safety glasses or goggles and suitable protective gloves while performing the experiments.**

2. *Concentrated* **hydrobromic acid (47–49%) is a** *highly corrosive* **and** *toxic* **material. Working at a hood, measure out the amount of acid required and avoid inhaling its vapors. If the acid comes in contact with your skin, flood the affected area immediately and thoroughly with water and rinse it with 5% sodium bicarbonate solution.**

3. **The 30% hydrogen peroxide used is a** *strong oxidant* **and may blister the skin on contact. If you accidentally spill some on your skin, wash the affected area with copious amounts of water.**

MINISCALE PROCEDURE

Preparation Refer to the online resources to answer Pre-Lab Exercises, access videos, and read the MSDSs for the chemicals used or produced in this procedure. Review Sections 2.5, 2.9, 2.11, 2.17, and 2.21.

Apparatus A 50-mL round-bottom flask, 1- and 2-mL graduated pipets, ice-water bath, apparatus for heating under reflux, magnetic stirring, vacuum filtration, and *flameless* heating.

Setting Up Place 0.5 g of (*E*)-stilbene and 12 mL of 95% ethanol in the round-bottom flask containing a stirbar and set up the apparatus for heating under reflux. Heat the mixture under reflux until most of the solid has dissolved.

Bromination and Isolation Using a graduated pipet, add 1.2 mL of *concentrated* hydrobromic acid dropwise to the boiling mixture; this may cause some precipitation. Continue heating and add 0.8 mL of 30% hydrogen peroxide dropwise, using a graduated pipet, and note the change in color of the mixture. Continue stirring and heating under reflux until the color is discharged. Allow the reaction mixture to cool to room temperature and adjust its pH to 5–7 by adding saturated aqueous sodium bicarbonate. Cool the resulting mixture in an ice-water bath to complete precipitation of the product, and collect it by vacuum filtration. Wash the filter cake with water, transfer the solid to a watch glass, a piece of filter paper, or weighing paper, and allow it to air-dry. Recrystallize the crude product from xylenes.

Analysis Weigh the product and determine the yield. Measure the melting point to determine the stereochemistry of the bromination. *Caution:* Do *not* use mineral oil as the heating fluid for this determination! Obtain IR and ^{1}H NMR spectra of your starting material and product and compare them with those of authentic samples (Figs. 10.34–10.37).

MICROSCALE PROCEDURE

Preparation Refer to the online resources to answer Pre-Lab Exercises, access videos, and read the MSDSs for the chemicals used or produced in this procedure. Review Sections 2.5, 2.9, 2.11, 2.17, and 2.21.

Apparatus A 5-mL conical vial, 1-mL graduated pipet, ice-water bath, apparatus for heating under reflux, magnetic stirring, vacuum filtration, Craig tube filtration, and *flameless* heating.

Setting Up Place 0.13 g of (*E*)-stilbene and 3 mL of 95% ethanol in the conical vial containing a stirbar and set up the apparatus for heating under reflux. Heat the mixture under reflux until most of the solid has dissolved.

Bromination and Isolation Using the pipet, add 0.3 mL of *concentrated* hydrobromic acid dropwise to the refluxing mixture; this may cause some precipitation to occur. Continue heating and add 0.2 mL of 30% hydrogen peroxide dropwise, noting the change in color of the mixture. Continue stirring and heating under reflux until the color is discharged. Allow the reaction mixture to cool to room temperature and adjust its pH to 5–7 by addition of saturated aqueous sodium bicarbonate. Cool the resulting mixture in an ice-water bath to complete precipitation of product, and collect it by vacuum filtration using a Hirsch funnel. Wash the filter cake with drops of water, transfer it to a watch glass or a piece of filter or weighing paper, and allow it to air-dry. Using a Craig tube, recrystallize the crude product from xylenes.

Analysis Weigh the product and determine the yield. Measure the melting point to determine the stereochemistry of the bromination. *Caution:* Do *not* use mineral oil as the heating fluid for this determination! Obtain IR and ^{1}H NMR spectra of your starting material and product and compare them with those of authentic samples (Figs. 10.34–10.37).

 Discovery Experiment *Bromination of (Z)-Stilbene*

Apply the procedure given for bromination of (*E*)-stilbene to study this reaction with (*Z*)-stilbene. Develop a protocol whereby you could demonstrate the stereochemical outcome of the dibromide(s) produced. Consult with your instructor before undertaking any experimental procedures.

Discovery Experiment   *Solvent Effects on the Stereochemistry of Bromination*

According to the literature (Buckles, R. E.; Bader, J. M.; Thurmaier, R. J., *J. Org. Chem.* **1962**, *27*, 4523–4527), the stereochemical outcome of the bromination of (Z)-stilbene is dependent upon the polarity of the solvent used in the reaction. Develop a protocol for confirming this report, using NMR spectroscopy (see Hartshorn, M. P.; Opie, M. C. A.; Vaughan, J., *Aust. J. Chem.* **1973**, *26*, 917–920) as the analytical tool for analyzing the stereochemistry of the reaction as a function of solvents of differing polarities. Consult with your instructor before undertaking any experimental procedures for evaluating the reported solvent effect.

Discovery Experiment   *Substituent Effects on the Stereochemistry of Bromination*

Apply the Microscale Procedure given for bromination of (*E*)-stilbene to study this reaction with one or more substituted (*E*)-stilbenes bearing electron-withdrawing or electron-donating substituents. Develop a protocol whereby you could demonstrate the stereochemical outcome of the dibromide(s) produced. Guidance for determining the stereochemical result is available in the following references: Buckles, R. E.; Bader, J. M.; Thurmaier, R. J., *J. Org. Chem.* **1962**, *27*, 4523–4527; Hartshorn, M. P.; Opie, M. C. A.; Vaughan, J., *Aust. J. Chem.* **1973**, *26*, 917–920. Compare your result(s) with those of others who have explored the reaction with substituted stilbenes different from yours. Consult with your instructor before undertaking any experimental procedures.

WRAPPING IT UP

Decolorize any *solutions* in which the color of bromine is visible by the dropwise addition of cyclohexene; then discard the resulting solutions together with *all other non-aqueous solutions* in a container for halogenated organic liquids. Pour all aqueous solutions down the drain.

B. Bromination of (E)-Cinnamic Acid

Discovery Experiment   **Purpose** To demonstrate the stereochemistry of bromination of an alkene.

SAFETY ALERT

1. **Wear safety glasses or goggles and suitable protective gloves while performing the experiment.**

2. **Pyridinium tribromide releases molecular bromine, so it should be considered a hazardous chemical. Do not let it come into contact with your skin because it may cause severe burns. If it does get on your skin, wash the area immediately with soap and warm water, then soak the affected area in a 0.6 *M aqueous* solution of sodium thiosulfate.**

3. **Bromine reacts with acetone to produce a powerful lachrymator, α-bromoacetone, $BrCH_2COCH_3$. Therefore, do *not* rinse residual pyridium tribromide out of glassware with acetone!**

4. *Glacial* acetic acid is caustic and can cause severe chemical burns. If it comes in contact with your skin, rinse the area immediately with cool water, followed by 5% *aqueous* solution of sodium bicarbonate.

MINISCALE PROCEDURE

Preparation Refer to the online resources to read the MSDS for the chemicals used in this experiment. Answer the Pre-Lab Exercises. Review Sections 2.7, 2.10, 2.11, 2.17, 2.22, and 3.2.

Apparatus A 50-mL round-bottom flask, ice-water bath, apparatus for heating under reflux, magnetic stirring, and vacuum filtration.

Setting Up Equip the flask with a stirbar and add 0.8 g of (*E*)-cinnamic acid and 10 mL of *glacial* acetic acid to it. Stir the mixture to effect dissolution.

Bromination and Isolation Add 1.76 g of pyridinium tribromide to the flask and stir the mixture; the pyridinium tribromide may not completely dissolve. Equip the flask with a reflux condenser and heat the reaction mixture under reflux for about 45 min. Allow the reaction mixture to cool to room temperature and slowly add 20 mL of water to the mixture with continued stirring. Cool the reaction mixture in an ice-water bath and then isolate the product by vacuum filtration. Wash the filter cake with 5 mL of cold 10% *aqueous* sodium bisulfite solution to remove color, followed by 5 mL of cold water. Recrystallize the crude product from a 1:1 mixture of 95% ethanol and water.

Analysis Weigh the product and determine the yield. Measure the melting point to determine the stereochemistry of bromination. *Caution*: Do *not* use mineral oil as the heating fluid for this determination! Obtain IR and ^{1}H NMR spectra of your starting material and product and compare them with those of authentic samples (Figs. 10.38–10.41)[1].

MICROSCALE PROCEDURE

Preparation Refer to the online resources to read the MSDS for the chemicals used in this experiment. Answer the Pre-Lab Exercises. Review Sections 2.7, 2.10, 2.11, 2.17, 2.22, and 3.2.

Apparatus A 5-mL conical vial, apparatus for heating under reflux, magnetic stirring, vacuum filtration, and Craig tube filtration.

Setting Up Equip the vial with a spinvane and add 150 mg of (*E*)-cinnamic acid and 1.5 mL of *glacial* acetic acid to it. Stir the mixture to effect dissolution.

Bromination and Isolation Add 330 mg of pyridinium tribromide to the vial and stir the contents of the flask; the pyridinium tribromide may not completely dissolve. Set up the apparatus for reflux and heat the reaction mixture under reflux for about

45 min. Allow the reaction mixture to cool to room temperature and slowly add 3 mL of water to the mixture with continued stirring. Cool the reaction mixture in an ice-water bath and then isolate the product by vacuum filtration using a Hirsch funnel. Wash the filter cake with a few drops of cold 10% aqueous sodium bisulfite solution to remove color, followed by a few drops of cold water. Using a Craig tube, recrystallize the crude product from a 1:1 mixture of 95% ethanol and water.

Analysis Weigh the product and determine the yield. Measure the melting point to determine the stereochemistry of bromination. *Caution*: Do *not* use mineral oil as the heating fluid for this determination! Obtain IR and ^{1}H NMR spectra of your starting material and product and compare them with those of authentic samples (Figs. 10.38–10.41).

WRAPPING IT UP

If necessary, add 10% aqueous sodium bisulfite to decolorize the *filtrate* and discard the mixture in a container for *aqueous acids*.

EXERCISES

General Questions

1. What does the "E" in (*E*)-stilbene or (*E*)-cinnamic acid stand for and what does it mean?

2. Write the expected products for the addition of bromine to the following alkenes.

   a. $CH_3CH_2CH{=}CH_2$ b. [structure: CH₃ and H on one carbon, H and CH₃ on the other]   c. [cyclohexene structure with D, D]

3. Write the structure of the product obtained when cyclohexene is used to decolorize bromine-containing solutions.

4. Draw structural formulae for the alkenes that give the following products upon bromination.

   a. [cyclopentane structure with CH₃, Br, Br]   b. [structure with two phenyl groups, Br, Br]   c. [structure: CH₃, Br, CH₃, Br, CH₃]

Questions for Part A

5. Is *meso*-stilbene dibromide optically active? What symmetry element is present in this molecule that precludes the existence of an enantiomer?

6. Explain why, in contrast to *meso*-stilbene dibromide, the dibromide **49** exists in two enantiomeric forms.

7. For the procedure in which a solution of bromine is used, why does solid not separate immediately after you begin adding the solution to that containing (*E*)-stilbene?

8. In the procedure involving *in situ* generation of bromine, what is the solid that appears when the *concentrated* hydrogen bromide is added?

9. What is the source of color that is produced upon addition of aqueous hydrogen peroxide to the reaction mixture?

10. Why should mineral oil not be used as the heating fluid for determining the melting point of *meso*-stilbene dibromide (**48**) or of the 2,3-dibromopropanoic acid (**53**) or (**54**)?

11. Using suitable stereochemical structures, write the mechanism for the addition of bromine to (*E*)-stilbene to give *meso*-stilbene dibromide via the intermediate cyclic bromonium ion **50**. Use curved arrows to symbolize the flow of electrons.

12. To what general class of mechanisms does the attack of bromide ion, Br⁻, on the cyclic bromonium ion **50** belong? Does this ring-opening step follow first- or second-order kinetics?

13. How many centers of chirality are present in the bromonium ion **50** and in the isomeric carbocation **51**?

14. Draw suitable three-dimensional structures of the two products obtained by attack of bromide ion at both of the carbon atoms in the cyclic bromonium ion **50** and show that they are identical.

15. Based upon the mechanism you provided in Exercise 11, predict the major product of the addition of bromine to (*Z*)-stilbene.

16. Demonstrate that attack of bromide ion, Br⁻, on the carbocation **51** can provide *both dl-* and *meso*-stilbene dibromide.

17. The reaction of bromine with cyclopentene according to Equation 10.18 is stereospecific and proceeds by *anti* addition. On the other hand, the addition of bromine to (*E*)-stilbene gives both *meso*-stilbene dibromide (**48**) and *dl*-stilbene dibromide (**49**). Rationalize this difference by comparing the relative stabilities of the carbocations resulting from ring-opening of the cyclic bromonium ions **38** and **50**.

18. How do the results of this experiment support the hypothesis that the addition of bromine to (*E*)-stilbene proceeds primarily through **50** rather than **51**?

Refer to the online resources and use the spectra viewer and Tables 8.1–8.8 as needed to answer the blue-numbered questions on spectroscopy.

19. Consider the spectral data for (*E*)-stilbene (Figs. 10.34 and 10.35).

    a. In the IR spectrum, identify the absorption(s) consistent with the *trans* relationship of the vinylic hydrogen atoms and explain the absence of a significant absorption in the range of 1600–1700 cm⁻¹ normally characteristic of alkenes.

    b. In the ¹H NMR spectrum, assign the various resonances to the hydrogen nuclei responsible for them.

    c. For the ¹³C NMR data, assign the various resonances to the carbon nuclei responsible for them.

20. Consider the ¹H NMR spectrum of *meso*-stilbene dibromide (Fig. 10.37a).

    a. In the ¹H NMR spectrum, assign the various resonances to the hydrogen nuclei responsible for them.

    b. For the ¹³C NMR data, assign the various resonances to the carbon nuclei responsible for them.

21. Discuss the differences observed in the IR (Fig. 10.36) and NMR spectra of (*E*)-stilbene and *meso*-stilbene dibromide that are consistent with addition of Br₂ occurring in this experiment.

Questions for Part B

22. Using curved arrows to symbolize the flow of electrons, propose a mechanism for the reaction of (*E*)-cinnamic acid and bromine that accounts for the stereochemistry of the product you obtained.

23. In light of the mechanism you wrote in question 1,

   a.   Draw the structure of the stereoisomers expected from the reaction of bromine with 1-methylcyclohexene.

   b.   Is the stereochemistry of the product(s) you drew *like* or *unlike*?

24. If the solid you isolated is red-orange, what is the most likely mistake made in the procedure?

25. For each of the following alkenes, determine if the major product will have *like* or *unlike* stereochemistry and if it is chiral or *meso*.

   a.   (cyclohexene)   b. CH_3 —CH=CH—C(=O)OH   c.   (1-methylcyclopentene with CH_3)

   d. CH_3 —CH=CH— CH_3 e.   (cinnamic acid structure with —OH)   f. CH_3 —CH=CH— CH_3

26. Why is a reflux condenser used in this reaction?

27. Consider the reaction of bicycloalkene **55** with bromine in CCl_4. Two diastereomers, **56** and **57**, are possible from *anti* addition, but only **56** is formed. Explain why this reaction is diastereoselective.

   55 $\xrightarrow[\text{CCl}_4]{\text{Br}_2}$ **56**   or   **57**

28. Consider the spectral data for (*E*)-cinnamic acid (Figs. 10.38 and 10.39).

   a.   In the IR spectrum, identify the absorption(s) in the functional group region for the carboxylic acid and alkene functions and any absorption(s) in the fingerprint region that symbolize the presence of a *trans* carbon-carbon double bond.

   b.   In the ^{1}H NMR spectrum, assign the various resonances to the hydrogen nuclei responsible for them.

   c.   How would you use the resonances from the vinylic hydrogens in the ^{1}H NMR spectrum to establish if the alkene was *E* or *Z*?

   d.   For the ^{13}C NMR data, assign the resonances to the carbon nuclei responsible for them.

29. Consider the spectral data for 2,3-dibromo-3-phenylpropanoic acid (Figs. 10.40 and 10.41).

    a. In the IR spectrum, identify the absorption(s) in the functional group region for the carboxylic acid function.

    b. In the ^{1}H NMR spectrum, assign the various resonances to the hydrogen nuclei responsible for them.

    c. For the ^{13}C NMR data, assign the resonances to the carbon nuclei responsible for them.

30. Discuss the differences observed in the IR and NMR spectra of (*E*)-cinnamic acid and 2,3-dibromo-3-phenylpropanoic acid that are consistent with addition of bromine to the double bond occurring in this experiment.

SPECTRA

Starting Materials and Products

Figure 10.34
IR spectrum of (E)-stilbene (IR card).

Figure 10.35
NMR data for (E)-stilbene (CDCl₃).

(a) ^{1}H NMR spectrum (300 MHz).
(b) ^{13}C data: δ 126.3, 127.8, 128.9, 129.0, 137.6.

Figure 10.36
IR spectrum of meso-*stilbene dibromide (IR card).*

Figure 10.37
NMR data for meso-*stilbene dibromide (CDCl₃).*

(a) ¹H NMR spectrum (300 MHz).
(b) ¹³C data: δ 57.0, 129.1, 129.6, 129.7, 141.7.

Figure 10.38
IR spectrum of (E)-*cinnamic acid.*

Figure 10.39
NMR data for (E)-cinnamic **(a)** ^{1}H NMR (400 MHz).

Figure 10.40
IR of 2,3-dibromo-3-phenyl-propanoic acid

Figure 10.41
NMR data for 2,3-dibromo-3-phenylpropanoic acid

10.7 ACID-CATALYZED HYDRATION OF ALKENES

The acid-catalyzed addition of the elements of water across a carbon-carbon π-bond to give an alcohol is referred to as **hydration** of an alkene (Eq. 10.17). Mechanistically, this process is simply the reverse of the acid-catalyzed **dehydration** of alcohols (Sec. 10.3). The position of the equilibrium for these two competing processes depends upon the reaction conditions. Hydration of a double bond requires the presence of excess water to drive the reaction to completion, whereas the dehydration of an alcohol requires removing water to complete the reaction. In the experiment that follows, you will examine the acid-catalyzed hydration of norbornene (**58**) to give *exo*-norborneol (**59**) (Eq. 10.27) as the exclusive product.

$$\tag{10.27}$$

58
Norbornene

59
exo-Norborneol

The mechanism of this reaction and the structure of the intermediate carbocation were once the subject of intense controversy. When the double bond of norbornene is protonated, a carbocation is produced (Eq. 10.28). The question of whether this cation is **60a** and is in rapid equilibrium with the isomeric cation **60b** or whether **60a** and **60b** are simply two contributing resonance structures to the resonance hybrid **61** has been the focus of the debate. Note that the contributing structures **60a** and **60b** *differ in the location of an σ-bond*, not in the location of π-bonds as is more commonly encountered in **resonance structures**. The delocalized structure **61** is thus referred to as a **nonclassical carbocation**, and most chemists now accept this formulation as the more likely representation of this intermediate. It is evident from empirical results that the sterically more accessible side of the carbocation is the one away from or *exo* to the bridging carbon atom as shown. Nucleophilic attack of water solely from this side leads to *exo*-norborneol (**59**) rather than *endo*-norborneol (**62**).

58   **60a**   **60b**

$$\tag{10.28}$$

62
endo-Norborneol

61

59
exo-Norborneol

Hydration of Norbornene

Purpose To study the acid-catalyzed addition of water to an alkene.

SAFETY ALERT

1. Wear safety glasses or goggles and suitable protective gloves while performing the experiment.
2. Norbornene is a volatile solid. Perform all weighings in a hood if possible.
3. If the acidic solution used in the first part of the experiment comes in contact with your skin, immediately flood the affected area with water and thoroughly rinse it with 5% sodium bicarbonate solution.
4. If the basic solution used in this experiment comes in contact with your skin, immediately flood the affected area with water and thoroughly rinse the area with a dilute solution (1%) of acetic acid.

MINISCALE PROCEDURE

Preparation Refer to the online resources to answer Pre-Lab Exercises, access videos, and read the MSDSs for the chemicals used or produced in this procedure. Review Sections 2.7, 2.20, 2.21, and 2.29.

Apparatus A 25-mL Erlenmeyer flask, separatory funnel, ice-water bath, apparatus for simple distillation, magnetic stirring, sublimation, and *flameless* heating.

Setting Up Slowly add 2 mL of *concentrated* sulfuric acid to 1 mL of water in a 25-mL Erlenmeyer flask containing a stirbar and cool the solution to room temperature.

Hydration Working at the hood, add 1.0 g of norbornene in small pieces to the stirred solution of aqueous sulfuric acid. Continue stirring the mixture until all of the norbornene dissolves. Cool the flask *briefly* in an ice-water bath if the mixture becomes noticeably warm, but *do not cool the contents of the flask below room temperature*. Stir this solution for 15–30 min. Prepare a solution of 1.5 g of potassium hydroxide in 7.5 mL of water.

Work-Up and Isolation Cool both solutions in an ice-water bath and slowly add the base to the acidic reaction mixture to partially neutralize the acid. Be sure that the resulting mixture is at room temperature or below, then transfer it to a small separatory funnel, and add 15 mL of diethyl ether. If any solid is present in the bottom of the separatory funnel, add 1–2 mL of water to dissolve it. After shaking the funnel, *periodically venting it to release pressure*, separate the aqueous layer (*Save!*) and pour the ethereal solution into an Erlenmeyer flask. Return the aqueous layer to the separatory funnel, and extract again with a fresh 10-mL portion of diethyl ether. Separate the aqueous layer, and combine the two ethereal extracts in the separatory funnel.

Wash the combined extracts with 5 mL of water. Separate the aqueous layer, and wash the ethereal solution sequentially with 5-mL portions of saturated sodium bicarbonate solution and saturated sodium chloride solution; *periodically vent the separatory funnel* to release pressure that may develop. Separate the layers, place the ethereal solution in a dry 50-mL Erlenmeyer flask, and dry it over several spatula-tips full of *anhydrous* sodium sulfate with occasional swirling for 10–15 min.★ Add additional portions of *anhydrous* sodium sulfate if the liquid remains cloudy.

Transfer the dried ethereal solution to a 50-mL round-bottom flask equipped for simple distillation and concentrate the solution by distillation. Alternatively, rotary evaporation or other techniques may be used to concentrate the solution. Transfer the residue to a 25- or 50-mL filter flask.

Sublimation Equip the filter flask with a filter adapter or rubber stopper containing a centrifuge tube or a 13-mm × 120-mm test tube that extends to within about 2 cm of the bottom of the flask (Fig. 2.58b). The adapter or stopper should be lightly greased to prevent water from entering the sublimation chamber. Evacuate the flask using a water aspirator equipped with a trap and *then* half-fill the centrifuge tube or test tube with tightly packed chipped ice. *Gently* heat the flask using a steam bath or the sweeping motion of a heat gun or small flame. Add ice as necessary to the cold finger on which the sublimed alcohol collects. When the alcohol has completely sublimed, cool the flask, break the vacuum to the aspirator, and disconnect the flask from the vacuum hose. Carefully decant the water out of the cold finger. Carefully remove the cold finger from the flask, while maintaining the cold finger in a horizontal or slightly inverted position. Scrape the soft crystalline product from the test tube.

Analysis Weigh the purified solid and determine the yield. Measure the melting point by placing the sample in a capillary tube that has been sealed about 2 cm from the open end. Obtain IR and ^{1}H NMR spectra of your starting material and product and compare them with those of authentic samples (Figs. 10.42–10.45).

WRAPPING IT UP

Slowly combine the *aqueous layers and washes*. If necessary, neutralize them with sodium carbonate, and flush them down the drain with excess water. Pour any *diethyl ether* obtained during the concentration step into a container for flammable organic solvents. After the diethyl ether has evaporated from the *sodium sulfate* in the hood, place the solid in a container for nonhazardous solids.

EXERCISES

1. What purpose is served by washing the ethereal solution of product with bicarbonate solution? Saturated brine?

2. How would the yield of *exo*-norborneol be affected if this alcohol were left under vacuum for an excessive period of time to remove all traces of solvent?

3. Why is it necessary to seal the capillary tube before taking the melting point of *exo*-norborneol?

4. Using curved arrows to symbolize the flow of electrons, show the mechanism by which the classical cations **60a** and **60b** can be interconverted.

5. What is the significance of the dashed bonds in the structure of the nonclassical cation **61**? What is "nonclassical" about this cation (i.e., how is **61** different from a classical carbocation)?

6. *Exo*-Norborneol (**59**) contains three centers of chirality, but the product obtained from the hydration of norbornene (**58**) in this experiment is optically inactive.

   a. Is norbornene chiral? Explain your answer by showing why it is or is not.

   b. Identify the three chiral centers in **59** with asterisks.

   c. Write the chemical structure for the stereoisomer of **59** that is produced by the hydration of **58** and show how it is formed from intermediate **61**. Are **59** and this stereoisomer enantiomers or diastereomers?

7. Why is sulfuric acid a superior acid to HCl for effecting the acid-catalyzed hydration of alkenes?

8. *Exo*-norborneol undergoes reaction with *concentrated* HBr to form a single diastereomer. Write the structure of this bromide and propose a stepwise mechanism for its formation, using curved arrows to symbolize the flow of electrons. Why is only one diastereomer produced?

9. For each of the alkenes **a–e**, write the structure of the alcohol that should be the major product of acid-catalyzed hydration.

10. Consider the spectral data for norbornene (Figs. 10.42 and 10.43).

   a. In the functional group region of the IR spectrum, identify any absorptions associated with the carbon-carbon double bond.

   b. In the ^{1}H NMR spectrum, assign the various resonances to the hydrogen nuclei responsible for them.

   c. For the ^{13}C NMR data, assign the various resonances to the carbon nuclei responsible for them. Explain why there are only four resonances in the spectrum.

11. Consider the spectral data for *exo*-norborneol (Figs. 10.44 and 10.45).

   a. In the functional group region of the IR spectrum, specify any absorptions associated with the alcohol function.

   b. In the ^{1}H NMR spectrum, assign the various resonances to the hydrogen nuclei responsible for them.

   c. For the ^{13}C NMR data, assign the various resonances to the carbon atom responsible for them.

12. Discuss the differences observed in the IR and NMR spectra of norbornene and *exo*-norborneol that are consistent with the addition of water occurring in this experiment.

SPECTRA

Starting Materials and Products

Figure 10.42
*IR spectrum of norbornene
(IR card).*

Figure 10.43
*NMR data for norbornene
(CDCl₃).*

(a) ¹H NMR spectrum (300 MHz).
(b) ¹³C data: 24.8, 42.0, 48.8, 135.4.

Figure 10.44
*IR spectrum of exo-norborneol
(IR card).*

Figure 10.45

NMR data for exo-norborneol (CDCl3).

(a) ^{1}H NMR spectrum (300 MHz).

(b) ^{13}C data: δ 24.7, 28.4, 34.5, 35.5, 42.2, 44.2, 74.6.

10.8 HYDROBORATION-OXIDATION OF ALKENES

The **regiochemistry** of acid-catalyzed hydration of alkenes to give alcohols is that predicted by Markovnikov's rule (Sec. 10.4) because the more stable intermediate carbocation is preferentially formed. Sometimes it is desirable to add the elements of water across a carbon-carbon π-bond in the opposite regiochemical sense to provide the **anti-Markovnikov** product (see the Historical Highlight *Additions Across Carbon-Carbon π-Bonds*, which is available online). To accomplish this goal, a process termed **hydroboration-oxidation** was developed that involves the reaction of an alkene sequentially with diborane, B_2H_6, and basic hydrogen peroxide (Eq. 10.29).

$$\underset{R^1}{\overset{R^2}{\diagup}}C{=}CHR^3 \xrightarrow[\text{2. }H_2O_2/HO^-]{\text{1. }B_2H_6} R^1{-}\underset{\underset{H}{|}}{\overset{\overset{R^2}{|}}{C}}{-}\underset{\underset{OH}{|}}{\overset{\overset{R^3}{|}}{CH}} \qquad (10.29)$$

Diborane is a dimer of borane, BH_3. The bonding in diborane is unusual because the hydrogen atoms bridge the two boron atoms with the two monomeric BH_3 subunits being bound by *two-electron, three-center bonds*. Because the boron atom in borane possesses an empty *p*-orbital, borane is a Lewis acid, and it forms stable complexes upon reaction with tetrahydrofuran (THF) and other ethers, which function as Lewis bases, as illustrated by the formation of a borane-THF complex (Eq. 10.30).

$$\qquad (10.30)$$

Diborane   Tetrahydrofuran   A borane-THF complex

As you may have learned from reading Section 10.4, alkenes may serve as Lewis bases. Hence, alkenes can react with borane by the net addition of the boron-hydrogen bond across the carbon-carbon π-bond to give a monoalkylborane as the initial product. This process, known as **hydroboration**, is illustrated for 1-butene (**63**) in Scheme 10.4. The intermediate monoalkylborane **64**, then adds successively to additional molecules of **63** to form a dialkylborane **65** or a trialkylborane **66**.

The extent to which dialkyl- and trialkylboranes will be formed is dependent upon steric factors and the degree of substitution on the double bond.

Scheme 10.4

$$CH_3CH_2\overset{H-BH_2}{CH=CH_2} \longrightarrow CH_3CH_2\overset{H\quad BH_2}{CH-CH_2} \xrightarrow{\text{1-butene}} [CH_3(CH_2)_2CH_2]_2BH \xrightarrow{\text{1-butene}} [CH_3(CH_2)_2CH_2]_3B$$

| **63** | **64** | **65** | **66** |
| 1-Butene | A monoalkylborane | A dialkylborane | A trialkylborane |

As observed in the electrophilic additions to alkenes depicted in Equations 10.16 and 10.17, the addition of borane to alkenes is *regioselective* (Sec. 10.4) and occurs so the *electrophilic boron atom* preferentially attaches to the *less substituted carbon atom* of the carbon-carbon π-bond. This selectivity may be understood by considering the mechanism of hydroboration. As borane approaches the double bond, a Lewis acid–Lewis base interaction occurs in which the π-electrons of the double bond begin to interact with the empty *p*-orbital on the Lewis acidic boron atom, resulting in the development of a partial positive charge on the more highly substituted carbon atom. The boron atom then assumes a partial negative charge that facilitates the transfer of the hydride ion, $H:^-$, from the boron to the carbon atom bearing the partial positive charge. This process occurs through the four-centered **transition state 67**, in which all four atoms simultaneously undergo changes in bonding.

$$\left[\begin{array}{c} \overset{\delta^-}{H\text{-}\text{-}BH_2} \\ \overset{\delta^+}{C_2H_5}\rule{0pt}{0pt} \end{array} \right]^{\ddagger}$$

67

Thus, the regiochemistry of hydroboration is predicted by the same general rule that applies to all electrophilic additions to alkenes: *The reaction of an electrophile with a carbon-carbon double bond occurs preferentially via the transition state in which a partial positive charge develops on that carbon atom better able to accommodate it.* Geometric constraints inherent in the cyclic transition state **67** require that the addition of borane to the alkene proceed so that both the boron and the hydrogen add from the same face of the double bond, a process called *syn*-**addition**.

Treating a trialkylborane with alkaline hydrogen peroxide promotes oxidative cleavage to form three moles of an alcohol and boric acid, as exemplified in the conversion of **66** into 1-butanol (**71**), shown in Scheme 10.5. The first step of the reaction involves nucleophilic attack of the hydroperoxide anion on the Lewis acidic boron atom of **66** to give **68**. In the next step, migration of the alkyl group from boron to oxygen proceeds with displacement of hydroxide ion and cleavage of the weak oxygen-oxygen bond to give the monoalkoxyborane **69**. An important consequence of the *intramolecularity* of this rearrangement is that it proceeds with *retention* of configuration at the carbon atom. The sequence is repeated until all of the boron-carbon bonds have been converted into oxygen-carbon bonds, whereupon the intermediate trialkoxyborane **70** undergoes alkaline hydrolysis to give 1-butanol and boric acid.

Scheme 10.5

$$[CH_3(CH_2)_2CH_2]_3B \ + \ ^-\ddot{O}-OH \longrightarrow [CH_3(CH_2)_2CH_2]_2\bar{B}\overset{CH_2(CH_2)_2CH_3}{\underset{O-OH}{|}} \xrightarrow{-HO^-}$$

66   **68**

$$[CH_3(CH_2)_2CH_2]_2B-O-CH_2(CH_2)_2CH_3 \xrightarrow{^-\ddot{O}-OH} CH_3(CH_2)_2CH_2B\overset{O(CH_2)_3CH_3}{\underset{O(CH_2)_3CH_3}{\big\langle}}$$

69

A monoalkoxyborane   A dialkoxyborane

$$\xrightarrow{^-\ddot{O}-OH} [CH_3(CH_2)_3O]_3B \xrightarrow[H_2O]{HO^-} 3 \ CH_3(CH_2)_2CH_2-OH \ + \ B(OH)_3$$

70   **71**

A trialkoxyborane   1-Butanol   Boric acid

By considering the reactions depicted in Schemes 10.4 and 10.5, you can see how the hydroboration-oxidation of an alkene gives an alcohol that is the product of overall anti-Markovnikov addition of the elements of H–OH to the carbon-carbon double bond. Remember, however, that the key step determining the regiochemistry of the reaction is the Markovnikov addition of the hydrogen-boron bond across the π-bond.

In this experiment, you will examine the regiochemistry and stereochemistry of the sequence of hydroboration-oxidation using $(+)$-α-pinene (**72**), a terpene, as the alkene and a solution of borane-tetrahydrofuran complex as the hydroborating reagent (Eq. 10.31). Using this complex avoids the need to handle diborane, B_2H_6, a potentially hazardous reagent. Since the double bond in **72** is trisubstituted and sterically hindered, only *two* molecules of this alkene react with each molecule of borane, thus forming the intermediate dialkylborane **73**. Subsequent oxidation of **73** with alkaline hydrogen peroxide then gives $(-)$-isopinocampheol (**74**).

$$\xrightarrow{H_3B:OC_4H_8} \qquad \xrightarrow{aq.\ NaOH,\ H_2O_2} \qquad\qquad (10.31)$$

72   **73**   **74**

$(+)$-α-Pinene $(-)$-Isopinocampheol

You will be able to verify that $(-)$-isopinocampheol (**74**) is the sole product of the hydroboration-oxidation of $(+)$-α-pinene (**72**) by determining the melting points of the product of your reaction and that of a suitable derivative, such as an α-naphthylurethane (Sec. 10.8B). Comparing these data with those in Table 10.1 for the alternative products **75–77**, which could also arise from adding borane in a Markovnikov sense, and the possible products **78** and **79**, which would be obtained from an anti-Markovnikov hydroboration, provides compelling evidence that the structure of the product is indeed **74**. Hence, by performing the hydroboration-oxidation of $(+)$-α-pinene (**72**), you will find that the addition of borane to alkenes occurs in a highly regioselective fashion with exclusive

Table 10.1 *Melting Points of Isomeric Campheols and Pinanols*

Structure	°C	°C, of α-Naphthylurethane Derivative
74	51–53 (55–56)	87.5–88
75	45–47 (48)	88
76	27	148
77	67	91
78	78–79	Not formed
79	58–59	Not formed

syn-addition. Moreover, you will discover that the hydroboration of (+)-α-pinene is an example of a **stereoselective** reaction because only one of several possible stereoisomeric products is formed by addition of the borane reagent to only one face of the π-bond of **65**.

75

(+)-Neoisopinocampheol

76

(−)-Neopinocampheol

77

(+)-Pinocampheol

78

(+)-*cis*-2-Pinanol

79

(+)-*trans*-2-Pinanol

EXPERIMENTAL PROCEDURES

A. Hydroboration-Oxidation of (+)-α-Pinene

Discovery Experiment   **Purpose** To study the hydration of an alkene by means of hydroboration-oxidation and to discover the regiochemistry and stereochemistry of the process.

SAFETY ALERT

1. **Wear safety glasses or goggles and suitable protective gloves while performing the experiments.**

2. **Tetrahydrofuran is a flammable and potentially toxic solvent. Do not use this solvent near open flames.**

3. **The 30% hydrogen peroxide used is a *strong oxidant* and may blister the skin on contact. If you accidentally spill some on your skin, wash the affected area with copious amounts of water.**

4. **The complex of borane and tetrahydrofuran reacts *vigorously* with water. Use care when disposing of any excess reagent and when quenching the reaction.**

MINISCALE PROCEDURE

Preparation Refer to the online resources to answer Pre-Lab Exercises, access videos, and read the MSDSs for the chemicals used or produced in this procedure. Review Sections 2.5, 2.10, 2.11, 2.13, 2.21, 2.27, 2.28, and 2.29.

Apparatus A 50-mL round-bottom flask, Claisen adapter, separatory funnel, two 5-mL glass syringes with needles, drying tube, water bath, ice-water bath, apparatus for simple distillation, magnetic stirring, and *flameless* heating.

Setting Up Equip the round-bottom flask with the Claisen adapter, and prepare a drying tube filled with calcium chloride for attachment to the *sidearm* of the adapter. Fit a rubber septum onto the joint of the adapter that is directly in line with the lower male joint (Fig. 10.46). Dry the glass parts of the syringes in an oven at 110 °C for at least 15 min. Using a small flame or a heat gun, gently heat the flask to remove traces of water from the apparatus; do *not* overheat the joint with the septum. After the flask has been heated for a few minutes, discontinue the heating, add a stirbar, attach the drying tube, and allow the flask to *cool to room temperature*. Remove the glass syringes from the oven and allow them to *cool to room temperature*, preferably in a desiccator. Prepare a solution of 1.6 mL of (+)-α-pinene in 2 mL of dry tetrahydrofuran.

Rubber septum

Figure 10.46
Miniscale hydroboration apparatus.

Hydroboration Draw 5 mL of a 1.0 *M* solution of borane-tetrahydrofuran complex in tetrahydrofuran into the *dry* 5-mL glass syringe. Pierce the rubber septum with the needle of the syringe, being careful not to press down on the plunger, and add the solution directly to the round-bottom flask. Place the apparatus in a water bath made from cold tap water and located on a magnetic stirrer. Use another dry 5-mL glass syringe to transfer the solution of (+)-α-pinene slowly and dropwise to the stirred solution of borane-tetrahydrofuran complex; this addition should require about 10–15 min. Remove the water bath and stir the reaction mixture at room temperature for about 1.5 h. Remove the septum from the reaction flask, and place the flask in an ice-water bath on a magnetic stirrer. Slowly add 0.5 mL of water with stirring to the reaction mixture to decompose any unreacted borane-tetrahydrofuran complex.

Oxidation Prepare 3 mL of an alkaline solution of hydrogen peroxide by mixing equal parts of 30 weight % aqueous hydrogen peroxide and 3 *M* aqueous sodium hydroxide. Using a Pasteur pipet, *slowly* add this solution to the stirred reaction mixture. Perform the addition *cautiously*, as vigorous gas evolution may occur. Once the addition is complete, stir the resulting mixture for 5–10 min.

Work-Up and Isolation Transfer the reaction mixture to a small separatory funnel. Add 5 mL of water and separate the layers. Extract the aqueous layer with two separate 10-mL portions of diethyl ether. Wash the combined organic layers sequentially with two 10-mL portions of water and two 10-mL portions of saturated brine, and dry them over several spatula-tips full of *anhydrous* sodium sulfate.★ Add additional portions of *anhydrous* sodium sulfate if the liquid remains cloudy. Filter or decant the dried solution into a dry 50-mL round-bottom flask and remove the volatile solvents by simple distillation. Alternatively, rotary evaporation or other techniques may be used to remove the solvents. The final traces of solvents may be removed by attaching the flask to a vacuum source and gently swirling the contents as the vacuum is applied.★ Transfer the solid residue to a tared container.

Analysis Weigh your product and determine the yield. Obtain IR and NMR spectra of your starting material and product and compare them with those of authentic samples (Figs. 10.48–10.51). You may also verify the identity of the product by comparing its properties and the melting point of its α-naphthylurethane derivative, prepared according to the procedure in Section 10.8B, with the data summarized in Table 10.1.

Discovery Experiment   *Regiochemistry of Hydroboration/Oxidation of an Acyclic Alkene*

Devise an experimental procedure for hydroboration/oxidation of an alkene, the results of which will test the net "anti-Markovnikov" orientation of the overall reaction. One possible substrate would be 1-hexene. Be certain the spectral data are available in the literature for the possible isomers that could be formed, so that you can analyze the product(s) you obtain. Check with your instructor before undertaking the procedure.

Discovery Experiment   *Regio- and Stereochemistry of Hydroboration/ Oxidation of a Cyclopentene*

Devise an experimental procedure for hydroboration/oxidation of 1-methylcyclopentene, the results of which will test the net "anti-Markovnikov" orientation of the overall reaction and its stereochemistry. Be certain the spectral data are available in the literature for the possible isomers that could be formed, so that you can analyze the product(s) you obtain. Check with your instructor before undertaking the procedure.

MICROSCALE PROCEDURE

Preparation Refer to the online resources to answer Pre-Lab Exercises, access videos, and read the MSDSs for the chemicals used or produced in this procedure. Review Sections 2.5, 2.10, 2.11, 2.13, 2.27, 2.28, and 2.29.

Apparatus A 3- and 5-mL conical vial, two 1-mL glass syringes with needles, screw-cap centrifuge tube, Claisen adapter, drying tube, water bath, ice-water bath, apparatus for magnetic stirring, simple distillation, and *flameless* heating.

Setting Up Dry the 5-mL conical vial, the Claisen adapter, and the glass parts of the two syringes in an oven at 110 °C for at least 15 min. Remove the vial and adapter from the oven, and after they have cooled to room temperature, preferably in a desiccator, place a spinvane in the vial, attach the adapter, and fit its sidearm with a drying tube. Seal the other opening of the adapter with a rubber septum (Fig. 10.47). Place the apparatus in a water bath made from cold tap water and located on a magnetic stirrer. Remove the glass syringes from the oven and allow them to *cool to room temperature*, preferably in a desiccator. Prepare a solution of 0.3 mL of (+)-α-pinene in 0.5 mL of dry tetrahydrofuran.

Figure 10.47
Microscale hydroboration apparatus.

Hydroboration Draw 1 mL of a 1.0 *M* solution of borane-tetrahydrofuran complex in tetrahydrofuran into a *dry* 1-mL glass syringe. Pierce the rubber septum with the needle of the syringe, being careful not to press down on the plunger, and add the solution directly to the conical vial. Use the other dry 1-mL glass syringe to transfer the solution of (+)-α-pinene slowly and *dropwise* to the stirred solution of borane-tetrahydrofuran complex; this addition should require about 5–10 min. The rate of addition may be controlled by slowly twisting the plunger down the body of the syringe. Remove the water bath and stir the reaction mixture at room temperature for about 1.5 h. Remove the septum from the reaction vial and place the vial in an ice-water bath on a magnetic stirrer. Slowly add 2–3 drops of water with stirring to the reaction mixture to decompose any unreacted borane-tetrahydrofuran complex.

Oxidation Prepare 0.6 mL of an alkaline solution of hydrogen peroxide by mixing equal parts of 30 weight % aqueous hydrogen peroxide and 3 *M* aqueous sodium hydroxide. Using a Pasteur pipet, *slowly* add this solution to the stirred reaction mixture. Perform this addition *cautiously*, as vigorous gas evolution may occur. Once the addition is complete, stir the resulting mixture for about 5 min and then add 1 mL of water followed by 1 mL of diethyl ether.

Work-Up and Isolation Stir the mixture in the vial for about 5 min to mix the liquid and solid materials. Remove the lower aqueous layer using a Pasteur pipet and transfer it to a 3-mL conical vial; transfer the organic layer to a screw-cap centrifuge tube. Extract the aqueous layer three times with separate 1-mL portions of diethyl ether, each time transferring the organic layer to the centrifuge tube. Vent the tube as necessary to relieve pressure. Wash the combined organic layers sequentially with two 2-mL portions each of water and saturated brine. Transfer the organic layer to a dry test tube using a Pasteur pipet, and add several microspatula-tips full of *anhydrous* sodium sulfate. Allow the solution to stand for 10–15 min, with occasional swirling to hasten drying.★ Add additional portions of *anhydrous* sodium sulfate if the liquid remains cloudy.

Using a Pasteur pipet, transfer the organic solution to a clean 5-mL conical vial and add a stirbar to the vial. Remove the volatile solvents by simple distillation. Alternatively, rotary evaporation or other techniques may be used to remove the solvent. Following removal of solvents, transfer the solid residue to a tared container.

Analysis Weigh your product and determine the yield. Obtain IR and NMR spectra of your starting material and product and compare them with those of authentic samples (Figs. 10.48–10.51). You may also verify the identity of the product by comparing its properties and the melting point of its α-naphthylurethane derivative, prepared according to the procedure in Section 10.8B, with the data summarized in Table 10.1.

Discovery Experiment *Regiochemistry of Hydroboration/Oxidation of an Acyclic Alkene*

Devise an experimental procedure for hydroboration/oxidation of an alkene, the results of which will test the net "anti-Markovnikov" orientation of the overall reaction. One possible substrate would be 1-hexene. Be certain the spectral data are available in the literature for the possible isomers that could be formed, so that you can analyze the product(s) you obtain. Check with your instructor before undertaking the procedure.

Discovery Experiment   *Regio- and Stereochemistry of Hydroboration/Oxidation of a Cyclopentene*

Devise an experimental procedure for hydroboration/oxidation of 1-methylcyclopentene, the results of which will test the net "anti-Markovnikov" orientation of the overall reaction and its stereochemistry. Be certain the spectral data are available in the literature for the possible isomers that could be formed, so that you can analyze the product(s) you obtain. Check with your instructor before undertaking the procedure.

WRAPPING IT UP

Dilute the *aqueous layer* with water, neutralize it with acetic acid, and then flush it down the drain with excess water. Pour the *distillate* containing THF and ether in the container for nonhalogenated organic solvents.

B. Derivatization

Urethanes

When an alcohol is allowed to react with an aryl-substituted isocyanate, $ArN=C=O$, addition of the alcohol to the carbon-nitrogen double bond occurs to give a urethane (Eq. 10.32). Some commonly used isocyanates are α-naphthyl (1-naphthyl), 4-nitrophenyl, and phenyl.

A major side reaction is hydrolysis of the isocyanate to an amine, which then reacts with more isocyanate to give a disubstituted urea (Eqs. 10.33 and 10.34). Because of their molecular symmetry, these ureas are high melting, and their presence can make purifying the desired urethane difficult. Therefore, you should take precautions to ensure that the alcohol is *anhydrous* when using this procedure. The method works best for water-insoluble alcohols because they are more easily obtained in anhydrous form.

$$ArN=C=O \ + \ ROH \longrightarrow \underset{\text{A urethane}}{ArNH-\overset{\overset{\displaystyle O}{\|}}{C}-OR} \qquad (10.32)$$

An aryl isocyanate   An alcohol

$$ArN=C=O \ + \ H_2O \longrightarrow \underset{\substack{\text{A carbamic acid} \\ \text{(unstable)}}}{ArNH-\overset{\overset{\displaystyle O}{\|}}{C}-OH} \xrightarrow{-CO_2} \underset{\text{An arylamine}}{ArNH_2} \qquad (10.33)$$

An aryl isocyanate

$$ArN=C=O \ + \ ArNH_2 \longrightarrow \underset{\text{A disubstituted urea}}{ArNH-\overset{\overset{\displaystyle O}{\|}}{C}-NHAr} \qquad (10.34)$$

An aryl isocyanate   An arylamine

This type of derivative is also useful for phenols; the procedure given here can thus be used for both alcohols and phenols.

EXPERIMENTAL PROCEDURE

Preparation of Urethanes

1. **Wear safety glasses or goggles and suitable protective gloves while performing the reaction.**
2. **Aryl isocyanates used for preparing urethanes are toxic. Take normal precautions in handling them. If cyanates contact the skin, wash the affected area with soap and warm water.**

Preparation Refer to the online resources to read the MSDSs for the chemicals used or produced in this procedure.

Apparatus A conical vial or 5-mL round-bottom flask, drying tube, and ice-water bath.

Setting Up Place 1 g of the *anhydrous* alcohol or phenol in the round-bottom flask and add 0.5 mL of phenyl isocyanate or α-naphthyl isocyanate; be sure to recap the bottle of the isocyanate tightly to minimize exposure of the reagent to atmospheric moisture. If you are preparing the derivative of a phenol, also add 2 or 3 drops of *dry* pyridine as a catalyst.

Attach the drying tube to the flask and warm the reaction mixture for 5 min. Cool the mixture in an ice-water bath, and scratch the mixture with a stirring rod at the air-liquid interface to induce crystallization. Recrystallize the crude derivative from petroleum ether. *Note:* 1,3- Di(α-naphthyl) urea has a mp of 293 °C, and 1,3-diphenylurea (carbanilide) has a mp of 237 °C; if your product has one of these melting points, repeat the preparation, taking greater care to maintain *anhydrous* conditions.

WRAPPING IT UP

Flush any *filtrates* that contain the starting isocyanate down the drain. Pour the *filtrates* from the recrystallization of the product into the container for nonhalogenated organic liquids. Put the *desiccant* used in the drying tube in the container for nonhazardous solids.

EXERCISES

1. Why is it necessary to maintain the pH of the reaction mixture above 7 during the oxidation of an alkyl borane with hydrogen peroxide?
2. Why is it necessary to use dry apparatus and solvents when using a solution of borane-THF complex?
3. The hydroborating agent used in this procedure is a complex of borane and tetrahydrofuran.

a. Why do you think this reagent was used rather than diborane, B_2H_6, itself?

b. Provide a Lewis dot structure for $BH_3\bullet THF$ showing the direction of each bond dipole with the symbol $\rightarrow$.

c. Why is uncomplexed BH_3 an unstable compound?

4. $(+)$-α-Pinene is the limiting reagent in the hydroboration reaction in this procedure. How many molar equivalents of BH_3 and of hydride ion, H^-, are used?

5. Label each center of chirality in $(+)$-α-pinene with an asterisk. Draw the chemical structure for $(-)$-α-pinene.

6. Why is the hydroboration-oxidation of an alkene considered an *addition* reaction? What small molecule has been "added" to the alkene in the overall reaction?

7. Provide explanations based on the mechanism of the hydroboration-oxidation process that would account for why compounds **75–79** are not formed from $(+)$-α-pinene.

8. Predict the structure of the major product that would be formed upon hydroboration-oxidation of each of the alkenes **a–e.** Include the correct stereochemistry wherever appropriate.

9. For each of the alkenes in Exercise 8, write the structure of the alcohol that should be the major product upon *acid-catalyzed* hydration.

10. An important advantage of hydroboration-oxidation for hydration of alkenes is that rearrangements of the carbon skeleton do not occur, whereas they do in acid-catalyzed hydration. For example, acid-catalyzed hydration of $(+)$-α-pinene (**72**) proceeds with skeletal rearrangements to produce a number of isomeric alcohols.

a. Draw the structure(s) of the product(s) you would expect to obtain upon acid-catalyzed hydration of $(+)$-α-pinene if no skeletal rearrangements occurred.

b. Draw the structures of the isomeric alcohols having the formula $C_{10}H_{18}O$ that you might expect to obtain upon treatment of $(+)$-α-pinene with aqueous acid.

11. The hydroboration of **72** is a stereoselective reaction in which borane approaches from only one face of the carbon-carbon double bond to give the dialkylborane **73**. Provide a rationale for this observed stereoselectivity.

12. Write the structure of the α-naphthylurethane derivative of **74**.

13. Neither of the alcohols **78** or **79** forms an α-naphthylurethane derivative, whereas the isomers **74–77** each do. Explain this difference in reactivities.

Refer to the online resources
and use the spectra viewer and
Tables 8.1–8.8 as needed to answer
the blue-numbered questions on
spectroscopy.

14. Consider the spectral data for (+)-α-pinene (Figs. 10.48 and 10.49).

   a. In the functional group region of the IR spectrum, identify any absorptions associated with the carbon-carbon double bond.

   b. In the ^{1}H NMR spectrum, assign the various resonances to the hydrogen nuclei responsible for them.

   c. For the ^{13}C NMR data, assign the resonances at δ 116.2 and 144.1 to the carbon nuclei responsible for them.

15. Consider the spectral data for (−)-isopinocampheol (Figs. 10.50 and 10.51).

   a. In the functional group region of the IR spectrum, specify any absorptions associated with the alcohol function.

   b. In the ^{1}H NMR spectrum, assign the various resonances to the hydrogen nuclei responsible for them.

   c. For the ^{13}C NMR data, assign the resonance at δ 71.7 to the carbon atom responsible for it.

16. Discuss the differences observed in the IR and NMR spectra of (+)-α-pinene and (−)-isopinocampheol that are consistent with overall addition of water to the double bond occurring in this experiment.

SPECTRA

Starting Materials and Products

Figure 10.48
IR spectrum of (+)-α-pinene (neat).

Figure 10.49
NMR data for (+)-α-pinene (CDCl₃).

(a) ¹H NMR spectrum (300 MHz).
(b) ¹³C data: δ 20.8, 22.8, 26.8, 31.4, 31.5, 38.0, 41.5, 42.2, 116.2, 144.1.

Figure 10.50
IR spectrum of (−)-isopinocampheol (IR card).

Figure 10.51
NMR data for (−)-isopinocampheol (CDCl₃).

(a) ¹H NMR spectrum (300 MHz).
(b) ¹³C data: δ 20.7, 23.7, 27.7, 34.4, 38.1, 39.1, 41.8, 47.8, 47.9, 71.7.

CHAPTER 11

Alkynes

You may have been introduced to the simplest alkyne, acetylene, because it is frequently used as a fuel in welding. Indeed, when acetylene is burned with oxygen, the flame reaches temperatures of about 3300 °C (6000 °F), hotter than all but two or three other mixtures of combustible gases. You already know that flames and organic labs can be a dangerous combination, so we will explore tamer properties of alkynes, avoiding studies of their combustibility. Like the alkenes discussed in Chapter 10, alkynes are unsaturated hydrocarbons, and their chemistry is also dominated by the presence of carbon-carbon multiple or π-bonds. Whereas alkenes have a double bond, alkynes are characterized by a triple bond, which is composed of two orthogonal carbon-carbon π-bonds. Hence, if you understand the reactions that lead to the formation of alkenes, you will be able to apply this knowledge to preparing alkynes. Similarly, you will be able to extend your knowledge of the reactions of alkenes to predicting products of reactions of alkynes. Because alkynes have two double bonds, however, they basically just do everything twice. Well, it's *almost* that simple.

11.1 INTRODUCTION

Unsaturated organic compounds that contain a carbon-carbon triple bond as the **functional group** are called **alkynes**. Acetylene (ethyne), H–C≡C–H, is the simplest alkyne and is widely used in industry as a fuel and as a chemical feedstock for the preparation of other organic compounds such as acetic acid (CH_3CO_2H), vinyl chloride (CH_2=CHCl), a monomer used in the manufacture of polyvinyl chloride, and chloroprene (CH_2=CCl–CH=CH_2), which polymerizes to give neoprene, the material from which your protective gloves may be made. The value of acetylene to Germany during World War II is described in the Historical Highlight *Acetylene: A Valuable Small Molecule*, which is available online.

The carbon-carbon triple bond in an alkyne is formed by the overlap of two *orthogonal* pairs of *p*-orbitals on adjacent *sp*-hybridized carbon atoms. Because the functional group in alkynes is related to the carbon-carbon double bond in alkenes (Chap. 10), the chemistry of these two classes of compounds is similar. For example, in much the same manner as one **elimination reaction** may be used to form the double bond in alkenes, *two* sequential elimination reactions yield the triple bond in alkynes (Eq. 11.1).

$$\underset{\text{A 1,2-dihaloalkane}}{R-\overset{\overset{\displaystyle X}{|}}{\underset{\underset{\displaystyle H}{|}}{C}}-\overset{\overset{\displaystyle X}{|}}{\underset{\underset{\displaystyle H}{|}}{C}}-R} \xrightarrow[\Delta]{\text{strong base}} \underset{\text{An alkyne}}{R-C\equiv C-R} \qquad (11.1)$$

The π-electrons in both alkenes and alkynes provide a Lewis-base site for interaction with electrophilic reagents, which are Lewis acids, so one of the typical reactions of alkenes *and* alkynes is **electrophilic addition** as illustrated in Equations 11.2 and 11.3. Such reactions of alkynes can be stopped after the addition of one equivalent of a reagent, but the use of excess reagent leads to the formation of saturated products in which a second equivalent has been added. Some aspects of the chemistry of alkynes will be explored in the experiments in this chapter.

$$\underset{\text{An alkyne}}{\text{R}-\text{C}\equiv\text{C}-\text{R}} + \underset{\text{A hydrogen halide}}{2\text{ H}-\text{X}} \longrightarrow \underset{\text{A 1,1-dihaloalkane}}{\text{R}-\overset{\overset{\text{X}}{|}}{\underset{\underset{\text{X}}{|}}{\text{C}}}-\overset{\overset{\text{H}}{|}}{\underset{\underset{\text{H}}{|}}{\text{C}}}-\text{R}} \qquad (11.2)$$

$$\underset{\text{An alkyne}}{\text{R}-\text{C}\equiv\text{C}-\text{R}} + \underset{\text{A halogen}}{2\text{ X}_2} \longrightarrow \underset{\text{A 1,1,2,2-tetrahaloalkane}}{\text{R}-\overset{\overset{\text{X}}{|}}{\underset{\underset{\text{X}}{|}}{\text{C}}}-\overset{\overset{\text{X}}{|}}{\underset{\underset{\text{X}}{|}}{\text{C}}}-\text{R}} \qquad (11.3)$$

11.2 DEHYDROHALOGENATION OF 1,2-DIHALOALKANES

The dehydrohalogenation of alkyl halides (Sec. 10.2) is a general method to produce a double bond (Eq. 11.4). By analogy, the sequential elimination of two molecules of a hydrogen halide will lead to the formation of a triple bond as shown by the general reaction in Equations 11.5 and 11.6. The first stage of the reaction involves the base-induced elimination of a hydrogen halide to give a haloalkene, also referred to as a **vinyl halide**. This step proceeds by a **concerted E2 mechanism**, with the base abstracting the proton on the carbon atom β to the leaving halide ion. This reaction involves a **transition state** in which the hydrogen atom and the departing halogen atom are at a dihedral angle of 180° to one another, the **anti-periplanar** orientation, as shown in Equation 11.5. The second step of the sequence (Eq. 11.6) requires a strong base and more forcing conditions such as heating. Although the relationship between the hydrogen and the leaving halide ion in the reaction shown in Equation 11.6 is anti-periplanar, elimination may also occur when the orientation of these two groups on the vinyl halide is **syn-periplanar**, wherein the dihedral angle between the groups is 180°. However, under these circumstances, even higher temperatures are required to effect dehydrohalogenation.

$$\underset{\text{A haloalkane}}{\text{RCH}-\overset{\overset{\text{H}}{|}}{\underset{\underset{\text{X}}{|}}{\text{CHR}}}} \xrightarrow{\text{base}} \underset{\text{An alkene}}{\text{RCH}=\text{CHR}} \qquad (11.4)$$

$$\underset{\text{A 1,2-dihaloalkane}}{\text{R}-\overset{\text{X}}{\underset{\text{H}}{\text{C}}}-\overset{\text{X}}{\underset{\text{H}}{\text{C}}}-\text{R}} \longrightarrow \underset{\text{A haloalkene}}{\overset{\text{X}}{\underset{\text{R}}{}}\text{C}=\overset{\text{R}}{\underset{\text{H}}{}}\text{C}} + \text{B}-\text{H} + \text{X}:^{-} \qquad (11.5)$$

B:⁻

$$X \overset{\frown}{\underset{R}{C}} = \overset{R}{\underset{H}{C}} \xrightarrow{\Delta} R-C≡C-R + B-H + X:^- \qquad (11.6)$$

A haloalkene   An alkyne

In the experiment that follows, you will perform the dehydrobromination of *meso*-stilbene dibromide (**1**), which can be prepared by the addition of bromine to (*E*)-stilbene (Sec. 10.6), to give diphenylacetylene (**2**) according to Equation 11.7. Thus, the combination of the bromination of an alkene followed by the double dehydrobromination of the intermediate *vicinal* dibromide represents a useful means of converting an alkene into an alkyne. Because the temperature required to effect elimination of hydrogen bromide from the intermediate vinyl bromide is nearly 200 °C, it is necessary to use a high-boiling solvent such as triethylene glycol so the reaction may be carried out without special apparatus. However, if a microwave apparatus is available for heating (Sec. 2.9), the reaction can be effected in a pressure-rated tube at 150 °C using methanol as solvent. It is instructive to compare the reaction times required and the results obtained using conventional versus microwave heating for promoting the reaction.

$$\text{Br} \overset{H}{\underset{}{C}} - \overset{}{\underset{H}{C}} \text{Br} + 2 \text{ KOH} \xrightarrow[190\ °C]{HO(CH_2CH_2O)_2CH_2CH_2OH} \text{Ph}-C≡C-\text{Ph} \qquad (11.7)$$

1   **2**
meso-Stilbene dibromide   Diphenylacetylene

EXPERIMENTAL PROCEDURES

Dehydrobromination of Meso-Stilbene Dibromide

Purpose To demonstrate the preparation of an alkyne by a double dehydrohalogenation.

MINISCALE PROCEDURE

Preparation Refer to the online resources to answer Pre-Lab Exercises, access videos, and read the MSDSs for the chemicals used or produced in this procedure. Review Sections 2.9, 2.10, and 2.17.

Apparatus A sand bath, hot plate, 25-mL Erlenmeyer flask, ice-water bath, apparatus for vacuum filtration.

Setting Up Preheat the sand bath to about 190–200 °C. Place 800 mg of *meso*-stilbene dibromide and 5 pellets (about 400 mg) of commercial potassium hydroxide in the Erlenmeyer flask. Add 4 mL of triethylene glycol and a *carborundum* boiling stone to the flask.

Dehydrobromination and Isolation Place the flask in the sand bath and heat the mixture. After potassium bromide begins to separate from solution, heat the mixture for an additional 5 min. Remove the flask from the sand bath and allow it to cool to room temperature. Add 10 mL of water and place the flask in an ice-water bath for 5 min. Collect the diphenylacetylene that precipitates by vacuum filtration. Wash the solid with about 2–3 mL of cold water. Recrystallize the product from a small quantity of 95% ethanol or an ethanol-water mixture. If the solution is allowed to cool slowly undisturbed, you should obtain large, sparlike, colorless crystals.

Analysis Determine the melting point, weight, and percent yield of the recrystallized product. Test the product for unsaturation using the bromine and Baeyer tests (Secs. 4.7A1 and 4.7A2, respectively). Obtain IR and ^{1}H NMR spectra of your starting material and product and compare them with those of authentic samples (Figs. 10.36, 10.37, 11.1, and 11.2).

MICROSCALE PROCEDURE

Preparation Refer to the online resources to answer Pre-Lab Exercises, access videos, and read the MSDSs for the chemicals used or produced in this procedure. Review Sections 2.9, 2.10, and 2.17.

Apparatus A sand bath, hot plate, 13-mm × 120-mm Pyrex test tube, ice-water bath, apparatus for vacuum filtration.

Setting Up Preheat the sand bath to about 190–200 °C. Place 150 mg of *meso*-stilbene dibromide and 1 pellet (about 80 mg) of commercial potassium hydroxide in the Pyrex test tube. Add 1 mL of triethylene glycol and a carborundum boiling stone to the test tube.

Dehydrobromination and Isolation Place the test tube in the sand bath and heat the mixture. After potassium bromide begins to separate from solution, heat the mixture for an additional 5 min. Remove the test tube from the sand bath and allow it to cool to room temperature. Add 2 mL of water and place the test tube in an ice-water bath for 5 min. Collect the diphenylacetylene that precipitates by vacuum filtration. Wash the solid with about 1 mL of cold water. Recrystallize the product from a small quantity of 95% ethanol or an ethanol-water mixture.

Analysis Determine the melting point, weight, and percent yield of the recrystallized product. Test the product for unsaturation using the bromine and Baeyer tests (4.7A1 and 4.7A2, respectively). Obtain IR and ^{1}H NMR spectra of your starting material and product and compare them with those of authentic samples (Figs. 10.36, 10.37, 11.1, and 11.2).

MINISCALE PROCEDURE FOR MICROWAVE OPTION

Preparation Refer to the online resources to answer Pre-Lab Exercises, access videos, and read the MSDSs for the chemicals used or produced in this procedure. Review Sections 2.9, 2.10, and 2.17.

Apparatus A 10-mL pressure-rated tube with cap, stirbar, ice-water bath, apparatus for microwave heating with magnetic stirring and vacuum filtration.

Setting Up Equip the 10-mL pressure-rated tube with the stirbar and add 0.40 g of *meso*-stilbene dibromide, about 0.2 g (2 pellets) of solid potassium hydroxide, and 2 mL of methanol. Cap the pressure-rated tube and gently shake it or place it on a magnetic stirrer to facilitate initial mixing of its contents. Place the tube in the cavity of the microwave apparatus.

Dehydrobromination and Isolation Program the unit to heat the reaction mixture with stirring according to the directions provided by your instructor. Generally, the reaction temperature should be set at 150 °C and the power set at a maximum of 25 W with a 1-min ramp time and a 5-min hold time; the pressure limit should be set at 275 psi. Allow the mixture to cool to room temperature and remove the tube from the microwave apparatus. Add 5 mL of water and place the tube in an ice-water bath for 5 min. Collect the diphenylacetylene that precipitates by vacuum filtration. Wash the solid with about 2 mL of cold water. Recrystallize the product from a small quantity of 95% ethanol or an ethanol-water mixture. If the solution is allowed to cool slowly undisturbed, you should obtain large, sparlike, colorless crystals.

Analysis Determine the melting point, weight, and percent yield of the recrystallized product. Test the product for unsaturation using the bromine and Baeyer tests (Secs. 4.7A1 and 4.7A2, respectively). Obtain IR and ^{1}H NMR spectra of your starting material and product and compare them with those of authentic samples (Figs. 10.36, 10.37, 11.1, and 11.2).

WRAPPING IT UP

Combine the *filtrate* from the reaction mixture with the *mother liquor* from the recrystallization, dilute with water, and neutralize with 10% aqueous hydrochloric acid; then flush the solution down the drain. Place the *dichloromethane solution* from the bromine test for unsaturation in a container for halogenated organic liquids; put the *manganese dioxide* from the Baeyer test for unsaturation in a container for heavy metals.

EXERCISES

1. Write three-dimensional structures of *meso*-stilbene dibromide and one enantiomer of *dl*-stilbene dibromide.

2. The solvent used in the preparation of diphenylacetylene (**2**) is triethylene glycol.

   a. Write the structure of triethylene glycol; circle and label each of the functional groups in this molecule.

   b. What structural features account for the high boiling point of this solvent?

3. The functional group in alkynes is the carbon-carbon triple bond.

   a. Using a suitable drawing, show how the atomic *p*-orbitals in acetylene overlap to form π-molecular orbitals.

   b. What is the angle between these two π-orbitals?

   c. What is the angle between each of the π-orbitals and the carbon-carbon σ-bond?

4. Give a stepwise mechanism showing the base-induced formation of diphenylacetylene from *meso*-stilbene dibromide. Use curved arrows to symbolize the flow of electrons.

5. The E2 reaction of most compounds is known to proceed preferentially by removal of a proton anti-periplanar to the leaving group. Based upon this generalization, predict the geometry of the 1-bromo-1,2-diphenylethylene that is produced as the intermediate in the double dehydrobromination of *meso*-stilbene dibromide.

6. Why do you think the enthalpy of activation, $\Delta H^{\ddagger}$, for a syn-periplanar elimination is higher than that for an anti-periplanar elimination?

7. If *meso*-stilbene dibromide is treated with KOH in ethanol, it is possible to isolate the 1-bromo-1,2-diphenylethylene that is formed from the first dehydrobromination. The E2 elimination of the second molecule of hydrogen bromide from this intermediate alkene to give diphenylacetylene has a higher activation enthalpy than the first elimination and thus requires a higher reaction temperature. Explain.

8. Provide an explanation for why neither ethanol nor ethylene glycol is a suitable solvent for the second dehydrobromination.

Refer to the online resources and use the spectra viewer and Tables 8.1–8.8 as needed to answer the blue-numbered questions on spectroscopy.

9. Consider the NMR spectral data for *meso*-stilbene dibromide (Fig. 10.37).

   a. In the ^{1}H NMR spectrum, assign the various resonances to the hydrogen nuclei responsible for them.

   b. For the ^{13}C NMR data, assign the various resonances to the carbon nuclei responsible for them.

10. Consider the spectral data for diphenylacetylene (Figs. 11.1 and 11.2).

   a. In the functional group region of the IR spectrum, identify the absorptions associated with the aromatic rings. Why is there no absorption for the carbon-carbon triple bond?

   b. In the ^{1}H NMR spectrum, assign the various resonances to the hydrogen nuclei responsible for them.

   c. For the ^{13}C NMR data, assign the various resonances to the carbon nuclei responsible for them.

11. Discuss the differences observed in the NMR spectra of *meso*-stilbene dibromide and diphenylacetylene that are consistent with the double dehydrobromination in this experiment.

SPECTRA

Starting Material and Product

The IR and NMR spectra of **meso**-*stilbene dibromide are provided in Figures 10.36 and 10.37, respectively.*

Figure 11.1

IR spectrum of diphenylacetylene (IR card).

Figure 11.2

NMR data for diphenylacetylene (CDCl₃).

(a) ¹H NMR spectrum (300 MHz).
(b) ¹³C NMR data: δ 89.6, 123.3, 128.1, 128.2, 131.6.

11.3 ADDITION REACTIONS OF ALKYNES

One of the general classes of reactions that characterizes the chemistry of alkynes as well as alkenes (Sec. 10.4) is the **electrophilic addition** of reagents generalized as **E–Nu** (Eq. 11.8) to the carbon-carbon multiple bond. This reaction typically proceeds in two stages. Under carefully controlled reaction conditions in which only one equivalent of electrophilic reagent **E–Nu** is used, it is sometimes possible to stop the addition reaction at the alkene stage. However, if more than one equivalent of the reagent is present, further addition occurs, and substituted alkanes are formed (Sec. 11.1). Typical reagents that add to alkynes include the halogens chlorine and bromine and acids such as hydrogen chloride and hydrogen bromide.

$$
R^1\!-\!C\!\equiv\!C\!-\!R^2 \xrightarrow[\text{1 mole}]{\text{E–Nu}}
\underset{E}{\overset{R^1}{}}C\!=\!C\underset{R^2}{\overset{Nu}{}}
\xrightarrow[\text{1 mole}]{\text{E–Nu}}
R^1\!-\!\underset{\underset{E}{|}}{\overset{\overset{E}{|}}{C}}\!-\!\underset{\underset{Nu}{|}}{\overset{\overset{Nu}{|}}{C}}\!-\!R^2 \qquad (11.8)
$$

An alkyne   A substituted alkene   A substituted alkane

The *regiochemistry* of the electrophilic addition of unsymmetrical reagents to terminal alkynes follows *Markovnikov's rule* in the same fashion as observed for alkenes (Sec. 10.4). Thus, the Lewis acid component, E^+, of the reagent E–Nu adds to the electron-rich triple bond to form the more stable, or more substituted, of the two possible intermediate vinyl carbocations. This carbocation then undergoes nucleophilic attack by the Lewis base component, Nu^-, of the reagent (Eq. 11.9). The addition of the second mole of E–Nu to the resulting alkene occurs with the same regiochemistry as the first (Eqs. 11.8 and 11.9).

$$
R\!-\!C\!\equiv\!C\!-\!H \xrightarrow[\text{1 mole}]{\text{E–Nu}}
\underset{Nu}{\overset{R}{}}C\!=\!C\underset{H}{\overset{E}{}}
\xrightarrow[\text{1 mole}]{\text{E–Nu}}
R\!-\!\underset{\underset{Nu}{|}}{\overset{\overset{Nu}{|}}{C}}\!-\!\underset{\underset{E}{|}}{\overset{\overset{E}{|}}{C}}\!-\!H \qquad (11.9)
$$

A terminal alkyne   A substituted alkene   A substituted alkane

Alkynes undergo electrophilic addition reactions with hydrogen halides and bromine, but these reactions have limited synthetic utility. However, one reaction of alkynes that is commonly used in organic chemistry is **hydration** of the carbon-carbon triple bond to give a **ketone**, a transformation that is catalyzed by mercuric ion in the presence of sulfuric acid (Eq. 11.10).

$$
\underset{RCH_3}{\overset{O}{\underset{\|}{}}} \qquad\qquad (11.10)
$$

The role that the mercuric ion, a polarizable Lewis acid, plays in this reaction is not completely understood. However, a reasonable hypothesis is that it first coordinates with the triple bond to generate a vinyl cation that may be stabilized by conversion to a cyclic mercurinium ion (Eq. 11.11). Attack by water on the cyclic ion then occurs at the more substituted carbon. Following several proton transfers, mercuric ion is lost to give a substituted vinyl alcohol, which is unstable and tautomerizes to give the **carbonyl group**, C=O, of the product ketone (Eq. 11.12).

$$ R-C{\equiv}C-H \xrightarrow{Hg^{2+}} R-\overset{+}{C}{=}C{\overset{\ddot{H}g^+}{\underset{H}{}}} \rightleftharpoons \text{(cyclic mercurinium ion)} \quad (11.11) $$

An alkyne A vinyl cation A cyclic mercurinium ion

$$ (11.12) $$

A vinyl alcohol (enol) A ketone
(*unstable; not isolated*)

In the experiment that follows, the hydration of a terminal alkyne is illustrated by the conversion of 2-methyl-3-butyn-2-ol (**3**) to 3-hydroxy-3-methyl-2-butanone (**4**), as shown in Equation 11.13. The presence of a hydroxyl group in **3** has little effect on the chemical properties of the carbon-carbon triple bond. Rather, the main effect of the polar hydroxyl group is on the physical properties of the molecule, with the boiling point of **3** being considerably higher than those of other acetylenic hydrocarbons having the same molecular weight.

$$ CH_3-\underset{CH_3}{\overset{OH}{C}}-C{\equiv}C-H + H_2O \xrightarrow[Hg^{2+}]{H_2SO_4} CH_3-\underset{CH_3}{\overset{OH}{C}}-\overset{O}{C}-CH_3 \quad (11.13) $$

3 **4**
2-Methyl-3-butyn-2-ol 3-Hydroxy-3-methyl-2-butanone

5
(not formed)

This experiment provides an opportunity to verify experimentally that the hydration of a terminal alkyne occurs in accordance with Markovnikov's rule (see the Historical Highlight *Additions Across Carbon-Carbon π-Bonds*, which is available online). For example, that the structure of the product of the hydration of **3** is **4** rather than **5** may be confirmed by comparing the IR, ^{1}H, and ^{13}C NMR spectra of the product of the reaction with known standard samples of **4** and **5**, the spectra of which are clearly distinct. When *all* of the spectral properties of a compound are superimposable with those of a standard sample, the two are identical.

Another common method of assigning a structure to the product of a reaction involves preparing a solid derivative and comparing its melting point with the melting point of the same derivative of a series of known compounds. If the melting points of the derivatives of the unknown and a known sample are identical, then a mixed melting point (Sec. 3.3) is taken to confirm the identity of the two samples. In the present case, a solid derivative of the product of the reaction can be made for comparison with the same derivatives of authentic samples of **4** and **5**. For example, forming the semicarbazone **6** from **4** proceeds smoothly, but converting the aldehyde **5** to its semicarbazone occurs with concomitant dehydration to give **7**, which is the semicarbazone of the unsaturated aldehyde $(CH_3)_2C=CHCHO$.

$$\underset{\textbf{6}}{CH_3-\overset{\overset{\displaystyle OH}{|}}{\underset{\underset{\displaystyle CH_3}{|}}{C}}-\overset{\overset{\displaystyle NNHCONH_2}{\nearrow}}{\underset{\underset{\displaystyle CH_3}{\diagdown}}{C}}} \qquad\qquad \underset{\textbf{7}}{\overset{CH_3}{\underset{CH_3}{}}C=CH-\overset{\overset{\displaystyle NNHCONH_2}{\nearrow}}{\underset{\underset{\displaystyle H}{\diagdown}}{C}}}$$

EXPERIMENTAL PROCEDURE

Preparation of 3-Hydroxy-3-methyl-2-butanone

Discovery Experiment   **Purpose** To demonstrate the hydration of a terminal alkyne and to determine the regiochemistry of the addition.

SAFETY ALERT

1. **Wear safety glasses or goggles and suitable protective gloves while performing the experiment.**

2. ***Concentrated* sulfuric acid may cause severe chemical burns, so do not allow it to come into contact with your skin. Wash any affected area immediately with water and then with 5% sodium bicarbonate solution.**

3. **Be sure that you add the concentrated acid slowly *to water* and *not* the reverse.**

MINISCALE PROCEDURE

Preparation Refer to the online resources to answer Pre-Lab Exercises, access videos, and read the MSDSs for the chemicals used or produced in this procedure. Review Sections 2.9, 2.10, 2.11, 2.13, 2.16, 2.21, 2.22, and 2.29.

Apparatus A 10-mL, a 50-mL, and a 100-mL round-bottom flask, separatory funnel, ice-water bath, 13-mm × 120-mm test tube, apparatus for heating under reflux, simple distillation, magnetic stirring, steam distillation with internal steam generation, vacuum filtration, and *flameless* heating.

Setting Up *Carefully* add 3 mL of *concentrated* sulfuric acid *to* 20 mL of water contained in the 100-mL round-bottom flask equipped with a stirbar. Dissolve 0.2 g of reagent-grade mercuric oxide in the resulting warm solution and cool the flask to about 50 °C. Attach a reflux condenser to the flask, and add 3.6 mL of 2-methyl-3-butyn-2-ol in one portion through the top of the condenser. The cloudy white precipitate that forms is presumably the mercury complex of the alkyne.

Reaction Stir the reaction flask to mix the contents thoroughly, whereupon a mildly exothermic reaction may ensue. Allow the reaction to proceed by itself for about 2 min. Heat the mixture under reflux for 30 min, and then allow it to cool to room temperature.★

Isolation While the reaction mixture is cooling, prepare an apparatus for steam distillation *with internal steam generation.* Add about 35–40 mL of water to the reaction flask. Heat the flask and steam-distill the product until about 30–35 mL of distillate has been collected.★

Transfer the distillate to a small separatory funnel, add 2–3 g of potassium carbonate, and then carefully saturate the solution with sodium chloride; avoid adding more sodium chloride than will dissolve readily. If a second layer forms at this point, do not separate it, but continue with the extraction. Extract the mixture sequentially with two 10-mL portions of dichloromethane and combine the extracts. Dry the organic extracts over several spatula-tips full of *anhydrous* sodium sulfate. Swirl the flask occasionally for a period of 10–15 min to facilitate drying; add further portions of *anhydrous* sodium sulfate if the solution does not become clear.

Decant the solution into a 50-mL round-bottom flask and remove most of the dichloromethane by simple shortpath distillation; cool the receiving flask in an icewater bath. Alternatively, use rotary evaporation or other techniques to remove most of the dichloromethane. Using a Pasteur pipet to minimize material loss, transfer the crude product into a 10-mL round-bottom flask. Remove the last traces of dichloromethane from the crude product by connecting the flask to a vacuum source. Then continue the distillation using the shortpath apparatus fitted with a clean, tared receiving flask. Since the volume of product is likely to be only 1–2 mL, an accurate boiling point may be difficult to obtain, and all material that distills above 110–115 °C should be collected to obtain a reasonable yield.

Derivatization Add 0.5 mL of **4** to a solution prepared from 0.5 g of semicarbazide hydrochloride and 0.8 g of sodium acetate dissolved in 2.5 mL of water contained in a test tube. Shake the mixture vigorously. Collect the crude solid semicarbazone that forms by vacuum filtration; use a glass rod to scratch at the air-liquid interface if necessary to initiate crystallization. Dissolve the solid product in a minimum volume of hot 2-propanol, cool the mixture to room temperature and then in an icewater bath, and collect the crystals by vacuum filtration. Air-dry the crystals.

Analysis Determine the weight and percent yield of the distilled product. Determine the melting point of the semicarbazone of your product. Use this information to determine the identity of the product of hydration of 2-methyl-3-butyn-2-ol. The melting point of **6** is 162–163 °C, whereas **7** melts at 222–223 °C. Obtain IR and ¹H NMR spectra of your starting material and product and compare them with those of authentic samples (Figs. 11.3–11.6).

WRAPPING IT UP

Since the *residue in the stillpot* from the steam distillation contains soluble mercury salts, place it in a special container for solutions of mercuric salts; do *not* pour it down the drain. If directed to do so by your instructor, you may convert the *soluble mercuric salts* into insoluble mercuric sulfide by reaction with sodium sulfide or hydrosulfide; collect the *mercuric sulfide* by vacuum filtration for recycling. Flush the *other aqueous solutions* down the drain with water. Pour the *dichloromethane* recovered as the forerun in the distillation into a container for halogenated organic solvents. Dilute the *filtrate* from the preparation of the semicarbazone derivative and the *mother liquor* from the recrystallization with water, and flush the solutions down the drain.

EXERCISES

1. 3-Hydroxy-3-methyl-2-butanone (**4**) was produced by the hydration of 2-methyl-3-butyn-2-ol (**3**) in the experiment you performed.

   a. Does the formation of **4** from **3** follow Markovnikov's rule?

   b. Draw the structure of **4** and a tautomer of **4**.

   c. Draw a resonance structure of **4**.

   d. Using your answers to Parts **b** and **c**, describe how two tautomeric structures and two resonance structures differ.

2. Using curved arrows to symbolize the flow of electrons, give a stepwise mechanism for the hydration of 2-methyl-3-butyn-2-ol with sulfuric acid in the presence of mercuric oxide.

3. How is the ketone **4** isolated from the reaction mixture, and what advantages does this technique offer over direct extraction?

4. Give the structures for the products that you would expect to obtain on hydration of 1-octyne and of 2-octyne.

5. Consider the compounds hexane, 1-hexene, and 1-hexyne (see Chap. 9 and 10 for a review of reactions of alkanes and alkenes). What similarities and what differences would you expect upon treatment of these compounds with

   a. bromine in dichloromethane at room temperature in the dark?

   b. an aqueous solution of potassium permanganate?

   c. an aqueous solution of sulfuric acid?

   Give the structures of the products, if any, that would be obtained from each of these reactions.

6. The reaction of 1-butyne with hydrogen bromide gives 2,2-dibromobutane, as shown in Equation 11.14. Write a stepwise mechanism for this process, using curved arrows to symbolize flow of electrons. Also provide a rationale for the observed regioselectivity for addition of *each* mole of hydrogen bromide.

$$CH_3CH_2C{\equiv}C{-}H \ + \ 2\ H{-}Br \longrightarrow CH_3CH_2CBr_2CH_3 \qquad (11.14)$$

Refer to the online resources and use the spectra viewer and Tables 8.1–8.8 as needed to answer the blue-numbered questions on spectroscopy.

7. Consider the spectral data for 2-methyl-3-butyn-2-ol (Figs. 11.3 and 11.4).

   a. In the functional group region of the IR spectrum, identify the absorptions associated with the carbon-carbon triple bond and the hydroxyl group.

   b. In the ^{1}H NMR spectrum, assign the various resonances to the hydrogen nuclei responsible for them.

   c. For the ^{13}C NMR data, assign the various resonances to the carbon nuclei responsible for them.

8. Consider the spectral data for 3-hydroxy-3-methyl-2-butanone (Figs. 11.5 and 11.6).

   a. In the functional group region of the IR spectrum, identify the absorptions associated with the carbonyl and hydroxyl groups.

   b. In the ^{1}H NMR spectrum, assign the various resonances to the hydrogen nuclei responsible for them.

   c. For the ^{13}C NMR data, assign the various resonances to the carbon nuclei responsible for them.

9. ^{1}H NMR spectroscopy could be used to determine whether **4** or **5** is produced by the hydration of **3**. Sketch the ^{1}H NMR spectrum you would expect from **5**, and discuss the differences that you would expect in the ^{1}H NMR spectra of **4** and **5**.

10. IR spectroscopy may also be used to determine whether **4** or **5** is produced by the hydration of **3**. Describe the important difference(s) you would expect in the IR spectra of the two compounds.

SPECTRA

Starting Material and Product

Figure 11.3

IR spectrum of 2-methyl-3-butyn-2-ol (neat).

Figure 11.4
NMR data for 2-methyl-3-butyn-2-ol (CDCl₃).

(a) ¹H NMR spectrum (300 MHz).
(b) ¹³C NMR data: δ 31.1, 64.8, 70.4, 89.1.

Figure 11.5
IR spectrum of 3-hydroxy-3-methyl-2-butanone (neat).

Figure 11.6
NMR data for 3-hydroxy-3-methyl-2-butanone (CDCl₃).

(a) ¹H NMR spectrum (300 MHz).
(b) ¹³C NMR data: δ 23.8, 26.4, 76.7, 213.2.

11.4 TERMINAL ALKYNES AS ACIDS

The acidity of the acetylenic hydrogen of terminal alkynes (R–C≡C–H), pK_a ≈ 25, provides the basis of a simple and specific qualitative test for such compounds. Treating a terminal alkyne with a solution containing silver ammonia complex, which is prepared by dissolving silver nitrate in ammonia, provides a solid silver salt, as illustrated by the general reaction shown in Equation 11.15. This test differentiates terminal alkynes from alkenes, internal alkynes, dienes, and allenes.

$$R\text{—}C\equiv C\text{—}H + [Ag(NH_3)_2]^+ \longrightarrow \underline{R\text{—}C\equiv C\text{—}Ag} + NH_3 + \overset{+}{N}H_4 \quad (11.15)$$

A terminal alkyne   A silver acetylide

This unique reaction of terminal alkynes may also be used to separate them from non-terminal alkynes, which do *not* form insoluble silver salts and remain in solution. The silver acetylide salts can be reconverted to the terminal alkynes on treatment with hydrochloric acid, as shown in Equation 11.16. *Dry silver salts of this type are quite sensitive to shock and tend to decompose explosively.* Therefore, they should *never* be allowed to dry out before being decomposed by hydrochloric acid.

$$\underline{R\text{—}C\equiv C\text{—}Ag} + HCl \longrightarrow R\text{—}C\equiv C\text{—}H + \underline{AgCl} \quad (11.16)$$

EXPERIMENTAL PROCEDURE

Formation of a Silver Acetylide and Its Decomposition

Purpose To demonstrate the formation and decomposition of the silver acetylide of a terminal alkyne.

SAFETY ALERT

1. **Wear safety glasses or goggles and suitable protective gloves while performing the experiment.**

2. **Avoid spilling solutions of silver salts on your skin. If you do come into contact with the solutions, wash the affected area immediately with copious amounts of water.**

3. **Take special care to destroy *all* of the silver salt acetylide with hydrochloric acid *before* discarding it.**

Procedure

Preparation Refer to the online resources to answer Pre-Lab Exercises, access videos, and read the MSDSs for the chemicals used or produced in this procedure. Review Section 2.17.

Apparatus Two 13-mm × 120-mm test tubes, apparatus for vacuum filtration.

Setting Up In a 10-mL Erlenmeyer flask, prepare a solution of silver ammonia complex from 2.5 mL of 0.1 *M* silver nitrate solution by adding ammonium hydroxide solution *dropwise*. Brown silver oxide forms first; add *just enough* ammonium hydroxide to dissolve the silver oxide. Dilute the solution by adding 1.5 mL of water.

Formation and Decomposition Add 3 mL of the diluted silver ammonia complex solution to about 0.1 mL of 2-methyl-3-butyn-2-ol in a test tube, and note the formation of the silver acetylide salt. Collect the silver salt by vacuum filtration of the aqueous solution; *be careful not to let the silver salt dry, because the dry salt is explosive.* Transfer the *wet* silver salt to a clean test tube, and add a small amount of dilute hydrochloric acid. Observe what changes occur, especially in the color and form of the precipitate.

WRAPPING IT UP

Dilute the remaining *silver ammonia complex test solution* with water and flush it down the drain with water. Destroy all *solid silver acetylide salt* by treatment with hydrochloric acid. Collect the *silver chloride* and put it in a container for heavy metals, and then flush the *filtrate* down the drain.

EXERCISES

1. The pK_a of a terminal alkyne is about 25 and the pK_a of ammonium ion, $\overset{+}{N}H_4$, is about 9.

   a. Use these values to predict the K_{eq} for Equation 11.17 to prove that the acid-base equilibrium lies to the left as shown.

$$R—C\equiv C—H + NH_3 \rightleftharpoons R—C\equiv \overset{-}{C} + \overset{+}{N}H_4 \qquad (11.17)$$

   b. Provide an explanation for the observation that the equilibrium for the reaction of a terminal alkyne with silver ammonia complex (Eq. 11.15) favors the formation of silver acetylide.

2. How might you separate, and obtain in pure form, 1-octyne and 2-octyne from a mixture containing both?

3. Explain the dramatic decrease in acidity of the boldfaced "terminal" hydrogen atom in the following series of hydrocarbons. (*Hint:* Consider the difference in the nature of the respective carbon-hydrogen σ-bonds and the anticipated relative stabilities of the anions resulting from deprotonation.)

$$H_3C—C\equiv C—\mathbf{H} > \underset{H_3C}{\overset{H}{>}}C=C\underset{H}{\overset{\mathbf{H}}{}} > \underset{H_3C}{\overset{H}{}}H—C—C\underset{H}{\overset{\mathbf{H}}{—H}}$$

4. Consider the acetylide anion $CH_3C{\equiv}C^-$, which can be prepared by deprotonation of propyne.

    a.   Explain why this anion is a good nucleophile.

    b.   Write the structure of the product of the reaction of the anion $CH_3C{\equiv}C^-$ with ethyl bromide and indicate to what mechanistic class this reaction belongs.

    c.   The reaction of the anion $CH_3C{\equiv}C^-$ with a tertiary halide does *not* yield a substitution product. Write the structure of the organic product(s) formed from the reaction of $CH_3C{\equiv}C^-$ and a tertiary alkyl halide of your choosing and explain why substitution does not occur.

5. Carbon-carbon bond-forming reactions are very important in synthetic organic chemistry. Explain how you could prepare heptane from 1-butyne using any other organic and inorganic reagents you desire.

CHAPTER 12

Dienes: The Diels-Alder Reaction

In Chapters 10 and 11, you can discover that both alkenes and alkynes undergo a number of different electrophile addition reactions. These transformations typically proceed by stepwise mechanisms, involving reactions of a π-bond first with an electrophile to give an intermediate cation that then reacts with a nucleophile. In this chapter, you will discover that alkenes and alkynes also undergo a unique cycloaddition reaction with conjugated 1,3-dienes in an interesting process known as the Diels-Alder reaction. Because two new carbon-carbon bonds *and* a new six-membered ring are formed in a single step, the Diels-Alder reaction is one of the most powerful reactions in organic chemistry, and it has been widely used as a key step in the preparation of complex molecules of biological interest. Indeed, there is now good evidence that the Diels-Alder reaction is even used to prepare compounds that occur in living organisms.

12.1 INTRODUCTION

In synthetic organic chemistry, reactions that produce carbocyclic rings from acyclic precursors are of great importance because they lead to the formation of complex molecular structures. Rings may be produced from the cyclization of an acyclic starting material by forming a *single* carbon-carbon bond. However, one of the most useful methods for constructing six-membered rings from acyclic starting materials involves the reaction of a **1,3-diene** with an alkene to give a derivative of cyclohexene by forming *two* new carbon-carbon bonds (Eq. 12.1). In this reaction, the alkene partner is generally referred to as a **dienophile**. This **cycloaddition** reaction is called the **Diels-Alder reaction** in honor of Otto Diels and Kurt Alder, the two German chemists who recognized its importance and shared the 1950 Nobel Prize in Chemistry for their extensive development of this reaction. A more detailed account of their elegant work is included in the Historical Highlight *Discovery of the Diels-Alder Reaction*, which is available online. The Diels-Alder reaction also belongs to a class of reactions termed **1,4-additions,** because the two new carbon-carbon σ-bonds are formed between the 1- and 4-carbon atoms of the diene and the two π-bonded carbon atoms of the dienophile.

$$\text{A 1,3-diene} \qquad \text{A dienophile} \qquad \text{A cycloadduct} \tag{12.1}$$

The scope of the Diels-Alder reaction is broad, and many combinations of dienes and dienophiles are known to furnish cycloadducts in good yields. The presence of electron-withdrawing substituents such as cyano, C≡N, and carbonyl, C=O, groups on the dienophile increases the rate and yield of the reaction. The Diels-Alder reaction is remarkably free of complicating side reactions, and yields of the desired product are often high. Probably the single most important side reaction that may be encountered is dimerization or polymerization of the diene and/or dienophile.

12.2 MECHANISTIC AND STEREOCHEMICAL ASPECTS

The mechanism of the Diels-Alder reaction is now fairly well understood. The accumulated evidence favors a picture in which reorganization of the π-electrons in the transition state for the reaction occurs so the two new σ-bonds are formed in a simultaneous, **concerted** fashion; little charge or free-radical character is developed at any of the terminal carbon atoms (Eq. 12.2). The reaction is one of a number of **pericyclic reactions** that are controlled by **orbital symmetry**. Detailed discussions of orbital symmetry and its ramifications may be found in all modern organic chemistry textbooks.

$$\text{Reactants} \qquad \begin{array}{c}\text{Six-membered} \\ \text{transition state}\end{array} \qquad \text{Product} \tag{12.2}$$

A consequence of the concerted nature of the Diels-Alder reaction is that the diene must be able to adopt a *planar conformation* in which the dihedral angle between the two double bonds is 0°. Such a conformation is designated as *s-cis*, meaning **cisoid** about the single bond. Dienes that have difficulty achieving this conformation are expected to undergo Diels-Alder reactions slowly or not at all. A concerted reaction of a dienophile with a diene in its more stable *s-trans* conformation would lead to a six-membered ring containing a highly strained *trans* double bond, an improbable structure.

s-cis 1,3-diene conformation   *s-trans* 1,3-diene conformation

A consequence of the highly ordered relationship between the diene and dienophile in the transition state depicted in Equation 12.2 is that Diels-Alder reactions are highly **stereoselective**, with the formation of the new carbon-carbon σ-bonds occurring on the *same* face of *both* the diene and the dienophile. Such addition reactions are termed *syn-additions*. Thus, the cycloadduct resulting from reaction of *trans, trans*-2,4-hexadiene and a dienophile is *exclusively cis*-3,6-dimethylcyclohexene (Eq. 12.3), whereas the product obtained from *cis, trans*-2,4-hexadiene and a dienophile is *trans*-3,6-dimethylcyclohexene. Both of these reactions proceed by *syn*-1,4-addition to the diene, with the stereochemical relationship between the two methyl groups on the diene being maintained. Similarly, the reaction between 1,3-butadiene and dimethyl maleate gives solely the *cis*-diester **1** (Eq. 12.4). This example illustrates that the Diels-Alder reaction proceeds by exclusive *syn*-1,2-addition to the double bond of the dienophile; no isomerization to the more stable *trans*-diester is observed. The results of the reactions depicted in Equations 12.3 and 12.4 are in accord with the hypothesis that the Diels-Alder reaction is concerted, with both new carbon-carbon bonds forming simultaneously, although not necessarily at the same rate.

$$(12.3)$$

trans, trans-2,4-Hexadiene   An alkene   A *cis*-3,6-dimethylcyclohexene

$$(12.4)$$

1

s-cis-1,3-Butadiene   Dimethyl maleate

Dimethyl 4-cyclohexene-*cis*-1,2-dicarboxylate

There is a second type of stereoselectivity that is characteristic of the Diels-Alder reaction. The addition of a dienophile such as maleic anhydride to a *cyclic* diene like 1,3-cyclopentadiene could provide two products, the *endo*-adduct **2** and the *exo*-adduct **3** (Eq. 12.5). However, only the *endo*-cycloadduct **2**, in which the two boldfaced hydrogens are *syn* to the one-carbon bridge, is observed experimentally, and its preferential formation follows what is now commonly termed the **Alder rule**. The basis for this result is believed to be stabilization of the **transition state 4** by **secondary orbital interactions** that occur through space between the *p*-orbitals on the internal carbons of the diene and the carbonyl carbon atoms of the dienophile, as shown by the dashed lines in **4**. Analogous stabilization is not possible in transition state **5**. Structure **4** is thus characterized as the one being stabilized by *maximum orbital overlap*. It should be noted that not all Diels-Alder reactions are as stereoselective as the one between 1,3-cyclopentadiene and maleic anhydride; mixtures of *endo*- and *exo*-products are sometimes obtained.

$$\text{(12.5)}$$

Maleic anhydride	1,3-Cyclo-pentadiene	Bicyclo[2.2.1]hept-5-ene-*endo*-2,3-dicarboxylic anhydride	Bicyclo[2.2.1]hept-5-ene-*exo*-2,3-dicarboxylic anhydride
		2	**3**

4   **5**

12.3 APPLICATIONS OF DIELS-ALDER REACTIONS

Two experimental procedures using maleic anhydride, a highly reactive dienophile, nicely illustrate various features of the Diels-Alder reaction. In Part **A**, the Diels-Alder reaction of 1,3-butadiene with maleic anhydride to give 4-cyclohexene-*cis*-1,2-dicarboxylic anhydride (**6**) is performed (Eq. 12.6). Because 1,3-butadiene is a gas at room temperature (bp −4.5 °C, 760 torr), Diels-Alder reactions involving this diene are normally performed in sealed steel pressure vessels into which the diene is introduced under pressure. However, 1,3-butadiene can be conveniently generated *in situ* by the thermal decomposition of 3-sulfolene that occurs with the extrusion of sulfur dioxide (Eq. 12.7), and this strategy is used in this experiment. The decomposition can be effected by heating a solution of 3-sulfolene under reflux in the presence of the diene. This process can also be promoted through microwave heating (Sec. 2.9) if the appropriate apparatus is available. It is instructive to compare the reaction times required and the results obtained using conventional versus microwave heating for promoting the reaction.

$$\text{(12.6)}$$

1,3-Butadiene	Maleic anhydride	4-Cyclohexene-*cis*-1,2-dicarboxylic anhydride
		6

$$\text{(12.7)}$$

3-Sulfolene   1,3-Butadiene

In Part **B**, the cycloaddition of 1,3-cyclopentadiene with maleic anhydride is examined (Eq. 12.5). This experiment provides an opportunity to investigate the stereoselectivity of the Diels-Alder reaction. Monomeric 1,3-cyclopentadiene cannot be purchased because it readily dimerizes at room temperature by a Diels-Alder reaction to give dicyclopentadiene (Eq. 12.8), which is commercially available. Fortunately, the equilibrium between the monomer and the dimer can be established at the boiling point of the dimer (170 °C, 760 torr) by a process commonly called **cracking,** and the lower-boiling 1,3-cyclopentadiene may then be isolated by fractional distillation. The diene *must* be kept cold to prevent its redimerization via a Diels-Alder reaction prior to use in the desired Diels-Alder reaction.

$$(12.8)$$

1,3-Cyclopentadiene   Dicyclopentadiene

The Diels-Alder adducts **2** and **6** retain the anhydride function that was originally present in the dienophile, maleic anhydride. As a rule, this functionality is unstable in the presence of water even at pH 7 and hydrolyzes to a dicarboxylic acid according to the general sequence outlined in Scheme 12.1. This process is initiated by nucleophilic attack of water on the carbonyl function, followed by cleavage of a C–O bond of the anhydride. Because anhydrides are reactive toward water, it is important that the apparatus and reagents you use are dry to maximize the yield of cycloadduct. You may be instructed to prepare the dicarboxylic acids **7** and **8** by intentional hydrolysis of the corresponding anhydrides. Protocols for doing so are included in Part **C** of the experimental procedures.

Scheme 12.1

4-Cyclohexene-*cis*-
1,2-dicarboxylic acid
7

Bicyclo[2.2.1]hept-5-ene-
endo-2,3-dicarboxylic acid
8

EXPERIMENTAL PROCEDURE

Diels-Alder Reaction

Purpose To demonstrate the formation of six-membered rings by a cycloaddition reaction.

SAFETY ALERT

1. **Wear safety glasses or goggles and suitable protective gloves while performing the experiments.**

2. **Be certain that all joints in the apparatus are tight and well lubricated before heating the reaction mixture so that sulfur dioxide, a toxic and foul-smelling gas, does not escape into the laboratory.**

3. **The organic solvents are highly flammable, so use *flameless* heating.**

A ▪ *Reaction of 1,3-Butadiene and Maleic Anhydride*

MINISCALE PROCEDURE

Preparation Refer to the online resources to answer Pre-Lab Exercises, access videos, and read the MSDSs for the chemicals used or produced in this procedure. Review Sections 2.9, 2.11, 2.17, 2.22, and 2.23.

Apparatus A 25-mL round-bottom flask, gas trap, ice-water bath, apparatus for heating under reflux, magnetic stirring, vacuum filtration, and *flameless* heating.

Setting Up Place 2.5 g of 3-sulfolene, 1.5 g of finely pulverized maleic anhydride, and 1 mL of dry xylenes in the flask and equip it with a stirbar. Assemble the apparatus for heating under reflux. Attach the condenser to the gas trap specified by your instructor.

Cycloaddition and Isolation Gently warm the flask with stirring to effect *complete* dissolution of all solids; failure to do so will lower the yield of the reaction *significantly.* Then heat the mixture under gentle reflux for about 0.5 h.★ Allow the reaction mixture to cool below the boiling point and add 10 mL of xylenes. Reheat the stirred mixture to dissolve all of the solids, and then pour the hot solution into an Erlenmeyer flask. If any product crystallizes during this operation, heat the mixture gently until all of the solid redissolves.★ Carefully add petroleum ether (bp 60–80 °C, 760 torr) to the hot solution until cloudiness develops, and then set the solution aside to cool to room temperature to complete the crystallization.★ Collect the product by vacuum filtration, transfer it to a piece of filter or weighing paper, and air-dry it thoroughly.

Analysis Weigh your product to determine the yield and measure its melting point. Test the product for unsaturation using the bromine in dichloromethane and Baeyer tests (Secs. 4.7A1 and 4.7A2, respectively). Obtain IR and ^{1}H NMR spectra of your starting materials and product and compare them with those of authentic samples (Figs. 12.1–12.6).

MICROSCALE PROCEDURE

Preparation Refer to the online resources to answer Pre-Lab Exercises, access videos, and read the MSDSs for the chemicals used or produced in this procedure. Review Sections 2.9, 2.11, 2.17, 2.22, and 2.23.

Apparatus A 3-mL conical vial, gas trap, ice-water bath, apparatus for heating under reflux, magnetic stirring, Craig tube filtration, and *flameless* heating.

Setting Up Equip the conical vial with a spinvane and add 250 mg of 3-sulfolene, 150 mg of finely pulverized maleic anhydride, and 0.5 mL of dry xylenes. Assemble the apparatus for heating under reflux using the gas trap specified by your instructor.

Cycloaddition and Isolation Gently warm the vial with stirring to effect *complete* dissolution of all solids; failure to do so will lower the yield of the reaction *significantly.* Then heat the mixture under gentle reflux for about 0.5 h. Allow the reaction mixture to cool below the boiling point and add 1 mL of xylenes.★ Reheat the mixture to effect dissolution of all solids and then rapidly transfer the contents of the vial to a Craig tube. If any product crystallizes during this operation, heat the mixture gently until all of the solid redissolves. Carefully add petroleum ether (bp 60–80 °C, 760 torr) to the hot solution until cloudiness develops, and then set the solution aside to cool to room temperature.★ Complete the crystallization by cooling the mixture in an ice-water bath, equip the Craig tube for filtration, and isolate the product by centrifugation.

Analysis Weigh your product to determine the yield and measure its melting point. Test the product for unsaturation using the bromine in dichloromethane and Baeyer tests (Secs. 4.7A1 and 4.7A2, respectively). Obtain IR and ^{1}H NMR spectra of your starting materials and product and compare them with those of authentic samples (Figs. 12.1–12.6).

MINISCALE PROCEDURE FOR MICROWAVE OPTION

Preparation Refer to the online resources to answer Pre-Lab Exercises, access videos, and read the MSDSs for the chemicals used or produced in this procedure. Review Sections 2.9, 2.11, 2.17, 2.22, and 2.23.

Apparatus A 25-mL round-bottom flask, stirbar, condenser, gas trap, ice-water bath, apparatus for microwave heating with magnetic stirring and vacuum filtration.

Setting Up Place 2.5 g of 3-sulfolene, 1.5 g of pulverized maleic anhydride, and 1 mL of dry xylenes in the flask and equip it with a stirbar. Place the flask in the cavity of the microwave apparatus on top of the vessel stand, secure the microwave door, and attach the condenser and gas trap to the flask.

Cycloaddition and Isolation Program the microwave apparatus for heating open reaction vessels with stirring according to the directions provided by your instructor. Generally, the reaction temperature should be set at 150 °C and the power set at 300 W with a 1-min ramp time and a 2-min hold time; the pressure limit should be set at 250 psi. Allow the mixture to cool to room temperature, remove the flask from the microwave apparatus, and add 10 mL of xylenes. Return the apparatus to the microwave and, using the same directions as before, heat the solution to dissolve all solids. Transfer the hot solution to an Erlenmeyer flask. If any product crystallizes during this operation, heat the mixture gently until all of the solid redissolves.★ Carefully add petroleum ether (bp 60–80 °C, 760 torr) to the hot solution until cloudiness develops, and then set the solution aside to cool to room temperature to complete the crystallization. Collect the product by vacuum filtration, transfer it to a piece of filter or weighing paper, and air-dry it thoroughly.

Analysis Weigh your product to determine the yield and measure its melting point. Test the product for unsaturation using the bromine in dichloromethane and Baeyer tests (Secs. 4.7A1 and 4.7A2, respectively). Obtain IR and ¹H NMR spectra of your starting materials and product and compare them with those of authentic samples (Figs. 12.1–12.6).

WRAPPING IT UP

Pour the *mixture of xylenes and petroleum ether* recovered from the reaction into a container for nonhalogenated organic solvents. Pour the *dichloromethane solution* from the bromine test for unsaturation in a container for halogenated organic liquids, and put the *manganese dioxide* from the Baeyer test for unsaturation in a container for heavy metals.

B ▪ *Reaction of 1,3-Cyclopentadiene and Maleic Anhydride*

SAFETY ALERT

1. **Wear safety glasses or goggles and suitable protective gloves while performing the experiments.**

2. **1,3-Cyclopentadiene is a mildly toxic, foul-smelling, and volatile substance. Prepare and use it in a hood if possible; avoid inhaling its vapors. Keep the diene cold at all times to minimize vaporization and dimerization.**

3. **Be certain that all joints in the apparatus are tight before heating dicyclopentadiene to produce the monomer.**

4. **The 1,3-cyclopentadiene and organic solvents used are highly flammable, so use *flameless* heating.**

MINISCALE PROCEDURE

Preparation Refer to the online resources to answer Pre-Lab Exercises, access videos, and read the MSDSs for the chemicals used or produced in this procedure. Review Sections 2.9, 2.10, 2.11, 2.14, and 2.17.

Apparatus A 10- and a 25-mL round-bottom flask, 25-mL Erlenmeyer flask, drying tube, ice-water bath, apparatus for fractional distillation, magnetic stirring, vacuum filtration, and *flameless* heating.

Setting Up Place a stirbar and 7 mL of dicyclopentadiene in a 25-mL round-bottom flask, and attach the flask to a fractional distillation apparatus that is equipped with an *unpacked* fractionating column and the 10-mL round-bottom flask as the receiver. Cool the receiver in an ice-water bath and fit the vacuum adapter with a drying tube.

Preparing Reactants Heat the dimer until the solution is briskly refluxing (*Caution*: occasional foaming) and the monomer begins to distill in the range 40–42 °C. Distill the monomer as *rapidly* as possible, but do *not* permit the head temperature to exceed 43–45 °C. About 2.5 mL of 1,3-cyclopentadiene should be obtained from the distillation, which will require about 0.5 h. If the distillate is cloudy because of condensation of moisture in the cold receiver, add about 0.5 g of *anhydrous* calcium chloride.

While the distillation is in progress, dissolve 1.5 g of maleic anhydride in 5 mL of ethyl acetate contained in a 25-mL Erlenmeyer flask. Gently heat the mixture to hasten dissolution and then cool the solution thoroughly in an ice-water bath.

Cycloaddition and Isolation Add 1.5 mL of dry 1,3-cyclopentadiene in one portion to the cooled solution of maleic anhydride, and swirl the resulting solution gently until the exothermic reaction subsides. Add 5 mL of petroleum ether (bp 60–80 °C, 760 torr) to cause precipitation of the cycloadduct as a white solid.★ Heat the mixture until the solid has redissolved.

To prevent the desired product from oiling out of solution rather than crystallizing, the reaction mixture *must* cool slowly. To accomplish this, place the Erlenmeyer flask in a water bath warmed to about 60 °C, and let the bath and flask cool to room temperature.★ Seeding the solution to effect crystallization may be required and can be performed according to the procedure described in Section 3.2. Once the solution has cooled to room temperature and no more crystallization is apparent, cool the flask in an ice-water bath for a few minutes to complete the process. Then isolate the product by vacuum filtration, and air-dry it.

Analysis Weigh your product to determine the yield and measure its melting point. Test the product for unsaturation using the bromine in dichloromethane and Baeyer tests (Secs. 4.7A1 and 4.7A2, respectively). Obtain IR and ¹H NMR spectra of your starting materials and product and compare them with those of authentic samples (Figs. 12.1, 12.2, and 12.7–12.10).

WRAPPING IT UP

Pour the *pot residue of dicyclopentadiene*, any unused *1,3-cyclopentadiene*, and the *mother liquors* from the filtration into a container for recovered dicyclopentadiene or a container for nonhalogenated organic solvents. Allow any volatiles on the *calcium chloride* that was used to dry the 1,3-cyclopentadiene to evaporate in the hood, and then place the *residual solid* in a container for nonhazardous solids. Pour the *dichloromethane solution* from the bromine test for unsaturation in a container for halogenated organic liquids, and put the *manganese dioxide* from the Baeyer test for unsaturation in a container for heavy metals.

MICROSCALE PROCEDURE

Preparation Refer to the online resources to answer Pre-Lab Exercises, access videos, and read the MSDSs for the chemicals used or produced in this procedure. Review Sections 2.9, 2.10, and 2.17.

Apparatus Water bath, ice-water bath, apparatus for Craig tube filtration.

Setting Up Place 100 mg of maleic anhydride and 0.4 mL of ethyl acetate in the Craig tube and mix the contents of the tube carefully to effect dissolution; warming may be helpful. Then add 0.4 mL of petroleum ether (bp 60–90 °C, 760 torr), and mix the contents of the tube by gentle swirling.

Cycloaddition and Isolation In one portion, add 0.1 mL of dry 1,3-cyclopentadiene, which may be obtained from your instructor, to the Craig tube and mix the contents by swirling the tube until a homogeneous solution results. The exothermicity of the reaction may keep the desired product in solution, but if not, heat the tube gently to dissolve any solids that form.

To prevent the desired product from oiling out of solution rather than crystallizing, the reaction mixture *must* cool slowly. To accomplish this, place the Craig tube in a water bath warmed to about 60 °C, and let the bath and tube cool to room temperature.★ Seeding the solution to effect crystallization may be required and can be performed according to the procedure described in Section 3.2. Once the solution has cooled to room temperature and no more crystallization is apparent, cool the Craig tube in an ice-water bath for a few minutes to complete the process. Then equip the Craig tube for filtration and isolate the product by centrifugation.

Analysis Weigh your product to determine the yield and measure its melting point. The reported melting point is 164–165 °C. Test the product for unsaturation using the bromine in dichloromethane and Baeyer tests (Secs. 4.7A1 and 4.7A2, respectively). Obtain IR and ^{1}H NMR spectra of your starting materials and product, and compare them with those of authentic samples (Figs. 12.1, 12.2, and 12.7–12.10).

WRAPPING IT UP

Pour the *mother liquors* from the filtration into a container for nonhalogenated organic solvents. Pour the *dichloromethane solution* from the bromine test for unsaturation in a container for halogenated organic liquids, and put the *manganese dioxide* from the Baeyer test for unsaturation in a container for heavy metals.

C ▪ *Hydrolysis of Anhydrides*

1. 4-Cyclohexene-*cis*-1,2-dicarboxylic Acid

MINISCALE PROCEDURE

Preparation Refer to the online resources to answer Pre-Lab Exercises, access videos, and read the MSDSs for the chemicals used or produced in this procedure. Review Sections 2.9, 2.10, and 2.17.

Apparatus A 25-mL Erlenmeyer flask, ice-water bath, apparatus for vacuum filtration and *flameless* heating.

Setting Up Place 1.0 g of anhydride **6** and 5 mL of distilled water in the Erlenmeyer flask and add a stirbar.

Hydrolysis and Isolation Heat the mixture to boiling and continue heating until all the oil that forms initially dissolves. Allow the solution to cool to room temperature and then induce crystallization by scratching the flask at the air-liquid interface.★ After crystallization begins, cool the flask in an ice-water bath to complete the process, collect the solid by vacuum filtration, and air-dry it. Recrystallize the diacid from water.

Analysis Weigh the product to determine the yield, and measure its melting point. Test the product for unsaturation using the bromine in dichloromethane and Baeyer tests (Secs. 4.7A1 and 4.7A2, respectively) to determine whether hydrolysis has affected the carbon-carbon π-bond. Also test a saturated aqueous solution of the diacid with pHydrion paper and record the result. Obtain IR and ^{1}H NMR spectra of your product and compare them with those of an authentic sample (Figs. 12.11 and 12.12).

Optional Procedure Hydrolysis of the anhydride **6** to the diacid **8** without isolating **6** is possible as follows. Modify the procedure in Part **A** such that after the initial 0.5 h period of reflux the reaction mixture is allowed to cool below the boiling point. Then add about 5 mL of water to the flask through the top of the condenser and heat the resulting mixture under gentle reflux with stirring for at least 0.5 h. Allow the mixture to cool to room temperature.★ If no crystals form, acidify the solution with about 0.3 mL (6 drops) of *concentrated* sulfuric acid, and cool the resulting mixture in an ice-water bath. Isolate and analyze the diacid according to the directions given above.

MICROSCALE PROCEDURE

Preparation Refer to the online resources to answer Pre-Lab Exercises, access videos, and read the MSDSs for the chemicals used or produced in this procedure. Review Sections 2.9, 2.10, and 2.17.

Apparatus A 10-mm × 75-mm test tube, ice-water bath, apparatus for Craig tube filtration and *flameless* heating.

Setting Up, Hydrolysis, and Isolation Place 100 mg of the anhydride **6** in the test tube and add 0.5 mL of water and a boiling stone. Heat the mixture to boiling and continue heating until all the oil that forms initially dissolves. Using a preheated Pasteur pipet, transfer the hot solution to a Craig tube. Allow the solution to cool to room temperature. If necessary, induce crystallization by gently scratching the tube at the air-liquid interface.★ Complete crystallization by cooling the mixture in an ice-water bath.

Equip the Craig tube for filtration, isolate the crystalline product by centrifugation, and transfer it to a piece of filter or weighing paper for air-drying. Recrystallize the diacid from water.

Analysis Weigh the product to determine the yield and measure its melting point. Test the product for unsaturation using the bromine in dichloromethane and Baeyer tests (Secs. 4.7A1 and 4.7A2, respectively) to determine whether hydrolysis has affected the carbon-carbon π-bond. Also test a saturated aqueous solution of the diacid with pHydrion paper and record the result. Obtain IR and ^{1}H NMR spectra of your product and compare them with those of an authentic sample (Figs. 12.11 and 12.12).

2. Bicyclo[2.2.1]hept-5-ene-*endo*-2,3-dicarboxylic Acid

MINISCALE PROCEDURE

Preparation Refer to the online resources to answer Pre-Lab Exercises, access videos, and read the MSDSs for the chemicals used or produced in this procedure. Review Sections 2.9, 2.10, and 2.17.

Apparatus A 25-mL Erlenmeyer flask, ice-water bath, apparatus for vacuum filtration and *flameless* heating.

Setting Up Place 1.0 g of anhydride **2** and 5 mL of distilled water in the Erlenmeyer flask and add a stirbar.

Hydrolysis and Isolation Follow the Miniscale Procedure of Part **1** to effect hydrolysis of **2** and isolation of the diacid **8**.

Analysis Perform the same analysis as that described in the Miniscale Procedure of Part **1**. The reported melting point of bicyclo[2.2.1]hept-5-en-*endo*-2,3-dicarboxylic acid is 180–182 °C. Obtain IR and ^{1}H NMR spectra of your product, and compare them with those of an authentic sample (Figs. 12.13 and 12.14).

MICROSCALE PROCEDURE

Preparation Refer to the online resources to answer Pre-Lab Exercises, access videos, and read the MSDSs for the chemicals used or produced in this procedure. Review Sections 2.9, 2.10, and 2.17.

Apparatus A 10-mm × 75-mm test tube, ice-water bath, apparatus for Craig tube filtration and *flameless* heating.

Setting Up, Hydrolysis, and Isolation Follow the Microscale Procedure of Part **1** to effect hydrolysis of **2** and isolation of the diacid **8**.

Analysis Perform the same analysis as that described in the Microscale Procedure of Part **1**. The reported melting point of bicyclo[2.2.1]hept-2-ene-*endo*-5,6-dicarboxylic acid is 180–182 °C. Obtain IR and ^{1}H NMR spectra of your product, and compare them with those of an authentic sample (Figs. 12.13 and 12.14).

Discovery Experiment   *Hydrolysis of Anhydrides*

Investigate the possible catalysis of the hydrolysis of anhydrides by designing a protocol for determining whether the hydrolysis of an anhydride may be accelerated by performing the reaction at pH > 7 and pH < 7 as well as at pH 7. The results test whether the hydrolysis is subject to base and acid catalysis. Consult with your instructor before undertaking any experimental procedures.

WRAPPING IT UP

Using sodium carbonate, neutralize the *aqueous mother liquors* from hydrolysis of the anhydride, dilute the solution, and flush it down the drain. Pour the *dichloromethane solution* from the bromine test for unsaturation in a container for halogenated organic liquids, and put the *manganese dioxide* from the Baeyer test for unsaturation in a container for heavy metals.

EXERCISES

General Questions

1. Explain why a *cyclic* 1,3-diene like 1,3-cyclopentadiene dimerizes much more readily than does an *acyclic* one like *trans*-1,3-pentadiene.

2. Write structures for the products expected in the following possible Diels-Alder reactions. If no reaction is anticipated, write "N.R."

a.   [structure] + [maleic anhydride structure] ⟶   b.   [2,3-dimethyl-1,3-butadiene structure] + [CH₂=CH−CHO structure] ⟶

c.   [diene structure] + $\begin{array}{c} CH_2 \\ \| \\ CH_2 \end{array}$ ⟶   d.   [cyclopentadiene structure] + $CH_3O_2C-C{\equiv}C-CO_2CH_3$ ⟶

3. When producing Diels-Alder adducts using maleic anhydride as the dienophile, why is it important that the reagents and apparatus be dry?

4. Write the structure of the product from the Diels-Alder reaction between dimethyl maleate (Sec. 7.3) and 1,3-cyclopentadiene.

5. Provide an explanation for why the stereoisomers shown below would not be produced in the Diels-Alder reaction of Exercise 4.

a.   [bicyclic structure with CO_2CH_3, CO_2CH_3, H, H]   b.   [bicyclic structure with CO_2CH_3, H, H, CO_2CH_3]

6. Consider the structure of bicyclo[2.2.1]hept-4-ene-*endo*-2,3-dicarboxylic anhydride (**2**), which is produced by the Diels-Alder reaction of maleic anhydride and 1,3-cyclopentadiene.

   a. Draw the structure of **2** and label the bridgehead carbons with an asterisk.

   b. Highlight the two new carbon-carbon bonds that were formed in the Diels-Alder reaction.

   c. What structural features of **2** lead to its classification as a bridged bicyclic compound?

   d. The nomenclature terms *endo* and *exo* are used to indicate the relative stereochemistry of a substituent on a bridged bicyclic structure. Circle the larger of the two unsubstituted carbon bridges in **2** and define the term *endo* as it relates to the orientation of substituents relative to the larger carbon bridge of **2**.

   e. Circle the carbon atoms in **2** that are chiral and determine whether **2** is optically active.

7. The primary orbital interactions in the transition state of normal Diels-Alder reactions occur between the highest occupied molecular orbital (the HOMO)

of the diene and the lowest unoccupied molecular orbital (the LUMO) of the dienophile; however, in so-called inverse demand Diels-Alder reactions, the primary orbital interactions occur between the LUMO of the diene and the HOMO of the dienophile.

a. The four π-molecular orbitals for 1,3-cyclopentadiene are shown below. Show how each of the orbitals is occupied with electrons and label the HOMO and the LUMO.

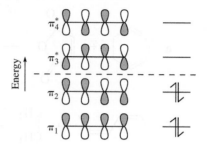

b. Write the two π-molecular orbitals for ethylene, CH_2=CH_2, and label the HOMO and the LUMO.

c. Is the HOMO or the LUMO an antibonding orbital?

d. Show the two possible HOMO and LUMO primary orbital interactions between ethylene and 1,3-butadiene.

e. Which of the HOMO-LUMO interactions in Part **d** are the orbital interactions between the carbon atoms in the bonds that are being formed in-phase (i.e., constructive orbital overlap)?

8. Write the structures of the diene and the dienophile that would form each of the products shown via a Diels-Alder reaction.

a.   **b.**

9. Consider the spectral data for maleic anhydride (Figs. 12.1 and 12.2).

a. In the functional group region of the IR spectrum, specify the absorptions associated with the carbon-carbon and carbon-oxygen double bonds.

b. The ^{1}H NMR spectrum shows the chemical shift of the vinylic protons to be δ 7.10. Explain why this resonance is downfield of the normal range observed for vinyl hydrogen atoms.

c. For the ^{13}C NMR data, assign the various resonances to the carbon nuclei responsible for them.

Refer to the online resources and use the spectra viewer and Tables 8.1–8.8 as needed to answer the blue-numbered questions on spectroscopy.

Questions for Part A

10. In this experiment, heating 3-sulfolene produces 1,3-butadiene and a gas.

a. What gas is evolved and is it toxic?

b. How did you avoid releasing this gas into the laboratory?

    c. Using curved arrows to symbolize the flow of electrons, give a mechanism for this reaction and indicate whether it is a concerted reaction or not.

    d. Is this reaction reversible under the conditions you used? If not, can you describe a set of conditions under which the reaction might be reversible?

11. Which of the two reactants in this procedure was the dienophile and which was the diene?

12. Draw the structures of the three isomeric xylenes and give two reasons why a mixture of these is a good solvent for this reaction.

13. Why should 3-sulfolene and maleic anhydride be *completely* dissolved in xylenes before heating the mixture to effect reaction?

14. Write the structure, including stereochemistry, of the expected addition product of bromine to the Diels-Alder adduct obtained by this procedure.

15. Consider the spectral data for 3-sulfolene (Figs. 12.3 and 12.4).

    a. In the IR spectrum, specify the absorption associated with the carbon-carbon double bond.

    b. In the ^{1}H NMR spectrum, assign the various resonances to the hydrogen nuclei responsible for them.

    c. For the ^{13}C NMR data, assign the various resonances to the carbon nuclei responsible for them.

16. Consider the spectral data for 4-cyclohexene-*cis*-1,2-dicarboxylic anhydride (Figs. 12.5 and 12.6).

    a. In the functional group region of the IR spectrum, specify the absorptions associated with the carbon-carbon double bond and the carbonyl groups of the anhydride function.

    b. In the ^{1}H NMR spectrum, assign the various resonances to the hydrogen nuclei responsible for them.

    c. For the ^{13}C NMR data, assign the various resonances to the carbon nuclei responsible for them.

17. a. Given that sulfones typically show two strong absorptions near 1300 and 1150 cm^{-1}, discuss the differences observed in the IR and NMR spectra of 3-sulfolene and of 4-cyclohexene-*cis*-1,2-dicarboxylic anhydride that are consistent with the loss of sulfur dioxide in this experiment.

    b. Discuss the differences observed in the IR and NMR spectra of maleic anhydride and of 4-cyclohexene-*cis*-1,2-dicarboxylic anhydride that are consistent with the involvement of the carbon-carbon π-bond of maleic anhydride in this experiment.

Questions for Part B

18. Why should the head temperature be maintained below about 45 °C when dicyclopentadiene is being cracked?

19. Why is it good technique to induce crystallization of the diacid derived from the Diels-Alder adduct *before* cooling the solution to 0 °C?

20. Why does 1,3-cyclopentadiene react more rapidly with maleic anhydride than with another molecule of itself?

Refer to the online resources and use the spectra viewer and Tables 8.1–8.8 as needed to answer the blue-numbered questions on spectroscopy.

21. The "cracking" of dicyclopentadiene to two moles of 1,3-cyclopentadiene (Eq. 12.8) is an example of a retro-Diels-Alder reaction. Predict the products to be anticipated from an analogous reaction with the compounds **a–d.**

    a.   **b.**   **c.**   **d.**

22. Write an equation that shows the equilibrium process involving dimerization of 1,3-cyclopentadiene to give dicyclopentadiene and cracking of dicyclopentadiene to give 1,3-cyclopentadiene. Using curved arrows to symbolize the flow of electrons, give a mechanism for each reaction.

23. Describe the process of "seeding" a solution and explain how it can facilitate crystallization of a solid.

24. What experimental evidence do you have that the Diels-Alder reaction of 1,3-cyclopentadiene with maleic anhydride is a stereoselective reaction?

25. Explain why maleic anhydride is a much better dienophile than ethylene, $CH_2{=}CH_2$.

26. Consider the spectral data for 1,3-cyclopentadiene (Figs. 12.7 and 12.8).

    a. In the functional group region of the IR spectrum, specify the absorption associated with the carbon-carbon π-bond.

    b. In the ^1H NMR spectrum, assign the various resonances to the hydrogen nuclei responsible for them. Explain why the resonance associated with the two C5 hydrogen atoms is not a triplet.

    c. For the ^{13}C NMR data, assign the various resonances to the carbon nuclei responsible for them.

Refer to the online resources and use the spectra viewer and Tables 8.1–8.8 as needed to answer the blue-numbered questions on spectroscopy.

27. Consider the spectral data for bicyclo[2.2.1]hept-5-ene-*endo*-2,3-dicarboxylic anhydride (Figs. 12.9 and 12.10).

    a. In the functional group region of the IR spectrum, specify the absorption associated with the carbon-carbon and carbon-oxygen double bonds.

    b. In the ^1H NMR spectrum, assign the various resonances to the hydrogen nuclei responsible for them.

    c. For the ^{13}C NMR data, assign the various resonances to the carbon nuclei responsible for them.

28. a. Discuss the differences observed in the IR and NMR spectra of 1,3-cyclopentadiene and of bicyclo[2.2.1]hept-5-en-*endo*-2,3-dicarboxylic anhydride that are consistent with the loss of the 1,3-diene function in this experiment.

    b. Discuss the differences observed in the IR and NMR spectra of maleic anhydride and of bicyclo[2.2.1]hept-2-ene-*endo*-5,6-dicarboxylic anhydride that are consistent with the involvement of the carbon-carbon π-bond of maleic anhydride in this experiment.

Questions for Part C

Refer to the online resources and use the spectra viewer and Tables 8.1–8.8 as needed to answer the blue-numbered questions on spectroscopy.

29. In this experiment, you prepared a 1,4-dicarboxylic acid by hydrolyzing an anhydride with water. However, it is known that this hydrolysis is facilitated by the addition of a catalytic amount of acid. Using curved arrows to symbolize the flow of electrons, give a mechanism for the acid-catalyzed hydrolysis of 4-cyclohexene-*cis*-1,2-dicarboxylic anhydride (**6**) and explain why acid increases the rate of the reaction.

30. Suggest a way to prepare 4-cyclohexene-*cis*-1,2-dicarboxylic anhydride (**6**) from the corresponding dicarboxylic acid **7**.

31. Amides and esters, which are also derivatives of carboxylic acids, are stable in water under neutral conditions, whereas anhydrides react readily with water by nucleophilic acyl substitution. Explain the basis for the difference in reactivity of anhydrides, amides, and esters toward water at neutral pH.

32. Consider the spectral data for 4-cyclohexene-*cis*-1,2-dicarboxylic acid (Figs. 12.11 and 12.12).

    a. In the functional group region of the IR spectrum, specify the absorptions associated with the carbon-carbon double bond and the carboxylic acid functions.

    b. In the ^{1}H NMR spectrum, assign the various resonances to the hydrogen nuclei responsible for them.

    c. For the ^{13}C NMR data, assign the various resonances to the carbon nuclei responsible for them. Does the number of resonances observed in the ^{13}C NMR spectrum allow you to prove that the *cis*- rather than the *trans*-dicarboxylic acid was produced by the hydrolysis of anhydride **6**? Explain.

33. Discuss the differences observed in the IR and NMR spectra of 4-cyclohexene-*cis*-1,2-dicarboxylic anhydride (Figs. 12.5 and 12.6) and of 4-cyclohexene-*cis*-1,2-dicarboxylic acid that *are* consistent with the conversion of an anhydride to a diacid function in this experiment.

34. Consider the spectral data for bicyclo[2.2.1]hept-5-ene-*endo*-2,3-dicarboxylic acid (Figs. 12.13 and 12.14).

    a. In the functional group region of the IR spectrum, specify the absorptions associated with the carbon-carbon double bond and the carboxylic acid functions.

    b. In the ^{1}H NMR spectrum, assign the various resonances to the hydrogen nuclei responsible for them.

    c. For the ^{13}C NMR data, assign the various resonances to the carbon nuclei responsible for them. Does the number of resonances observed in the ^{13}C NMR spectrum allow you to prove that the *cis*- rather than the *trans*-dicarboxylic acid was produced by the hydrolysis of anhydride **2**? Explain.

35. Discuss the differences observed in the IR and NMR spectra of bicyclo[2.2.1]hept-5-ene-*endo*-2,3-dicarboxylic anhydride and of bicyclo[2.2.1]hept-2-ene-*endo*-5,6-dicarboxylic acid that are consistent with the conversion of an anhydride to a diacid function in this experiment.

Starting Materials and Products

Figure 12.1
IR spectrum of maleic anhydride (IR card).

Figure 12.2
NMR data for maleic anhydride (DMSO-d₆).

(a) ¹H NMR spectrum (300 MHz).
(b) ¹³C NMR data: δ 137.0, 165.0.

Figure 12.3
IR spectrum of 3-sulfolene (IR card).

Figure 12.4
NMR data for 3-sulfolene
(CDCl₃).

(a) ¹H NMR spectrum (300 MHz).
(b) ¹³C NMR data: δ 55.7, 124.7.

Figure 12.5
IR spectrum of 4-cyclohexene-
cis-dicarboxylic anhydride
(IR card).

Figure 12.6
NMR data for 4-cyclohexene-
cis-dicarboxylic anhydride
(CDCl₃).

(a) ¹H NMR spectrum (300 MHz).
(b) ¹³C NMR data. Chemical shifts: δ 26.1, 39.3, 125.5, 174.7.

Figure 12.7

IR spectrum of 1,3-cyclopenta-diene (neat).

Figure 12.8

NMR data for 1,3-cyclopentadiene (CDCl₃).

(a) ¹H NMR spectrum (300 MHz).

(b) ¹³C NMR data. Chemical shifts: δ 42.2, 133.0, 134.4.

Figure 12.9

IR spectrum of bicyclo[2.2.1]hept-5-ene-endo-2,3-dicarboxylic anhydride (IR card).

Figure 12.10

NMR data for bicyclo[2.2.1]hept-5-ene-endo-2,3-dicarboxylic anhydride (CDCl₃).

(a) 1H NMR spectrum (300 MHz).
(b) 13C NMR data: δ 47.6, 53.5, 136.1, 171.9.

Figure 12.11

IR spectrum of 4-cyclohexene-cis-1,2-dicarboxylic acid (IR card).

Figure 12.12

NMR data for 4-cyclohexene-cis-1,2-dicarboxylic acid (DMSO-d₆).

(a) ¹H NMR spectrum (300 MHz).
(b) ¹³C NMR data: δ 25.9, 39.2, 125.2, 174.9.

Figure 12.13

IR spectrum of bicyclo[2.2.1]hept-5-ene-endo-2,3-dicarboxylic acid (IR card).

Figure 12.14

NMR data for bicyclo[2.2.1]hept-5-ene-endo-2,3-dicarboxylic acid (DMSO-d₆).

(a) ¹H NMR spectrum (300 MHz).
(b) ¹³C NMR data: δ 46.8, 48.7, 49.1, 135.6, 174.4.

Kinetic and Thermodynamic Control of a Reaction

Only one product is formed in many of the reactions you will study in organic chemistry. However, sometimes two (or more) different products may be formed from competing reactions. If two products are formed, the inquiring scientist wants to know which is formed faster and which is more stable. While the product that forms faster is frequently also the more stable, this is not always the case. Sometimes it is possible to favor the formation of one product or another by modifying the conditions under which the reaction is conducted. The two experimental parameters that are most commonly varied to favor formation of one product over another are reaction time and the temperature at which the reaction is conducted. By performing the experimental procedures in this chapter, you will have an unusual opportunity to discover different means of controlling the course of chemical reactions using some of the same techniques that are practiced in industry for the production of compounds of practical interest.

13.1 INTRODUCTION

Organic chemists frequently need to predict which of several possible competing reactions will predominate when a specified set of reactants is used. The problem is illustrated by considering the general situation where reagents **W** and **X** can combine to produce either **Y** (Eq. 13.1) or **Z** (Eq. 13.2). Predicting which product will be formed preferentially under a particular set of conditions requires developing an **energy diagram** for the two reactions. Strictly speaking, such a diagram should be based on the **free energy changes**, ΔG, that occur during the conversion of reactants to products. To generate the needed diagram would require knowledge of both the **enthalpic**, ΔH, and **entropic**, ΔS, changes involved. If the reactions being considered are similar, as we assume is the case for the processes of Equations 13.1 and 13.2, entropic changes are not too important, and using values for ΔH gives a close approximation to the changes occurring in ΔG. This is the basis on which our potential energy diagram will be constructed.

$$W + X \rightleftharpoons Y \tag{13.1}$$

$$W + X \rightleftharpoons Z \tag{13.2}$$

443

We will start our analysis by assuming that both of the reactions are *exothermic and* that **Y** is *thermodynamically less stable* than **Z.** These assumptions mean that (1) the energy levels of both products are below those of the starting materials and (2) the heat of reaction for formation of **Z** is greater than that for **Y,** as symbolized by ΔH_Z and ΔH_Y in Figure 13.1. The activation barriers for converting the reactants to the two transition states **Y‡** and **Z‡** for forming **Y** and **Z,** respectively, then correspond to $\Delta H^{\ddagger}{}_Y$ and $\Delta H^{\ddagger}{}_Z$. The latter barrier has been set *lower* on the diagram, so $\Delta H^{\ddagger}{}_Z < \Delta H^{\ddagger}{}_Y$. This ordering of the relative energies of the two transition states is arbitrary but is consistent with the *empirical* observation that when comparing two *similar* organic reactions, the *more exothermic process often has the lower enthalpy of activation,* $\Delta H^{\ddagger}$. In other words, the relative energies of the transition states reflect the relative energies of the products that are produced from those transition states. This generalization is useful because it allows us to complete the potential energy diagram and to predict that because less energy is required for reactants **W** and **X** to reach the transition state leading to **Z,** this product will be formed faster than **Y.**

For all practical purposes, most organic reactions are either irreversible or are performed under conditions such that equilibrium between products and starting materials is *not* attained. The ratio of products obtained from reactions run under these circumstances is said to be subject to **kinetic control.** This means that the relative amounts of the various products are determined by the relative rates of their formation rather than by their relative thermodynamic stabilities. The major product of such a reaction is often called the **kinetically controlled product** even though other compounds may be formed. In the example of Figure 13.1, **Z** would represent the kinetically controlled product; note that $\Delta H^{\ddagger}{}_Z < \Delta H^{\ddagger}{}_Y$.

Organic reactions that are performed under conditions in which *equilibrium* is established give ratios of products, and perhaps reactants too, that are subject to **thermodynamic control.** In such cases, the predominant product is called the **thermodynamically controlled product** and would be the more stable **Z** under the circumstances illustrated in Figure 13.1. Note that the ratio **Y:Z** would probably *not* be the same under both sets of reaction conditions because it is unlikely that

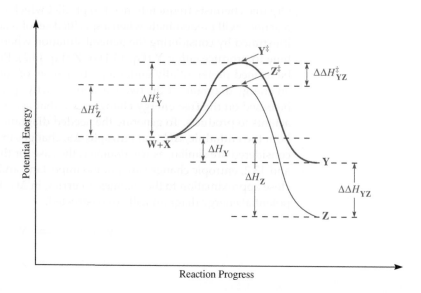

Figure 13.1
Typical reaction profile for competing reactions.

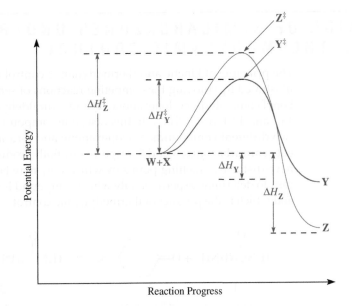

Figure 13.2

Reaction profile that predicts different products from kinetic and thermodynamic control of competing reactions.

the difference in enthalpies of activation, $\Delta\Delta H^{\ddagger}_{YZ}$, between the two transition states would be equal to the enthalpy difference, $\Delta\Delta H_{YZ}$, between the two ground states of the products.

The situation shown in Figure 13.1 is the one most frequently encountered in organic chemistry; namely, the major product of kinetic control and thermodynamic control is the same. There are exceptions to this, however, and the reaction profiles of Figure 13.2 depict just such a case. To generate these profiles, it is again assumed that the reactions of Equations 13.1 and 13.2 are exothermic and that **Z** is more stable than **Y**, but the activation barrier, $\Delta H^{\ddagger}_{Y}$, for forming **Y** is now *less* than that for forming **Z**, $\Delta H^{\ddagger}_{Z}$. This lower barrier means that **Y** would now be the product of kinetic control, although **Z** would remain the product of thermodynamic control.

Note that in developing Figures 13.1 and 13.2, conclusions about the preferred products of kinetic control and thermodynamic control do not depend on the exothermicity or endothermicity of the overall reactions. *All that matters are the relative energies of the transition states leading to the products* if *the reaction is under kinetic control, or the relative energies of the products themselves* if *the conditions are those of thermodynamic control.* Whether different major products are produced under reversible or irreversible reaction conditions is determined by measuring the ratios of products formed under the two different circumstances. If the products of a reaction can be separated, there is another way to ascertain whether the reaction was performed under conditions of kinetic or thermodynamic control. Each product is resubjected separately to the original reaction conditions. If each is recovered unchanged, the original reaction was conducted under conditions of kinetic control, but if each product gives a mixture of products identical to that of the original reaction, then the reaction was subject to thermodynamic control. In the procedures that follow, you will have the opportunity to determine whether reactions are being conducted under thermodynamic or kinetic control using several experimental techniques.

13.2 FORMATION OF SEMICARBAZONES UNDER KINETIC AND THERMODYNAMIC CONTROL

The principle of kinetic and thermodynamic control in organic chemistry may be illustrated by studying the competing reactions of semicarbazide (**1**) with two carbonyl compounds, cyclohexanone (**2**) and 2-furaldehyde (**4**), as shown in Equations 13.3 and 13.4, respectively. In this case, one compound, semicarbazide, reacts with two different compounds, cyclohexanone and 2-furaldehyde, to give two different **semicarbazones, 3 and 5,** respectively. Both products are crystalline solids that have distinctive melting points by which they may be easily identified. Thus, it is easy to determine experimentally which compound is the product of kinetic control and which is the product of thermodynamic control.

$$
\underset{\substack{\textbf{1}\\\text{Semicarbazide}}}{H_2NCNHNH_2} + \underset{\substack{\textbf{2}\\\text{Cyclohexanone}}}{O=} \; \rightleftharpoons \; \underset{\substack{\textbf{3}\\\text{Cyclohexanone semicarbazone}}}{H_2NCNHN=} + H_2O \qquad (13.3)
$$

$$
\underset{\textbf{1}}{H_2NCNHNH_2} + \underset{\substack{\textbf{4}\\\text{2-Furaldehyde}}}{O=} \; \rightleftharpoons \; \underset{\substack{\textbf{5}\\\text{2-Furaldehyde semicarbazone}}}{H_2NCNHN} + H_2O \qquad (13.4)
$$

The rates and equilibrium constants of reactions of carbonyl compounds with nucleophiles such as semicarbazide are affected significantly by variations in the pH of the reaction medium. This may be understood by examining the results of protonating the reactants.

First, consider the effect that protonation of the carbonyl function has on the rate of reaction of a carbonyl compound with nucleophiles. The addition of a proton to the oxygen atom of a carbonyl group produces **6** (Eq. 13.5). The carbonyl carbon atom in **6** is now more *electrophilic* than in the unprotonated compound because the partial positive charge on the carbon atom is increased, as shown by consideration of the two resonance structures for **6**. Because the rate-determining step for the formation of semicarbazones is the nucleophilic attack of semicarbazide (**1**) on the carbonyl carbon atom of the substrate, the enhanced electrophilicity of **6** *increases* the rate of formation of the semicarbazone.

$$
\underset{\substack{\text{Carbonyl}\\\text{compound}}}{R_2C=\ddot{O}} + H^+ \; \rightleftharpoons \; \underset{\textbf{6}}{\left[R_2C=\overset{+}{O}H \longleftrightarrow R_2\overset{+}{C}-\ddot{O}H \right]} \qquad (13.5)
$$

Now assess what happens if the nucleophile is protonated. Adding a proton to **1** produces its conjugate acid **7** (Eq. 13.6), which is *not* nucleophilic and does not add to a carbonyl group. Hence, if the medium in which the reaction is performed is sufficiently acidic, essentially all of **1** will be converted to **7**, thereby lowering the concentration of **1**, the nucleophile in the reaction. The formation of the semicarbazone would then become extremely slow or might not occur at all. Thus, both the rate and equilibrium constant for the reaction of Equation 13.7 are affected significantly by the pH at which the process is conducted.

$$H_2NCNHNH_2 + H^+ \rightleftharpoons H_2NCNHNH_3^+ \tag{13.6}$$

$$\text{1} \qquad\qquad\qquad \text{7}$$

$$H_2NCNHNH_2 \quad R_2C=\overset{+}{O}H \rightleftharpoons \left[H_2NCNHN-CR_2\right] \rightleftharpoons \left[H_2NCNHN-CR_2\right] \tag{13.7}$$

$$\text{1} \qquad \text{6}$$

$$H_3O^+ + H_2NCNHN=CR_2 \rightleftharpoons H_2NCNHN=CR_2 + H_2O$$

The preceding discussion makes clear that with respect to Equation 13.7, the concentration of the nucleophile **1** is reduced at low pH, whereas that of the activated electrophile **6** is reduced at high pH. It follows that there is an optimum pH at which the mathematical product of the concentrations of the nucleophilic reagent and conjugate acid of the carbonyl compound, [**1**] × [**6**], is maximized.

$$\text{—NHNH}_2 \qquad\qquad \text{HONH}_2$$

$$\text{8} \qquad\qquad\qquad \text{9}$$

For reactions of aldehydes and ketones with reagents such as semicarbazide (**1**), phenylhydrazine (**8**), and hydroxylamine (**9**), the desired pH is produced and maintained by **buffer solutions**. A buffer solution resists changes in its pH. In general, a buffer is comprised of a pair of substances in the solution, one that is a weak acid and neutralizes hydroxide ions and another that is a weak base and neutralizes protons. For example, the $H_2PO_4^-/HPO_4^{2-}$ **buffer system** is produced by the addition of dibasic potassium phosphate, K_2HPO_4, a weak base, to semicarbazide hydrochloride. The weakly acidic $H_2PO_4^-$ component of the buffer system is produced as shown in Equation 13.8. The function of $H_2PO_4^-$ and HPO_4^{2-} in neutralizing hydroxide and hydrogen ions is illustrated in Equations 13.9 and 13.10, respectively.

Each different buffer system functions to maintain the pH within a rather narrow range characteristic of the weak acid and weak base of that particular system. In the case of the phosphate system, the range is pH 6.1–6.2. By comparison, a carbonate/bicarbonate buffer system (CO_3^{2-}/HCO_3^-) maintains a pH of 7.1–7.2.

$$H_2NCNHNH_3^+ Cl^- + HPO_4^{2-} \rightleftharpoons H_2NCNHNH_2 + H_2PO_4^- + Cl^- \tag{13.8}$$

$$\text{1}$$

$$H_2PO_4^- + HO^- \rightleftharpoons HPO_4^{2-} + H_2O \tag{13.9}$$

$$HPO_4^{2-} + H_3O^+ \rightleftharpoons H_2PO_4^- + H_2O \tag{13.10}$$

The maximum rates of the reactions of most aldehydes and ketones with semicarbazide occur in the pH range of 4.5–5.0. For the purpose of making derivatives of carbonyl compounds (Sec. 25.7), semicarbazide is best used in an acetate buffer ($CH_3CO_2H/CH_3CO_2^-$) solution, which maintains a pH in the maximum rate range of 4.5–5.0. However, to demonstrate the principle of kinetic and thermodynamic control of reactions, buffers that maintain higher pHs, and thus produce lower rates, are more desirable. Parts A–C of the experimental procedure involve a phosphate buffer system, whereas the bicarbonate system is used in Part D. It is then possible to compare how the difference in rates in the two buffer systems affects the product ratio. Analysis of the products from the various parts of these experiments provides strong clues as to which of the semicarbazones is the product of kinetic control and which is the product of thermodynamic control.

In Section 13.1, it was noted that implicit in the theory of kinetic and thermodynamic control is the fact that the *kinetic product* (**Z** in Fig. 13.1 and **Y** in Fig. 13.2), which is produced more rapidly, is also reconverted to starting material more rapidly. The *thermodynamic product,* on the other hand, being more stable, is not so readily reconverted to starting material. Experimental Procedure E provides tests of the relative stabilities of the two semicarbazone products toward the reverse reaction. The results of these experiments should provide additional evidence as to which of the semicarbazones is the product of kinetic control and which is the product of thermodynamic control.

EXPERIMENTAL PROCEDURES

Kinetic and Thermodynamic Control of a Reaction

Discovery Experiment   **Purpose** To study the phenomenon of kinetic and thermodynamic control of competing reactions by determining which product is formed faster and which is more stable.

SAFETY ALERT

1. **Wear safety glasses or goggles and suitable protective gloves while performing the experiments.**

2. **Because the melting point of one of the compounds is near 200 °C, do *not* use a liquid-filled apparatus such as a Thiele tube *unless* the heating medium is silicone oil. A metal block melting-point apparatus is satisfactory for determining the melting points of the semicarbazones in this experiment.**

Preparation Refer to the online resources to answer Pre-Lab Exercises, access videos, and read the MSDSs for the chemicals used or produced in this procedure. Review Sections 2.9, 2.10, and 2.17.

You may be assigned only some parts of these procedures. If so, plan to share results with other students who perform the other parts. This will enable you to draw your own conclusions and report them accordingly.

Apparatus Two 13-mm × 100-mm test tubes, two 25-mL, seven 5-mL, and two 125-mL Erlenmeyer flasks, 1-mL graduated pipet, ice-water bath, apparatus for vacuum filtration and *flameless* heating.

A ■ *Preparation of Cyclohexanone Semicarbazone*

Dissolve 0.5 g of semicarbazide hydrochloride and 1.0 g of dibasic potassium phosphate (K_2HPO_4) in 6 mL of water contained in a 25-mL Erlenmeyer flask. Using a 1-mL graduated pipet, deliver 0.5 mL of cyclohexanone into a test tube containing 2.5 mL of 95% ethanol. Pour the ethanolic solution into the aqueous semicarbazide solution, and swirl or stir the mixture immediately. Allow 5–10 min for crystallization of the semicarbazone to reach completion. Collect the crystals by vacuum filtration and wash them with a little cold water. *Thoroughly* air-dry the crystals★ and determine their weight and melting point. Obtain IR and NMR spectra of your starting material and product. Compare your spectra with those of authentic samples (Figs. 13.3–13.6).

B ■ *Preparation of 2-Furaldehyde Semicarbazone*

For best results, the 2-furaldehyde should be freshly distilled immediately before use. Prepare the semicarbazone of 2-furaldehyde by following the procedure of Part A exactly, except use 0.4 mL of 2-furaldehyde instead of 0.5 mL of cyclohexanone.★ Recrystallization of the product is unnecessary. Obtain IR and NMR spectra of your starting material and product. Compare your spectra with those of authentic samples (Figs. 13.7–13.10).

C ■ *Reactions of Semicarbazide with Cyclohexanone and 2-Furaldehyde in Phosphate Buffer Solution*

Dissolve 3.0 g of semicarbazide hydrochloride and 6.0 g of dibasic potassium phosphate in 75 mL of water. Because this is an aqueous solution, it is referred to as *solution W.* Prepare a solution of 3.0 mL of cyclohexanone and 2.5 mL of 2-furaldehyde in 15 mL of 95% ethanol. Because this is an ethanolic solution, it is referred to as *solution E.*★

1. Cool a 25-mL portion of solution W and a 5-mL portion of solution E separately in an ice-water bath to 0–2 °C. Add solution E to solution W and swirl the mixture; crystals should form almost immediately. Place the mixture in an ice-water bath for 3–5 min; then collect the crystals by vacuum filtration and wash them on the filter with 2–3 mL of cold water. *Thoroughly* air-dry the crystals★ and determine their weight and melting point.

2. Add a 5-mL portion of solution E to a 25-mL portion of solution W at room temperature; crystals should be observed in 1–2 min. Allow the mixture to stand at room temperature for 5 min, cool it in an ice-water bath for about 5 min, and then collect the crystals by vacuum filtration and wash them on the filter with 2–3 mL of cold water. *Thoroughly* air-dry the crystals★ and determine their weight and melting point.

3. Warm a 25-mL portion of solution W and a 5-mL portion of solution E *separately* to 80–85 °C; add solution E to solution W and swirl the mixture. Continue to heat the solution for 10–15 min, cool it to room temperature, and then place it in an ice-water

bath for about 5–10 min. Collect the crystals by vacuum filtration, and wash them on the filter with 2–3 mL of cold water. *Thoroughly* air-dry the crystals★ and determine their weight and melting point.

D ▪ Reactions of Semicarbazide with Cyclohexanone and 2-Furaldehyde in Bicarbonate Buffer Solution

Dissolve 2.0 g of semicarbazide hydrochloride and 4.0 g of sodium bicarbonate in 50 mL of water. Prepare a solution of 2.0 mL of cyclohexanone and 1.6 mL of 2-furaldehyde in 10 mL of 95% ethanol. Divide each of these solutions into two equal portions.

1. Mix half of the aqueous solution and half of the ethanolic solution and allow the mixture to stand at room temperature for 5 min. Collect the crystals by vacuum filtration and wash them on the filter with 2–3 mL of cold water. *Thoroughly* air-dry the crystals★ and determine their weight and melting point.

2. Warm the other portions of the aqueous and ethanolic solutions *separately* to 80–85 °C. Combine them and continue heating the mixture for 10–15 min. Cool the solution to room temperature, and place it in an ice-water bath for 5–10 min. Collect the crystals by vacuum filtration and wash them on the filter with 2–3 mL of cold water. *Thoroughly* air-dry the crystals★ and determine their weight and melting point.

E ▪ Tests of Reversibility of Semicarbazone Formation

1. Place 0.3 g of cyclohexanone semicarbazone, prepared in Part A, 0.3 mL of 2-furaldehyde, 2 mL of 95% ethanol, and 10 mL of water in a 25-mL Erlenmeyer flask. Warm the mixture until a homogeneous solution is obtained (about 1 or 2 min should suffice), and continue warming for an additional 3 min. Cool the mixture to room temperature and then place it in an ice-water bath. Collect the crystals on a filter, and wash them with 2–3 mL of cold water. *Thoroughly* air-dry the crystals★ and determine their weight and melting point.

2. Repeat the preceding experiment, but use 0.3 g of 2-furaldehyde semicarbazone (prepared in Part B) and 0.3 mL of cyclohexanone in place of the cyclohexanone semicarbazone and 2-furaldehyde.

On the basis of your results from experiments C, D, and E, deduce which semicarbazone is the product of kinetic control and which is the product of thermodynamic control. To do this, first use the observed melting points of the crystals produced in parts C1, C2, and C3 to deduce whether the product in each part is the semicarbazone of cyclohexanone, of 2-furaldehyde, or of a mixture of the two. Note that in C1 the crystals of product separate almost immediately, in C2 after 1 or 2 min, and in C3 only after 10–15 min at a higher temperature; thus the reaction time is shortest in C1, intermediate in C2, and longest in C3. Then compare the product of D1 with that of C2 and the product of D2 with that of C3.

When analyzing the results of the experiments in Part E, remember that the thermodynamic product, being the more stable, is not easily transformed into the less stable kinetic product. In contrast, the kinetic product is readily converted into the more stable thermodynamic product. These circumstances arise because the reverse reaction from the less stable kinetic product has a lower activation enthalpy than the reverse reaction of the thermodynamic product.

Your completed laboratory report should include the diagram specified in Exercise 2 and answers to Exercises 3 and 4 as well, unless you were instructed to omit some parts of the experiment.

Discovery Experiment   *Effect of pH on Kinetic vs. Thermodynamic Control*

Repeat experiments C1–3 using a buffer that will provide a pH different from that obtained using dibasic potassium phosphate and sodium bicarbonate. You may be instructed to prepare a buffer having a specific pH, or you may be instructed to use sodium citrate, potassium hydrogen phosphate, sodium carbonate, potassium phosphate, or a salt of another acid to serve as the buffer for the reaction. Use the same quantities of semicarbazide hydrochloride, 2-furaldehyde, cyclohexanone, water, and 95% ethanol as were used in experiments C1–3; the molar ratio of the salt to semicarbazide hydrochloride should be about 2:1. Depending upon the buffer you use, crystals may form faster or slower than under the conditions of experiments C1–3; record the length of time required for crystals to form with the different buffers under the different conditions. On the basis of your results from these experiments, deduce which semicarbazone is the product of kinetic control and which is the product of thermodynamic control at different pHs. Do this by using the observed melting points of the crystals produced in Parts C1, C2, and C3 to deduce the nature of your product(s). If possible, compare the rates of forming the kinetic or thermodynamic product at different pHs. Your instructor may have you work in groups with several students using different buffers, so you may compare your results with theirs. As always, consult with your instructor before undertaking any experimental work.

WRAPPING IT UP

Neutralize the various *filtrates* and then flush them down the drain.

EXERCISES

1. Why is it particularly important in this experiment that the semicarbazones be dry before their melting points are determined?

2. On the basis of the results from experiments **C, D,** and **E,** draw a diagram similar to Figure 13.2 and clearly label the products corresponding to those obtained in experiments **A** and **B.**

3. On the basis of the results from the experiments of Part **D,** explain the effect of the higher pH on the reactions between semicarbazide and the two carbonyl compounds.

4. What results would be expected if sodium acetate buffer, which provides a pH of approximately 5, were used in experiments analogous to those of Part **C?** Explain.

5. What different result, if any, is expected if the heating period at 80–85 °C is extended to one hour in experiment D2?

6. There are two NH_2 groups in semicarbazide (**1**), yet only one of them reacts with the carbonyl group to form a product. Explain.

7. Based upon your observations, describe the impact that temperature has upon whether a reaction is subject to kinetic or thermodynamic control.

8. The change in the free energy, ΔG, for a chemical reaction is mathematically related to the change in enthalpy, ΔH, and the change in entropy, ΔS. What is this mathematical relationship?

9. For a chemical reaction at equilibrium, the difference in the free energy, ΔG_{eq}, between the reactants and the products is mathematically related to the equilibrium constant, K_{eq}, for the reaction. What is this mathematical relationship?

10. To draw the energy diagrams in Figures 13.1 and 13.2 and in Exercise 2, we assumed that entropic changes were not important.

    a. Define entropy.

    b. Explain why it was appropriate to ignore entropy changes in comparing the reactions of cyclohexanone and 2-furaldehyde with semicarbazide.

11. Using the energy diagram in Figure 13.1 to illustrate your answer, describe how the terms below are related.

    a. ΔH and K_{eq}

    b. $\Delta H^{\ddagger}$ and k

12. Formation of a semicarbazone is the basis for a qualitative test to determine whether an aldehyde or a ketone functional group is present in a molecule (Sec. 25.7). For the compounds listed below, indicate whether they will yield a positive or negative semicarbazone test. For those compounds giving a positive test result, write the structure of the semicarbazone produced.

13. Using curved arrows to symbolize the flow of electrons, provide a stepwise mechanism showing all proton transfer steps for the reaction between cyclohexanone and semicarbazide at pH < 7.

14. Which step of the mechanism for semicarbazone formation occurs at a reduced rate when the solution pH is

    a. very high?

    b. very low?

15. Explain how a strong acid and a weak acid differ in terms of their respective K_as and pK_as.

16. Which of the following pairs contain the prerequisites for a good buffer? (More than one answer may be correct.)

    a. strong acid/conjugate base

    b. weak acid/conjugate base

    c. weak base/conjugate acid

    d. strong base/conjugate acid

17. The carbonate/bicarbonate buffer system is an important biological buffer that is critical in maintaining blood pH.

    a. Write the acid-base reaction that shows how this buffer prevents *acidosis*, or a decrease in blood pH.

    b. Write the acid-base reaction that shows how this buffer system prevents an increase in blood pH.

18. The Henderson-Hasselbach equation, shown below, is used by biochemists and chemists to calculate the pH of a buffer solution.

$$pH = pK_a + \log \left(\frac{[A^-]_{initial}}{[HA]_{initial}} \right)$$

    a. Calculate the pH of a buffer solution that is $0.02\ M\ CH_3CO_2Na$ and $0.04\ M\ CH_3CO_2H$. The pK_a of acetic acid is 4.75.

    b. Using the Henderson-Hasselbach equation, show how $pH = pK_a$ when the concentration of the acid and the conjugate base are the same.

19. Protonation of an enolate ion can occur via two competing exothermic reactions, as shown. Protonation on the oxygen atom of the enolate ion to form the enol is faster than protonation on the carbon atom to form the ketone; however, the ketone is thermodynamically more stable than the enol.

Assuming that the ΔSs for the two reactions are similar, draw an energy diagram similar to that shown in Figure 13.1 or 13.2 for the protonation of an enolate ion to give an enol and a ketone. Clearly label the curve with the structures of the reactants and the products. Your curve should show the relative enthalpies of the reactants and products as well as the relative enthalpies of activation, $\Delta H^{\ddagger}$, and the relative enthalpies, ΔH, for the two reactions.

20. Consider the spectral data for cyclohexanone (Figs. 13.3 and 13.4).

    a. In the functional group region of the IR spectrum, specify the absorption associated with the carbon-oxygen double bond.

    b. In the ^1H NMR spectrum, assign the various resonances to the hydrogen nuclei responsible for them and explain why the multiplet integrating for four protons is downfield of that integrating for six.

    c. For the ^{13}C NMR data, assign the various resonances to the carbon nuclei responsible for them.

21. Consider the NMR data for cyclohexanone semicarbazone (Fig. 13.6).

    a. In the ^1H NMR spectrum, assign the various resonances to the hydrogen nuclei responsible for them.

    b. For the ^{13}C data, assign the various resonances to the carbon nuclei responsible for them.

Refer to the online resources and use the spectra viewer and Tables 8.1–8.8 as needed to answer the blue-numbered questions on spectroscopy.

22. Consider the spectral data for 2-furaldehyde (Figs. 13.7 and 13.8).

    a. In the functional group region of the IR spectrum, specify the absorptions associated with the carbon-oxygen and carbon-carbon double bonds and with the aldehydic C–H bond.

    b. In the ^{1}H NMR spectrum, assign the various resonances to the hydrogen nuclei responsible for them.

    c. For the ^{13}C NMR data, assign the various resonances to the carbon nuclei responsible for them.

23. a. The IR spectra portrayed in Figures 13.5 and 13.9 contain a sharp absorption at approximately 3500 cm^{-1}. Specify the functional group responsible.

    b. Both of these spectra have complex absorptions in the range of 3100–3400 cm^{-1}. What functional group is responsible?

    c. In Figure 13.9, what functionality is responsible for the absorption near 1700 cm^{-1}?

24. Consider the NMR data for 2-furaldehyde semicarbazone in Figure 13.10.

    a. In the ^{1}H NMR spectrum, assign the various resonances to the hydrogen nuclei responsible for them.

    b. For the ^{13}C data, assign the various resonances to the carbon nuclei responsible for them.

25. Discuss the differences observed in the IR and NMR spectra of cyclohexanone and of its semicarbazone that are consistent with the replacement of the carbonyl function by a carbon-nitrogen double bond in this experiment.

26. Discuss the differences observed in the IR and NMR spectra of 2-furaldehyde and of its semicarbazone that are consistent with the replacement of the carbonyl function by a carbon-nitrogen double bond in this experiment.

SPECTRA

Starting Materials and Products

Figure 13.3

IR spectrum of cyclohexanone (neat).

Figure 13.4
NMR data for cyclohexanone (CDCl₃).

(a) ¹H NMR spectrum (300 MHz).
(b) ¹³C NMR data: δ 25.1, 27.2, 41.9, 211.2.

Figure 13.5
IR spectrum of cyclohexanone semicarbazone (IR card).

Figure 13.6
NMR data for cyclohexanone semicarbazone (CDCl₃).

(a) ¹H NMR spectrum (300 MHz).
(b) ¹³C NMR data: δ 25.6, 25.8, 26.5, 27.0, 35.4, 153.4, 158.8.

Figure 13.7
IR spectrum of 2-furaldehyde (neat).

Figure 13.8
NMR data for 2-furaldehyde (CDCl₃).

(a) ¹H NMR spectrum (300 MHz).
(b) ¹³C NMR data: δ 112.9, 121.6, 148.6, 153.3, 178.1.

Figure 13.9
IR spectrum of 2-furaldehyde semicarbazone (IR card).

Figure 13.10
NMR data for 2-furaldehyde semicarbazone (DMSO-d₆).

(a) ¹H NMR spectrum (300 MHz).
(b) ¹³C NMR data: δ 110.6, 111.8, 129.8, 143.8, 149.9, 156.4.

CHAPTER 14

Nucleophilic Aliphatic Substitution
Preparation of Alkyl Halides

Nucleophiles, Nu:, may effect substitutions at aliphatic carbon atoms by a process that reveals many interesting principles of organic chemistry, including the roles steric and solvent effects play in substitutions. For the synthetic chemist, these types of reactions are very important because they frequently allow a wide variety of functionalities to be prepared from a single starting material simply by varying the nature of the nucleophilic partner. You might say that "variety is the spice of life," and this certainly applies to the range of transformations available through substitution reactions. In this chapter we'll review the general features of these types of reactions, and you will conduct experiments to study the conversion of alcohols into alkyl halides, a representative substitution reaction.

14.1 GENERAL CONCEPTS

The substitution of one group for another at a saturated, sp^3-hybridized carbon atom is a reaction commonly used to interconvert different functional groups, and the process is called a **nucleophilic aliphatic substitution**. This conversion is exemplified in Equation 14.1, where **Nu:** is a symbol for a generalized **nu**cleophile and represents a neutral molecule or an anion that has Lewis basic character, and **L:** represents a **l**eaving group. The reaction may be considered to be a Lewis acid-base reaction because the carbon atom bearing the electronegative leaving group has Lewis acidic character. This results from the polarization of the C–L bond, which produces a partial positive charge on the carbon atom.

$$\text{Nu:}^- + \text{R}-\overset{|}{\underset{|}{\text{C}}}\overset{\delta^+ \; \delta^-}{-\text{L}} \longrightarrow \text{Nu}-\overset{|}{\underset{|}{\text{C}}}-\text{R} + \text{L:}^- \qquad (14.1)$$

Nucleophiles have the property in common of bearing *at least one pair of nonbonding electrons* and being either *neutral or negatively charged.* During a substitution reaction, the nonbonding pair of electrons of the nucleophile is donated to an electrophilic, Lewis acidic carbon atom with concomitant formation of a new covalent

459

bond. Examples of some typical nucleophiles include Cl^-, Br^-, I^-, HO^-, $N\equiv C^-$, $R-C\equiv C^-$, H_2O, N_3^-, R_3N, and RS^-.

The leaving group, L:, which may also be neutral or negatively charged, must accept the pair of bonding electrons from the carbon atom as the C–L bond breaks. The ease with which various groups leave in nucleophilic substitution reactions has been determined experimentally by studying the rates of reaction of a specific nucleophile with substrates having different leaving groups. These investigations show that the leaving ability of a particular group L: correlates with the strength of its conjugate acid: The better leaving groups are those that are **conjugate bases** of the stronger acids. For example, a leaving group such as Cl^- is the conjugate base of hydrochloric acid, a strong acid, so Cl^- is a good leaving group. On the other hand, HO^- is a poor leaving group because it is the conjugate base of the weak acid water; however, H_2O itself is a good leaving group because it is the conjugate base of hydronium, H_3O^+, another strong acid.

The procedures in this chapter allow you to perform the conversion of alcohols to alkyl halides (Secs. 14.4 and 14.5) and to study the kinetics of the reverse reaction (Sec. 14.6), as seen in Equation 14.2. You will also have the opportunity to explore two different types of competition associated with nucleophilic aliphatic substitution reactions: that between nucleophiles when two or more of them are present (Sec. 14.7), illustrated in Equation 14.3 and that between substitution and elimination (Sec. 14.8), shown in Equation 14.4.

$$R-OH + HX \rightleftharpoons R-X + H_2O \tag{14.2}$$
$$\text{where } X = Cl, Br, I$$

(14.3)

(14.4)

14.2 CLASSIFICATION OF NUCLEOPHILIC SUBSTITUTION REACTIONS

Nucleophilic substitution is a general reaction for aliphatic compounds in which the leaving group is attached to an sp^3-hybridized carbon (Eq. 14.1). However, the mechanism for a given transformation depends upon the structure of the alkyl group bearing the leaving group. The two different mechanistic pathways that apply to such substitutions are designated by the symbols S_N1 (S for substitution, N for nucleophilic, and 1 for unimolecular) and S_N2 (2 for bimolecular). These two mechanisms are depicted in general form in Equations 14.5 and 14.6.

S$_N$1 Mechanism

$$R_3C\!-\!L \xrightleftharpoons[\text{slow}]{\text{rds}} L\!:^- + R_3\overset{+}{C} \xrightarrow{\text{Nu:}^-}_{\text{fast}} R_3C\!-\!Nu \qquad (14.5)$$

S$_N$2 Mechanism

$$Nu\!:\overset{R^1}{\underset{R^2}{\overset{|}{\underset{|}{C}}}}\!\!\!-\!L \xrightarrow{\text{rds}} \left[\overset{R^1}{\underset{H\;\;R^2}{\overset{\delta^-}{Nu\!:\!-\!-\!-\!C\!-\!-\!-\!L}}}\right]^{\ddagger} \longrightarrow Nu\!-\!\overset{R^1}{\underset{R^2}{C}}\!\!\!\!\text{''H} + L\!:^- \qquad (14.6)$$

Transition state

When nucleophilic substitution occurs by an S$_N$1 mechanism, the reaction proceeds in two successive steps, as illustrated in Equation 14.5. The first one involves the **heterolytic cleavage**, or **ionization**, of the bond between the carbon atom and the leaving group. This step is assisted by polar interactions between solvent molecules and the incipient cationic and anionic centers. Because the leaving group acquires the pair of bonding electrons, the organic fragment becomes a **carbocation**, with the carbon atom formerly bound to L now bearing a positive charge. The intermediate carbocation may then undergo the usual reactions of carbocations: (1) rearrangement to a more stable carbocation (Sec. 10.3); (2) loss of a proton to give an alkene by net **elimination** of the elements of **H–L** (Sec. 10.3); or (3) combination with a nucleophile to form the substitution product (Eq. 14.5). Normally, the concentration of the nucleophile, Nu:$^-$, is high compared to that of the L:$^-$ that has been produced, so the reaction of the carbocation with L:$^-$ to give the starting material is relatively unimportant. If the nucleophile is the solvent, the reaction is known as a **solvolysis**.

The first step of an S$_N$1 reaction is much slower than the second because it involves breaking the C–L bond to generate an unstable carbocation, an endothermic process. The second step is a fast, exothermic process involving bond formation. Thus, the first step of an S$_N$1 reaction is the **rate-determining step (rds)** of the reaction, and the rate of the reaction depends *only* upon the concentration of the substrate, R–L. Such a reaction is termed **unimolecular**. This is expressed mathematically in Equation 14.7, where k_1 is the **first-order rate constant**.

$$\text{Rate} = k_1[\text{R}\!-\!\text{L}] \qquad (14.7)$$

When substitution occurs by an S$_N$2 mechanism, the nucleophile directly attacks the substrate, with the angle of approach being 180° to the C–L bond. This is called "backside attack," and the reaction proceeds with inversion of stereochemistry, the so-called "Walden inversion." The C–L bond is being broken concurrently with the formation of the C–Nu bond, so both the substrate, R–L, *and* the nucleophile are involved in the transition state of the rate-determining step. Reactions in which two reactants are involved in the transition state of the rate-determining step are termed **bimolecular**, and the rate of such processes depends on the concentration of the substrate and the nucleophile, as shown in Equation 14.8, where k_2 is the **second-order rate constant**.

$$\text{Rate} = k_2[\text{R}\!-\!\text{L}][\text{Nu:}] \qquad (14.8)$$

We can summarize some of the important factors that dictate whether a particular substrate undergoes substitution preferentially by an S_N1 or S_N2 mechanism as follows.

1. As more alkyl groups are attached to the carbon atom **C–L** undergoing substitution, it becomes sterically more difficult for the nucleophile to attack from the backside because of the bulk of these groups, thereby decreasing the ease with which the S_N2 process can occur.

2. With increasing substitution of alkyl groups on the carbon atom **C–L**, the incipient carbocation in the S_N1 reaction becomes more stable, thereby increasing its ease of formation along the S_N1 pathway.

These two effects reinforce one another and yield the following general trends:

$$CH_3\!-\!L \quad RCH_2\!-\!L \quad R_2CH\!-\!L \quad R_3C\!-\!L$$

Primary substrates (1°, RCH_2–L) undergo nucleophilic substitution predominantly by an S_N2 mechanism, whereas tertiary substrates (3°, R_3C–L) react via an S_N1 mechanism. Secondary substrates (2°, R_2CH–L) may react by both mechanisms, and the specific pathway is dictated by various factors such as solvent, reaction conditions, and nature of the nucleophile. Such generalizations must always be applied cautiously, and the mechanism by which a particular reaction occurs *must* be confirmed *experimentally*. You may be able to deduce the mechanism by which a particular substitution occurred if you know the structures of the starting substrate and the product(s). For example, because carbocations are involved in the S_N1 mechanism, the formation of a product arising from a rearranged carbocation points to this mechanism; the lack of rearrangement does *not* necessarily exclude an S_N1 process, however. Determining whether the role of the substitution reaction follows a first- or a second-order rate law (Eqs. 14.7 and 14.8, respectively) also allows you to characterize the process as S_N1 or S_N2.

14.3 COMPETITION BETWEEN SUBSTITUTION AND ELIMINATION

Elimination reactions to produce alkenes may compete with reactions in which nucleophilic aliphatic substitution is the desired process. Unimolecular elimination reactions, **E1**, compete with S_N1 substitutions, and bimolecular elimination processes, **E2** (E stands for elimination and 2 for bimolecular), compete with S_N2 transformations. These competitions are shown in Equations 14.9 and 14.10. The nature of E1 reactions is discussed in detail in Section 10.3 and that of E2 processes in Section 10.2.

S$_N$2 Versus E2

$$\text{(14.9)}$$

| Substitution (S$_N$2) | Elimination (E2) |

S$_N$1 Versus E1

$$\text{(14.10)}$$

| Substitution (S$_N$1) | Elimination (E1) |

The course of the reaction may be influenced by the nature of the nucleophile that is present. *Substitution is favored* with weakly basic and highly polarizable nucleophiles such as I^-, Br^-, Cl^-, H_2O, and $CH_3CO_2^-$, whereas *elimination is favored* when strongly basic and only slightly polarizable nucleophiles such as RO^-, H_2N^-, H^-, and HO^- are used. **Polarizability** is a measure of the ease with which the electron cloud of the Lewis base is distorted by a nearby center that bears a partial or full positive charge. Furthermore, bulky nucleophiles tend to favor elimination because the hydrogen atom is more sterically accessible than is the carbon atom bearing the leaving group.

14.4 PREPARATION OF 1-BROMOBUTANE: AN S$_N$2 REACTION

A common technique for converting a primary alcohol to an alkyl halide involves treating the alcohol with a hydrogen halide H–X (X = Cl, Br, or I) as shown in Equation 14.11. This reaction is *reversible*, and displacing the equilibrium to the right normally involves using a large excess of the acid, a strategy in accord with the **Le Chatelier principle**.

$$H\!-\!X \ + \ R\!-\!OH \ \underset{\Delta}{\overset{}{\rightleftharpoons}} \ R\!-\!X \ + \ H_2O \qquad (14.11)$$
$$\text{where X = Cl, Br, I}$$

For example, 1-bromobutane may be prepared by heating 1-butanol with hydrobromic acid, H–Br, in the presence of sulfuric acid (Eq. 14.12). The mechanism for this reaction is known to occur in two steps, the first being protonation of the alcohol to give the oxonium ion **1** via a Lewis acid-base reaction (Eq. 14.13). This oxonium ion then undergoes displacement by the bromide ion to form 1-bromobutane and water. This process is an S_N2 reaction in which *water is the leaving group and bromide ion is the nucleophile.* The sulfuric acid serves two important purposes: (1) It is a *dehydrating* agent that reduces the activity of water and shifts the position of equilibrium to the right and (2) it provides an added source of hydrogen ions to increase the concentration of oxonium ion **1**. The use of *concentrated* hydrobromic acid also helps to establish a favorable equilibrium.

$$CH_3CH_2CH_2CH_2\!-\!OH \ + \ HBr \ \underset{\Delta}{\overset{H_2SO_4}{\rightleftharpoons}} \ CH_3CH_2CH_2CH_2\!-\!Br \ + \ H_2O \qquad (14.12)$$

$$\begin{array}{cc} \text{1-Butanol} & \text{1-Bromobutane} \\ (n\text{-butyl alcohol}) & (n\text{-butyl bromide}) \end{array}$$

$$n\text{-}C_3H_7CH_2\!-\!\ddot{O}H \ + \ \overset{..}{H}{}^{+} \ \rightleftharpoons \ n\text{-}C_3H_7CH_2\!-\!\overset{+}{O}\!\!\begin{smallmatrix}H\\ \\H\end{smallmatrix} \ \overset{Br:^-}{\rightleftharpoons} \ n\text{-}C_3H_7CH_2\!-\!Br \ + \ H_2O \qquad (14.13)$$
$$\mathbf{1}$$

No reaction occurs between 1-butanol and NaBr in the absence of strong acid because leaving groups in nucleophilic substitution reactions must be *weakly* basic (Sec. 14.1), as is water in Equation 14.13. If the reaction of 1-butanol and NaBr were to occur (Eq. 14.14), the leaving group would necessarily be the *strongly* basic hydroxide ion, a poor leaving group. Thus, the forward reaction depicted in Equation 14.14 does *not* occur. On the other hand, the reverse reaction between 1-bromobutane and hydroxide proceeds readily because the leaving group in this reaction is the weakly basic bromide ion.

$$CH_3CH_2CH_2CH_2\text{–OH} \ + \ NaBr \ \overset{}{\underset{}{\rlap{\,/}{\rightleftharpoons}}} \ CH_3CH_2CH_2CH_2\text{–Br} \ + \ NaOH \qquad (14.14)$$

The mixture of hydrobromic acid and sulfuric acid may be prepared by either adding *concentrated* sulfuric acid to *concentrated* hydrobromic acid or by generating the hydrobromic acid *in situ* by adding *concentrated* sulfuric acid to aqueous sodium bromide (Eq. 14.15). Both of these methods work well and give good yields of the alkyl bromide from low-molar-mass alcohols. The method of generating H–Br *in situ* is not effective with higher-molar-mass alcohols because of their low solubility in concentrated salt solutions, so *concentrated* hydrobromic acid is used instead.

$$NaBr \ + \ H_2SO_4 \ \rightleftharpoons \ HBr \ + \ NaHSO_4 \qquad (14.15)$$

Although the presence of *concentrated* sulfuric acid promotes the formation of the alkyl bromide, several side reactions involving sulfuric acid and the alcohol can occur. One of these is esterification of the alcohol by sulfuric acid to form an alkyl hydrogen sulfate **2** (Eq. 14.16).

$$RO\text{—}H + H_2SO_4 \rightleftharpoons \underset{\textbf{2}}{RO\text{—}SO_3H} + H_2O \qquad (14.16)$$

An alkyl hydrogen sulfate

This reaction is reversible, and the position of the equilibrium is shifted to the left, regenerating the alcohol from **2** as the alkyl bromide is produced. The formation of **2** itself does not directly decrease the yield of alkyl bromide; rather, **2** undergoes other reactions to give undesired by-products. For example, it suffers elimination on heating to give a mixture of alkenes (Eq. 14.17). It may also react with another molecule of alcohol to give a dialkyl ether by an S_N2 reaction in which the nucleophile is ROH (Eq. 14.18). Both of these side reactions consume alcohol, so the yield of alkyl bromide is decreased. Fortunately, these side reactions may be minimized for *primary* alcohols by controlling the temperature of the reaction and the concentration of sulfuric acid.

$$RO\text{—}SO_3H \xrightarrow{\text{heat}} \text{alkenes} + H_2SO_4 \qquad (14.17)$$

$$RO\text{—}SO_3H + ROH \xrightarrow{\text{heat}} R\text{—}O\text{—}R + H_2SO_4 \qquad (14.18)$$

It is necessary to use different procedures to prepare secondary alkyl bromides from *secondary* alcohols because such alcohols are easily dehydrated by *concentrated* sulfuric acid to give alkenes by way of Equations 14.16 and 14.17. In fact, the acid-catalyzed dehydration of secondary and tertiary alcohols is a common method for synthesizing alkenes (Sec. 10.3). This problem may be circumvented by using *concentrated* hydrobromic acid; however, it is better to prepare secondary alkyl bromides by the reaction of secondary alcohols with phosphorus tribromide, PBr_3 (Eq. 14.19).

$$3\,R\text{—}OH + \underset{\substack{\text{Phosphorus} \\ \text{tribromide}}}{PBr_3} \longrightarrow 3\,R\text{—}Br + H_3PO_3 \qquad (14.19)$$

EXPERIMENTAL PROCEDURES

Preparation of 1-Bromobutane

Purpose To demonstrate the conversion of a primary alcohol to a primary bromoalkane using hydrobromic acid.

SAFETY ALERT

1. **Wear safety glasses or goggles and suitable protective gloves while performing the experiments.**
2. **Examine your glassware for cracks and chips. This procedure involves heating concentrated acids, and defective glassware could break under these conditions, spilling hot corrosive chemicals on you and those working around you.**

3. Be very careful when handling *concentrated* sulfuric acid. If any acid comes in contact with your skin, immediately wash it off with copious amounts of cold water and then with dilute sodium bicarbonate solution.

4. *Concentrated* sulfuric acid and water mix with the evolution of lots of heat. *Always add the acid to the water*, a technique that disperses the heat through warming of the water. Add the acid slowly and with swirling to ensure continuous and thorough mixing.

MINISCALE PROCEDURE

Preparation Refer to the online resources to answer Pre-Lab Exercises, access videos, and read the MSDSs for the chemicals used or produced in this procedure. Review Sections 2.9, 2.10, 2.11, 2.13, 2.21, and 2.22.

Apparatus A 25-mL and a 100-mL round-bottom flask, separatory funnel, ice-water bath, apparatus for heating under reflux, simple distillation, magnetic stirring, and *flameless* heating.

Setting Up Place 11.1 g of sodium bromide in the 100-mL round-bottom flask, and add about 10 mL of water and 10 mL of 1-butanol. Mix the contents of the flask thoroughly by swirling; then cool the flask in an ice-water bath. *Slowly* add 10 mL of *concentrated* sulfuric acid to the cold mixture with swirling and continuous cooling. Remove the flask from the ice-water bath, add a stirbar, and set up the apparatus for heating under reflux.

Reaction, Work-Up, and Isolation Warm the flask gently until most of the salts have dissolved and then heat the mixture under gentle reflux for about 45 min.★ Equip the flask for simple distillation. Distill the mixture rapidly and collect the distillate of water and 1-bromobutane in an ice-cooled receiver. Continue distilling until the distillate is clear; the head temperature should be around 115 °C at this point because sulfuric acid and hydrobromic acid are now co-distilling with water.★

Transfer the distillate to a separatory funnel and add about 10 mL of water. Shake the funnel gently with venting. Separate the layers and determine which of these is the organic layer. Wash the organic layer sequentially with two 5-mL portions of cold 2 *M* aqueous sodium hydroxide solution, 10 mL of water, and 10 mL of saturated aqueous sodium chloride. Transfer the cloudy 1-bromobutane layer to an Erlenmeyer flask, and dry it over several spatula-tips full of *anhydrous* sodium sulfate.★ Swirl the flask occasionally for a period of 10–15 min until the crude 1-bromobutane is clear; add further small portions of *anhydrous* sodium sulfate if the solution appears cloudy.

Using a Pasteur pipet, carefully transfer the crude 1-bromobutane to a clean, dry 25-mL round-bottom flask. Add a stirbar, equip the flask for shortpath distillation, and carefully distill the product into a tared receiver. Because of the relatively small quantity of product expected, it may be difficult to obtain an accurate boiling point, so you should collect the fraction having a boiling point greater than about 90 °C (760 torr).

Analysis Weigh the product and compute the yield. Perform the alcoholic silver nitrate and sodium iodide/acetone classification tests on your product (Sec. 10.5C). Obtain IR and ^{1}H NMR spectra of your starting material and product and compare them with those of authentic samples (Figs. 8.28 and 14.1–14.4).

MICROSCALE PROCEDURE

Preparation Refer to the online resources to answer Pre-Lab Exercises, access videos, and read the MSDSs for the chemicals used or produced in this procedure. Review Sections 2.9, 2.10, 2.11, 2.13, 2.21, and 2.22.

Apparatus A 3-mL and a 5-mL conical vial, ice-water bath, apparatus for heating under reflux, simple distillation, magnetic stirring, and *flameless* heating.

Setting Up Place 1.1 g of sodium bromide in the 5-mL conical vial and add 1.0 mL of water, 1.0 mL of 1-butanol, and a spinvane. Mix the contents of the vial thoroughly by stirring and then cool the vial in an ice-water bath. *Slowly* add 1.0 mL of *concentrated* sulfuric acid to the cold mixture with stirring and continuous cooling. Remove the vial from the bath and set up the apparatus for heating under reflux.

Reaction, Work-Up, and Isolation Warm the vial gently until most of the salts have dissolved and then heat the mixture under gentle reflux for 45 min.★ Equip the vial for simple distillation. Distill the mixture rapidly and collect the distillate water and 1-bromobutane in the Hickman stillhead. Continue the distillation, increasing the temperature of the heating source as necessary until there is about 1.5–2.0 mL remaining in the vial and the distillate is clear.★

Using a Pasteur pipet, transfer the distillate to the 3-mL conical vial and add 1.0 mL of water. Cap the vial and shake it gently with venting to mix the contents. Separate the layers and determine which of these is the organic layer. Wash the organic layer sequentially with two 0.5-mL portions of cold 2 *M* aqueous sodium hydroxide solution, 1.0 mL of water, and 1.0 mL of saturated aqueous sodium chloride. Transfer the cloudy 1-bromobutane layer to a screw-cap centrifuge tube, and dry it over several microspatula-tips full of *anhydrous* sodium sulfate.★ Swirl the contents in the centrifuge tube occasionally for a period of 10–15 min until the crude 1-bromobutane is clear; add further small portions of *anhydrous* sodium sulfate if the solution appears cloudy. Using a Pasteur pipet, carefully transfer the crude 1-bromobutane to a tared sample vial.

Analysis Weigh the product and compute the yield. Perform the alcoholic silver nitrate and sodium iodide/acetone classification tests on your product (Sec. 10.5B). Obtain IR and ¹H NMR spectra of your starting material and product and compare them with those of authentic samples (Figs. 8.28 and 14.1–14.4).

Discovery Experiment *Analysis of S$_N$ Reactions as a Function of Substrate*

In Section 14.2, we state that the relative rate of an S$_N$ reaction for a substrate R–L is a function of the degree of the carbon atom bearing the leaving group. This proposition is subject to experimental verification using simple qualitative tests. By referring to the procedures described in Section 14.6, propose and execute experiments that explore the relative rates of S$_N$1 and S$_N$2 reactions for substrates in which the leaving group is on carbon atoms of various degrees. You may consider using both alkyl and aryl halides in your study, but be certain to consult with your instructor before undertaking any experiments.

WRAPPING IT UP

Carefully dilute the *stillpot residue from the reaction* with water and then slowly combine this with the *water and the sodium hydroxide washes.* Neutralize the *combined aqueous mixture* with sodium carbonate and flush the solution down the drain with water. The *sodium sulfate* used as the drying agent is contaminated with product, so place it in the container for halogenated solids. Place the *residue in the stillpot from the distillation in the Miniscale Procedure* into the container for halogenated liquids. Dilute the *silver nitrate test solution* with water and flush it down the drain unless instructed otherwise. Place the *sodium iodide/acetone test solution* in the container for halogenated liquids.

EXERCISES

1. Some water was added to the initial reaction mixture in the procedure you performed.

   a. How might the yield of 1-bromobutane be affected if the water were not added, and what product(s) would be favored?

   b. How might the yield of product be affected by adding twice as much water as is specified, while keeping the quantities of the other reagents the same?

2. In the purification process, the organic layer is washed sequentially with 2 *M* NaOH, water, and saturated aqueous sodium chloride. What is the purpose of each of these washes and why is *cold* 2 *M* NaOH recommended?

3. After the washes described in Exercise 2, the 1-bromobutane is treated with *anhydrous* sodium sulfate.

   a. Why is this done?

   b. Could solid sodium hydroxide or potassium hydroxide be used for this purpose? Explain.

4. The final step of the purification process in the Miniscale Procedure involves a simple distillation. What impurities are removed by this distillation?

5. Using curved arrows to symbolize the flow of electrons, propose a mechanism for formation of the by-product(s) from elimination that could be produced when 1-butanol is treated with HBr. Specify whether your mechanism is E1, E2, or both.

6. List two factors, one in the substrate R–L and the other in the nucleophile/base Nu: being used, that should increase the yield of E2 relative to S_N2 products in the reaction of R–L with Nu:.

7. Treating 1-butanol with sulfuric acid establishes an equilibrium with 1-butyl bisulfate and water as shown.

$$CH_3(CH_2)_2CH_2OH + H_2SO_4 \rightleftharpoons CH_3(CH_2)_2CH_2OSO_3H + H_2O$$
$$\text{1-Butanol} \qquad\qquad\qquad\qquad \text{1-Butyl bisulfate}$$

   a. Using curved arrows to symbolize the flow of electrons, propose a mechanism for the conversion of 1-butanol to 1-butyl bisulfate.

   b. 1-Butyl bisulfate may serve as the precursor to an ether and an alkene, as well as the desired 1-bromobutane.

      i. Provide the structures expected for the ether and the alkene that might be produced from 1-butyl bisulfate.

    ii. Using curved arrows to symbolize the flow of electrons, provide reaction mechanisms for the formation of each of these by-products and specify whether the mechanisms you write are S_N1, S_N2, E1, E2, or none of these.

    iii. Explain why you would expect bisulfate to be a good leaving group for substitution or elimination reactions.

8. Consider the mechanistic step in the conversion of 1-butanol to 1-bromobutane in which bromide ion displaces water from the oxonium ion **1** (Eq. 14.13).

    a. What evidence, if any, is there that this step involves backside attack by the nucleophile?

    b. Propose an experiment that would allow you to prove that backside attack of the nucleophile on the substrate was indeed the mechanism for the reaction.

9. How would doubling the concentration of the nucleophile affect the rate of an S_N2 reaction?

10. 1-Butanol does not undergo S_N2 reactions in the absence of acids. If the alcohol were converted to the corresponding *p*-toluenesulfonate ester shown below, would you expect this ester to undergo an S_N2 reaction with NaBr in the absence of acid? Give your reasoning.

$$CH_3(CH_2)_2CH_2O-\overset{\displaystyle O}{\underset{\displaystyle O}{S}}-\langle\!\!\langle\;\rangle\!\!\rangle-CH_3$$

1-Butyl *p*-toluenesulfonate

11. Consider the structure of 1-butanol and explain why this alcohol does not undergo S_N1 reactions.

12. Specify the most electrophilic carbon atom in 1-butanol and rationalize your choice.

13. Neopentyl chloride, $CH_3C(CH_3)_2CH_2Cl$, is a primary alkyl halide that undergoes S_N2 reactions at extremely slow rates. Offer an explanation for this fact that is consistent with the mechanism for this reaction. Draw a suitable illustration of the transition state for the rate-determining step as part of your explanation.

14. A student attempted to prepare *tert*-butyl ethyl ether by a substitution reaction between 2-chloro-2-methylpropane (*tert*-butyl chloride) and sodium ethoxide. None of the desired ether was obtained. What product was formed and why did this route to the ether fail? Suggest a different substitution reaction that should yield the desired product.

$$(CH_3)_3C-Cl + NaOCH_2CH_3 \xrightarrow{\;\;/\!\!/\;\;} (CH_3)_3C-O-CH_2CH_3$$

2-Chloro-2-methylpropane	Sodium ethoxide	*tert*-Butyl ethyl ether

15. The reaction of 3-methyl-2-butanol with *concentrated* hydrobromic acid gives the two isomeric bromoalkanes shown.

$$(CH_3)_2CHCHCH_3 \xrightarrow{\text{HBr}} (CH_3)_2CHCHCH_3 + (CH_3)_2CCH_2CH_3$$
$$\qquad\quad|\qquad\qquad\qquad\qquad\qquad|\qquad\qquad\qquad\qquad|$$
$$\qquad\;\;OH\qquad\qquad\qquad\qquad\;\;Br\qquad\qquad\qquad\;\;Br$$

3-Methyl-2-butanol	2-Bromo-3-methylbutane	2-Bromo-2-methylbutane
	10%	90%

Using curved arrows to symbolize the flow of electrons, suggest two reasonable mechanisms for this reaction that account for the formation of each of the products. Indicate whether each is an S_N1 or an S_N2 reaction.

Refer to the online resources and use the spectra viewer and Tables 8.1–8.8 as needed to answer the blue-numbered questions on spectroscopy.

16. Consider the spectral data for 1-butanol (Figs. 8.28, 14.1, and 14.2).

    a. In the functional group region of the IR spectrum, specify the absorption associated with the hydroxyl group, and explain why this peak is broad.

    b. In the ^1H NMR spectrum, assign the various resonances to the hydrogen nuclei responsible for them.

    c. If the NMR sample of 1-butanol in $CDCl_3$ were shaken with D_2O *prior* to taking the ^1H NMR spectrum, what differences might you expect to observe?

    d. For the ^{13}C NMR data, assign the various resonances to the carbon nuclei responsible for them.

17. Consider the NMR spectral data for 1-bromobutane (Fig. 14.4).

    a. In the ^1H NMR spectrum, assign the various resonances to the hydrogen nuclei responsible for them.

    b. For the ^{13}C NMR data, assign the various resonances to the carbon nuclei responsible for them.

18. Discuss the differences in the IR and NMR spectra of 1-butanol and 1-bromobutane that are consistent with the conversion of an alcohol into a bromoalkane in this procedure.

19. Discuss the differences observed in the ^1H and ^{13}C NMR spectra of 1-bromobutane, which appear in Figure 14.4a and b, with those of 1-chlorobutane, which appear in Figure 9.3a and b. To what factor do you attribute these differences?

SPECTRA

Starting Material and Product

The ^1H NMR spectrum of 1-butanol is provided in Figure 8.28.

Figure 14.1
IR spectrum of 1-butanol (neat).

Figure 14.2
^{13}C NMR data for 1-butanol (CDCl$_3$).
Chemical shifts: Δ 13.9, 19.2, 35.0, 62.2.

$CH_3(CH_2)_2 CH_2OH$

Figure 14.3

IR spectrum of 1-bromobutane (neat).

Figure 14.4

NMR data for 1-bromobutane (CDCl$_3$).

(a) ^1H NMR spectrum (300 MHz).
(b) ^{13}C NMR data: δ 13.2, 21.5, 33.1, 35.0.

14.5 PREPARATION OF 2-CHLORO-2-METHYLBUTANE: AN S$_N$1 REACTION

Different reagents such as HX and PX$_3$ may be used to prepare alkyl halides from primary and secondary alcohols. However, because elimination reactions predominate when tertiary alcohols are treated with phosphorous trihalides, preparing tertiary alkyl halides from tertiary alcohols proceeds with good yields only if *concentrated* hydrogen halides, HX, are used. The reaction of 2-methyl-2-butanol with hydrochloric acid to produce 2-chloro-2-methylbutane (Eq. 14.20) illustrates this transformation.

$$CH_3CH_2-\overset{\overset{\displaystyle CH_3}{|}}{\underset{\underset{\displaystyle CH_3}{|}}{C}}-OH + HCl \rightleftharpoons CH_3CH_2-\overset{\overset{\displaystyle CH_3}{|}}{\underset{\underset{\displaystyle CH_3}{|}}{C}}-Cl + H_2O \qquad (14.20)$$

2-Methyl-2-butanol   2-Chloro-2-methylbutane

The conversion of tertiary alcohols to the corresponding tertiary alkyl halide using hydrogen halides proceeds via an S_N1 mechanism (Eqs. 14.21–14.23). The first step of the process involves protonation of the hydroxyl group of the alcohol, a Lewis acid-base reaction (Eq. 14.21). Ionization then occurs in the second step, and a molecule of water is lost (Eq. 14.22). Owing to steric hindrance at the tertiary carbon atom, attack of chloride ion on 3 in an S_N2 process (Sec. 14.4) does not occur. This difference in reactivity between the oxonium ions derived from tertiary alcohols and those of secondary or primary alcohols reflects the relative stabilities of the three types of carbocations ($3° > 2° > 1°$) that would be formed upon the loss of a molecule of water. The ionization is the slow, rate-determining (rds) step in the sequence. In the final step of the reaction, chloride ion attacks the intermediate carbocation to give 2-chloro-2-methylbutane (Eq. 14.23).

$$CH_3CH_2-\overset{\overset{\displaystyle CH_3}{|}}{\underset{\underset{\displaystyle CH_3}{|}}{C}}-\underset{\ddot{}}{O}\overset{H}{} + H^+ \overset{fast}{\rightleftharpoons} CH_3CH_2-\overset{\overset{\displaystyle CH_3}{|}}{\underset{\underset{\displaystyle CH_3}{|}}{C}}-\overset{+}{O}\overset{H}{\underset{H}{}} \qquad (14.21)$$

3

$$CH_3CH_2-\overset{\overset{\displaystyle CH_3}{|}}{\underset{\underset{\displaystyle CH_3}{|}}{C}}-\overset{+}{O}\overset{H}{\underset{H}{}} \overset{rds}{\rightleftharpoons} CH_3CH_2-\overset{\overset{\displaystyle CH_3}{|}}{\underset{\underset{\displaystyle CH_3}{|}}{C}}{}^+ + H_2O \qquad (14.22)$$

Relatively stable
tertiary carbocation

$$CH_3CH_2-\overset{\overset{\displaystyle CH_3}{|}}{\underset{\underset{\displaystyle CH_3}{|}}{C}}{}^+ + :\ddot{Cl}:^- \overset{fast}{\rightleftharpoons} CH_3CH_2-\overset{\overset{\displaystyle CH_3}{|}}{\underset{\underset{\displaystyle CH_3}{|}}{C}}-Cl \qquad (14.23)$$

The principal side reaction in S_N1 reactions of this type is E1 elimination, which results from the loss of a proton from the tertiary carbocation to give 2-methyl-1-butene together with 2-methyl-2-butene (Eqs. 14.24 and 14.25), as discussed in more detail in Section 10.3. Under the reaction conditions you will use, however, elimination is reversible through Markovnikov addition of HCl to the 2-methyl-1-butene or the 2-methyl-2-butene that is produced, and this addition gives the desired 2-chloro-2-methylbutane (Eq. 14.26). A more extensive discussion of the ionic addition of hydrogen halides to alkenes is presented in Sections 10.4 and 10.5.

$$CH_3CH_2-\overset{\overset{\displaystyle CH_2-H}{|}}{\underset{\underset{\displaystyle CH_3}{|}}{C}}{}^+ + :\ddot{Cl}:^- \longrightarrow CH_3CH_2-\overset{CH_2}{\underset{CH_3}{C}} + H-Cl \qquad (14.24)$$

2-Methyl-1-butene

$$\text{CH}_3\text{CH}{\overset{\displaystyle \text{CH}_3}{\underset{\displaystyle \underset{\displaystyle :\ddot{\text{Cl}}:^-}{\text{H}}}{-\text{C}^+}}}\text{CH}_3 \longrightarrow \text{CH}_3\text{CH}=\text{C}{\overset{\displaystyle \text{CH}_3}{\underset{\displaystyle \text{CH}_3}{}}} + \text{H}-\text{Cl} \qquad (14.25)$$

2-Methyl-2-butene

$$\text{CH}_3\text{CH}_2-\text{C}{\overset{\displaystyle \text{CH}_2}{\underset{\displaystyle \text{CH}_3}{}}} + \text{H}-\ddot{\text{Cl}}: \longrightarrow \text{CH}_3\text{CH}_2-\overset{\displaystyle \text{CH}_3}{\underset{\displaystyle \text{CH}_3}{\text{C}^+}} + :\ddot{\text{Cl}}:^- \longrightarrow \text{CH}_3\text{CH}_2-\overset{\displaystyle \text{CH}_3}{\underset{\displaystyle \text{CH}_3}{\text{C}}}-\text{Cl}$$

or
$$(14.26)$$

$$\text{CH}_3\text{CH}=\text{C}{\overset{\displaystyle \overset{\displaystyle \text{H}-\text{Cl}}{\text{CH}_3}}{\underset{\displaystyle \text{CH}_3}{}}}$$

EXPERIMENTAL PROCEDURES

Preparation of 2-Chloro-2-Methylbutane

Purpose To demonstrate the conversion of a tertiary alcohol into a tertiary alkyl chloride using hydrochloric acid.

SAFETY ALERT

1. **Wear safety glasses or goggles and suitable protective gloves while performing the experiments.**

2. **If *concentrated* hydrochloric acid spills on your skin, wash it off with large volumes of water and then with dilute sodium bicarbonate solution.**

MINISCALE PROCEDURE

Preparation Refer to the online resources to answer Pre-Lab Exercises, access videos, and read the MSDSs for the chemicals used or produced in this procedure. Review Sections 2.10, 2.13, and 2.21.

Apparatus A 25-mL round-bottom flask, separatory funnel, ice-water bath, apparatus for simple distillation, magnetic stirring, and *flameless* heating.

Setting Up Place 10 mL of 2-methyl-2-butanol and 25 mL of *concentrated* (12 *M*) hydrochloric acid in the separatory funnel.

Reaction, Work-Up, and Isolation Swirl the contents of the separatory funnel gently *without* the stopper on the funnel to mix the reactants. After swirling the funnel for about 1 min, stopper and then carefully invert it. Release the excess pressure by opening the stopcock with the funnel inverted. *Do not shake the funnel until the pressure has been equalized.* Now close the stopcock and shake the funnel for several minutes, with intermittent venting. Allow the mixture to separate into two distinct layers.

Separate the layers and determine which is the organic layer. Wash the organic layer sequentially with 10-mL portions of saturated aqueous sodium chloride and *cold* saturated aqueous sodium bicarbonate. On initial addition of the bicarbonate solution, vigorous gas evolution will normally occur; gently swirl the *unstoppered* separatory funnel until this stops. Stopper the funnel and *carefully* invert it; vent the funnel immediately to release gas pressure. Shake the funnel gently with *frequent* venting and then shake it vigorously with frequent venting. Separate the organic layer, and again wash it sequentially with 10-mL portions of water and saturated aqueous sodium chloride. Carefully remove the aqueous layer, transfer the 2-chloro-2-methylbutane to an Erlenmeyer flask, and dry the product over several spatula-tips full of *anhydrous* sodium sulfate.★ Swirl the flask occasionally for 10–15 min until the product is dry; add further small portions of *anhydrous* sodium sulfate if the liquid appears cloudy.

Using a Pasteur or a filter-tip pipet, carefully transfer the crude product to the round-bottom flask, add a stirbar, and equip the flask for shortpath distillation. Distill the product, collecting it in a tared receiver cooled in an ice-water bath. Because of the relatively small quantity of product, it may be difficult to obtain an accurate boiling point, so you should collect the fraction having a boiling point greater than about 75 °C (760 torr).

Analysis Determine the mass of 2-chloro-2-methylbutane isolated and calculate the percent yield. Perform the alcoholic silver nitrate and sodium iodide/acetone classification tests on your product (Sec. 10.5B). Obtain IR and ^{1}H NMR spectra of your starting material and product and compare them with those of authentic samples (Figs. 14.5–14.8).

MICROSCALE PROCEDURE

Preparation Refer to the online resources to answer Pre-Lab Exercises, access videos, and read the MSDSs for the chemicals used or produced in this procedure. Review Sections 2.10, 2.13, and 2.21.

Apparatus A 5-mL conical vial.

Setting Up Place 1.0 mL of 2-methyl-2-butanol and 2.5 mL of *concentrated* (12 *M*) hydrochloric acid in the conical vial.

Reaction, Work-Up, and Isolation Using a small glass rod, stir the contents of the conical vial gently to mix the reactants. After about 1 min of stirring, cap and invert the vial, and then return it to its original upright position. Release the excess pressure in the vial by *carefully* venting. *Do not shake the vial until the pressure has been equalized.* Replace the screw-cap and shake the vial for several minutes with intermittent venting. Allow the mixture to separate into two distinct layers.

Separate the layers and determine which is the organic layer. Wash the organic layer sequentially with 1-mL portions of saturated aqueous sodium chloride and *cold* saturated aqueous sodium bicarbonate. On initial addition of the bicarbonate solution, vigorous gas evolution will normally occur. Using a glass rod, gently stir the contents of the vial until this stops. Cap and invert the vial and return it to its original upright position; release the excess pressure by venting *carefully.* Shake the vial gently with *frequent* venting and then shake it vigorously with frequent venting. Separate the organic layer, and again wash it sequentially with 1-mL portions of water and saturated sodium chloride solution. Carefully remove the aqueous layer.

Using a filter-tip pipet, transfer the 2-chloro-2-methylbutane to a centrifuge tube, and dry the liquid over several microspatula-tips full of *anhydrous* sodium sulfate.★ Swirl the tube occasionally for 10–15 min until the product is dry. Add several small portions of *anhydrous* sodium sulfate if the liquid is cloudy. Using a Pasteur or a filter-tip pipet, transfer the crude product to a tared sample vial.

Analysis Determine the mass of the 2-chloro-2-methylbutane isolated and calculate the percent yield. Perform the alcoholic silver nitrate and sodium iodide/acetone classification tests on your product (Sec. 10.5B). Obtain IR and ^{1}H NMR spectra of your starting material and product and compare them with those of authentic samples (Figs. 14.5–14.8).

Discovery Experiment   *Analysis of S_N Reactions as a Function of Substrate*

In Section 14.2, we stated that the relative rate of an S_N reaction for a substance R–L is a function of the degree of the carbon atom bearing the leaving group. This proposition is subject to experimental verification using simple qualitative tests. By referring to the procedures described under "Alkyl Halides" in Section 10.5B, propose and execute experiments that explore the relative rates of S_N1 and S_N2 reactions for substrates in which the leaving group is on carbon atoms of various degrees. You may consider using both alkyl and aryl halides in your study, but be certain to consult with your instructor before undertaking any experiments.

WRAPPING IT UP

Carefully dilute the *aqueous layer* from the first separation with water and then neutralize it with sodium carbonate. Combine this solution with the other *aqueous washes (water, saturated sodium bicarbonate, and saturated sodium chloride) and layers,* and flush them down the drain with water. The *sodium sulfate* used as the drying agent is contaminated with product, so place it in the container for halogenated solids. Pour the *residue in the stillpot from the Miniscale Procedure* into the container for halogenated liquids. Dilute the *silver nitrate test solution* with water and flush it down the drain unless instructed otherwise. Place the *sodium iodide/acetone test solution* in the container for halogenated liquids.

EXERCISES

1. The work-up procedure calls for washing the crude 2-chloro-2-methylbutane with *cold* sodium bicarbonate solution.

   a.   What purpose does this wash serve?

b. This washing procedure is accompanied by vigorous gas evolution, which increases the difficulty of handling and requires considerable caution. Alternatively, one might consider using a dilute solution of sodium hydroxide instead of the sodium bicarbonate. Discuss the relative advantages and disadvantages of using these two basic solutions in the work-up.

c. On the basis of your answer to **b**, why were you instructed to use sodium bicarbonate, even though it is more difficult to handle?

2. What is the purpose of the final wash of the organic layer with saturated sodium chloride solution in the purification process?

3. The 2-chloro-2-methylbutane is dried with *anhydrous* sodium sulfate in this procedure. Could solid sodium hydroxide or potassium hydroxide be used for this purpose? Explain your answer.

4. Draw a reaction profile to illustrate the conversion of 2-methyl-2-butanol and HCl to 2-chloro-2-methylbutane and water (Eqs. 14.21–14.23). Label enthalpy of activation, $\Delta H^{\ddagger}$, for the rate-determining step and the enthalpy, ΔH_{rxn}, for the overall reaction.

5. Explain why polar solvents would be expected to increase the rate of the reaction in Exercise 4.

6. Consider the by-products that might be formed by E1 and/or E2 processes in the reaction of 2-methyl-2-butanol with HCl.

$$\overset{\overset{\displaystyle CH_3}{\displaystyle |}}{\underset{\underset{\displaystyle OH}{\displaystyle |}}{CH_3CCH_2CH_3}} + HCl_{aq} \longrightarrow \text{?}$$

2-Methyl-2-butanol

a. Provide structures for these products.

b. Which alkene would you expect to be favored and why?

7. Draw the structures of all the alcohols that are isomeric with 2-methyl-2-butanol.

a. Arrange these alcohols in order of increasing reactivity toward *concentrated* hydrochloric acid.

b. Which, if any, of these other alcohols would you expect to give a reasonable yield of the corresponding alkyl chloride under such reaction conditions?

8. Consider the reaction of (*S*)-3-methyl-3-hexanol with *concentrated* HCl.

$$\overset{\overset{\displaystyle CH_3}{\displaystyle |}}{\underset{\underset{\displaystyle OH}{\displaystyle |}}{CH_3CH_3CCH_2CH_2CH_3}} + HCl_{conc} \longrightarrow \overset{\overset{\displaystyle CH_3}{\displaystyle |}}{\underset{\underset{\displaystyle Cl}{\displaystyle |}}{CH_3CH_3CCH_2CH_2CH_3}}$$

3-Methyl-3-hexanol   3-Chloro-3-methylhexane

a. Draw a three-dimensional representation for this enantiomer.

b. Predict whether the 3-chloro-3-methylhexane formed in the reaction would be *R*-, *S*-, or racemic, and explain your prediction.

9. Why does ionization of a tertiary substrate $R_3C{-}L$ proceed at a rate faster than that of a secondary substrate $R_2CH{-}L$?

10. How would doubling the concentration of the nucleophile affect the rate of an S_N1 reaction?

11. Benzyl chloride, $C_6H_5CH_2Cl$, readily undergoes an S_N1 reaction, yet it is a primary substrate. Explain this seemingly anomalous result.

12. For each of the following pairs, predict which one will undergo an S_N1 reaction faster. Explain your predictions.

13. On prolonged heating with *concentrated* hydrochloric acid, 2,2-dimethyl-1-propanol, a primary alcohol, reacts to give 2-chloro-2-methylbutane. Provide a stepwise mechanism for this reaction using curved arrows to symbolize the flow of electrons. Also account for the fact that the observed product is a tertiary alkyl chloride even though the starting alcohol is primary.

2,2-Dimethyl-1-propanol   2-Chloro-2-methylbutane

Refer to the online resources and use the spectra viewer and Tables 8.1–8.8 as needed to answer the blue-numbered questions on spectroscopy.

14. Consider the spectral data for 2-methyl-2-butanol (Figs. 14.5 and 14.6).

    a. In the functional group region of the IR spectrum, identify the absorption associated with the hydroxyl group, and explain why this peak is broad.

    b. In the ^1H NMR spectrum, assign the various resonances to the hydrogen nuclei responsible for them.

    c. If the NMR sample of 2-methyl-2-butanol in $CDCl_3$ were shaken with D_2O *prior* to taking the ^1H NMR spectrum, what differences might you expect to observe?

    d. For the ^{13}C NMR data, assign the various resonances to the carbon nuclei responsible for them.

15. Consider the NMR spectral data for 2-chloro-2-methylbutane (Fig. 14.8).

    a. In the ^1H NMR spectrum, assign the various resonances to the hydrogen nuclei responsible for them.

    b. For the ^{13}C NMR data, assign the various resonances to the carbon nuclei responsible for them.

16. Discuss the differences in the **IR** and **NMR** spectra of 2-methyl-2-butanol and 2-chloro-2-methylbutane that are consistent with the conversion of an alcohol into an alkyl chloride in this procedure.

SPECTRA

Starting Material and Product

Figure 14.5
*IR spectrum of 2-methyl-
2-butanol (neat).*

Figure 14.6
*NMR data for 2-methyl-2-butanol
(CDCl₃).*

(a) ¹H NMR spectrum (300 MHz).
(b) ¹³C NMR data: δ 8.7, 28.6, 36.5, 71.1.

Figure 14.7
*IR spectrum of 2-chloro-
2-methylbutane (neat).*

Figure 14.8
NMR data for 2-chloro-2-methylbutane (CDCl₃).

(a) ¹H NMR spectrum (300 MHz).
(b) ¹C NMR data: δ 9.5, 32.0, 39.0, 70.9.

14.6 CHEMICAL KINETICS: EVIDENCE FOR NUCLEOPHILIC SUBSTITUTION MECHANISMS

Introduction

Continuous changes in the concentrations of reactants and products are observed during the course of chemical reactions. Thus, as an *irreversible* reaction proceeds, we find that the concentration of each reactant decreases until that of the limiting reagent becomes zero, at which point the reaction stops. Simultaneously, the concentration of the product increases from zero to its maximum value when the reaction is complete. How fast these concentrations change *as a function of time* is determined by the **reaction rate** for a chemical transformation.

The field of **chemical kinetics** involves investigation of the interplay of factors and variables influencing the rates of reactions. Because chemical kinetics provide great insight into the nature and details of reaction mechanisms, it is hard to overstate the importance of this area of research. Indeed, kinetic studies have provided some of the most important evidence in support of the S_N1 and S_N2 mechanisms (Sec. 14.2).

To simplify our discussion of kinetics, let's first assume that the reactions are *irreversible* and then consider the mathematical expressions or **rate laws** governing these classes of reaction. To do so, we need only examine the *rate-determining step* for the particular type of substitution. In the case of an S_N1 reaction, this step is the formation of a carbocation from the precursor $R-L$ (Eq. 14.5). The rate of the overall reaction is then proportional only to the concentration of substrate, as expressed in Equation 14.27. We see that the rate is **first order** in the concentration of $R-L$, expressed as $[R-L]^1$, and **zeroth order** in that of Nu:, that is, $[Nu:]^0$, which means that the rate is *independent* of its concentration. Adding the two exponents for the concentration gives the overall order of the S_N1 reaction, which is seen to be **first order**. A simplified version of the rate law is seen in Equation 14.28, and this is the form in which it is normally written.

$$\text{Rate of reaction} = k_1[R-L]^1[Nu:]^0 \tag{14.27}$$

$$\text{Rate of reaction} = k_1[R-L] \tag{14.28}$$

For S_N2 reactions, the rate-determining step involves attack by the nucleophile, Nu:, on the substrate R—L (Eq. 14.6). The corresponding rate law is shown in Equation 14.29, wherein the rate is seen to be proportional to the concentrations of R—L *and* Nu:. Both concentrations are taken to the first power, so the rate of the reaction is said to be **second order** overall.

$$\text{Rate of reaction} = k_1[\text{R}-\text{L}][\text{Nu:}] \tag{14.29}$$

Both k_1 and k_2 (Eqs. 14.27–14.29) are **rate constants** or proportionality factors that relate the rate of reaction and the concentrations of reactants; the subscripts "1" and "2" indicate that they are for unimolecular and bimolecular processes, respectively. Rate constants may have different units: that for k_1 is $(\text{time})^{-1}$, whereas that for k_2 is $(\text{concentration})^{-1}\,(\text{time})^{-1}$.

As compared to their S_N2 relatives, S_N1 reactions are somewhat easier to study experimentally, and calculations of their rate constants are easier. Consequently, you will be performing quantitative measurements of the kinetics of an S_N1 rather than an S_N2 reaction in the procedure of this section.

Kinetic Study of an S_N1 Reaction

Because an S_N1 reaction is a first-order reaction, the rate of the reaction is linearly dependent on the concentration of the reactant R–L (Eq. 14.27). For example, doubling the concentration doubles the rate. A graph of *rate* versus *concentration* thus yields a straight line whose slope is k_1.

When the R–L is being consumed in a first-order reaction, its concentration decreases exponentially with time. If C_0 is the *initial* concentration of the substrate at time $t = 0$ (t_0) and C is its concentration at any elapsed time t, where t is measured in a unit of time, typically seconds or minutes, their relationship is given by Equation 14.30. This equation may be rewritten as shown in Equations 14.31 and 14.32.

$$C_t = C_0 e^{-\text{k}/\text{t}} \tag{14.30}$$

$$k_1 t = \ln\,(C_0/C_t) \tag{14.31}$$

$$k_1 t = 2.303 \log\,(C_0/C_t) \tag{14.32}$$

If the initial concentration, C_0, of R–L is known and if its concentration, C_t, is measured at various time intervals, t, while the reaction is proceeding, the rate constant may be determined in the following ways.

1. The values of C_0, C_t, and t measured at each point during the reaction are substituted into Equation 14.31 or 14.32, which is then solved for k_1. This produces several values of k_1, which may then be averaged. The *correct* rate constant is not easily obtained by this method because the average value will be determined without bias, that is, without compensating for any measurements that may be incorrect due to experimental error.

2. A better method for calculating k_1 from experimental data is to plot either (C_0/C_t) or $\log\,(C_0/C_t)$ versus t. A straight line is then drawn so that it lies closest to the largest number of points on the graph and is called the "best fit." This line is drawn *with bias* in that it purposely gives more "weight" to the majority of points lying close to the line; those lying farther from the line are taken to be less reliable and thus are given less "weight." The slope of the line is the

rate constant k_1 if natural logarithms are used (Eq. 14.31) or $k_1/2.303$ if $\log_{10}$ are used (Eq. 14.32). Alternatively, a least-squares analysis may be performed on the data using commercially available software such as the "LINEST" option provided in Microsoft® Excel to obtain the slope of the line; this eliminates the human element in defining where the line should be drawn. You should consult with your instructor if this technique is to be used.

The discussion and experiments presented below illustrate methods of studying chemical kinetics and determining the effects of structure on reactivity, as exemplified by the **solvolysis** of tertiary alkyl halides. The term "solvolysis" describes a substitution reaction in which the solvent, HOS, functions as the nucleophile (Eq. 14.33). In principle, solvolyses may be performed in any nucleophilic solvent such as water (hydrolysis), alcohols (alcoholysis), and carboxylic acids (e.g., acetolysis with acetic acid). However, a practical limitation in choosing a solvent is the solubility of the substrate in the solvent because the reaction mixture must be homogeneous; if it is not, surface effects at the interface of the phases will make the kinetic results difficult to interpret and probably nonreproducible as well. In the experiment described here, you will explore solvolyses in mixtures of 2-propanol and water.

$$R-X + H-OS \rightarrow R-OS + H-X \tag{14.33}$$

You will need to measure the quantities C_0, C_i, and t to determine k_1. The procedure we'll use to measure C_t depends on the fact that a molecule of acid, H–X, is produced for each molecule of alkyl halide that reacts (Eq. 14.33). Thus, you may monitor the progress of the reaction by determining the concentration of hydrogen ion, $[H^+]$, produced as a function of time: At any time t, $[H^+]_t = C_0 - C_t$, so that C_t is defined by Equation 14.34.

$$C_t = C_0 - [H^+]_t \tag{14.34}$$

The value of $[H^+]_t$ is determined experimentally by withdrawing an accurately measured sample, referred to as an **aliquot**, from the reaction mixture and "quenching" it in a quantity of 98% 2-propanol sufficient to prevent further solvolysis; the solvolysis does not proceed at a measurable rate when so little water is present. The elapsed time t is simply the time that the aliquot was removed minus the time t_0 that the reaction was started. The aliquot is titrated with base to yield $[H^+]$. If C_0 is known, C_t may then be calculated from Equation 14.34.

The value of C_0 can be determined in two different ways. (1) A solution of defined molarity of the alkyl halide in the solvent mixture to be used may be prepared by accurately measuring the mass of alkyl halide and the volume of solvent. (2) A more reliable and easier procedure is to allow the solvolysis to go to completion, at which time all the alkyl halide will have reacted. Accurate titration of an aliquot of this mixture is the "infinity point," the point at which $[H^+]_\infty = C_0$. Equation 14.34 may then be transformed to Equation 14.35.

$$C_t = [H^+]_\infty - [H^+] \tag{14.35}$$

It is *not* necessary to use a standardized solution of base for the titrations so long as the *same* basic solution is used for *all* the titrations you perform and the volumes of the aliquots withdrawn are *identical*. Furthermore, it is unnecessary to calculate $[H^+]$ for each aliquot, because the value is directly proportional to the *volume* of basic solution required to titrate the acid that has formed according to

Equation 14.33. Using this fact, and Equation 14.35, allows Equation 14.32 to be rewritten as Equation 14.36.

$$2.303 \log \frac{(mL\ of\ NaOH)_q}{(mL\ of\ NaOH)_q - (mL\ of\ NaOH)_t} = k_1 t \qquad (14.36)$$

A variety of factors influence the rate of S_N1 reactions, and the experimental procedures you perform will allow you to investigate some of them. These include the influence of solvent and the structure of the alkyl halide undergoing solvolysis, as discussed below. Additional factors affecting reaction rates include temperature and the nature of L:, the leaving group; exploration of these is possible in the **Optional Discovery Experiments** provided.

1. **Solvent Composition.** The nature of the solvent is expected to affect the rate of an S_N1 reaction because formation of ions occurs in the rate-determining step (Eq. 14.5). Based on your understanding of the principles of solvation, you would predict that more polar solvents would accelerate the rate of the reaction. You will test this hypothesis with experiments that use various mixtures of 2-propanol and water for the solvolysis.

2. **Alkyl Group.** Tertiary alkyl halides may be used in solvolyses with confidence that the reactions are proceeding by the S_N1 mechanism. Two compounds that allow you to examine the possible effects of the structure of the alkyl group on reaction rates are 2-chloro-2-methylpropane (*tert*-butyl chloride) and 2-chloro-2-methylbutane (*tert*-pentyl chloride).

The composition of the reaction mixture formed in an S_N1 reaction affects neither the rate of the reaction nor the value of the rate constant because the products all arise from a common intermediate, namely a carbocation, that is formed in the rate-determining step (Eq. 14.5). Thus, the four products that are likely to be formed in the solvolysis of 2-chloro-2-methylpropane in aqueous 2-propanol may individually be formed at different rates because of the differing values of the rate constants, $k_a - k_d$, and of the concentrations of the reagents that react with the carbocation. Nevertheless, the overall rate of the reaction is defined by the ionization that produces the carbocation and chloride ion and has the rate constant k_1 (Scheme 14.1).

Scheme 14.1

EXPERIMENTAL PROCEDURES

Kinetics of Solvolysis of 2-Chloro-2-Methylbutane

Discovery Experiment   **Purpose** To investigate the effect of various factors on the rates of S_N1 reactions.

> ### SAFETY ALERT
>
> 1. **Wear safety glasses or goggles and suitable protective gloves while performing the experiments.**
> 2. **Avoid contact of the aqueous solutions with your skin, as they may dry it out and/or cause chemical burns. Wash any affected areas with copious amounts of water.**

MINISCALE PROCEDURE

Buret

Clamp

Preparation Refer to the online resources to answer Pre-Lab Exercises, access videos, and read the MSDSs for the chemicals used or produced in this procedure. Review Sections 2.5 and 2.11. Prepare a table in your notebook for recording (1) the solvent(s) that have been assigned to you and their volume(s); (2) the mass or volume of alkyl halide to be used; (3) the time, t_0, at which the kinetic run is initiated; (4) the temperature of the reaction mixture; (5) the results of a "blank" titrimetric determination; (6) a series of times, t, at which aliquots are withdrawn; and (7) the initial and final buret readings observed in the titration of each aliquot.

Apparatus A 10-mL and a 100-mL graduated cylinder, 50-mL buret, a 25-mL beaker, 10-mL volumetric pipet, thermometer, one 125-mL and three 250-mL Erlenmeyer flasks, a 10-mm × 75-mm test tube, disposable pipet, timer, apparatus for magnetic stirring.

Setting Up *Use the same pipet and buret, as well as the same solution of sodium hydroxide, throughout this experiment.* Using a graduated cylinder, accurately measure 100 mL of your assigned solvent into a 250-mL Erlenmeyer flask equipped with a well-fitting rubber stopper. Measure and record the temperature of this solution. Place 80 mL of 98% 2-propanol in a second flask for use in quenching the solvolysis. Prepare 125–150 mL of approximately 0.04 *M* sodium hydroxide solution in a third flask. Stopper the flask with a well-fitting rubber stopper.

Set up the buret, and after rinsing it with a small amount of the 0.04 *M* sodium hydroxide solution, fill it with this solution, ensuring that all air bubbles are out of the tip, and cover the top with a test tube or beaker to minimize absorption of carbon dioxide from the air.

Put about 2 mL of phenolphthalein indicator solution in the test tube and have it and a dropper available for use in each titration. Connect a short length of rubber tubing to the nearest aspirator or vacuum line. This will be used to draw air through the volumetric pipet for a minute or two after each sampling to dry it before taking the next sample. Have a timer available that may be read to at least the nearest minute.

This experiment requires that a series of quantitative measurements be performed in a relatively short time. You should be prepared to work *rapidly* and *carefully* to maintain a high standard of accuracy. Buret readings should be made to the nearest 0.02 mL if possible, although precision within 0.05 mL will normally be satisfactory. Time measurements should be made *at least* to the nearest minute.

Reaction   Initiate the kinetic run by adding about 1 g of the alkyl halide to the solvent mixture. The alkyl halide may either be weighed out or be measured with a 1-mL pipet. Swirl the mixture gently to obtain homogeneity. Record the time of addition as t_0. Keep the flask tightly stoppered to avoid evaporation that would change the concentration of the mixture.

While waiting to make the first measurement, determine a "blank" correction for the solvent as follows. Using a graduated cylinder, measure a 10-mL portion of your assigned solvent into the 125-mL Erlenmeyer flask. Next add 10 mL of 98% 2-propanol and 4–5 drops of phenolphthalein to the flask and titrate the stirred solution with aqueous base until a faint pink color persists for 30 sec. In this and all other titrations, use a white background below the titration flask to assist you in detecting color. You may also wish to accentuate the lower edge of the meniscus in the buret by holding dark paper or some other dark object just below it to make the graduations on the buret easier to read. The blank correction will probably be no more than 0.05–0.15 mL. Dispose of the titrated solution as specified in "Wrapping It Up." Rinse the flask with water and then with a small amount of 98% 2-propanol so that it is ready for further use.

At regular intervals, use the volumetric pipet to remove a 10-mL aliquot from the reaction mixture, and quench the reaction by adding the aliquot to 10 mL of 98% 2-propanol contained in the 125-mL Erlenmeyer flask. Be sure to note the time of addition of the aliquot, probably best taken as the time at which one-half of it has been added. Titrate the solution with base to the phenolphthalein endpoint just as you did in the blank determination.

The suggested *approximate* times for taking aliquots under various conditions are as follows:

1. 50% 2-propanol/water and 2-chloro-2-methylpropane: 10, 20, 35, 50, 75, and 100 min.

2. 55% 2-propanol/water and 2-chloro-2-methylpropane: 15, 30, 50, 75, 100, and 135 min.

3. 60% 2-propanol/water and 2-chloro-2-methylpropane: 20, 40, 70, 100, 130, and 170 min.

4. 50% 2-propanol/water and 2-chloro-2-methylbutane: 10, 20, 30, 40, 50, and 60 min.

5. 55% 2-propanol/water and 2-chloro-2-methylbutane: 15, 30, 45, 60, 80, 110, and 140 min.

6. 60% 2-propanol/water and 2-chloro-2-methylbutane: 20, 40, 60, 80, 100, and 120 min.

At room temperature, the *fastest* of these solvolyses requires about 4 h to reach 99.5% completion, and the slowest requires over 12 h. Therefore, it is easiest to wait until the next laboratory period to perform the infinity titration that is necessary to obtain C_0. Stopper the reaction flask *tightly* to avoid evaporation and store it in your desk. *Be sure to save at least 30 mL of the sodium hydroxide solution in a tightly stoppered flask so that it will be available for the infinity titration.*

Analysis Obtain IR and NMR spectra of the starting materials and compare them with those of authentic samples (Figs. 14.7–14.10). Compute the desired rate constants, k_1, by the following sequence of steps.

1. Using the buret readings, determine by difference the number of milliliters of sodium hydroxide solution used in each titration. Apply the blank correction to all values by subtracting it from each volume and use these corrected values in your calculations. Determine the elapsed time at which each aliquot was withdrawn from the reaction mixture. Apply Equation 14.36 by calculating the log term for each kinetic point, multiplying that value by 2.303, and *plotting* the resulting number (ordinate) versus elapsed time t (abscissa) in hours. Draw the best straight line through the points. Determine the slope of the line; this slope is the rate constant k_1. Alternatively, analyze your experimental data using least-squares techniques to obtain the rate constant.

2. Using the same data and Equation 14.34, calculate the value of k_1 separately for each kinetic point. Compare the *average* of these values with the rate constant obtained graphically. Also compare this average with each of the values that were averaged. In your laboratory report for this experiment, specify which procedure, graphical or averaging, allows you most easily to identify a point that is likely in error.

3. Equation 14.37 provides the half-life, $t_{1/2}$, the time necessary for one-half of the original alkyl halide to react. Calculate the half-life of your reaction using the value of k_1 obtained from the graph. Then reexamine your experimental data to determine whether about one-half of the total volume of NaOH used in the infinity titration had been consumed by the end of the first half-life of the reaction. If not, an error has been made in the calculations, and they should be rechecked.

$$t_{1/2} = 0.693/k_1 \qquad (14.37)$$

Discovery Experiment *Effect of Temperature on Rates of Solvolysis*

Temperature changes affect the rates of reactions, the *rough* rule of thumb being that the rate doubles for each rise of 10 °C. You may explore this effect by comparing solvolysis rates at room temperature and at 0 °C. Based on the experimental procedure of this section, design and execute a protocol for measuring the rate of solvolysis of either 2-chloro-2-methylpropane or 2-chloro-2-methylbutane at 0 °C. You should consider the effect of solvent polarity on the rate of the reaction so that the conditions you propose to use will allow measurable amounts of solvolysis to occur during the time period available to you for removing aliquots for titration. Consult with your instructor before performing your proposed procedure.

Discovery Experiment *Effect of Leaving Group on Rates of Solvolysis*

Because the bond between the carbon atom and the leaving group, L, in R–L is broken in the rate-determining step of an S_N reaction, you might expect that the rate of the reaction would be dependent on the nature of L. You may investigate this possibility by comparing the rates of solvolysis of alkyl bromides with those of alkyl chlorides. Based on the experimental procedure of this section, design and execute a protocol for measuring the rate of solvolysis of 2-bromo-2-methylpropane at room temperature. Consult with your instructor before performing your proposed procedure.

WRAPPING IT UP

Although you normally should not dispose of solutions containing organic halides by pouring them down the drain, it is acceptable to do so in this case because of the small quantities of the alkyl halides being used and their facile solvolysis. Make the solvolysis solution strongly basic and allow it to remain in your desk until the next laboratory period. Neutralize the basic solution and flush it down the drain using copious amounts of water.

EXERCISES

1. Why does the titration endpoint color fade after 30–60 sec?

2. You covered the top of the buret with a test tube or beaker to protect its contents from air. Why was a rubber or cork stopper not used instead?

3. In the kinetics experiment you performed, individual aliquots were quenched by adding them to 98% 2-propanol. Why does this stop the solvolysis?

4. Suppose the reaction flask was left unstoppered until the following laboratory period when the infinity titration was performed. Would the calculated rate constant have been larger or smaller than the "correct" value if some evaporation of solvent had occurred during this time?

5. Derive Equation 14.36 from Equation 14.32. (*Hint:* Set up the concentrations C_0 and C_t in terms of moles/liter for the volume of NaOH used in each titration, and recognize the fact that $[OH^-] = [H^+]$. Substitute these values into Equation 14.32, and cancel out constants that appear in the new equation.)

6. Give the possible advantages for using an infinity titration and Equation 14.36 for analyzing your data instead of calculating concentrations of the alkyl halide and using Equation 14.32 for the analysis. Be sure to consider the alternatives for obtaining C_0.

7. Equation 14.37 gives the *half-life, $t_{1/2}$,* of a reaction, which is defined as the time required for a reaction to reach 50% completion. This equation applies to any first-order reaction, including radioactive disintegration. Show how Equation 14.37 may be derived from Equation 14.32. (*Hint:* When the reaction is 50% complete, $C_t = 1/2\ C_0$.)

8. List the possible errors involved in the determination of rate constants by the procedure you used and state the relative importance of each.

9. Refer to Scheme 14.1 and predict the ratio of 2-methyl-2-butanol:*tert*-amyl isopropyl ether expected if the solvent for the reaction were equimolar in water and 2-propanol and $k_c = 5k_d$.

Refer to the online resources and use the spectra viewer and Tables 8.1–8.8 as needed to answer the blue-numbered questions on spectroscopy.

10. Consider the NMR spectral data for 2-chloro-2-methylbutane (Fig. 14.8).

    a. In the ^{1}H NMR spectrum, assign the various resonances to the hydrogen nuclei responsible for them.

    b. For the ^{13}C NMR data, assign the various resonances to the carbon nuclei responsible for them.

11. Consider the NMR spectral data for 2-chloro-2-methylpropane (Fig. 14.10).

    a. In the ^{1}H NMR spectrum, assign the resonance to the hydrogen nuclei responsible for it.

    b. For the ^{13}C NMR data, assign the resonance to the carbon nuclei responsible for it.

12. Consider the spectral data for 2-methyl-2-butanol (Figs. 14.5 and 14.6).

    a. In the functional group region of the IR spectrum, identify the absorption associated with the hydroxyl group, and explain why this peak is broad.

    b. In the ^{1}H NMR spectrum, assign the various resonances to the hydrogen nuclei responsible for them.

    c. If the NMR sample of 2-methyl-2-butanol in CDCl$_3$ were shaken with D$_2$O *prior* to taking the ^{1}H NMR spectrum, what differences might you expect to observe?

    d. For the ^{13}C NMR data, assign the various resonances to the carbon nuclei responsible for them.

13. Consider the spectral data for 2-methyl-1-butene (Figs. 10.4 and 10.5).

    a. In the functional group region of the IR spectrum, specify the absorptions associated with the carbon-carbon double bond and the hydrogens attached to the C1 carbon atom.

    b. In the ^{1}H NMR spectrum, assign the various resonances to the hydrogen nuclei responsible for them.

    c. For the ^{13}C NMR data, assign the various resonances to the carbon nuclei responsible for them.

14. Consider the spectral data for 2-methyl-2-methylbutene (Figs. 10.6 and 10.7).

    a. In the functional group region of the IR spectrum, specify the absorptions associated with the carbon-carbon double bond and the hydrogens attached to the C3 carbon atom.

    b. In the ^{1}H NMR spectrum, assign the various resonances to the hydrogen nuclei responsible for them.

    c. For the ^{13}C NMR data, assign the various resonances to the carbon nuclei responsible for them.

SPECTRA

Starting Materials and Products

The IR spectrum and NMR spectral data for 2-chloro-2-methylbutane, 2-methyl-2-butanol, 2-methyl-1-butene, and 2-methyl-2-methylbutene are provided in Figures 14.7, 14.8, 14.5, 14.6, and 10.4–10.7, respectively.

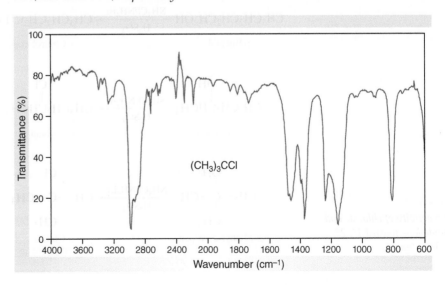

Figure 14.9

IR spectrum of 2-chloro-2-methylpropane (neat).

Figure 14.10

NMR data for 2-chloro-2-methylpropane (CDCl$_3$).

(a) ^{1}H NMR spectrum (300 MHz).

(b) ^{13}C NMR data: δ 34.5, 66.5.

14.7 COMPETING NUCLEOPHILES IN S$_N$ REACTIONS

The nature of the nucleophile, the structure of the alcohol, and the solvent are all important factors in the S$_N$1 and S$_N$2 reactions of alcohols. The role of the leaving group in such substitution reactions is also important. As noted in Section 14.1, leaving group ability is positively correlated with the acidity of the conjugate acid of the leaving group. For example, hydroxide ion is a poor leaving group because as the conjugate base of water, a weak acid, it is a strong base. Therefore, protonating the hydroxyl group of the alcohol converts it to an oxonium species so that the leaving group becomes water, the conjugate base of hydronium ion, H$_3$O$^+$, a strong acid (Eq. 14.38). The reactions in this experiment are all performed in acidic media, so water is the leaving group; your focus will be on exploring the reactivity of chloride and bromide ions toward 1°, 2°, and 3° butanols (Fig. 14.11).

$$R\!-\!OH + HA \rightleftharpoons R\!-\!\overset{+}{O}H_2 + A^- \qquad (14.38)$$

$$CH_3CH_2CH_2CH_2OH \xrightarrow[\text{H}_2\text{SO}_4]{\text{NH}_4\text{Cl}/\text{NH}_4\text{Br}} CH_3CH_2CH_2CH_2Cl \;+\; CH_3CH_2CH_2CH_2Br$$

1-Butanol   1-Chlorobutane   1-Bromobutane

Figure 14.11

The reactivity of chloride and bromide ions toward 1°, 2°, and 3° butanols.

2-Butanol   2-Chlorobutane   2-Bromobutane

2-Methyl-2-propanol   2-Chloro-2-methylpropane   2-Bromo-2-methylpropane

To evaluate the reactivity of these nucleophile toward each class of alcohol, chloride and bromide are used in equimolar amounts, thereby creating a "competition" between the halides. The mechanism of the reaction may be S_N1 or S_N2 depending on the structure of the alcohol, so that the results you obtain provide data regarding the relative reactivity of these halide ions as nucleophiles in such substitution reactions. You will analyze the ratio of products by gas chromatography. From the ratio of products obtained, you will be able to formulate an order of reactivity of halides toward 1°, 2°, and 3° alcohols and gain insight into the mechanism of the reaction for each.

EXPERIMENTAL PROCEDURES

Competing Nucleophiles in S_N Reactions

Discovery Experiment

Purpose To investigate the relative reactivity of chloride and bromide ions in acid-catalyzed nucleophilic substitution reactions of 1°, 2°, and 3° butanols.

SAFETY ALERT

1. **Wear safety glasses or goggles and suitable protective gloves while performing the experiment.**

2. **Examine your glassware for cracks and chips. This procedure involves heating *concentrated* acids, and defective glassware could break under these conditions, spilling hot corrosive chemicals on you and those working around you.**

3. **Use care when handling *concentrated* sulfuric acid. If any acid comes in contact with your skin, immediately wash it off with copious amounts of cold water and then with 10% sodium bicarbonate solution.**

4. **Mixing sulfuric acid and water can be highly exothermic. *Always* add the acid *to* the water so the heat is dispersed through warming of the water. Add the acid slowly and with swirling to ensure continuous and thorough mixing.**

Preparation Refer to the online resources to answer Pre-Lab Exercises, access videos, and read the MSDSs for the chemicals used or produced in this procedure. Review Sections 2.9, 2.11, 2.21, 2.22, 2.25, and 6.4.

MICROSCALE PROCEDURE FOR 1-BUTANOL AND 2-BUTANOL

Apparatus A 10-mL conical vial, apparatus for heating under reflux, magnetic stirring, and *flameless* heating.

Setting Up Place 1.0 mL of either 1-butanol or 2-butanol and a spinvane in the 10-mL vial. Slowly add 1.0 mL of a 1.5 *M* solution of ammonium chloride in 9 *M* sulfuric acid and 1.0 mL of a 1.5 *M* solution of ammonium bromide in 9 *M* sulfuric acid to the 10-mL vial equipped with a spinvane. Set up the apparatus for heating under reflux.

Reaction Heat the solution under gentle reflux for about 60 min.

Work-Up and Isolation Allow the reaction mixture to cool to room temperature and *slowly* pour it into a 100-mL beaker containing 5 g of ice. Add *saturated* aqueous sodium carbonate solution *slowly* with stirring until the pH of the mixture is about 8. Add 5 mL of hexane. Transfer the liquids to a 25-mL centrifugation tube, remove the *aqueous* layer using a pipet, and wash the organic layer with 5 mL of *saturated* sodium chloride solution. Remove the *aqueous* layer and dry the organic layer over several spatula-tips full of 0.5 g of *anhydrous* sodium sulfate. Cap the vial tightly for later analysis.

MICROSCALE PROCEDURE FOR 2-METHYL-2-PROPANOL

Apparatus A 10-mL conical vial, apparatus for magnetic stirring.

Setting Up *Slowly* add 1.0 mL of a 1.5 *M* solution of ammonium chloride in 9 *M* sulfuric acid and 1.0 mL of a 1.5 *M* solution of ammonium bromide in 9 *M* sulfuric acid to the 10-mL vial equipped with a spinvane. *Slowly* add 1.0 mL of 2-methyl-2-propanol.

Reaction Stir the reaction mixture at room temperature for about 30 min.

Work-Up and Isolation *Slowly* pour the reaction mixture into a 100-mL beaker containing 5 g of ice. Add *saturated* aqueous sodium carbonate solution *slowly* with stirring until the pH of the mixture is about 8. Add 5 mL of hexane. Transfer the liquids to a 25-mL centrifugation tube, remove the *aqueous* layer using a pipet, and wash the organic layer with 5 mL of *saturated* sodium chloride solution. Remove the *aqueous* layer and dry the organic layer over several spatula-tips full of *anhydrous* sodium sulfate. Cap the vial tightly for later analysis.

Analysis Following the directions of your instructor, perform a gas chromatographic analysis of your sample and compare your result to the known standards. Calculate the relative ratio of products for each of the starting alcohols that you used in the reaction.

WRAPPING IT UP

Dilute the *combined aqueous layers* with water and flush down the drain. Place the *hexane solution of alkyl halides* in the container for halogenated organic liquids.

EXERCISES

1. What ratio of products did you observe in the acid-catalyzed reaction of 1-butanol with the mixture of ammonium halides? What conclusions can you draw from those results?

2. Using curved arrows to symbolize the flow of electrons, propose *two* mechanisms for the conversion of 1-butanol to 1-chlorobutane.

   a. Identify which of these is S_N1 and which is S_N2.

   b. Based on your results from this experiment, can you conclude which mechanism is operating? Why is the other mechanistic pathway disfavored?

   c. What ratio of products would you expect, if the concentration of ammonium halides were doubled in this reaction? If they were reduced to one half?

3. What ratio of products did you observe in the reaction of 2-butanol with the mixture of ammonium halides? What conclusions can you draw from those results?

4. Using curved arrows to symbolize the flow of electrons, propose *two* mechanisms for the conversion of 2-butanol to 2-bromobutane.

   a. Identify which is S_N1 and S_N2.

   b. Based on your results from this experiment, can you differentiate which mechanism is operating?

   c. What ratio of products would you expect, if the concentration of ammonium halides were doubled in this reaction? If they were reduced to one-half?

   d. Suggest an experiment that would enable you to differentiate between an S_N1 and an S_N2 reaction.

5. What was the ratio of products for the 2-methyl-2-propanol reaction? What conclusions can you draw from those results?

6. Based on your results for 2-methyl-2-propanol and using curved arrows to symbolize the flow of electrons, propose a mechanism for this substitution reaction.

7. In the reaction of 2-methyl-2-propanol, what ratio of products would you expect if the concentration of ammonium halides were doubled? If they were reduced to one-half?

8. Based upon the results of these experiments, predict the ratios of products you might obtain from the reactions of 1-butanol, 2-butanol, and 2-methyl-2-propanol if a mixture of equimolar amounts of ammonium chloride, ammonium bromide, and ammonium iodide were used.

Refer to the online resources and use the spectra viewer and Tables 8.1–8.8 as needed to answer the blue-numbered questions on spectroscopy.

9. Consider the spectral data for 1-butanol (Figs. 8.28, 14.1, and 14.2).

   a. In the functional group region of the IR spectrum, specify the absorption associated with the hydroxyl group, and explain why this peak is broad.

   b. In the ^1H NMR spectrum, assign the various resonances to the hydrogen nuclei responsible for them.

   c. If the NMR sample of 1-butanol in $CDCl_3$ were shaken with D_2O *prior* to taking the ^1H NMR spectrum, what differences might you expect to observe?

   d. For the ^{13}C NMR data, assign the various resonances to the carbon nuclei responsible for them.

10. Consider the NMR spectral data for 1-bromobutane (Fig. 14.4).

    a. In the ^1H NMR spectrum, assign the various resonances to the hydrogen nuclei responsible for them.

    b. For the ^{13}C NMR data, assign the various resonances to the carbon nuclei responsible for them.

11. Discuss the differences in the IR and ^1H NMR spectra of 1-butanol and 1-bromobutane that are consistent with the conversion of an alcohol into a bromoalkane in this procedure.

12. Consider the NMR data for 1-chlorobutane (Fig. 9.3).

    a. In the ^{1}H NMR spectrum, assign the various resonances to the hydrogen nuclei responsible for them.

    b. For the ^{13}C NMR data, assign the various resonances to the carbon nuclei responsible for them.

13. Discuss the differences observed in the ^{1}H and ^{13}C NMR spectra of 1-bromobutane (Fig. 14.4) with those of 1-chlorobutane (Fig. 9.3). To what factor(s) do you attribute these differences?

14. Consider the spectral data for 2-butanol (Figs. 14.12 and 14.13).

    a. In the functional group region of the IR spectrum, specify the absorption associated with the hydroxyl group, and explain why this peak is broad.

    b. In the ^{1}H NMR spectrum, assign the various resonances to the hydrogen nuclei responsible for them.

    c. If the NMR sample of 2-butanol in CDCl$_3$ were shaken with D$_2$O *prior* to taking the ^{1}H NMR spectrum, what differences might you expect to observe?

    d. For the ^{13}C NMR data, assign the various resonances to the carbon nuclei responsible for them.

15. Consider the spectral data for 2-methyl-2-propanol (Figs. 14.14 and 14.15).

    a. In the functional group region of the IR spectrum, specify the absorption associated with the hydroxyl group, and explain why this peak is broad.

    b. In the ^{1}H NMR spectrum, assign the various resonances to the hydrogen nuclei responsible for them.

    c. If the NMR sample of 2-methyl-2-propanol in CDCl$_3$ were shaken with D$_2$O *prior* to taking the ^{1}H NMR spectrum, what differences might you expect to observe?

SPECTRA

Starting Materials and Products

The IR and NMR spectra of 1-butanol, 1-bromobutane, and 1-chlorobutane are provided in Figures 14.1, 8.28, 14.2–14.4, and 9.2 and 9.3, respectively.

Figure 14.12

IR spectrum of 2-butanol (ATR).

Figure 14.13
NMR data for 2-butanol (CDCl₃).

(a) ¹H NMR spectrum (MHz).
(b) ¹³C NMR data: δ 10.0, 22.8, 32.1, 69.3.

Figure 14.14
IR spectrum of 2-methyl-2-propanol (ATR).

Figure 14.15
NMR data for 2-methyl-2-propanol (CDCl₃).

(a) ¹H NMR spectrum (MHz).
(b) ¹³C NMR data: δ 30.8, 69.4.

14.8 COMPETITION BETWEEN SUBSTITUTION AND ELIMINATION

As noted in Section 14.3, the course of nucleophilic aliphatic substitution may be influenced by the nature of the nucleophile that is present. To review, *substitution is favored* with weakly basic and highly polarizable nucleophiles such as I^-, Br^-, Cl^-, H_2O, and $CH_3CO_2^-$, whereas *elimination is favored* when strongly basic and only slightly polarizable nucleophiles such as RO^-, H_2N^-, H^-, and HO^- are used. Furthermore, elimination is favored when sterically bulky nucleophiles are used because the hydrogen atom is more sterically accessible than is the carbon atom bearing the leaving group as described in Section 10.2.

The procedures in this section are designed to allow comparison of the effect of different nucleophiles in defining the competition between substitution and elimination reactions. The reactions you perform will involve treating 1-bromohexane (**4**) and 2-bromohexane (**5**), 1° and 2° bromides, respectively, with sodium methoxide and potassium *tert*-butoxide functioning as the nucleophile or base.

The potential products of nucleophilic substitution and of elimination are shown in Equations 14.39 and 14.40, respectively. Both of the alkoxides that you will be using are strong bases, but *tert*-butoxide is sterically bulkier than methoxide. Hence, steric hindrance tends to disfavor nucleophilic substitution to produce the ether **9**, so preferential formation of the elimination product **8** by way of an E2 reaction might be expected. On the other hand, the less sterically demanding methoxide should be more likely to produce **7** by an S_N2 process. Determining the ratio of products from the transformations of Equation 14.39 will allow you to evaluate the effect of steric bulk on S_N2 and E2 reactions of 1° alkyl halides.

$$\text{(14.39)}$$

5
1-Bromohexane

7
1-Methoxyhexane

8
1-Hexene

9
1-(*tert*-Butoxy)hexane

8
1-Hexene

$$(14.40)$$

Extending this study of steric effects to 2-bromohexane (**6**) is informative because now the steric environment of the C-2 carbon atom of **6** is more hindered than that of the C-1 carbon atom of 1-bromohexane (**6**). You might therefore predict that the S_N2 process for **6** is less likely than the E2 reaction. Analysis of the ratio of **8** and **12** formed in this reaction will allow you to determine whether the thermodynamically more stable alkenes are formed preferentially according to Zaitsev's rule (Sec. 10.2) and how the steric bulk of the base may affect the product ratio. To obtain all the data necessary to analyze the effect of the structures of alkyl halides and nucleophiles/bases on the competition between S_N2 and E2 reactions, your instructor may have you work in teams for these discovery experiments.

EXPERIMENTAL PROCEDURES

Competition between Substitution and Elimination

Discovery Experiment   **Purpose** To demonstrate the effect of steric hindrance in competing S_N2 and E2 reactions.

SAFETY ALERT

1. **Wear safety glasses or goggles and suitable protective gloves while performing the experiment.**

2. **Most of the chemicals used in this experiment are *flammable*, so do *not* handle these liquids near open flames.**

3. **The solutions used in this experiment are highly caustic. Do *not* allow them to come in contact with your skin. If this should occur, flood the affected area with water and then thoroughly rinse the area with a solution of dilute acetic acid.**

A. Reactions of 1-Bromohexane with Sodium Methoxide and Potassium tert-Butoxide

Preparation Refer to the online resources to answer Pre-Lab Exercises, access videos, and read the MSDSs for the chemicals used or produced in this procedure. Review Sections 2.3, 2.4, 2.5, 2.9, 2.11, 2.21, 2.22, 2.25, and 6.4.

Apparatus A 25-mL round-bottom flask, apparatus for heating under reflux, magnetic stirring, and *flameless* heating.

Setting Up Add 1 mL of 1-bromohexane and a stirbar to the 25-mL round-bottom flask. Slowly add 7 mL of a 1.5 *M* solution of sodium methoxide in methanol or 7 mL of a 1.5 *M* solution of potassium *tert*-butoxide in 2-methyl-2-propanol to the flask. Set up the apparatus for heating under reflux.

Reaction Heat the solution under gentle reflux for 1 h and then allow the mixture to cool to room temperature.

Work-Up and Isolation Pour the reaction mixture into a beaker containing 5 mL of water. Add about 3 mL of diethyl ether and transfer the liquids to a 25-mL screw-cap centrifuge tube. Rinse the beaker with 1–2 mL of diethyl ether and transfer to the centrifuge tube. Screw the cap on the centrifuge tube and shake it briefly. Remove the *aqueous* layer, and wash the organic layer with 5 mL of *saturated* sodium chloride solution. Again remove the *aqueous* layer and dry the organic layer over several spatula-tips full of *anhydrous* sodium sulfate. Cap the vial tightly for later analysis.

Analysis Analyze your solutions by GLC or submit a sample of it for analysis. Compare your product mixture with the known standards. After obtaining the results, calculate the relative percentages of the products formed, assuming that the GLC response factors are identical for the products.

B. Reactions of 2-Bromohexane with Sodium Methoxide and Potassium tert-Butoxide

Apparatus A 25-mL round bottom flask, apparatus for heating under reflux, magnetic stirring, and *flameless* heating.

Setting Up Add 1 mL of 2-bromohexane to the 25-mL round-bottom flask equipped with a stirbar. Slowly add 7 mL of a 1.5 *M* solution of sodium methoxide in methanol or 7 mL of a 1.5 *M* solution of potassium *tert*-butoxide in 2-methyl-2-propanol to the flask. Set up the apparatus for heating under reflux.

Reaction Heat the solution under gentle reflux for 1 h and then allow the mixture to cool to room temperature.

Work-Up and Isolation Pour the reaction mixture into a beaker containing 5 mL of water. Add about 3 mL of diethyl ether and transfer the liquids to a 25-mL screw-cap centrifuge tube. Rinse the beaker with 1–2 mL of diethyl ether and transfer to the centrifuge tube. Screw the cap on the centrifuge tube and shake it briefly. Remove the *aqueous* layer, and wash the organic layer with 5 mL of *saturated* sodium chloride solution. Again remove the *aqueous* layer and dry the organic layer over several spatula-tips full of *anhydrous* sodium sulfate. Cap the vial tightly for later analysis.

Analysis Analyze your solutions by GLC or submit a sample of it for analysis. Compare your product mixture with the known standards. After obtaining the results, calculate the relative percentages of the products formed, assuming that the GLC response factors are identical for the products.

WRAPPING IT UP

Neutralize the *aqueous layer* with 10% *aqueous* hydrochloric acid, then flush the solution down the drain.

EXERCISES

1. Each of the reactions in these procedures uses an alcohol as a solvent. What effect, if any, would you expect if water were the solvent?

2. An ether is formed from the reaction of 1-bromohexane and sodium methoxide. Suggest another similar reaction to prepare the same compound.

3. Does the reaction of 1-bromohexane and potassium *tert*-butoxide produce the corresponding ether? If not, why not?

4. How might the ratio of alkenes obtained in the *tert*-butoxide-promoted elimination change if the 2-methyl-2-propanol contains water?

5. Why does the excess of base used in these eliminations favor the E2 over the E1 mechanism for elimination?

6. If all the elimination reactions in the experimental section had proceeded by the E1 mechanism, would the results have been different from those actually obtained? Why?

Refer to the online resources and use the spectra viewer and Tables 8.1–8.8 as needed to answer the blue-numbered questions on spectroscopy.

7. Consider the NMR data for 1-bromohexane (Fig. 10.31).

    a. In the ^{1}H NMR spectrum, assign the various resonances to the hydrogen nuclei responsible for them.

    b. For the ^{13}C NMR data, assign the various resonances to the carbon nuclei responsible for them.

8. Consider the NMR data for 2-bromohexane (Fig. 10.33).

    a. In the ^{1}H NMR spectrum, assign the various resonances to the hydrogen nuclei responsible for them.

    b. For the ^{13}C NMR data, assign the various resonances to the carbon nuclei responsible for them.

9. Consider the spectral data for 1-hexene (Figs. 10.28 and 10.29).

    a. In the IR spectrum, specify the absorptions associated with the carbon-carbon double bond and the vinylic hydrogen atoms.

    b. In the ^{1}H NMR spectrum, assign the various resonances to the hydrogen nuclei responsible for them.

    c. For the ^{13}C NMR data, assign the various resonances to the carbon nuclei responsible for them.

CHAPTER 15

Electrophilic Aromatic Substitution

Something chemically remarkable happens when three conjugated double bonds are incorporated into a six-membered ring to produce a so-called **aromatic compound**. In sharp contrast to the addition reactions that typify alkenes (Secs. 10.4 and 17.2 and Chap. 12) and alkynes (Sec. 11.3), the chemistry of aromatic molecules is characterized by substitutions. Namely, aromatic compounds undergo chemical transformations in which a proton on the ring is replaced by some other atom or functional group. We'll explore these fascinating and useful reactions in this chapter. You will have the opportunity not only to execute reactions in which functional groups are introduced onto the aromatic ring but also to study both qualitatively and quantitatively the effects that substituents on the ring have on the rates at which these substitutions occur. These experiments will help you understand the extraordinary chemistry of aromatic compounds.

15.1 INTRODUCTION

Electrophilic aromatic substitution is an important reaction that allows the introduction of many different functional groups onto an aromatic ring. A general form of the reaction is given by Equation 15.1, where Ar–H is an aromatic compound, an **arene**, and E^+ represents an **electrophile** that replaces an H on the ring. This equation is oversimplified because the electrophile is usually generated during the reaction, and a Lewis base assists in the removal of H^+.

$$\text{Ar}\!-\!\text{H} \ + \ \text{E}^+ \ \longrightarrow \ \text{Ar}\!-\!\text{E} \ + \ \text{H}^+ \qquad (15.1)$$

An arene   An electrophile   A substituted arene

The rates of electrophilic aromatic substitutions are generally *second-order* overall, being *first-order* in both the aromatic component *and* the electrophile (Eq. 15.2). A general mechanism that is consistent with this kinetic expression is depicted in Equations 15.3–15.5.

$$\text{Rate} = k_2\,[\text{Ar}\!-\!\text{H}][\text{E}^+] \qquad (15.2)$$

499

Step 1: Formation of electrophile:

$$E\!-\!Nu \; \underset{}{\overset{\text{catalyst}}{\rightleftharpoons}} \; E^+ + Nu^- \tag{15.3}$$

Step 2: Reaction with electrophile:

$$Ar\!-\!H + E^+ \rightleftharpoons \left[\overset{+}{Ar}\underset{E}{\overset{H}{<}} \right] \tag{15.4}$$

Step 3: Loss of proton to give product:

$$\left[\overset{+}{Ar}\underset{E}{\overset{H}{<}} \right] \overset{\text{fast}}{\rightleftharpoons} Ar\!-\!E + H^+ \tag{15.5}$$

The electrophile is usually produced by the reaction between a catalyst and a compound containing a potential electrophile (Eq. 15.3). The second-order nature of the reaction arises from the step shown in Equation 15.4 in which one molecule each of arene and electrophile react to give a cationic intermediate. The formation of this cation is oftentimes the **rate-determining step (rds)** in the overall reaction; the subsequent deprotonation of the cation (Eq. 15.5) is fast relative to the previous step. The bimolecular nature of the transition state for the rate-limiting step and the fact that an electrophile is involved in attacking the aromatic substrate classifies the reaction as S_E2 (**S**ubstitution **E**lectrophilic **B**imolecular). Experiments involving four different such reactions are given in this chapter: Friedel-Crafts alkylation and acylation, nitration, and bromination.

15.2 FRIEDEL-CRAFTS ALKYLATION OF *p*-XYLENE WITH 1-BROMOPROPANE

The **Friedel-Crafts alkylation reaction** using alkyl halides (Eq. 15.6) is a classic example of electrophilic aromatic substitution (see the Historical Highlight *Discovery of the Friedel-Crafts Reaction,* which is available online). As a versatile method for directly attaching alkyl groups to aromatic rings, it is a process of great industrial importance. The three main limitations to this reaction as a synthetic tool are: (1) the substitution fails if the aromatic ring carries strongly **ring-deactivating groups**, such as NO_2, R_3N^+, C(=O)R, and CN; (2) more than one alkyl group may be introduced onto the aromatic ring, a process termed polyalkylation; and (3) mixtures of isomeric products may be formed if the alkyl group rearranges. The first restriction is associated with the fact that aromatic rings bearing electron-withdrawing substituents are not sufficiently nucleophilic, or Lewis basic, to react with the electrophiles, or Lewis acids, generated under the reaction conditions. The second problem arises because addition of the first alkyl group activates the ring toward further substitution. This side reaction is minimized by using a large excess of the arene, as is done in our experiment. The final limitation results from structural rearrangements within the alkyl group, R.

$$\underset{\substack{\text{An arene}}}{Ar\!-\!H} + \underset{\substack{\text{An alkyl halide} \\ (X = \text{halogen})}}{R\!-\!X} \; \overset{\text{catalyst}}{\rightleftharpoons} \; \underset{\substack{\text{An alkylarene}}}{Ar\!-\!R} + HX \tag{15.6}$$

A generally accepted mechanism for Friedel-Crafts alkylation using alkyl halides is illustrated in Equations 15.7–15.9. In the first step, the active electrophile, written as R^+ for the present purpose, is produced from an alkyl halide by reaction with a Lewis acid such as $AlCl_3$ (Eq. 15.7). The product-determining step involves attack of the arene, which functions as a Lewis base, on the electrophile to yield a positively charged, resonance-stabilized **σ-complex** (Eq. 15.8). Finally, a base such as a halide ion from $X\overline{A}lCl_3$ deprotonates this complex to give HX, reforming the aromatic system and regenerating the catalyst (Eq. 15.9).

$$R\text{—}X + \overline{A}lCl_3 \rightleftharpoons R^+ + X\overline{A}lCl_3 \tag{15.7}$$

$$\tag{15.8}$$

σ-Complex

$$\tag{15.9}$$

Representing the electrophile as a carbocation, as in Equation 15.7, is reasonable when *secondary* and *tertiary* alkyl halides are used. However, this is probably *not* the case with the electrophile derived from a *primary* alkyl halide, where the reactive species is better written as a *polarized complex* such as $R\text{–}X^+\text{–}\overline{A}lCl_3$.

Molecular rearrangements involving hydride, H^-, and alkyl, R^-, shifts become possible when a carbocation or a polarized cation-like complex is an intermediate in a reaction. Thus, the aluminum chloride-promoted reaction of 1-bromopropane with benzene gives *n*-propylbenzene (**1**) and isopropylbenzene (**2**) in a ratio of about 1:2 (Eq. 15.10). As shown in Scheme 15.1, the formation of **2** is explained by a 1,2-hydride rearrangement of the *primary carbocation-like* complex **3** to produce the *secondary carbocation* **4**. The formation of comparable amounts of **1** and **2** in the reaction means that the rate of addition of **3** to benzene followed by deprotonation to give **1** must be competitive with that for isomerization of **3** to the secondary ion **4**, which is the precursor of **2**. In mathematical terms, $k_2[\mathbf{3}][C_6H_6]$ must be similar to $k_1[\mathbf{3}]$, since these expressions measure the rates of formation of **1** and **2**, respectively.

$$\tag{15.10}$$

Benzene   1-Bromopropane   *n*-Propylbenzene   Isopropylbenzene

Scheme 15.1

CH$_3$CH$_2$CH$_2$—Br: $\xrightarrow{\text{AlCl}_3}$ CH$_3$CHCH$_2$—Br—AlCl$_3$ $\xrightarrow[k_1]{\sim\text{H}^-}$ CH$_3$CHCH$_2$

1-Bromopropane   *n*-Propyl carbocation-like complex   Isopropyl carbocation

3   **4**

k_2 | C$_6$H$_6$ C$_6$H$_6$

CH$_2$CH$_2$CH$_3$ CH(CH$_3$)$_2$

1   **2**

33%   67%

The present experiment demonstrates Friedel-Crafts alkylation and allows you to examine the following hypothesis: The amount of alkyl rearrangement that occurs during a Friedel-Crafts alkylation of a series of arenes depends on the *nucleophilicity* of the arene, Ar–R. This hypothesis is derived as follows. The rate constant, k_1, is for a *unimolecular* process of **3** that does *not* involve the arene (Scheme 15.1), so its value is *independent* of the nucleophilicity of the arene. In contrast, k_2 is the rate constant for a *bimolecular* reaction between **3** *and* the arene, so its value is *dependent* on the nucleophilicity of the arene. Substituents, R, that make the arene more nucleophilic yield higher values of the bimolecular rate constant k_2, thereby increasing the overall rate of the electrophilic substitution. Consequently, formation of the product derived from rearrangement becomes a relatively less important reaction pathway.

The importance of the nucleophilicity of the arene in determining the ratio of unrearranged products to rearranged products may be studied by comparing the Friedel-Crafts reactions of benzene and *p*-xylene with 1-bromo-propane (Eq. 15.11). Because methyl groups activate the ring toward the S$_E$2 reaction and make k_2 (*p*-xylene) > k_2 (benzene), we expect from our hypothesis that the ratio of unrearranged product **5** to its rearranged isomer **6** is *greater* than that of 1:2. In the experiment that follows, you will be able to test this prediction.

CH$_3$CH$_2$CH$_2$—Br + [p-Xylene] $\xrightarrow{\text{AlCl}_3}$ [n-Propyl-p-xylene] + [Isopropyl-p-xylene]

(15.11)

1-Bromopropane   *p*-Xylene   *n*-Propyl-*p*-xylene   Isopropyl-*p*-xylene

5   **6**

Friedel-Crafts Alkylation of p-*Xylene*

Discovery Experiment

Purpose To demonstrate alkylation by electrophilic aromatic substitution and to assess the relationship between the extent of carbocation rearrangement and arene nucleophilicity.

SAFETY ALERT

1. **Wear safety glasses or goggles and suitable protective gloves while performing the experiments.**

2. ***Anhydrous*** **aluminum chloride is** ***extremely*** **hygroscopic and reacts rapidly with water, even the moisture on your hands, producing highly corrosive fumes of hydrogen chloride.** *Do not allow aluminum chloride to come in contact with your skin.* **If it does, flush the affected area with copious amounts of water.** *Minimize exposure of this chemical to the atmosphere!*

3. **You may be provided with a weighed amount of aluminum chloride. If not,** *quickly* **weigh it into a dry vial in the hood and then transfer it to the reaction flask. The success of this experiment is highly dependent on the quality of the aluminum chloride that is used, so obtain it from a** *freshly* **opened bottle. In handling this chemical, be aware that it is a powdery solid that easily becomes airborne.**

4. ***p*-Xylene is flammable. Assemble the apparatus carefully and be sure that all joints are tightly mated. Have your instructor inspect your set-up before you begin the distillation and use** *flameless* **heating.**

MINISCALE PROCEDURE

Preparation Refer to the online resources to answer Pre-Lab Exercises, access videos, and read the MSDSs for the chemicals used or produced in this procedure. Review Sections 2.9, 2.10, 2.11, 2.13, 2.14, 2.23, 2.27, 4.4, and 6.4.

Apparatus A 25- and a 50-mL round-bottom flask, drying tube, Claisen adapter, separatory funnel, gas trap, ice-water bath, apparatus for simple distillation, fractional distillation, magnetic stirring, and *flameless* heating.

Setting Up Equip the 50-mL round-bottom flask for simple distillation and add a stirbar and 15 mL of *p*-xylene. Dry the *p*-xylene through azeotropic distillation by distilling it into a receiver until the distillate is not cloudy; this should occur after no more than 3 mL of distillate has been obtained.

Allow the residual *p*-xylene in the stillpot to cool to room temperature, and equip the flask with a Claisen adapter, a water-cooled condenser, the gas trap specified by your instructor for removing fumes of hydrogen chloride, and a separatory funnel (Fig. 2.66b); fit the funnel with a drying tube rather than a stopper. *Briefly* remove the condenser from the reaction flask, *quickly* pour 0.7 g of *anhydrous* powdered aluminum chloride into the flask, and *immediately* reassemble the apparatus.

Using a graduated cylinder, measure 8.5 mL of 1-bromopropane and pour it into the funnel (stopcock *closed!*). Prepare an ice-water bath in case it is needed to cool the reaction mixture.

Addition and Work-Up Add the 1-bromopropane dropwise to the stirred mixture of *p*-xylene and aluminum chloride. If the evolution of gas becomes too vigorous, reduce the rate of addition and raise the cooling bath to lower the temperature of the reaction mixture. The addition should take about 10 min. After completing the addition, continue stirring the reaction mixture for about 0.5 h. *Working at the hood* because of the hydrogen chloride that will be evolved, pour the mixture into a beaker containing about 10 g of crushed ice. Stir the mixture until all the ice has melted, pour it into a separatory funnel, and remove the aqueous layer. Pour the organic layer into an Erlenmeyer flask containing *anhydrous* sodium sulfate and swirl it for about 1 min; if the solution appears cloudy, add additional sodium sulfate with swirling.★

Isolation Decant the dried solution into the 25-mL round-bottom flask containing a stirbar and equip the flask for fractional distillation using an *air-cooled* condenser. Make sure that all connections are tight and then distill. Collect the excess *p*-xylene and forerun in a tared, labeled receiver until the head temperature reaches 180 °C (760 torr). Discontinue heating and allow any liquid in the column to drain into the stillpot.★ Refit the stillpot for shortpath distillation and resume the distillation, collecting any additional distillate boiling below 180 °C (760 torr) in the original receiver. Use a second tared, labeled receiver to collect the fraction boiling between 180 °C and 207 °C (760 torr). Isomers **5** and **6** have bp 204 °C and 196 °C (760 torr), respectively.

Analysis Weigh the two fractions and analyze each by GLC; a typical trace of the second fraction is provided in Figure 15.1. Use the GLC analysis to calculate the yield of propylxylenes obtained and their proportion to each other; assume that *p*-xylene and the propylxylenes have the same density and GLC response factor. Obtain IR and ^{1}H NMR spectra of your starting materials and products and compare them with those of authentic samples (Figs. 15.4–15.11).

Figure 15.1

GLC analysis of reaction mixture from miniscale alkylation of p-xylene with 1-bromopropane. Numbers in parentheses are retention times in minutes and peak areas, respectively. Peak 1: solvent; peak 2: p-xylene (1.7, 14417); peak 3: isopropyl-p-xylene (1.8, 71721); peak 4: n-propyl-p-xylene (2.0, 229275). Column and conditions: 0.25-mm × 25-m AT-1 with 100% dimethylsiloxane as stationary phase; flowrate 1.2 mL/min; temperature 185 °C.

MICROSCALE PROCEDURES

Preparation Refer to the online resources to answer Pre-Lab Exercises, access videos, and read the MSDSs for the chemicals used or produced in this procedure. Review Sections 2.9, 2.11, 2.13, 2.17, 2.23, 4.4, and 6.4.

Apparatus A 3- and a 5-mL conical vial, 1-mL syringe, Claisen adapter, gas trap, ice-water bath, apparatus for distillation, magnetic stirring, and *flameless* heating.

Setting Up Equip the 5-mL vial for simple distillation and add a spinvane and 2 mL of *p*-xylene. Dry the *p*-xylene through azeotropic distillation by distilling it until the distillate is not cloudy; this should occur after no more than 0.5 mL of distillate has been obtained.

Allow the residual *p*-xylene in the vial to cool to room temperature. Replace the condenser on the vial with a Claisen adapter fitted with a rubber septum on the opening directly above the vial; on the other arm of the adapter, put a water-cooled condenser equipped with the gas trap specified by your instructor for removing fumes of hydrogen chloride (Fig. 15.2). *Briefly* remove the vial from the adapter, *quickly* pour 0.1 g of *anhydrous* powdered aluminum chloride into it, and *immediately* reassemble the apparatus. Draw 0.8 mL of 1-bromopropane into the syringe and pierce the rubber septum with its needle.

Addition and Work-Up Add the 1-bromopropane dropwise to the stirred mixture of *p*-xylene and aluminum chloride over a period of 5–10 min. If the evolution of gas becomes too vigorous, reduce the rate of addition. After completing the addition, continue stirring the reaction mixture for about 20 min. *Working at the hood* because of the hydrogen chloride that will be evolved, pour the mixture into a beaker containing about 1 g of crushed ice. Stir the mixture until all the ice has melted, pour it into a centrifuge tube, and remove the aqueous layer. Add a microspatula-tip full of *anhydrous* sodium sulfate to the organic layer and swirl it for about 1 min; add additional sodium sulfate with swirling if the solution appears cloudy.★

Isolation Using a filter-tip pipet, transfer the dried solution into the 3-mL conical vial containing a spinvane and equip the vial for distillation, with internal monitoring of the temperature. Make sure that all connections are tight and then distill. Collect the excess *p*-xylene and forerun until the bath temperature reaches 180 °C (760 torr). Discontinue heating and allow the apparatus to cool for a few minutes.★ Transfer the distillate to a tared, labeled screw-cap vial and resume the distillation, collecting as one fraction all volatiles that distill with a bath temperature of up to 240 °C. Isomers **5** and **6** have bp 204 °C and 196 °C (760 torr), respectively.

Analysis Weigh the two fractions and analyze each by GLC; a typical trace of the second fraction is provided in Figure 15.3. Obtain IR and ^{1}H NMR spectra of your starting materials and products and compare them with those of authentic samples (Figs. 15.4–15.11).

To gas trap

Water out

Water in

Figure 15.2
Microscale apparatus for alkylation of p-xylene.

Figure 15.3

GLC trace of reaction mixture from microscale alkylation of p-xylene with 1-bromopropane (numbers in parentheses are retention times in minutes and peak areas, respectively). Peak 1: solvent; peak 2: p-xylene (1.8, 74565); peak 3: isopropyl-p-xylene (4.3, 92699); peak 4: n-propyl-p-xylene (5.0, 236258). Column and conditions: 0.25-mm × 25-m AT-1 column with 100% dimethylpolysiloxane as stationary phase; flowrate 1.2 mL/min; temperature 110 °C.

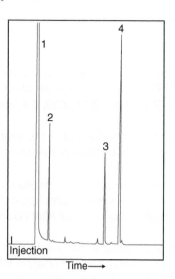

WRAPPING IT UP

Neutralize the *aqueous layer* from hydrolysis of the reaction mixture with sodium carbonate and flush it down the drain. Spread the *sodium sulfate* used for drying on a tray in the hood to allow evaporation of the volatiles; then put spent *drying agent* in the container for nontoxic solids. Place the *liquid collected* as the *low-boiling fraction* and the *pot residue* from the distillation in the container for nonhalogenated organic liquids.

EXERCISES

1. What purpose is served by distilling a small quantity of *p*-xylene prior to combining the reagents in this experiment?

2. What compound(s) might be present in the higher-boiling residue remaining in the stillpot after you collect the isomeric propylxylenes?

3. Calculate the molar ratio of 1-bromopropane to *p*-xylene used in the experiment. What would be the effect on the experimental results of using a ratio twice that actually used? Consider this in connection with your answer to Exercise 2.

4. Use either your own GLC analysis or that given in Figure 15.1 or 15.3 to calculate the yield of propylxylenes obtained and their proportion to each other; assume that *p*-xylene and the propylxylenes have the same densities and GLC response factor. Discuss whether these calculations support the hypothesis regarding the relative amounts of unrearranged product versus rearranged product to be expected from *p*-xylene as compared to benzene.

5. Account for the higher boiling point of **5** as compared to that of **6**. Are the boiling points consistent with the relative retention times of the two isomers as seen in the GLC trace of Figure 15.1 or 15.3? Explain.

6. Why do you think it is recommended that the distillation fraction containing the mixture of **5** and **6** be taken over such a wide temperature range?

7. Consider aluminum chloride, $AlCl_3$, the catalyst used in this procedure.

   a. Provide a Lewis dot structure for $AlCl_3$.

   b. Define the term *Lewis acid*.

c. What aspect of the Lewis structure is consistent with this compound being a Lewis acid?

d. How does addition of $AlCl_3$ promote formation of carbocations from alkyl halides?

e. Provide an example of a Lewis acid that does not contain aluminum.

8. Using curved arrows to symbolize the flow of electrons, write the stepwise mechanism for the reaction between *p*-xylene and *n*-propyl cation. Show all resonance structures where applicable.

9. On the assumption that relative cost is not an overriding consideration, give a reason why *p*-xylene was used in preference to *o*-xylene as the arene in this experiment.

10. Outline a synthesis of *p*-xylene starting from benzene. Assume that the isomeric xylenes that might be formed could be separated from one another through a purification procedure.

11. Nitrobenzene is sometimes used as the solvent for Friedel-Crafts alkylation of other arenes. Why is this possible?

12. Alkylation of toluene, $C_6H_5CH_3$, with 1-bromopropane gives a mixture of four isomeric propyltoluenes. Write formulas for these compounds, and predict the relative amounts in which they would be formed, based on the results of the alkylation of *p*-xylene.

13. Consider the Friedel-Crafts alkylation of toluene and anisole, $C_6H_5OCH_3$, with 1-chloropropane/$AlCl_3$. Given your understanding of substituent effects on S_E2 reactions, how would you expect the ratio of *n*-propyl alkylation to isopropyl alkylation to differ for these two arenes? Explain your answer.

Refer to the online resources and use the spectra viewer and Tables 8.1–8.8 as needed to answer the blue-numbered questions on spectroscopy.

14. Consider the spectral data for *p*-xylene (Figs. 15.4 and 15.5).

a. In the functional group region of the IR spectrum, specify the absorption associated with the π-bonds of the aromatic ring. Indicate with what structural feature the strong band at about 800 cm^{-1} is associated.

b. In the ^1H NMR spectrum, assign the various resonances to the hydrogen nuclei responsible for them.

c. For the ^{13}C NMR data, assign the various resonances to the carbon nuclei responsible for them.

15. Consider the NMR spectral data for 1-bromopropane (Fig. 15.7).

a. In the ^1H NMR spectrum, assign the various resonances to the hydrogen nuclei responsible for them.

b. For the ^{13}C NMR data, assign the various resonances to the carbon nuclei responsible for them.

16. Consider the spectral data for *n*-propyl-*p*-xylene (Figs. 15.8 and 15.9).

a. In the functional group region of the IR spectrum, specify the absorption associated with the π-bonds of the aromatic ring. Indicate with what structural feature the strong band at about 810 cm^{-1} is associated.

b. In the ^1H NMR spectrum, assign the various resonances to the hydrogen nuclei responsible for them.

c. For the ^{13}C NMR data, assign the various resonances to the carbon nuclei responsible for them.

17. Consider the spectral data for isopropyl-*p*-xylene (Figs. 15.10 and 15.11).

   a. In the functional group region of the IR spectrum, specify the absorption associated with the π-bonds of the aromatic ring. Indicate with what structural feature the strong band at about 800 cm^{-1} is associated.

   b. In the ^{1}H NMR spectrum, assign the various resonances to the hydrogen nuclei responsible for them.

   c. For the ^{13}C NMR data, assign the various resonances to the carbon nuclei responsible for them.

18. What differences in the IR and NMR spectra of *p*-xylene and of **5** and **6** are consistent with the incorporation of a C$_3$H$_7$ group on the ring in this experiment?

19. What differences in the IR and NMR spectra of *p*-xylene, and of **5** and **6**, are consistent with *mono*substitution rather than *di*substitution of *p*-xylene having occurred in the Friedel-Crafts alkylation you performed?

20. *Ipso* substitution, as given in the equation below, is sometimes observed in electrophilic aromatic substitution reactions. Explain how the data provided in Figures 15.8–15.11 are inconsistent with the structures of the *ipso* products shown.

p-Xylene   1-Methyl-4-*n*-propylbenzene   1-Methyl-4-isopropylbenzene

Ipso substitution products

SPECTRA

Starting Materials and Products

Figure 15.4
IR spectrum of p-xylene *(neat).*

Figure 15.5
NMR data for p-xylene (CDCl$_3$).

(a) ^{1}H NMR spectrum (300 MHz).
(b) ^{13}C NMR data: δ 20.9, 129.0, 134.6.

Figure 15.6
IR spectrum of 1-bromopropane (neat).

Figure 15.7
NMR data for 1-bromopropane (CDCl$_3$).

(a) ^{1}H NMR spectrum (300 MHz).
(b) ^{13}C NMR data: δ 13.0, 26.2, 35.9.

Figure 15.8

IR spectrum of n-*propyl-*p-*xylene (neat).*

Figure 15.9

NMR data for n-*propyl-*p-*xylene (contains 4% of isopropyl isomer) (CDCl₃).*

(a) ¹H NMR spectrum (300 MHz).
(b) ¹³C NMR data: δ 14.2, 18.8, 21.0, 23.5, 35.4, 126.4, 129.7, 129.9, 132.7, 135.1, 140.7.

Figure 15.10

*IR spectrum of isopropyl-*p-*xylene (neat).*

Figure 15.11

NMR data for isopropyl-p-xylene (CDCl₃).

(a) ¹H NMR spectrum (300 MHz).

(b) ¹³C NMR data: δ 18.8, 21.2, 23.2, 29.1, 125.4, 126.2, 130.1, 131.8, 135.4, 146.6.

15.3 FRIEDEL-CRAFTS ACYLATION OF ANISOLE

As described in Section 15.2 and in the Historical Highlight *Discovery of the Friedel-Crafts Reaction,* which is available online, Friedel and Crafts discovered that alkyl groups could be introduced onto an aromatic ring by reaction of arenes with alkyl halides in the presence of aluminum chloride, $AlCl_3$, and other Lewis acids (Eq. 15.1, E^+ = alkyl). The role of aluminum chloride, a strong Lewis acid, in this reaction is to convert the alkyl halide into a reactive electrophilic intermediate, in the form of a carbocation or a highly polarized carbon-halogen bond (Eq. 15.8). The electrophile then undergoes attack by an arene, which functions as a Lewis base, resulting in aromatic substitution (Eqs. 15.8 and 15.9). The overall transformation is an example of an S_E2 reaction, as noted in Section 15.1.

They also investigated the reaction of arenes with acid chlorides, **7**, in the presence of Lewis acids and discovered that this produced aryl ketones, **9**, as shown in Equation 15.12. The first step in this process is reaction of the Lewis acid with the acid chloride to form an **acylonium ion**, **8**, which serves as the electrophile. Subsequently **8** adds to the arene, introducing an acyl group on the ring to provide an aryl ketone, **9**. In honor of its inventors, the reaction is named the **Friedel-Crafts acylation**.

The resonance structures shown for **8** in Equation 15.12 symbolize the distribution of the positive charge in acylonium ions. This charge delocalization stabilizes the cation so that acylonium ions such as **8** do *not* undergo the structural rearrangements that are seen for alkyl carbocations in Friedel-Crafts alkylations (Sec. 15.2). The increased stability of acylonium ions also makes them less electrophilic and thus less reactive than alkyl carbocations. Nevertheless, cation **8** is sufficiently reactive to undergo the S_E2 reaction with arenes, *provided* the aromatic ring does not bear strongly deactivating groups such as nitro, NO_2, ammonium, R_3N^+, acyl, C(=O)R, and cyano, C≡N.

$$
\begin{array}{ccccc}
\underset{\substack{\text{R} \quad \text{Cl} \\ \textbf{7} \\ \text{An acid chloride}}}{\text{R—C(=O)—Cl}} & \underset{\text{Acid}}{\overset{\text{Lewis}}{\rightleftharpoons}} & \left[\; \overset{\cdot\cdot}{\underset{\text{R}}{\text{O}}}{\equiv}\overset{+}{\underset{\text{R}}{\text{C}}} \;\right]^{+} & \xrightarrow{\text{Ar—H}} & \underset{\substack{\textbf{9} \\ \text{An aryl ketone}}}{\text{Ar—C(=O)—R}}
\end{array}
\qquad (15.12)
$$

7   **8**   **9**

An acid chloride   An acylonium ion   An aryl ketone

The Friedel-Crafts acylation performed in this section involves the reaction of benzoyl chloride (**10**) with anisole (**12**) in the presence of zinc oxide, a mild Lewis acid that is much easier to handle than aluminum chloride, the strong Lewis acid used in the Friedel-Crafts alkylation experiment (Sec. 15.2), to furnish 4-methoxybenzophenone (**14**), as shown in Equation 15.13. The first step in this transformation involves formation of the acylonium ion **11**, which is a Lewis acid. The next step is the reaction of **11** with anisole (**12**), which serves as the Lewis base, to form a σ-complex that is depicted as the resonance hybrid **13**. The subsequent reaction of **13** with a base such as ClZnO⁻ leads to loss of a proton and formation of **14**.

| **10** | **11** | **12** |
| Benzoyl chloride | An acylonium ion | Anisole |

(15.13)

| **13** | **14** |
| σ-Complex | 4-Methoxybenzophenone |

Anisole (**12**) is a monosubstituted arene, so the question of the regioselectivity of the acylation arises. The structures of the two possible resonance hybrids other than **13** from the reaction of the acylonium ion **11** with **12** are **15** and **16**, the precursors to the corresponding diaryl ketones **17** and **18**. The activation enthalpies, $\Delta H^{\ddagger}$, for attack at the positions *ortho* and *para* to the methoxy group of **12** are lower than for attack at the *meta* position because the σ-complexes **13** and **15** are more stable than **16**. This results from the additional stabilizing interaction of the non-bonding electrons on the oxygen atom with the positive charge in the complexes **13** and **15**. Thus, like an alkyl group (Sec. 15.2), a methoxy group is an *ortho-para* director because it favors formation of these intermediates over the *meta* isomer **16** so that acylation is expected to produce the *ortho-* and *para*-substituted products **14** and **17** selectively. Furthermore, even though substitution at the *ortho* position of anisole (**12**) is statistically favored by a factor of two over the *para* position, **14** rather than **17** is the observed product of the reaction, presumably because of steric effects.

| **15** | **16** |

| **17** | **18** |

Although solvents are used in many Friedel-Crafts reactions, you will perform this reaction *without* solvent, and it is possible to recover and reuse the zinc oxide catalyst. This "green" variant of the traditional procedure is more environmentally friendly because there is no need to dispose of organic solvents. However, whenever doing a reaction without solvent, you must be careful because many reactions are exothermic, so it is important to add one reactant *slowly* to the other. Otherwise, the reaction may quickly spiral out of control, spewing the contents of the flask onto you and the bench.

EXPERIMENTAL PROCEDURES

Friedel-Crafts Acylation of Anisole

Green Experiment   **Purpose** To demonstrate the acylation of anisole by electrophilic aromatic substitution in a solvent free reaction.

SAFETY ALERT

1. **Wear safety glasses or goggles and suitable protective gloves while performing the experiment.**
2. **Benzoyl chloride is hygroscopic and reacts rapidly with water, even the moisture on your hands, producing highly corrosive fumes of hydrogen chloride. Do not allow benzoyl chloride to come in contact with your skin. If it does, flush the affected area with copious amounts of water. Minimize exposure of this chemical to the atmosphere.**

MINISCALE PROCEDURE

Rubber septum

Preparation Refer to the online resources to answer Pre-Lab Exercises, access videos, and read the MSDSs for the chemicals used or produced in this procedure. Review Sections 2.5, 2.11, 2.17, 2.19, 2.21, 2.25, 2.27, and 2.29.

Apparatus A dry 25-mL round-bottom flask, Claisen adapter, drying tube, two 3-mL syringes with needles, small separatory funnel, apparatus for magnetic stirring, simple distillation, and vacuum filtration.

Setting Up Weigh 0.4 g of zinc oxide into a dry 25-mL round-bottom flask equipped with a stirbar. Assemble the apparatus shown in Figure 10.46. Using a 3-mL syringe, add 1.2 mL of benzoyl chloride and begin stirring the mixture.

Reaction Using a 3-mL syringe, *slowly* add 1.1 mL of anisole dropwise over a period of about 5 min to the stirred slurry. *Caution:* Reaction is very exothermic. Continue stirring for 1 h.

Work-Up and Isolation Add 5 mL of dichloromethane to the reaction flask and decant the liquid into the separatory funnel. Rinse the residual zinc oxide and the flask with 5 mL of dichloromethane and again decant the liquid into the funnel. Add 5 mL of water and separate the layers. Wash the organic layer sequentially with a 5-mL portion of *saturated* sodium bicarbonate solution and a 5 mL portion of *saturated* sodium chloride solution. Transfer the organic layer to a dry flask. Dry the solution over several spatula-tips full of *anhydrous* sodium sulfate. Swirl the flask occasionally for a period of 10–15 min to facilitate drying; add further small portions of *anhydrous* sodium sulfate if the solution remains cloudy.★

Filter or carefully decant the dried organic solution into a round-bottom flask equipped for simple distillation, and remove the solvents by simple distillation. Alternatively, use rotary evaporation or other techniques to concentrate the solution. The final traces of solvent may be removed by attaching the flask to a vacuum source and gently swirling the contents as the vacuum is applied. Recrystallize the crude product from hexane and air-dry the purified product.

Analysis Weigh the product and determine the percent yield of crude 4-methoxybenzophenone; determine the melting point. Obtain IR and ¹H NMR spectra of your starting materials and product, and compare them with those of authentic samples (Figs. 8.15, 15.12–15.16).

MICROSCALE PROCEDURES

Preparation Refer to the online resources to answer Pre-Lab Exercises, access videos, and read the MSDSs for the chemicals used or produced in this procedure. Review Sections 2.5, 2.11, 2.17, 2.19, 2.21, 2.25, and 2.29.

Apparatus A 5-mL conical vial, Claisen adapter, drying tube, two 5-mL centrifuge tubes with caps, two 1-mL syringes with needles, apparatus for magnetic stirring, simple distillation, vacuum filtration, and Craig tube filtration.

Setting Up Weigh 0.04 g of zinc oxide into the 5-mL conical vial equipped with a spinvane. Assemble the apparatus shown in Figure 10.47. Using a 1-mL syringe, add 0.12 mL of benzoyl chloride and begin stirring the mixture.

Reaction Using a 1-mL syringe, *slowly* add 0.11 mL of anisole dropwise over a period of about 1 min to the stirred slurry. *Caution:* Reaction is exothermic. Continue stirring for 1 h.

Work-up and Isolation Add 1 mL of dichloromethane to the reaction vial and decant the liquid into a 5-mL screw-cap centrifuge tube. Rinse the residual zinc oxide and the conical vial with 1 mL of dichloromethane and again decant the liquid into the 5-mL centrifuge tube. Wash the organic layer sequentially with a 1-mL portion of *saturated* sodium bicarbonate solution and a 1-mL portion of *saturated* sodium chloride solution. Using a Pasteur pipet, transfer the organic layer to a dry screw-cap centrifuge tube and add several microspatula-tips full of *anhydrous* sodium sulfate. Allow the solution to stand for 15 min, with occasional shaking to hasten drying. Add further small portions of *anhydrous* sodium sulfate if the solution remains cloudy.★

Using a Pasteur pipet, transfer the organic solution to a clean 5-mL conical vial, add a spinvane to the vial, and remove the solvents by simple distillation. Alternatively, use rotary evaporation or other techniques to concentrate the solution. The final traces of solvent may be removed by attaching the flask to a vacuum source and gently swirling the contents as the vacuum is applied. Recrystallize the crude product from hexane and air-dry the purified product.

Analysis Weigh the product and determine the percent yield of crude 4-methoxybenzophenone; determine the melting point. Obtain IR and ^{1}H NMR spectra of your starting materials and product, and compare them with those of authentic samples (Figs. 8.15, 15.12–15.16).

WRAPPING IT UP

Dilute the combined *aqueous layers* with water, neutralize the resulting solution with sodium carbonate, and flush the mixture down the drain with excess water. After the dichloromethane evaporates from the *sodium sulfate* in the hood, place the solid in the container for nonhazardous solids. Place the *recovered dichloromethane* that was used as the extraction solvent in the container for halogenated organic solvents.

EXERCISES

1. What is the *final* form of the zinc oxide used in this experiment?

2. What is an advantage of using a "solvent-free" reaction? What precautions must be taken when performing such reactions?

3. What is the purpose of washing the solution with *saturated* sodium carbonate solution?

4. After the washes, the solution is treated with *anhydrous* sodium sulfate.

   a. Why is this done?

   b. Could solid sodium hydroxide or potassium hydroxide be used for this purpose? Explain.

5. Why is loss of carbon monoxide (decarbonylation) from the acylonium ion **11** an unlikely reaction?

6. Anisole is considered an "activated" aromatic compound with respect to Friedel-Crafts acylation. What effect on the rate of the reaction would you expect to observe if benzaldehyde were the reactant? If nitrobenzene were the reactant?

7. Why is it important that the apparatus used for the Friedel-Crafts acylation be dry?

8. What is the product of the reaction between benzoyl chloride and water? Using curved arrows to symbolize the flow of electrons, write a mechanism for this reaction.

9. The reaction of benzoyl chloride (**10**) with anisole (**12**) in this experiment gives 4-methoxybenzophenone (**14**) rather than 3-methoxybenzophenone (**18**) as the product. Write the important contributing resonance structures for the σ-complexes for **13** and **16** and briefly explain why **13** is more stable than **16**.

Refer to the online resources
and use the spectra viewer and
Tables 8.1–8.8 as needed to answer
the blue-numbered questions on
spectroscopy.

10. Consider the spectral data for benzoyl chloride (Figs. 8.15 and 15.12).

   a. In the IR spectrum, specify the absorptions associated with the acid chloride group and the π-bonds of the aromatic ring.

   b. In the ¹H NMR spectrum, assign the various resonances to the hydrogen nuclei responsible for them.

   c. For the ¹³C NMR data, assign the various resonances to the carbon nuclei responsible for them.

11. Consider the spectral data for anisole (Figs. 15.13 and 15.14).

   a. In the functional group region of the IR spectrum, specify the absorption associated with the π-bonds of the aromatic ring.

   b. In the ¹H NMR spectrum, assign the various resonances to the hydrogen nuclei responsible for them.

   c. For the ¹³C NMR data, assign the various resonances to the carbon nuclei responsible for them.

12. Consider the spectral data for 4-methoxybenzophenone (Figs. 15.15 and 15.16).

   a. In the IR spectrum, specify the absorptions associated with the carbonyl group and the π-bonds of the aromatic ring.

   b. In the ¹H NMR spectrum, assign the various resonances to the hydrogen nuclei responsible for them.

   c. For the ¹³C NMR data, assign the various resonances to the carbon nuclei responsible for them.

13. Discuss the differences observed in the IR and NMR spectra of anisole, benzoyl chloride, and 4-methoxybenzophenone that are consistent with a Friedel-Crafts acylation reaction having occurred.

SPECTRA

Starting Materials and Products

The IR spectrum of benzoyl chloride is provided in Figure 8.15.

Figure 15.12
NMR data for benzoyl chloride (CDCl₃).

(a) ¹H NMR spectrum (400 MHz).
(b) ¹³C-NMR data. Chemical shifts: δ 129.2, 131.6, 133.3, 135.7, 168.5.

Figure 15.13
IR spectrum of anisole (neat).

Figure 15.14
NMR data for anisole (CDCl₃).

(a) ¹H NMR spectrum (400 MHz).
(b) ¹³C-NMR data. Chemical shifts: δ 54.8, 114.1, 120.7, 159.9.

Figure 15.15
IR spectrum of 4-methoxybenzo-phenone (ATR).

Figure 15.16
NMR data for 4-methoxybenzo-phenone (CDCl₃).

(a) ¹H NMR spectrum (400 MHz).
(b) ¹³C-NMR data. Chemical shifts: δ 55.5, 113.5, 128.2, 129.7, 130.1, 131.9. 132.6, 138.3, 163.2, 195.5.

15.4 NITRATION OF BROMOBENZENE

Reaction of an aromatic compound such as benzene with a mixture of *concentrated* sulfuric and nitric acids introduces a nitro group on the ring by way of electrophilic aromatic substitution (Sec. 15.1) as depicted in Equation 15.14. The electrophilic species is the **nitronium ion**, NO_2^+, which is produced by reaction of sulfuric acid with nitric acid (Eq. 15.15). It is interesting to note that nitric acid, the *weaker* of these two strong mineral acids, is serving as a *base* rather than as an acid in this equilibrium.

$$\text{Benzene} \xrightarrow[\text{H}_2\text{SO}_4]{\text{HNO}_3} \text{Nitrobenzene} + \text{H}_2\text{O} \qquad (15.14)$$

$$\text{HÖ—NO}_2 \;\xrightleftharpoons[]{\text{H—OSO}_3\text{H}}\; \text{H}_2\overset{+}{\text{O}}\text{—NO}_2 \;\xrightleftharpoons[]{}\; \text{H}_2\text{O} + {}^+\text{NO}_2 \qquad (15.15)$$

Nitric acid   Nitronium ion

The product-determining step in the nitration reaction involves nucleophilic attack of the aromatic ring, a Lewis base, on the nitronium ion, a Lewis acid, to form the delocalized σ-complex **19** (Eq. 15.16). In the final step of the reaction, a base such as water or bisulfate deprotonates this complex to regenerate the aromatic ring (Eq. 15.17).

$$ \qquad (15.16)$$

19
σ-Complex

$$\text{(19)} \quad \xrightarrow{:\ddot{O}H_2} \quad \text{NO}_2 \quad + \quad H_3O^+ \qquad (15.17)$$

The mechanism of the reaction of a nitronium ion with a *substituted* benzene such as bromobenzene (**20**), the arene used in this experiment, is similar to that with benzene. Now, however, *three* different products can be formed because of the presence of the bromine atom on the ring (Eq. 15.18). The activation enthalpies, $\Delta H^{\ddagger}$, for attack at the positions *ortho* and *para* to the substituent are lower than that for attack

$$\text{(15.18)}$$

20	**21**	**22**	**23**
Bromobenzene	2-Bromo-nitrobenzene	4-Bromo-nitrobenzene	3-Bromo-nitrobenzene

at the *meta* position. This is because the σ-complexes **24** and **25**, which are the precursors to the *ortho* and *para* products **21** and **22**, respectively, are more stable than complex **26**, which gives the *meta* derivative **23**. The increased stability of **24** and **25** relative to **26** is associated with the participation of nonbonding electrons on the bromine atom in delocalizing the positive charge in the complexes **24** and **25**. Thus, like an alkyl group, a bromine atom is an **ortho-para director**, because it favors formation of these isomers over the *meta* isomer. The preference is found to be about 100:1 for this nitration reaction.

24 **25** **26**

The *o:p* ratio itself is of interest because isolating pure disubstituted products requires separation of these isomers. Assuming that attack of the electrophile at the two positions occurs on a purely statistical basis, this ratio is predicted to be 2:1, because there are two *ortho* carbon atoms and only one *para* carbon atom in bromobenzene. This prediction neglects the possibility that the steric bulk and the inductive electron-withdrawing effect of the bromine atom might inhibit approach of the electrophile to the *ortho* relative to the *para* position. By determining the relative amounts of **21** and **22** produced, you will be able to evaluate this hypothesis.

Another product may result if further nitration of **21** or **22** occurs. Consideration of the directing effects of the bromo and nitro substituents leads to the prediction that the major product from both isomers would be 1-bromo-2,4-dinitrobenzene (**27**), as shown in Equation 15.19. The possible formation of 1-bromo-2,6-dinitrobenzene (**28**) from **21** is less likely because of the steric and inductive effects of the bromo substituent. Fortunately, it is relatively easy to minimize dinitration by performing the reaction at temperatures below 60 °C. Controlling the temperature suppresses dinitration because the strongly deactivating nitro substituent in **21** and **22** raises the activation enthalpy for their nitration above that of **20**. Another factor decreasing the dinitration of **22** is its precipitation during the course of the reaction, thereby removing it from the nitrating medium.

$$\textbf{21 or 22} \quad \xrightarrow[\text{H}_2\text{SO}_4]{\text{HNO}_3} \quad \text{(structure)} \quad + \quad \text{H}_2\text{O} \qquad (15.19)$$

27
1-Bromo-2,4-dinitrobenzene

28
1-Bromo-2,6-dinitrobenzene

Based on the preceding discussion, the major obstacle to isolating pure 4-bromonitrobenzene (**22**) is separating it from 2-bromonitrobenzene (**21**). The strategy for doing this relies on the greater polarity of **21,** which makes this isomer more soluble in polar solvents than **22**. This difference is dramatically demonstrated by their relative solubilities in ethanol. At room temperature, the *ortho* isomer **21** is very soluble, whereas the *para* isomer is only slightly soluble, dissolving to the extent of only 1.2 g/100 mL. This large difference in solubilities allows separation of the isomers by the technique of **fractional crystallization**. The mixture of products obtained from nitration of bromobenzene is dissolved in hot 95% ethanol, and the solution is allowed to cool. The less soluble *para* isomer selectively crystallizes from solution and is isolated by filtration. Concentration of the filtrate allows isolation of a second crop of **22**.

In a fractional crystallization, it is commonly possible to induce crystallization of the more soluble component, in this case the *ortho* isomer **21,** once the less soluble material has been mostly removed. The low melting point of **21** means that this isomer is difficult to crystallize in the presence of impurities. Consequently, isolating this isomer involves column chromatography, however (Sec. 6.3).

EXPERIMENTAL PROCEDURES

Nitration of Bromobenzene

Discovery Experiment   **Purpose** To demonstrate nitration by electrophilic aromatic substitution and to test the directing effects of a bromo substituent.

SAFETY ALERT

1. **Wear safety glasses or goggles and suitable protective gloves while performing the experiments.**

2. Because *concentrated* **sulfuric and nitric acids may cause severe chemical burns,** *do not allow them to contact your skin.* **Wipe off any drips and runs on the outside surface of reagent bottles and graduated cylinders** *before* **picking them up. Wash any affected area immediately and thoroughly with cold water, and apply 5% sodium bicarbonate solution.**

A ■ *Nitration*

MINISCALE PROCEDURE

Figure 15.17
Apparatus for brominating nitrobenzene.

Preparation Refer to the online resources to answer Pre-Lab Exercises, access videos, and read the MSDSs for the chemicals used or produced in this procedure. Review Sections 2.9, 2.10, 2.11, and 2.29.

Apparatus A 25-mL round-bottom flask, Claisen adapter, thermometer, water-cooled condenser, ice-water bath, apparatus for magnetic stirring, simple distillation, vacuum filtration, and *flameless* heating.

Setting Up Prepare a solution of 4.0 mL of *concentrated* nitric acid and 4.0 mL of *concentrated* sulfuric acid in the round-bottom flask and cool it to room temperature with a water bath. Equip the flask with a stirbar and a Claisen adapter fitted with the condenser and a thermometer that extends into the flask (Fig. 15.17).

Reaction and Work-Up In portions of approximately 0.5 mL, add 4.5 mL of bromobenzene to the stirred mixture through the top of the condenser over a period of about 10 min. Do *not* allow the temperature of the reaction mixture to exceed 50–55 °C during the addition. Control the temperature by allowing more time between the addition of successive portions of bromobenzene and by cooling the reaction flask with an ice-water bath.

After the addition is complete and the exothermic reaction has subsided, heat the stirred mixture for 15 min, keeping its temperature below 60 °C.★ Cool the reaction mixture to room temperature and then pour it carefully and with stirring into 40 mL of cold water contained in a beaker.

Isolation and Purification Isolate the mixture of crude bromonitrobenzenes by vacuum filtration. Wash the filter cake thoroughly with cold water until the washes are neutral to pHydrion paper; allow the solid to drain under vacuum until nearly dry.★

Transfer the filter cake to an Erlenmeyer flask and recrystallize the crude product from 95% ethanol. Allow the residual solution to cool slowly to room temperature; then cool it to 0 °C in an ice-water bath. Isolate the crystalline product by vacuum filtration. Wash the product with a little *ice-cold* 95% ethanol, allowing the washes to drain into the filter flask with the mother liquors. Transfer the product to a watchglass or a piece of filter paper for air-drying.

Concentrate the mother liquors to a volume of about 10 mL by simple distillation. Perform this operation in a hood to prevent release of vapors into the laboratory. Alternatively, use rotary evaporation or other techniques to concentrate the solution. Allow the residual solution to cool to room temperature to produce a second crop of 4-bromonitrobenzene. Isolate it by vacuum filtration and, after air-drying, put it in a separate vial from the first crop.

Further concentrate the mother liquors from the second crop to a volume of 3–4 mL. The resulting oil contains crude 2-bromonitrobenzene. Separate the oil from the two-phase mixture by means of a Pasteur pipet and weigh and reserve it for chromatographic analysis (Parts B and C).

Analysis Weigh both crops of product and calculate the yield. Measure the melting points of both crops. Obtain IR and NMR spectra of your starting material and product and compare them with those of authentic samples (Figs. 8.31, 15.18–15.21).

MICROSCALE PROCEDURES

Preparation Refer to the online resources to answer Pre-Lab Exercises, access videos, and read the MSDSs for the chemicals used or produced in this procedure. Review Sections 2.9, 2.10, 2.11, and 2.29.

Apparatus A 3-mL conical vial, water-cooled condenser, ice-water bath, apparatus for magnetic stirring, vacuum and Craig tube filtration, simple distillation, and *flameless* heating.

Setting Up Prepare a solution of 0.5 mL of *concentrated* nitric acid and 0.5 mL of *concentrated* sulfuric acid in the vial, and cool it to room temperature with a water bath. Equip the vial with a spinvane and the condenser.

Reaction and Work-Up In portions of approximately 0.1 mL, add 0.5 mL of bromobenzene to the stirred mixture through the top of the condenser over a period of about 5 min. Although the temperature of the reaction mixture may become *warm* to the touch, do not allow the reaction mixture to become *hot* to the touch during the addition. Control the temperature by allowing more time between the addition of successive portions of bromobenzene and by cooling the vial with an ice-water bath if necessary.

After the addition is complete and the exothermic reaction has subsided, heat the stirred reaction mixture for 15 min, keeping the bath temperature below 70 °C.★ Cool the vial to room temperature, and then pour the reaction mixture carefully and with stirring into 5 mL of cold water contained in a beaker.

Isolation and Purification Isolate the mixture of crude bromonitrobenzenes by vacuum filtration. Wash the filter cake with small portions of cold water until the washes are neutral, and allow the solid to drain under vacuum until nearly dry.★

Transfer the filter cake to a Craig tube and recrystallize the crude product from 95% ethanol. Allow the solution to cool slowly to room temperature; then cool it to 0 °C in an ice-water bath. Isolate the crystalline product by vacuum or Craig tube filtration.

Concentrate the mother liquors to a volume of about 1 mL by simple distillation. Perform this operation in a hood to prevent release of vapors into the laboratory. Alternatively, use rotary evaporation or other techniques to concentrate the solution. Allow the residual solution to cool to room temperature to produce a second crop of 4-bromonitrobenzene. Isolate it by Craig tube filtration and, after air-drying, put it in a separate vial from the first crop.

Further concentrate the mother liquors from the second crop to a volume of about 0.5 mL. The resulting oil contains crude 2-bromonitrobenzene. Separate the oil from the two-phase mixture by means of a Pasteur pipet and weigh and reserve it for TLC analysis (Part B).

Analysis Weigh both crops of product and calculate the yield. Measure the melting points of both crops. Obtain IR and NMR spectra of your starting material and product and compare them with those of authentic samples (Figs. 8.31, 15.18–15.21).

WRAPPING IT UP

Neutralize the *aqueous filtrates* with sodium carbonate and flush them down the drain. Pour any *oil* obtained by concentration of the *ethanolic filtrate* into the container for halogenated liquids, *unless* the oil is to be subjected to chromatography.

B ▪ *Thin-Layer Chromatography*

Preparation Refer to the online resources to answer Pre-Lab Exercises, access videos, and read the MSDSs for the chemicals used in this procedure. Review Section 6.3.

Apparatus Capillary pipets, developing chamber.

Setting Up In two small vials, separately prepare solutions of 4-bromonitrobenzene and of the oil containing 2-bromonitrobenzene in about 0.5 mL of dichloromethane. Apply spots of each solution to the TLC plate and allow the spots to dry.★

Chromatography and Analysis Develop the chromatogram in a TLC chamber with 9:1 (*v*:*v*) hexane:ethyl acetate as the eluting solvent. When the solvent is within about 0.5 cm of the top of the plate, remove the developed chromatogram from the chamber, *quickly* mark the solvent front with a pencil, and allow the plate to dry.

Make the spots visible by placing the dry plate under an ultraviolet (UV) lamp or in a chamber whose atmosphere is saturated with iodine vapor. Calculate the R_f-values of the spots observed, and identify them as either 2- or 4-bromonitrobenzene. A small orange spot may be observed very near the origin of the chromatogram; this spot is 1-bromo-2,4-dinitrobenzene.

WRAPPING IT UP

Put the unused *eluant* in the container for nonhalogenated organic liquids. Discard the dry *TLC plates* in the container for halogenated organic compounds.

C ▪ *Column Chromatography*

MINISCALE PROCEDURE

Preparation Refer to the online resources to answer Pre-Lab Exercises, access videos, and read the MSDSs for the chemicals used in this procedure. Review Sections 6.2 and 6.3.

Apparatus Silica gel chromatography column, eight 13-mm × 100-mm test tubes, capillary pipets, developing chamber, apparatus for simple distillation and *flameless* heating.

Setting Up Obtain a 25- or 50-mL buret and 5 g of silica gel. Pack the column according to the procedure described in the experimental procedure of Section 6.2 with the exception that a 9:1 (*v:v*) solution of hexane and ethyl acetate rather than petroleum ether is to be used for packing the column. With a Pasteur pipet, apply a 0.2-g sample of the oil containing the 2-bromonitrobenzene to the head of the column, and rinse the inside of the buret with 1 mL of ethyl acetate. Open the stopcock and allow the liquid to drain just to the top of the sand. Close the stopcock and fill the buret with 9:1 (*v:v*) hexane:ethyl acetate.

Chromatography and Analysis Elute the column until a total of 40 mL of the solvent has passed through the column. Do not allow the level of liquid to drain below the sand at the top of the column. Collect the eluate in 5- to 8-mL fractions in separate, labeled test tubes.

Analyze each fraction by TLC using 9:1 (*v:v*) hexane:ethyl acetate as the eluting solvent. With careful spotting, it should be possible to analyze at least three different fractions on each 3-cm × 8-cm strip of a silica gel chromatogram. To introduce sufficient sample onto the strip, respot two or three times.

Combine the fractions containing only 2-bromonitrobenzene and then remove the solvent by simple distillation. Alternatively, use rotary evaporation or other techniques to concentrate the solution. Characterize the residue obtained by either melting point determination, TLC, or spectroscopic analysis. 2-Bromonitrobenzene has a low reported melting point and thus may be isolated as an oil. Obtain IR and NMR spectra of your product and compare them with those of an authentic sample (Figs. 15.22 and 15.23).

WRAPPING IT UP

After first allowing any residual solvent to evaporate from the silica gel in the hood, put the *silica gel* in the container for nontoxic solids. Pour all *eluates containing bromonitrobenzenes* into the container for halogenated organic compounds; put *all other eluates* and any unused eluant in the container for nonhalogenated solvents.

EXERCISES

1. Precipitation of the mononitration product of bromobenzene prevents dinitration from occurring. Explain how this experimental result is an application of the Le Chatelier principle.

2. How does maintaining the reaction temperature below 60 °C help suppress formation of dinitration by-products?

3. The pK_as of sulfuric and nitric acids are −3 (estimated value) and −1.3, respectively.

    a. Which of the two acids is stronger?

    b. Given your prediction in Part **a**, write a chemical equation for the equilibrium involving reaction of sulfuric and nitric acids.

    c. What is the value of K_{eq} for the acid-base reaction in Part **b**? Assume that the pK_a of the conjugate acid of nitric acid is −13.

4. Explain why 4-bromonitrobenzene (**22**) is *less* polar than 2-bromonitrobenzene (**21**).

5. Explain why the melting point of 4-bromonitrobenzene (**22**) is considerably higher than that of the 2-isomer **21**.

6. a. Provide the resonance structures that contribute to the σ-complexes **25** and **26**.

    b. Use these resonance structures to explain why the formation of 3-bromonitrobenzene (**23**) is disfavored relative to 4-bromonitrobenzene.

7. Why does 4-bromonitrobenzene (**22**) have a larger R_f-value in the TLC analysis than does the 2-isomer **21**?

8. Which isomer, 4-bromonitrobenzene or 2-bromonitrobenzene, will elute from a silica gel chromatography column first, and what physical property accounts for this order?

9. Why should a chromatography column never be allowed to go dry?

10. It is hard to measure the yield of 2-bromonitrobenzene accurately in the reaction you performed because this isomer is not purified in the procedure. However, the yield may be crudely approximated by assuming that the oil obtained from concentrating the mother liquors from recrystallization is comprised entirely of this isomer.

    a. Using the assumption that the oil is the 2-bromonitrobenzene, determine the *o:p* ratio of isomers formed in the mononitration of bromobenzene.

    b. Explain whether the result of your determination in Part **a** supports the hypothesis that the steric effect of the bromo substituent suppresses or augments the formation of the *o*-isomer relative to that expected statistically.

    c. A better way to assess the *o:p* ratio of the two bromonitro isomers would be to analyze the crude reaction mixture prior to any purification steps. Propose a way by which you might perform such an analysis. Be specific in your answer.

    d. The experimentally observed *o:p* ratio is reported to be 38:62.

        i. What does this ratio indicate with regard to the steric effect of the bromo substituent?

        ii. Use this ratio and the amount of 4-bromonitrobenzene actually isolated to estimate the experimental yield of mononitration in the reaction.

        iii. What errors might attend using this method to calculate the extent of mononitration?

11. The *o:p* ratio in the mononitration of bromobenzene has been reported to be 38:62. Use this ratio and the amount of 4-bromonitrobenzene actually isolated to estimate the experimental yield of mononitration in the reaction. What errors are there in using this method to calculate the extent of mononitration?

12. Explain why 4-bromonitrobenzene cannot be prepared efficiently by the bromination of nitrobenzene.

13. Would nitration of bromobenzene or nitration of 4-bromonitrobenzene be expected to have the higher enthalpy of activation? Explain your answer.

14. Consider the spectral data for bromobenzene (Figs. 15.18 and 15.19).

    a. In the IR spectrum, specify the absorption associated with the π-bonds of the aromatic ring. Indicate with what structural feature the strong absorptions at about 740 cm^{-1} and 690 cm^{-1} are associated.

Refer to the online resources and use the spectra viewer and Tables 8.1–8.8 as needed to answer the blue-numbered questions on spectroscopy.

    b. In the ^1H NMR spectrum, assign the various resonances to the hydrogen nuclei responsible for them.

    c. For the ^{13}C NMR data, assign the various resonances to the carbon nuclei responsible for them.

15. Consider the spectral data for 4-bromonitrobenzene (Figs. 8.31, 15.20, and 15.21).

    a. In the functional group region of the IR spectrum, specify the absorption associated with the π-bonds of the aromatic ring and indicate with what structural feature the strong absorption at 825 cm^{-1} is associated.

    b. In the ^1H NMR spectrum, assign the various resonances to the hydrogen nuclei responsible for them.

    c. For the ^{13}C NMR data, assign the various resonances to the carbon nuclei responsible for them.

16. Consider the spectral data for 2-bromonitrobenzene (Figs. 15.22 and 15.23).

    a. In the functional group region of the IR spectrum, specify the absorption associated with the π-bonds of the aromatic ring and indicate with what structural feature the strong absorption at about 750 cm^{-1} is associated.

    b. In the ^1H NMR spectrum, assign the various resonances to the hydrogen nuclei responsible for them.

    c. For the ^{13}C NMR data, assign the various resonances to the carbon nuclei responsible for them.

17. What differences in the IR and NMR spectra of bromobenzene and 4-bromonitrobenzene are consistent with the introduction of a nitro group onto the ring in this experiment?

Starting Materials and Products

The ^1H NMR spectrum of 4-bromonitrobenzene is provided in Figure 8.31.

Figure 15.18
IR spectrum of bromobenzene (neat).

Figure 15.19

NMR data for bromobenzene (CDCl₃).

(a) ¹H NMR spectrum (300 MHz).
(b) ¹³C NMR data: δ 122.5, 126.7, 129.8, 131.4.

Figure 15.20

IR spectrum of 4-bromonitrobenzene (IR card).

Figure 15.21

NMR data for 4-bromonitrobenzene.

¹³C NMR data: δ 124.9, 129.8, 132.5, 147.0.

Figure 15.22

IR spectrum of 2-bromonitrobenzene (neat).

Figure 15.23
NMR data for 2-bromonitrobenz-ene (CDCl₃).

(a) ¹H NMR spectrum (300 MHz).
(b) ¹³C NMR data: δ 114.1, 125.4, 128.2, 133.2, 134.8, 136.4.

15.5 SUBSTITUENT EFFECTS ON ELECTROPHILIC AROMATIC SUBSTITUTION

Electrophilic aromatic substitution reactions (Sec. 15.1) are among the best understood of all organic reactions. The *qualitative* aspects of the reactions that are discussed in textbooks include the effect substituents have on the reactivity of arenes toward electrophiles and the orientation, *ortho, meta,* or *para,* of their attack on the ring. To explore how the *rates* of electrophilic substitution are affected by the nature of a substituent, the experimental procedures in Part A of this section provide both *semiquantitative* and *quantitative* measures of the differences in reactivity of a series of arenes toward the bromonium ion, Br^+, to produce the corresponding aryl bromides (Eq. 15.20). The electronic and steric effects that substituents have on the orientation or regiochemistry of substitution will be revealed by the experimental procedures in Parts B and C of this section. These procedures involve determining the distribution of products, *ortho, meta,* and *para,* obtained from bromination and nitration of monosubstituted arenes (Eq. 15.21, G = Br and NO_2, respectively). They will also allow you to make a *semi-qualitative* assessment of relative reactivities. The discussion that follows will first focus on the issue of relative rates of substitution and then on that of orientation in the reaction.

$$Ar-H \ + \ Br_2 \ \xrightarrow{\text{HOAc}} \ Ar-Br \ + \ HBr \qquad (15.20)$$

An arene   Bromine   An aryl bromide
         (red-brown)   *(colorless)*

$$(15.21)$$

A mono-   An   An *ortho-*   A *meta-*   A *para-*
substituted arene   electrophile   substituted arene   substituted arene   substituted arene

Studies of electrophilic substitutions on arenes are reported in which the experimental conditions allow a direct comparison of the relative reaction rates. For example, the relative reactivities of benzene and toluene toward halogenation, acetylation, sulfonation, nitration, and methylation have been determined. In all cases, electrophilic aromatic substitution was more rapid with toluene. For example, bromination of toluene is some 600 times faster than that of benzene. Such studies have led to the classification of substituents as **ring activators** or **deactivators**, depending on whether the substituted arene reacts *faster* or *slower* than benzene itself. Thus, the methyl group of toluene is a ring activator.

The effect substituents have on the relative rate of a specific type of S_E2 reaction may exceed several orders of magnitude, and the range of reactivity when comparing substituents may be enormous. For example, phenol reacts 10^{12} times faster and nitrobenzene reacts 10^5 times *slower* than benzene in electrophilic bromination. The hydroxy group is thus a powerful *ring activator,* and the nitro group is a potent *ring deactivator.* The range of difference in reactivity due to these two substituents is a remarkable 10^{17}!

You will be measuring the relative rates of bromination of six monosubstituted arenes **12** and **29–33**. This particular electrophilic aromatic substitution is selected because the relative rates may be measured both qualitatively and quantitatively. The results should allow you to determine the order of ring activation associated with the various substituents that you will investigate.

29	**12**	**30**	**31**	**32**	**33**
Anisole	Phenol	Diphenyl ether	Acetanilide	4-Bromophenol	1-Naphthol

The qualitative measurements are easier to perform because they simply rely on your observing the disappearance of the characteristic reddish color of molecular bromine as a function of time. You should be able to rank-order the substituent effects, but the results will be subject to the vagaries of human judgment with respect to knowing when all of the color has been discharged from the reaction mixture.

Quantitative measurements also depend on determining the rate of disappearance of the color of bromine, but a spectrophotometer rather than the human eye is used to measure changes in the visible portion of the electromagnetic spectrum (Sec. 8.1). None of the substrates **12**, **30**, and **31** that we are studying absorb at this wavelength; only molecular bromine does, so a decrease in light absorption reflects a decrease in the concentration of molecular bromine. It is not possible to extend quantitative analyses to **29**, **32**, and **33** because they react so fast that obtaining accurate measurements of absorbance versus time requires use of more expensive spectrophotometers than are normally available in undergraduate laboratories.

To analyze the data from the quantitative measurements, we start by considering the rate expression for an S_E2 reaction that is given in Equation 15.22; this equation is identical to Equation 15.2 (Sec. 15.1) except that $[Br_2]$, the precursor to Br^+, has replaced $[E^+]$. This expression could be more complex if a Lewis acid catalyst were required to promote formation of the electrophile, which is the case when the arene is less reactive toward electrophilic aromatic substitution than the ones we've selected. The rate is seen to be first-order in both reagents and therefore is second-order overall (Sec. 14.6).

$$\text{Rate} = k_2[\text{ArH}][\text{Br}_2] \tag{15.22}$$

Calculations for second-order reactions are more complex than those for first-order processes, so we've designed our experimental protocol so that the kinetics will be pseudo-first-order by using a sufficiently large excess of the aromatic substrate so that its concentration does not change appreciably during the reaction. Equation 15.21 then may be simplified to Equation 15.23, in which the concentration of the arene has been incorporated into the rate constant, k_1; k_1 is then called a *pseudo-first-order rate constant*.

$$\text{Rate} = k_1[\text{Br}_2] \tag{15.23}$$

The *integrated* form of Equation 15.23 is given in Equation 15.24, which can be rearranged to provide Equations 15.25 and 15.26, where $[\text{Br}_2]_0$ and $[\text{Br}_2]_t$ are the initial concentration of bromine and the concentration of bromine at any time t, respectively. By plotting the natural log, ln, or $\log_{10}$ of the ratio of concentrations versus time, a straight line should result if the reaction is indeed pseudo-first-order; the slope of the line is k_1 if the natural log is used, but is $k_1/2.303$ if the common log is used.

$$[\text{Br}_2]_t = [\text{Br}_2]_0 e^{-k/t} \tag{15.24}$$

$$k_1 t = \ln \frac{[\text{Br}_2]_0}{[\text{Br}_2]_t} \tag{15.25}$$

$$k_1 t = 2.303 \log \frac{[\text{Br}_2]_0}{[\text{Br}_2]_t} \tag{15.26}$$

You will be using an ultraviolet-visible (UV-Vis) spectrophotometer to measure the concentration of molecular bromine as a function of time. Fortunately, a relatively inexpensive nonrecording instrument is sufficient for our purposes because all measurements will be made at a single wavelength of 400 nm. You personally will be serving as the recorder, noting the absorbance, A (Eq. 15.27), of Br_2 at this wavelength as a function of time. Equation 15.27 was originally discussed in Section 8.4 (Eq. 8.14), and you should review that discussion. Because the same cell will be used for all measurements, l is constant. The molar extinction coefficient, ε, is also constant because only molecular bromine is absorbing at the selected wavelength. Consequently, Equations 15.26 and 15.27 may be combined to give Equation 15.28, since $A_0 = \varepsilon[\text{Br}_2]_0 l$ and $A_t = \varepsilon[\text{Br}_2]_t l$. This is a useful form of Equation 15.25 because it relates absorbance to the rate constant k_1.

$$A = \log \frac{I_0}{I} = \varepsilon c l \tag{15.27}$$

$$k_1 t = 2.303 \log \frac{A_0}{A_t} \tag{15.28}$$

The presence of a substituent on an aromatic ring also affects the orientation of electrophilic aromatic substitutions because three different regioisomers may be formed, as illustrated in Equation 15.21. As described in Sections 15.2–15.4, the S_E2 process involves nucleophilic attack of π-electrons from the aromatic ring onto an electrophile to form a σ-complex (Scheme 15.2). This is often the rate-determining step of the reaction. The rate of the reaction can be increased by the addition of a Lewis or Brønsted acid, A, to form the electrophile, E^+ (Eq. 15.3), or to activate an electrophilic reagent through complexation, as portrayed for the activation of bromine in the Scheme.

Scheme 15.2

As noted above, the regiochemical outcome and the rate of electrophilic substitution reactions of monosubstituted benzenes depend on the nature of the substituent, G, that is on the aromatic ring. Arenes bearing electron-donating substituents, G = alkyl (R), amido (NHC(=O)R) and alkoxy (OR), for example, preferentially give *ortho*- and/or *para*-substituted products, and those with electron-withdrawing groups, G = nitro (NO_2) and acyl (RC=O), for instance, mainly yield *meta*-substituted products.

The basis for the orienting effects of substituents is understandable by considering the resonance contributors **34–38** in Scheme 15.3. When addition of the electrophile occurs *ortho* or *para* to the substituent G, partial positive charge in the σ-complex resides on the carbon atom bearing G, as shown in **34** and **36**, as well as on the carbon atoms having the symbol, (+). When addition occurs *meta* to G, the partial positive charge is not found on the carbon atom bearing the substituent; rather it resides at least one carbon atom away, as seen in **35**.

Considering possible inductive effects, G may *stabilize* the positive charge in contributing resonance structures, especially **34** and **36**, through electron release, as is the case for alkyl substituents such as methyl or *tert*-butyl. Such stabilization will favor formation of *ortho*- and *para*-substituted products. On the other hand, G groups such as alkoxy, amido, halogen, and nitro groups that are inductively electron withdrawing will *destabilize* all of the contributing resonance structures, but especially those represented by **34** and **36**. If this inductive effect were the only factor defining the regiochemistry of reaction of the electrophile, *meta* substitution would be favored because the partial positive charge in the σ-complex is further removed from the location of G, as seen in **35**.

Scheme 15.3

Ortho-addition	*Meta*-addition	*Para*-addition	*Ortho*-addition	*Para*-addition
34	**35**	**36**	**37**	**38**

However, an additional factor must be considered when G bears non-bonded electrons on the atom attached to the aromatic ring, as is the case for alkoxy, amido, and halogen substituents. Donation of these electrons, as shown in the contributing resonance forms **37** and **38**, further delocalizes and stabilizes the positive charge in the σ-complex. Moreover, resonance structures such as **37** and **38** are particularly stable because all atoms have an octet of electrons. It would be difficult for you to predict whether the inductive effect or delocalization by resonance ultimately dominates in determining the regioselectivity of substitution when G is an electron-withdrawing group having non-bonding electrons. The procedures of Parts B and C of this section will enable you to explore this question. In addition, you will be able to determine how substituents affect the relative rates of the S_E2 reaction, and you will evaluate the roles that Lewis acids and steric factors play in determining the regioselectivity of the substitution. Finally, analysis of the ratio of *ortho:para* products will allow you to assess the possible role of statistics in defining the relative amounts of the two regioisomers produced. Because there are two *ortho* positions in a monosubstituted arene and only one *para* position, statistics would predict the ratio of the two regioisomers to be 2:1, so you should consider this additional factor when assessing the relative amounts of products formed.

In the experiments of Part B, two sets of reaction conditions for the electrophilic bromination of monosubstituted arenes will be investigated. The first set involves reaction with anisole (**12**), bromobenzene (**18**), toluene (**39**), or methyl benzoate (**40**) and in the *absence* of a catalyst. For the second, a catalytic amount of ferric bromide, $FeBr_3$, will be used to assess the effect the *presence* of a Lewis acid has on the reaction. Analysis of the crude reaction mixtures by ^1H NMR spectroscopy will enable you to determine the structures of the products as well as the ratio of products if multiple products are produced. The results you obtain will be compared with those of your classmates to determine the overall patterns of reactivity and orientation as a function both of the substituent on the ring and whether or not a catalyst is used.

Anisole	Bromobenzene	Toluene	Methyl benzoate
12	**18**	**39**	**40**

In the experiments of Part C, you will study how the presence of substituents on an aromatic ring affect the orientation of nitration (Eq. 15.29). Specifically, the *ortho*- and *para*-directing effects of alkyl groups will be explored by the nitration of toluene (**39**) and *tert*-butylbenzene (**41**). Importantly, these two alkyl groups have very different steric requirements, so this experiment will allow you to assess whether steric factors can affect the ratio of regioisomers produced. The analysis of directing effects of other substituents on the aromatic ring will be extended by performing the nitration of chlorobenzene (**42**) and acetanilide (**43**). Both of these groups are also *ortho* and *para* directing, but one of them yields all three possible regioisomers of the S_E2 nitration product, whereas the other provides only one. Finally, nitration of methyl benzoate (**40**) allows you to assess how an electron-withdrawing substituent, G, affects the orientation of electrophilic aromatic substitution. The product ratios for each arene may be determined by either gas chromatographic analyses or ^1H NMR spectroscopy, allowing you to deduce the pattern of the directing effects of the groups.

$$\text{(15.29)}$$

ortho-	*meta*-	*para*-
Substitution	Substitution	Substitution

EXPERIMENTAL PROCEDURES

A. Relative Rates of Electrophilic Aromatic Bromination

Discovery Experiment

Purpose To evaluate the relative reactivities of different arenes in electrophilic aromatic bromination.

SAFETY ALERT

1. **Wear safety glasses or goggles and suitable protective gloves while performing the experiments.**

2. *Bromine is a hazardous chemical that may cause serious chemical burns.* **Do not breathe its vapors or allow it to come into contact with your skin. Perform all operations involving the transfer of the pure liquid or its solutions in a hood. Even though the solutions used in the experiments are dilute, take proper precautions. If you get bromine on your skin, wash the area** *immediately* **with soap and warm water and soak any affected area in 0.6 *M* sodium thiosulfate solution, for up to 3 h if the burn is particularly serious.**

3. Do not use your mouth to fill pipets when transferring any of the solutions in this experiment; this applies to both the bromine solutions and the acetic acid solutions of the substrates.

1 ▪ *Qualitative Measurements*

Note It is convenient to work in pairs for this experiment.

MINISCALE PROCEDURE

Preparation Refer to the online resources to answer Pre-Lab Exercises, access videos, and read the MSDSs for the chemicals used or produced in this procedure. Review Section 2.5.

Apparatus A 1-L beaker, three calibrated Pasteur pipets, six 13-mm × 100-mm test tubes, ice-water bath, and *flameless* heating.

Setting Up As accurately as possible, calibrate several Pasteur pipets to contain 1.5 mL. In separate labeled test tubes, place 1.5 mL of a 0.2 M stock solution of each of the following substrates in 15 M (90%) acetic acid: (a) phenol, (b) anisole, (c) diphenyl ether, (d) acetanilide, (e) 4-bromophenol, and (f) 1-naphthol. Use special care to ensure that one stock solution is not contaminated by another during any transfer operations you perform.

Use the 1-L beaker to make a water bath and establish a bath temperature of 35 ± 2 °C. Suspend the test tubes partially in the water bath by looping a piece of copper wire around the neck of the test tube and over the rim of the beaker. Transfer about 25 mL of a solution containing 0.05 M bromine in 15 M acetic acid to an Erlenmeyer flask, and equilibrate this solution in the water bath for a few minutes.

Bromination Add 1.5 mL of the bromine solution to a test tube containing one of the substrates. Make the addition *rapidly,* mix the solution *quickly,* and *record the exact time* of addition. Monitor the reaction mixture and note the elapsed time for the bromine color to become faint yellow or to disappear. If decoloration has not occurred within 5 min, which can happen because some substrates react slowly, proceed to a second substrate while waiting for the endpoint of the first to be reached. Record the reaction times. Repeat this procedure with each of the arenes, using the same endpoint color for each one.

To confirm your conclusions regarding the relative reactivities of the series of arenes, repeat the experiment at 0 °C in an ice-water bath, using only those substrates that were not clearly differentiated at 35 °C.

Analysis On the basis of your observations, list the arenes in order of *decreasing* reactivity toward bromine. Obtain IR and NMR spectra of the starting materials and compare them with those of authentic samples (Figs. 8.7, 15.13, 15.14, and 15.24–15.32).

2 ▪ *Quantitative Measurements*

Note It is convenient to work in pairs for this experiment. In preparing for and executing the kinetic runs, you must be well organized so that you can work quickly. You may need to perform several kinetic runs before you are able to work rapidly enough to get acceptable results.

MINISCALE PROCEDURE

Preparation Refer to the online resources to answer Pre-Lab Exercises, access videos, and read the MSDSs for the chemicals used or produced in this procedure. Review Section 2.5.

Apparatus UV-Vis spectrophotometer or colorimeter, three capped cuvettes, three calibrated Pasteur pipets, glass stirring rod, timer, and thermometer.

Setting Up Have at your disposal 0.5 *M* stock solutions of anisole, diphenyl ether, and acetanilide in 15 *M* (90%) *aqueous* acetic acid and a 0.02 *M* solution of bromine in 15 *M* acetic acid. Calibrate the spectrophotometer to zero absorbance at 400 nm using a solution of the arene substrate in acetic acid. Prepare this solution by using a 2-mL pipet to combine 2 mL of 15 *M* acetic acid and 2 mL of the stock solution of the substrate arene in a clean and dry cuvette. Stir the solution to ensure homogeneity and place the cuvette in the spectrophotometer. Adjust the absorbance reading of the spectrometer to zero. Record the laboratory temperature.

Reaction Once you have zeroed the spectrophotometer for a particular substrate, clean and dry the cuvette used for the calibration, and continue with the following operations. (1) Add 2 mL of 0.02 *M* bromine solution to the cuvette; (2) add 2 mL of the 0.5 *M* solution of substrate; (3) quickly stir the solution one time only, and have your partner record the exact time of mixing; (4) place the cuvette in the spectrophotometer; and (5) record the absorbance of the solution at 400 nm as a function of time. Obtain as many readings as possible before the absorbance drops below about 0.5, although readings may be taken below 0.5. Record the time at which the needle on the meter crosses a line on the absorbance scale rather than attempting interpolation between those lines.

Repeat the above procedure for each of the other substrates. For those runs that are particularly fast, it is advisable to perform duplicate runs.

Analysis Obtain IR and NMR spectra of the starting materials and compare them with those of authentic samples (Figs. 8.7, 15.13, 15.14, 15.26–28). Compute the desired rate constants, k_l, according to the following procedure.

1. The data recorded above constitute a series of values A_t, and it is necessary to extrapolate these data to the initial time, t_0, of mixing to obtain A_0. Plot log A_t as the ordinate versus time in seconds as the abscissa and extrapolate the straight line obtained to zero time to obtain log A_0. Determine A_0 for each substrate.

2. Using the value of A_0 and the recorded values of A_t, calculate a series of values of 2.303 log A_0/A_t for each substrate. Plot these values as the ordinate versus time in seconds as the abscissa. Draw the best straight line possible through these points. Calculate the slope of that line to obtain k_1 (sec^{-1}) for that substrate. You may also wish to determine the slope of the line using a least-squares analysis.

3. Divide all values of k_1 by the *smallest* value of this rate constant to obtain the *relative* reactivities of the substrate toward electrophilic aromatic bromination.

WRAPPING IT UP

Neutralize the combined *acetic acid solutions* by adding saturated *aqueous* sodium bicarbonate and filter the resulting mixture under vacuum. Flush the *filtrate* down the drain; put the *filter cake,* which contains aryl bromides, in the container for halogen-containing solid waste.

EXERCISES

General Questions

1. Write the structures for the major monobromination products that would be formed from each of the substrates used. Explain your predictions.

2. Using curved arrows to symbolize the flow of electrons, write the stepwise mechanism for the bromination of anisole to provide the major product you proposed in Exercise 1. Write the resonance structures for any charged, delocalized intermediates that are involved.

3. Explain how a nonpolar molecule like bromine can act as an electrophile.

4. Why does performing some of the S_E2 reactions at 0 °C allow you to identify the order of reactivity more definitively?

5. What problem might there be if you attempted to perform the kinetic experiment at −20 °C?

6. Would you expect the relative rate of S_E2 reactions to be faster or slower if molecular chlorine, Cl_2, were used instead of molecular bromine, Br_2? Explain your reasoning.

7. Why is the first step in an electrophilic aromatic bromination reaction so much more endothermic than the first step in an electrophilic addition to an alkene?

8. Do any of the substrates in this experiment contain substituents that deactivate the aromatic ring toward electrophilic aromatic substitution?

9. Which compound would you expect to undergo bromination at a faster rate, aniline, $C_6H_5NH_2$, or acetanilide? Anilinium ion, $C_6H_5NH_3^+$, or aniline? Explain your reasoning.

10. Explain why a methyl group is a mildly activating group, whereas a methoxy group is strongly activating.

11. Using inductive and/or resonance effects, explain the reactivity order you observed.

12. **a.** Determine the *initial* concentrations of bromine and of substrate arene in each reaction mixture.

    b. Which reagent, bromine or arene, is present in excess in each case?

    c. Explain why the experimental procedure calls for an excess of the specified reactant in **b** over the other.

    d. Discuss whether you would be able to determine the rates of reaction if the concentrations of arene and bromine had been reversed over what is specified in the experimental procedure.

Questions for Part A1

13. Provide a mechanistically based explanation for the relative order of reactivity that you obtained.

14. What might be a major source of error in determining relative rates of reaction by the procedure that you used?

Refer to the online resources and use the spectra viewer and Tables 8.1–8.8 as needed to answer the blue-numbered questions on spectroscopy.

Questions for Part A2

15. Derive Equation 15.28 from Equations 15.26 and 15.27. On the basis of this derivation, explain why it is unnecessary to use either the extinction coefficient, ε, or the cell length, l, for analyzing the data.

16. Specify sources of error in this procedure for determining the relative rates of electrophilic aromatic substitution.

17. Consider the spectral data for anisole (Figs. 15.13 and 15.14).

    a. In the functional group region of the IR spectrum, specify the absorption associated with the π-bonds of the aromatic ring.

    b. In the ^{1}H NMR spectrum, assign the various resonances to the hydrogen nuclei responsible for them.

    c. For the ^{13}C NMR data, assign the various resonances to the carbon nuclei responsible for them.

18. Consider the spectral data for phenol (Figs. 15.24 and 15.25).

    a. In the functional group region of the IR spectrum, specify with what structural feature the broad band centered at about 3280 cm^{-1} is associated, and explain why the band is broad.

    b. In the ^{1}H NMR spectrum, assign the various resonances to the hydrogen nuclei responsible for them.

    c. For the ^{13}C NMR data, assign the various resonances to the carbon nuclei responsible for them.

19. Consider the spectral data for diphenyl ether (Figs. 15.26 and 15.27).

    a. In the functional group region of the IR spectrum, specify the absorption associated with the π-bonds and with the C–H bonds of the aromatic rings.

    b. In the ^{1}H NMR spectrum, assign the various resonances to the hydrogen nuclei responsible for them.

    c. For the ^{13}C NMR data, assign the various resonances to the carbon nuclei responsible for them.

20. Consider the spectral data for acetanilide (Figs. 8.7 and 15.28).

    a. In the functional group region of the IR spectrum, specify the absorptions associated with the N–H bond, with the carbon-oxygen double bond, and with the aromatic ring.

    b. In the ^{1}H NMR spectrum, assign as many as possible of the various resonances to the hydrogen nuclei responsible for them.

    c. For the ^{13}C NMR data, assign the various resonances to the carbon nuclei responsible for them.

21. Consider the spectral data for 4-bromophenol (Figs. 15.29 and 15.30).

    a. In the functional group region of the IR spectrum, specify the absorption associated with the π-bonds of the aromatic ring; indicate with what structural feature the broad band centered at about 3400 cm^{-1} is associated, and explain why the band is broad.

    b. In the ^{1}H NMR spectrum, assign the various resonances to the hydrogen nuclei responsible for them.

    c. For the ^{13}C NMR data, assign the various resonances to the carbon nuclei responsible for them.

22. Consider the spectral data for 1-naphthol (Figs. 15.31 and 15.32).

    a. In the functional group region of the IR spectrum, specify the absorption associated with the π-bonds of the aromatic ring. Indicate with what structural feature the broad band centered at about 3300 cm^{-1} is associated, and explain why the band is broad.

    b. In the ^{1}H NMR spectrum, assign the various resonances to the hydrogen nuclei responsible for them.

    c. For the ^{13}C NMR data, assign the various resonances to the carbon nuclei responsible for them.

23. When the NMR solutions of phenol, 4-bromophenol, and 1-naphthol were each shaken with a drop of D_2O, the resonances assigned to the hydrogen of the hydroxyl group disappeared. Explain this observation.

SPECTRA

Starting Materials and Products

The IR spectrum of acetanilide is provided in Figure 8.7. The IR and NMR spectra of anisole are given in Figures 15.13 and 15.14.

Figure 15.24
IR spectrum of phenol (neat).

Figure 15.25
NMR data for phenol (CDCl₃).

(a) ^{1}H NMR spectrum (300 MHz).
(b) ^{13}C NMR data: δ 115.4, 121.0, 129.7, 154.9.

Figure 15.26
IR spectrum of diphenyl ether (IR card).

Figure 15.27
NMR data for diphenyl ether (CDCl₃).

(a) ¹H NMR spectrum (300 MHz).
(b) ¹³C NMR data: δ 119.0, 123.2, 129.8, 157.6.

Figure 15.28
NMR data for acetanilide (CDCl₃).

(a) ¹H NMR spectrum (300 MHz).
(b) ¹³C NMR data: δ 24.0, 119.7, 123.3, 128.5, 139.2, 168.7.

Figure 15.29
IR spectrum of 4-bromophenol (IR card).

Figure 15.30
NMR data for 4-bromophenol (CDCl$_3$).

(a) ^{1}H NMR spectrum (300 MHz).
(b) ^{13}C NMR data: δ 113.2, 117.2, 132.5, 153.9.

Figure 15.31
IR spectrum of 1-naphthol (IR card).

Figure 15.32
NMR data for 1-naphthol
(CDCl₃).

(a) ¹H NMR spectrum (300 MHz).
(b) ¹³C NMR data: δ 108.8, 119.8, 122.1, 124.9, 126.2, 127.6, 134.9, 152.4.

B. Electrophilic Aromatic Bromination of Monosubstituted Arenes

MINISCALE PROCEDURE

Discovery Experiment

Purpose To explore how the electronic nature of substituents and the use of a catalyst influence the regiochemistry and the rate of reaction in the bromination of monosubstituted arenes.

SAFETY ALERT

1. **Wear safety glasses or goggles and suitable protective gloves while performing the experiment.**

2. **Bromine is corrosive. Do not let it come into contact with your skin because it may cause severe burns. If it does get on your skin, wash the area immediately with soap and cool water, and then soak the affected area in a 0.6 *M* *aqueous* solution of sodium thiosulfate.**

3. **Iron (III) bromide causes skin and eye irritation. Do not let it come into contact with your skin. If it does get on your skin, wash the area immediately with soap and cool water.**

4. **Bromine reacts with acetone to produce a powerful lachrymator, α-bromoacetone, BrCH₂COCH₃. Therefore, do *not* rinse residual bromine out of glassware with acetone!**

5. **The hydrogen bromide generated during the reaction is volatile and corrosive. Do not breath the vapors because it may cause severe respiratory irritation. Do not let crude reaction mixtures come into contact with your skin because it may cause severe burns. If it does get on your skin, wash the area immediately with soap and water, and then soak the affected area in a 5% *aqueous* sodium bicarbonate solution.**

6. **Do not let crude reaction mixtures come into contact with your skin. If they do, wash the affected area with soap and water and then soak the area in a 5%** *aqueous* **sodium bicarbonate solution. If irritation persists, seek medical attention.**

Preparation Refer to the online resources to answer Pre-Lab Exercises, access videos, and read the MSDSs for the chemicals used or produced in this procedure. Review Sections 2.5, 2.6, 2.11, 2.17, 2.21, 2.23, 2.25, and 2.29.

Apparatus A 25-mL round-bottom flask, gas trap, apparatus for magnetic stirring and simple distillation.

UNCATALYZED REACTION

Setting Up Equip the flask with a stirbar and add 1 mmol of the substituted aromatic compound that is assigned to you to the flask. Add 5 mL of dichloromethane to the flask and stir the mixture to effect dissolution. Then add 1 mL of a 1 *M* solution of bromine in dichloromethane with stirring and attach the gas trap to the apparatus.

Reaction Stir the mixture for 1 h at room temperature. During this time, observe the reaction, noting whether the color of bromine disappears; if it does, record the time required for the color of bromine to disappear. Add 5 mL of 10% *aqueous* sodium bisulfite solution, followed by 5 mL of *saturated aqueous* sodium bicarbonate solution, and stir this mixture for about 5 min. If the color due to bromine persists, add additional 10% *aqueous* sodium bisulfite solution until the color disappears.

Work-Up and Isolation Transfer the mixture to a separatory funnel and separate the layers. Extract the aqueous layer with two 5-mL portions of dichloromethane and combine the organic layers. Transfer the organic layer to a dry flask. Dry the solution over several spatula-tips full of *anhydrous* magnesium sulfate. Swirl the flask occasionally for a period of 10–15 min to facilitate drying; add further small portions of *anhydrous* magnesium sulfate if the solution remains cloudy.★ Filter or carefully decant the dried organic solution into a round-bottom flask equipped for simple distillation, and remove the solvents by simple distillation. Alternatively, use rotary evaporation or other techniques to concentrate the solution. The final traces of solvent may be removed by attaching the flask to a vacuum source and gently swirling the contents as the vacuum is applied.

CATALYZED REACTION

Setting Up Equip the flask with a stirbar and add 30 mg of ferric bromide and 1 mmol of the substituted aromatic compound that is assigned to you to the flask. Add 5 mL of dichloromethane to the flask and stir the mixture to effect dissolution. Then add 1 mL of a 1 *M* solution of bromine in dichloromethane with stirring and attach the gas trap to the apparatus.

Reaction, Work-Up, and Isolation Follow the procedures used for the uncatalyzed reaction. You will not be able determine the time required for disappearance of color in the reaction, however.

Analysis Obtain a ^{1}H NMR spectrum of your crude reaction mixture and compare it to those of authentic samples (Figs. 15.34–15.58). Determine which isomers were formed in the reaction you performed and the ratio in which they were produced. Provide details in your notebook that explain how you arrived at your conclusions.

EXERCISES

Refer to the online resources and use the spectra viewer and Tables 8.1–8.8 as needed to answer the blue-numbered questions on spectroscopy.

1. Write the structures and names of the possible isomers from monobromination of the arene(s) you used in this experiment.

2. Combining your results with those of others, what products are present in the crude reaction mixture and in what ratio for each of the arenes?
   a. without FeBr$_3$
   b. with FeBr$_3$

3. a. Compare your results with those of others and identify which substrate reacted fastest *in the absence of* Lewis acid, based on the time required for disappearance of the bromine color in the reaction mixture?
   b. Which substrate underwent the reaction the next most rapidly?

4. What difference is observed between the non catalyzed and the FeBr$_3$-catalyzed reaction of toluene? Explain the difference in reactivity.

5. What difference is observed between the non catalyzed and the ferric bromide-catalyzed reaction of bromobenzene? Explain the difference in reactivity.

6. Draw the resonance contributors for the σ-complexes arising from the reaction of bromine with the arene you used.

7. Consider the spectral data for anisole (Figs. 15.13 and 15.14).
   a. In the functional group region of the IR spectrum, specify the absorption(s) associated with the π-bonds of the aromatic ring.
   b. In the ^{1}H NMR spectrum, assign the various resonances to the hydrogen nuclei responsible for them.
   c. For the ^{13}C NMR spectrum, assign the various resonances to the carbon nuclei responsible for them.

8. Consider the spectral data for 2-bromoanisole (Figs. 15.35 and 15.36).
   a. In the functional group region of the IR spectrum, specify the absorption(s) associated with the π-bonds and C—H bonds of the aromatic ring and any absorption(s) in the fingerprint region that correspond to a 1,2-disubstituted aromatic ring.
   b. In the ^{1}H NMR spectrum, assign the various resonances to the hydrogen nuclei responsible for them.
   c. For the ^{13}C NMR spectrum, assign the various resonances to the carbon nuclei responsible for them.

9. Consider the spectral data for 3-bromoanisole (Figs. 15.37 and 15.38).
   a. In the functional group region of the IR spectrum, specify the absorption(s) associated with the π-bonds and C—H bonds of the aromatic ring and any absorption(s) in the fingerprint region that correspond to a 1,3-disubstituted aromatic ring.

    b. In the ^{1}H NMR spectrum, assign the various resonances to the hydrogen nuclei responsible for them.

    c. For the ^{13}C NMR spectrum, assign the various resonances to the carbon nuclei responsible for them.

10. Consider the spectral data for 4-bromoanisole (Figs. 15.39 and 15.40).

    a. In the ^{1}H NMR spectrum, assign the various resonances to the hydrogen nuclei responsible for them.

    b. For the ^{13}C NMR spectrum, assign the various resonances to the carbon nuclei responsible for them.

11. Consider the spectral data for toluene (Figs. 9.6 and 9.7).

    a. In the functional group region of the IR spectrum, specify the absorption(s) associated with the π-bonds of the aromatic ring.

    b. In the ^{1}H NMR spectrum, assign the various resonances to the hydrogen nuclei responsible for them.

    c. For the ^{13}C NMR spectrum, assign the various resonances to the carbon nuclei responsible for them.

12. Consider the spectral data for 2-bromotoluene (Figs. 15.41 and 15.42).

    a. In the functional group region of the IR spectrum, specify the absorption(s) associated with the π-bonds and C−H bonds of the aromatic ring and any absorption(s) in the fingerprint region that correspond to a 1,2-disubstituted aromatic ring.

    b. In the ^{1}H NMR spectrum, assign the various resonances to the hydrogen nuclei responsible for them.

    c. For the ^{13}C NMR spectrum, assign the various resonances to the carbon nuclei responsible for them.

13. Consider the spectral data for 3-bromotoluene (Figs. 15.43 and 15.44).

    a. In the functional group region of the IR spectrum, specify the absorption(s) associated with the π-bonds and C−H bonds of the aromatic ring and any absorption(s) in the fingerprint region that correspond to a 1,3-disubstituted aromatic ring.

    b. In the ^{1}H NMR spectrum, assign the various resonances to the hydrogen nuclei responsible for them.

    c. For the ^{13}C NMR spectrum, assign the various resonances to the carbon nuclei responsible for them.

14. Consider the spectral data for 4-bromotoluene (Figs. 15.45 and 15.46).

    a. In the functional group region of the IR spectrum, specify the absorption(s) associated with the π-bonds and C−H bonds of the aromatic ring and any absorption(s) in the fingerprint region that correspond to a 1,4-disubstituted aromatic ring.

    b. In the ^{1}H NMR spectrum, assign the various resonances to the hydrogen nuclei responsible for them.

    c. For the ^{13}C NMR spectrum, assign the various resonances to the carbon nuclei responsible for them.

15. Consider the spectral data for bromobenzene (Figs. 15.19 and 15.20).

    a. In the functional group region of the IR spectrum, specify the absorption associated with the π-bonds of the aromatic ring. Indicate with what structural feature the strong absorptions at about 740 cm^{-1} and 690 cm^{-1} are associated.

    b. In the ^{1}H NMR spectrum, assign the various resonances to the hydrogen nuclei responsible for them.

    c. For the ^{13}C NMR spectrum, assign the various resonances to the carbon nuclei responsible for them.

16. Consider the spectral data for 1,2-dibromobenzene (Figs. 15.47 and 15.48).

    a. In the functional group region of the IR spectrum, specify the absorption associated with the π-bonds and the C$-$H bonds of the aromatic ring. Indicate with what structural feature the strong absorption at about 741 cm^{-1} is associated.

    b. In the ^{1}H NMR spectrum, assign the various resonances to the hydrogen nuclei responsible for them.

    c. For the ^{13}C NMR spectrum, assign the various resonances to the carbon nuclei responsible for them.

17. Consider the spectral data for 1,3-dibromobenzene (Figs. 15.49 and 15.50).

    a. In the functional group region of the IR spectrum, specify the absorption associated with the π-bonds and the C$-$H bonds of the aromatic ring. Indicate with what structural feature the strong absorptions at 766 cm^{-1} and 723 cm^{-1} are associated.

    b. In the ^{1}H NMR spectrum, assign the various resonances to the hydrogen nuclei responsible for them.

    c. For the ^{13}C NMR spectrum, assign the various resonances to the carbon nuclei responsible for them.

18. Consider the spectral data for 1,4-dibromobenzene (Figs. 15.51 and 15.52).

    a. In the functional group region of the IR spectrum, specify the absorption associated with the π-bonds and the C$-$H bonds of the aromatic ring. Indicate with what structural feature the strong absorption at 803 cm^{-1} is associated.

    b. In the ^{1}H NMR spectrum, assign the various resonances to the hydrogen nuclei responsible for them.

    c. For the ^{13}C NMR spectrum, assign the various resonances to the carbon nuclei responsible for them.

19. Consider the spectral data for methyl benzoate (Figs. 15.33, 15.34, and 8.48).

    a. In the functional group region of the IR spectrum, identify the absorptions associated with the ester group and the absorption(s) associated with the π-bonds of the aromatic ring.

    b. In the ^{1}H NMR spectrum, assign the various resonances to the hydrogen nuclei responsible for them.

    c. For the ^{13}C NMR spectrum, assign the various resonances to the carbon nuclei responsible for them.

20. Consider the spectral data for methyl 2-bromobenzoate (Figs. 15.53 and 15.54).

    a. In the functional group region of the IR spectrum, specify the absorption(s) associated with the ester, the π-bonds, and C$-$H bonds of the aromatic ring, and any absorption(s) in the fingerprint region that correspond to a 1,2-disubstituted aromatic ring.

    b. In the ^{1}H NMR spectrum, assign the various resonances to the hydrogen nuclei responsible for them.

    c. For the ^{13}C NMR spectrum, assign the various resonances to the carbon nuclei responsible for them.

21. Consider the spectral data for methyl 3-bromobenzoate (Figs. 15.55 and 15.56).

    a. In the functional group region of the IR spectrum, identify the absorptions associated with the ester group, the absorption(s) associated with the π-bonds of the aromatic ring, and any absorption(s) in the fingerprint region that correspond to a 1,3-disubstituted aromatic ring.

    b. In the ^{1}H NMR spectrum, assign the various resonances to the hydrogen nuclei responsible for them.

    c. For the ^{13}C NMR spectrum, assign the various resonances to the carbon nuclei responsible for them.

22. Consider the spectral data for methyl 4-bromobenzoate (Figs. 15.57 and 15.58).

    a. In the functional group region of the IR spectrum, identify the absorptions associated with the ester group, the absorption(s) associated with the π-bonds of the aromatic ring, and any absorption(s) in the fingerprint region that correspond to a 1,4-disubstituted aromatic ring.

    b. In the ^{1}H NMR spectrum, assign the various resonances to the hydrogen nuclei responsible for them.

    c. For the ^{13}C NMR spectrum, assign the various resonances to the carbon nuclei responsible for them.

SPECTRA

Starting Materials and Products

The IR and NMR spectra for anisole, toluene, bromobenzene, and methyl benzoate are provided in Figures 15.13 and 15.14, 9.6 and 9.7, 15.19 and 15.20, and 15.33 and 15.34, respectively. The ^{13}C data for methyl benzoate are found in Figure 8.48.

Figure 15.33
IR spectrum of methyl benzoate.

Figure 15.34
¹H NMR spectrum of methyl benzoate.

Figure 15.35
IR spectrum of 2-bromoanisole (ATR).

Figure 15.36
NMR data for 2-bromoanisole (CDCl₃).

(a) ¹H NMR (400 MHz).
(b) ¹³C NMR: δ 56.1, 111.6, 112.0, 121.8, 128.5, 133.3, 155.8.

Figure 15.37
IR spectrum of 3-bromoanisole (ATR).

Figure 15.38
NMR data for 3-bromoanisole (CDCl₃).

(a) ¹H NMR (400 MHz).
(b) ¹³C NMR: δ 55.3, 113.0, 117.1, 122.8, 123.7, 130.5, 160.3.

Figure 15.39
IR spectrum of 4-bromoanisole (ATR).

Figure 15.40

NMR data for 4-bromoanisole (CDCl₃).

(a) ¹H NMR (400 MHz).
(b) ¹³C NMR: δ 55.3, 112.7, 115.6, 132.2, 158.6.

Figure 15.41

IR spectrum of 2-bromotoluene (ATR).

Figure 15.42

NMR data for 2-bromotoluene (CDCl₃).

(a) ¹H NMR (400 MHz).
(b) ¹³C NMR: δ 22.9, 124.9, 127.2, 127.3, 130.8, 132.3, 137.8.

Figure 15.43
IR spectrum of 3-bromotoluene (ATR).

Figure 15.44
NMR data for 3-bromotoluene (CDCl₃).

(a) ^{1}H NMR (400 MHz).
(b) ^{13}C NMR: δ 21.1, 122.2, 127.6, 128.4, 129.7, 132.0, 140.1.

Figure 15.45
IR spectrum of 4-bromotoluene (ATR).

Figure 15.46

NMR data for 4-bromotoluene (CDCl₃)

(a) ¹H NMR (400 MHz).
(b) ¹³C NMR: δ 136.7, 131.2, 130.8, 119.0, 20.9.

Figure 15.47

IR spectrum of 1,2-dibromobenzene (ATR).

Figure 15.48

NMR data for 1,2-dibromobenzene (CDCl₃).

(a) ¹H NMR (400 MHz).
(b) ¹³C NMR: δ 124.8, 128.4, 133.7.

Figure 15.49
IR spectrum of 1,3-dibromo-benzene (ATR).

Figure 15.50
NMR data for 1,3-dibromo-benzene (CDCl₃).

(a) ¹H NMR (400 MHz).
(b) ¹³C NMR: δ 123.0, 130.2, 131.1, 134.2.

Figure 15.51
IR spectrum of 1,4-dibromobenz-ene (ATR).

Figure 15.52
NMR data for 1,4-dibromo-benzene (CDCl₃).

(a) ¹H NMR (400 MHz).
(b) ¹³C NMR: δ 121.1, 133.1.

Figure 15.53
IR spectrum of methyl 2-bromobenzoate (ATR).

Figure 15.54
NMR data for methyl 2-bromo-benzoate (CDCl₃).

(a) ¹H NMR (400 MHz).
(b) ¹³C NMR: δ 53.3, 121.4, 127.0, 131.1, 131.9, 132.4, 134.1, 166.3.

Figure 15.55
IR spectrum of methyl 3-bromobenzoate (ATR).

Figure 15.56
NMR data for methyl 3-bromobenzoate (CDCl₃).

(a) ¹H NMR (400 MHz).
(b) ¹³C NMR: δ 52.3, 122.4, 128.1, 129.9, 132.0, 132.5, 135.8, 165.6.

Figure 15.57
IR spectrum of methyl 4-bromobenzoate (ATR).

Figure 15.58
*NMR data for methyl
4-bromobenzoate (CDCl₃).*

(a) ¹H NMR (400 MHz).
(b) ¹³C NMR: δ 52.2, 128.0, 129.0, 131.1, 131.7, 166.3.

C. Nitration of Monosubstituted Arenes

Discovery Experiment

Purpose To demonstrate the directing effects of four different substituents on the course of electrophilic nitration reactions.

SAFETY ALERT

1. **Wear safety glasses or goggles and suitable protective gloves while performing the experiment.**

2. **Concentrated sulfuric and nitric acids may cause severe chemical burns. Do not allow them to contact your skin. Wipe off any drips and runs on the outside surface of reagent bottles and other glassware before handling them. Wash any affected area immediately and thoroughly with cold water and apply 5% sodium bicarbonate solution.**

Preparation Refer to the online resources to answer Pre-Lab Exercises, access videos, and read the MSDSs for the chemicals used or produced in this procedure. Review Sections 2.9, 2.10, 2.11, 2.17, 2.21, 2.29, and 6.4.

Nitration of Arenes — General Procedure for Setting Up and Reaction except for Acetanilide

Apparatus A 5-mL conical vial, ice-water bath, apparatus for magnetic stirring.

Setting Up Add 1.0 mL of *concentrated* sulfuric acid and 1.0 mL of *concentrated* nitric acid to the conical vial equipped with a spinvane or small stirbar. Place the vial in the ice-water bath and stir the solution of mixed acids for 10 min.

Reaction In portions of approximately 0.1 mL, add 1.0 mL of the arene assigned to you to the stirred mixture dropwise over a period of about 5 min. Maintain the reaction temperature between 0 °C and 10 °C during the course of this addition.

After addition is complete, remove the ice-water bath and continue to stir the mixture. Once the mixture has warmed to room temperature, continue stirring for the indicated time for the arene you used as the starting material:

> toluene — 15 min; *tert*-butylbenzene — 1 h; chlorobenzene — 1 h; methyl benzoate — 15 min.

Nitration of Acetanilide — Setting Up and Reaction

Setting Up Add 1.0 mL of *concentrated* sulfuric acid and 1.0 mL of *concentrated* nitric acid to the conical vial equipped with a spinvane or small stirbar. Place the vial in the ice-water bath and begin stirring the solution of mixed acids. Dissolve 1.2 g of acetanilide in 2–3 mL of *glacial* acetic acid—gentle heating may be required to dissolve all of the solid—and set the solution aside.

Reaction In portions of approximately 0.1 mL, add the solution of acetanilide to the stirred mixture dropwise over a period of about 5 min. Maintain the reaction temperature between 0 °C and 10 °C during the course of this addition. After addition is complete, remove the ice-water bath and continue to stir the mixture. Once the mixture has warmed to room temperature, continue stirring for 1 h.

General Work-Up and Analysis

General Work-Up Pour the reaction mixture from each of the nitration reactions onto about 2 g of ice contained in a 100-mL beaker. Bring the mixture to a pH of about 8 by *slowly* adding saturated *aqueous* sodium carbonate solution; no more than 10 mL should be required. Use care because some foaming occurs during this step. Add about 3 mL of dichloromethane to the beaker and transfer the liquids to a 25-mL screw-cap centrifuge tube. Rinse the beaker with an additional 2 mL of dichloromethane and add this rinse to the centrifuge tube. Cap the tube and shake it briefly. Using a pipet, remove the aqueous layer and wash the organic layer with 5 mL of *saturated* sodium chloride solution. Using a pipet, remove the *aqueous* layer and add several spatula-tips full of *anhydrous* sodium sulfate. Shake the contents of the tube to facilitate drying and carefully decant the solution into a dry screw-cap vial. Cap the vial tightly for analysis by GLC. If you are instructed to analyze the mixture of products by ^{1}H NMR spectroscopy, filter or carefully decant the dried organic solution into a round-bottom flask equipped for simple distillation, and remove the solvents by simple distillation. Alternatively, use rotary evaporation or other techniques to concentrate the solution. The final traces of solvent may be removed by attaching the flask to a vacuum source and gently swirling the contents as the vacuum is applied.

Analysis Consult with your instructor to analyze the solution by GLC or by ^{1}H NMR spectroscopic analysis and determine the ratio of regioisomers as well as the percentage of any unreacted arene present.

WRAPPING IT UP

Neutralize the *aqueous filtrates* with sodium carbonate and flush them down the drain with excess water. Spread the *sodium sulfate* on a tray in the hood, and after the volatiles have evaporated, place it in the container for nonhazardous solids. Transfer solutions of *dichloromethane* into the container for halogenated liquids.

EXERCISES

1. Using curved arrows to symbolize the flow of electrons, show the mechanism by which the nitronium ion, NO_2^+, is formed.

2. The pK_a values of sulfuric and nitric acids are −3 (estimated value) and −1.3, respectively.

   a. Which of the two acids is stronger?

   b. Given your prediction in Part **a**, write a chemical equation for the equilibrium involving reaction of sulfuric and nitric acids.

   c. What is the value of K_{eq} for the acid-base reaction in Part **b**? Assume that the pK_a of the conjugate acid of nitric acid is −13.

3. What is the purpose for maintaining lower temperatures during these nitration reactions?

4. What is the approximate ratio of regioisomers for the nitration of toluene? Provide a rationale for the preferential formation of the major product.

5. What is the approximate ratio of regioisomers for the nitration of *tert*-butyl-benzene? Explain why the ratio of *ortho*- and *para*-products obtained from the nitration of toluene and *tert*-butylbenzene is so different.

6. Given the ratio of regioisomers for all these experiments, what conclusions can you draw about the directing effects of the substitutents: methyl, *tert*-butyl, chloro, carbomethoxy, and acetamido?

7. Why are *concentrated* acids used in this experiment?

8. Why are no di- or trinitrated products formed in significant amounts in these reactions? If you wanted to produce greater quantities of di- and trinitrated products, what changes could you make to the experimental conditions?

9. Using curved arrows to symbolize the flow of electrons, write the stepwise mechanism for the nitration of toluene with nitronium ion to provide *o*-nitrotoluene. Write the contributing resonance structures for any charged, delocalized intermediates that are involved.

10. Using curved arrows to symbolize the flow of electrons, write the stepwise mechanism for the nitration of chlorobenzene to provide *o*- and *p*-chloronitrobenzene. Write the contributing resonance structures for any charged, delocalized intermediates that are involved.

11. Using curved arrows to symbolize the flow of electrons, write the stepwise mechanism for the nitration of methyl benzoate to provide methyl *m*-nitrobenzoate.

    a. Write the contributing resonance structures for any charged, delocalized intermediates that are involved.

    b. Write the contributing resonance structures for the formation of methyl *o*-nitrobenzoate.

    c. Why is this regioisomer a minor product? Provide other resonance structures to support your answer.

Refer to the online resources and use the spectra viewer and Tables 8.1–8.8 as needed to answer the blue-numbered questions on spectroscopy.

12. Consider the spectral data for toluene (Figs. 9.6 and 9.7).

    a. In the IR spectrum, specify the absorption associated with the π-bonds of the aromatic ring. Indicate with what structural feature the strong absorptions at about 750 cm^{-1} and 700 cm^{-1} are associated.

    b. In the ^{1}H NMR spectrum, assign the various resonances to the hydrogen nuclei responsible for them.

    c. For the ^{13}C NMR data, assign the various resonances to the carbon nuclei responsible for them.

13. Consider the spectral data for *tert*-butylbenzene (Figs. 9.12 and 9.13).

    a. In the functional group region of the IR spectrum, specify the absorptions associated with the aromatic ring.

    b. In the ^{1}H NMR spectrum, assign the various resonances to the hydrogen nuclei responsible for them.

    c. For the ^{13}C NMR data, assign the various resonances to the carbon nuclei responsible for them.

14. Consider the spectral data for methyl benzoate (Figs. 15.33, 15.34, and 8.48).

    a. In the functional group region of the IR spectrum, specify the absorptions associated with the aromatic ring and the ester functional group.

    b. In the ^{1}H NMR spectrum, assign the various resonances to the hydrogen nuclei responsible for them.

    c. For the ^{13}C NMR data, assign the various resonances to the carbon nuclei responsible for them.

15. Consider the spectral data for acetanilide (Figs. 15.27 and 15.28).

    a. In the functional group region of the IR spectrum, specify the absorptions associated with the aromatic ring, the nitrogen-hydrogen bond, and the carbon-oxygen double bond.

    b. In the ^{1}H NMR spectrum, assign the various resonances to the hydrogen nuclei responsible for them.

    c. For the ^{13}C NMR data, assign the various resonances to the carbon nuclei responsible for them.

16. Consider the spectral data for chlorobenzene (Figs. 15.59 and 15.60).

    a. In the functional group region of the IR spectrum, specify the absorptions associated with the aromatic ring.

    b. In the ^{1}H NMR spectrum, assign the various resonances to the hydrogen nuclei responsible for them.

    c. For the ^{13}C NMR data, assign the various resonances to the carbon nuclei responsible for them.

17. Consider the spectral data for *o*-nitrotoluene (Figs. 15.61 and 15.62).

    a. In the functional group region of the IR spectrum, specify the absorptions associated with the aromatic ring and the nitro functional group.

    b. In the ^{1}H NMR spectrum, assign the various resonances to the hydrogen nuclei responsible for them.

    c. For the ^{13}C NMR data, assign the various resonances to the carbon nuclei responsible for them.

18. Consider the spectral data for *p*-nitro-*tert*-butylbenzene (Figs. 15.69 and 15.70).

    a. In the functional group region of the IR spectrum, specify the absorptions associated with the aromatic ring and the nitro functional group.

    b. In the ^{1}H NMR spectrum, assign the various resonances to the hydrogen nuclei responsible for them.

    c. For the ^{13}C NMR data, assign the various resonances to the carbon nuclei responsible for them.

19. Consider the spectral data for *p*-chloronitrobenzene (Figs. 15.75 and 15.76).

    a. In the functional group region of the IR spectrum, specify the absorptions associated with the aromatic ring and the nitro functional group.

    b. In the ^{1}H NMR spectrum, assign the various resonances to the hydrogen nuclei responsible for them.

    c. For the ^{13}C NMR data, assign the various resonances to the carbon nuclei responsible for them.

20. Consider the spectral data for *p*-nitroacetanilide (Figs. 15.81 and 15.82).

    a. In the functional group region of the IR spectrum, specify the absorptions associated with the aromatic ring, the nitro functional group, the nitrogen-hydrogen bond, and the carbon-oxygen double bond.

    b. In the ^{1}H NMR spectrum, assign the various resonances to the hydrogen nuclei responsible for them.

    c. For the ^{13}C NMR data, assign the various resonances to the carbon nuclei responsible for them.

21. Consider the spectral data for methyl *m*-nitrobenzoate (Figs. 15.85 and 15.86).

    a. In the functional group region of the IR spectrum, specify the absorptions associated with the aromatic ring, the nitro functional group, and the ester functional group.

    b. In the ^{1}H NMR spectrum, assign the various resonances to the hydrogen nuclei responsible for them.

    c. For the ^{13}C NMR data, assign the various resonances to the carbon nuclei responsible for them.

SPECTRA

Starting Materials and Products

*The IR and NMR spectra for toluene, **tert-butylbenzene**, acetanilide, and methyl benzoate are provided in Figures 9.6 and 9.7, 9.12 and 9.13, 8.7 and 15.28, and 15.33 and 15.34, respectively.*

Figure 15.59
IR spectrum of chlorobenzene (ATR).

Figure 15.60
¹H NMR spectrum of chlorobenzene (CDCl₃).

(a) ¹H NMR (400 MHz).
(b) ¹³C NMR: δ 126.4, 128.6, 129.7, 134.3.

Figure 15.61
IR spectrum of o-nitrotoluene (ATR).

Figure 15.62
¹H NMR spectrum of o-nitrotoluene (CDCl₃).

(a) ¹H NMR (400 MHz).
(b) ¹³C NMR: δ 20.3, 124.5, 126.9, 132.8, 133.0, 133.5, 149.4.

Figure 15.63
IR spectrum of m-*nitrotoluene (ATR).*

Figure 15.64
¹H NMR *spectrum of*
m-*nitrotoluene (CDCl₃).*

(a) ¹H NMR (400 MHz).
(b) ¹³C NMR: δ 21.2, 120.7, 123.8, 129.2, 135.4, 139.9, 148.4.

Figure 15.65
IR spectrum of p-*nitrotoluene (ATR).*

Figure 15.66
¹H NMR *spectrum of*
p-*nitrotoluene (CDCl₃).*

(a) ¹H NMR (400 MHz).
(b) ¹³C NMR: δ 21.6, 123.5, 129.8, 145.9, 146.3.

Figure 15.67
IR spectrum of m-*nitro-tert-butylbenzene (ATR).*

Figure 15.68
¹H NMR *spectrum of* m-*nitro-tert-butylbenzene (CDCl₃).*

(a) ¹H NMR (400 MHz).
(b) ¹³C NMR: δ 31.3, 36.3, 119.5, 122.4, 132.4, 133.6, 149.5, 152.9.

Figure 15.69
IR spectrum of p-*nitro-tert-butylbenzene (ATR).*

Figure 15.70
¹H NMR spectrum of p-*nitro-tert-butylbenzene (CDCl₃).*

(a) ¹H NMR (400 MHz).
(b) ¹³C NMR: δ 31.3, 34.2, 126.3, 129.2, 146.8, 158.5.

Figure 15.71
IR spectrum of o-*chloronitrobenzene (ATR).*

Figure 15.72
¹H NMR spectrum of o-chloro-nitrobenzene (CDCl₃).

(a) ¹H NMR (400 MHz).
(b) ¹³C NMR: δ 125.5, 126.9, 127.7, 131.9, 133.3, 148.1.

Figure 15.73
IR spectrum of m-chloronitrobenz-ene (ATR).

Figure 15.74
¹H NMR spectrum of m-chloro-nitrobenzene (CDCl₃).

(a) ¹H NMR (400 MHz).
(b) ¹³C NMR: δ 121.7, 123.8, 130.5, 134.7, 135.5, 148.9.

Figure 15.75
IR spectrum of p-*chloronitrobenzene (ATR).*

Figure 15.76
¹H NMR *spectrum of* p-*chloronitrobenzene (CDCl₃).*

(a) ¹H NMR (400 MHz).
(b) ¹³C NMR: δ 124.9, 129.6, 141.4, 146.6.

Figure 15.77
IR spectrum of o-*nitroacetanilide (ATR).*

Figure 15.78
^{1}H NMR *spectrum of* o-*nitroacet-anilide (CDCl$_3$).*

(a) ^{1}H NMR (400 MHz).
(b) ^{13}C NMR: δ 25.5, 122.3, 123.3, 125.6, 134.7, 135.8, 136.5, 169.2.

Figure 15.79
IR spectrum of m-*nitroacetanilide (ATR).*

Figure 15.80
^{1}H NMR *spectrum of* m-*nitroac-etanilide (DMSO-d$_6$).*

(a) ^{1}H NMR (400 MHz).
(b) ^{13}C NMR: δ 24.3, 114.5, 116.7, 125.1, 129.5, 141.2, 147.4, 169.5.

Figure 15.81
IR spectrum of p-*nitroacetanilide (ATR).*

Figure 15.82
¹H NMR spectrum of p-*nitroacetanilide (DMSO-d₆).*

(a) ¹H NMR (400 MHz).
(b) ¹³C NMR: δ 24.2, 118.5, 124.8, 142.0, 145.4, 169.3.

Figure 15.83
IR spectrum of methyl o-*nitrobenzoate (ATR).*

Figure 15.84
¹H NMR *spectrum of methyl*
o-nitrobenzoate (CDCl₃).

(a) ¹H NMR (400 MHz).
(b) ¹³C NMR: δ 53.2, 123.9, 127.5, 129.9, 131.9, 133.0, 148.3, 165.8.

Figure 15.85
IR spectrum of methyl
m-nitrobenzoate (ATR).

Figure 15.86
¹H NMR *spectrum of methyl*
m-nitrobenzoate (CDCl₃).

(a) ¹H NMR (400 MHz).
(b) ¹³C NMR: δ 52.8, 124.6, 127.4, 129.7, 131.9, 135.3, 148.3, 164.9.

Figure 15.87

IR spectrum methyl p-nitrobenz-oate (ATR).

Figure 15.88

^{1}H NMR *spectrum methyl p-nitrobenzoate (CDCl₃).*

(a) ^{1}H NMR (400 MHz).

(b) ^{13}C NMR: δ 52.8, 123.6, 130.8, 135.6, 150.6, 165.2.

15.6 AZO DYES AND THE CHEMISTRY OF DYEING FABRICS

Humans have used colored pigments extracted from plants and animals to dye fabrics for over 5000 years. Tyrian purple, which is also known as royal purple, was prized for its rich reddish-purple color. It could only be obtained from a predatory sea snail found in the Mediterranean Sea, and some 12,000 snails are needed to produce 1.4 g of pure dye, enough to color the trim on only one garment. Accordingly, it was expensive and its use was reserved exclusively for royalty. The principal compound in Tyrian purple is 6,6'-dibromoindigo (**44**).

44

6,6'-Dibromoindigo

The first synthetic organic dye was inadvertently prepared in 1856 by William Henry Perkin when he was trying to synthesize quinine, (**45**), an antimalarial drug that was not actually synthesized until 1944. Perkin noticed a rich mauve color very-similar to that of Tyrian purple in the alcohol extract of a tarry sludge left in a reaction flask that he was cleaning after a failed reaction. Mauveine, which chemists now call the mixture he made, actually comprises two closely related compounds, mauveine A (**46**) and mauveine B (**47**). Perkin quickly patented the mixture and opened a dye works for producing and using it. Since Perkin's discovery of mauveine, literally thousands of new synthetic dyes have been made.

45	**46**	**47**
Quinine	Mauveine A	Mauveine B

Synthetic dyes quickly replaced dyes from natural sources because they are less expensive and have a broader range of color. For example, the class of **azo dyes**, which can have a wide range of colors, can be conveniently prepared by the reaction of aromatic diazonium salts with aromatic compounds bearing activating groups such as amino and hydroxy on the rings. The diazonium salts themselves are easily prepared from substituted anilines by oxidation with nitrous acid (Eq. 15.30).

$$(15.30)$$

The color of an organic molecule depends upon the energy gap between the π (or n) and $\pi^\star$ orbitals (see Section 8.4). When energy is absorbed, an electron is promoted from the π to the $\pi^\star$ orbital. For most organic molecules, this energy gap usually corresponds to the ultraviolent region of the electromagnetic spectrum, so we see all of the visible wavelengths either transmitted or reflected. Increasing conjugation reduces the energy difference between the π and $\pi^\star$ orbitals, and substituting electron-withdrawing and electron-donating groups on either the aniline precursor to the diazonium compound or the activated arene can also change the magnitude of this energy gap. When white light passes through a compound, the radiation with energy corresponding to the difference between the π and $\pi^\star$ orbitals is absorbed, and the light that either passes through the material or is reflected from its surface corresponds to the color complement of the absorbed light, or in other words, the wavelengths that are *not* absorbed. A simplified color wheel showing the

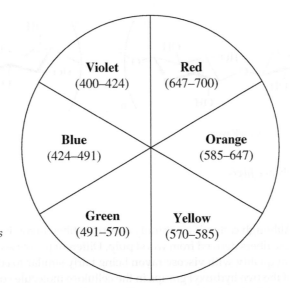

Figure 15.89
Color wheel showing wavelengths in nm. Colors opposite to each other on the wheel are complementary.

wavelengths in nanometers and where complementary colors are opposite to each other is shown in Figure 15.89.

A solution of Tyrian purple, for example, absorbs light at $\lambda_{max} = 590$ nm. When exposed to white light, the dye in the solution will absorb light in the orange region of the visible spectrum and allow other wavelengths to be transmitted. As a result, we see solutions of Tyrian purple as blue. When measured on fabric that has been dyed, the λ_{max} shifts to 520 nm, which corresponds to a yellow-green absorption, resulting in a red-violet or purple color. The same principle applies to azo dyes: The reflected color we perceive is the color of the complement to the wavelength they most strongly absorb.

The structure of the colored pigments is only one aspect of the chemistry associated with dyeing fabrics. The manner in which the pigment interacts with the fabric is also important. Two types of dyes can be directly applied to fabrics without further chemical action. These are **direct dyes**, which are water-soluble pigments, and **disperse dyes**, which are suspensions of insoluble pigments rather than aqueous solutions. By contrast, **vat dyes** are characteristically applied to fabric in a soluble, reduced form of the pigments and then chemically "fixed" through an oxidation process that converts the soluble pigment into an insoluble pigment. There are advantages and disadvantages to each type of dyeing process, and the type of fabric to be dyed is an important factor in determining how well a particular process works.

The "fastness" of dyes, or how well they stay in the fabric, depends upon the kinds of interactions the pigment molecules have with the molecules in the fabric. For direct dyes, these are typically ionic, dipole-dipole, or hydrogen-bonding interactions. Acidic dyes, such as those having a sulfonic acid, SO_3H, group, will form salts with fabric molecules containing basic functional groups such as an amino, NH_2, group. Accordingly, acidic pigments are applied to fabric from an acidic dyeing solution and basic pigments from an alkaline solution. Azo dyes are typically considered as direct dyes, so the functional groups on both the pigment and the fabric dictate how well the dye is retained.

Fabrics have a variety of functional groups, but they tend to fall within a few general structural types. For example, cotton and rayon are cellulose fibers (Figure 15.90).

Cotton/Rayon   Acetate rayon

Figure 15.90
Structures of cellulosic fibers.

Although cotton is a naturally occurring fiber, rayon is made by reconstructing cellulose fibers derived from wood pulp. Different processes provide rayon fibers of different quality, with **viscose rayon** being fairly similar to cotton. **Acetate rayon** has some of the free hydroxyl groups in the cellulose molecule converted into acetate esters.

Wool and silk are primarily strands of protein. Wool contains a wide range of amino acids (such as lysine and glutamic acid) in its primary structure, and the predominant secondary structure is α-helical. By contrast, silk fibers are made of fibroin protein, a specific repeating sequence of glycine-serine-glycine-alanine-glycine-alanine in its primary structure, with a secondary structure consisting of anti-parallel β-sheets.

Fibroin (silk)

Nylon is a synthetic polyamide fiber formed by reacting a diacid with a diamine. Different nylons are distinguished by the number of carbons in each of the diacid and diamine component. Nylon 6,6, for example, is made from sebacic acid and hexamethylene diamine (Eq. 22.15, Sec. 22.3). Polyester fibers are similar, except that a diol is used rather than a diamine. Dacron, for instance, is a fabric made from fibers of polyethylene terephthalate (Eq. 22.14, Sec. 22.3).

Nylon 6,6   Dacron

Acrylic fibers are polymers of acrylonitrile formed through a free-radical polymerization reaction. Orlon was the first so-called acrilan fabric, but a variety of modified acrylic fibers have been developed. The modifications typically include vinyl acetate or methyl acrylate as a co-polymer, though vinyl chloride and vinyl bromide are used in "self-extinguishing fabric" acrilans.

Acrylic fiber

To explore the way that azo dyes interact with different fabrics, it is necessary to synthesize a variety of dyes. **Combinatorial chemistry** is a common strategy for rapidly producing a collection of molecules, commonly referred to as a **compound library**, that differ from each other in the nature, position, and orientation of substituents. A common goal of such studies is to maximize the diversity of molecules being studied. Using such a library, it is possible to assess the relationships between structural variations of the molecules in the library and their properties or "activities." For example, the activities of a drug-like compound would be a measure of how effective it is in inhibiting an enzyme. Investigations in which structure and properties of a series of molecules are correlated are often called **structure-activity relationship** (SAR) studies.

Two combinatorial approaches are commonly used to build libraries. The first is **parallel synthesis**, in which each compound is prepared individually, and activity is assayed separately for each compound. The second is **split synthesis**, in which the compounds are prepared as mixtures, and the activity of the mixture is evaluated. Parallel synthesis has the advantage of easy analysis of the assays because the identity of the compound being tested is known, but it has the disadvantage that preparing and isolating each member of the library is time consuming. Split synthesis requires **deconvolution**, or figuring out which compound is active when the mixture is tested, but the synthesis of the library is much faster than with parallel synthesis.

It is possible to extend this combinatorial approach to the synthesis of azo dyes because they are formed from coupling an aniline-derived diazonium salt with an activated arene, each of which can be independently substituted in a variety of ways. Each coupling component thus represents a "point of diversification" for the library. In a combinatorial approach to a library of azo dyes, the number of compounds possible in the library is the number of substituted anilines times the number of activated arenes that are used as starting materials. For example, if five different anilines and five different activated arenes are used, the library will have 25 compounds, each with a different combination of substituents.

In this experiment, the class will generate a library from the three isomers of aminobenzenesulfonic acid (**47–49**) and the four activated arenes, 1-naphthol (**50**), 2-naphthol (**51**), ammonium 8-anilino-1-naphthalenesulfonate (**52**), and salicylic acid (**53**). Each student will select one combination and synthesize one member of the compound library. There are two "activities" of interest related to the substituents on the pigment molecules that you will test: the color of the dye and how well a dye is retained by different fabrics. By comparing your results with those of others in the laboratory, you can begin to see emerging patterns that you can correlate to structures of both the pigment molecules and the fabric different fibers.

47

4-Aminobenzene-
sulfonic acid
(Sulfanilic acid)

48

3-Aminobenzene-
sulfonic acid
(Metanilic acid)

49

2-Aminobenzene-
sulfonic acid
(Orthanilic acid)

50

1-Naphthol

51

2-Naphthol

52

Ammonium 8-anilino-
1-naphthalenesulfonate

53

Salicylic acid

EXPERIMENTAL PROCEDURES

Combinatorial Synthesis of Azo Dyes

Discovery Experiment

Purpose To explore the relationship between substitution patterns in an azo dye, the color of the dye, and the way the dye interacts with fibers of different fabrics.

SAFETY ALERT

1. **Wear safety glasses or goggles and suitable protective gloves while performing the experiment.**

2. **Aromatic amines, as a class of compounds, are toxic. Do not allow them to come into contact with your skin. If this should occur, immediately wash the area with soap and cool water. If irritation persists, seek medical attention.**

3. **Naphthol derivatives are irritants. Do not allow them to come into contact with your skin. If this should occur, immediately wash the area with soap and cool water. If irritation persists, seek medical attention.**

4. **Sodium nitrite is an oxidizing agent and a strong irritant, particularly to eyes. Do not allow it to come into contact with your skin or eyes. If this should occur, immediately wash the area with soap and cool water. Flush eyes with water for 15 minutes. If irritation persists, seek medical attention.**

5. **Concentrated hydrochloric acid is a strong acid. It is corrosive and can cause severe burns. Do not allow it to come into contact with your skin. If this should occur, immediately rinse the area with cool water and then rinse with 5% aqueous sodium bicarbonate solution.**

6. **Sodium hydroxide is caustic. Do not allow it to come into contact with your skin. If this should occur, immediately rinse the area with cool water and then rinse with *dilute aqueous* acetic acid.**

7. **Diazonium salts are explosive in the solid state and should be kept in solution. They should be used immediately after preparation.**

Preparation Refer to the online resources to answer Pre-Lab Exercises, access videos, and read the MSDSs for the chemicals used or produced in this procedure. Review Sections 2.5, 2.6, 2.9, 2.10, 2.11, and 2.17.

Apparatus Two 10-mm × 75-mm test tubes, ice-water bath, Hirsch funnel, multi-fiber strip, and apparatus for *flameless* heating.

Setting Up Mix 0.49 g of aminobenzenesulfonic acid, 0.13 g of sodium carbonate, and 5 mL of water in a test tube.

Reaction Gently warm the test tube in a hot-water bath until the solids are dissolved. Remove the test tube from the water bath and add a solution of 0.2 g of sodium nitrite in 0.5 mL of water. To a second test tube containing about 3 g of ice, add 0.53 mL of *concentrated* hydrochloric acid. Using a Pasteur pipet, transfer the solution dropwise from the first test tube to the second test tube, and place the resulting mixture in an ice-water bath. The diazonium salt should precipitate, and the resulting suspension is used in the next step.

Weigh 2.6 mmol of one of the arene coupling reagents, selected from among 1-naphthol, 2-naphthol, salicylic acid, or 8-anilino-1-naphthalenesulfonic acid ammonium salt, into a 25-mL Erlenmeyer flask, and add 2 mL of 2.5 *M aqueous* sodium hydroxide. Cool the Erlenmeyer flask in an ice-water bath. Using a Pasteur pipet, transfer the suspension of the diazonium salt prepared above to the Erlenmeyer flask in several portions. After each addition, stir the reaction mixture with a glass rod. The color of the solution in the Erlenmeyer flask should change as you add the diazonium salt. After all of the salt has been added, allow the reaction to proceed for an additional 10 min, stirring the mixture occasionally. Remove the flask from the ice-water bath and heat the suspension to dissolve the solid. Some solutions will be very dark, and it may be difficult to determine when the solids are dissolved. If this is the case, heat the mixture until it boils for about 2 min. Remove the flask from the heating source, allow the flask to cool for 2–3 min, and then add 1 g of sodium chloride to the solution of your dye. Reheat the mixture for a few minutes, and remove the flask from the heat after all of the sodium chloride has dissolved. Allow the solution to cool for 10–15 min and then place it in an ice-water bath to cool further. If a solid separates, collect it by vacuum filtration in a Hirsch funnel. If you do not see any solid, you will use about 1 mL of the reaction mixture in the fabric dying procedure. Wash the solid with 2 mL of saturated *aqueous* sodium chloride, and let it dry on the Hirsch funnel for 2–3 min.

Dyeing Obtain a strip that has a variety of fibers on it. One version has a black thread followed in order by (a) acetate rayon; (b) acrilan; (c) Arnel; (d) bleached cotton; (e) Creslan 61; (f) Dacron 54; (g) Dacron 64; (h) Nylon 6,6; (i) Orlon 75; (j) silk; (k) polypropylene; (l) viscose rayon; (m) wool. Mix 50 mg of the azo dye prepared in the previous step and 20 mL of water in a 100-mL beaker. Submerge the multi-fiber strip in the solution of the azo dye and boil the solution for 10 min, keeping the strip immersed with the aid of a glass stirring rod. Add water as necessary to maintain the volume of the dye solution. Remove the strip from the solution with forceps and allow it to cool for 5 min. Rinse the dyed strip with tap water and pat it dry.

Analysis Compare the colors of your strip with those of others that were produced in your lab. Obtain IR and ^{1}H NMR spectra of your starting material and compare them with those of authentic samples (Figs. 15.34, 15.35, and 15.91–15.102).

OPTIONAL EXPERIMENTS

UV/Vis spectroscopy Acquire the UV/Vis spectrum of your compound and find the λ_{max} for your dye.

pH dependence of color Repeat the procedure for dyeing the multi-fiber strip and add a few drops of *concentrated* hydrochloric acid or 2.5 *M aqueous* sodium hydroxide to the dye solution before submersing the multi-fiber strip.

WRAPPING IT UP

Place any *filter papers* and the *fabric strip* in a container for nontoxic solid waste. Pour the *filtrate* in a container for acids. Pour the *aqueous dye solutions* in the appropriate container for *aqueous* acids or bases.

EXERCISES

1. Consider the fabric strips dyed by all the pigments synthesized in your lab section. Which fabrics best absorbed the various dyes in the library generated by your section?

2. What is the mechanism by which the dye molecules interact with the fabrics that best retained the dye.

3. Using curved arrows to symbolize the electron flow, propose a mechanism for the formation of the diazonium salt in the reaction below.

4. Using curved arrows to symbolize the flow of electrons, propose a mechanism for the diazo-coupling reaction below.

5. Why is sodium carbonate added to the aminobenzenesulfonic acid in the first step for forming the diazonium salt?

6. Why add solid sodium chloride to the solution of pigment before cooling and filtering?

7. Why is it necessary to keep the diazonium salt cold (0 °C)?

8. Recall that solutions of **48** absorb light at a different λ_{max} than the dye adsorbed onto fabric. The combination of **48** and **53** produces different color hues (as opposed to just different saturation of color) when adsorbed onto different fabrics (e.g., wool and silk versus rayon and nylon). Explain this phenomenon in terms of differing molecular interactions.

9. Is the method used to generate the library of azo dyes an example of parallel or split synthesis? Explain your answer.

10. Why do compounds that appear blue generally have more conjugation than compounds that appear red?

11. The pigment Sudan IV has $\lambda_{max} = 516$. What is the color of this dye? Explain why.

12. From what compounds would you synthesize "American Flag Red" (also known as para-red)

Para red

13. Draw the azo dye expected to be produced upon coupling of the diazonium salt derived from 2-aminobenzenesulfonic acid and salicylic acid.

Refer to the online resources and use the spectra viewer and Tables 8.1–8.8 as needed to answer the blue-numbered questions on spectroscopy.

14. Consider the spectral data for 4-aminobenzenesulfonic acid (Figs. 15.91 and 15.92).

    a. In the functional group region of the spectrum, specify the absorptions associated with the aromatic ring.

    b. In the ^{1}H NMR spectrum, assign the various resonances to the hydrogen nuclei responsible for them.

    c. For the ^{13}C NMR data, assign the various resonances to the carbon nuclei responsible for them.

15. Consider the spectral data for 3-aminobenzenesulfonic acid (Figs. 15.93 and 15.94).

    a. In the functional group region of the spectrum, specify the absorptions associated with the aromatic ring.

    b. In the ^{1}H NMR spectrum, assign the various resonances to the hydrogen nuclei responsible for them.

    c. For the ^{13}C NMR data, assign the various resonances to the carbon nuclei responsible for them.

16. Consider the spectral data for 2-aminobenzenesulfonic acid (Figs. 15.95 and 15.96).

    a. In the functional group region of the spectrum, specify the absorptions associated with the aromatic ring.

    b. In the ^{1}H NMR spectrum, assign the various resonances to the hydrogen nuclei responsible for them.

    c. For the ^{13}C NMR data, assign the various resonances to the carbon nuclei responsible for them.

17. Consider the spectral data for 1-naphthol (Figs. 15.34 and 15.35).

    a. In the functional group region of the spectrum, specify the absorptions associated with the phenol.

    b. In the ^{1}H NMR spectrum, assign the various resonances to the hydrogen nuclei responsible for them.

    c. For the ^{13}C NMR data, assign the various resonances to the carbon nuclei responsible for them.

18. Consider the spectral data for 2-naphthol (Figs. 15.97 and 15.98).

    a. In the functional group region of the spectrum, specify the absorptions associated with the phenol.

    b. In the ^{1}H NMR spectrum, assign the various resonances to the hydrogen nuclei responsible for them.

    c. For the ^{13}C NMR data, assign the various resonances to the carbon nuclei responsible for them.

19. Consider the spectral data for ammonium 8-anilino-1-naphthalene sulfonate (Figs. 15.99 and 15.100).

    a. In the functional group region of the spectrum, specify the absorptions associated with the aromatic rings.

    b. In the ^{1}H NMR spectrum, assign the various resonances to the hydrogen nuclei responsible for them.

    c. For the ^{13}C NMR data, assign the various resonances to the carbon nuclei responsible for them.

20. Consider the spectral data for salicylic acid (Figs. 15.101 and 15.102).

    a. In the functional group region of the spectrum, specify the absorptions associated with the carboxylic acid.

    b. In the ^{1}H NMR spectrum, assign the various resonances to the hydrogen nuclei responsible for them.

    c. For the ^{13}C NMR data, assign the various resonances to the carbon nuclei responsible for them.

SPECTRA

Starting Materials

The IR and NMR spectra of 1-naphthol are given in Figures 15.34 and 15.35, respectively.

Figure 15.91

IR spectrum of 4-aminobenzene-sulfonic acid (ATR).

Figure 15.92

NMR data for 4-aminobenzene-sulfonic acid (DMSO-d₆).

(a) ¹H NMR spectrum (400 MHz).
(b) ¹³C NMR data: δ 115.4, 127.0, 132.2, 149.7.

Figure 15.93
IR spectrum of 3-aminobenzene-sulfonic acid (ATR).

Figure 15.94
NMR data for 3-amino-benzenesulfonic acid (DMSO-d₆).

(a) ^{1}H NMR spectrum (400 MHz).
(b) ^{13}C NMR data: δ 112.5, 115.8, 119.0, 129.9, 143.1, 146.9.

Figure 15.95
IR spectrum of 2-aminobenzene-sulfonic acid (ATR).

Figure 15.96
NMR data for 2-aminobenzene-sulfonic acid (DMSO-d₆).

(a) ¹H NMR spectrum (400 MHz).
(b) ¹³C NMR data: δ 117.8, 117.9, 126.4, 127.2, 132.7, 143.9.

Figure 15.97
IR spectrum of 2-naphthol (ATR).

Figure 15.98
NMR data for 2-naphthol (DMSO-d₆).

(a) ¹H NMR spectrum (400 MHz).
(b) ¹³C NMR data: δ 109.5, 117.7, 123.7, 126.4, 126.6, 127.8, 129.0, 129.9, 134.6, 153.3.

Figure 15.99

IR spectrum of ammonium-8-anilino-1-naphthalene sulfonate (ATR).

Figure 15.100

NMR data for ammonium-8-anilino-1-naphthalene sulfonate (DMSO-d$_6$ and with D$_2$O).

(a) ^{1}H NMR spectrum (400 MHz).

(b) ^{13}C NMR data: δ 114.9, 116.7, 119.1, 120.9, 122.3, 124.1, 125.8, 126.6, 128.9, 131.5, 136.8, 139.5, 142.9, 144.6.

Figure 15.101

IR spectrum of salicylic acid (ATR).

Figure 15.102
NMR data for salicylic acid (DMSO-d₆).

(a) ¹H NMR spectrum (400 MHz).
(b) ¹³C NMR data: δ 113.0, 117.1, 119.2, 130.3, 135.7, 161.2, 172.0.

CHAPTER 16

Oxidation of Alcohols and Carbonyl Compounds

In this and the following chapter, we'll explore oxidation and reduction reactions in organic chemistry. Such reactions are extremely important as they are commonly used to convert one functional group into another during the course of preparing more complex materials from simpler ones. In the simplest sense, oxidation in organic chemistry involves increasing the number of carbon-oxygen bonds. For example, the complete combustion of hydrocarbons or other organic compounds to produce carbon dioxide, water, and heat is an oxidation, but because such oxidations destroy the organic molecule, they are not useful in synthesis. Hence, in the experiments that follow, we'll explore the controlled oxidations of alcohols and aldehydes to give carbonyl compounds and carboxylic acids, respectively, as these transformations are widely used in contemporary synthetic organic chemistry.

16.1 INTRODUCTION

Oxidation, often represented by the symbol [O], is a fundamental type of reaction in chemistry and is the opposite of **reduction.** The general concept of oxidation is typically introduced in beginning chemistry courses as an electron transfer process involving a *loss* of electrons from an ion or a neutral atom. Such definitions are difficult to apply to reactions in organic chemistry, however, as carbon forms covalent bonds and does not normally lose electrons. Nevertheless, oxidations of organic compounds usually do involve a loss of electron density at carbon as a consequence of forming a new bond between a carbon atom and a more electronegative atom such as nitrogen, oxygen, or a halogen. Reactions that result in breaking carbon-hydrogen bonds are also oxidations.

Oxidations are frequently used in organic chemistry to effect functional group transformations. For example, oxidations involving the conversions of an alkene into a 1,2-diol (Eq. 16.1 and Sec. 10.4), of a primary or secondary alcohol into an aldehyde or ketone, respectively (Eq. 16.2), and of an aldehyde into a carboxylic acid (Eq. 16.3) are common laboratory reactions. In processes involving the degradation of organic molecules, oxidations may be used to cleave carbon-carbon double bonds, as illustrated by the ozonolysis of an alkene to give the corresponding carbonyl compounds (Eq. 16.4).

$$R^1 \diagup \xrightarrow{\text{KMnO}_4} R^1 \underset{\text{OH}}{\overset{\text{OH}}{\diagdown}} \text{OH} \qquad (16.1)$$

An alkene   A 1,2-diol

$$R^1\text{—CH}_2\text{—OH} \quad \text{or} \quad R^1\text{—CH—OH} \xrightarrow{\text{[O]}} \underset{R^1}{\overset{O}{\overset{\|}{C}}}\diagdown_H \quad \text{or} \quad \underset{R^1}{\overset{O}{\overset{\|}{C}}}\diagdown_{R^2} \qquad (16.2)$$
                                     $\underset{R^2}{|}$

A primary   A secondary   An aldehyde   A ketone
alcohol   alcohol

$$\underset{R^1}{\overset{O}{\overset{\|}{C}}}\diagdown_H \xrightarrow{\text{[O]}} \underset{R^1}{\overset{O}{\overset{\|}{C}}}\diagdown_{\text{OH}} \qquad (16.3)$$

An aldehyde   A carboxylic acid

$$R^1 \diagup \overset{R^2}{\underset{R^3}{\diagdown}} \xrightarrow[\text{(2) Zn, HOAc}]{\text{(1) O}_3} R^1\text{—CHO} + \underset{R^2}{\overset{O}{\overset{\|}{C}}}\diagdown_{R^3} \qquad (16.4)$$

An alkene   An aldehyde   A ketone

The processes of life also depend on oxidation of organic substrates, whereby metabolic energy is derived from the overall oxidation of carbohydrates, fats, and proteins to carbon dioxide and water, among other products, as illustrated in Equation 16.5. Potentially poisonous substances are commonly detoxified by biological oxidation to more benign substances. Thus, nicotine, which is toxic to humans if present in sufficiently high concentration, is oxidized in the liver to cotinine (Eq. 16.6), a substance of low toxicity.

$$\xrightarrow{\text{[O]}} 6\,CO_2 + 6\,H_2O + \text{energy} \qquad (16.5)$$

A carbohydrate

$$\xrightarrow[\text{Liver}]{\text{[O]}} \qquad (16.6)$$

Nicotine   Cotinine

As noted previously, oxidation does *not* necessarily result in introducing oxygen into an organic molecule. In a general sense, and in analogy to concepts of oxidation as applied to inorganic compounds, oxidation of an organic substance involves an increase in the **oxidation number,** or **oxidation state,** of carbon. There are several ways of defining the oxidation numbers of carbon atoms, but the method that follows is a useful one.

1. *Select* the carbon atom whose oxidation number is to be determined.

2. *Assign* oxidation numbers to the atoms attached to this carbon atom using the following values:

   a. +1 for hydrogen

   b. −1 for halogen, nitrogen, oxygen, and sulfur

   c. 0 for carbon

   Simply put, with respect to the atoms bound to the carbon atom, those that are *more* electronegative are assigned a value of −1, whereas those that are *less* electronegative are given a value of +1.

3. *Sum* the oxidation numbers of these atoms. If a heteroatom (see 2b) is multiply bound to a carbon atom, multiply its oxidation number by the number of bonds linking it to the carbon atom involved.

4. *Determine* that the sum of the number from step 3 and the oxidation number of the carbon atom under consideration equals the charge on the carbon atom, which is zero unless it bears a positive or negative charge.

Application of this method shows that the conversion of an alkene to the corresponding dibromide (Eq. 16.7) and that of an alkane to an alkyl halide (Eq. 16.8), reactions discussed in Section 10.6 and Chapter 9, respectively, are oxidations even though oxygen is not incorporated in these reactions. Determining the changes in the oxidation number of the various carbon atoms during the conversion of acetic acid to carbon dioxide and water (Eq. 16.9) illustrates how the method is used when heteroatoms are multiply bound to a carbon atom.

$$\begin{array}{ccc} \text{Cyclopentene} & \xrightarrow[\text{CCl}_4]{\text{Br}_2} & \textit{trans}\text{-1,2-Dibromo-}\\ & & \text{cyclopentane} \end{array} \left.\begin{array}{c}\\ \\ \\ \\ \end{array}\right\} \begin{array}{c}\text{Net increase in}\\ \text{oxidation number of 2}\\ \text{(oxidation at carbon atom)}\end{array}$$

(16.7)

$$\overset{-3}{\text{CH}_3}\text{CH}_2\text{—H} \xrightarrow{\text{Cl}_2} \overset{-1}{\text{CH}_3}\text{CH}_2\text{—Cl}$$

Ethane   1-Chloroethane

(16.8)

$$\text{R—C}\equiv\text{C—R(H)} \xrightarrow[\substack{\text{H}_2\text{SO}_4\\ \text{H}_2\text{O}}]{\text{HgSO}_4} \text{R—}\overset{\displaystyle\overset{\text{O}}{\|}}{\text{C}}\text{—CH}_2\text{—R(H)}$$

An alkyne   A ketone

(16.9)

16.2 PREPARATION OF ALDEHYDES AND KETONES BY OXIDATION OF ALCOHOLS

Aldehydes and ketones have a central role in organic synthesis, and efficient procedures for their preparation are of great importance. Such compounds are synthesized in a number of ways, including hydration or hydroboration-oxidation of alkynes (Eqs. 16.10 and 16.11, respectively, and Chap. 11) and reaction of carboxylic acids or their derivatives with organometallic reagents or reducing agents

(Eqs. 16.12 and 16.13). The latter two reactions involve net **reduction** of the carbonyl carbon atom. However, the transformations shown in Equations 16.10 and 16.11 are neither *net* oxidation nor reduction of the alkyne. You should determine the net change in oxidation state at the carbon atoms involved in these transformations to convince yourself that this is true.

$$R-C{\equiv}C-R(H) \xrightarrow[\substack{H_2SO_4 \\ H_2O}]{HgSO_4} R-\overset{\overset{\displaystyle O}{\|}}{C}-CH_2-R(H) \qquad (16.10)$$

An alkyne   A ketone or aldehyde

$$R-C{\equiv}C-R(H) \xrightarrow[\substack{(2)\ H_2O_2/ \\ HO^-}]{(1)\ B_2H_6} R-CH_2-\overset{\overset{\displaystyle O}{\|}}{C}-R(H) \qquad (16.11)$$

An alkyne   Á ketone or aldehyde

$$R^1-\overset{\overset{\displaystyle O}{\|}}{C}-OH \xrightarrow[\substack{(2)\ H_3O^+}]{(1)\ 2\ R^2Li} R^1-\overset{\overset{\displaystyle O}{\|}}{C}-R^2 \qquad (16.12)$$

A carboxylic acid   A ketone

$$R^1-\overset{\overset{\displaystyle O}{\|}}{C}-OR \xrightarrow[\substack{(2)\ H_3O^+}]{(1)\ (i\text{-}Bu)_2AlH} R^1-\overset{\overset{\displaystyle O}{\|}}{C}-H \qquad (16.13)$$

An ester   An aldehyde

One of the most common synthetic methods for preparing aldehydes and ketones is the oxidation of primary and secondary alcohols, a reaction involving a two-electron change in the oxidation number of the functionalized carbon atom (Eq. 16.14). Chromic ion, Cr^{6+}, frequently in the form of chromic acid, H_2CrO_4, is a common oxidant, although other oxidants, such as halonium ion, X^+, as found in hypohalous acids, HOX, and permanganate ion, Mn^{7+}, as in potassium permanganate, $KMnO_4$, can be used.

$$R-\overset{\overset{\displaystyle OH}{|}}{CH}-R(H) \xrightarrow[\substack{or\ HOX \\ or\ KMnO_4}]{H_2CrO_4} R-\overset{\overset{\displaystyle O}{\|}}{C}-R(H) \qquad (16.14)$$

Oxidation number 0 (2 alcohol)   +2 (ketone)
                        −1 (1 alcohol)   +1 (aldehyde)

Chromic acid and potassium permanganate are excellent oxidizing agents, but these oxidants are derived from **heavy metals**, a class of elements that are commonly toxic and therefore environmentally hazardous, as exemplified in the film "Erin Brockovich." The safe disposal of these metals and their derivatives is of considerable concern, and avoiding their use is desirable from an environmental perspective. Fortunately alternatives to chromate and permanganate ions for oxidizing alcohols are the hypohalites, XO^- (X = halogen), which are made by reaction of the corresponding halogen with aqueous base (Eq. 16.15). The most familiar sources of this type of oxidizing agent are household bleach, which is an aqueous solution of sodium hypochlorite, NaOCl, and granular $Ca(OCl)_2$, which is a chlorinating agent for swimming pools.

$$X\text{—}X + HO^- \rightleftharpoons X^- + XO\text{—}H \xrightarrow{HO^-} XO^- + H_2O \qquad (16.15)$$

(X = halogen)   Hypohalous   Hypohalite
                       acid   ion

The mechanism by which hypohalites oxidize alcohols probably involves initial formation of an alkyl hypohalite (Eq. 16.16). This product arises from reaction of the alcohol with the hypohalous acid that is in equilibrium with hypohalite ion in aqueous medium (Eq. 16.15). Base-promoted E2 elimination of the elements of H–X from the alkyl hypochlorite leads directly to either an aldehyde or ketone. The advantage of using hypohalite as an oxidant is immediately obvious upon examining Equation 16.16. The inorganic by-product derived from the oxidant is a halide salt that can be safely flushed down the drain. Reactions that do not produce toxic by-products are environmentally friendly and are now commonly referred to as "Green Chemistry." (See the Historical Highlight *Green Chemistry*, which is available online.)

$$R\text{—}\underset{\underset{H}{|}}{\overset{R}{\underset{|}{C}}}\text{—}\ddot{O}H + X\text{—}OH \rightleftharpoons R\text{—}\underset{\underset{H}{|}}{\overset{R}{\underset{|}{C}}}\text{—}O \xrightarrow{-OAc} \underset{R'}{\overset{R}{C}}{=}O + X^- \qquad (16.16)$$

Alkyl hypohalite   Halide ion

The stoichiometry of the reaction may be determined from changes in oxidation numbers in the reactants and products. During the oxidation, the carbon atom of the alcohol undergoes a two-electron change in oxidation state, from 0 to +2 in the case of a secondary alcohol, and the halogen also undergoes a two-electron change in oxidation number from +1 to −1 (Eq. 16.16). Consequently, a one-to-one ratio of the two reactants is required stoichiometrically.

Some side reactions may complicate the oxidation of a primary alcohol to an aldehyde using hypohalous acid. For example, an aldehyde may undergo reaction with an additional equivalent of hypohalite to form a carboxylic acid (Eq. 16.17), a process that may be initiated by acid-catalyzed formation of the hydrate of the intermediate aldehyde as shown in Equation 16.18. Subsequent steps in the oxidation are then analogous to those for converting an alcohol to an aldehyde or ketone (Eq. 16.16). It may be difficult to suppress this further oxidation, so unless carboxylic acids are the desired products, the use of hypohalite as an oxidant is limited to the conversion of *secondary* alcohols to ketones.

$$R^1\text{—}\overset{O}{\overset{||}{C}}\text{—}H + X\text{—}OH + H_2O \longrightarrow R^1\text{—}\overset{O}{\overset{||}{C}}\text{—}OH \qquad (16.17)$$

An aldehyde   A carboxylic acid

$$R\text{—}\overset{O}{\overset{||}{C}}\text{—}H \underset{H^+}{\rightleftharpoons} R\text{—}\overset{\overset{H}{\underset{|}{\overset{+}{O}}}}{\overset{||}{C}}\text{—}H \rightleftharpoons R\text{—}\underset{\underset{+}{\overset{|}{O}H_2}}{\overset{\overset{OH}{|}}{C}}\text{—}H \underset{-H^+}{\rightleftharpoons} R\text{—}\underset{\underset{OH}{|}}{\overset{\overset{OH}{|}}{C}}\text{—}H \qquad (16.18)$$

An aldehyde   A hydrate

Another side reaction that may occur when hypohalite is used as an oxidizing agent is the α-halogenation of the desired carbonyl compound (Eq. 16.19). This process involves the rapid, *base-catalyzed* formation of an enolate ion, followed by its reaction with hypohalous acid or some other source of halonium ion, such as X_2 (X = halogen). When oxidations of alcohols to give carbonyl compounds are conducted in acidic media, α-halogenation of the product is not a significant problem for two reasons: the rate of enolization of a ketone is slower in acidic than in basic media, and the *enols* that are produced in acid are less reactive nucleophiles than the enolates that are formed in base.

$$
\underset{\substack{\text{A carbonyl} \\ \text{compound}}}{\text{R}-\overset{\text{O}}{\overset{\|}{\text{C}}}-\overset{\text{H}}{\underset{|}{\text{CHR}}} \;\underset{\text{HO}^-}{\rightleftharpoons}\; \underset{\text{An enolate}}{\text{R}-\overset{\text{O}^-}{\overset{|}{\text{C}}}=\text{CHR}} \;\longleftrightarrow\; \text{R}-\overset{\text{O}}{\overset{\|}{\text{C}}}-\overset{-}{\text{CHR}} \;\xrightarrow{\text{X}-\text{OH}}\; \underset{\substack{\text{An } \alpha\text{-halocarbonyl} \\ \text{compound}}}{\text{R}-\overset{\text{O}}{\overset{\|}{\text{C}}}-\overset{\text{X}}{\underset{|}{\text{CHR}}}} \qquad (16.19)
$$

In two of the experiments that follow, several representative oxidations of alcohols using hypochlorous acid will be performed. In the first experiment, cyclododecanol (**1**), a secondary alcohol, is oxidized to cyclododecanone (**2**) using sodium hypochlorite or commercial household bleach (Eq. 16.20). In the second experiment, 4-chlorobenzyl alcohol (**3**) is oxidized directly to 4-chlorobenzoic acid (**4**) using calcium hypochlorite (Eq. 16.21).

Because the concentration of sodium hypochlorite in bleach depends on the age of the solution, it is important to monitor the oxidation of **1** with starch/iodide test paper to ensure that excess oxidant is present. The test paper should turn blue-black owing to the formation of the complex of starch and iodine that is produced upon oxidation of iodide ion by the hypochlorite ion. On the other hand, commercially available calcium hypochlorite is quite stable and may be stored at room temperature without decomposition. Therefore, the amount of oxidant used is easy to determine, but it is still good practice to ensure that an excess of oxidant is present by applying the starch/iodide test, especially if the reaction mixture is heated. Excess oxidant may be reduced at the end of the experiment using sodium bisulfite, $NaHSO_3$.

$$
\underset{\substack{\textbf{1} \\ \text{Cyclododecanol}}}{} \xrightarrow[\text{CH}_3\text{CO}_2\text{H}]{\text{NaOCl, H}_2\text{O}} \underset{\substack{\textbf{2} \\ \text{Cyclododecanone}}}{} \qquad (16.20)
$$

$$
\underset{\substack{\textbf{3} \\ \text{4-Chlorobenzyl alcohol}}}{\text{Cl}-\bigcirc-\text{CH}_2\text{OH}} \xrightarrow[\text{CH}_3\text{CO}_2\text{H}]{\text{Ca(OCl)}_2\text{, H}_2\text{O}} \underset{\substack{\textbf{4} \\ \text{4-Chlorobenzoic acid}}}{\text{Cl}-\bigcirc-\text{CO}_2\text{H}} \qquad (16.21)
$$

Many oxidation reactions require the use of a stoichiometric amount of oxidant. For example, sodium or calcium hypochlorite are used in stoichiometric quantities in the oxidation of cyclododecanol and 4-chlorobenzyl alcohol in other experiments in this chapter, but hypochlorite salts are not always suitable as the oxidant. For example, the production and requisite disposal of stoichiometric quantities of sodium chloride or calcium chloride can be problematic in largescale processes, and the hypochlorite oxidant can promote undesirable side reactions, such as formation of halohydrins or oxidation of the desired aldehyde to the corresponding carboxylic acid (Eq. 16.17).

Molecular oxygen is arguably the ideal stoichiometric oxidant from a "green chemistry" perspective, because it is abundant, nontoxic, and renewable, and it does not generate hazardous waste products. The process of using atmospheric oxygen to oxidize a functional group is termed *aerobic oxidation*. Historically, such reactions have been rare, except in the production of commodity chemicals on a largescale, for which oxygen is the only viable oxidant for economic and/or environmental reasons. An important goal of modern organic chemistry, however, is to identify ways to perform chemical reactions in a more efficient and environmentally benign manner using a catalyst. A catalyst, of course, only enables an otherwise slow or unselective reaction to proceed readily on a practical timescale by providing a selective, lower-energy pathway. Because the catalyst is regenerated during the reaction, only a small amount (e.g., 0.01–0.05 equivalents) with respect to the starting materials is required. It is thus significant that efficient catalytic methods that enable molecular oxygen to be used as the stoichiometric oxidant in alcohol oxidation reactions have been discovered recently.

Recall that reduction and oxidation (redox) processes involve the donation or acceptance of electrons to or from a molecule. **Transition metals** such as Ni, Pd, and Cu are able to donate and accept electrons to and from other molecules, and these metals are commonly used in catalysts. Many important industrial processes are based upon transition-metal catalysis. When the metal catalyst is present in the same phase as the reagents, it is referred to as a **homogeneous catalyst.** Homogeneous catalysis frequently involves use of organic solvents, and the catalyst often contains organic molecules with Lewis-basic heteroatoms, such as O, N, P, or S, that bond to the metal atom or ion to make it soluble in the chosen solvent. These molecules, which are called **ligands**, also modify the steric and electronic properties of the metal or metallic ion involved.

A copper-based catalyst system that has been recently developed enables the aerobic oxidation of alcohols under very mild conditions. The net reaction for the oxidation of a benzylic alcohol **5** to the corresponding aldehyde **6** is shown in Equation 16.22. The catalytic process is based on the copper(I)/copper(II) redox cycle and employs 2,2,6,6-tetramethyl-1-piperidinyloxyl (TEMPO, **7**), a stable nitroxyl radical, as a co-catalyst along with cuprous bromide, and 2,2-bipyridyl (bpy, **8**) and *N*-methylimidazole (NMI, **9**) as ligands.

$$\text{ArCH}_2\text{OH} + 1/2\,\text{O}_2 \xrightarrow{\text{Catalyst}} \text{ArCHO} + \text{H}_2\text{O} \qquad (16.22)$$
$$\underset{\textbf{5}}{\phantom{\text{ArCH}_2\text{OH}}} \qquad\qquad\qquad \underset{\textbf{6}}{\phantom{\text{ArCHO}}}$$

The important chemical changes that take place during a catalytic process can be described via a **catalytic cycle**. A simplified catalytic cycle for the Cu/TEMPO aerobic oxidation of **5** is shown in Scheme 16.1.

Scheme 16.1

The overall reaction can be divided into a catalyst oxidation stage (Step I) and a substrate oxidation stage (Steps II and III). The 2,2-bipyridyl (bpy, **8**) and *N*-methylimidazole (NMI, **9**) ligands bind to the cuprous ion and enter the catalytic cycle as the Cu^I coordination complex [(bpy)(NMI)Cu]Br, represented as L$_n$Cu^I (**10**) in Scheme 16.1. In Step I of the catalytic cycle, **10** is oxidized by atmospheric O$_2$ to the Cu^{II}-hydroxide species [(bpy)(NMI)Cu(OH)]Br, written as L$_n$Cu^{II}(OH) (**11**). During the initial cycle, **11** is generated by direct O$_2$ oxidation of **10** because TEMPOH (R$_2$NOH, **12**) is not present at the start of the reaction. Following the first complete cycle, the co-catalyst TEMPO (**7**) is produced by oxidation of **12** in Step I, with concomitant oxidation of **10** to **11**.

In Steps II and III of the cycle, the benzylic alcohol **5** binds to **11** producing the Cu^{II}–alkoxide species L$_n$Cu^{II}(OCH$_2$Ar) (**13**) and water. The alkoxide ligand is then oxidized to the aldehyde by **7**, which results in the reformation of **10** and **12**. The aldehyde does not re-enter the catalytic cycle, but **10** and **12** repeat the cycle until the benzyl alcohol is fully converted to the aldehyde. The remaining copper species is likely a cupric hydroxide complex analogous to **11**.

The substrate oxidation sequence (Steps II and III) can be presented as a pair of simplified mechanisms as seen in Scheme 16.2 and Equation 16.23. The benzylic alcohol initially coordinates to **11** and undergoes an acid-base-mediated ligand-substitution reaction to form the alkoxide complex **13** (Scheme 16.2).

Scheme 16.2

The Cu^{II}-bound alkoxide **13** then reacts with **7** to afford the aldehyde **6** (Step III). The mechanism of this process is not known, but it could occur via a homolytic mechanism as shown in Equation 16.23, where the cupric center is reduced to cuprous ion, resulting in the formation of **10** and **12**. Overall, the conversion of the alcohol to the aldehyde is a net two-electron oxidation reaction, with the cupric atom and TEMPO (**7**) serving as complementary one-electron oxidants.

(16.23)

In the experiment described in this section, a substituted benzylic alcohol **5** will be oxidized under aerobic conditions to the corresponding benzaldehyde **6** using the Cu/TEMPO catalyst system. For additional information, you may wish to read the original research paper from which this experiment is adapted: see Hill, N. J.; Hoover, J. M.; Stahl, S. S., *J. Chem. Educ.* **2013**, *90*, 102–105.

EXPERIMENTAL PROCEDURES

Oxidation of Alcohols

A ▪ *Oxidation of Cyclododecanol to Cyclododecanone*

Green Experiment   **Purpose** To demonstrate the oxidation of a secondary alcohol to the corresponding ketone using hypochlorous acid.

SAFETY ALERT

1. **Wear safety glasses or goggles and suitable protective gloves while performing the experiment.**

2. **Do not allow the solution of sodium hypochlorite to come in contact with your skin or eyes. If it does, flush the affected area immediately with copious amounts of water. Sodium hypochlorite will also bleach clothing.**

MINISCALE PROCEDURE

Preparation Refer to the online resources to answer Pre-Lab Exercises, access videos, and read the MSDSs for the chemicals used or produced in this procedure. Review Sections 2.9, 2.11, 2.13, 2.21, 2.22, and 2.29.

Apparatus A 25-mL round-bottom flask, separatory funnel, apparatus for heating under reflux, magnetic stirring, simple distillation, and *flameless* heating.

Setting Up Place 0.5 g of cyclododecanol, 1.2 mL of acetone, and 0.4 mL of *glacial* acetic acid in the round-bottom flask. Equip the flask with a stirbar, and set up the apparatus for heating under reflux.

Oxidation Stir the mixture and warm it to approximately 45 °C; maintain this temperature within ±5 °C throughout the course of the reaction. You may wish to monitor the temperature by suspending a thermometer through the top of the condenser using a copper wire to hold the thermometer in place. Using a Pasteur pipet, add 4.5 mL of commercial bleach (ca. 5.3% sodium hypochlorite) dropwise to the stirred mixture through the top of the condenser over a period of about 0.5 h. Upon completing the addition, stop stirring and heating the mixture so the layers may separate. Using a Pasteur pipet, remove a small portion of the aqueous layer, and place a drop or two of this solution on a dampened piece of starch/iodide test paper to determine whether sufficient hypochlorite has been added. The indicator paper immediately turns blue-black in color if sufficient bleach has been added. If this color does not develop, add an additional 0.4 mL of bleach to the reaction mixture. Stir the resulting mixture with heating for 2–3 min and repeat the test for excess hypochlorite. Add additional 0.4-mL portions of bleach until a positive test for oxidant is observed. Then stir the reaction mixture with heating for an additional 10 min and retest for hypochlorite. If the test is negative, add a final 0.4 mL of bleach. Whether this last test is positive or negative, stir the mixture with heating for 10 min more to complete the reaction.

Work-Up Allow the reaction mixture to cool to room temperature, and transfer it to a separatory funnel using a Pasteur pipet. Rinse the round-bottom flask with 5 mL of diethyl ether, and use a filter-tip pipet to transfer this wash to the separatory funnel. Shake the two-phase mixture, and separate the layers. Extract the aqueous layer with an additional 5-mL portion of diethyl ether, and add this extract to the original one. Wash the combined organic extracts with 5 mL of saturated sodium bicarbonate. Before shaking this mixture, swirl the *unstoppered* funnel until the evolution of carbon dioxide ceases. Shake the mixture, venting the funnel frequently to relieve any pressure that might develop. Wash the organic solution sequentially with 5-mL portions of saturated aqueous sodium bisulfite and saturated aqueous sodium chloride. Transfer the organic solution to an Erlenmeyer flask, and dry it over several spatula-tips full of *anhydrous* sodium sulfate.* Swirl the flask occasionally for a period of 10–15 min to facilitate drying; add further small portions of *anhydrous* sodium sulfate if the solution does not become clear.

Isolation and Purification Using a filter-tip pipet, transfer the dried ethereal solution to a *tared* 25-mL round-bottom flask, and equip it for simple distillation. Remove the diethyl ether by simple distillation. Alternatively, use rotary evaporation or other techniques to concentrate the solution. The final traces of solvent may be removed by attaching the flask to a vacuum source and gently swirling the contents as the vacuum is applied.★ The oil that is initially formed after removal of the solvents should solidify. Recrystallize the cyclododecanone from aqueous methanol.

Analysis Weigh the flask and calculate the yield of solid cyclododecanone. Determine the melting point of the product. Prepare the semicarbazone (mp 218–219 °C) or oxime (mp 131–132 °C) according to the procedures given in Section 16.2C. If necessary, recrystallize the derivatives from methanol or some other appropriate solvent. Obtain IR and ^{1}H NMR spectra of your starting material and product, and compare them with those of authentic samples (Figs. 16.1–16.4).

WRAPPING IT UP

Flush all *aqueous solutions* down the drain. Spread the *sodium sulfate* on a tray in the hood to evaporate residual solvent from it, and then put the used drying agent in the nonhazardous solid waste container. Place the *diethyl ether* recovered by distillation in the container for nonhalogenated organic solvents.

EXERCISES

1. Prove that no net oxidation occurs at the carbon atoms in the transformations of Equations 16.10 and 16.11 by comparing the oxidation numbers for carbon in the reactants and products.

2. Consider the oxidation of cyclododecanol (**1**) to cyclododecanone (**2**) using hypochlorous acid to answer the following questions.

   a. What is the oxidation number of the carbon atom bearing the alcohol functional group in **1**?

   b. What is the oxidation number of this carbon atom in **2**?

   c. What is reduced in this reaction, and to what is it reduced?

   d. Suggest a reagent other than bleach that would oxidize **1** into **2**.

   e. Write the chemical structure of a side-product that might be formed by the α-halogenation of **2**.

3. Would it be possible to convert cyclododecanone (**2**) to cyclododecanol (**1**)? If so, what type of reaction would this be and what reagent(s) might you use?

4. Why is *glacial* acetic acid used in this reaction?

5. What is the function of sodium bisulfite in the procedure for isolating cyclododecanone?

Refer to the online resources for Chapter 25 to answer Question 6.

6. Would cyclododecanone be expected to give a positive iodoform test (Sec. 25.7E)? Explain.

7. Determine whether the conversion of iodide ion to iodine is an oxidation or a reduction.

8. What was the purpose of testing the reaction mixture with starch/iodide paper? Illustrate your answer by showing the chemical reaction that transforms iodide ion into iodine in this test.

9. For each of the reactions shown below, indicate whether it is a net reduction, an oxidation, or neither and calculate the change in oxidation number for any carbon being reduced or oxidized.

a.

b.

c.

d.

e.

Refer to the online resources and use the spectra viewer and Tables 8.1–8.8 as needed to answer the blue-numbered questions on spectroscopy.

10. Consider the spectral data for cyclododecanol (Figs. 16.1 and 16.2).

   a.   In the IR spectrum, there is a broad absorption at about 3250 cm^{-1}. What functional group is responsible for this absorption, and why is the absorption broad?

   b.   In the ^{1}H NMR spectrum, assign the various resonances to the hydrogen nuclei responsible for them.

   c.   For the ^{13}C NMR data, assign the various resonances to the carbon nuclei responsible for them.

11. Consider the spectral data for cyclododecanone (Figs. 16.3 and 16.4).

   a.   In the functional group region of the IR spectrum, specify the absorption associated with the carbon-oxygen double bond.

   b.   In the ^{1}H NMR spectrum, assign the various resonances to the hydrogen nuclei responsible for them.

   c.   For the ^{13}C NMR data, assign the various resonances to the carbon nuclei responsible for them.

12. Discuss the differences observed in the IR and NMR spectra of cyclododecanol and cyclododecanone that are consistent with the conversion of an alcohol to a ketone in this experiment.

SPECTRA

Starting Material and Product

Figure 16.1
IR spectrum of cyclododecanol (IR card).

Figure 16.2
NMR data for cyclododecanol (CDCl₃).

(a) ¹H NMR spectrum (300 MHz).
(b) ¹³C NMR data: δ 21.0, 23.5, 24.0, 24.4, 32.5, 69.1.

Figure 16.3
IR spectrum of cyclododecanone (IR card).

Figure 16.4
NMR data for cyclododecanone (CDCl₃).

(a) ¹H NMR spectrum (300 MHz).
(b) ¹³C NMR data: δ 22.7, 24.4, 24.9, 40.2, 211.4.

B ■ *Oxidation of 4-Chlorobenzyl Alcohol to 4-Chlorobenzoic Acid*

Green Experiment

Purpose To demonstrate the oxidation of an alcohol to the corresponding carboxylic acid using hypochlorous acid.

SAFETY ALERT

1. **Wear safety glasses or goggles and suitable protective gloves while performing the experiment.**

2. **Do not allow the solution of calcium hypochlorite to come in contact with your skin or eyes. If it does, flush the affected area immediately with copious amounts of water. Calcium hypochlorite will also bleach clothing.**

MINISCALE PROCEDURE

Preparation Refer to the online resources to answer Pre-Lab Exercises, access videos, and read the MSDSs for the chemicals used or produced in this procedure. Review Sections 2.9, 2.11, 2.17, and 2.21.

Apparatus A 50-mL round-bottom flask, separatory funnel, apparatus for heating under reflux, magnetic stirring, vacuum filtration, and *flameless* heating.

Setting Up Place 2.4 g of commercial calcium hypochlorite (65%) in the round-bottom flask containing a stirbar. Add 20 mL of water to the flask and stir the mixture while adding 2 mL of *glacial* acetic acid dropwise. Weigh 0.5 g of 4-chlorobenzyl alcohol, and add it to a small Erlenmeyer flask containing 5 mL of acetonitrile; swirl the contents to dissolve the alcohol. Set up the apparatus for heating under reflux.

Oxidation Using a Pasteur pipet, transfer the solution of 4-chlorobenzyl alcohol in acetonitrile dropwise to the stirred solution of calcium hypochlorite through the top of the condenser. Stir the mixture vigorously for 1.5–2 h at 45–50 °C; *do not heat above 50 °C*. You may wish to monitor the temperature by suspending a thermometer through the top of the condenser using a copper wire to hold the thermometer in place. Remove a small aliquot of the reaction mixture using a Pasteur pipet, and place a drop or two of this solution on a dampened piece of starch/iodide paper to determine whether sufficient hypochlorite has been added. The indicator paper immediately turns blue-black in color if excess oxidizing agent is present. If this color does not develop, add an additional 150 mg of commercial calcium hypochlorite to the reaction mixture. Whether or not additional calcium hypochlorite is required, continue stirring the mixture for an additional 1 h at 45–50 °C, and then cool it to room temperature.★

Work-Up, Isolation, and Purification Transfer the cooled reaction mixture to the separatory funnel and add 10 mL of water. Rinse the reaction flask with 10 mL of diethyl ether, and transfer this rinse to the separatory funnel. Shake the funnel and separate the layers. Extract the aqueous layer sequentially with two additional 10-mL portions of diethyl ether. Extract the *combined* diethyl ether layers with two separate 10-mL portions of saturated aqueous sodium bicarbonate. Slowly add *concentrated* hydrochloric acid dropwise to the *combined* bicarbonate washings until the solution is slightly acidic (pH = 3). Collect the precipitate by vacuum filtration, wash the filter cake with 5–10 mL of water, and air-dry the crude product. 4-Chlorobenzoic acid may be recrystallized from methanol or some other appropriate solvent.

Analysis Weigh the product and calculate the percent yield. Determine the melting point of the recrystallized 4-chlorobenzoic acid. Obtain IR and ¹H NMR spectra of your starting material and product, and compare them with those of authentic samples (Figs. 16.5–16.8).

MICROSCALE PROCEDURE

Preparation Refer to the online resources to answer Pre-Lab Exercises, access videos, and read the MSDSs for the chemicals used or produced in this procedure. Review Sections 2.7, 2.9, 2.11, 2.17, and 2.21.

Apparatus A 5-mL conical vial, apparatus for heating under reflux, magnetic stirring, vacuum filtration, Craig tube filtration, and *flameless* heating.

Setting Up Weigh 0.5 g of commercial calcium hypochlorite (65%), and transfer it to the conical vial containing a spinvane. Add 3.0 mL of water to the vial, and stir the mixture while adding 0.3 mL of *glacial* acetic acid dropwise. Weigh 0.1 g of 4-chlorobenzyl alcohol, and add it to a small test tube containing 1 mL of acetonitrile; swirl the contents to dissolve the alcohol. Set up the apparatus for heating under reflux.

Oxidation Using a Pasteur pipet, transfer the solution of 4-chlorobenzyl alcohol in acetonitrile dropwise to the stirred solution of calcium hypochlorite through the top

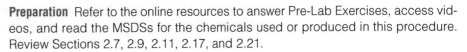

of the condenser. Stir the mixture vigorously for 1.5–2 h at 45–50 °C; *do not heat above 50 °C.* You may wish to monitor the temperature by suspending a thermometer through the top of the condenser using a copper wire to hold the thermometer in place. Remove a small aliquot of the reaction mixture using a Pasteur pipet, and place a drop or two of this solution on a dampened piece of starch/iodide paper to determine whether sufficient hypochlorite has been added. The indicator paper immediately turns blue-black in color if there is excess oxidizing agent present. If this color does not develop, add an additional 50 mg of commercial calcium hypochlorite to the reaction mixture. Whether or not additional calcium hypochlorite is required, continue stirring the mixture for an additional 1 h and then cool it to room temperature.★

Work-Up, Isolation, and Purification   Transfer the cooled reaction mixture to a screw-cap centrifuge tube and add 1.5 mL of water. Rinse the reaction vial with 2 mL of diethyl ether, and add this rinse to the centrifuge tube. Cap the centrifuge tube, and shake the contents to mix the layers. Separate the layers and extract the aqueous layer sequentially with two additional 2-mL portions of diethyl ether. Extract the *combined* diethyl ether layers with two separate 2-mL portions of saturated aqueous sodium bicarbonate. Slowly add *concentrated* hydrochloric acid dropwise to the *combined* bicarbonate washings until the solution is slightly acidic (pH = 3). Collect the precipitate by vacuum filtration, wash the filter cake with 1–2 mL of water, and air-dry the crude product. 4-Chlorobenzoic acid may be recrystallized in a Craig tube from methanol or some other appropriate solvent.

Analysis   Weigh the product and calculate the percent yield. Determine the melting point of the recrystallized 4-chlorobenzoic acid. Obtain IR and ^{1}H NMR spectra of your starting material and product, and compare them with those of authentic samples (Figs. 16.5–16.8).

WRAPPING IT UP

Neutralize the *combined aqueous solutions* with solid sodium carbonate and flush them down the drain. Because there may be some residual starting alcohol or intermediate aldehyde in the *diethyl ether extract,* transfer it to the container for halogenated organic liquids.

EXERCISES

1. Prove that no net oxidation occurs at the carbon atoms in the transformations of Equations 16.10 and 16.11 by comparing the oxidation numbers for carbon in the reactants and products.

2. Consider the oxidation of 4-chlorobenzyl alcohol (**3**) to 4-chlorobenzoic acid (**4**) using hypochlorous acid to answer the following questions.

   a. What is the oxidation number of the carbon atom bearing the alcohol functional group in **3**?

   b. What is the oxidation number of this carbon atom in **4**?

   c. What reactant is reduced in this reaction, and what is the reduced product?

3. Why is *glacial* acetic acid used in this reaction?

4. Write a balanced equation for the oxidation of **3** into **4** to determine how many equivalents of hypochlorous acid must be used.

5. Why is 4-chlorobenzaldehyde not produced in this reaction?

6. There are reagents that may be used to convert 4-chlorobenzoic acid (**4**) into 4-chlorobenzyl alcohol (**3**). Suggest one such reagent, and indicate whether this reaction is an oxidation, a reduction, or neither.

7. What was the purpose of testing the reaction mixture with starch/iodide paper? Illustrate your answer by showing the chemical reaction that transforms iodide ion into iodine in this test.

8. Determine whether the conversion of iodide ion to molecular iodine is an oxidation or a reduction.

9. What is the reduced form of $Ca(OCl)_2$ that remains at the end of the experiment?

10. For each of the reactions shown below, indicate whether it is a reduction, an oxidation, or neither.

11. The equilibrium between an aldehyde and its hydrate greatly favors the aldehyde, as shown.

    a. Using curved arrows to symbolize the flow of electrons, show the mechanism for the acid-catalyzed conversion of the hydrate into the aldehyde.

    b. If only the hydrate is capable of undergoing oxidation, explain why the aldehyde is oxidized efficiently to the carboxylic acid, even though only small amounts of the hydrate are present at any given time.

    c. Based upon the principle that an alcohol such as methanol, CH_3OH, is a derivative of water, what compound might be formed upon mixing an aldehyde, RCHO, with methanol?

    d. What product would be formed if the intermediate in Part **c** were oxidized?

Refer to the online resources and use the spectra viewer and Tables 8.1–8.8 as needed to answer the blue-numbered questions on spectroscopy.

12. One of the side reactions that may occur when hypohalite is used as an oxidizing agent for benzylic alcohols is halogenation of the aromatic ring. Although this side reaction is a problem when benzyl alcohol is oxidized with calcium hypochlorite, it is not observed in the oxidation of 4-chlorobenzyl alcohol. Explain this observation.

13. Consider the spectral data for 4-chlorobenzyl alcohol (Figs. 16.5 and 16.6).

   a. In the IR spectrum, there is a broad absorption at about 3400 cm^{-1}. What functional group is responsible for this absorption, and why is the absorption broad?

   b. In the ^{1}H NMR spectrum, assign the various resonances to the hydrogen nuclei responsible for them.

   c. For the ^{13}C NMR data, assign the various resonances to the carbon nuclei responsible for them.

14. Consider the spectral data for 4-chlorobenzoic acid (Figs. 16.7 and 16.8).

   a. In the functional group region of the IR spectrum, specify the absorptions associated with the carbonyl component of the carboxyl group. What functional group is responsible for the broad absorption in the region of about 2800 cm^{-1}, and why is it broad?

   b. In the ^{1}H NMR spectrum, assign the various resonances to the hydrogen nuclei responsible for them.

   c. For the ^{13}C NMR data, assign the various resonances to the carbon nuclei responsible for them.

15. Discuss the differences observed in the IR and NMR spectra of 4-chlorobenzyl alcohol and 4-chlorobenzoic acid that are consistent with the oxidation of an alcohol to a carboxylic acid rather than to an aldehyde in this experiment.

SPECTRA

Starting Material and Product

Figure 16.5

IR spectrum of 4-chlorobenzyl alcohol (IR card).

Figure 16.6

NMR data for 4-chlorobenzyl alcohol (CDCl₃).

(a) ¹H NMR spectrum (300 MHz).
(b) ¹³C NMR data: δ 62.5, 128.0, 131.6, 141.4.

Figure 16.7

IR spectrum of 4-chlorobenzoic acid (IR card).

Figure 16.8

NMR data for 4-chlorobenzoic acid (DMSO-d₆).

(a) ¹H NMR spectrum (300 MHz).
(b) ¹³C NMR data: δ 128.5, 129.8, 131.1, 138.3, 166.8.

C ▪ *Aerobic Oxidation of Benzylic Alcohols*

Green Experiment   **Purpose** To demonstrate the aerobic oxidation of a benzyl alcohol to the corresponding aldehyde using a CuBr/TEMPO catalyst system.

> ### SAFETY ALERT
>
> 1. **Wear safety glasses or goggles and suitable protective gloves while performing the experiment.**
> 2. **Dispense pentane in a well-ventilated work area.**

MINISCALE PROCEDURE

Preparation Refer to the online resources and read the MSDSs for the chemicals used or produced in this procedure. Review Sections 2.6, 2.7, 2.11, 2.17, 2.21, 2.24, 2.25, 2.26, and 2.29.

Apparatus A 50-mL Erlenmeyer flask, 250-mL separatory funnel, 250-mL round-bottom flask, apparatus for magnetic stirring and simple distillation.

Setting Up Place 1.0 g of the benzylic alcohol selected from among 4-nitrobenzyl alcohol, 3-nitrobenzyl alcohol, and 4-chlorobenzyl alcohol, 15 mL of reagent-grade acetone, and a stirbar in the Erlenmeyer flask and stir the solution to dissolve the solid. Weigh 85 mg of cuprous bromide, 100 mg of bpy (**8**), and 100 mg of TEMPO (**7**) into separate glass vials.

Oxidation Sequentially add the solid CuBr, **8**, and then **7** to the *rapidly* stirred solution of the benzylic alcohol, noting any color changes during the addition of each solid. Rinse each vial with 2 mL of reagent-grade acetone and add this wash to the Erlenmeyer flask. Using a Pasteur pipet, add 10 drops of NMI (**9**) to the stirred solution and again note any color change. Continue stirring the reaction mixture rapidly until you observe a significant color change (~2 h) and then continue stirring for an additional 5 min.

Work-Up Transfer the reaction mixture to the separatory funnel. Use a Pasteur pipet to rinse the Erlenmeyer flask with 2 mL of reagent-grade acetone and add this wash to the funnel. Add 30 mL of water to the funnel, swirl the mixture, and then add 20 mL of pentane. If a precipitate appears, add an additional 20 mL of reagent-grade acetone. Extract the organic products into the pentane layer. Allow the layers to separate and extract the aqueous layer with an additional 20-mL portion of pentane. Transfer the combined organic extracts to an Erlenmeyer flask and dry the solution by stirring over several spatula-tips full of *anhydrous* magnesium sulfate. Add further portions of *anhydrous* magnesium sulfate if the solution does not become clear.

Isolation and Purification Remove the magnesium sulfate from the solution by gravity filtration, collecting the filtrate in a tared round-bottom flask. Equip the flask for simple distillation, and remove the solvent by simple distillation. Alternatively, remove the solvent by rotary evaporation or other appropriate techniques. Add 5 mL of cold pentane to the flask, stir the mixture briefly, and isolate the solid product by vacuum filtration. Rinse the solid twice with 2-mL portions of cold pentane.

Analysis Weigh the product aldehyde and calculate the percent yield. Determine the melting point of the aldehyde. Obtain IR and ^{1}H NMR spectra of your starting material and product, and compare them with those from authentic samples (Figs. 16.5, 16.6, and 16.9–16.18).

MICROSCALE PROCEDURE

Preparation Refer to the online resources and read the MSDSs for the chemicals used or produced in this procedure. Review Sections 2.6, 2.7, 2.11, 2.17, 2.21, 2.24, 2.25, 2.26, and 2.29.

Apparatus A 50-mL Erlenmeyer flask, 250-mL separatory funnel, 100-mL round-bottom flask, apparatus for magnetic stirring and simple distillation.

Setting Up Place 0.40 g of the benzylic alcohol selected from among 4-nitrobenzyl alcohol, 3-nitrobenzyl alcohol, and 4-chlorobenzyl alcohol, 10 mL of *reagent-grade* acetone, and a stirbar in the Erlenmeyer flask and stir the solution to dissolve the solid. Weigh 35 mg of cuprous bromide, 40 mg of bpy (**8**), and 40 mg of TEMPO (**7**) into separate glass vials.

Oxidation Sequentially add the solid CuBr, **8**, and **7** to the *rapidly* stirred solution of the benzylic alcohol, noting any color changes during the addition of each solid. Rinse each vial with 1 mL of *reagent-grade* acetone and add this wash to the Erlenmeyer flask. Using a Pasteur pipet, add 4 drops of NMI (**9**) to the stirred solution and again note any color change. Continue stirring the reaction mixture rapidly until you observe a significant color change (~45 min) and then continue stirring for an additional 5 min.

Work-Up Transfer the reaction mixture to the separatory funnel. Use a Pasteur pipet to rinse the Erlenmeyer flask with 2 mL of *reagent-grade* acetone and add this wash to the funnel. Add 30 mL of water to the funnel, swirl the mixture, and then add 20 mL of pentane. If a precipitate appears, add an additional 10 mL of *reagent-grade* acetone. Extract the organic products into the pentane layer. Allow the layers to separate and extract the aqueous layer with an additional 10-mL portion of pentane. Transfer the combined organic extracts to an Erlenmeyer flask and dry the solution by stirring over several spatula-tips full of *anhydrous* magnesium sulfate.★ Add further portions of *anhydrous* magnesium sulfate if the solution does not become clear.

Isolation and Purification Remove the magnesium sulfate from the solution by gravity filtration, collecting the filtrate in a tared round-bottom flask. Equip the flask for simple distillation, and remove the solvent by simple distillation. Alternatively, remove the solvent by rotary evaporation or other appropriate techniques. Add 5 mL of cold pentane to the flask, stir the mixture briefly, and isolate the solid product by vacuum filtration. Rinse the solid twice with 1-mL portions of cold pentane.

Analysis Weigh the product aldehyde and calculate the percent yield. Determine the melting point of the aldehyde. Obtain IR and ^{1}H NMR spectra of your starting material and product, and compare them with those from authentic samples (Figs. 16.5, 16.6, and 16.9–16.18).

WRAPPING IT UP

Flush all *aqueous solutions* down the drain. Spread the magnesium sulfate on a tray or filter paper in the fume hood to evaporate any solvent from it, and then place it in the nonhazardous solid waste container. Transfer the *acetone* and *pentane* recovered from the reaction mixture to the container for nonhalogenated organic liquids.

EXERCISES

1. The CuBr/TEMPO aerobic oxidation is highly selective toward oxidizing primary alcohols. Draw the organic product obtained from CuBr/TEMPO aerobic oxidation of each molecule shown below.

2. A student chose to monitor the progress of the aerobic oxidation reaction by IR spectroscopy. Sketch the functional group region of the IR spectrum you would expect at 50% completion of the reaction.

3. A student chose to monitor the progress of the aerobic oxidation reaction by TLC on silica plates. Sketch the TLC plate you would expect at 50% completion of the reaction.

4. The compounds *N*-methyl imidazole (NMI) and 2,2'-bipyridyl (bpy) function as mono- and bidentate ligands, respectively, in the CuBr/TEMPO-catalyzed aerobic oxidation reaction. Define the terms "ligand," "monodentate," and "bidentate."

5. Acetone was used as the solvent for the oxidation reaction. Briefly explain why pentane was added to the reaction mixture during the work-up process.

6. What was the purpose of rinsing the substituted benzaldehyde product with 5 mL of pentane?

7. Calculate the amount of each catalyst component (CuBr, bpy, and TEMPO) you used in the aerobic oxidation. Express your answer in mmol, equivalents, and mol % with respect to the benzylic alcohol starting material.

8. Briefly list some reasons why the Cu/TEMPO aerobic oxidation is regarded as a "green" reaction in comparison to classic oxidation methods using Cr- and Mn-based oxidizing agents.

Refer to the online resources and use the spectra viewer and Tables 8.1–8.8 as needed to answer the blue-numbered questions on spectroscopy.

9. Consider the spectra of your starting benzylic alcohol (Figs. 16.5 and 16.6, or 16.11 and 16.12, or 16.15 and 16.16).

   a. In the functional group region of the IR spectrum, identify any absorption bands associated with the alcohol function.

   b. In the ^{1}H NMR spectrum, assign the various resonances to the hydrogen nuclei responsible for them.

   c. For the ^{13}C NMR data, assign the various resonances to the carbon nuclei responsible for them.

10. Consider the IR and NMR spectra of your benzaldehyde product (Figs. 16.9 and 16.10, or 16.13 and 16.14, or 16.17 and 16.18)

   a. In the IR spectrum, identify the absorption band due to the carbonyl group.

   b. In the ^{1}H NMR spectrum, assign the various resonances to the hydrogen nuclei responsible for them.

   c. For the ^{13}C NMR data, assign the various resonances to the carbon nuclei responsible for them.

11. Discuss the differences observed in the IR and NMR spectra of your starting alcohol and product aldehyde that are consistent with oxidation having occurred.

12. What evidence would appear in the IR, ^{1}H NMR, and ^{13}C NMR spectra indicating partial over-oxidation of the substituted benzaldehyde to the corresponding benzoic acid?

SPECTRA

Starting Materials and Products

The IR and NMR spectra of 4-chlorobenzyl alcohol are given in Figures 16.5 and 16.6, respectively.

Figure 16.9

IR spectrum of 4–chlorobenzaldehyde (IR card).

Figure 16.10

NMR data for 4-chlorobenzalde-hyde (CDCl₃).

(a) ¹H NMR spectrum (300 MHz).
(b) ¹³C NMR data: δ 129.4, 130.8, 134.9, 140.7, 190.5.

Figure 16.11

IR spectrum of 3-nitrobenzyl alcohol (ATR).

Figure 16.12

NMR data for 3-nitrobenzyl alcohol (CDCl₃).

(a) ¹H NMR spectrum (300 MHz)
(b) ¹³C NMR data: δ 64.0, 121.6, 122.6, 129.7, 132.9, 143.1, 148.5

Figure 16.13
IR spectrum of 3-nitrobenzalde-hyde (ATR).

Figure 16.14
NMR data for 3-nitrobenzalde-hyde (CDCl₃).

(a) ¹H NMR spectrum (300 MHz)
(b) ¹³C NMR data: δ 124.3, 128.6, 130.5, 134.9, 137.5, 148.9, 189.8

Figure 16.15
IR spectrum of 4-nitrobenzyl alcohol (ATR).

Figure 16.16

NMR data for 4-nitrobenzyl alcohol (CDCl₃).

(a) ¹H NMR spectrum (300 MHz)
(b) ¹³C NMR data: δ 64.2, 123.9, 127.2, 147.4, 148.5

Figure 16.17

IR spectrum of 3-nitrobenzalde-hyde (ATR).

Figure 16.18

NMR data for 4-nitrobenzalde-hyde (CDCl₃).

(a) ¹H NMR spectrum (300 MHz)
(b) ¹³C NMR data: δ 124.5, 130.7, 140.3, 151.3, 190.7

D ▪ *Derivative Preparation*

There are several reactions that allow the preparation of solid derivatives of aldehydes and ketones. Two such procedures are provided here.

1. Semicarbazones

Semicarbazide reacts with aldehydes and ketones to produce a derivative that is called a semicarbazone (Eq. 16.24)

$$
\underset{\substack{\text{An aldehyde} \\ \text{or ketone}}}{\overset{\displaystyle O}{\underset{R^1}{\overset{\|}{C}}}\!\!{}_{R^2}} + \underset{\text{Semicarbazide}}{H_2N-NHCNH_2} \xrightarrow{\ H^+\ } \underset{\text{A semicarbazone}}{\overset{\displaystyle O}{\underset{R^1}{\overset{\underset{\displaystyle \|}{N}}{\overset{\displaystyle \|}{C}}}\!\!{}_{R^2}}} \tag{16.24}
$$

via a reaction that is discussed in Chapter 13. Because semicarbazide is unstable as the free base, it is usually stored in the form of its hydrochloric acid salt. In the procedure that follows, the free base is liberated from the salt by the addition of sodium acetate.

EXPERIMENTAL PROCEDURE

Preparation of Semicarbazones

SAFETY ALERT

1. Wear safety glasses or goggles and suitable protective gloves while performing the experiment.
2. Semicarbazide and its hydrochloride may have carcinogenic, teratogenic, and mutagenic effects. Should either of these substances contact your skin, immediately flood the affected area with water.

Preparation Refer to the online resources and read the MSDSs for the chemicals used or produced in this procedure.

Apparatus Test tubes, graduated cylinder, calibrated Pasteur pipets, ice-water bath, apparatus for heating.

Setting Up Dissolve 0.5 g of semicarbazide hydrochloride and 0.8 g of sodium acetate in 5 mL of water in a test tube, and then add about 0.5 g of the carbonyl compound. If the carbonyl compound is insoluble in water, dissolve it in 5 mL of ethanol. Add water until the solution becomes turbid, then add a little ethanol until the turbidity disappears. Add the semicarbazide hydrochloride and sodium acetate to this solution, and continue as described below.

Reaction and Isolation Stopper and shake the tube vigorously to effect mixing. Remove the stopper, and place the test tube in a beaker of boiling water. Discontinue heating the water, and allow the test tube to cool to room temperature in the beaker of water. Transfer the test tube to an ice-water bath, and if crystals have not formed, scratch the side of the tube with a glass rod at the interface between the liquid and air to induce crystallization. Recrystallize the semicarbazone from water or aqueous ethanol.

WRAPPING IT UP

Flush *excess reagent* and *filtrates* down the drain.

2. Oximes

Hydroxylamine reacts with aldehydes or ketones to yield an oxime (Eq. 16.25).

$$\underset{\substack{\text{An aldehyde} \\ \text{or ketone}}}{\overset{\overset{\displaystyle O}{\|}}{R^1 \diagup C \diagdown R^2}} + \underset{\text{Hydroxylamine}}{H_2N-OH} \xrightarrow{H^+} \underset{\text{An oxime}}{\overset{\overset{\displaystyle N^{\diagup OH}}{\|}}{R^1 \diagup C \diagdown R^2}} + H_2O \qquad (16.25)$$

Hydroxylamine is usually stored as the hydrochloric acid salt because it is not stable as the free base; it is liberated from its salt by addition of base. The use of an oxime as a derivative has the limitation that such compounds frequently are *not* crystalline solids.

EXPERIMENTAL PROCEDURE

Preparation of Oximes

SAFETY ALERT

1. Wear safety glasses or goggles and suitable protective gloves while performing the experiment.

2. Hydroxylamine and its hydrochloride may have mutagenic effects. Should either of these substances contact your skin, immediately flood the affected area with water.

Preparation Refer to the online resources and read the MSDSs for the chemicals used or produced in this procedure.

Apparatus A 25-mL Erlenmeyer flask, graduated cylinder, Pasteur pipets, ice-water bath, apparatus for heating.

Setting Up Dissolve 0.5 g of hydroxylamine hydrochloride in 5 mL of water and 3 mL of 3 *M* sodium hydroxide solution and add 0.5 g of the aldehyde or ketone.

If the carbonyl compound is insoluble in water, add just enough ethanol to give a clear solution.

Reaction and Isolation Warm the mixture for 10 min, then cool it in an ice-water bath. If crystals do not form immediately, scratch the side of the tube with a glass rod at the air-liquid interface to induce crystallization. Recrystallize the oxime from water or aqueous ethanol.

In some cases, the use of 3 mL of pyridine and 3 mL of absolute ethanol instead of the 3 mL of 3 *M* sodium hydroxide solution and 5 mL of water is more effective. A longer heating period is often necessary. After finishing heating, pour the mixture into an evaporating dish, and remove most of the solvent with a current of air in a hood. Grind the mostly solid residue with 3–4 mL of cold water, and collect the solid by vacuum filtration. Recrystallize the oxime from water or aqueous ethanol.

WRAPPING IT UP

Flush *excess reagent* and *filtrates* down the drain.

16.3 BASE-CATALYZED OXIDATION-REDUCTION OF ALDEHYDES: THE CANNIZZARO REACTION

Aldehydes having no hydrogens on the **α-carbon atom,** the carbon atom adjacent to the carbonyl group, undergo mutual oxidation and reduction in the presence of strong alkali (Eq. 16.26). In contrast, aldehydes with hydrogen atoms on the **α-carbon atom** preferentially undergo other types of base-promoted reactions such as the **aldol condensation** (Eq. 16.27), which is described in Section 18.3. The mutual oxidation-reduction, called the **Cannizzaro reaction,** is a consequence of the fact that an aldehyde is intermediate in oxidation state between an alcohol and a carboxylic acid. It may be converted into either one by a decrease or gain of 2, respectively, in the oxidation number (Sec. 16.1) of its carbonyl carbon atom, as given in Equation 16.26.

$$
2\ \underset{\text{An aldehyde}}{R-\underset{R}{\overset{R}{|}}C-\overset{O}{\overset{\|}{C}}{}_{+1}-H} \xrightarrow[\text{(2) H}_3\text{O}^+]{\text{(1) HO}^-} \underset{\text{An alcohol}}{R-\underset{R}{\overset{R}{|}}C-\underset{H}{\overset{O-H}{|}}C{}_{-1}-H} + \underset{\text{A carboxylic acid}}{R-\underset{R}{\overset{R}{|}}C-\overset{O}{\overset{\|}{C}}{}_{+3}-OH} \qquad (16.26)
$$

$$
2\ \underset{\text{An aldehyde}}{R-\underset{H}{\overset{H}{|}}C-\overset{O}{\overset{\|}{C}}-H} \xrightarrow[\text{(2) H}_3\text{O}^+]{\text{(1) HO}^-} \underset{\substack{\text{An aldol}\\ (\beta\text{-hydroxycarbonyl compound})}}{R-\underset{H}{\overset{H}{|}}C-\underset{H}{\overset{OH}{|}}C-\underset{R}{\overset{}{|}}CH-\overset{O}{\overset{\|}{C}}-H} \qquad (16.27)
$$

The accepted mechanism of the reaction follows logically from the ease with which nucleophiles add to the carbonyl group, particularly that of an aldehyde.

The first step is attack of hydroxide ion on the carbonyl group to give a tetrahedral intermediate (Eq. 16.28), and this is followed by transfer of a hydride ion to the carbonyl group of another aldehyde function (Eq. 16.29). The step depicted in Equation 16.29 is the one in which both oxidation and reduction occur. You should verify this fact by determining the oxidation numbers of the carbon atoms involved. The remaining steps (Eqs. 16.30 and 16.31) simply illustrate the types of acid-base chemistry expected to occur in the strongly basic medium in which the reaction is performed. Summing of Equations 16.28–16.31 gives Equation 16.26.

$$\text{(16.28)}$$

$$\text{(16.29)}$$

$$\text{(16.30)}$$

$$\text{(16.31)}$$

Aromatic aldehydes, formaldehyde, and α-trisubstituted acetaldehydes all undergo the Cannizzaro reaction. In the experiment of this section, 4-chlorobenzaldehyde, an aromatic aldehyde, is converted to 4-chlorobenzyl alcohol and potassium 4-chlorobenzoate, which gives 4-chlorobenzoic acid upon acidification of the reaction mixture (Eq. 16.32).

$$\text{(16.32)}$$

| 4-Chlorobenzaldehyde | 4-Chlorobenzyl alcohol | Potassium 4-chlorobenzoate | 4-Chlorobenzoic acid |

EXPERIMENTAL PROCEDURES

Base-Catalyzed Oxidation-Reduction of Aldehydes by the Cannizzaro Reaction

Purpose To demonstrate the oxidation-reduction of an aromatic aldehyde to give the corresponding alcohol and carboxylic acid salt by the Cannizzaro reaction.

SAFETY ALERT

1. **Wear safety glasses or goggles and suitable protective gloves while performing the experiment.**

2. **The concentrated solution of potassium hydroxide is *highly* corrosive and caustic. *Do not allow it to come into contact with skin.* Should it do so, flood the affected area immediately with water and then thoroughly rinse it with 1% acetic acid.**

MINISCALE PROCEDURE

Preparation Refer to the online resources to answer Pre-Lab Exercises, access videos, and read the MSDSs for the chemicals used or produced in this procedure. Review Sections 2.9, 2.11, 2.13, 2.17, 2.21, 2.22, and 2.29.

Apparatus A 25-mL round-bottom flask, thermometer, separatory funnel, apparatus for heating under reflux, magnetic stirring, simple distillation, vacuum filtration, and *flameless* heating.

Setting Up Dissolve 5.0 g of potassium hydroxide in 5 mL of distilled water contained in an Erlenmeyer flask. Weigh 1.0 g of 4-chlorobenzaldehyde and transfer it to the round-bottom flask containing a stirbar. Add 2.5 mL of methanol and stir to dissolve the 4-chlorobenzaldehyde. Using a Pasteur pipet, transfer 1.5 mL of the 50% aqueous potassium hydroxide solution to the round-bottom flask; do not get any of this solution on the ground-glass joint. Assemble the apparatus for heating under reflux; *be sure that the joint of the flask is well greased so that it does not freeze.*

Reaction and Work-Up Place the round-bottom flask in a heating bath that has been preheated to about 75 °C, and heat the reaction mixture with stirring at this temperature for 1.5 h. Remove the flask from the bath, cool the mixture to room temperature, and add 15 mL of water. Transfer the mixture to a separatory funnel. Extract the aqueous mixture with three separate 5-mL portions of dichloromethane; shake the funnel *gently* to avoid forming an emulsion. Test the lower layer to verify that it is the organic layer, and combine all of the organic extracts. *Save both the organic and the aqueous layers, as each must be worked up separately to isolate the two products.*

Isolation and Purification Wash the combined *dichloromethane extracts* with two separate 5-mL portions of saturated aqueous sodium chloride. Dry the organic layer over several spatula-tips full of *anhydrous* sodium sulfate.★ Swirl the flask occasionally for a period of 10–15 min to facilitate drying; add further small portions of *anhydrous* sodium sulfate if the solution does not become clear. Filter or decant the dried solution into a clean, tared round-bottom flask, rinse the sodium sulfate with about 1 mL of fresh dichloromethane, and transfer the rinse to the round-bottom flask. Remove the dichloromethane by simple distillation. Alternatively, use rotary evaporation or other techniques to concentrate the solution. The final traces of solvent may be removed by attaching the flask to a vacuum source and gently swirling the contents as the vacuum is applied.★ The crude 4-chlorobenzyl alcohol may be recrystallized by mixed-solvent techniques from acetone and hexane (Sec. 3.2). Air-dry the purified product.

Cool the aqueous phase saved from the initial dichloromethane extraction in an ice-water bath. Acidify the solution by slowly adding 2.5 mL of concentrated hydrochloric acid. If the solution is not acidic (pH ~3), continue adding concentrated hydrochloric acid *dropwise* until it is. Collect the precipitate by vacuum filtration, wash the filter cake with 5–10 mL of water, and air-dry the product. The crude 4-chlorobenzoic acid may be recrystallized from methanol or some other appropriate solvent.

Analysis Weigh the products and calculate percent yields. Determine the melting points of the recrystallized products. Perform the chromic acid test (Sec. 4.7B) on the 4-chlorobenzyl alcohol. Obtain IR and ^{1}H NMR spectra of your starting material and product and compare them with those of authentic samples (Figs. 16.5–16.8, 16.19, and 16.20).

MICROSCALE PROCEDURE

Preparation Refer to the online resources to answer Pre-Lab Exercises, access videos, and read the MSDSs for the chemicals used or produced in this procedure. Review Sections 2.9, 2.11, 2.13, 2.17, 2.21, 2.22, and 2.29.

Apparatus A 5-mL conical vial, thermometer, screw-cap centrifuge tubes, apparatus for heating under reflux, magnetic stirring, simple distillation, vacuum filtration, Craig tube filtration, and *flameless* heating.

Setting Up Dissolve 1.0 g of potassium hydroxide in 1 mL of distilled water contained in a test tube. Weigh 200 mg of 4-chlorobenzaldehyde and transfer it to the 5-mL conical vial containing a spinvane. Add 0.5 mL of methanol and stir to dissolve the 4-chlorobenzaldehyde. Using a Pasteur pipet, transfer 0.3 mL of the 50% aqueous potassium hydroxide solution to the vial; do not get any of this solution on the ground-glass joint. Assemble the apparatus for heating under reflux; *be sure that the joint of the vial is well greased so it does not freeze.*

Reaction and Work-Up Place the base of the 5-mL conical vial in a heating bath that has been preheated to about 75 °C, and heat the reaction mixture with stirring at this temperature for 1.5 h. Remove the vial from the bath, cool the mixture to room temperature, and add 2.5 mL of water. Extract the aqueous mixture with three separate 0.7-mL portions of dichloromethane. For each extraction, you should cap the vial and shake it gently to avoid forming an emulsion; carefully vent the vial by slowly loosening the cap. Remove the lower layer using a Pasteur pipet and verify that it is the organic layer; combine all of the organic extracts in a centrifuge tube.

Save both the organic and the aqueous layers, as each must be worked up separately to isolate the two products.

Isolation and Purification Wash the combined *dichloromethane extracts* with two separate 0.5-mL portions of saturated aqueous sodium chloride. Transfer the dichloromethane layer to a dry centrifuge tube, and add several microspatula-tips full of *anhydrous* sodium sulfate.★ Swirl the tube occasionally for a period of 10–15 min to facilitate drying; add further small portions of *anhydrous* sodium sulfate if the solution does not become clear. Using a filter-tip pipet, transfer the dried solution into a clean 5-mL conical vial, rinse the sodium sulfate with about 0.2 mL of dichloromethane, and transfer the rinse to the conical vial. Remove the dichloromethane by simple distillation. Alternatively, use rotary evaporation or other techniques to concentrate the solution. The final traces of solvent may be removed by attaching the vial to a vacuum source and gently swirling the contents as the vacuum is applied.★ The crude 4-chlorobenzyl alcohol may be recrystallized from 4% acetone in hexane in a Craig tube. Air-dry the purified product.

Cool the aqueous phase saved from the initial dichloromethane extraction in an ice-water bath. Acidify the solution by slowly adding 0.5 mL of *concentrated* hydrochloric acid. If the solution is not acidic (pH ~3), continue adding *concentrated* hydrochloric acid dropwise until it is. Collect the precipitate by vacuum filtration, wash the filter cake with 1–2 mL of water, and air-dry the product. The crude 4-chlorobenzoic acid may be recrystallized in a Craig tube from methanol or some other appropriate solvent.

Analysis Weigh the products and calculate the percent yields. Determine the melting points of the recrystallized products. Perform the chromic acid test (Sec. 4.7B) on the 4-chlorobenzyl alcohol. Obtain IR and ^{1}H NMR spectra of your starting material and product, and compare them with those of authentic samples (Figs. 16.5–16.10).

WRAPPING IT UP

Flush the *saturated sodium chloride solution* used during the extraction process to wash the dichloromethane down the drain. Neutralize the *aqueous filtrate* from isolation of 4-chlorobenzoic acid with solid sodium carbonate, and flush it down the drain. Place the *dichloromethane* obtained by distillation in the container for halogenated organic liquids. Dispose of the *sodium sulfate* in the container for nontoxic solids after allowing residual solvent to evaporate from it in the hood.

EXERCISES

1. Determine the oxidation number for the carbonyl carbon in 4-chlorobenzaldehyde. Then confirm that oxidation and reduction of the aldehyde group occurs in the Cannizzaro reaction by calculating the oxidation numbers for the corresponding carbon atoms in 4-chlorobenzyl alcohol and 4-chlorobenzoic acid.

2. After completion of the Cannizzaro reaction of 4-chlorobenzaldehyde, the mixture was transferred into a separatory funnel and extracted with CH_2Cl_2. Write the structures of the products that were dissolved in the organic and aqueous layers, respectively.

3. After separating the organic and aqueous layers, the dichloromethane layer is first washed with two portions of saturated aqueous sodium chloride prior to drying with *anhydrous* sodium sulfate. What is the purpose of these washes?

4. What is the maximum yield of 4-chlorobenzoic acid that can be expected in this experiment?

5. When the Cannizzaro reaction is performed on 4-chlorobenzaldehyde in D_2O solution, no deuterium is found on the benzylic carbon atom in the 4-chlorobenzyl alcohol formed. How does this support the mechanism given in Equations 16.28–16.31?

6. What is the solid that is formed after 4-chlorobenzaldehyde has been allowed to react with aqueous potassium hydroxide?

7. The Cannizzaro reaction occurs much more slowly in dilute than in concentrated potassium hydroxide solution. Why is this?

8. How would 2,2-dimethylpropanal react with potassium hydroxide solution under the conditions of this experiment?

9. A *crossed* Cannizzaro reaction involving two different aldehydes may be performed if neither of the aldehydes contains α-hydrogens. Write an equation for the crossed Cannizzaro reaction of a mixture of 4-chlorobenzaldehyde and formaldehyde with concentrated potassium hydroxide solution, followed by acidification of the reaction mixture. Show all organic products.

10. Heterolytic cleavage of a C–H bond is typically a very endothermic step. Explain why hydride transfer can occur in the Cannizzaro reaction.

11. The Cannizzaro reaction is an example of a general class of reactions known as *disproportionation* reactions. Explain how this term applies to the Cannizzaro reaction.

12. A carbon-based hydride reducing reagent in biological systems is nicotinamide adenine dinucleotide, NADH, which reduces carbonyl compounds by a mechanism related to the Cannizzaro reaction. In these reactions, NADH transfers a hydride to a carbonyl compound to yield an alcohol and NAD^+, as shown.

a. Using curved arrows to symbolize the flow of electrons, write the mechanism for the hydride transfer from NADH to a carbonyl compound. Your mechanism should show how NADH becomes NAD^+.

b. What is the driving force for this reaction? (*Hint:* Consider why NAD^+ is more stable than NADH.)

c. Does the conversion of NADH to NAD^+ represent an oxidation or a reduction of NADH?

13. Consider the spectral data for 4-chlorobenzaldehyde (Figs. 16.9 and 16.10).

Refer to the online resources and use the spectra viewer and Tables 8.1–8.8 as needed to answer the blue-numbered questions on spectroscopy.

a. In the functional group region of the IR spectrum, specify the absorptions associated with the carbonyl group and with the aromatic ring.

b. In the ^1H NMR spectrum, assign the various resonances to the hydrogen nuclei responsible for them.

c. For the ^{13}C NMR data, assign the various resonances to the carbon nuclei responsible for them.

14. Consider the spectral data for 4-chlorobenzyl alcohol (Figs. 16.5 and 16.6).

   a. In the functional group region of the IR spectrum, there is a broad absorption at about 3400 cm^{-1}. What functional group is responsible for this absorption, and why is it broad?

   b. In the ^{1}H NMR spectrum, assign the various resonances to the hydrogen nuclei responsible for them.

   c. For the ^{13}C NMR data, assign the various resonances to the carbon nuclei responsible for them.

15. Consider the spectral data for 4-chlorobenzoic acid (Figs. 16.7 and 16.8).

   a. In the functional group region of the IR spectrum, specify the absorptions associated with the carbonyl component of the carboxyl group. What functional group is responsible for the broad absorption in the region of about 2800 cm^{-1}, and why is the absorption broad?

   b. In the ^{1}H NMR spectrum, assign the various resonances to the hydrogen nuclei responsible for them.

   c. For the ^{13}C NMR data, assign the various resonances to the carbon nuclei responsible for them.

16. Discuss the differences observed in the IR and NMR spectra of 4-chlorobenzaldehyde, 4-chlorobenzyl alcohol, and 4-chlorobenzoic acid that are consistent with the formation of the two products from 4-chlorobenzaldehyde by the Cannizzaro reaction.

SPECTRA

Starting Material and Products

The IR and NMR spectra of 4-chlorobenzyl alcohol are given in Figures 16.5 and 16.6, respectively; those of 4-chlorobenzoic acid are presented in Figures 16.7 and 16.8, respectively; those of 4-chlorobenzaldehyde are provided in Figures 16.9 and 16.10, respectively.

CHAPTER 17

Reduction Reactions of Double Bonds: Alkenes, Carbonyl Compounds, and Imines

In this chapter, we'll examine the reductions of a number of different functional groups. A reduction in organic chemistry frequently involves increasing the number of carbon-hydrogen bonds by simply adding a molecule of dihydrogen across a double or triple bond. For example, a reduction that is commonly performed in the food industry is the catalytic hydrogenation of the polyunsaturated triglycerides that are found in peanut, soybean, cottonseed, and other oils. In this process, which is commonly known as hardening, oils are converted into fats like shortening, margarine, and butter substitutes. You will have an opportunity to conduct a catalytic hydrogenation similar to the reaction used to make fats from oils in one of the experiments in this chapter. In other experiments, you will be able to investigate the reduction of one functional group selectively in the presence of others, a common and often challenging transformation in modern research laboratories.

17.1 INTRODUCTION

Reduction, often represented by the symbol [H], is a basic type of chemical reaction and is the opposite of **oxidation** (Chap. 16). You were probably introduced to reductions as electron-transfer processes in which an ion or a neutral atom *gained* electrons. In the reactions that are typically encountered in organic chemistry, however, carbon usually forms covalent bonds and hence does not really *gain* electrons. Nevertheless, reductions of organic compounds often do involve an increase in the electron density at a carbon atom as a consequence of replacing a bond between the carbon atom and a more electronegative atom, for example, nitrogen, oxygen, or halogen, with a carbon-hydrogen bond. Reactions in which new carbon-hydrogen bonds are formed by adding one or more hydrogen atoms to a functional group are also reductions.

The term **oxidation number** was defined in Section 16.1, and an organic reaction was considered an oxidation whenever the oxidation number on a carbon atom *increased.* This same concept may be applied to ascertain if an organic reaction is a reduction, because a reduction results in a *decrease* in the oxidation number of a carbon atom. For example, from the discussion in Section 16.1, we know that the oxidation number of each carbon atom in ethylene, $CH_2=CH_2$, is −2; however, in ethane, CH_3-CH_3, the corresponding number is −3. Thus, conversion of ethylene to ethane involves a *decrease* in the oxidation number of each carbon atom by 1 and a net decrease in the oxidation numbers of both carbon atoms by 2.

Organic functional groups containing double and triple bonds undergo reduction by the *net* addition of the elements of one or two molecules of dihydrogen across the π-bond(s). The starting compounds in these processes are termed **unsaturated**, whereas the final products are **saturated** with respect to hydrogen atoms because they contain only single bonds and can absorb no more hydrogen atoms. Equations 17.1–17.4 illustrate some types of reductions that are commonly performed by organic chemists. In addition to the functional groups shown in these equations, many other functional groups containing π-bonds can be reduced to form products containing only single σ-bonds. These include carboxylic acids and their derivatives (esters, amides, anhydrides, and nitriles), nitro groups, and aromatic rings.

$$\text{An alkene} \quad \xrightarrow{[H]} \quad \text{An alkane} \tag{17.1}$$

$$\text{An alkyne} \quad \xrightarrow[\text{1 mol}]{[H]} \quad \text{An alkene} \quad \xrightarrow[\text{1 mol}]{[H]} \quad \text{An alkane} \tag{17.2}$$

$$\text{A carbonyl compound} \quad \xrightarrow{[H]} \quad \text{An alcohol} \tag{17.3}$$

$$\text{An imine} \quad \xrightarrow{[H]} \quad \text{An amine} \tag{17.4}$$

Reductions like those shown in Equations 17.1–17.4 are commonly effected by either **catalytic** or **chemical** methods. Sometimes both techniques can be used, but one of them will usually be preferred for a given substrate, depending upon what other functional groups might be present in the molecule. A brief discussion of these methods will help enable you to understand the basic differences.

The technique of **catalytic hydrogenation**, which is discussed in more detail in Section 17.2, is widely used in organic chemistry to reduce the π-bonds in a variety of functional groups. This reaction involves the addition of a molecule of dihydrogen across a carbon-carbon or carbon-heteroatom double or triple bond in the presence of a metal **catalyst** such as platinum, palladium, nickel, or rhodium that is often adsorbed on an inert solid support like carbon. These catalysts have different properties and selectivities toward reducible groups, so it is sometimes necessary to try several in order to optimize a particular catalytic hydrogenation. Preparative catalytic hydrogenation is frequently used in research and industrial laboratories to synthesize compounds of interest, and analytical methods have been developed in which the amount of dihydrogen consumed and the heat, ΔH_{hydrog}, evolved may be accurately measured. In the experimental procedures provided in Section 17.2, you will have an opportunity to perform the catalytic hydrogenation of a carbon-carbon double bond in the presence of other reducible functional groups and with and without the use of gaseous dihydrogen.

Selective reactions are an integral part of organic synthesis. A chemist planning the synthesis of a complex molecule must be able to rely on certain reactions to give a specific isomer of the desired product. This could require that the reaction be **regioselective** (Sec. 10.4) or **stereoselective** (Sec. 10.8), or both. Many molecules of interest have multiple different functional groups, so practical challenge in contemporary synthetic organic chemistry is the ability to effect the selective reaction of one functional group in the presence of all others. Achieving such **chemoselective reactions**, which is often critical to the success of a synthesis, may require considerable experimentation or the development of new reagents or methods.

In the case of catalytic hydrogenation, chemoselectivity can be accomplished by the proper choice of the reaction conditions. You will see in Section 17.2 that preferential catalytic reduction of a carbon-carbon double bond (Eq. 17.1) in the presence of a carbon-oxygen double bond (Eq. 17.3) is possible because reducing a carbonyl function by hydrogenation typically requires that the reaction be performed above room temperature and at pressures of dihydrogen greater than 1 atm. Other functional groups such as aryl halides, esters, and nitro groups may be reduced by catalytic hydrogenation, and a procedure described in Section 17.2 will allow you to test the selective reduction of carbon-carbon double bonds in the presence of these functional groups.

Chemical reduction differs from catalytic hydrogenation because the **metal hydride** reducing agents that are used in the latter have differing reactivities toward specific functional groups. Such reagents selectively reduce functional groups that are susceptible to nucleophilic attack by hydride ion. This is an important property because metal hydrides may be thus used to reduce a carbon-oxygen or a carbon-nitrogen double bond in the presence of a carbon-carbon double bond, a reaction that is not easily accomplished by catalytic hydrogenation. As a general rule, the carbon atoms of an isolated, carbon-carbon double bond are not susceptible to nucleophilic attack, whereas those of carbon-oxygen and carbon-nitrogen double bonds are. One such selective reduction is accomplished using the experimental procedure of Section 17.3.

There are a large number of hydride-containing reducing agents that exhibit a broad range of selectivities toward various unsaturated functional groups. For example, lithium aluminum hydride, $LiAlH_4$, is a very reactive, and hence unselective, hydride donor that rapidly reduces aldehydes, ketones, and esters with ease. Sodium borohydride, $NaBH_4$, on the other hand, is less reactive and thus more selective. It reduces aldehydes and ketones but *not* esters, for example. Not only is chemical selectivity a concern in choosing the appropriate reagent, but experimental factors such as ease of handling and safety often play important roles. Thus, if lithium aluminum hydride and sodium borohydride could both be used for a specific reduction, the latter would be preferred because it is safer and easier to handle. Sections 17.3 and 17.4 include experiments for several chemical reductions that utilize sodium borohydride. In one of these, you will also explore the consequences of using chiral additives upon the stereochemistry of hydride reductions.

The nitro group, $-NO_2$, is reduced to the amino group, $-NH_2$, by catalytic hydrogenation or by using lithium aluminum hydride. However, this conversion may also be accomplished by an alternative method of chemical reduction known as **dissolving metal reduction**. These reductions are typically performed either using finely divided iron or tin with hydrochloric acid or an alkaline earth metal such as lithium, sodium, and potassium in liquid ammonia. The reduction of nitrobenzene, $C_6H_5NO_2$, to aniline, $C_6H_5NH_2$, using tin and hydrochloric acid (Sec. 20.2) is an example of such a reduction.

17.2 CATALYTIC HYDROGENATION OF THE CARBON-CARBON DOUBLE BOND

Catalytic hydrogenation of alkenes belongs to the general class of reactions known as **addition reactions**, which were introduced in Section 10.4. In these reactions, one mole of dihydrogen is *stereospecifically* added to the *same* side of the carbon-carbon π-bond, as exemplified by the hydrogenation of 1,2-dimethylcyclopentene (**1**) to give *cis*-1,2-dimethylcyclopentane (**2**), as shown in Equation 17.5. Such processes are termed *syn*-**additions**. Although the detailed mechanism by which this reaction occurs is not fully understood, it is generally believed that dihydrogen gas is adsorbed on the surface of the finely divided metal catalyst to produce hydrogen atoms. The π-electrons of the double bond also complex with the surface of the catalyst, and the hydrogen atoms are transferred from the surface to the carbon atoms to form the saturated product. Catalytic hydrogenation is the method of choice for reducing alkenes to alkanes and alkynes to *cis*-alkenes, although a catalyst poison is generally used to prevent overreduction.

$$\text{(17.5)}$$

1
1,2-Dimethylcyclopentene

2
cis-1,2-Dimethylcyclopentane

Catalytic hydrogenation is commonly performed by first suspending the appropriate metallic catalyst in a solution containing the reactant in a suitable solvent. The resulting mixture is then stirred or shaken under an atmosphere of dihydrogen gas at pressures typically ranging from 15 to 60 psi. Dihydrogen, a highly flammable gas, is normally purchased in pressurized gas cylinders, and special apparatus and precautions must be used to deliver the dihydrogen from the cylinder to the apparatus. However, it is more convenient in simple experiments to generate dihydrogen gas *in situ*. Combining sodium borohydride and *concentrated* hydrochloric acid provides a ready source of dihydrogen via the reaction of H^+ from HCl with the hydride ions, $H:^-$, contained in $NaBH_4$ (Eq. 17.6). The catalyst, platinum on carbon, may be prepared in the same flask by reducing chloroplatinic acid, H_2PtCl_6, with dihydrogen in the presence of decolorizing carbon, which serves as a solid support for the finely divided metallic platinum produced (Eq. 17.7).

$$NaBH_4 + HCl + 3\,H_2O \rightarrow 3\,H_2 + B(OH)_3 + NaCl \qquad (17.6)$$

$$H_2PtCl_6 + 2\,H_2 \rightarrow Pt^0 + 6\,HCl \qquad (17.7)$$

The first experiment that follows involves the conversion of 4-cyclohexene-*cis*-1,2-dicarboxylic acid (**3**) to cyclohexane-*cis*-1,2-dicarboxylic acid (**4**), as shown in Equation 17.8. The catalytic hydrogenation of **3** is considered to be selective because the carboxylic acid groups, $-CO_2H$, are *not* reduced under the reaction conditions. This classifies this reduction as chemoselective because the reaction is occurring with one functional group in the preference to others. The starting material for this reduction may be readily prepared by the hydrolysis of 4-cyclohexene-*cis*-1,2-dicarboxylic acid anhydride, which is the product of the Diels-Alder reaction between butadiene and maleic anhydride, according to the procedure outlined in Section 12.3, Part C1.

$$\text{(17.8)}$$

3

4-Cyclohexene-*cis*-
1,2-dicarboxylic acid

4

Cyclohexane-*cis*-
1,2-dicarboxylic acid

In the second experiment in this section, you will explore the chemoselectivity of palladium-catalyzed hydrogenation using cinnamic acid derivatives **6–9** (Fig. 17.1). You will also use a different technique for generating the dihydrogen needed to effect the reduction, as seen in Equations 17.9 and 17.10.

$$\text{(17.9)}$$

5

Ammonium formate

$$\text{(17.10)}$$

A cinnamic acid derivative

A 3-phenylpropanoic
acid derivative

In contrast to a reduction with dihydrogen gas, which can be both hazardous and inconvenient to use, alternative methods for hydrogenation have been developed that involve transferring an equivalent of dihydrogen from other molecules, a process termed **transfer hydrogenation**. One common hydrogen atom donor is

6

4-Fluorocinnamic acid

7

4-Chlorocinnamic acid

8

Benzyl cinnamate

9

4-Nitrocinnamic acid

Figure 17.1
Cinnamic acid derivatives.

ammonium formate (**5**), which generates dihydrogen, ammonia, and carbon dioxide in the presence of palladium on carbon (Eq. 17.9). The dihydrogen thus produced is then available for reducing various functional groups, including carbon-carbon double bonds, resulting in a net transfer of two hydrogen atoms from ammonium formate to an alkene as illustrated in Equation 17.10.

A goal of the procedure you will perform is defining the chemoselectivity associated with the reduction. Thus, not only is the desired reduction of the carbon-carbon double bond of the cinnamic acid derivatives **6–9** to give **10–13** possible, but the **hydrogenolysis** of carbon-heteroatom bonds or the reduction of a nitro group may also occur to yield the undesired products **14** and **15** (Fig. 17.2). In short, you will explore the reduction of a multifunctional molecule and determine whether the reaction conditions used effect chemoselective reaction with the carbon-carbon double bond. Melting point data provided in Figure 17.2 and/or spectral analysis will allow you to assess the chemoselectivity for the particular substrate you are assigned.

10
3-(4-Fluorophenyl)-
propanoic acid
mp = 95 °C

11
3-(4-Chlorophenyl)-
propanoic acid
mp = 93 °C

12
Benzyl 3-phenylpropanoate

mp < 25 °C

Figure 17.2
Some possible reduction products of cinnamic acid derivatives.

13
3-(4-Nitrophenyl)
cinnamic acid
mp = 164 °C

14
3-Phenylpropanoic acid

mp = 46 °C

15
3-(4-Aminophenyl)
propanoic acid
mp = 131 °C

EXPERIMENTAL PROCEDURE

Catalytic Hydrogenation of Alkenes

A. Hydrogenation of 4-Cyclohexene-cis-1,2-dicarboxylic Acid

Purpose To demonstrate the selective reduction of an alkene by catalytic hydrogenation.

SAFETY ALERT

1. **Wear safety glasses or goggles and suitable protective gloves while performing the experiment.**

2. ***Dihydrogen gas is extremely flammable.*** **Use no flames in this experiment, and make certain that there are no open flames nearby.**

3. **The platinum catalyst prepared in this experiment may be** *pyrophoric* **and combust spontaneously in air. Do** *not* **allow it to become dry!**

4. **Use** *flameless* **heating when removing the diethyl ether.**

5. **Avoid skin contamination by the solutions you are using. If you do get these solutions on your hands, wash them thoroughly with water. In the case of contact with the sodium borohydride solution, rinse the affected areas with 1% acetic acid solution. In the case of acid burns, apply a paste of sodium bicarbonate to the area for a few minutes and then rinse with large amounts of water.**

MINISCALE PROCEDURE

Preparation Refer to the online resources to answer Pre-Lab Exercises, access videos, and read the MSDSs for the chemicals used or produced in this procedure. Review Sections 2.9, 2.11, 2.13, 2.17, 2.21, and 2.29.

Apparatus A 50-mL filter flask, a 2-mL plastic syringe, separatory funnel, a 50-mL round-bottom flask, ice-water bath, apparatus for magnetic stirring, simple distillation, vacuum filtration, and *flameless* heating.

Setting Up Prepare a reaction vessel for hydrogenation by using rubber bands or wire to attach a balloon to the sidearm of the filter flask. Place a stirbar in the flask. Prepare a 1 *M* aqueous solution of sodium borohydride by dissolving 0.2 g of sodium borohydride in 5 mL of 1% aqueous sodium hydroxide. Place 5 mL of water, 0.5 mL of a 5% solution of chloroplatinic acid ($H_2PtCl_6 \cdot 6\ H_2O$), and 0.2 g of decolorizing carbon in the reaction flask, and slowly add 1.5 mL of the 1 *M* sodium borohydride solution with stirring. Stir the resulting slurry for 5 min to allow the active catalyst to form. During this time, dissolve 0.5 g of 4-cyclohexene-*cis*-1,2-dicarboxylic acid (Sec. 12.3, Part C1) in 5 mL of water by heating. Pour 2 mL of *concentrated* hydrochloric acid into the reaction flask containing the catalyst, and add the hot aqueous solution of the diacid to the flask. Seal the flask with a rubber septum that is securely wired in place, and stir the aqueous suspension.

Reaction Draw 0.8 mL of the 1 *M* sodium borohydride solution into the syringe, carefully push the needle of the syringe through the rubber septum, and inject the solution dropwise into the stirred solution of reactants in the filter flask. If the balloon on the sidearm of the flask begins to inflate *rapidly,* add the solution of sodium borohydride more slowly. When the syringe is empty, remove it from the flask, refill it with an additional 0.8 mL of 1 *M* sodium borohydride solution, and add this second portion of sodium borohydride to the stirred reaction mixture. The balloon should inflate somewhat, indicating a positive pressure of dihydrogen in the system. Remove the syringe from the rubber septum, and continue stirring for 5 min. Then heat the flask at 70–80 °C for 1.5–2 h with continued stirring to complete the reaction.

Work-Up and Isolation Release the pressure from the reaction flask by pushing a syringe needle through the rubber septum. *Do not allow the reaction mixture to cool.* Remove the catalyst by vacuum filtration of the *hot* reaction mixture using a Hirsch funnel, and *immediately* treat the catalyst according to the protocol given in the Wrapping It Up section.★ Cool the filtrate and use pHydrion paper to determine its pH. If the filtrate is not acidic, add *concentrated* hydrochloric acid dropwise until the pH is below 5. Transfer the filtrate to the separatory funnel. Rinse the filter flask with 15 mL of diethyl ether to dissolve the solid product that may have separated upon cooling, and add this rinse to the separatory funnel. Based upon the amount of water used in the experiment, calculate the weight of sodium chloride that would be needed to saturate the filtrate, and add this amount of solid sodium chloride to the separatory funnel. Shake the separatory funnel, venting as necessary to relieve pressure. After separating the layers, extract the aqueous layer sequentially with two 10-mL portions of diethyl ether. Dry the *combined* ethereal extracts by swirling them over several spatula-tips full of *anhydrous* sodium sulfate.★ Swirl the flask occasionally for a period of 10–15 min to facilitate drying; add further small portions of *anhydrous* sodium sulfate if the solution remains cloudy.

Carefully decant the dried solution into a tared round-bottom flask containing a stirbar. Equip the flask for simple distillation and remove the diethyl ether by simple distillation. Alternatively, use rotary evaporation or other techniques to concentrate the solution. The final traces of solvent may be removed by attaching the flask to a vacuum source and gently swirling the contents as the vacuum is applied. The product remains as a solid residue in the flask.★ Air-dry the crude product and determine its weight and melting point.

Purification The crude product may be recrystallized from a *minimum* amount of water as follows. Carefully scrape the bulk of the crude diacid from the round-bottom flask into a *small* Erlenmeyer flask. Add *no more* than 1 mL of water to the round-bottom flask, heat the contents to dissolve any residual diacid, and pour the hot solution into the Erlenmeyer flask. Heat the aqueous mixture to boiling. If necessary, *slowly* add water *dropwise* to the boiling mixture to dissolve all of the diacid. Do *not* add more than an additional 1 mL of water, because the diacid is rather soluble in water. Determine the pH of the aqueous solution to ascertain whether hydrogenation has affected the acidic nature of the molecule. When all of the diacid has dissolved, add 2 drops of *concentrated* hydrochloric acid to decrease its solubility. Allow the mixture to cool to room temperature, and then place the flask in an ice-water bath to complete crystallization of the product. Collect the product by vacuum filtration and air-dry it completely.★ The reported melting point of cyclohexane-*cis*-1,2-dicarboxylic acid is 192–193 °C.

Analysis Weigh the product and calculate the percent yield; determine its melting point. Perform the qualitative tests for unsaturation using the bromine and Baeyer tests (Secs. 4.7A1 and B, respectively) on your product, and record your observations. Obtain IR and ^{1}H NMR spectra of your starting material and product, and compare them with those of authentic samples (Figs. 12.11, 12.12, 17.1, and 17.2).

WRAPPING IT UP

Wet the *catalyst* and the *filter paper* immediately with water, and remove them from the Hirsch funnel; place them in the container reserved for recovered catalyst, which should be kept wet with water at all times. Neutralize the *combined aqueous filtrates and washes* with sodium carbonate, and then flush them down the drain with excess water. Place the *recovered ethereal distillate* in the container for nonhalogenated organic solvents. After allowing the diethyl ether to evaporate from the *sodium sulfate* on a tray in the hood, place the used drying agent in the container for nonhazardous solids. Place the *dichloromethane solution* from the bromine test for unsaturation in a container for halogenated organic liquids; put the *manganese dioxide* from the Baeyer test for unsaturation in a container for heavy metals.

EXERCISES

1. Why is the sodium borohydride solution prepared in basic rather than acidic aqueous media?

2. Why is decolorizing carbon used in the procedure for the catalytic hydrogenation of 4-cyclohexene-*cis*-1,2-dicarboxylic acid (**3**)?

3. Why does adding hydrochloric acid to an aqueous solution of a carboxylic acid decrease its solubility in water?

4. What is the purpose of swirling the ethereal extracts over sodium sulfate?

5. How did you determine that the two carboxylic acid groups were not reduced by catalytic hydrogenation of **3**?

6. Determine the oxidation numbers (Sec. 16.1) of the carbon atoms undergoing change in the following reactions, and indicate the net increase or decrease of oxidation number in each.

   a. $CH_3Cl \rightarrow CH_4$

   b. $CH_3CHO \rightarrow CH_3CH_2OH$

   c. acetylene $\rightarrow$ ethylene

   d. $CH_3CO_2H \rightarrow CH_3CHO$

   e. $CH_3CH = NH \rightarrow CH_3CH_2NH_2$

   f. $H_2C = CH_2 \rightarrow BrCH_2CH_2Br$

7. Draw the structure of the product of catalytic hydrogenation of:

   a. 1,2-dimethylcyclohexene

   b. *cis*-2,3-dideutero-2-butene

   c. *trans*-2,3-dideutero-2-butene

   d. *cis*-3,4-dimethyl-3-hexene

8. Consider the conversion of nitrobenzene to aniline as shown.

   Nitrobenzene   Aniline

   a. Determine the oxidation number of the nitrogen atom in the starting material and the product. What is the net change in the oxidation number of this atom?

   b. Is this reaction an oxidation or a reduction?

   c. Suggest a method, including necessary reagents, that may be used to accomplish this transformation.

9. The heats of hydrogenation, ΔH_{hydrog}, for three isomeric butenes have been determined quantitatively by catalytic hydrogenation.

1-butene	$\Delta H_{hydrog} = -30.3 \text{ kcal/mol}$
cis-2-butene	$\Delta H_{hydrog} = -28.6 \text{ kcal/mol}$
trans-2-butene	$\Delta H_{hydrog} = -27.6 \text{ kcal/mol}$

   a. Give the reagent(s) that would effect these catalytic hydrogenations.

   b. Write the structure(s) of the products(s) formed from each of these reductions.

c. Rank these butenes in order of *decreasing* stability.

d. Draw a reaction profile for each of the three hydrogenations on the same diagram. For each reaction, label the starting alkene, the product, and ΔH_{hydrog}, and show the relative stabilities of the starting material and the product.

Refer to the online resources and use the spectra viewer and Tables 8.1–8.8 as needed to answer the blue-numbered questions on spectroscopy.

10. Consider the spectral data for 4-cyclohexene-*cis*-1,2-dicarboxylic acid (Figs. 12.11 and 12.12).

   a. In the functional group region of the IR spectrum, specify the absorptions associated with the hydroxyl and the carbonyl components of the carboxyl group.

   b. In the ^{1}H NMR spectrum, assign the various resonances to the hydrogen nuclei responsible for them.

   c. For the ^{13}C NMR data, assign the various resonances to the carbon nuclei responsible for them.

11. Consider the spectral data for cyclohexane-*cis*-1,2-dicarboxylic acid (Figs. 17.3 and 17.4).

   a. In the functional group region of the IR spectrum, specify the absorptions associated with the hydroxyl and the carbonyl components of the carboxyl group.

   b. In the ^{1}H NMR spectrum, assign the various resonances to the hydrogen nuclei responsible for them.

   c. For the ^{13}C NMR data, assign the various resonances to the carbon nuclei responsible for them.

12. Discuss the differences observed in the IR and NMR spectra of 4-cyclohexene-*cis*-1,2-dicarboxylic acid and cyclohexane-*cis*-1,2-dicarboxylic acid that are consistent with the selective reduction of the carbon-carbon double bond in this experiment.

SPECTRA

Starting Material and Product

*The IR and NMR spectra of 4-cyclohexene-*cis*-1,2-dicarboxylic acid are given in Figures 12.11 and 12.12, respectively.*

Figure 17.3

IR spectrum of cyclohexane-cis-1,2-dicarboxylic acid (IR card).

Figure 17.4

NMR data for cyclohexane-cis-1,2-dicarboxylic acid (DMSO-d₆).

(a) ¹H NMR spectrum (300 MHz).

(b) ¹³C NMR data: δ 23.8, 26.3, 42.3, 175.6.

B. Transfer Hydrogenation of Cinnamic Acid Derivatives

Purpose To determine the chemoselectivity of transfer hydrogenation.

SAFETY ALERT

1. Wear safety glasses or goggles and suitable protective gloves while performing the experiment.

2. *The dihydrogen gas generated in this procedure is extremely flammable.* Do not use flames in this experiment and make certain that there are no open flames nearby.

3. Add the palladium on carbon catalyst to your reaction flask first, *before* adding the methanol. If the dry catalyst comes into contact with vapors of methanol, the vapors may be ignited.

4. Clean up any spilled catalyst with a damp paper towel and then rinse the paper towel with plenty of water before discarding it in the *solid waste* container.

5. Ammonia gas is generated in these reactions. Ammonia vapors cause severe respiratory tract burns. Avoid inhalation.

6. All of the substrates and their reaction products are mild irritants. Avoid skin contact. If any of them do come in contact with your skin, wash the area immediately with soap and cool water.

7. 1 *M* aqueous hydrochloric acid is corrosive. If it comes in contact with your skin, rinse the area immediately with copious amounts of cool water.

MINISCALE PROCEDURE

Preparation Refer to the online resources to answer Pre-Lab Exercises, access videos, and read the MSDSs for the chemicals used or produced in this procedure. Review Sections 2.7, 2.9, 2.13, 2.17, 2.21, 2.24, 2.29, 3.2, and 6.2.

Apparatus Separatory funnel, 25- and 50-mL round-bottom flasks, filtering pipet containing 1/4 of a glass-fiber filter disk, apparatus for *flameless* heating under reflux, magnetic stirring, simple distillation, and thin-layer chromatography.

Setting Up Mark two tracks on each of three TLC plates and label one track as "t = 0" (time = zero) on all three plates. The second track should be labeled as "t = 15" on one of the plates, as "t = 30" on a second plate, and as "t = 45" on the third plate. Spot the reaction mixture on each TLC plate in the t = 0 track.

Equip the round-bottom flask with a stirbar. Add 70 mg of 10% palladium on carbon and 300 mg of one of the cinnamic acid derivatives **6–9** to the flask. Add 5 mL of methanol and swirl the flask to effect dissolution. Add 7.5 molar equivalents of ammonium formate.

Reaction Equip the flask for heating under reflux and gently heat the mixture under reflux for 45 min. Monitor the progress of the reaction at 15-min intervals by spotting a small sample of the reaction mixture on a TLC plate in the appropriate track and developing the plate with a mixture (70:30) of ethyl acetate and hexane as the eluent. Add 1 or 2 drops of *glacial* acetic acid to the eluent if the spots streak. Visualize the plate under UV light and compare the two tracks, noting that t = 0 is starting material. If after 45 min it does not seem like the reaction is complete, add another 2 equivalents of ammonium formate and continue heating. When the reaction is complete, as determined by TLC analysis, allow the mixture to cool to room temperature.

Isolation Filter the reaction mixture into a 50-mL round-bottom flask, using the filter-tip pipet, rinse the round-bottom flask with about 0.5 mL of methanol, and filter the rinse into the Erlenmeyer flask. Remove the methanol by simple distillation. Alternatively, rotary evaporation or other techniques may be used.

If your starting material was 4-nitrocinnamic acid, omit the following aqueous work-up and proceed to the recrystallization. For any other starting materials, add 10 mL of 1 *M aqueous* hydrochloric acid and 20 mL of *technical* diethyl ether to the round-bottom flask. Swirl the flask to effect dissolution and then transfer the mixture into a separatory funnel. Shake the mixture in the funnel and separate the layers. Extract the aqueous layer with an additional 10 mL of ether and combine the organic layers. Dry the organic solution over *anhydrous* sodium sulfate and filter it into 50-mL round-bottom flask. Remove the ether by simple distillation. Alternatively, rotary evaporation or other techniques may be used to remove it. Recrystallize the crude product using petroleum ether (bp 35–60 °C) unless your starting material was 4-nitrocinnamic acid, in which case use 2-propanol as the solvent. Isolate the product by vacuum filtration.

Analysis Weigh the product and calculate the percent yield; determine its melting point and compare it to the reported melting point (see Fig. 17.2). Obtain IR and ¹H NMR spectra of your starting material and product, and compare them with those of authentic samples (Figs. 17.5–17.24).

WRAPPING IT UP

Discard *aqueous waste* in a container for aqueous acid and the *filtrate* from recrystallization in a container for organic solvents. Place *filtering pipets* containing any Pd/C in a container for palladium-containing waste.

EXERCISES

1. Using curved arrows to symbolize the flow of electrons, propose a mechanism for the palladium-catalyzed reaction of cinnamic acid with dihydrogen.

2. What functional groups other than the carbon-carbon double bond were also transformed by the hydrogenation?

3. Why was treatment with 1 *M aqueous* hydrochloric acid omitted in the work-up procedure when 4-nitrocinnamic acid was the starting material?

4. Why is 1 *M aqueous* hydrochloric acid necessary in the work-up of the reactions with the other substrates used?

5. Why was a large excess of ammonium formate used in the procedure?

6. What are the driving forces that favor the palladium-catalyzed decomposition of ammonium formate to give dihydrogen gas?

7. Based upon your results and those of your labmates, which of the following transformations could *not* be accomplished using ammonium formate and palladium on carbon?

8. Why is hexane a poor choice as a recrystallization solvent for the products of both benzyl cinnamate and 4-chlorocinnamic acid?

9. Why do a few drops of acetic acid need to be added to the eluent if the spots streak when performing the TLC analysis?

Refer to the online resources and use the spectra viewer and Tables 8.1–8.8 as needed to answer the blue-numbered questions on spectroscopy.

10. Consider the spectral data for 4-fluorocinnamic acid (Figs. 17.5 and 17.6).

   a. In the IR spectrum, identify the absorption(s) in the functional group region for the carboxylic acid and alkene groups and any absorption(s) in the fingerprint region that symbolize the presence of a *trans* carbon-carbon double bond.

   b. In the ^{1}H NMR spectrum, assign the various resonances to the hydrogen nuclei responsible for them.

   c. For the ^{13}C NMR spectrum, assign the various resonances to the carbon nuclei responsible for them.

11. Consider the spectral data for 3-(4-fluorophenyl)propanoic acid (Figs. 17.7 and 17.8).

   a. In the functional group region of the IR spectrum, specify the absorption(s) associated with the carboxylic acid and the π-bonds of the aromatic ring.

   b. In the ^{1}H NMR spectrum, assign the various resonances to the hydrogen nuclei responsible for them.

   c. For the ^{13}C NMR spectrum, assign the various resonances to the carbon nuclei responsible for them.

12. Consider the spectral data for 4-chlorocinnamic acid (Figs. 17.9 and 17.10).

   a. In the IR spectrum, identify the absorption(s) in the functional group region for the carboxylic acid and alkene groups and any absorption(s) in the fingerprint region that symbolize the presence of a *trans* carbon-carbon double bond.

   b. In the ^{1}H NMR spectrum, assign the various resonances to the hydrogen nuclei responsible for them.

   c. For the ^{13}C NMR spectrum, assign the various resonances to the carbon nuclei responsible for them.

13. Consider the spectral data for 3-(4-chlorophenyl)propanoic acid (Figs. 17.11 and 17.12).

   a. In the functional group region of the IR spectrum, specify the absorption(s) associated with the carboxylic acid function.

   b. In the ^{1}H NMR spectrum, assign the various resonances to the hydrogen nuclei responsible for them.

   c. For the ^{13}C NMR spectrum, assign the various resonances to the carbon nuclei responsible for them.

14. Consider the spectral data for benzyl cinnamate (Figs. 17.13 and 17.4).

   a. In the IR spectrum, identify the absorption(s) in the functional group region for the ester and alkene groups and any absorption(s) in the fingerprint region that symbolize the presence of a *trans* carbon-carbon double bond.

   b. In the ^{1}H NMR spectrum, assign the various resonances to the hydrogen nuclei responsible for them.

   c. For the ^{13}C NMR spectrum, assign the various resonances to the carbon nuclei responsible for them.

15. Consider the spectral data for benzyl 3-phenylpropanoate (Figs. 17.15 and 17.16).

    a. In the functional group region of the IR spectrum, specify the absorption(s) associated with the ester and the π-bonds of the aromatic ring.

    b. In the ^{1}H NMR spectrum, assign the various resonances to the hydrogen nuclei responsible for them.

    c. For the ^{13}C NMR spectrum, assign the various resonances to the carbon nuclei responsible for them.

16. Consider the spectral data for 4-nitrocinnamic acid (Figs. 17.17 and 17.18).

    a. In the IR spectrum, identify the absorption(s) in the functional group region for the carboxylic acid and alkene groups and any absorption(s) in the fingerprint region that symbolize the presence of a *trans* carbon-carbon double bond.

    b. In the ^{1}H NMR spectrum, assign the various resonances to the hydrogen nuclei responsible for them.

    c. For the ^{13}C NMR spectrum, assign the various resonances to the carbon nuclei responsible for them.

17. Consider the spectral data for 3-(4-nitrophenyl)propanoic acid (Figs. 17.19 and 17.20).

    a. In the functional group region of the IR spectrum, specify the absorption(s) associated with the carboxylic acid and nitro groups and the π-bonds of the aromatic ring.

    b. In the ^{1}H NMR spectrum, assign the various resonances to the hydrogen nuclei responsible for them.

    c. For the ^{13}C NMR spectrum, assign the various resonances to the carbon nuclei responsible for them.

18. Consider the spectral data for 3-(4-aminophenyl)propanoic acid (Figs. 17.20 and 17.21).

    a. In the functional group region of the IR spectrum, specify the absorption(s) associated with the amino and carboxylic acid groups and the π-bonds of the aromatic ring.

    b. In the ^{1}H NMR spectrum, assign the various resonances to the hydrogen nuclei responsible for them.

    c. For the ^{13}C NMR spectrum, assign the various resonances to the carbon nuclei responsible for them.

19. Consider the spectral data for 3-phenylpropanoic acid (Figs. 17.23 and 17.24).

    a. In the functional group region of the IR spectrum, specify the absorptions associated with the carboxylic acid and the π-bonds of the aromatic ring.

    b. In the ^{1}H NMR spectrum, assign the various resonances to the hydrogen nuclei responsible for them.

    c. For the ^{13}C NMR spectrum, assign the various resonances to the carbon nuclei responsible for them.

20. Discuss the differences observed in the IR and NMR spectra of the starting material you used and the product(s) obtained that are consistent with reduction occurring in this experiment.

SPECTRA

Starting Materials and Products

Figure 17.5
IR spectrum of 4-fluorocinnamic acid (ATR).

Figure 17.6
NMR data for 4-fluorocinnamic acid (DMSO-d₆).

(a) ¹H NMR spectrum (400 MHz).
(b) ¹³C NMR data: δ 116.2, 116.4, 119.6, 130.9, 131.0, 131.3, 131.4, 143.1, 162.4, 164.8, 168.0.

Figure 17.7
IR spectrum of 3-(4-fluoro-phenyl)propanoic acid (ATR).

Figure 17.8
NMR data for 3-(4-fluorophenyl)-propanoic acid (DMSO-d$_6$).

(a) ^{1}H NMR spectrum (400 MHz).
(b) ^{13}C NMR data: δ 29.5, 35.3, 114.8, 115.0, 130.0, 130.1, 136.9, 137.0, 159.5, 162.0, 173.7.

Figure 17.9
IR spectrum of 4-chlorocinnamic acid (ATR).

Figure 17.10
NMR data for 4-chloro-cinnamic acid (DMSO-d$_6$).

(a) ^{1}H NMR spectrum (400 MHz).
(b) ^{13}C NMR data: δ 120.5, 129.4, 130.4, 133.7, 135.2, 142.9, 167.9.

Figure 17.11
IR spectrum of 3-(4-chloro-phenyl)propanoic acid (ATR).

Figure 17.12
NMR data for 3-(4-chloro-phenyl)propanoic acid (DMSO-d$_6$).

(a) ^{1}H NMR spectrum (400 MHz).
(b) ^{13}C NMR data: δ 29.6, 34.9, 128.1, 130.1, 130.5, 140.0, 173.5.

Figure 17.13
IR spectrum of benzyl cinnamate (ATR).

Figure 17.14
NMR data for benzyl cinnamate (DMSO-d₆).

(a) ¹H NMR spectrum (400 MHz).
(b) ¹³C NMR data: δ 66.1, 118.2, 128.6, 128.9, 129.0, 129.4, 131.0, 134.4, 136.7, 145.4, 166.6.

Figure 17.15
IR spectrum of benzyl 3-phenylpropanoate (ATR).

Figure 17.16
NMR data for benzyl 3-phenylpropanoate (DMSO-d₆).

(a) ¹H NMR spectrum (400 MHz).
(b) ¹³C NMR data: δ 30.2, 35.0, 65.4, 126.0, 127.9, 128.2, 128.3, 128.4, 136.1, 140.4, 172.1.

Figure 17.17

IR spectrum of 4-nitrocinnamic acid (ATR).

Figure 17.18

NMR data for 4-nitrocinnamic acid (DMSO-d₆).

(a) ¹H NMR spectrum (400 MHz).
(b) ¹³C NMR data: δ 124.1, 124.4, 129.7, 141.2, 141.8, 148.4, 167.5.

Figure 17.19

IR spectrum of 3-(4-nitrophenyl)-propanoic acid (ATR).

Figure 17.20

NMR data for 3-(4-nitrophenyl)-propanoic acid (DMSO-d$_6$).

(a) ^{1}H NMR spectrum (400 MHz).
(b) ^{13}C NMR data: δ 30.1, 34.3, 123.3, 129.6, 146.0, 149.3, 173.4.

Figure 17.21

IR spectrum of 3-(4-aminophenyl)-propanoic (ATR).

Figure 17.22

NMR data for 3-(4-aminophenyl)-propanoic acid (DMSO-d$_6$).

(a) ^{1}H NMR spectrum (400 MHz).
(b) ^{13}C NMR data: δ 29.7, 36.0, 113.9, 127.8, 128.5, 146.6, 174.0.

Figure 17.23
IR spectrum of 3-phenyl propanoic acid (ATR).

Figure 17.24
NMR data for 3-phenylpropanoic acid (DMSO-d_6).

(a) ^{1}H NMR spectrum (400 MHz).

(b) ^{13}C NMR data: δ 30.4, 35.2, 126.0, 128.2, 128.3, 140.9, 173.8.

17.3 REDUCTION OF IMINES; PREPARATION OF AMINES

Imines are readily formed by the condensation of primary amines with carbonyl compounds such as aldehydes and ketones according to Equation 17.11. The stability of an imine depends upon the nature of the substituents, R^1–R^3, on the carbon and nitrogen atoms of the imine functional group. For example, imines that bear only alkyl groups are less stable toward hydrolysis and polymerization than those having one or more aromatic substituents on the carbon-nitrogen double bond; the most stable imines are those having only aryl substituents.

$$
\begin{array}{c}
R^1 \\
\diagdown \\
C=O \\
\diagup \\
R^2
\end{array}
+\; R^3{-}NH_2 \longrightarrow
\begin{array}{c}
R^1 \\
\diagdown \\
C=N \diagup\!^{R^3} \\
\diagup \\
R^2
\end{array}
+\; H_2O \qquad (17.11)
$$

An aldehyde or   A primary amine   An imine
a ketone

Imines undergo addition of one mole of dihydrogen in the presence of a catalyst such as palladium, platinum, or nickel to produce secondary amines (Eq. 17.12).

This process is analogous to the hydrogenation of alkenes to alkanes and of carbonyl compounds to alcohols.

$$\underset{\substack{R^2 \\ \text{An imine}}}{\overset{R^1}{C}}=\overset{R^3}{N} \; + \; H_2 \quad \xrightarrow{\text{Catalyst}} \quad \underset{\substack{R^2 \qquad \overset{|}{NR^3} \\ \qquad \overset{|}{H} \\ \text{A secondary amine}}}{\overset{R^1 \qquad H}{C}} \qquad\qquad (17.12)$$

Other functional groups, such as C=C, C≡C, N=O (as in NO$_2$), and C–X, undergo catalytic hydrogenation under the same conditions required to reduce imines. Hence, it may not be possible to reduce an imine selectively by catalytic hydrogenation in the presence of one or more of these functional groups. This problem may be readily overcome, however, by using a metal hydride, such as sodium borohydride, NaBH$_4$, to reduce the imine, as depicted in Equation 17.13. Sodium borohydride also reacts with methanol, which is frequently used as the solvent, but it reacts much faster with the imine. Thus, when performing small-scale reactions, it is generally more convenient to use an excess of NaBH$_4$ to allow for its reaction with the solvent than to use a less-reactive solvent in which this reagent is less soluble.

$$\underset{\substack{R^2 \\ \text{An imine}}}{\overset{R^1}{C}}=\overset{R^3}{N} \quad \xrightarrow[\text{(2) } H_2O]{\text{(1) NaBH}_4,\text{CH}_3\text{OH}} \quad \underset{\substack{R^2 \qquad \overset{|}{N}-R^3 \\ \qquad \overset{|}{H} \\ \text{A secondary amine}}}{\overset{R^1 \qquad H}{C}} \qquad\qquad (17.13)$$

In this section, you will prepare *N*-cinnamyl-*m*-nitroaniline (**20**) by a sequence beginning with the condensation of cinnamaldehyde (**16**) with *m*-nitroaniline (**17**), followed by reduction of the intermediate imine **18** with sodium borohydride, as shown in Equations 17.14–17.16. The formation of the imine is reversible, but the reaction is driven to completion by **azeotropic distillation**. Because cyclohexane and water form a **minimum-boiling azeotrope** (Sec. 4.4), the water generated by the condensation of **16** and **17** is continuously removed by distilling the cyclohexane-water azeotrope throughout the course of the reaction.

$$\underset{\substack{| \\ H \\ \textbf{16} \\ \text{Cinnamaldehyde}}}{C_6H_5CH=CH-\overset{}{C}=O} \; + \; \underset{\substack{ \\ NO_2 \\ \textbf{17} \\ \textit{m}\text{-Nitroaniline}}}{H_2N-} \quad \underset{\text{Cyclohexane}}{\rightleftharpoons} \quad \underset{\substack{| \\ H \\ \textbf{18} \\ \textit{N}\text{-Cinnamylidene-}\textit{m}\text{-nitroaniline}}}{C_6H_5CH=CH-\overset{}{C}=N-} \; + \; H_2O \qquad (17.14)$$

$$4 \; \underset{\substack{| \\ H \\ NO_2 \\ \textbf{18}}}{C_6H_5CH=CH-\overset{}{C}=N-} \; + \; NaBH_4 \quad \longrightarrow \quad \left(\underset{\substack{| \\ O_2N \\ \textbf{19}}}{C_6H_5CH=CH-\overset{H}{\underset{|}{CH}}-N-} \right)_4 B^- Na^+ \qquad (17.15)$$

$$\left(C_6H_5CH=CH-CH_2-N \underset{O_2N}{\underset{|}{\bigcirc}} \right)_4 B^- Na^+ + 3\ H_2O \longrightarrow 4\ C_6H_5CH=CH-CH_2-NH \underset{O_2N}{\underset{|}{\bigcirc}} + NaH_2BO_3 \qquad (17.16)$$

19

20

N-Cinnamyl-*m*-nitroaniline

The reaction of sodium borohydride with the imine **18** (Eq. 17.15) is analogous to the addition of sodium borohydride (Sec. 17.4) or a Grignard reagent (Secs. 19.1 and 19.3) to a carbonyl compound. A nucleophilic hydride ion (H:) is transferred from the borohydride anion (BH$_4^-$) to the electrophilic carbon atom of the carbon-nitrogen double bond, and the electron-deficient boron atom becomes attached to the nitrogen atom. All four hydrides of the borohydride anion may be transferred to the imine carbon to produce the organoborate anion **19**, which is subsequently decomposed with water to yield the secondary amine **20** as shown in Equation 17.16. In this experiment, you will have an opportunity to determine whether sodium borohydride selectively reduces the imine group in **18** but not the carbon-carbon double bond, the nitro group, or the benzene rings that are also present.

EXPERIMENTAL PROCEDURES

Formation and Reduction of N-Cinnamylidene-m-Nitroaniline

Discovery Experiment   **Purpose** To demonstrate the reductive alkylation of an amine by the sequence of forming and reducing an imine and to evaluate the selectivity of a hydride reducing agent.

SAFETY ALERT

1. **Wear safety glasses or goggles and suitable protective gloves while performing the experiment.**

2. **Avoid contact of methanolic solutions of sodium borohydride with your skin; they are *highly caustic*. If these solutions accidentally get on your skin, wash the affected area first with a 1% acetic acid solution and then with copious quantities of water.**

3. **Do *not* stopper flasks containing methanolic sodium borohydride; the solution slowly evolves dihydrogen gas, and a dangerous buildup of pressure could occur in a stoppered flask.**

4. **Cyclohexane is flammable; use *flameless* heating in this experiment.**

MINISCALE PROCEDURE

Preparation Refer to the online resources to answer Pre-Lab Exercises, access videos, and read the MSDSs for the chemicals used or produced in this procedure. Review Sections 2.9, 2.11, 2.13, 2.17, and 2.22.

Apparatus A 25-mL round-bottom flask, ice-water bath, apparatus for heating under reflux, simple distillation, magnetic stirring, vacuum filtration, and *flameless* heating.

Setting Up Place 0.6 g of *freshly distilled* cinnamaldehyde, 0.6 g of *m*-nitroaniline, 5 mL of cyclohexane, and a stirbar in the round-bottom flask. Assemble the apparatus for simple distillation.

Imine Formation Heat the reaction flask with stirring, and distill until most of the cyclohexane is removed; about 4 mL of distillate should be obtained in approximately 5–10 min. *Discontinue* heating when the distillation rate decreases. Add another 5 mL of cyclohexane, and resume the distillation. Stop heating when the rate of distillation again decreases, and remove the heat source.

Isolation and Characterization of the Imine (Optional) Take about 0.2 mL of the residual liquid from the stillpot with a Pasteur pipet. Transfer this aliquot to a small test tube, and add 1 mL of methanol. Swirl the tube to effect solution and place the test tube in an ice-water bath. Collect any crystals that separate, air-dry them, and determine their melting point. The crude *N*-cinnamylidene-*m*-nitroaniline may be purified by recrystallization from a small volume of methanol or some other suitable solvent. Pure *N*-cinnamylidene-*m*-nitroaniline has a melting point of 92–93 °C.

Reduction Add 4 mL of methanol to the stillpot, stir to dissolve the crude imine, and set up the apparatus for heating under reflux. Prepare a solution of 0.15 g of sodium borohydride in 3 mL of methanol. Because sodium borohydride reacts slowly with methanol to evolve dihydrogen gas, this solution should be prepared in an *unstoppered* vessel *immediately* before use. Using a Pasteur pipet or a syringe, transfer the methanolic solution of sodium borohydride dropwise through the top of the reflux condenser to the stirred solution of the imine; the addition should be completed within 1 min. Try to add the solution *directly* into the flask without touching the walls of the condenser. After all of the borohydride solution is added, heat the reaction mixture under reflux for 15 min.

Work-Up, Isolation, and Purification Cool the reaction mixture to room temperature and add 10 mL of water.★ Stir the mixture for 10–15 min. Collect the crystals that separate, and wash them with water.★ Recrystallize the crude *N*-cinnamyl-*m*-nitroaniline from 95% ethanol or some other solvent. Air-dry the product.

Analysis Weigh the product and calculate the percent yield; determine its melting point. The melting point of pure *N*-cinnamyl-*m*-nitroaniline is 106–107 °C. Perform the qualitative test for unsaturation using the bromine test (Sec. 4.7A1) on your product, and record your observations. Obtain IR and ^{1}H NMR spectra of your starting materials, the intermediate imine if isolated, and the product, and compare them with those of authentic samples (Figs. 17.25–17.32).

MICROSCALE PROCEDURE

Preparation Refer to the online resources to answer Pre-Lab Exercises, access videos, and read the MSDSs for the chemicals used or produced in this procedure. Review Sections 2.9, 2.11, 2.13, 2.17, and 2.22.

Apparatus A 5-mL conical vial, 1-mL syringe, apparatus for heating under reflux, simple distillation, vacuum filtration, magnetic stirring, Craig tube filtration, and *flameless* heating.

Setting Up Place 120 mg of *freshly distilled* cinnamaldehyde, 120 mg of *m*-nitroaniline, 1 mL of cyclohexane, and a spinvane in the conical vial. Assemble the apparatus for simple distillation.

Imine Formation Heat the reaction vial with stirring and distill until most of the cyclohexane is removed; this should require about 3–5 min. *Discontinue* heating when the distillation rate decreases. Add another 1 mL of cyclohexane, and resume the distillation; stop heating when the rate of distillation again decreases, and remove the heat source.

Reduction Add 0.8 mL of methanol to the conical vial, stir to dissolve the crude imine, and set up the apparatus for heating under reflux. Prepare a solution of 30 mg of sodium borohydride in 0.6 mL of methanol. Because sodium borohydride slowly reacts with methanol to evolve dihydrogen gas, this solution should be prepared in an *unstoppered* vessel *immediately* before use. Using a Pasteur pipet or a syringe, transfer the methanolic solution of sodium borohydride *dropwise* through the top of the reflux condenser to the stirred solution of the imine; the addition should be completed within 1 min. Try to add the solution *directly* into the vial without touching the walls of the condenser. After all of the borohydride solution is added, heat the reaction mixture under reflux for 15 min.

Work-Up, Isolation, and Purification Cool the reaction mixture to room temperature and add 2 mL of water.★ Stir the mixture for 10–15 min. Collect the crystals that separate, and wash them with water.★ Recrystallize the crude *N*-cinnamyl-*m*-nitroaniline from 95% ethanol using a Craig tube. Air-dry the product.

Analysis Weigh the product and calculate the percent yield; determine its melting point. The melting point of pure *N*-cinnamyl-*m*-nitroaniline is 106–107 °C. Perform the qualitative test for unsaturation using the bromine test (Sec. 4.7A1) on your product, and record your observations. Obtain IR and ^{1}H NMR spectra of your starting materials and product, and compare them with those of authentic samples (Figs. 17.19–17.22, 17.25, and 17.26).

WRAPPING IT UP

Pour the *cyclohexane distillates* in the container for nonhalogenated organic solvents. Dilute the *aqueous methanolic filtrate* with water, neutralize it with acetic acid to destroy any excess sodium borohydride, and flush the mixture down the drain with excess water. Place the *dichloromethane solution* from the bromine test for unsaturation in a container for halogenated organic liquids.

EXERCISES

1. What causes the turbidity in the distillate collected while heating the carbonyl compound and primary amine in cyclohexane solution?

2. Why was it not necessary to use *dry* cyclohexane?

3. How does the fact that water forms a minimum-boiling azeotrope with cyclohexane allow water to be removed during the course of the reaction to produce **18**?

4. Explain how Le Chatelier's principle is being applied in this experiment to give **18**.

5. Draw a Lewis dot structure for $NaBH_4$, and indicate the direction of the B–H bond dipoles.

6. Determine the molar ratio of $NaBH_4$ to imine that you actually used in the experiment. Why is it necessary to use a greater molar ratio than theoretical?

7. After the reaction between sodium borohydride and the imine is complete, the reaction mixture is treated with water to produce the desired secondary amine. Explain this reaction by indicating the source of the hydrogen atom that ends up on the nitrogen atom.

8. Although sodium borohydride is fairly unreactive toward methanol, adding a mineral acid to this solution results in the rapid destruction of the $NaBH_4$. Explain this observation.

9. Suppose that each of the following pairs of compounds were subjected to the reactions provided in the experimental procedures. Draw the structures of the intermediate imines that would be formed, as well as those of the final amine product.

   a.   cinnamaldehyde and aniline

   b.   benzaldehyde and *m*-nitroaniline

   c.   benzaldehyde and aniline

   d.   cyclohexanone and aniline

10. Propose a mechanism to show how imine **18** can be hydrolyzed by water at pH 6 to produce **16** and **17**. Symbolize the flow of electrons with curved arrows.

11. Explain why aliphatic imines are much less stable than those having at least one aromatic group attached to the carbon-nitrogen double bond.

12. Write the structure(s) of the product(s) that you would expect from reducing *N*-cinnamylidene-*m*-nitroaniline by *catalytic* hydrogenation.

13. A nitro group, a carbon-carbon double bond, a carbon-nitrogen double bond, and an aromatic ring may each be reduced under certain conditions. Rank these functional groups in the order of *increasing* reactivity toward hydride reducing agents such as sodium borohydride.

14. Why do you think an imine bearing an aryl group on the nitrogen atom is more stable than one bearing an alkyl group?

15. The optimal pH for condensing an amine with a carbonyl group to give an imine according to Equation 17.11 is about 5–6.

   a.   Using curved arrows to symbolize the flow of electrons, provide a stepwise mechanism showing all proton transfers for this reaction.

b. In the mechanism you proposed in Part **a**, what step(s) would be slower at *higher* pH? Explain your reasoning.

c. In the mechanism you proposed in Part **a**, what steps(s) would be slower at *lower* pH? Explain your reasoning.

16. What fundamental characteristic of alkenes makes them unreactive toward hydride reagents?

Refer to the online resources and use the spectra viewer and Tables 8.1–8.8 as needed to answer the blue-numbered questions on spectroscopy.

17. Consider the spectral data for cinnamaldehyde (Figs. 17.25 and 17.26).

a. In the functional group region of the IR spectrum, specify the absorptions associated with the α, β-unsaturated aldehyde and with the aromatic ring.

b. In the ^{1}H NMR spectrum, assign the various resonances to the hydrogen nuclei responsible for them.

c. For the ^{13}C NMR data, assign the various resonances to the carbon nuclei responsible for them.

18. Consider the spectral data for *m*-nitroaniline (Figs. 17.27 and 17.28).

a. In the functional group region of the IR spectrum, specify the absorptions associated with the nitro group and the aromatic ring. There is a sharp absorption at about 3400 cm^{-1}. What functional group is responsible for this absorption?

b. In the ^{1}H NMR spectrum, assign the various resonances to the hydrogen nuclei responsible for them.

c. For the ^{13}C NMR data, assign the various resonances to the carbon nuclei responsible for them.

19. Consider the spectral data for *N*-cinnamylidene-*m*-nitroaniline (Figs. 17.29 and 17.30).

a. In the functional group region of the IR spectrum, specify the absorptions associated with the nitro group, the carbon-nitrogen double bond, and the carbon-carbon π-bonds of the aromatic ring and alkene.

b. In the ^{1}H NMR spectrum, assign the various resonances to the hydrogen nuclei responsible for them.

c. For the ^{13}C NMR data, assign the various resonances to the carbon nuclei responsible for them.

20. Consider the spectral data for *N*-cinnamyl-*m*-nitroaniline (Figs. 17.31 and 17.32).

a. In the functional group region of the IR spectrum, specify the absorptions associated with the amino group, the nitro group, and the aromatic ring.

b. In the ^{1}H NMR spectrum, assign the various resonances to the hydrogen nuclei responsible for them.

c. For the ^{13}C NMR data, assign the various resonances to the carbon nuclei responsible for them.

21. Discuss the differences observed in the IR and NMR spectra of cinnamaldehyde and *N*-cinnamylidene-*m*-nitroaniline that are consistent with the formation of the carbon-nitrogen double bond of the imine in this experiment.

22. Discuss the differences observed in the IR and NMR spectra of *N*-cinnamylidene-*m*-nitroaniline and *N*-cinnamyl-*m*-nitroaniline that are consistent with the *selective* reduction of the carbon-nitrogen double bond in this experiment.

Starting Materials and Products

Figure 17.25
IR spectrum of cinnamaldehyde (neat).

Figure 17.26
NMR data for cinnamaldehyde (CDCl₃).

(a) ¹H NMR spectrum (300 MHz).
(b) ¹³C NMR data: δ 128.5, 129.0, 131.1, 134.1, 152.3, 193.2.

Figure 17.27
IR spectrum of m-*nitroaniline (IR card).*

Figure 17.28

NMR data for m-*nitroaniline (CDCl₃).*

(a) ¹H NMR spectrum (300 MHz).
(b) ¹³C NMR data: δ 107.8, 110.5, 120.2, 129.6, 149.1, 149.8.

Figure 17.29

IR spectrum of N-*cinnamylidene-*
m-nitroaniline (IR card).

Figure 17.30

NMR data for N-*cinnamylidene-*
m-nitroaniline (CDCl₃).

(a) ¹H NMR spectrum (300 MHz).
(b) ¹³C NMR data: δ 114.8, 120.1, 127.4, 128.6, 129.5, 129.8, 129.9, 130.2, 134.8, 145.7, 148.6, 152.6, 163.4.

Figure 17.31

IR spectrum of N-*cinnamyl-m-nitroaniline (IR card).*

Figure 17.32

NMR data for N-*cinnamyl-m-nitroaniline (CDCl$_3$).*

(a) ^{1}H NMR spectrum (300 MHz).

(b) ^{13}C NMR data: δ 45.7, 106.5, 111.9, 118.8, 125.4, 126.3, 127.7, 128.6, 129.6, 132.1, 136.4, 148.7, 149.3.

17.4 REDUCTION OF CARBONYL COMPOUNDS; PREPARATION OF ALCOHOLS

Carbonyl compounds are commonly reduced to alcohols by catalytic hydrogenation or with metal hydrides. When applied to aldehydes, reduction, represented by the symbol [H], provides a convenient route to primary alcohols (Eq. 17.17), whereas the reduction of ketones gives secondary alcohols (Eq. 17.18). Although catalytic hydrogenation of carbonyl groups is frequently the method of choice in industrial processes, lithium aluminum hydride, sodium borohydride, and their derivatives are generally used in the research laboratory. Sodium borohydride may be used in alcoholic and even aqueous solutions, because it reacts much more rapidly with the carbonyl group than with the solvent. On the other hand, lithium aluminum hydride reacts rapidly with protic solvents, so it *must* be used in *anhydrous* ethereal solvents such as diethyl ether or tetrahydrofuran.

$$\underset{\substack{\text{An aldehyde}\\ \text{R = alkyl or aryl}}}{\overset{\displaystyle O}{\underset{R}{\overset{\parallel}{\underset{}{C}}}}\diagdown H} \xrightarrow{\text{[H]}} \underset{\text{A primary alcohol}}{R\!-\!CH_2OH} \qquad (17.17)$$

$$\underset{\substack{\text{A ketone}\\ \text{R}^1 \text{ and } \text{R}^2 = \text{alkyl or aryl}}}{\overset{\displaystyle O}{\underset{R^1}{\overset{\parallel}{\underset{}{C}}}}\diagdown R^2} \xrightarrow{\text{[H]}} \underset{\substack{\text{A secondary alcohol}}}{\overset{R^1\!-\!CHOH}{\underset{\displaystyle R^2}{|}}} \qquad (17.18)$$

In this experiment, you will examine the reduction of 9-fluorenone (**21**) using sodium borohydride to give 9-fluorenol (**23**), as shown in Equation 17.19. This reaction is mechanistically analogous to the reduction of imines with sodium borohydride (Sec. 17.3) and involves the transfer of hydride ion (H:$^-$) from borohydride ion, BH_4^-, to the electrophilic carbonyl carbon with concomitant transfer of the electron-deficient boron atom to the carbonyl oxygen. Theoretically, all four of the hydrogen atoms attached to boron may be transferred in this way to produce the intermediate borate salt **22**, which is decomposed upon addition of water and acid to yield 9-fluorenol (**23**).

(17.19)

21
9-Fluorenone
(*yellow*)

22

23
9-Fluorenol
(*colorless*)

EXPERIMENTAL PROCEDURES

Reduction of 9-Fluorenone

Discovery Experiment   **Purpose** To demonstrate the reduction of a ketone to an alcohol using sodium borohydride.

SAFETY ALERT

1. **Wear safety glasses or goggles and suitable protective gloves while performing the experiment.**

2. **Avoid contact of methanolic solutions of sodium borohydride with your skin; they are *highly caustic*. If these solutions accidentally get on your skin, wash the affected area first with a 1% acetic acid solution and then with copious quantities of water.**

3. **Do *not* stopper flasks containing methanolic sodium borohydride; the solution slowly evolves dihydrogen gas, and a dangerous buildup of pressure could occur in a stoppered flask.**

MINISCALE PROCEDURE

Preparation Refer to the online resources to answer Pre-Lab Exercises, access videos, and read the MSDSs for the chemicals used or produced in this procedure. Review Sections 2.9 and 2.17.

Apparatus A 25-mL Erlenmeyer flask, ice-water bath, apparatus for vacuum filtration and *flameless* heating.

Setting Up Add 0.6 g of 9-fluorenone to the Erlenmeyer flask containing 6 mL of methanol, and swirl the flask with slight warming to dissolve the ketone. Allow the solution to cool to room temperature. *Quickly* weigh 0.05 g of sodium borohydride into a dry test tube, and stopper the test tube *immediately* to avoid undue exposure of the hygroscopic reagent to atmospheric moisture.

Reduction Add the sodium borohydride in *one* portion to the solution of 9-fluorenone in methanol, and swirl the mixture vigorously to dissolve the reagent. After all of the sodium borohydride dissolves, allow the solution to stand at room temperature for 20 min; swirl the solution occasionally. If the solution does not become colorless during this time, add an additional small portion of sodium borohydride with swirling to complete the reaction.★

Work-Up Add 2 mL of 3 *M* sulfuric acid to the reaction mixture. Heat the contents in the flask gently and intermittently for 5–10 min. Stir the mixture occasionally with a glass rod to help dissolve the solid; maintain the internal temperature just below the reflux point to *minimize the loss of solvent.* If all of the solids do not dissolve, *gradually* add methanol in about 0.5- to 1-mL portions with continued heating until a solution is obtained. When all of the precipitated solids redissolve, allow the solution to cool to room temperature, and then cool in an ice-water bath for 10–15 min.

Isolation and Purification Collect the solid product by vacuum filtration, *wash it thoroughly* with water until the filtrate is *neutral,* and air-dry the product.★ Recrystallize the 9-fluorenol by a mixed solvent recrystallization (Sec. 3.2) using methanol and water or some other suitable solvent. After cooling the solution first to room temperature and then in an ice-water bath for 10–15 min, collect and air-dry the crystals.

Analysis Weigh the product and calculate the percent yield; determine its melting point. Obtain IR and ^{1}H NMR spectra of your starting material and product, and compare them with those of authentic samples (Figs. 17.33–17.36).

MICROSCALE PROCEDURE

Preparation Refer to the online resources to answer Pre-Lab Exercises, access videos, and read the MSDSs for the chemicals used or produced in this procedure. Review Sections 2.9 and 2.17.

Apparatus A 5-mL conical vial, ice-water bath, apparatus for magnetic stirring, heating under reflux, vacuum filtration, Craig tube filtration, and *flameless* heating.

Setting Up Add 100 mg of 9-fluorenone to the conical vial containing a spinvane and 1 mL of methanol, and stir the contents of the vial with slight warming to dissolve the ketone. Allow the solution to cool to room temperature. *Quickly* weigh 10 mg of sodium borohydride into a dry test tube, and stopper the test tube *immediately* to avoid undue exposure of the hygroscopic reagent to atmospheric moisture.

Reduction Add the sodium borohydride in *one* portion to the stirred solution of 9-fluorenone in methanol, and then stir the mixture rapidly to dissolve the reagent. After all of the sodium borohydride dissolves, continue stirring at room temperature for 20 min. If the solution does not become colorless during this time, add an additional small portion of sodium borohydride to complete the reaction.★

Work-Up Add 0.35 mL of 3 *M* sulfuric acid to the reaction mixture. Set up the apparatus for heating under reflux, and heat the vial gently with stirring for 5–10 min to dissolve the solid; adjust the internal temperature to maintain a *gentle* reflux. If all of the solids do not dissolve, *gradually* add methanol *dropwise* to the vial with continued heating and stirring until a solution is obtained. When all of the precipitated solids redissolve, allow the solution to cool to room temperature, and then cool it in an ice-water bath for 10–15 min.

Isolation and Purification Collect the solid product by vacuum filtration, *wash it thoroughly* with water until the filtrate is *neutral,* and air-dry the product.★ Recrystallize the 9-fluorenol by a mixed solvent recrystallization (Sec. 3.2) using methanol and water or some other suitable solvent using a Craig tube. After cooling the solution first to room temperature and then in an ice-water bath for 10–15 min, collect and air-dry the crystals.

Analysis Weigh the product and calculate the percent yield; determine its melting point. Obtain IR and ^{1}H NMR spectra of your starting material and product, and compare them with those of authentic samples (Figs. 17.33–17.36).

Discovery Experiment   *Reduction of 4-tert-Butylcyclohexanone*

Consult with your instructor before performing this experiment, in which you will determine the stereoselectivity of the reduction of 4-*tert*-butylcyclohexanone with sodium borohydride. You might also be asked to use other hydride reducing agents and compare their stereoselectivities with that of sodium borohydride. Follow either the Miniscale or the Microscale Procedure for the reduction of 9-fluorenone to reduce 4-*tert*-butylcyclohexanone using sodium borohydride. Modify the work-up by using diethyl ether to extract the aqueous mixture obtained after acidification. Dry the extracts over *anhydrous* sodium sulfate, and remove the solvent. Obtain the ^{1}H NMR spectrum of your product and compare it with the ^{1}H NMR spectra of authentic samples of *cis*- and *trans*-4-*tert*-butylcyclohexanol. Determine the ratio of *cis*- and *trans*-4-*tert*-butylcyclohexanols in your product, using the integrations of the multiplets for the methine protons at 4.0 and 3.5 ppm, respectively, on the carbon atoms bearing the hydroxyl groups. Perform a GLC analysis of your reaction mixture to determine the ratio of *cis*- and *trans*-4-*tert*-butylcyclohexanols in your product, and compare the results of these two determinations.

Discovery Experiment   *Reduction of Benzoin*

Consult with your instructor before performing this experiment, in which you will determine the stereoselectivity of the reduction of benzoin with sodium borohydride. Follow either the Miniscale or Microscale Procedure described for reducing 9-fluorenone to reduce benzoin with sodium borohydride, but use ethanol rather than methanol as the solvent. After slowly adding 3 *M* HCl to decompose the excess borohydride, add enough water to adjust the solvent composition to 50% (*v:v*) ethanol and water. You may recrystallize the crude product from 50% (*v:v*) ethanol and water. Obtain the melting point and the IR, ^{1}H, and ^{13}C NMR spectra of the purified product for characterization. Compare these data with those for racemic and *meso*-hydrobenzoin to determine the identity of the product and the stereochemistry of the reduction. If authentic samples of racemic and *meso*-hydrobenzoin are available, determine mixed melting points to support your assignment.

WRAPPING IT UP

Dilute the combined *aqueous methanol filtrates* with water, neutralize the resulting solution with sodium carbonate, and flush the mixture down the drain with excess water.

EXERCISES

1. Determine the molar ratio of sodium borohydride to 9-fluorenone that you used in the experiment. Why is it necessary to use a greater molar ratio than theoretical?

2. After the reaction between sodium borohydride and the ketone is complete, the reaction mixture is treated with water and acid to produce the desired secondary alcohol. Explain this reaction by indicating the source of the hydrogen atom that ends up on the oxygen atom.

3. Sodium borohydride is fairly unreactive toward methanol, but adding a mineral acid to this solution results in the rapid destruction of the sodium borohydride. Explain.

4. How many molar equivalents of hydride does sodium borohydride contain?

5. Using curved arrows to symbolize the flow of electrons, write the mechanism for the steps involved in the conversion of 9-fluorenone to 9-fluorenol with sodium borohydride followed by aqueous acid according to Equation 17.19.

6. Suggest a structure for the white precipitate formed in the reaction of 9-fluorenone with sodium borohydride.

7. What gas is evolved when sulfuric acid is added to the reaction mixture?

8. 9-Fluorenone is colored, but 9-fluorenol is not. What accounts for this difference?

9. Draw the structure of the product that results from complete reduction of the following compounds by sodium borohydride.

   a. cyclohexanone   c. 1, 4-butanediol   e. acetophenone

   b. 3-cyclohexen-1-one   d. 4-oxohexanal

10. Draw the structure of the product that would be formed from allowing each of the compounds in Exercise 9 to react with excess dihydrogen gas in the presence of a nickel catalyst.

11. A nitro group, a carbon-carbon double bond, a carbon-nitrogen double bond, and an aromatic ring may each be reduced under certain conditions. Rank these functional groups in order of *increasing* reactivity toward hydride reducing agents such as sodium borohydride.

12. Esters are normally unreactive toward sodium borohydride, but they react readily with lithium aluminum hydride to produce alcohols.

    a. Propose a rationale for the lack of reactivity of esters toward sodium borohydride.

    b. Propose a rationale for the greater reactivity of lithium aluminum hydride over that of sodium borohydride.

    c. Write the structures of the organic products that will be formed when methyl benzoate is treated with lithium aluminum hydride and the reaction mixture is worked up with aqueous acid.

Methyl benzoate

13. Consider the spectral data for 9-fluorenone (Figs. 17.33 and 17.34).

    a. In the functional group region of the IR spectrum, specify the absorptions associated with the carbonyl group and with the aromatic rings.

    b. In the ^{1}H NMR spectrum, assign the various resonances to the hydrogen nuclei responsible for them.

    c. For the ^{13}C NMR data, assign the various resonances to the carbon nuclei responsible for them.

14. Consider the spectral data for 9-fluorenol (Figs. 17.35 and 17.36).

    a. In the functional group region of the IR spectrum, specify the functional group that is associated with the broad absorption at about 3100–3250 cm^{-1}; why is this absorption broad?

    b. In the ^{1}H NMR spectrum, assign the various resonances to the hydrogen nuclei responsible for them.

    c. For the ^{13}C NMR data, assign the various resonances to the carbon nuclei responsible for them.

15. Discuss the differences observed in the IR and NMR spectra of 9-fluorenone and 9-fluorenol that are consistent with the reduction of the ketone functional group in this experiment.

Refer to the online resources and use the spectra viewer and Tables 8.1–8.8 as needed to answer the blue-numbered questions on spectroscopy.

Starting Materials and Products

Figure 17.33
*IR spectrum of 9-fluorenone
(IR card).*

Figure 17.34
*NMR data for 9-fluorenone
(CDCl₃).*

(a) ¹H NMR spectrum (300 MHz).
(b) ¹³C NMR data: δ 120.1, 123.8, 128.8, 133.9, 134.4, 144.1, 193.1.

Figure 17.35
*IR spectrum of 9-fluorenol
(IR card).*

Figure 17.36
NMR data for 9-fluorenol (CDCl₃).

(a) ¹H NMR spectrum (300 MHz).
(b) ¹³C NMR data: δ 73.8, 119.6, 125.0, 127.2, 128.2, 139.5, 146.8.

17.5 ENANTIOSELECTIVE REDUCTIONS: A CHIRAL ALCOHOL FROM A KETONE

In modern synthetic organic chemistry, it is often necessary to use optically pure starting materials to prepare a target molecule, such as a drug, in enantiomerically pure form. However, reduction by catalytic hydrogenation or with hydride reducing agents such as sodium borohydride or lithium aluminum hydride of a **prochiral** ketone **24** that bears two *different* alkyl or aryl residues on the carbonyl carbon atom gives equal amounts of the chiral alcohol **25** and its enantiomer *ent*-25 (Eq. 17.20). This **racemate** is produced because these *achiral* reducing reagents attack *both* faces *a* and *b* of the planar prochiral carbonyl function with *equal* probability.

$$\tag{17.20}$$

However, if the reducing agent is chiral, it is possible to reduce a prochiral ketone to give a chiral alcohol in enantiomerically pure or enriched form. For example, the reduction of such ketones by catalytic hydrogenation in the presence of chiral catalysts or by the use of chiral hydride reducing agents may produce secondary alcohols with high levels of **enantiomeric purity**. There are also certain enzymes, which are polypeptides composed of l-amino acids and thus have chiral active sites, that can perform such **enantioselective** reductions. Reactions resulting in preferential formation of one enantiomer are examples of **asymmetric synthesis**.

When a chiral product is formed from an achiral starting material, the **optical yield or optical purity**, not just the chemical yield, of the product is important.

Optical purity is commonly evaluated by calculating the **enantiomeric excess (ee)**. For example, reduction of a prochiral ketone **24** with sodium borohydride produces 50% (*R*)- and 50% (*S*)-alcohol **25** and **ent-25**; the ee for this process is 0%. If 90% of the (*S*)- and 10% of the (*R*)-isomer were produced by the reaction, the ee would be 80%.

In the first of the two experimental procedures that follow, you will perform an enantioselective reduction by reaction of methyl acetoacetate (**26**) with sodium borohydride in the presence of (2*R*,3*R*)-(+)-tartaric acid (**27**) to produce methyl (*R*)-(−)-3-hydroxybutanoate (**28**) preferentially (Eq. 17.21). Alternatively, you may use (2*S*,3*S*)-(−)-tartaric acid (**29**), the enantiomer of **27**, to produce methyl (*S*)-(+)-3-hydroxybutanoate (**30**), the enantiomer of **28**, preferentially (Eq. 17.22). The active reducing agent in both cases is presumably either the chiral complex **31** or **32**, which result from the reaction between **27** or **29**, respectively, and the borohydride anion, which is achiral (Eq. 17.23). The enantiomeric excess that results from this reduction can be determined in several ways, as described in the experimental procedure. Note that your instructor may divide the class into two groups, so you and your fellow students can compare the stereochemical outcome of the two reactions.

$$(17.21)$$

26
Methyl acetoacetate

28
Methyl (*R*)-(−)-3-hydroxybutanoate

$$(17.22)$$

26
Methyl acetoacetate

30
Methyl (*S*)-(+)-3-hydroxybutanoate

27 or **29** + NaBH$_4$ $\longrightarrow$ ... + H$_2$ (17.23)

31   **32**

The hydroxyesters **28** and **30** are key building blocks and may be used as starting materials in a variety of syntheses of optically pure products having useful biological activities. These include Trusopt (**33**), an anti-glaucoma medication, and **34**, which is a precursor to the carbopenams, a family of antibiotics that includes penicillin G (**35**).

The importance of being able to produce pure enantiomers of drugs is discussed in the Historical Highlight *Chiral Drugs*, which is available online.

33

Trusopt
(Dorzpolamide)

34

35

Penicillin G

In the second experimental procedure provided in this section, methyl aceto-acetate (**26**) is reduced to methyl (S)-(+)-3-hydroxybutanoate (**30**) using one of the reducing enzymes found in baker's yeast (Eq. 17.24). The enzyme is one of many that are involved in the metabolism of D-glucose to ethanol. Enantiomeric excesses ranging from 70% to 97% have been reported for this reaction, with higher optical purities being reported when oxygen is excluded during the fermentation. Such fermentations are termed **anaerobic**; an aerobic process is described in Section 16.2. The hydroxy ester **30** is a building block that has been used as a starting material in a number of syntheses of optically pure natural products having useful biological activities.

$$\underset{\textbf{26}}{\ce{CH3COCH2CO2CH3}} \xrightarrow[\text{sucrose}]{\text{baker's yeast}} \underset{\textbf{30}}{\ce{CH3CH(OH)CH2CO2CH3}} \qquad (17.24)$$

26
Methyl acetoacetate

30
Methyl (S)-(+)-3-hydroxybutanoate

EXPERIMENTAL PROCEDURES

A. Tartaric Acid-Mediated Enantioselective Reduction of Methyl Acetoacetate

Discovery Experiment   **Purpose** To demonstrate the enantioselective reduction of a ketone using chiral reducing agents.

SAFETY ALERT

1. **Wear safety glasses or goggles and suitable protective gloves while performing the experiment.**

2. **Tetrahydrofuran is a flammable and potentially toxic solvent. Do not use this solvent near open flames.**

3. Avoid contact of solutions of sodium borohydride with your skin; they are highly caustic. If any of these solutions accidentally get on your skin, wash the affected area first with a 1% acetic acid solution and then with copious amounts of water.

4. Do not stopper flasks containing solutions of sodium borohydride; the solution slowly evolves dihydrogen gas, and a dangerous buildup of pressure could occur in a stoppered flask.

5. Diethyl ether is *extremely* flammable and volatile, and its vapors can easily travel several feet along the bench top or the floor and then be ignited. Consequently, *be certain there are no open flames anywhere in the laboratory* whenever you are working with ether. Use a *flameless* heating source whenever heating is required.

MINISCALE PROCEDURE

Preparation Refer to the online resources to answer Pre-Lab Exercises, access videos, and read the MSDSs for the chemicals used or produced in this procedure. Review Sections 2.5, 2.9, 2.10, 2.11, 2.17, 2.21, 2.25, 2.29, 6.4, and 7.5.

Apparatus A 100-mL round-bottom flask, ice-water bath separatory funnel, 1-mL syringe with needle, apparatus for magnetic stirring, simple distillation, and *flameless* heating.

Setting Up Add 0.5 g of sodium borohydride and 2.0 g of the enantiomer of tartaric acid that is assigned to you by your laboratory instructor to the 100-mL round-bottom flask containing 30 mL of tetrahydrofuran and a stirbar. Stir the mixture vigorously for 15 min.

Reaction Cool the flask in an ice-water bath and using the 1-mL syringe, transfer 0.40 mL of methyl acetoacetate to the mixture with stirring. Stir the mixture for 2 h in the ice-water bath. Using a Pasteur pipet, add 10 mL of 1 *M* hydrochloric acid *dropwise* to the mixture with stirring (*Caution:* vigorous foaming!). After this addition is complete, remove the flask from the ice-water bath and continue stirring the suspension for 10 min.

Work-Up and Isolation Transfer the reaction mixture to a separatory funnel, and add 25 mL of diethyl ether. Shake the funnel vigorously with periodic venting to relieve pressure. Separate the layers, and wash the ethereal solution sequentially with one 20-mL portion of saturated *aqueous* sodium bicarbonate (*Caution:* evolution of gas!) and one 20-mL portion of saturated *aqueous* sodium chloride. Dry the ethereal solution over several spatula-tips full of *anhydrous* sodium sulfate with occasional swirling for 10–15 min.* Add additional portions of *anhydrous* sodium sulfate if the liquid remains cloudy.

Filter or carefully decant the dried organic solution into a dry 100-mL round-bottom flask equipped for simple distillation and concentrate the solution by distillation (*Foaming!*). Alternatively, use rotary evaporation or other techniques to concentrate the solution. Remove the last traces of solvent by attaching the flask to a vacuum source and gently swirling the contents as the vacuum is applied.★

Analysis Weigh the product and determine the percent yield of crude methyl 3-hydroxybutanoates. Obtain IR and ^{1}H NMR spectra of your starting material and product and compare them with those of authentic samples (Figs. 17.37–17.40). If starting material is present, calculate the ratio of starting material and product in the mixture.

The enantioselectivity of the reaction can be determined in several different ways. One option is to use the procedure outlined in Section 17.6. A second approach is first to prepare a solution of the crude reaction material in chloroform at a concentration of 20 mg/mL, analysis of which by ^{1}H NMR spectrum will allow you to calculate the concentration of product in your sample. Knowing the concentration and using a polarimeter (Sec. 7.5) to measure the optical rotation of your solution provides the data necessary for calculating the enantiomeric excess of the product. A third option is to determine the enantiomeric excess by analyzing your product by chiral gas chromatography using a Chiraldex capillary column (Sec. 6.4), if it is available.

WRAPPING IT UP

Dilute the combined *aqueous layers* with water, neutralize the resulting solution with sodium carbonate, and flush the mixture down the drain with excess water. After the diethyl ether evaporates from the *sodium sulfate* in the hood, place the solid in the container for nonhazardous solids. Place the *ether distillate* in the container for nonhalogenated organic solvents. Put the dried TLC plates in the container for solid waste.

SPECTRA

Starting Materials and Products

Figure 17.37
IR spectrum of methyl acetoacetate (ATR).

Figure 17.38

NMR data for methyl acetoacetate (CDCl₃).

(a) ¹H NMR spectrum (300 MHz).
(b) ¹³C NMR data: δ.

Figure 17.39

IR spectrum of methyl (S)-(+)-3-hydroxybutanoate (ATR).

Figure 17.40

NMR data for methyl (S)-(+)-3-hydroxybutanoate (CDCl₃).

(a) ¹H NMR spectrum (300 MHz).
(b) ¹³C NMR data: δ.

B. Enzymatic Reduction of Methyl Acetoacetate

Purpose To demonstrate the enantioselective reduction of a ketone with an enzyme.

SAFETY ALERT

Wear safety glasses or goggles and suitable protective gloves while performing the experiment.

MINISCALE PROCEDURE

Preparation Refer to the online resources to answer Pre-Lab Exercises, access videos, and read the MSDSs for the chemicals used or produced in this procedure. Review Sections 2.11, 2.17, 2.21, and 2.29.

Apparatus A 250-mL Erlenmeyer flask, separatory funnel, apparatus for magnetic stirring, anaerobic fermentation, vacuum filtration, simple distillation, and *flameless* heating.

Setting Up Dissolve 40 g of sucrose and 0.25 g of disodium hydrogen phosphate, which buffers the mixture to maintain an optimal pH, in 75 mL of warm (40 °C) tap water contained in an Erlenmeyer flask. Add a stirbar and one packet (8 g) of dry, *active* baker's yeast, and assemble the apparatus for anaerobic fermentation shown in Figure 17.41. The 3% barium hydroxide solution should be protected from atmospheric carbon dioxide by a layer of mineral oil. A precipitate of barium carbonate will form as carbon dioxide is produced during the course of the fermentation. Stir the mixture *vigorously* for about 1 h in a warm location to suspend all of the yeast and to initiate the fermentation.

Reduction Remove the stopper in the Erlenmeyer flask, and add 2.5 mL of methyl acetoacetate. Replace the stopper and continue to stir the contents of the flask *vigorously,* under anaerobic conditions, and in a warm place, ideally at 30–35 °C, for at least 48 h, although one week is better.★

Figure 17.41
Apparatus for the anaerobic fermentation of methyl acetoacetate.

Work-Up and Isolation Prepare a uniform bed of a filter-aid by filtering a slurry made from 10 g of filter-aid in 50 mL of water through a 10-cm Büchner funnel containing a piece of filter paper; do not allow the bed to dry completely. Discard the filtrate according to the procedure described in the Wrapping It Up section. Add another 10 g of filter-aid to the reaction mixture containing the yeast cells, and swirl the mixture thoroughly to mix the filter-aid with the cells. Allow the solids to settle for a few minutes. Then *carefully* decant the clear supernatant fluid through the bed of filter-aid while applying *gentle* suction to remove the suspended yeast cells. After filtering the supernatant liquid, pour the residual solids and liquid through the bed of filter-aid. Because the tiny yeast particles tend to clog the filter-bed, it will be necessary to increase the vacuum during the filtration.

Wash the filter-aid with 25 mL of water and saturate the combined filtrates with solid sodium chloride. Extract the resulting solution three times with 25-mL portions of dichloromethane by *gently* shaking the separatory funnel to mix the layers. Do *not* shake the funnel too vigorously, because a bad emulsion will form. If this occurs, pass the contents of the separatory funnel through a *fresh* bed of filter-aid by vacuum filtration; this filtration will remove most of the fine particles that cause the emulsion. Wash the bed with 15 mL of dichloromethane. Transfer the filtrate and washing to the separatory funnel, and continue the extractions; any fine emulsion that remains may be included with the aqueous phase. Dry the combined dichloromethane layers with several spatula-tips full of *anhydrous* sodium sulfate.★ Swirl the flask occasionally for a period of 10–15 min to facilitate drying; add further small portions of *anhydrous* sodium sulfate if the solution remains cloudy.

Decant the dry dichloromethane solution into a tared distilling flask, and remove most of the dichloromethane by simple distillation. Alternatively, use rotary evaporation or other techniques to concentrate the solution.★ Cool the stillpot to room temperature and remove the last traces of solvent by attaching the flask to a vacuum source and gently swirling the contents as the vacuum is applied.★

Analysis Weigh the product and determine the percent yield of crude methyl (*S*)-(+)-3-hydroxybutanoate. Use the ferric ion test (Sec. 17.5C) to ascertain whether starting ketoester remains; record the result. Also analyze the product by TLC using dichloromethane as the eluting solvent. If starting material is present, the optical yield of methyl (*S*)-(+)-3-hydroxybutanoate as measured by specific rotation will be lower. Obtain IR and ^{1}H NMR spectra of your starting material and product, and compare them with those of authentic samples (Figs. 17.37–17.40).

WRAPPING IT UP

Flush the *filtrate* from preparing the filter-aid bed down the drain. Dilute the *aqueous layer* with water, and flush it down the drain. Dry the *filter-aid and yeast cells*, and place them in the container for nonhazardous solids. After the dichloromethane evaporates from the *sodium sulfate* in the hood, place the solid in the container for nonhazardous solids. In the container for halogenated organic solvents, place the *recovered dichloromethane* that was used as the extraction solvent and as the eluting solvent for the TLC.

C. Qualitative Analysis

Ferric Chloride Test Most enols as well as phenols react with ferric chloride to give colored complexes. The colors vary, depending on the nature of the phenol or enol and also on the solvent, concentration, and time of observation. Some phenols that do not give coloration in aqueous or alcoholic solutions do so in chloroform solution, especially after addition of a drop of pyridine. The nature of the colored complexes is still uncertain, but they are likely ferric phenoxide salts.

EXPERIMENTAL PROCEDURE

Ferric Chloride Test for Phenols and Enols

SAFETY ALERT

Wear safety glasses or goggles and suitable protective gloves while performing the experiment.

Preparation Refer to the online resources to read the MSDSs for the chemicals used or produced in this procedure.

Apparatus Test tubes, calibrated Pasteur pipets.

Setting Up Dissolve 30–50 mg of the unknown compound in 1–2 mL of water, or a mixture of water and 95% ethanol if the compound is not water soluble.

Reaction Add several drops of a 0.2 M aqueous solution of ferric chloride. Enols give red, violet, or tan coloration.

WRAPPING IT UP

Flush all *test solutions* down the drain.

EXERCISES

General Questions

1. Define the term *enantioselective*.

2. For applying the rules for naming centers of chirality as S or R, which substituent in **28** or **30** has the highest priority? Which has the second highest?

3. The keto group of methyl acetoacetate (**26**) may also be reduced selectively with sodium borohydride. Describe how the product of this reaction would differ from the product of reduction of (**26**) with either sodium borohydride/tartaric acid or baker's yeast.

4. Even though an excess of reducing agent is used in this procedure, starting material may be recovered. Postulate why this might be the case.

Questions for Part A

5. The enantioselectivity in this reaction is not particularly good, but improved levels of enantioselectivity can be obtained if the reaction is performed at a lower temperature. Explain why this might be the case.

6. Sodium borohydride is relatively unreactive toward methanol, but adding a mineral acid to this solution results in the rapid destruction of the sodium borohydride. Explain why.

7. How many molar equivalents of hydride does sodium borohydride contain?

8. In the experiment, the chiral reducing agent is allowed to react with methyl acetoacetate, and hydrochloric acid is added at the end of the reaction. What is the purpose of adding hydrochloric acid at this point?

9. Draw the structure of the product resulting from reduction of the following compounds by sodium borohydride. If no reaction is expected, indicate this by "NR."

   a. cyclohexanone   **c.** 1,4-butanediol   **e.** acetophenone

   b. 3-cyclohexen-1-one   **d.** 4-oxohexanal

10. Esters are normally unreactive toward sodium borohydride, but they react readily with lithium aluminum hydride to produce alcohols.

    a. Propose a rationale for the lack of reactivity of esters toward sodium borohydride.

    b. Propose a rationale for the greater reactivity of lithium aluminum hydride over that of sodium borohydride.

    c. Write the structures of the organic products that will be formed when methyl acetoacetate is treated with lithium aluminum hydride and the reaction mixture is worked up with aqueous acid.

11. If the reduction of methyl acetoacetate with sodium borohydride were performed in the presence of *meso*-tartaric acid instead of one of the enantiomers of tartaric acid, what product would you expect to be formed?

Refer to the online resources and use the spectra viewer and Tables 8.1–8.8 as needed to answer the blue-numbered questions on spectroscopy.

12. Consider the spectral data for methyl acetoacetate (Figs. 17.37 and 17.38).

    a. In the functional group region of the IR spectrum, specify the absorptions associated with the two carbonyl groups. What is the source of the weak absorption at about 3700–3300 cm^{-1} and why is it broad?

    b. In the ^{1}H NMR spectrum, assign the various resonances to the hydrogen nuclei responsible for them.

    c. For the ^{13}C NMR data, assign the various resonances to the carbon nuclei responsible for them.

13. Consider the spectral data for methyl (*S*)-(+)-3-hydroxybutanoate (Figs. 17.39 and 17.40).

    a. In the functional group region of the IR spectrum, specify the absorptions associated with the hydroxyl and ester groups. Why is the absorption at about 3600–3200 cm^{-1} broad?

    b. In the ^{1}H NMR spectrum, assign the various resonances to the hydrogen nuclei responsible for them.

    c. For the ^{13}C NMR data, assign the various resonances to the carbon nuclei responsible for them.

14. Discuss the differences observed in the IR and NMR spectra of methyl acetoacetate and methyl (*S*)-(+)-3-hydroxybutanoate that are consistent with the reduction of the carbonyl group in this experiment.

15. Would any of your answers to Questions 12 and 13 been different if spectral data for methyl (*R*)-(−)-3-hydroxybutanoate had been provided? If so, why?

Questions for Part B

16. What nucleophile is supplied by the yeast?

17. What is the purpose of using sucrose in the baker's yeast reduction of methyl acetoacetate?

18. What gas is evolved during the reduction of methyl acetoacetate with baker's yeast?

19. What is an emulsion, and what substances present in the reaction mixture could be responsible for the formation of an emulsion during the work-up?

20. *Alcohol dehydrogenase* is one of the active enzymes in yeast. The active site in *alcohol dehydrogenase* contains a zinc ion, Zn^{2+}, that is coordinated to the sulfur atoms of two cysteine residues of the enzyme. The hydride reducing reagent in alcohol dehydrogenase is nicotinamide adenine dinucleotide, NADH, which transfers a hydride ion to a carbonyl compound to yield an alcohol and NAD^+, in a mechanism that is related to the Cannizzaro reaction (Sec. 16.3).

a. Explain how the zinc ion might assist the addition of hydride ion to a ketone such as methyl acetoacetate.

b. Using curved arrows to symbolize the flow of electrons, write the mechanism for the steps involved in the hydride transfer from NADH to a carbonyl compound. Your mechanism should show how NADH becomes NAD^+.

c. Describe what general structural features are characteristic of enzymes that enable them to catalyze enantioselective reactions.

21. Why is it necessary to saturate the aqueous mixture with sodium chloride before extracting the product with dichloromethane?

22. Based upon the stereochemistry of the reduction of **26** with baker's yeast, predict the stereochemistry of the alcohols obtained by reducing the ketones **a–c** with baker's yeast.

Refer to the online resources and use the spectra viewer and Tables 8.1–8.8 as needed to answer the blue-numbered questions on spectroscopy.

23. Consider the spectral data for methyl acetoacetate (Figs. 17.37 and 17.38).

a. In the functional group region of the IR spectrum, specify the absorptions associated with the two carbonyl groups. What is the source of the weak absorption at about 3700–3300 cm^{-1} and why is it broad?

b. In the ^1H NMR spectrum, assign the various resonances to the hydrogen nuclei responsible for them.

c. For the ^{13}C NMR data, assign the various resonances to the carbon nuclei responsible for them.

24. Consider the spectral data for methyl (*S*)-(+)-3-hydroxybutanoate (Figs. 17.39 and 17.40).

   a. In the functional group region of the IR spectrum, specify the absorptions associated with the hydroxyl and ester groups. Why is the absorption at about 3600–3200 cm^{-1} broad?

   b. In the ^{1}H NMR spectrum, assign the various resonances to the hydrogen nuclei responsible for them.

   c. For the ^{13}C NMR data, assign the various resonances to the carbon nuclei responsible for them.

25. Discuss the differences observed in the IR and NMR spectra of methyl acetoacetate and methyl (*S*)-(+)-3-hydroxybutanoate that are consistent with the reduction of the carbonyl group in this experiment.

26. Would any of your answers to Questions **23** and **24** been different if spectral data for methyl (*R*)-(−)-3-hydroxybutanoate had been provided? If so, why?

17.6 DETERMINING OPTICAL PURITY

There are several ways to determine the optical purity of a compound obtained as the product of an enantioselective synthesis. The simplest technique involves measuring the optical rotation using a polarimeter (Sec. 7.5). The reported specific rotation, $[\alpha]_D^{25}$, of pure methyl (*S*)-(−)-3-hydroxybutanoate is +38.5° (*c* = 1.80, chloroform), but the specific rotation of the product you obtain might range from +27° to +36°. For example, a measured specific rotation of 32.7° (*c* = 1.3, chloroform) for the product of this enzymatic reduction would correspond to an ee of 85%.

When only small quantities of a material are available, the use of ^{1}H NMR and a chiral shift reagent is a more accurate method for determining optical purity. In an *achiral* environment, the ^{1}H NMR spectra of methyl (*S*)-(+)-3-hydroxybutanoate and methyl (*R*)-(−)-3-hydroxybutanoate are *identical,* and each corresponds to that shown in Figure 17.40a. However, a Lewis acid having chiral ligands can form *diastereomeric complexes* with the two enantiomers, and these complexes often give different ^{1}H NMR spectra. In the present case, the Lewis acid is the chiral shift reagent *tris*[3-(heptafluoropropylhydroxymethylene)-(+)-camphorato]europium (III) or Eu(hfc)$_3$ (**36**); the europium atom coordinates with the hydroxyl group in methyl 3-hydroxybutanoate and its camphor-derived ligand to form diastereomeric complexes, each of which gives a different ^{1}H NMR spectrum.

36

An important property of europium shift reagents is that they spread the peaks in the ^{1}H NMR spectrum over a wider range, generally moving all of the peaks *downfield*. The magnitude of the shift of each peak depends on the concentration of the shift reagent and the proximity of the europium ion to the protons corresponding to that peak. The complexes that are formed between **36** and methyl (*S*)-(+)-3-hydroxybutanoate and methyl (*R*)-(+)-3-hydroxybutanoate are *diastereomeric*, so the peaks for the protons on one enantiomer move downfield to different extents than the corresponding peaks for the other. Thus, the two diastereomeric complexes may be easily distinguished in the ^{1}H NMR spectrum, as shown in Figure 17.42. It is then possible to integrate the peak areas corresponding to each of the enantiomers and determine the enantiomeric excess. Determining the ee using this NMR method gives results that are accurate to within approximately 2–3%.

In this experiment, you will use the chiral europium shift reagent **36** to determine the enantiomeric excess of the product that you obtained upon enantioselective reduction of methyl acetoacetate with baker's yeast.

EXPERIMENTAL PROCEDURE

Determining Optical Purity of Methyl (S)-(+)-3-Hydroxybutanoate

Discovery Experiment

Purpose To determine the optical purity of methyl (*S*)-(+)-3-hydroxybutanoate by ^{1}H NMR spectroscopy.

SAFETY ALERT

Wear safety glasses or goggles and suitable protective gloves while performing the experiment.

Preparation Refer to the online resources to answer Pre-Lab Exercises, access videos, and read the MSDSs for the chemicals used in this procedure.

Apparatus Two NMR tubes.

Preparing the Samples Dissolve 25–30 mg of racemic methyl 3-hydroxybutanoate in 0.75 mL of deuterochloroform and transfer the solution to an NMR tube using a Pasteur or a filter-tip pipet. Dissolve 25–30 mg of enantiomerically enriched methyl (*S*)-(+)-3-hydroxybutanoate, prepared by the enzymatic reduction in the first part of this experiment, in 0.75 mL of deuterochloroform and transfer the solution to another NMR tube, using a Pasteur or a filter-tip pipet.

Measurements Measure the ^{1}H NMR spectrum of the racemic methyl 3-hydroxybutanoate. Add 20 mg of *tris*[3-(heptafluoropropylhydroxymethylene)-(+)-camphorato]europium shift reagent to this solution, shake the mixture to effect dissolution, and allow the solution to stand for about 20 min. Again measure the ^{1}H NMR spectrum. The peaks of primary interest for determining the optical purity are the methoxy hydrogen atoms of the ester. Continue the procedure of adding shift reagent and measuring the spectrum until these peaks are separated and accurate integrals may be obtained as shown in Figure 17.42; no more than about 80 mg of shift reagent should be required.

Now perform the same series of measurements on the sample of enantiomerically enriched methyl (S)-(+)-3-hydroxybutanoate. A typical spectrum is shown in Figure 17.43. Determine the ratio of the two enantiomers in both of the samples by comparing the heights of the singlets that correspond to the methoxy protons of the ester, and calculate the percentages of each enantiomer in the two samples based upon this ratio. Calculate the enantiomeric excess (ee) of methyl (S)-(+)-3-hydroxybutanoate.

WRAPPING IT UP

Place the *solution of deuterochloroform and shift reagent* in the container for halogenated organic liquids.

EXERCISES

1. What are the advantages of using a chiral shift reagent instead of polarimetry to determine the optical purity of a sample?

2. Which of the following terms may be applied to the nature of the complex formed between Eu^{3+} and the alcohol functional group in methyl 3-hydroxybutanoate: Covalent? Ionic? Lewis acid/Lewis base? Van der Waals?

3. What is the meaning of the word "shift" in the term "chiral shift reagent"?

4. What properties of a chiral shift reagent allow you to determine the ratio of enantiomeric alcohols formed by an enantioselective reduction?

5. The ^1H NMR spectrum in Figure 17.43 is of a mixture of the chiral shift reagent **36** and a sample of methyl (S)-(+)-3-hydroxybutanoate that was produced by the enzymatic reduction of methyl acetoacetate using baker's yeast. Determine the enantiomeric excess of the methyl (S)-(+)-3-hydroxybutanoate contained in this sample.

6. Why is it necessary to obtain a ^1H NMR spectrum of *racemic* methyl 3-hydroxybutanoate in the presence of **36**?

7. Which of the following compounds will form diastereomeric complexes with the chiral europium shift reagent **36**?

SPECTRA

Chiral Complexes of Products

Figure 17.42
¹H NMR spectrum (300 MHz, CDCl₃) of racemic methyl 3-hydroxybutanoate in the presence of chiral shift reagent 36.

Figure 17.43
¹H NMR spectrum (300 MHz, CDCl₃) of enriched methyl (S)-(+)-3-hydroxybutanoate in the presence of chiral shift reagent 36.

CHAPTER 18

Reactions of Carbonyl Compounds

The chemistry of carbon-carbon multiple bonds is illustrated by the experiments with alkenes (Chaps. 10 and 12, and Sec. 17.2), alkynes (Sec. 11.3), and arenes (Chaps. 15 and 21). The reactions of these classes of organic compounds typically involve electrophilic processes, with the transformations being initiated by addition of an electrophile or a Lewis acid such as a proton, a bromonium ion, and so forth, to the π-system. What you will *not* find in your study of alkenes, alkynes, and arenes are reactions in which a nucleophile or a Lewis base attacks the carbon-carbon π-bond, although examples of such processes may be found when a carbon atom involved in the π-bond is appropriately substituted. The picture changes dramatically when you examine the chemistry of carbonyl compounds, which are characterized by the presence of a carbon-oxygen double bond. Now you will find that the carbonyl carbon atom is highly susceptible to attack by nucleophiles and that this leads to reactions having broad applications, including the formation of new carbon-carbon bonds. In this chapter, we can only give you a small sample of the diversity of reactions available to aldehydes and ketones, a class of compounds in which the carbonyl carbon atom, in addition to being doubly bound to an oxygen atom, is singly bound to other carbon or hydrogen atoms. Nonetheless, we think that this "taste" of carbonyl chemistry will suffice to whet your appetite for more.

18.1 INTRODUCTION

The **carbonyl group, 1,** is a rich source of many important reactions in organic chemistry, and two fundamental properties of this functionality are primarily responsible for its diverse chemistry. The first is the polarization of the carbon-oxygen π-bond, owing to the relatively high electronegativity of the oxygen atom. In terms of resonance theory, this polarization arises from the contribution of the dipolar resonance structure **1b** to the resonance hybrid for this functional group.

1a   **1b**

An aldehyde or ketone

Consequently, the carbonyl group undergoes a variety of reactions in which the electrophilic carbonyl carbon atom is attacked by **nucleophiles**, which are Lewis bases; the oxygen atom, in turn, reacts with **electrophiles** or **Lewis acids**. The net effect is *addition* of a reagent, symbolized as Nu–E, across the π-bond of the carbonyl function. The sequence by which the elements of Nu–E add to the carbonyl group varies according to the particular reagents and reaction conditions used but, in a general sense, follows the pathway of Equation 18.1 in basic media and that of Equation 18.2 in acidic media.

An aldehyde or ketone

(18.1)

An aldehyde or ketone

(18.2)

The second property of a carbonyl moiety is to increase the acidity of the **α-hydrogen atoms**, which are those on the carbon atoms directly attached to the carbonyl carbon atom. A result of the enhanced acidity of these α-hydrogens is that the **α-carbon atoms** can become **nucleophilic** either through deprotonation to form an **enolate ion** (Eq. 18.3), or by a keto–enol equilibration, called **tautomerization**, to give an **enol** (Eq. 18.4). As shown in Equation 18.3, an enolate, which is the resonance hybrid of the two contributing resonance structures **2a** and **2b**, can react with electrophiles, E$^+$, at an α-carbon atom to give net substitution of the electrophile for an α-hydrogen atom. A similar result attends the reaction of an enol **3** with an electrophile, E$^+$ (Eq. 18.4).

An aldehyde or ketone	**2a** **2b**	(18.3)
	An enolate ion	

An aldehyde or ketone   An enol

3

(18.4)

The experiments described in this chapter illustrate some of the representative reactions of carbonyl compounds according to the principles outlined in Equations 18.1–18.4. For example, you may study reactions in which the electrophilic carbonyl carbon atom of an aldehyde undergoes attack by a carbon nucleophile that is stabilized by an electron-withdrawing group such as a phosphorus-containing substituent, as found in the **Wittig reaction** and the **Horner-Wadsworth-Emmons modification** thereof (Sec. 18.2), or a carbonyl function, as found in the **aldol condensation** (Sec. 18.3). The **1,4-** or **conjugate addition** of the enol form of an aldehyde to an α,β-unsaturated ketone to give a dicarbonyl compound that cyclizes by an *intramolecular* aldol reaction is explored in Section 18.4.

Finally, the Biginelli reaction, a condensation involving a β-ketoester, an aldehyde, and urea that produces a heterocyclic compound and can be performed without the use of a solvent is described in Section 18.5.

18.2 THE WITTIG AND RELATED REACTIONS

Introduction

As noted in the preceding section, the ability of the carbonyl group to increase the acidity of the α-hydrogens makes it possible to generate enolate ions, carbon nucleophiles that are stabilized by resonance (Eq. 18.3). Certain phosphorus-containing groups are also known to increase the acidity of the hydrogen atoms bound to the same carbon atom, and deprotonation gives rise to another type of carbon nucleophile. For example, the bold-faced hydrogen atom of the phosphonium salt **4** exhibits enhanced acidity, and treatment of **4** with a strong base produces the nucleophilic species **5** (Eq.18.5), a **phosphorane**.

$$(C_6H_5)_3\overset{+}{P}\text{—}CHC_6H_5 \ Cl^- \longrightarrow \left[(C_6H_5)_3\overset{+}{P}\text{—} \ \bar{C}HC_6H_5 \longleftrightarrow (C_6H_5)_3P\text{=}CHC_6H_5 \right] \quad (18.5)$$

| **4** | **5a** | **5b** |

Benzyltriphenylphosphonium
chloride

Benzylidenetriphenylphosphorane

Phosphoranes like **5** are frequently referred to as **Wittig reagents** in recognition of their inventor, Georg Wittig, who showed that they react with aldehydes and ketones to produce alkenes. Wittig won the Nobel Prize in Chemistry in 1979 for this discovery. (See the Historical Highlight *The Wittig Reaction*, which is available online). An example of a Wittig reaction is illustrated in Scheme 18.1 for the synthesis of (*Z*)-stilbene (**9**) and (*E*)-stilbene (**10**) by the reaction of the phosphorane **5** with benzaldehyde (**6**). The reaction commences with the addition of the nucleophilic carbon atom of **5** to the electrophilic carbonyl carbon atom of **6** to give the **betaine 7** as a mixture of two diastereomers. The betaine **7** collapses to the **oxaphosphetane 8**, which then undergoes fragmentation to give a mixture of the isomeric alkenes **9** and **10,** together with triphenylphosphine oxide (**11**). Although phosphorane **5** may be written as the neutral species **5b**, the contribution of resonance structure **5a** imparts dipolar character to it, so it is commonly referred to as a phosphorus **ylide**.

Scheme 18.1

5a

6
Benzaldehyde

7
Betaine

8

An oxaphosphetane

9 + **10** + (C$_6$H$_5$)$_3$P=O

9
(Z)-Stilbene

10
(E)-Stilbene

11
Triphenylphosphine oxide

The net effect of the Wittig reaction is the conversion of a *carbon-oxygen* double bond to a *carbon-carbon* double bond. This transformation represents a general method for preparing alkenes that has two important advantages over other methods for preparing this functional group: (1) the carbonyl group is replaced *specifically* by a carbon-carbon double bond without forming isomeric alkenes having the π-bond at other positions, and (2) the reactions are carried out under mild conditions.

The Wittig reaction typically involves highly reactive phosphoranes and is therefore not normally suited for the introductory laboratory. For example, the very strong bases such as *n*-butyllithium that are normally used to produce the intermediate ylides require special handling techniques, and it is generally necessary to perform all experimental operations under a dry, inert atmosphere. However, it has been shown that the Wittig reaction of **5** with aromatic aldehydes such as benzaldehyde can be performed under **phase-transfer** conditions in which the phosphonium salt serves as the phase-transfer catalyst (Sec. 10.5). This simple expedient is used to prepare a mixture of (Z)-stilbene (**9**) and (E)-stilbene (**10**) in one of the experimental procedures in this section. You will be able to determine the ratio of **9** and **10** by examining an ^{1}H NMR spectrum of the mixture. It is also possible to isomerize (Z)-stilbene to (E)-stilbene by irradiation of a solution of (Z)-stilbene containing a small amount of molecular iodine (Eq. 18.6).

9
(Z)-Stilbene

10
(E)-Stilbene

(18.6)

The **Horner-Wadsworth-Emmons reaction** is an important variant of the Wittig reaction and involves using a **phosphonate ester** in place of a phosphonium salt. Like the phase-transfer Wittig reaction just discussed, these reactions may be easily performed in the undergraduate laboratory. In one of the procedures that follows, the phosphonate ester **12** is deprotonated with potassium *tert*-butoxide in the polar, *aprotic* solvent *N,N*-dimethylformamide, $(CH_3)_2NCHO$ (DMF), to provide the resonance-stabilized, nucleophilic phosphonate anion **13** (Eq. 18.7).

$$(18.7)$$

12		**13a**	**13b**
Diethyl benzylphosphonate	*tert*-Butoxide ion	Potassium diethyl benzylphosphate	

As shown in Scheme 18.2, the anion **13** reacts with benzaldehyde to give one of the stilbenes **9** or **10** with high selectivity by a sequence of reactions similar to that observed for the reaction of the ylide **5** with **6** (Scheme 18.1). The intermediate adduct **14** undergoes cyclization to form **15**, which then fragments to give the stilbene and potassium diethyl phosphate (**16**).

By obtaining spectral data for your product, you will be able to determine which of the stilbenes is produced. An explanation of why the Wadsworth-Emmons reaction mainly yields a single diastereomer of stilbene is beyond the scope of this discussion.

Scheme 18.2

13a	**6**	**14**

15

$C_6H_5CH{=}CHC_6H_5$ + $(C_2H_5O)_2\overset{O}{\overset{\|}{P}}{-}O^- \ K^+$

9 and/or **10**	**16**
Stilbene	Potassium diethyl phosphate

EXPERIMENTAL PROCEDURES

Wittig and Horner-Wadsworth-Emmons Reactions

A ■ *Preparation of (Z)- and (E)-Stilbenes by a Wittig Reaction*

Purpose To synthesize a mixture of stilbenes by a phase-transfer Wittig reaction, to determine the major product, and to perform their geometric isomerization.

SAFETY ALERT

1. **Wear safety glasses or goggles and suitable protective gloves while performing the experiment.**

2. **The 50% (by mass) aqueous sodium hydroxide solution is *highly* caustic. *Do not allow it to come in contact with your skin.* If this should happen, wash the affected area with dilute acetic acid and then copious amounts of water.**

MINISCALE PROCEDURE

Preparation Refer to the online resources to answer Pre-Lab Exercises, access videos, and read the MSDSs for the chemicals used or produced in this procedure. Review Sections 2.9, 2.11, 2.13, 2.17, 2.21, 2.22, and 2.29.

Apparatus A 25- and a 50-mL round-bottom flask, separatory funnel, ice-water bath, apparatus for heating under reflux, magnetic stirring, simple distillation, vacuum filtration, and *flameless* heating.

Setting Up Prepare a 50% (by mass) sodium hydroxide solution by dissolving 5.0 g of sodium hydroxide in 5 mL of water. The mixture may be heated gently to facilitate dissolution, but be sure to cool the solution to room temperature before proceeding. Add 3.8 g of benzyltriphenylphosphonium chloride and 1 mL of benzaldehyde to the 50-mL round-bottom flask containing 10 mL of dichloromethane and a stirbar. Set up the apparatus for heating under reflux. Heat the mixture with stirring until it refluxes *gently*.

Reaction Using a Pasteur pipet, add 5 mL of the 50% aqueous sodium hydroxide solution dropwise through the top of the reflux condenser to the solution, which should be stirred as *vigorously* as possible to ensure complete mixing of the phases. Try to add the solution *directly* into the flask without allowing it to touch the walls of the condenser. After the addition is complete, continue stirring under gentle reflux for 30 min.

Work-Up and Isolation of Isomeric Stilbenes Allow the reaction mixture to cool to room temperature and transfer it to a separatory funnel. Rinse the round-bottom flask with a 5-mL portion of dichloromethane and transfer the wash to the separatory funnel. Separate the layers and wash the organic layer sequentially with 10 mL of water and 15 mL of saturated aqueous sodium bisulfite. Finally, wash the organic layer with 10-mL portions of water until the pH of the wash is neutral. Transfer the organic solution to a dry Erlenmeyer flask and add several spatula-tips full of *anhydrous* sodium sulfate. Occasionally swirl the mixture over a period of 10–15 min to facilitate drying; add further small portions of *anhydrous* sodium sulfate if the solution remains cloudy.★

To determine the ratio of (*Z*)- and (*E*)-stilbene in the *crude* reaction mixture by ^{1}H NMR spectroscopy, use a filter-tip pipet to transfer a 0.5-mL portion of the dry organic solution to a small round-bottom flask and evaporate the dichloromethane by one of the techniques described in Section 2.29. The final traces of solvent may be removed by attaching the flask to a vacuum source and gently swirling the contents as vacuum is applied.★

Isomerization Filter or decant the remainder of the dry dichloromethane solution into the 25-mL round-bottom flask containing a stirbar, rinse the sodium sulfate with about 1 mL of fresh dichloromethane, and transfer this wash to the round-bottom flask. Add about 75 mg of iodine to the dichloromethane solution, fit the flask with a reflux condenser, and irradiate the solution with stirring for 1 h with a 150-W lightbulb.★

Work-Up and Isolation of Stilbene Decant the solution into a separatory funnel, and wash the dichloromethane solution with a 5-mL portion of saturated aqueous sodium bisulfite; shake the mixture vigorously to decolorize the dichloromethane layer. Wash the organic solution with a 5-mL portion of saturated aqueous sodium chloride, transfer it to a dry Erlenmeyer flask, and add several spatula-tips full of *anhydrous* sodium sulfate. Occasionally swirl the mixture over a period of 10–15 min to facilitate drying; add further small portions of *anhydrous* sodium sulfate if the solution remains cloudy.★

Filter or decant the dried solution into a tared round-bottom flask, equip the flask for simple distillation, and distil the dichloromethane. Alternatively, use rotary evaporation or other techniques to concentrate the solution. The final traces of solvent may be removed by attaching the flask to a vacuum source and gently swirling the contents as vacuum is applied.★

Remove about 50 mg of the residue for analysis by ^{1}H NMR spectroscopy. Dissolve the remaining residue in 10–12 mL of hot 95% ethanol. Cool the resulting solution first to room temperature and then in an ice-water bath for 15–20 min to complete crystallization. Collect the precipitate and air-dry it. Recrystallize the crude stilbene from 95% ethanol or some other suitable solvent.

Analysis Weigh the product and calculate the percent yield; determine its melting point. Perform the tests for unsaturation on the product using the bromine and Baeyer tests provided in Section 4.7A. Obtain IR and ^{1}H NMR spectra of your starting materials, the final product, and the crude product mixtures before and after irradiation with iodine. Compare these spectra with those of authentic samples (Figs. 10.34, 10.35, 18.1–18.6) to determine the ratio of the isomeric stilbenes **9** and **10** in the crude mixture and which one is the product after irradiation.

MICROSCALE PROCEDURE

Preparation Refer to the online resources to answer Pre-Lab Exercises, access videos, and read the MSDSs for the chemicals used or produced in this procedure. Review Sections 2.9, 2.11, 2.13, 2.17, 2.21, 2.22, and 2.29.

Apparatus A 5-mL conical vial, screw-cap centrifuge tube, ice-water bath, apparatus for heating under reflux, magnetic stirring, simple distillation, vacuum filtration, Craig tube filtration, and *flameless* heating.

Setting Up Prepare a 50% (by mass) sodium hydroxide solution by dissolving 1.0 g of sodium hydroxide in 1 mL of water. The mixture may be heated gently to facilitate dissolution, but be sure to cool the solution to room temperature before proceeding. Add 0.76 g of benzyltriphenylphosphonium chloride and 0.2 mL of benzaldehyde to the conical vial containing 2 mL of dichloromethane and a spinvane. Set up the apparatus for heating under reflux. Heat the solution with stirring until it refluxes *gently.*

Wittig Reaction Using a Pasteur pipet, add 1 mL of the 50% aqueous sodium hydroxide solution dropwise through the top of the reflux condenser to the solution, which should be stirred as *vigorously* as possible to ensure adequate mixing of the phases. Try to add the solution *directly* into the flask without allowing it to touch the walls of the condenser. After the addition is complete, continue stirring under *gentle* reflux for 30 min.

Work-Up and Isolation of Isomeric Stilbenes Allow the reaction mixture to cool to room temperature, and transfer it to a screw-cap centrifuge tube. Rinse the conical vial with 0.5 mL of dichloromethane, and transfer the rinse to the centrifuge tube. Separate the layers and wash the organic layer sequentially with 1.5 mL of water and then with 2.5 mL of saturated aqueous sodium bisulfite. Finally, wash the organic layer with 1.5-mL portions of water until the pH of the wash is neutral. Transfer the organic solution to a dry centrifuge tube and add several microspatula-tips full of *anhydrous* sodium sulfate. Occasionally shake the mixture over a period of 10–15 min to facilitate drying; add further small portions of *anhydrous* sodium sulfate if the solution remains cloudy.★ Consult with your instructor to determine whether you should obtain a ^{1}H NMR spectrum of the crude product mixture as in the Miniscale Procedure.

Isomerization Using a filter-tip pipet, transfer the dried dichloromethane solution into the 5-mL conical vial equipped with a spinvane. Rinse the sodium sulfate with 0.5 mL of dichloromethane and transfer the rinse to the conical vial. Add a *very small* crystal of iodine to the dichloromethane solution and irradiate the solution with stirring for 1 h with a 150-W lightbulb.★

Work-Up and Isolation of Stilbene Remove the spinvane, add 1 mL of saturated aqueous sodium bisulfite, and shake the mixture vigorously to decolorize the dichloromethane layer. Separate the layers and wash the organic layer with 1 mL of saturated aqueous sodium chloride. Transfer the organic solution to a dry centrifuge tube and add several microspatula-tips full of *anhydrous* sodium sulfate. Occasionally swirl the mixture over a period of 10–15 min to facilitate drying; add further small portions of *anhydrous* sodium sulfate if the solution remains cloudy.★ Using a filter-tip pipet, transfer the dried dichloromethane solution into a clean, dry 5-mL conical vial equipped with a spinvane. Remove the dichloromethane by simple distillation; withdraw solvent from the Hickman stillhead as needed. Alternatively, use rotary evaporation or other techniques to concentrate the solution. The final traces of solvents may be removed by attaching the vial to a vacuum source.★

Remove about 25 mg of the residue for analysis by ^{1}H NMR. Dissolve the remaining thick, cloudy residue in 2 mL of hot 95% ethanol. Cool the resulting solution first to room temperature and then in an ice-water bath for 15–20 min to complete crystallization. Collect the precipitate and air-dry it. Recrystallize the crude stilbene from 95% ethanol or some other suitable solvent.

Analysis Weigh the product and calculate the percent yield; determine its melting point. Perform the tests for unsaturation on the product using the bromine and Baeyer tests provided in Section 4.7. Obtain IR and ^{1}H NMR spectra of your starting materials and the final product, and compare them with those of authentic samples (Figs. 10.34, 10.35, 18.1–18.6) to determine which one of the isomeric stilbenes **9** and **10** has been formed.

Discovery Experiment   *Wittig Reaction of 9-Anthraldehyde*

The Wittig reaction of benzyltriphenylphosphorane (**5**) with benzaldehyde (**6**) produces a mixture of (*E*)- and (*Z*)-stilbene (Scheme 18.1). These isomers are expected to have differing thermodynamic stabilities owing to steric interactions that should make the *Z*-isomer **9** less stable than the *E*-isomer **10.** You might then predict that increasing the steric interactions would foster a larger preference for forming the thermodynamically favored diastereomer in a Wittig reaction. This hypothesis is testable by executing a base-promoted Wittig reaction of benzyltriphenylphosphonium chloride with 9-anthraldehyde rather than benzaldehyde (Eq. 18.8). Even if you don't perform the experiment, you may explore the proposal that the difference in thermodynamic stabilities is less for **9** and **10** than it is for the corresponding isomers derived from 9-anthraldehyde if you have access to a molecular mechanics program for calculating energies. Consult with your instructor to ascertain whether this is possible.

$$\text{(reaction scheme)} \tag{18.8}$$

Design and implement an experimental protocol for performing the reaction of Equation 18.8. Because 9-anthraldehyde is relatively expensive, you should plan to carry out the reaction at the microscale level unless instructed to do otherwise. You should be sure that your proposal includes a means to analyze for the possible presence of both isomeric products in the reaction mixture. Appropriate quantities of reagents are 0.97 g of benzyltriphenylphosphonium chloride (**4**), 0.57 g of 9-anthraldehyde, and 1.5 mL of 50% (by mass) of aqueous sodium hydroxide solution. Use 2-propanol as the solvent for recrystallizing the crude product.

WRAPPING IT UP

Neutralize the combined *aqueous layers and washes* with dilute hydrochloric acid, and flush them down the drain. Place the *recovered dichloromethane*, the *dichloromethane solution* from the bromine test for unsaturation, and any NMR solutions in a container for halogenated organic liquids; put the *manganese dioxide* from the Baeyer test for unsaturation in a container for heavy metals. Flush the *ethanolic filtrate* down the drain.

B ■ *Preparation of a Stilbene by the Horner-Wadsworth-Emmons Reaction*

Purpose To demonstrate the synthesis of a stilbene by the Horner-Wadsworth-Emmons modification of the Wittig reaction.

SAFETY ALERT

1. **Wear safety glasses or goggles and suitable protective gloves while performing the experiment.**

2. **Solutions of potassium *tert*-butoxide are *highly* caustic. *Do not allow them to come in contact with your skin.* If this should happen, wash the affected area with dilute acetic acid and then copious amounts of water.**

MINISCALE PROCEDURE

Preparation Refer to the online resources to answer Pre-Lab Exercises, access videos, and read the MSDSs for the chemicals used or produced in this procedure. Review Sections 2.10, 2.11, 2.17, 2.27, and 2.28.

Apparatus A 25-mL round-bottom flask, Claisen adapter, rubber septum, drying tube, two 1-mL syringes and a 5-mL syringe with needles, cold-water bath, apparatus for magnetic stirring, vacuum filtration, and *flameless* heating.

Setting Up Dry the glass parts of the syringes in an oven at 110 °C for 15 min. Equip the round-bottom flask with a stirbar and the Claisen adapter. Fit the rubber septum onto the joint of the Claisen adapter that is directly *in line* with the lower male joint (see Fig. 10.42). Using a small flame or a heat gun, gently heat the flask and Claisen adapter to remove traces of water from the apparatus; do *not* overheat the joint with the septum. After the flask has been heated for a few minutes, discontinue the heating, attach the drying tube to the *sidearm* of the adapter, and allow the apparatus to cool to room temperature. Remove the glass syringes from the oven and allow them to cool to room temperature, preferably in a desiccator.

Draw 5 mL of a 1 *M* solution of potassium *tert*-butoxide in dry *N,N*-dimethylformamide into the 5-mL syringe. Pierce the rubber septum with the needle of the syringe, being careful not to press down on the plunger, and add the solution directly to the round-bottom flask. Cool the flask below 20 °C in a cold-water bath. Using a 1-mL syringe, transfer 1 mL of diethyl benzylphosphonate through the rubber septum into the stirred solution of potassium *tert*-butoxide and continue stirring the solution for 5 min.

Reaction Using a 1-mL syringe, slowly add 0.5 mL of benzaldehyde through the rubber septum while continuing to stir the cooled solution. Once the addition is complete, remove the apparatus from the cooling bath and stir the mixture at room temperature for 30 min.★

Work-Up and Isolation Remove the Claisen adapter from the flask and add 5 mL of water with stirring. Collect the precipitated product by vacuum filtration, wash it with cold 1:1 methanol-water, and air-dry it. Reserve about 50 mg of the crude stilbene for analysis by ^{1}H NMR spectroscopy, and recrystallize the remainder from 95% ethanol or some other suitable solvent.

Analysis Weigh the product and calculate the percent yield; determine its melting point. Perform the tests for unsaturation on the product using the bromine and

Baeyer tests provided in Section 4.7A. Obtain IR and ^{1}H NMR spectra of your starting materials and product, and compare them with those of authentic samples (Figs. 10.34, 10.35, 18.1–18.4, 18.7, and 18.8). In the ^{1}H NMR spectrum of the crude product, determine whether there is any of the other diastereomeric stilbene present; if so, calculate the isomeric ratio.

MICROSCALE PROCEDURE

Preparation Refer to the online resources to answer Pre-Lab Exercises, access videos, and read the MSDSs for the chemicals used or produced in this procedure. Review Sections 2.10, 2.11, 2.17, 2.27, and 2.28.

Apparatus A 5-mL conical vial, Claisen adapter, rubber septum, drying tube, three 1-mL syringes with needles, cold-water bath, apparatus for magnetic stirring, vacuum filtration, Craig tube filtration, and *flameless* heating.

Setting Up Dry the conical vial, the Claisen adapter, and the glass parts of the syringes in an oven at 110 °C for 15 min. Remove the vial and adapter from the oven, and after they have cooled, preferably in a desiccator, place a spinvane in the vial and attach the Claisen adapter. Fit the rubber septum onto the joint of the Claisen adapter that is directly *in line* with the lower male joint and attach the drying tube to the *sidearm* of the adapter (see Fig. 10.43). Remove the glass syringes from the oven and allow them to cool to room temperature, preferably in a desiccator.

Draw 1 mL of a 1 *M* solution of potassium *tert*-butoxide in dry *N,N*-dimethylformamide into one of the 1-mL syringes. Pierce the rubber septum with the needle of the syringe, being careful not to press down on the plunger, and add the solution directly to the vial. Cool the vial below 20 °C in a cold-water bath. Using another 1-mL syringe, transfer 0.2 mL of diethyl benzylphosphonate through the rubber septum into the stirred solution of potassium *tert*-butoxide, and continue stirring the solution for 5 min.

Reaction Using a 1-mL syringe, slowly add 0.1 mL of benzaldehyde through the rubber septum while continuing to stir the cooled solution. Once the addition is complete, remove the apparatus from the cooling bath and stir the mixture at room temperature for 30 min.★

Work-Up and Isolation Remove the Claisen adapter from the vial, and add 1 mL of water with stirring. Collect the precipitated product by vacuum filtration, wash it with cold 1:1 methanol-water, and air-dry it. Recrystallize the crude stilbene from 95% ethanol or some other suitable solvent.

Analysis Weigh the product and calculate the percent yield; determine its melting point. Perform the tests for unsaturation on the product using the bromine and Baeyer tests provided in Section 4.7. Obtain IR and ^{1}H NMR spectra of your starting materials and product, and compare them with those of authentic samples (Figs. 10.34, 10.35, 18.1–18.4, 18.7, and 18.8). In the ^{1}H NMR spectrum of the product, determine whether there is any of the other diastereomeric stilbene present; if so, calculate the isomeric ratio and identify the major isomer.

WRAPPING IT UP

Place the *recovered dichloromethane* and the *dichloromethane solution* from the bromine test for unsaturation in a container for halogenated organic liquids; put the *manganese dioxide* from the Baeyer test for unsaturation in a container for heavy metals. Flush all *aqueous filtrates* down the drain.

EXERCISES

General Questions

1. Compare the mechanism of aldol addition (Sec. 18.3) to that of the Wittig synthesis, pointing out similarities and differences.

2. Why should the aldehydes used as starting materials in Wittig syntheses be free of contamination by carboxylic acids?

3. Considering the mechanism of the Wittig reaction, speculate what the driving force for the decomposition of the oxaphosphetane intermediate might be.

4. Ylides react readily with aldehydes and ketones but slowly or not at all with esters. Explain this difference in reactivity.

5. Explain why you would expect an anion of type **5** to be more stable if one of the R groups is cyano, $C≡N$, rather than alkyl.

6. An ylide like **5** is a stabilized carbanion. In what way(s) does the phosphorus atom provide stability to the carbanion?

7. Write equations for the preparation of the following alkenes by the Wittig reaction or the Horner-Wadsworth-Emmons modification of the Wittig reaction. Start with any carbonyl compound and Wittig or Horner-Wadsworth-Emmons reagent together with any other organic or inorganic reagents that you require.

   a. $C_6H_5CH=C(CH_3)C_6H_5$

   b. $CH_2=CH–CH=CH–C_6H_5$

   c. $(CH_3)_2C=CH–CO_2C_2H_5$

8. Suggest *two* different routes for preparing $C_6H_5CH=CHCH_3$ via a Wittig reaction, using a phosphorane as the ylide precursor in both cases.

Questions for Part A

9. Suggest a method for preparing benzyltriphenylphosphonium chloride (**4**) via an S_N2 reaction.

10. Define the term *phase-transfer catalyst*.

11. Consider the step in which you washed the crude reaction mixture obtained *immediately* following the Wittig reaction with saturated aqueous sodium bisulfite.

    a. What potential contaminant of the final product is removed in this step?

    b. Write the chemical equation by which the contaminant is removed and specify whether it is being oxidized, reduced, or neither in the process.

12. Propose a mechanism for the isomerization of (Z)-stilbene into (E)-stilbene using a trace of iodine and light (Eq. 18.6). (*Hint:* The iodine-iodine bond undergoes homolysis upon irradiation with light; see also Section 7.3.)

13. Based on your experimental results, what are you able to conclude about the iodine/light-promoted isomerization of (E)-stilbene? Which of the stilbene isomers appears to be the more stable, and how do you rationalize this difference in stabilities?

Refer to the online resources and use the spectra viewer and Tables 8.1–8.8 as needed to answer the blue-numbered questions on spectroscopy.

14. Consider the spectral data for benzaldehyde (Figs. 18.1 and 18.2).

    a. In the functional group region of the IR spectrum, specify the absorptions associated with the carbonyl group and the aromatic ring.

    b. In the ^1H NMR spectrum, assign the various resonances to the hydrogen nuclei responsible for them.

    c. For the ^{13}C NMR data, assign the various resonances to the carbon nuclei responsible for them.

15. Consider the NMR spectral data for benzyltriphenylphosphonium chloride (Fig. 18.6). Note that ^{31}P has $I_z = 1/2$ and therefore couples with neighboring hydrogen and carbon and carbon nuclei having $I_z = 1/2$.

   a. In the ^{1}H NMR spectrum, assign the various resonances to the hydrogen nuclei responsible for them.

   b. For the ^{13}C NMR data, assign the various resonances to the carbon nuclei responsible for them.

16. Consider the spectral data for (E)-stilbene (Figs. 10.34 and 10.35).

   a. In the functional group region of the IR spectrum, specify the absorptions associated with the aromatic rings.

   b. In the ^{1}H NMR spectrum, assign the various resonances to the hydrogen nuclei responsible for them.

   c. For the ^{13}C NMR data, assign the various resonances to the carbon nuclei responsible for them.

17. Consider the spectral data for (Z)-stilbene (Figs. 18.3 and 18.4).

   a. In the functional group region of the IR spectrum, specify the absorptions associated with the aromatic rings.

   b. In the ^{1}H NMR spectrum, assign the various resonances to the hydrogen nuclei responsible for them.

   c. For the ^{13}C NMR data, assign the various resonances to the carbon nuclei responsible for them.

18. Discuss the differences observed in the IR and NMR spectra of (Z)-stilbene and (E)-stilbene that enable you to distinguish between the two isomers. What differences in the IR and NMR spectra of benzaldehyde, (Z)-stilbene, and (E)-stilbene are consistent with the conversion of benzaldehyde to a mixture of the isomeric stilbenes in this experiment?

19. By analysis of the ^{1}H NMR spectra of the crude product before *and* after irradiation in the presence of iodine, determine the approximate ratios of (Z)-stilbene and (E)-stilbene.

Questions for Part B

20. Account for the fact that diethyl benzylphosphonate (**12**) is a stronger acid than benzyltriphenylphosphonium chloride (**4**).

21. Which phosphorus compound, **5** or **13**, is more nucleophilic? Explain your answer.

22. Why should the exposure of the solution of potassium *tert*-butoxide to the atmosphere be minimized?

23. *N,N*-Dimethylformamide (DMF) is very water-soluble. Why?

Refer to the online resources and use the spectra viewer and Tables 8.1–8.8 as needed to answer the blue-numbered questions on spectroscopy.

24. Consider the spectral data for benzaldehyde (Figs. 18.1 and 18.2).

   a. In the functional group region of the IR spectrum, specify the absorptions associated with the carbonyl group and the aromatic ring.

   b. In the ^{1}H NMR spectrum, assign the various resonances to the hydrogen nuclei responsible for them.

   c. For the ^{13}C NMR data, assign the various resonances to the carbon nuclei responsible for them.

25. Consider the spectral data for (*E*)-stilbene (Figs. 10.34 and 10.35).

   a. In the functional group region of the IR spectrum, specify the absorptions associated with the aromatic rings.

   b. In the ¹H NMR spectrum, assign the various resonances to the hydrogen nuclei responsible for them.

   c. For the ¹³C NMR data, assign the various resonances to the carbon nuclei responsible for them.

26. Consider the NMR spectral data for diethyl benzylphosphonate (Fig. 18.8). Note that ^{31}P has $I_z = 1/2$ and therefore couples with neighboring hydrogen and carbon and carbon nuclei having $I_z = 1/2$.

   a. In the ¹H NMR spectrum, assign the various resonances to the hydrogen nuclei responsible for them.

   b. For the ¹³C NMR data, assign the various resonances to the carbon nuclei responsible for them.

SPECTRA

Starting Materials and Products

The IR and NMR spectra of (E)-stilbene are provided in Figures 10.34 and 10.35, respectively.

Figure 18.1
IR spectrum of benzaldehyde (neat).

Figure 18.2
NMR data for benzaldeyde (CDCl₃).

(a) ¹H NMR spectrum (300 MHz).
(b) ¹³C NMR data: δ 129.0, 129.7, 134.4, 136.6, 192.0.

Figure 18.3
IR spectrum of (Z)-stilbene (neat).

Figure 18.4
NMR data for (Z)-stilbene (CDCl₃).

(a) ^{1}H NMR spectrum (300 MHz).
(b) ^{13}C NMR data: δ 127.3, 128.4, 129.1, 130.5, 137.5.

Figure 18.5
IR spectrum of benzyltriphe-nylphosphonium chloride (IR card).

Figure 18.6
NMR data for benzyltri-phenylphosphonium chloride (CDCl₃).

(a) ¹H NMR spectrum (300 MHz).
(b) ¹³C NMR data (³¹P-coupled): δ 29.8, 30.6, 116.6, 118.0, 126.8, 127.0, 127.9, 128.3, 128.4, 129.6, 129.8, 13.10, 131.1, 133.8, 133.9, 134.6.

Figure 18.7
IR spectrum of diethyl benzylphosphonate (neat).

Figure 18.8
NMR data for diethyl benzylphos-phonate (CDCl₃).

(a) ¹H NMR spectrum (300 MHz).
(b) ¹³C NMR data (³¹P-coupled): δ 15.9, 16.0, 32.2, 34.4, 61.5, 61.6, 126.3, 126.4, 128.0, 128.1, 129.3, 129.4, 131.1, 131.3.

18.3 REACTIONS OF STABILIZED CARBANIONS FROM CARBONYL COMPOUNDS

The presence of the carbonyl function in a molecule allows the α-carbon atom to become nucleophilic through formation of an **enolate ion** (Eq. 18.3) or an **enol** (Eq. 18.4). The nucleophilic property of an enol is the basis for the experiment in Section 18.4. The corresponding property of enolate ions, which are considerably more reactive nucleophiles than enols, is the subject of this section.

Formation of enolate ions by deprotonation (Eq. 18.3) involves the use of a base, and the strength of the base required will depend on the pK_a of the hydrogen atom attached to the α-carbon atom. The pK_a is defined as the negative log of K_a, the equilibrium constant for the ionization shown in Equation 18.9. Thus, $pK_a = -\log K_a$. Water is the usual reference solvent for reporting values of pK_a, which are known to be solvent-dependent. Experimentally, the pK_a values of aldehydes and ketones are in the range of 18–20. This is a remarkably low value when compared to those for saturated hydrocarbons, which are in excess of 50!

$$\underset{\text{Aldehyde or ketone}}{\overset{\text{O}}{\underset{}{\parallel}}\ce{\alpha-H_{(aq)}}} \underset{}{\overset{K_a}{\rightleftharpoons}} \ce{H^+_{(aq)}} + \underset{\underset{\textbf{17}}{\text{An enolate ion}}}{\overset{\text{O}}{\underset{}{\parallel}}\ce{-_{(aq)}}} \qquad (18.9)$$

The dramatic acidifying effect of the carbonyl group is due to its inductive effect and its ability to delocalize and thereby stabilize the negative charge in the enolate ion, as reflected in **17** (Eq. 18.9), which is the **resonance hybrid** of the **resonance structures 3a** and **3b** (Sec. 18.1). Because the pK_a values of aldehydes and ketones fall in the range of water (pK_a 15.7) and alcohols (pK_a 15.5–18), it is possible to generate enolate ions using anions such as hydroxide or alkoxide (Eq. 18.3, $B^- = HO^-$ and RO^-, respectively); these are the bases utilized in the experimental procedures of this section.

Aldol Condensations

An important general reaction of enolate ions involves **nucleophilic addition** to the electrophilic carbonyl carbon atom of the aldehyde or ketone from which the enolate is derived. A dimeric anion **18** results, which may then be neutralized by abstraction of a proton to produce a β-hydroxycarbonyl compound, **19** (Eq. 18.10). If the reaction is performed in hydroxylic solvents such as water or an alcohol, the source of the proton may be the solvent, whose deprotonation will regenerate the base required for forming the enolate ion. Thus, the overall process is *catalytic* in the base that is used.

$$\left[\underset{\textbf{3b}}{\overset{\text{O}^-}{\underset{}{\parallel}}} \longleftrightarrow \underset{\textbf{3a}}{\overset{\text{O}}{\underset{}{\parallel}}}_- \right] + \overset{\text{O}}{\underset{}{\parallel}} \rightleftharpoons \underset{\textbf{18}}{\overset{\text{O}}{\underset{}{\parallel}}\overset{\text{O}^-}{\underset{}{\parallel}}} \underset{}{\overset{\text{H}-\text{OR}}{\rightleftharpoons}} \underset{\underset{\textbf{19}}{}}{\overset{\text{O}}{\underset{}{\parallel}}\ce{\alpha}\overset{\beta}{\underset{}{|}}\overset{\text{OH}}{}} + \ce{RO^-} \qquad (18.10)$$

An enolate ion   Aldehyde or ketone   An aldol
(β-hydroxycarbonyl compound)

A β-hydroxycarbonyl compound such as **19** is commonly called an **aldol**, and the reaction leading to its formation is referred to as an **aldol addition**. The term "aldol" reflects the fact that use of an aldehyde in the reaction provides a product containing an *ald*ehyde function and an **alcoh**ol group. Isolating such addition products depends on the reaction conditions used and the nature of the carbonyl compound(s) involved. As shown in Equation 18.10, the aldol reaction involves an equilibrium, and in the case of self-condensation of ketones such as acetone, the equilibrium is to the left. Preparing an aldol product derived from a ketone in acceptable yields requires that the product be removed from the reaction continuously. In contrast, the equilibrium favors formation of the aldol product in the case of self-condensation of aldehydes, so these aldols are more readily isolated.

Another challenge to isolating aldols such as **19** is their potential dehydration to an α,β-unsaturated carbonyl compound **20**, which may occur *because* **19** still has a hydrogen atom on the α-carbon atom that served as the nucleophile in the addition reaction (Eq. 18.11). Indeed, dehydration can be used to drive aldol reactions because reversal of dehydration often requires forcing conditions. The overall reaction described by Equations 18.10 and 18.11 is called an **aldol condensation**.

$$+ \text{ROH} + \text{HO}^- \qquad (18.11)$$

19   **20**

An α,β-unsaturated
compound

The experiment presented in this section is an example of a **mixed-** or **crossed-aldol condensation**. This term describes cases in which two *different* carbonyl compounds are the reactants. Such reactions are synthetically practical under certain circumstances, selectively producing a single major condensation product. For example, a ketone may preferentially condense with an aldehyde rather than undergoing self-addition with another molecule of itself (Scheme 18.3). This is because the carbonyl carbon atom of ketones is sterically and electronically not as susceptible to nucleophilic attack as is that of aldehydes. The aldehydic partner in such a reaction generally has no α-hydrogen atoms, so it is unable to undergo an aldol reaction.

Scheme 18.3

You will perform the crossed-aldol condensation of acetophenone (**21**) with *p*-anisaldehyde (4-methoxybenzaldehyde, **23**) to give *trans-p*-anisalacetophenone (**25**) according to the sequence outlined in Scheme 18.4. For steric and electronic reasons, the enolate ion **22** reacts preferentially with **23** rather than **21**, leading to the aldol **24**. This product then dehydrates in the presence of base, in the manner outlined in Equation 18.11, to yield the condensation product **25**. Part of the thermodynamic driving force for this dehydration is associated with forming a new carbon-carbon π-bond that is **conjugated** with the aromatic ring as well as with the carbonyl function. As a general rule, formation of an extended conjugated system, as occurs here, increases the stability of a molecule.

Scheme 18.4

$$C_6H_5-\overset{\overset{\displaystyle O}{\|}}{C}-CH_2 \quad + \quad \overline{}OH \quad \underset{\xrightarrow{-H_2O}}{\rightleftharpoons} \quad \left[C_6H_5-\overset{\overset{\displaystyle O}{\|}}{C}-\overline{C}H_2 \quad \longleftrightarrow \quad C_6H_5-\overset{\overset{\displaystyle \overline{O}}{\|}}{C}=CH_2 \right]$$

21
Acetophenone

22

$$CH_3O-\left\langle\right\rangle-\overset{\displaystyle \overset{H}{|}}{C}$$
$$\overset{\displaystyle \|}{O}$$

23
p-Anisaldehyde

$$HO\overline{} \;+\; C_6H_5-\overset{\overset{\displaystyle O}{\|}}{C}-\underset{\underset{\displaystyle H}{|}}{C}H-\overset{\overset{\displaystyle C_6H_4OCH_3}{|}}{\underset{\underset{\displaystyle OH}{|}}{C}H} \quad \underset{HO-H}{\rightleftharpoons} \quad C_6H_5-\overset{\overset{\displaystyle O}{\|}}{C}-CH_2-\overset{\overset{\displaystyle C_6H_4OCH_3}{|}}{\underset{\underset{\displaystyle \overline{O}}{|}}{C}H}$$

24

$\updownarrow \quad \begin{array}{l} -H_2O \\ -HO^- \end{array}$

$$C_6H_5-\overset{\overset{\displaystyle O}{\|}}{C}\underset{\overset{\displaystyle |}{H}}{\overset{\displaystyle |}{C}}=\underset{\overset{\displaystyle |}{C}\!-\!\!\left\langle\right\rangle\!-\!OCH_3}{\overset{\displaystyle H}{}}$$

25
trans-p-Anisalacetophenone

Preparation of trans-p-*Anisalacetophenone*

Purpose To demonstrate the synthesis of an α,β-unsaturated carbonyl compound by a crossed-aldol condensation.

SAFETY ALERT

1. **Wear safety glasses or goggles and suitable protective gloves while performing the experiment.**

2. **The 50% (by mass) aqueous sodium hydroxide solution is *highly* caustic. *Do not allow it to come in contact with your skin.* If this should happen, wash the affected area with dilute acetic acid and then copious amounts of water.**

MINISCALE PROCEDURE

Preparation Refer to the online resources to answer Pre-Lab Exercises, access videos, and read the MSDSs for the chemicals used or produced in this procedure. Review Sections 2.9, 2.10, 2.11, and 2.17.

Apparatus A 13-mm × 100-mm test tube, two 1-mL syringes, ice-water bath, apparatus for vacuum filtration and *flameless* heating.

Setting Up Prepare a 50% (by mass) sodium hydroxide solution by dissolving 1.0 g sodium hydroxide in 1 mL of water. You may heat the mixture gently to hasten dissolution, but be sure to cool the solution to room temperature before proceeding. Place 1.0 mL of *p*-anisaldehyde and 1.0 mL of acetophenone in the test tube and add 3.0 mL of 95% ethanol. Shake the test tube gently to mix and dissolve the reactants.

Aldol Reaction, Work-Up, and Isolation Using a Pasteur pipet, transfer 5 drops of the 50% sodium hydroxide solution into the ethanolic solution of the carbonyl compounds, shake the mixture for a minute or two until a homogeneous solution results, and allow it to stand with occasional shaking at room temperature for 15 min.★ Cool the reaction mixture in an ice-water bath. If crystals do not form, scratch at the liquid-air interface to induce crystallization. Collect the product by vacuum filtration, wash the product with 1–2 mL of *cold* 95% ethanol, and air-dry the crystals.★ Recrystallize the crude *trans-p*-anisalacetophenone from methanol or some other suitable solvent.

See *Who was Baeyer?*

Analysis Weigh the product and calculate the percent yield; determine its melting point. Prepare the 2,4-dinitrophenylhydrazones of the product and of the two starting materials according to the procedure in Section 7.4B; compare the melting points of these derivatives. Perform the tests for unsaturation on the product using the bromine and Baeyer tests provided in Section 4.7A. Obtain IR and ¹H NMR spectra of your starting materials and product and compare them with those of authentic samples (Figs. 18.9–18.14).

MICROSCALE PROCEDURE

Preparation Refer to the online resources to answer Pre-Lab Exercises, access videos, and read the MSDSs for the chemicals used or produced in this procedure. Review Sections 2.9, 2.10, 2.11, and 2.17.

Apparatus A 3-mL conical vial, two 1-mL syringes, ice-water bath, apparatus for magnetic stirring, vacuum filtration, Craig tube filtration, and *flameless* heating.

Setting Up Prepare a 50% (by mass) sodium hydroxide solution by dissolving 0.5 g of sodium hydroxide in 0.5 mL of water. You may heat the mixture gently to hasten dissolution, but be sure to cool the solution to room temperature before proceeding. Place 0.2 mL of *p*-anisaldehyde and 0.2 mL of acetophenone in the conical vial containing a spinvane. Add 0.7 mL of 95% ethanol and stir the contents of the vial to mix and dissolve the reactants.

Aldol Reaction, Work-Up, and Isolation Using a Pasteur pipet, transfer 2 drops of the 50% NaOH solution into the ethanolic solution of the carbonyl compounds and stir the reaction mixture at room temperature for 15 min.★ Cool the reaction mixture in an ice-water bath. If crystals do not form, scratch at the liquid-air interface to induce crystallization.★ Collect the product by vacuum filtration, wash the product with about 5–10 drops of *cold* 95% ethanol, and air-dry the crystals. Recrystallize the crude *trans-p-* anisalacetophenone from methanol or some other suitable solvent.

Analysis Weigh the product and calculate the percent yield; determine its melting point. Prepare the 2,4-dinitrophenylhydrazones of the product and the two starting materials according to the procedure in Section 7.4B; compare the melting points of these derivatives. Perform the tests for unsaturation on the product using the bromine and Baeyer tests provided in Section 4.7A. Obtain IR and ¹H NMR spectra of your starting materials and product, and compare them with those of authentic samples (Figs. 18.9–18.14).

Discovery Experiment

Synthesis of **trans,trans-Dibenzylideneacetone**

Consult with your instructor before performing this experiment, in which you will prepare *trans,trans*-dibenzylideneacetone by the crossed-aldol condensation of benzaldehyde with acetone. Design an experiment using either a miniscale or microscale procedure to produce the product. Determine the relative amounts of benzaldehyde and acetone that should be used by considering the stoichiometry of the reaction and the potential problems that might be associated with side reactions and the use of large excesses of reactants. Freshly distilled benzaldehyde should be used. The product may be recrystallized from an 85% (*v:v*) ethanol-water mixture. Perform the tests for unsaturation on the product using the bromine and Baeyer tests provided in Section 4.7A. Obtain the melting point and the IR, ¹H, and ¹³C NMR spectra of the purified product for purposes of characterization, and compare these with the literature data. Determine the λ_{max} and ε_{max} in the UV spectrum for the product, and compare to the literature values.

Discovery Experiment

Solvent-Free Aldol Condensation

Some but not all organic reactions proceed well in the absence of solvent(s). Develop a solvent-free protocol for the aldol reaction using solid sodium or potassium hydroxide as the base. The reaction can be successfully performed on a 5–10 mmol scale, using an equivalent of base. Bear in mind that the base should be finely divided, so it should first be crushed with a mortar and pestle. In addition, the condensation partners should be mixed thoroughly before the base is added. The reaction should take no more than 20–25 min, and the mixture should be acidified to pH ~5 with 10% HCl prior to isolation of the product. Characterize the product by its melting point and using spectroscopic techniques.

Possible combinations that you might use include acetophenone with either 4-methylbenzaldehyde or 4-chlorobenzaldehyde, 1-indanone with either 4-phenylcyclohexanone or 3,4-dimethoxybenzaldehyde, 4-phenylcyclohexanone with 3,4-dimethoxybenzaldehyde, and 3,4-dimethoxybenzaldehyde with 3,4-dimethoxyacetophenone. Self-condensation of 4-methylbenzaldehyde is also a possibility. Consult with your instructor before undertaking any experiments.

WRAPPING IT UP

Neutralize the *ethanolic filtrates* with dilute hydrochloric acid, and flush them down the drain. Place the *dichloromethane solution* from the bromine test for unsaturation in a container for halogenated organic liquids.

EXERCISES

1. Compute the equilibrium constant, K_{eq}, for the reaction of equimolar amounts of acetophenone, $C_6H_5COCH_3$, and hydroxide ion to generate the enolate ion. The pK_a values of the ketone and of water are 19.0 and 15.7, respectively.

2. Explain why the main reaction between acetophenone and *p*-anisaldehyde is the mixed-aldol reaction rather than (a) self-condensation of acetophenone or (b) the Cannizzaro reaction (Sec. 16.3) of *p*-anisaldehyde.

3. Identify the nucleophile and electrophile involved in the rate-determining step of the synthesis of *trans-p*-anisalacetophenone.

4. Explain why *trans-p*-anisalacetophenone would be expected to be more stable than the corresponding *cis* isomer.

5. Propose a synthesis of cinnamaldehyde using a mixed-aldol condensation reaction.

Cinnamaldehyde

6. Write structures for the various aldol condensation products you expect from the aldol self-condensation of 2-butanone, $CH_3C(=O)CH_2CH_3$.

7. Predict the product of the *intramolecular* aldol reaction of the diketone shown below.

2,4-Pentanedione

8. α,β-Unsaturated carbonyl compounds such as *trans-p*-anisalacetophenone (**25**) undergo reactions that are typical of both carbonyl compounds and alkenes. Illustrate this by showing the reaction of **25** with bromine (Sec. 10.6) and 2,4-dinitrophenylhydrazine (Sec. 7.4B) to give the corresponding products.

9. The addition of bromine to *trans-p*-anisalacetophenone (**25**) gives a single diastereomer.

   a. Using curved arrows to symbolize the flow of electrons and suitable stereochemical drawings, show the mechanism of this reaction (*Hint*: See Sec. 10.6).

   b. How does the mechanism you proposed in Part **a** support the formation of the racemic form of a single diastereomer?

Refer to the online resources and use the spectra viewer and Tables 8.1–8.8 as needed to answer the blue-numbered questions on spectroscopy.

10. Consider the spectral data for acetophenone (Figs. 18.9 and 18.10).

    a. In the functional group region of the IR spectrum, specify the absorptions associated with the carbonyl group and the aromatic ring.

    b. In the ¹H NMR spectrum, assign the various resonances to the hydrogen nuclei responsible for them.

    c. For the ¹³C NMR data, assign the various resonances to the carbon nuclei responsible for them.

11. Consider the spectral data for *p*-anisaldehyde (Figs. 18.11 and 18.12).

    a. In the functional group region of the IR spectrum, specify the absorptions associated with the carbonyl group and the aromatic ring. Also specify the absorption in the fingerprint region that is associated with the *para* substitution on the aromatic ring.

    b. In the ¹H NMR spectrum, assign the various resonances to the hydrogen nuclei responsible for them.

    c. For the ¹³C NMR data, assign the various resonances to the carbon nuclei responsible for them.

12. Consider the spectral data for *trans-p*-anisalacetophenone (Figs. 18.13 and 18.14).

    a. In the functional group region of the IR spectrum, specify the absorptions associated with the unsaturated ketone group and the aromatic rings. Also specify the absorptions in the fingerprint region that are associated with the *para* substitution on the aromatic ring and the *trans* carbon-carbon double bond.

    b. In the ¹H NMR spectrum, assign the various resonances to the hydrogen nuclei responsible for them.

    c. For the ¹³C NMR data, assign the various resonances to the carbon nuclei responsible for them.

13. Discuss the differences observed in the IR and NMR spectra of acetophenone, *p*-anisaldehyde, and *trans-p*-anisalacetophenone that are consistent with the crossed-aldol condensation occurring in this experiment.

SPECTRA

Starting Materials and Products

Figure 18.9

IR spectrum of acetophenone (neat).

Figure 18.10
NMR data for acetophenone (CDCl₃).

(a) ¹H NMR spectrum (300 MHz).
(b) ¹³C NMR data: δ 26.3, 128.3, 128.8, 133.0, 137.3, 197.4.

Figure 18.11
IR spectrum of p-*anisaldehyde (neat).*

Figure 18.12
NMR data for p-*anisaldehyde (CDCl₃).*

(a) ¹H NMR spectrum (300 MHz).
(b) ¹³C NMR data: δ 55.5, 114.5, 130.2, 131.9, 164.6, 190.5.

Figure 18.13
IR spectrum of trans-p-anisalacetophenone (IR card).

Figure 18.14
NMR data for trans-p-anisalacetophenone (CDCl₃).

(a) ¹H NMR spectrum (300 MHz).

(b) ¹³C NMR data: δ 55.3, 114.4, 119.7, 127.6, 128.4, 128.5, 130.2, 132.5, 138.5, 144.6, 161.7, 190.4.

18.4 CONJUGATE ADDITION TO AN α,β-UNSATURATED KETONE

A carbon atom *alpha* to a carbonyl function may become nucleophilic either by forming an *enolate* (Eq. 18.3) or an *enol* (Eq. 18.4). In the crossed-aldol reaction described in the previous section, the nucleophilic character of the carbon atom *alpha* to a ketone function was provided by enolate formation. However, in the experiment that follows, nucleophilicity at the α-carbon atom is promoted by enol formation. The overall reaction is the *acid-catalyzed* conversion of 2-methylpropanal (**26**) and 3-buten-2-one (**27**) into 4,4-dimethyl-2-cyclohexen-1-one (**29**), as shown in Equation 18.12. For the first stage of the reaction, the enol form of **26** serves as the nucleophile in the **1,4-** or **conjugate addition** to the α,β-unsaturated ketone **27** to give **28**. In the second stage, an enol is the nucleophile in the cyclization of **28** by an **intramolecular aldol condensation** that is acid-catalyzed. The conjugate addition to give **28** is an example of a **Michael-type addition**, whereas the combination of this addition followed by an aldol reaction to form the six-membered ring of **29** is called the **Robinson annulation**. The names honor the chemists who discovered the two types of reactions.

26	**27**	**28**	**29**
2-Methylpropanal (isobutyraldehyde)	3-Buten-2-one (methyl vinyl ketone)	2,2-Dimethylhexanal-5-one	4,4-Dimethyl-2-cyclohexen-1-one

(18.12)

A plausible overall mechanism for the reactions leading to **29** is shown in Scheme 18.15. 2-Methylpropanal (**26**) first undergoes acid-catalyzed tautomerization to its enol form **30**, in which the α-carbon atom is now nucleophilic. Protonation of the 3-buten-2-one (**27**) converts it into the delocalized cation **31**, which is more electrophilic than the unprotonated form. Because **30** is a weak nucleophile relative to an enolate ion, the formation of **31** facilitates the next stage of the reaction, which results in a new carbon-carbon bond between **26** and **27** to give the enol **32**. The boldfaced atoms in structure **32** show that the α-C–H bond of **26** has been added in a conjugate-, or 1,4-, manner to **27**. Acid-catalyzed tautomerization of **32** leads to the thermodynamically more stable keto form **28**.

Although **28** is potentially isolable, it is not observed under our reaction conditions but rather undergoes an acid-catalyzed *intramolecular aldol addition*. This process is initiated by the acid-catalyzed tautomerization of **28** to the isomeric enol form **33**. Activating the remaining carbonyl function by protonation sets the stage for cyclization to give the β-hydroxycarbonyl compound **34** after deprotonation. The fact that **28** undergoes efficient cyclization rather than intermolecular condensation with the enolic form of 2-methylpropanal or other potential nucleophiles is mainly attributable to *entropic* considerations that favor *intra*molecular processes over *inter*molecular ones. In the final stage of the reaction, acid-catalyzed dehydration of **34** to give the desired **29** occurs; the mechanism of this reaction is left as the topic of Exercise 15 at the end of this section.

The dehydration of **34** to **29** is a reversible process that is driven to completion by removing the water from the reaction mixture. This is conveniently done by using toluene as the solvent for the reaction. Water and toluene form an **azeotrope** (Sec. 4.4), so azeotropic distillation allows the continuous separation of water as dehydration occurs. To minimize the amount of solvent that is required for distillations of this type, a Dean-Stark trap (Fig. 18.15) is commonly used. Because such traps are often not available in the undergraduate laboratory, an operational equivalent may be devised by assembling the apparatus in a way such that water, but not toluene, can be prevented from returning to the reaction flask (Fig. 18.16).

The choice of an aldehyde such as 2-methylpropanal (**26**) having only *one* α-hydrogen atom is important to the success of this crossed-aldol condensation because its dimerization is of no consequence. For example, aldehyde **26** can readily react with another molecule of itself to give **35** (Eq. 18.13), but this process is *reversible* under the reaction conditions. Moreover, unlike the aldol product **34**, **35** is incapable of dehydrating to give an α,β-unsaturated product and water, which would drive the equilibrium leading to the dimerization product. It is left to you to consider the possibility of self-condensation of 3-buten-2-one (**27**) and of the reaction of enol **30** with the carbon atom of the protonated carbonyl function in **31** (see Exercises 12 and 13 at the end of this section).

Tautomerize

Tautomerize

26

27

30

31

32

33

28

34

29

2 (CH$_3$)$_2$CH—CHO $\xrightleftharpoons{\text{H}_3\text{O}^+}$ (CH$_3$)$_2$CH—C(CH$_3$)$_2$—CHO (18.13)

26

35

2,2,4-Trimethyl-3-
hydroxypentanal

Figure 18.15
Dean-Stark trap for removal of water by azeotropic distillation.

Figure 18.16
Apparatus for reaction of 2-methylpropanal with 3-buten-2-one.

EXPERIMENTAL PROCEDURES

Preparation of 4,4-Dimethyl-2-Cyclohexen-1-One

Purpose To demonstrate the preparation of a cyclohexenone via sequential conjugate addition and an intramolecular aldol condensation.

SAFETY ALERT

1. **Wear safety glasses or goggles and suitable protective gloves while performing the experiment.**

2. **Because volatile and flammable solvents are used in this experiment, *flameless* heating should be used for heating the reaction under reflux and for distilling the product. However, if it is necessary to use a small microburner to distill the product, be sure that no one is using flammable solvents in the laboratory.**

3. **Be certain that all joints in the apparatus are properly lubricated and firmly mated to keep vapors from escaping.**

4. **2-Methylpropanal and 3-buten-2-one are highly volatile irritants with penetrating odors. Do *not* breathe their vapors. If possible, these chemicals should be handled in a hood and measured using syringes to minimize their release into the laboratory.**

MINISCALE PROCEDURE

Preparation Refer to the online resources to answer Pre-Lab Exercises, access videos, and read the MSDSs for the chemicals used or produced in this procedure. Review Sections 2.9, 2.11, 2.13, 2.21, 2.22, 2.29, and 6.4.

Apparatus A 125-mL and a 25-mL round-bottom flask, two 5-mL syringes, stillhead, reflux condenser, separatory funnel, microburner, apparatus for magnetic stirring, simple distillation, and *flameless* heating.

Setting Up Equip the 125-mL round-bottom flask with a stirbar and add 0.1 g of 2-naphthalenesulfonic acid and 25 mL of toluene. Fit the stillhead with a rubber septum or stopper firmly seated in the male joint that is normally used for connecting to a condenser. Secure the rubber septum or stopper to the joint with wire. Working at the hood, use two different syringes to measure 5.0 mL of 2-methylpropanal and 3.5 mL of 3-buten-2-one into an Erlenmeyer flask containing 10 mL of toluene. Transfer this solution quickly, but carefully to avoid spillage, to the round-bottom flask. Attach the stillhead to the flask and fit the stillhead with a reflux condenser. Finish assembling the apparatus shown in Figure 18.16 for heating under reflux with azeotropic removal of water. Tilt the flask at an angle of about 30° from the vertical so that the condensate will drip into the sidearm of the distillation head.

Robinson Annulation Bring the solution to a brisk reflux, and continue heating under reflux for 2–2.5 h. The reflux ring must rise above the sidearm of the stillhead so that water collects in the plugged sidearm as the reaction proceeds. Following the period of reflux, allow the reaction mixture to cool to room temperature. Carefully disconnect the apparatus, *making certain that the collected water is not spilled back into the reaction flask.* Pour the water and solvent contained in the sidearm into a small graduated cylinder to determine what percentage of the theoretical amount of water has been collected.★

Work-Up and Isolation Transfer the reaction mixture to a separatory funnel, and wash it with a 20-mL portion of saturated aqueous sodium bicarbonate. Transfer the organic solution to a dry Erlenmeyer flask, and add several spatula-tips full of *anhydrous* sodium sulfate. Occasionally shake the mixture over a period of 10–15 min to facilitate drying; add further small portions of *anhydrous* sodium sulfate if the solution remains cloudy.★

Filter or decant the dry organic solution into a dry 125-mL round-bottom flask and equip the flask for simple distillation. Carefully remove the solvent and any unreacted starting materials by simple distillation; (*Caution: foaming*) collect a single fraction boiling up to 110–115 °C (760 torr). Alternatively, use rotary evaporation or other techniques to concentrate the solution. The final traces of solvent may be removed by attaching the flask to a vacuum source and gently swirling the contents as the vacuum is applied.★

Purification Using a Pasteur pipet, transfer the *cooled* pot residue to a 25-mL round-bottom flask containing a stirbar. Rinse the original stillpot with 1–2 mL of diethyl ether and add the rinse to the 25-mL round-bottom flask. Remove the diethyl ether by attaching the flask to a vacuum source and gently swirling the contents as the vacuum is applied. Fit this flask for a shortpath distillation, insulating the top of the distillation flask and stillhead with cotton or glasswool wrapped with aluminum foil. Using a microburner, distill the 4,4-dimethyl-2-cyclohexen-1-one into a *tared* receiver. Collect it as a single fraction boiling above 130 °C (760 torr), and record the temperature range. Note that the head temperature may never reach the reported boiling point, owing to the high boiling temperature and the small amount of sample being distilled. Terminate heating when the dark pot residue becomes viscous and starts to evolve fumes.

Analysis Weigh the product and calculate the percent yield. The product may be contaminated with residual toluene, as illustrated in Figures 18.17 and 18.18, the amount of which will be dependent upon the efficiency with which toluene was removed in the first stage of the distillation. Determine the purity of your product by GLC or ^{1}H NMR spectroscopic analysis. Prepare a 2,4-dinitrophenylhydrazone derivative from a portion of your product according to the procedure in Section 7.4B; this derivative may be recrystallized from 95% ethanol or some other suitable solvent. Obtain IR and ^{1}H NMR spectra of your starting materials and product, and compare them with those of authentic samples (Figs. 18.19–18.24).

MINISCALE PROCEDURE

Preparation Refer to the online resources to answer Pre-Lab Exercises, access videos, and read the MSDSs for the chemicals used or produced in this procedure. Review Sections 2.9, 2.11, 2.13, 2.21, 2.22, 2.25, 2.29, and 6.4.

Apparatus A 5-mL conical vial, two 1-mL syringes, microburner, apparatus for magnetic stirring, simple distillation, and *flameless* heating.

Setting Up Equip the conical vial with a spinvane and add 10 mg of 2-naphthalenesulfonic acid and 2.5 mL of toluene. Working at the hood, use two different syringes to measure first 0.35 mL of 3-buten-2-one and then 0.50 mL of 2-methylpropanal directly into the conical vial. Set up the apparatus for simple distillation. Using a Pasteur pipet, fill the well of the Hickman stillhead with toluene, being careful not to allow any toluene to flow back into the conical vial.

Robinson Annulation Bring the solution to a brisk reflux with stirring and continue heating under reflux for 2–2.5 h. To remove water from the reaction mixture, it is important that the reflux ring rise *into the bulb* of the Hickman stillhead. Allow the reaction mixture to cool to room temperature.★

Work-Up and Isolation Remove the spinvane and add 0.5 mL of saturated aqueous sodium bicarbonate to the conical vial. Cap the vial, invert it, and return it to its original upright position; release the pressure by *carefully* venting the screw-cap. Now replace the cap and shake the vial with intermittent venting to relieve pressure. Separate the layers and transfer the organic layer to a dry, screw-cap centrifuge tube. Add several microspatula-tips full of *anhydrous* sodium sulfate. Shake the centrifuge tube occasionally for a period of 10–15 min to facilitate drying; add further small portions of *anhydrous* sodium sulfate if the solution remains cloudy.★

Using a Pasteur pipet, transfer the organic solution to a dry 5-mL conical vial and equip the vial for simple distillation. Carefully remove the solvent by simple distillation (*Caution:* foaming); withdraw solvent from the Hickman stillhead as needed. Alternatively, use rotary evaporation or other techniques to concentrate the solution. The final traces of solvents may be removed by attaching the vial to a vacuum source and gently swirling the contents as the vacuum is applied.★

Purification (Optional) Remove the source of heat and dissemble the apparatus. Clean and dry the Hickman stillhead. Reassemble the apparatus for simple distillation, insulating the upper portion of the conical vial and the lower portion of the stillhead with cotton or glasswool wrapped with aluminum foil. Using a micro-burner, distill the 4,4-dimethyl-2-cyclohexen-1-one. Collect it as a single fraction. Terminate heating when the dark pot residue becomes viscous and starts to evolve fumes. Transfer the product to a tared sample vial.

Analysis Weigh the product and calculate the percent yield. The product may be contaminated with residual toluene, as illustrated in Figures 18.17 and 18.18, the amount of which will be dependent upon the efficiency with which toluene was removed. Determine the purity of your product by GLC or ^{1}H NMR spectroscopic analysis. Prepare a 2,4-dinitrophenylhydrazone derivative of a portion of your product according to the procedure in Section 7.4B; this derivative may be recrystallized from 95% ethanol or some other suitable solvent. Obtain IR and ^{1}H NMR spectra of your starting materials and product, and compare them with those of authentic samples (Figs. 18.19–18.24).

WRAPPING IT UP

Place the *recovered toluene* that was obtained from the azeotropic distillation and from removal from the reaction mixture into the container for nonhalogenated organic liquids. Flush the *sodium bicarbonate wash* down the drain. Use a small amount of *recovered toluene* to rinse the *pot residue* from the simple distillation of the product into the container for nonhalogenated organic liquids. After residual solvent has evaporated from the *sodium sulfate* on a tray in the hood, place the drying agent in a container for nonhazardous waste.

EXERCISES

1. Although benzene could be used as the solvent for this experiment, toluene is the solvent of choice because it is not carcinogenic. Furthermore, the reaction appears to go faster in toluene than in benzene, as judged by the rate at which water is formed. How might you account for this rate acceleration?

2. What is the structure of 2-naphthalenesulfonic acid, and what function does it serve in the preparation of 4,4-dimethyl-2-cyclohexen-1-one?

3. Why is it important that water *not* return to the reaction vessel during the reflux period?

4. Why is the final product of this reaction washed with aqueous sodium bicarbonate?

5. **a.** Figure 18.23 is the GLC trace of a sample of product obtained in this experiment. Calculate the ratio of toluene to **29** that is present, assuming that the response factor of the detector (Sec. 6.4) is the same for the two compounds. Note that the GLC column used in this analysis separates substances on the basis of their relative boiling points.

   b. Figure 18.24 is the ^{1}H NMR spectrum of this same sample. Calculate the ratio of toluene to **29** that is present according to this method of analysis.

   c. Account for the discrepancy, if any, between the two ratios that you obtain.

   d. The distillate from which the samples for these analyses were drawn weighed a total of 2.8 g. Based on the ^{1}H NMR analysis of the mixture, calculate how many grams of the desired **29** were present in the distillate. Calculate the yield of the reaction, assuming that it was run on the scale described in the Miniscale Procedure of this section.

6. Why is an enolate more nucleophilic than an enol?

7. Write two resonance structures for the α,β-unsaturated ketone **27**, and discuss how these structures show that both the carbonyl carbon atom and the β-carbon atom have electrophilic character.

8. Refer to Equation 18.12 and write the stepwise mechanism for the self-condensation of **26** to form **35**, using curved arrows to symbolize the flow of electrons.

9. Why does compound **35** *not* undergo dehydration?

10. Why is compound **31** less likely to form an enol than compound **26**?

11. Product **36** could potentially be produced by aldol condensation of **28** but is not. Provide a stepwise reaction mechanism for the possible formation of **36** in the presence of an acid catalyst. Explain why this alternative product is not observed.

36

12. Write a product of self-condensation of 3-buten-2-one (**27**), and explain why this process might be less favorable than the observed reaction between **26** and **27**.

13. Write the product expected to result from attack of enol **30** on the carbonyl carbon atom of **31**. Explain why formation of this product should be thermodynamically *less* favorable than formation of **28**.

14. Why is protonation of the carbonyl group in **33** likely to be involved in the conversion of **33** to **34**?

15. Using curved arrows to symbolize the flow of electrons, write a stepwise reaction mechanism for the acid-catalyzed dehydration of **34** to **29**.

16. Attempted acid-catalyzed reaction of propanal, CH_3CH_2CHO, with 3-buten-2-one (**27**) fails to give a good yield of the desired product, **37**. Explain why this might be.

27		**37**
Propanal		4-Methyl-2-cyclohexen-1-one

17. In the Robinson ring annulation reaction, the formation of five- and six-membered rings is favored over formation of rings having other sizes or bicyclic rings. Given this information, predict the product of the following Robinson ring annulation.

1,3-Cyclohexanedione   3-Buten-2-one

18. Outline a synthesis of the compound below via a Michael reaction using cyclo-pentanone as one of your starting materials.

Refer to the online resources and use the spectra viewer and Tables 8.1–8.8 as needed to answer the blue-numbered questions on spectroscopy.

19. Consider the spectral data for 2-methylpropanal (Figs. 18.17 and 18.18).

     a. In the functional group region of the IR spectrum, specify the absorptions associated with the carbonyl component and the carbon-hydrogen bond of the aldehyde group.

     b. In the ^{1}H NMR spectrum, assign the various resonances to the hydrogen nuclei responsible for them.

     c. For the ^{13}C NMR data, assign the various resonances to the carbon nuclei responsible for them.

20. Consider the spectral data for 3-buten-2-one (Figs. 18.19 and 18.20).

     a. In the functional group region of the IR spectrum, specify the absorptions associated with the carbonyl group and the carbon-carbon double bond. Also specify the absorption in the fingerprint region that is characteristic for the terminal vinyl group.

     b. In the ^{1}H NMR spectrum, assign the various resonances to the hydrogen nuclei responsible for them.

     c. For the ^{13}C NMR data, assign the various resonances to the carbon nuclei responsible for them.

21. Consider the spectral data for 4,4-dimethyl-2-cyclohexen-1-one (Figs. 18.21 and 18.22).

     a. In the functional group region of the IR spectrum, specify the absorptions associated with the carbonyl group and the carbon-carbon double bond.

     b. In the ^{1}H NMR spectrum, assign the various resonances to the hydrogen nuclei responsible for them.

     c. For the ^{13}C NMR data, assign the various resonances to the carbon nuclei responsible for them.

22. Discuss the differences observed in the IR and NMR spectra of 2-methylpropanal, 3-buten-2-one, and 4,4-dimethyl-2-cyclohexen-1-one that are consistent with the formation of the latter by the Robinson annulation in this experiment.

SPECTRA

Starting Materials and Products

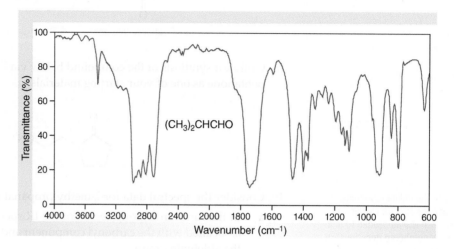

Figure 18.17
IR spectrum of 2-methylpropanal (neat).

Figure 18.18
NMR data for 2-methylpropanal (CDCl₃).

(a) ¹H NMR spectrum (300 MHz).
(b) ¹³C NMR data: δ 15.5, 41.2, 204.7.

Figure 18.19
IR spectrum of 3-buten-2-one (neat).

Figure 18.20

NMR data for 3-buten-2-one (CDCl₃).

(a) ¹H NMR spectrum (300 MHz).
(b) ¹³C NMR data: δ 26.3, 129.0, 137.5, 198.8.

Figure 18.21

IR spectrum of 4,4-dimethyl-2-cyclohexen-1-one (neat).

Figure 18.22

NMR data for 4,4-dimethyl-2-cyclohexen-1-one (CDCl₃).

(a) ¹H NMR spectrum (300 MHz).
(b) ¹³C NMR data: δ 27.7, 32.8, 34.4, 36.1, 126.8, 159.6, 199.0.

Figure 18.23
GLC analysis of reaction mixture for Exercise 5a. The retention times in minutes and the peak areas, respectively, are: Peak 1: 1.7, 10250; peak 2: 1.8, 37513.

Figure 18.24
¹H NMR spectrum (300 MHz) of 4,4-dimethyl-2-cyclohexen-1-one and toluene for Exercise 5b.

18.5 HETEROCYCLIC COMPOUNDS: SYNTHESIS OF A 1,2,3,4-TETRAHYDROPYRIMIDIN-2-ONE

Cyclic compounds that contain a **heteroatom** (non-carbon atom) as part of the ring are called **heterocyclic compounds**. The heterocyclic compounds you are most likely to encounter are those that have a nitrogen, oxygen, phosphorus, or sulfur atom in the ring. Heterocyclic compounds may be either aromatic or nonaromatic, and they are commonly found in naturally occurring molecules as well as in pharmaceutical agents that are clinically used to treat disease. There are many different heterocyclic ring systems, but the heterocyclic compound you will prepare in the experiment in this section belongs to the class of 1,2,3,4-tetrahydropyrimid-2-ones, **36**,whose ring structure is common to many biologically active substances. The derivative **37**, for example, has promise for treating hypertension, whereas **38** has potential as an anticancer agent.

36

37
Antihypertensive properties

38
Anticancer properties

Given the tremendous importance of many heterocyclic compounds, researchers must develop efficient methods for their synthesis. This is particularly true for pharmaceuticals and other fine chemicals that must be prepared on multi kilogram scales in high yield via procedures that ideally use inexpensive reagents and solvents and produce as little chemical waste as possible. In this context, **green chemistry** is a term that generally applies to efforts to develop new chemical processes and procedures that minimize waste and energy usage, use inexpensive chemicals, and limit the effects chemistry has on the environment. Green chemistry is becoming increasingly important in many large-scale industrial applications.

In order to introduce you to some of the principles of heterocyclic and green chemistry, you will prepare a simple derivative of **36** using a technique that simply involves grinding or mechanically stirring the reactants together *without* any solvent. Obviously, this process will not work for many reactions, especially those that are very exothermic, but when successful, it can be used in large-scale synthetic operations at lower costs than traditional solution-phase methods. Cost reductions result from reducing the amount of solvents used and limiting the cost of waste disposal, as well as sometimes decreasing the input of energy that is used to heat, or cool, reaction mixtures.

The specific reaction you will perform is the Biginelli reaction, named after Pietro Biginelli, who discovered the reaction in 1891. Overall, the reaction is a three-component process involving an aldehyde, a urea, and a β–ketoester as starting materials and produces the dihydropyrimidinone ring system, **36**. The transformation is catalyzed by acids, and *p*-toluenesulfonic acid, which is a common and inexpensive strong organic acid, is used in the procedure you will perform. Thoroughly mixing the two liquid reagents, benzaldehyde **(39)** and ethyl acetoacetate **(40)**, with urea, **(41)**, a solid, together with the catalyst allows the reaction to occur without using a solvent (Eq. 18.14). The product **42** has a formidable IUPAC name: ethyl 6-methyl-2-oxo-4-phenyl-1,2,3,4-tetrahydropyrimidine-5-carboxylate; of course, it is easier to label it as compound **42**.

39
Benzaldehyde

40
Ethyl acetoacetate

41
Urea

42
Ethyl 6-methyl-2-oxo-4-phenyl-1,2,3,4-tetra-hydropyrimidine-5-carboxylate

EXPERIMENTAL PROCEDURE

Synthesis of Ethyl 6-Methyl-2-Oxo-4-Phenyl-1,2,3,4-Tetrahydropyrimidine-5-Carboxylate

 Green Experiment   **Purpose** To prepare a tetrahydropyrimidin-2-one using a "green" chemistry technique.

SAFETY ALERT

1. **Wear safety glasses or goggles and suitable protective gloves while performing the experiment.**

2. ***p*-Toluenesulfonic acid is very corrosive and may cause serious chemical burns if allowed to come into contact with your skin. If any *p*-toluenesulfonic acid comes into contact with your skin, immediately flood the affected area with cold water and then with 5% sodium bicarbonate solution.**

MINISCALE PROCEDURE

Preparation Refer to the online resources to answer Pre-Lab Exercises, access videos, and read the MSDSs for the chemicals used or produced in this procedure. Review Sections 2.7, 2.9, 2.11, 2.17, 2.19, 2.26, and 6.2.

Apparatus A 25-mL round-bottom flask, oil bath, apparatus for magnetic stirring, *flameless* heating, and vacuum filtration.

Setting Up Preheat the oil bath to 70 °C, preferably on a magnetic stirrer hot plate. Add 1.33 g of benzaldehyde, 1.63 g of ethyl acetoacetate, 0.95 g of finely powdered urea, and 0.10 g of *p*-toluenesulfonic acid to the 25-mL round-bottom flask containing a stirbar.

Reaction Immerse the flask halfway into the hot oil bath and begin stirring the contents of the flask immediately. The mixture will begin to solidify in 3–5 min. Continue heating at 70°C for 1 h.

Work-Up and Isolation Remove the flask from the oil bath and allow it to cool to room temperature. Add 10 mL of water to the flask and stir its contents to break up the solid mass. Collect the solid by vacuum filtration using a Hirsch funnel, pressing the filter cake with a spatula to remove as much water as possible. Reserve a small of amount of crude product for TLC analysis. Rinse the filter cake with two 2- to 3-mL portions of ice-cold 95% ethanol, carefully stirring the rinse and the solid on the filter before reapplying the vacuum. Reserve a small amount of the ethanol-washed product for TLC analysis, and recrystallize the remaining crude product from 95% ethanol or some other suitable solvent. Isolate the product by vacuum filtration and air-dry it.

Analysis Weigh the product and calculate the percent yield. Determine the melting point of the recrystallized solid. Perform a TLC analysis comparing the purities of the crude product that was isolated before *and* after resuspending in 95% ethanol and after recrystallization. Use 95% ethanol to dissolve the crude and purified materials for spotting the plate and a 1:1 mixture of hexanes and ethyl acetate as the eluting solvent. Obtain the IR and NMR spectra of your starting materials and product and compare them with those of authentic samples (Figs.18.1, 18.2, 18.25–18.28).

MICROSCALE PROCEDURE

Preparation Refer to the online resources to answer Pre-Lab Exercises, access videos, and read the MSDSs for the chemicals used or produced in this procedure. Review Sections 2.7, 2.9, 2.11, 2.17, 2.19, 2.26, and 6.2.

Apparatus A 13-mm × 100-mm test tube, glass stir rod, oil bath, apparatus for *flameless* heating and vacuum filtration.

Setting Up Preheat the oil bath to 70°C. Add 0.27 g of benzaldehyde,0.33 g of ethyl acetoacetate, 0.19 g of finely powdered urea, and 0.02 g of *p*-toluenesulfonic acid to the test tube.

Reaction Immerse the test tube halfway into the hot oil bath and stir the contents of the flask using the glass rod. The mixture will begin to solidify in 3–5 min. Continue heating at 70 °C for 1 h with occasional stirring.

Work-Up and Isolation Remove the test tube from the oil bath and allow it to cool to room temperature. Add 1–2 mL of water to the test tube, stir the contents to break up the solid mass, and collect the solid by vacuum filtration using a Hirsch funnel, pressing the filter cake with a spatula to remove as much water as possible. Reserve a small amount of crude product for TLC analysis. Rinse the filter cake with two 0.5- to 1-mL portions of ice-cold 95% ethanol, carefully stirring the rinse and the solid on the filter before reapplying the vacuum. Reserve a small amount of the ethanol-washed product for TLC analysis, and recrystallize the remaining crude product from 95% ethanol or some other suitable solvent. Isolate the product by vacuum filtration and air-dry it.

Analysis Weigh your product and calculate the percent yield. Determine the melting point of the recrystallized solid. Perform a TLC analysis comparing the purities of the crude product that was isolated before *and* after resuspending in 95% ethanol and after recrystallization. Use 95% ethanol to dissolve the crude and purified materials for spotting the plate and a 1:1 mixture of hexanes and ethyl acetate as the eluting solvent. Obtain the IR and NMR spectra of your starting materials and product and compare them with those of authentic samples (Figs.18.1, 18.2, 18.25–18.28).

WRAPPING IT UP

Combine the *aqueous* washes and wash down the drain, and dispose the *ethanol* in the container for nonhalogenated compounds.

EXERCISES

1. Compare the melting point you determined for your product with the literature value. If your melting point is different, suggest reasons for the difference. For example, what, if any, effect might the presence of impurities have on the melting point of your compound?

2. This Biginelli reaction was performed without any solvent. Based on your experience with conducting this reaction, can you suggest one or more reasons one might want to use a solvent for this reaction? What solvent might you choose?

3. Using curved arrows to symbolize the flow of electrons, write a complete mechanism for the Biginelli reaction you performed.

4. If the Biginelli reaction is not heated, a solid product is formed, but it is not **42**. Based upon the mechanism you proposed to answer Exercise 3, suggest one or more possible structures for this material.

5. How many side products did you observe in the TLC of the crude product? Of the purified product? Assuming none of these are starting materials, propose possible structures of these compounds.

Refer to the online resources and use the spectra viewer and Tables 8.1–8.8 as needed to answer the blue-numbered questions on spectroscopy.

6. Consider the spectral data for benzaldehyde (Figs. 18.1 and 18.2).

   a. In the functional group region of the IR spectrum, specify the absorptions associated with the carbonyl group and the π-bonds of the aromatic ring.

   b. In the ^{1}H NMR spectrum, assign the various resonances to the hydrogen nuclei responsible for them.

   c. For the ^{13}C NMR data, assign the various resonances to the carbon nuclei responsible for them.

7. Consider the spectral data for ethyl acetoacetate (Figs. 18.25 and 18.26).

   a. In the functional group region of the IR spectrum, specify the absorptions associated with the carbonyl groups. Explain the source of the broad absorption at about $3700 - 3400$ cm^{-1}.

   b. In the ^{1}H NMR spectrum, assign the various resonances to the hydrogen nuclei responsible for them.

   c. For the ^{13}C NMR data, assign the various resonances to the carbon nuclei responsible for them.

8. Consider the spectral data for ethyl 6-methyl-2-oxo-4-phenyl-1,2,3,4-tetrahydropyrimidine-5-carboxylate (Figs. 18.27 and 18.28).

   a. In the functional group region of the IR spectrum, specify the absorptions associated with the carbonyl group and the π-bonds of the aromatic ring.

   b. In the ^{1}H NMR spectrum, assign the various resonances to the hydrogen nuclei responsible for them.

   c. For the ^{13}C NMR data, assign the various resonances to the carbon nuclei responsible for them.

9. What differences observed in the IR and NMR spectra of benzaldehyde, ethyl acetoacetate, and ethyl 6-methyl-2-oxo-4-phenyl-1,2,3,4-tetrahydropyrimidine-5-carboxylate are consistent with formation of the desired product in this reaction?

Starting Materials and Products

The IR and NMR spectra of benzaldehyde are provided in Figures 18.1 and 18.2, respectively.

Figure 18.25
IR spectrum of ethyl acetoacetate (ATR).

Figure 18.26
NMR data for ethyl acetoacetate (CDCl₃).

(a) ¹H NMR spectrum (MHz).

(b) ¹³C NMR data: δ (keto) 13.8, 29.8, 49.8, 61.0, 166.9, 200.4; (enol) 14.0, 20.8, 59.6, 89.5, 175.2.

Figure 18.27
IR spectrum of ethyl 6-methyl-2-oxo-4-phenyl-1,2,3,4-tetra-hydropyrimidine-5-carboxylate (ATR).

Figure 18.28
NMR data for ethyl 6-methyl-2-oxo-4-phenyl-1,2,3,4-tetrahydropyrimidine-5-carboxylate (DMSO-d_6).

(a) ^{1}H NMR spectrum (MHz).
(b) ^{13}C NMR data: δ 14.5, 18.2, 54.4, 59.6, 99.7, 126.7, 127.7, 128.8, 145.3, 148.8, 152.6, 165.7.

CHAPTER 19

Organometallic Chemistry

It should by now be no surprise to you that carbon atoms are not the least bit "happy" about bearing a charge. This reluctance to form ions is seen in the high energies of carbocations, for example, which are not formed at all at primary carbon atoms and are possible for tertiary carbon atoms only if a highly polar solvent is present to assist in stabilizing the ion (Chap. 14). The same aversion to accommodate a charge attends the formation of unstabilized carbanions, that is, anions that are not delocalized through electronegative substitutents like carbonyl groups (Chap. 18). Nonetheless, it's possible to generate chemical species that behave as though they are unstabilized carbanions. Their creation and reactions are the subject of this chapter.

19.1 INTRODUCTION

Organometallic compounds are substances that contain carbon-metal bonds, and they may be generally represented by the structure **1**, in which the metal, **M**, may be Li, Na, Mg, Cu, Zn, Hg, Pd, or other transition elements. Organomagnesium compounds **1**, M = MgBr, which are commonly called **Grignard reagents**, were the first organometallic substances to be extensively studied, and they are still among the most important. (See the Historical Highlight *Grignard and the Beginnings of Modern Organometallic Chemistry*, which is available online.) The polarization of the bond between the carbon atom and the electropositive metal in these reagents renders the carbon atom electron-rich, or Lewis basic, and the carbon atom bears a partial negative charge, δ^-, as shown. With this polarization in mind, you should not consider the metallated carbon atom as a true anion, because it retains some covalent bonding to its metallic partner, although the extent of this bonding remains a point of discussion among chemists.

One of the characteristic properties of organometallic reagents **1** is that the carbon atom serves as a **nucleophile** in chemical reactions. In contrast, when a carbon atom is bonded to more electronegative elements such as the halogen atom in the alkyl halide **2** (X = Cl, Br, or I) or oxygen atom in the carbonyl compound **3**, it is electron-deficient or Lewis acidic and possesses a partial positive charge. Such carbon atoms then serve as **electrophiles** in chemical reactions.

$$R^2-\underset{\underset{R^3}{|}}{\overset{\overset{R^1}{|}}{C}}{}^{\delta^-}-M^{\delta^+} \qquad R^2-\underset{\underset{R^3}{|}}{\overset{\overset{R^1}{|}}{C}}{}^{\delta^-}-X^{\delta^+} \qquad \underset{R^2}{\overset{R^1}{}}{}^{\delta^+}C=O^{\delta^-}$$

1	**2**	**3**
An organometallic reagent	An alkyl halide	A carbonyl-containing compound

715

Because of their nucleophilic character, organometallic compounds are widely utilized as reagents in reactions that produce new carbon-carbon bonds. For example, two typical reactions of organometallic reagents **1** with carbon electrophiles such as alkyl halides **2** and carbonyl compounds **3** are illustrated by the general transformations shown in Equations 19.1 and 19.2. In each of these reactions, the nucleophilic carbon atom of one reactant becomes attached to the electrophilic carbon atom of the other reactant with the resulting formation of a new carbon-carbon bond. Thus, like many bond-forming processes, these reactions may be viewed in the simple context of *combinations of Lewis bases with Lewis acids.*

$$\overset{\delta^-}{R^1CH_2}-\overset{\delta^+}{M} + \overset{\delta^-}{X}-\overset{\delta^+}{CH_2R^2} \longrightarrow R^1CH_2-CH_2R^2 + MX \qquad (19.1)$$

An organometallic   An alkyl halide   An alkane
reagent

$$\overset{\delta^-}{R^1CH_2}-\overset{\delta^+}{M} + \underset{R^3}{\overset{R^2}{\underset{\displaystyle}{}}}\overset{\delta^+}{C}=\overset{\delta^-}{O} \longrightarrow R^1CH_2-\underset{R^3}{\overset{R^2}{C}}-O^- M^+ \xrightarrow{H_3O^+} R^1CH_2-\underset{R^3}{\overset{R^2}{C}}-OH \qquad (19.2)$$

An organometallic   An aldehyde   A metal   An alcohol
reagent   or ketone   alkoxide

The following discussions focus on the preparation and reactions of two important classes of organometallic compounds, Grignard reagents **1** (M = MgX) and organozinc reagents **1** (M = ZnX). Of course, many of the principles that are presented may be applied to the chemistry of other organometallic reagents. The first series of experiments encompass the preparation of Grignard reagents from aryl and alkyl bromides, followed by their representative reactions with (1) an ester to produce a tertiary alcohol, (2) carbon dioxide to produce a carboxylic acid, and (3) an aldehyde to produce a secondary alcohol. The other set of experiments involves the preparation of an organozinc reagent and its reaction with a ketone to prepare a tertiary alcohol.

19.2 GRIGNARD REAGENTS: PREPARATION

Reactions of Organic Halides with Magnesium Metal

Grignard reagents, R–MgX or Ar–MgX, are typically prepared by the reaction of an alkyl halide, R–X, or an aryl halide, Ar–X, with magnesium metal in an *anhydrous* ethereal solvent (Eq. 19.3); the organometallic reagent dissolves as it is formed. You may note that carbon is transformed from an *electrophilic center* in the starting material R–X or Ar–X into a *nucleophilic center* in the product R–MgX or Ar–MgX in this process.

$$\overset{\delta^+}{R}-\overset{\delta^-}{X} \text{ or } \overset{\delta^+}{Ar}-\overset{\delta^-}{X} + Mg^\circ \xrightarrow[\text{or THF (solvent)}]{\text{dry diethyl ether}} \overset{\delta^-}{R}-\overset{\delta^+}{MgX} \text{ or } \overset{\delta^-}{Ar}-\overset{\delta^+}{MgX} \qquad (19.3)$$

An alkyl   An aryl   An alkyl   An aryl
halide   halide   magnesium halide   magnesium halide

Although it is customary to represent the Grignard reagent by the formula R–MgX or Ar–MgX, the structure of the organometallic species in solution is rather more complex. For example, with alkyl magnesium halides, there is an equilibrium between RMgX, R_2Mg, and MgX_2 that depends on the solvent, the halide ion, and the nature of the alkyl group. Moreover, the various organometallic species form aggregates in solution.

Grignard reagents are readily prepared from alkyl and aryl chlorides, bromides, and iodides but are rarely synthesized from organofluorides. The ease of formation of Grignard reagents from alkyl halides follows the order R–I > R–Br > R–Cl. Aryl halides are less reactive than their alkyl counterparts, and aryl bromides and chlorides are comparable in reactivity.

In the experiments in this chapter, you will prepare one or both of the Grignard reagents derived from bromobenzene (**4**) and 1-bromobutane (**6**) in an ethereal solvent, according to Equations 19.4 and 19.5. The preparation of these reagents theoretically requires equivalent amounts of the organic halide and magnesium, but a slight excess of magnesium is normally used.

$$\text{4} \quad + \quad Mg \quad \xrightarrow{\text{diethyl ether}} \quad \text{5} \qquad (19.4)$$

4
Bromobenzene

5
Phenylmagnesium bromide

$$CH_3CH_2CH_2CH_2{-}Br \; + \; Mg \; \xrightarrow{\text{diethyl ether}} \; CH_3CH_2CH_2CH_2{-}MgBr \qquad (19.5)$$

6
1-Bromobutane

7
1-Butylmagnesium bromide

Using an ethereal solvent is critical for the efficient preparation of the Grignard reagent because the basic oxygen atom of the ether complexes with the electropositive magnesium atom to help stabilize the organometallic species. The ethereal solvents most commonly used in this reaction are diethyl ether, $(C_2H_5)_2O$, and tetrahydrofuran, THF, $(CH_2)_4O$. Diethyl ether is often the solvent of choice because it is less expensive, may be purchased in anhydrous form, and is easily removed from the reaction mixture owing to its low boiling point (bp 36 °C, 760 torr). Tetrahydrofuran is a stronger Lewis base than diethyl ether and it also has better solvating ability; it may be used when the Grignard reagent does not form readily in diethyl ether.

You should be careful to use freshly opened containers of *anhydrous* diethyl ether or THF, because opening the container exposes its contents to atmospheric oxygen, which promotes the formation of **hydroperoxides**. For example, diethyl ether can be converted to the hydroperoxide, $CH_3CH(OOH)OCH_2CH_3$. Such peroxides are *explosive*, and large volumes of *anhydrous* diethyl ether or THF suspected to contain peroxides should *not* be evaporated to dryness. Moreover, emptied containers should be *thoroughly* rinsed with water to remove any peroxides before being discarded. Peroxides in *anhydrous* diethyl ether or THF may be conveniently destroyed by distillation from alkali metals such as sodium or potassium metal or from lithium aluminum hydride, $LiAlH_4$. The presence of peroxides in diethyl ether and THF may be detected by placing a drop of the solvent on a piece of *moistened* starch/iodide test paper. If the paper turns dark violet, owing to formation of a starch-iodine complex, then the ether contains peroxides.

The magnesium metal used for the preparation of the Grignard reagent is normally in the form of turnings—thin shavings that have a high surface area relative to chunks of the metal. This type of magnesium is generally suitable for the preparation of most Grignard reagents. However, if the turnings have been repeatedly exposed to atmospheric oxygen, their surface may be covered by a coating of magnesium oxide, which decreases their reactivity. Magnesium ribbon rather than turnings may also be used *after* the oxide coating on it has been scraped off with the edge of a spatula.

Initiating the reaction between the organic halide and magnesium may be difficult, especially with unreactive halides. In these cases, the reaction may often be initiated by adding a small crystal of iodine, I_2, to the reaction mixture. The iodine facilitates the reaction either by activating the metal through removal of some of its oxide coating or by converting a small amount of the R–X halide to the corresponding iodide, which is more reactive toward magnesium. Alternatively, the mixture can be placed in the bath of an ultrasonicator to initiate the reaction.

The formation of a Grignard reagent is an **exothermic** process. Since side reactions may occur if the reaction is allowed to proceed uncontrolled, it is important to regulate the rate at which the reaction proceeds. Grignard reactions often require an **induction period**, so it is *important* to be sure that the reaction of the alkyl or aryl halide with the magnesium has initiated *before* adding large quantities of the halide. Once the reaction is in progress, the halide should be added *dropwise and slowly* to the stirred suspension of the magnesium metal in the ethereal solvent; the halide may be added either neat or as a solution in the ethereal solvent. Adding the halide slowly keeps its concentration low, thus enabling better control of the rate of the reaction and the evolution of heat. The heat that is generated usually brings the solvent to its reflux temperature, so the heat of the reaction is ultimately transferred to the cooling water in the condenser. Should it appear that the reaction is getting out of control, as evidenced by vapors escaping from the top of the condenser, the reaction mixture must be immediately cooled with an ice-water bath. Indeed, as a general rule, it is prudent always to have such a cooling bath prepared when performing an exothermic reaction.

Side Reactions

Side reactions may be encountered when forming Grignard reagents. For example, as the **conjugate bases** of the exceedingly weak organic acids R–H (R = aryl and alkyl), Grignard reagents are *very* strong bases that react rapidly with water according to Equation 19.6. This reaction results in the destruction of the Grignard reagent and the formation of the corresponding hydrocarbon, RH, and a basic magnesium salt. Grignard reagents also cannot be prepared when carboxyl (CO_2H), hydroxyl (OH), or amino (NH_2) groups are present in the alkyl or aryl halide; the acidic hydrogens of these functional groups will simply protonate the highly basic carbon atom of the Grignard reagent, thus destroying it as it forms.

$$R\text{—}MgX \ + \ H_2O \longrightarrow R\text{—}H \ + \ HO\text{—}Mg\text{—}X \qquad (19.6)$$

The reaction of Grignard reagents with water (Eq. 19.6) dictates that all reagents, solvents, and apparatus used for their preparation must be *thoroughly* dry. Consequently, *anhydrous* ethereal solvents that typically contain less than 0.01% water must be used. *Anhydrous* diethyl ether rapidly absorbs atmospheric moisture, so opening a container of it a number of times over a period of several days renders the diethyl ether unsuitable for use in preparing a Grignard reagent. Only freshly opened cans should be used, and the cans should always be tightly sealed *immediately* after the necessary volume of ether has been removed. These operations will also minimize the formation of dangerous hydroperoxides.

Other side reactions that may occur during formation of a Grignard reagent are shown in Equations 19.7–19.9, but these can be minimized by taking certain precautions. For example, reaction of the Grignard reagent with oxygen and carbon dioxide (Eqs. 19.7 and 19.8) is typically avoided in the research laboratory by performing the reaction under an inert atmosphere such as nitrogen (N_2) or argon. However, this precaution is not essential in the undergraduate laboratory, because when diethyl ether is used as the solvent, its very high vapor pressure effectively excludes most of the air from the reaction vessel.

The coupling reaction (Eq. 19.9) is an example of a Wurtz-type reaction. Although this can be a useful process for preparing symmetrical hydrocarbons, R–R, it is normally desirable to minimize this side reaction by using dilute solutions, thereby avoiding high localized concentrations of the halide. This is accomplished by efficient stirring and by slowly adding the halide to the suspension of magnesium in the ethereal solvent.

$$RMgX + O_2 \longrightarrow R—O—O^- \: ^+MgX \tag{19.7}$$

$$RMgX + CO_2 \longrightarrow RCO_2^- \: ^+MgX \tag{19.8}$$

$$RMgX + RX \longrightarrow R—R + MgX_2 \tag{19.9}$$

In the experiments that follow, phenylmagnesium bromide (**5**) and 1-butylmagnesium bromide (**7**) are prepared according to Equations 19.4 and 19.5, respectively. The most important side reaction in these experiments involves the Wurtz-type coupling of the Grignard reagent with the organic halide. For example, during the preparation of phenylmagnesium bromide, small quantities of biphenyl (**8**) are formed according to Equation 19.10; however, the presence of **8** does not interfere with the subsequent reactions of phenylmagnesium bromide. Although coupling also occurs during the preparation of 1-butylmagnesium bromide, the *n*-octane that is produced is volatile and easily removed.

$$\tag{19.8}$$

| **5** | **4** | **8** |
| Phenylmagnesium bromide | Bromobenzene | Biphenyl |

EXPERIMENTAL PROCEDURES

Preparation of Grignard Reagents

Purpose To perform techniques required to prepare Grignard reagents from aryl and alkyl halides.

SAFETY ALERT

1. **Wear safety glasses or goggles and suitable protective gloves while performing the experiment.**

2. **Diethyl ether is *extremely* flammable and volatile, and its vapors can easily travel several feet along the bench top or the floor and then be ignited.**

Consequently, *be certain there are no open flames anywhere in the laboratory whenever you are working with ether. Use a flameless heating source whenever heating is required.*

3. The *anhydrous* diethyl ether used in this experiment is contained in metal cans, and the screw or plastic cap should *always* be in place when the can is not in use to prevent evaporation, absorption of atmospheric moisture and oxygen, and accidental fires.

4. *Open containers of diethyl ether must not be kept at your laboratory bench or stored in your laboratory drawer.* Estimate the total volume of ether you will need and measure it in the hood into a container that is *loosely* stoppered.

5. You should use ovens to dry your glassware if possible. However, if drying ovens are not available and it is necessary to dry the glass apparatus with a flame or a heat gun, be *certain* that no one in the laboratory is working with diethyl ether. *Consult with your instructor before using any open flame.* Avoid excessive heating in the vicinity of the ring seals in the condenser and near the stopcock in the addition funnel, particularly if the stopcock is made of plastic or Teflon.

6. Lubricate all ground-glass joints in the apparatus carefully and mate them tightly to prevent the escape of diethyl ether during the reaction.

7. On the small scale of these experiments, the exothermic formation and reaction of Grignard reagents rarely causes a problem. Nevertheless, it is still good laboratory practice for you to have an ice-water bath ready if the reaction proceeds too rapidly, as evidenced by an *excessively* rapid rate of reflux and the emission of vapors from the top of the condenser.

MINISCALE PROCEDURE

Preparation Refer to the online resources to answer Pre-Lab Exercises, access videos, and read the MSDSs for the chemicals used or produced in this procedure. Review Sections 2.9, 2.10, 2.11, 2.22, 2.27, and 2.28.

Apparatus A 50-mL round-bottom flask, 5-mL syringe, separatory funnel with a ground-glass joint, condenser, Claisen adapter, drying tube, ice-water bath, apparatus for magnetic stirring, and *flameless* heating.

Setting Up Weigh 0.5 g of magnesium turnings that have been freshly crushed with a spatula into the round-bottom flask, and add a stirbar. Place this flask and its contents, the separatory funnel, condenser, Claisen adapter, and drying tube in an oven at 110 °C for at least 30 min. If the separatory funnel has a plastic or Teflon stopcock and stopper, do *not* put the stopcock, its plastic retaining nut, and the stopper in the oven, as they may melt or soften. Using gloves or tongs, remove the glassware from the oven and let it cool, preferably in a desiccator. After the glassware is cool enough to handle, lubricate all the joints and quickly assemble the apparatus shown in Figure 2.66b. Attach the drying tube to the top of the condenser and place the stopper and stopcock in the separatory funnel. Allow the apparatus to cool to room temperature.

Optional Measures If an oven is not available, it will be necessary to dry the apparatus with a microburner or a heat gun. Assemble the apparatus as described above. *Be sure that no one in the laboratory is working with diethyl ether,* and then dry the assembled apparatus. Do not overheat any plastic parts of the apparatus. Allow the apparatus to cool to room temperature.

Verify that there are no *flames* in the laboratory before continuing. Prepare a solution of *either* 2.4 mL of bromobenzene or 2.5 mL of 1-bromobutane in 5 mL of *anhydrous* diethyl ether in a small, *dry* Erlenmeyer flask. Swirl the solution to achieve homogeneity. Add 5 mL of *anhydrous* diethyl ether to the round-bottom flask *through* the separatory funnel; *close the stopcock.* Be sure that water is running through the condenser. Transfer the ethereal solution of halide to the separatory funnel.

Reaction Add a 0.5-mL portion of the ethereal solution from the separatory funnel onto the magnesium turnings and stir the resulting mixture. If small bubbles form at the surface of the magnesium turnings or if the mixture becomes slightly cloudy or chalky, the reaction has started. The flask should become *slightly* warm. If the reaction has started, *disregard* the optional instructions in the next paragraph.

Optional Measures If the reaction does not start spontaneously, warm the mixture gently for several minutes and observe whether the mixture becomes slightly cloudy or chalky. If it does not, then obtain one or two additional magnesium turnings and crush them *thoroughly* with a heavy spatula or the end of a clamp. Remove the separatory funnel just long enough to add these broken pieces of magnesium to the flask and quickly replace the funnel. The clean, unoxidized surfaces of magnesium that are exposed should aid in initiating the reaction. If the reaction still has not started after an additional 3–5 min of warming, consult your instructor. The best remedy at this point is to add a small crystal of iodine to the mixture. Alternatively, a small amount of the preformed Grignard reagent may be added if it is available.

Once the reaction has started, *gently* heat the reaction mixture so that the solvent refluxes smoothly. Add another 5-mL portion of *anhydrous* diethyl ether to the reaction mixture through the top of the *condenser* and continue heating and stirring until the solvent is again refluxing. Add the remainder of the ethereal solution of the halide dropwise to the stirred reaction mixture at a rate that is just fast enough to maintain a gentle reflux. If the reaction becomes too vigorous, reduce the rate of addition and discontinue heating the flask, if necessary. If the spontaneous boiling of the mixture slows, increase the rate of addition slightly. If the rate of reflux still does not increase, heat the mixture as necessary to maintain gentle reflux during the remainder of addition. *It is important that reflux be maintained throughout the addition of the ethereal solution.* The addition should take about 5–10 min. Upon completing the addition, continue heating the mixture under gentle reflux for 15 min. If necessary, add *anhydrous* diethyl ether to the reaction flask so that there is no less than about 15 mL of solution. At the end of the reaction, the solution normally has a tan to brown, chalky appearance, and most of the magnesium will have disappeared, although residual bits of metal usually remain. Discontinue heating and allow the mixture to cool to room temperature.

Use the Grignard reagent as soon as possible after preparing it. Phenylmagnesium bromide is used in Parts A and B of Section 19.4, and 1-butylmagnesium bromide is used in Part C of that section.

MICROSCALE PROCEDURE

Preparation Refer to the online resources to answer Pre-Lab Exercises, access videos, and read the MSDSs for the chemicals used or produced in this procedure. Review Sections 2.9, 2.10, 2.11, 2.22, 2.27, and 2.28.

Apparatus A 3-mL and 5-mL conical vial, screw-cap centrifuge tube, two 1-mL plastic (or glass) syringes, Pasteur pipet with 0.5- and 1.0-mL calibration marks, Claisen adapter, drying tube, ice-water bath, apparatus for magnetic stirring, and *flameless* heating.

Setting Up Dry the conical vials, Claisen adapter, centrifuge tube, drying tube, and calibrated Pasteur pipet in an oven at 110 °C for at least 30 min. Do *not* put any plastic connectors or rubber O-rings in the oven, as they may melt or soften. Using gloves or tongs, remove the glassware from the oven and let it cool, preferably in a desiccator. Lubricate the joints and assemble the apparatus shown in Figure 19.1 by adding a spinvane to the 5-mL conical vial and then fitting the Claisen adapter to the vial. Finally, place a rubber septum and the drying tube on the Claisen adapter.

Optional Measures If an oven is not available, it will be necessary to dry the apparatus with a microburner or a heat gun. Assemble the apparatus as described above. *Be sure that no one in the laboratory is working with diethyl ether,* and then dry the assembled apparatus. Be careful heating the conical vial, as this heavy-walled vessel can easily crack from thermal shock. Also, do not overheat any plastic parts of the apparatus. Allow the apparatus to cool to room temperature.

Verify that there are no *flames* in the laboratory before continuing. Transfer about 3–4 mL of *anhydrous* diethyl ether to a screw-cap centrifuge tube and cap the tube. Use this during the remainder of the experiment whenever *anhydrous* diethyl ether is required. Weigh 0.05 g of magnesium turnings that have been freshly crushed with a heavy spatula or the end of a clamp. Remove the rubber septum and transfer the turnings to the reaction vial. Add one *small* crystal of iodine to the vial and replace the rubber septum. Add 0.2 mL of *anhydrous* diethyl ether

Figure 19.1
Microscale apparatus for preparing and reaction of Grignard reagents.

- Syringe
- Rubber septum
- Drying tube
- Claisen adapter
- Conical vial
- Spinvane
- Magnetic stirrer/hot plate
- Laboratory jack

to the vial using a dry 1-mL syringe inserted through the rubber septum. Prepare a solution of 0.24 mL of bromobenzene in 0.5 mL of *anhydrous* diethyl ether in a dry 3-mL conical vial. Swirl the solution to achieve homogeneity.

Reaction Stir the contents of the conical vial while warming them gently. Draw the ethereal solution of bromobenzene into the plastic syringe and cap the conical vial containing the bromobenzene solution. Insert the syringe needle through the rubber septum, and add a 0.1-mL portion of this solution onto the magnesium turnings; stir the mixture. *Hold* the plunger of the syringe to control the amount of solution added. If small bubbles form at the surface of the magnesium turnings or if the mixture becomes slightly cloudy or chalky, the reaction has started.

Optional Measures If the reaction does not start spontaneously, consult your instructor. The best remedy at this point is to continue warming the vial and add an additional crystal of iodine to the mixture. You may also add a small sample of phenylmagnesium bromide from a stock solution if it is available.

Once the reaction has started, continue heating the reaction mixture gently so that the solvent refluxes slightly. Using a second plastic syringe inserted through the rubber septum, add another 0.5-mL portion of *anhydrous* diethyl ether to the reaction mixture and continue heating and stirring until the solvent is again refluxing. Add the remainder of the bromobenzene-ether solution *dropwise* to the stirred reaction mixture at a rate that is just fast enough to maintain a gentle reflux. If the reaction becomes too vigorous, reduce the rate of adding the ethereal solution of aryl halide and, if necessary, discontinue heating the vial. If the spontaneous boiling of the mixture slows, increase the rate of addition slightly. If the rate of reflux still does not increase, increase heating and maintain gentle reflux during the remainder of addition. *It is important that reflux be maintained throughout the addition of the bromobenzene-ether solution.* The addition should take about 3–5 min. Upon completion of the addition, place about 0.5 mL of *anhydrous* diethyl ether in the vial that contained the bromobenzene solution. Draw this solution into a syringe and add it in one portion to the reaction mixture. Continue heating the mixture with stirring under gentle reflux for 15 min. If necessary, add *anhydrous* diethyl ether to maintain a level *no lower* than the 2-mL mark on the conical vial. At the end of the reaction, the solution normally has a tan to brown, chalky appearance, and most of the magnesium will have disappeared, although residual bits of metal usually remain. Discontinue heating and allow the mixture to cool to room temperature.

Use the Grignard reagent as soon as possible after preparing it, following one of the procedures given in Section 19.4.

EXERCISES

1. Answer the following questions for a Grignard reagent, R_3C-MgX.

   a. Indicate the direction of the dipole in the $C-Mg$ bond.

   b. Is the carbon atom associated with magnesium electrophilic or nucleophilic?

   c. Write the conjugate acid of a Grignard reagent, and explain why Grignard reagents are strong bases.

2. In the reaction of bromobenzene with magnesium to form phenylmagnesium bromide, C_6H_5MgBr, which reagent undergoes oxidation and which undergoes reduction? Write the half-reactions for the two processes.

3. Why should the glassware not be flame-dried in the vicinity of diethyl ether, the solvent used for preparing the Grignard reagent?

4. Why were you cautioned not to heat excessively in the vicinity of plastic parts, rubber O-rings, and the ring seals of the condenser when *flame*-drying the apparatus for this experiment?

5. Ethanol is often present in solvent-grade diethyl ether. If this grade rather than anhydrous were used, what effect would the ethanol have on the formation of the Grignard reagent?

6. Why is it necessary to lubricate the joints of the apparatus used to prepare a Grignard reagent?

7. What is the purpose of adding an additional portion of diethyl ether after the Grignard reaction is initiated and before the addition of halide is continued?

8. Why is it important to add the ethereal solution of the halide *slowly* to the magnesium turnings?

9. Why is it possible to use solvent-grade diethyl ether in the work-up procedure, whereas the more expensive anhydrous form must be used to prepare the Grignard reagent?

10. Explain how the undesired coupling product **8** is formed and how its formation can be minimized.

11. Why is it unwise to allow the solution of the Grignard reagent to remain exposed to air, even if it is protected from moisture by drying tubes?

12. Give a three-dimensional structure for the complex RMgBr · 2(C$_2$H$_5$)$_2$O. What is the nature of the bonding between the diethyl ether molecules and the Grignard reagent?

13. The intermediates in the mechanism for the conversion of an aryl halide into a Grignard reagent are shown below. Use curved arrows to symbolize the flow of electrons in the three steps given.

14. What problems might be encountered if bromocyclohexane were used as an additive to help initiate the formation of phenylmagnesium bromide and 1-butylmagnesium bromide? How does the use of 1,2- dibromoethane avoid such experimental difficulties?

15. The reaction of 5-bromo-1-pentanol with magnesium in *anhydrous* diethyl ether did not give the expected Grignard reagent, 5-hydroxy-1-pentylmagnesium bromide. Provide an explanation for the failure of this reaction. What organic product is formed instead?

$$H—OCH_2(CH_2)_3CH_2—Br \ + \ Mg^0 \xrightarrow{\text{dry Et}_2O} H—OCH_2(CH_2)_3CH_2—MgBr$$

5-Bromo-1-pentanol   5-Hydroxy-1-pentylmagnesium bromide

Refer to the online resources and use the spectra viewer and Tables 8.1–8.8 as needed to answer the blue-numbered questions on spectroscopy.

16. Consider the spectral data for bromobenzene (Figs. 15.19 and 15.20).

   a.   In the functional group region of the IR spectrum, specify the absorptions associated with the aromatic ring.

b. In the ^{1}H NMR spectrum, assign the various resonances to the hydrogen nuclei responsible for them.

c. For the ^{13}C NMR data, assign the various resonances to the carbon nuclei responsible for them.

17. Consider the NMR spectral data for 1-bromobutane (Fig. 14.4).

a. In the ^{1}H NMR spectrum, assign the various resonances to the hydrogen nuclei responsible for them.

b. For the ^{13}C NMR data, assign the various resonances to the carbon nuclei responsible for them.

SPECTRA

Starting Materials and Products

The IR and NMR spectra of 1-bromobutane and bromobenzene are given in Figures 14.3, 14.4, 15.19, and 15.20, respectively.

19.3 GRIGNARD REAGENTS: REACTIONS

Reaction of Phenylmagnesium Bromide with Methyl Benzoate

The aryl carbon atom bearing the magnesium in phenylmagnesium bromide is nucleophilic and reacts readily with electrophiles (Sec. 19.1). In Part A of the experimental procedures that follow, two equivalents of phenylmagnesium bromide (**5**, Eq. 19.4) are allowed to react with methyl benzoate (**9**), an ester, in an exothermic reaction to give triphenylmethanol (**11**) via the sequence of reactions depicted in Scheme 19.1. In the first step of the reaction, nucleophilic attack of phenylmagnesium bromide on the electrophilic carbon atom of the ester group of **9** leads to benzophenone (**10**) as an intermediate. Subsequent reaction of **10** with an additional mole of **5** then produces an alkoxide salt, which is converted to **11** upon protonation with acid. The reaction is worked up with acid rather than water alone to avoid precipitation of the basic magnesium salt, HOMgX.

Scheme 19.1

© 2016 Cengage Learning. All Rights Reserved. May not be scanned, copied or duplicated, or posted to a publicly accessible website, in whole or in part.

The principal organic products present after the aqueous work-up are the desired triphenylmethanol (**11**), benzene from any unreacted **5**, and biphenyl (**8**). Fortunately, it is possible to separate **11** and **8** easily owing to their relative solubilities in nonpolar hydrocarbon solvents. Biphenyl is considerably more soluble in cyclohexane than is triphenylmethanol, so recrystallization of the crude product mixture gives pure triphenylmethanol.

Reaction of Phenylmagnesium Bromide with Carbon Dioxide

Nucleophilic addition of phenylmagnesium bromide (**5**, Eq. 19.4) to carbon dioxide produces an intermediate carboxylate salt that may be converted into benzoic acid (**12**) by acidification of the reaction mixture (Eq. 19.11). This transformation, which is the subject of Part B of the following experimental procedures, may be accompanied by side reactions that lead to the formation of benzophenone (**10**) and/or triphenylmethanol (**11**), as shown in Equation 19.12. However, formation of these by-products can be minimized by controlling the reaction conditions. Thus, the bromomagnesium salt of benzoic acid is only slightly soluble in diethyl ether, so the salt precipitates from solution and cannot undergo further reaction when the reaction is performed in this solvent. A large excess of carbon dioxide is also used to increase the likelihood that **5** will react with carbon dioxide rather than with the magnesium salt of benzoic acid. If the phenylmagnesium bromide is added slowly *to* the dry ice, carbon dioxide is always present in excess, thus favoring the desired reaction. Finally, when dry ice is used as the source of carbon dioxide, the temperature of the reaction is maintained at −78 °C until all of the Grignard reagent has been consumed. At this low temperature, the reaction of phenylmagnesium bromide with the bromomagnesium salt of benzoic acid is slow.

$$(19.11)$$

$$(19.12)$$

After the addition of the Grignard reagent to carbon dioxide is complete, the excess dry ice is allowed to evaporate, and the mixture is acidified. Extraction of the aqueous mixture with diethyl ether then gives a solution containing benzoic acid, benzophenone, triphenylmethanol, benzene, and biphenyl. The benzoic acid is readily separated from the neutral side products in the crude mixture by extracting the solution with dilute aqueous sodium hydroxide, whereby benzoic acid is converted to its water-soluble sodium salt; the neutral by-products remain in the organic layer. Acidification of the aqueous layer regenerates the benzoic acid, which can be purified by recrystallization.

Reaction of 1-Butylmagnesium Bromide with 2-Methylpropanal

The reaction of 1-butylmagnesium bromide (**7**, Eq. 19.5) with 2-methylpropanal (**13**) provides an excellent example of the Grignard synthesis of secondary alcohols (Eq. 19.13). In this reaction, the Grignard reagent **7** adds to 2-methylpropanal to give the magnesium salt of 2-methyl-3-heptanol, and the alcohol **14** may be isolated after acidification of the mixture. This transformation is found in Part C of the following experimental procedures.

$$CH_3CH_2CH_2\overset{\delta^-}{CH_2}\!-\!\overset{\delta^+}{MgBr} \;+\; \underset{\underset{\displaystyle CH_3}{\overset{\displaystyle |}{CHCH_3}}}{\overset{\displaystyle \overset{\delta^-}{O}}{\underset{H}{\overset{\parallel}{C}}}}\!\!{}^{\delta^+} \longrightarrow$$

7   **13**

1-Butylmagnesium bromide   2-Methylpropanal   (19.13)

$$CH_3(CH_2)_3\!-\!\underset{\underset{\displaystyle CH_3}{\overset{\displaystyle |}{}}}{\overset{\displaystyle O^-\ {}^+MgBr}{\overset{\displaystyle |}{CH}}}\!-\!CHCH_3 \xrightarrow{H_3O^+} CH_3(CH_2)_3\!-\!\underset{\underset{\displaystyle CH_3}{\overset{\displaystyle |}{}}}{\overset{\displaystyle OH}{\overset{\displaystyle |}{CH}}}\!-\!CHCH_3$$

14

2-Methyl-3-heptanol

19.4 SPECIAL EXPERIMENTAL TECHNIQUES

Some of the side reactions that are encountered during the preparation of Grignard reagents and during their reaction with electrophiles can be controlled by prudent choice of solvent and reaction temperature. Another experimental technique that can be used to minimize undesired side reactions involves the **order of addition** of the Grignard reagent and the electrophilic reactant. For example, the electrophiles methyl benzoate and 2-methylpropanal are added to a solution of the appropriate Grignard reagent, which is a highly reactive nucleophile, to prepare triphenylmethanol (**11**) and 2-methyl-3-heptanol (**14**), respectively. This mode of addition is referred to as **normal addition**. On the other hand, in the preparation of benzoic acid, the nucleophilic Grignard reagent is added to the carbon dioxide, the electrophile. This procedure is called **inverse addition**.

Whether an addition is performed according to the *normal* or *inverse* mode is dictated by the specific nature of the reactants and the potential side reactions that may occur. For example, if carbon dioxide gas were bubbled into a solution of phenylmagnesium bromide (**5**) to prepare benzoic acid (normal addition), the by-products **10** and/or **11** (Eq. 19.12) would become significant, perhaps predominant. There would always be an excess of the Grignard reagent **5** present relative to carbon dioxide, so further reaction between **5** and the bromomagnesium salt of benzoic acid would be favored. In the preparation of triphenylmethanol (**11**), the combination of the Grignard reagent with methyl benzoate could be done equally well in either the normal or inverse fashion because two equivalents of **5** are necessary to complete the reaction and the magnesium alkoxide salt of **11** is unreactive toward **5**. Inverse addition is less convenient in this case, however, because it requires transferring the Grignard reagent **5** from one flask to a dropping funnel and then adding it to the reaction flask containing the methyl benzoate. Similarly, the reaction of 1-butylmagnesium bromide with 2-methylpropanal (Eq. 19.13) may be performed using either normal or inverse addition, but normal addition is again more convenient.

EXPERIMENTAL PROCEDURES

Reactions of Grignard Reagents

A ▪ *Preparation of Triphenylmethanol*

Purpose To demonstrate the preparation of a tertiary alcohol by the reaction of a Grignard reagent with an ester.

SAFETY ALERT

Review the Safety Alert for Preparation of Grignard Reagents (Sec. 19.2).

MINISCALE PROCEDURE

Preparation Refer to the online resources to answer Pre-Lab Exercises, access videos, and read the MSDSs for the chemicals used or produced in this procedure. Review Sections 2.10, 2.11, 2.13, 2.17, 2.21, 2.22, and 2.29.

Apparatus Glass apparatus from the miniscale experimental procedure of Section 19.2, separatory funnel, ice-water bath, apparatus for magnetic stirring, simple distillation, vacuum filtration, and *flameless* heating.

Setting Up While the reaction mixture for the preparation of phenylmagnesium bromide (Sec. 19.2) is cooling to room temperature, dissolve 1.2 mL of methyl benzoate in about 5 mL of *anhydrous* diethyl ether, and place this solution in the separatory funnel with the *stopcock closed.* Cool the reaction flask containing the phenylmagnesium bromide in the ice-water bath.

Reaction Begin the *slow, dropwise* addition of the solution of methyl benzoate to the *stirred* solution of phenylmagnesium bromide. This reaction is *exothermic,* so you should control the rate of reaction by adjusting the rate of addition *and* by occasionally cooling the reaction flask as needed with the ice-water bath. The ring of condensate should be allowed to rise no more than one-third of the way up the reflux condenser. A white solid may form during the reaction, but this is normal. After the addition is complete and the exothermic reaction subsides, you may complete the reaction in one of two ways. Consult with your instructor to determine whether you should (1) heat the reaction mixture at reflux for 30 min or (2) stopper the flask after cooling the contents to room temperature and place it in the *hood* until the next laboratory period (no reflux required).★

Work-Up, Isolation, and Purification Place about 10 mL of cold 6 *M* sulfuric acid and about 5–10 g of crushed ice in a beaker. If the reaction mixture solidified upon cooling, add a small quantity of solvent-grade diethyl ether to the reaction flask. Pour the reaction mixture gradually with stirring into the ice-acid mixture. Rinse the round-bottom flask with 2–3 mL of solvent-grade diethyl ether and add this wash to the beaker. Continue stirring until the heterogeneous mixture is completely free of undissolved solids. It may be necessary to add a small portion of solvent-grade diethyl

ether to dissolve all the organic material; the total volume of ether should be about 15–20 mL. Verify that the aqueous layer is acidic; if it is not, add cold 6 *M* sulfuric acid dropwise until the layer is acidic. If necessary, sequentially add 2- to 3-mL portions of solvent-grade diethyl ether and then water to dissolve all of the solids.

Transfer the entire mixture to a separatory funnel. Shake the funnel vigorously with venting to relieve pressure; separate the aqueous layer.★ Wash the organic layer sequentially with about 5 mL of 3 *M* sulfuric acid, two 5-mL portions of saturated aqueous sodium bicarbonate (*vent!*), and finally with one 5-mL portion of saturated sodium chloride solution. Dry the organic layer using several spatula-tips full of *anhydrous* sodium sulfate. Swirl the flask occasionally for a period of 10–15 min to facilitate drying; add further small portions of *anhydrous* sodium sulfate if the solution remains cloudy.★

Filter or decant the solution into a 50-mL round-bottom flask and equip the flask for simple distillation. Remove the diethyl ether by simple distillation. Alternatively, use rotary evaporation or other techniques to concentrate the solution. The final traces of solvent may be removed by attaching the flask to a vacuum source and gently swirling the contents as the vacuum is applied. After the crude solid residue has dried, determine its melting range, which may be wide.★

Purify the triphenylmethanol by dissolving it in a *minimum* amount of boiling cyclohexane (ca. 10 mL/g product). Perform this operation at the hood or use a funnel that is attached to a vacuum source and inverted over the flask (Fig. 2.71b). Once all the material is in solution, evaporate the solvent *slowly* until small crystals of triphenylmethanol start to form. Allow the crystallization to continue at room temperature and then in an ice-water bath until no more crystals form. Isolate the product by vacuum filtration and air-dry it.

Analysis Weigh the triphenylmethanol and calculate the percent yield; determine its melting point. Obtain IR and ¹H NMR spectra of your starting materials and product, and compare them with those of authentic samples (Figs. 8.48, 15.19, 15.20, 15.33, 15.34, 19.2, and 19.3). If possible, analyze your product by GC-MS to determine if it is contaminated with benzophenone (**10**).

MICROSCALE PROCEDURE

Preparation Refer to the online resources to answer Pre-Lab Exercises, access videos, and read the MSDSs for the chemicals used or produced in this procedure. Review Sections 2.10, 2.11, 2.13, 2.17, 2.21, 2.22, and 2.29.

Apparatus Glass apparatus from the microscale experimental procedure of Section 19.2, 3-mL conical vial, two screw-cap centrifuge tubes, 1-mL plastic syringe, Pasteur pipet with 0.5- and 1.0-mL calibration marks, ice-water bath, apparatus for magnetic stirring, simple distillation, vacuum filtration, Craig tube filtration, and *flameless* heating.

Setting Up While the reaction mixture for the preparation of phenylmagnesium bromide (Sec. 19.2) is cooling to room temperature, dissolve 120 μL of methyl benzoate in 0.5 mL of *anhydrous* diethyl ether contained in the 3-mL conical vial. Cool the conical vial containing the phenylmagnesium bromide in the ice-water bath.

Reaction Draw the solution of methyl benzoate into the syringe and recap the vial. Insert the syringe needle through the rubber septum and begin the *slow, dropwise* addition of the solution of methyl benzoate to the *stirred* solution of phenylmagnesium bromide. This reaction is *exothermic,* so you should control the rate of reaction by adjusting the rate of addition *and* by occasionally cooling the reaction vial with the ice-water bath if necessary. The reflux ring should not be allowed to rise into the Claisen adapter. A white solid may form during the reaction, but this is normal. Upon completion of the addition, place about 0.2–0.3 mL of *anhydrous* diethyl ether in the vial that contained the methyl benzoate solution. Draw this solution into a syringe and add it in one portion to the reaction mixture. After the exothermic reaction subsides, you may complete the reaction in one of two ways. Consult with your instructor to determine whether you should (1) heat the reaction mixture at *gentle* reflux for 30 min or (2) cap the vial after cooling the contents to room temperature and place it in the *hood* until the next laboratory period (no reflux required). The mixture may solidify on cooling.★

Work-Up, Isolation, and Purification   Remove the Claisen adapter and drying tube, and place the conical reaction vial in an ice-water bath. Using a Pasteur pipet, slowly add 1.5 mL of cold 3 *M* sulfuric acid dropwise to neutralize the reaction mixture. *Be careful* because the addition of acid may be accompanied by frothing. Use a small glass stirring rod to break up the solid during this addition. You may need to cap the vial tightly and shake it with venting to facilitate dissolution of the solids. Continue shaking the vial until the heterogeneous mixture is free of undissolved solids. You should obtain two distinct layers as the solid gradually dissolves. It may be necessary to add solvent-grade diethyl ether to dissolve all the organic material; the total volume of ether should be about 2.0–2.5 mL. Verify that the aqueous layer is acidic; if it is not, add cold 3 *M* sulfuric acid dropwise until the layer is acidic.

Optional Measures   If some solid material remains after the addition of aqueous acid is complete, the entire mixture may be transferred to a screw-cap centrifuge tube. Sequentially add solvent-grade diethyl ether and water in 0.5-mL portions to the centrifuge tube. After each addition, cap the tube and shake it to dissolve the solids. Continue the extraction as described, using appropriately sized conical vials or screw-cap centrifuge tubes.

    Remove the spinvane with forceps and, holding it above the vial, rinse it with several drops of solvent-grade ether. Cap the vial tightly, invert it gently several times with venting to mix the layers, and allow the layers to separate. Using a Pasteur pipet, transfer the lower aqueous layer to a conical vial. Wash this layer with one 0.5-mL portion of solvent-grade ether and combine this ethereal layer with the first.★ Wash the combined organic layers sequentially with a 0.5-mL portion of 3 *M* sulfuric acid, two 0.5-mL portions of saturated aqueous sodium bicarbonate (*vent!*), and finally with one 0.5-mL portion of saturated aqueous sodium chloride. Using a clean filter-tip pipet, transfer the ether layer to a screw-cap centrifuge tube containing several microspatula-tips full of *anhydrous* sodium sulfate. Swirl the tube occasionally for a period of 10–15 min to facilitate drying; add further small portions of *anhydrous* sodium sulfate if the solution remains cloudy.★

    With a filter-tip pipet, transfer the dried organic solution into a clean 5.0-mL conical vial, rinse the sodium sulfate with about 0.2 mL of diethyl ether, and add this rinse to the vial. Equip the vial for simple distillation. Remove the solvent by simple distillation; withdraw the distillate from the Hickman stillhead as needed. Alternatively, use rotary evaporation or other techniques to concentrate the solution. The final traces of solvents may be removed by attaching the vial to a vacuum source. After the crude solid residue has dried, determine its melting range, which may be wide.★

    Purify the triphenylmethanol by dissolving it in a *minimum* amount of boiling cyclohexane (ca. 1 mL/0.1 g product) in a Craig tube. Perform this operation at the hood. Once all the material is in solution, evaporate the solvent *slowly* until small crystals of triphenylmethanol start to form. Allow the crystallization to continue at room temperature and then in an ice-water bath until no more crystals form. Isolate and air-dry the product.

Analysis Weigh the triphenylmethanol and calculate the percent yield; determine its melting point. Obtain IR and ^{1}H NMR spectra of your starting materials and product and compare them with those of authentic samples (Figs. 8.48, 15.19, 15.20, 15.33, 15.34, 19.2, and 19.3). If possible, analyze your product by GC-MS to determine if it is contaminated with benzophenone (**10**).

WRAPPING IT UP

Dilute the *combined aqueous layers and washes* with water, neutralize the solution if necessary, and flush it down the drain with excess water. Place the *ether distillate* and the *cyclohexane mother liquor* in the container for nonhalogenated organic solvents. Spread the *calcium chloride* from the drying tube and the *sodium sulfate* on a tray in the hood and, after the ether has evaporated, place them and the *filter paper* in the container for nonhazardous solids.

B ■ *Preparation of Benzoic Acid*

Purpose To demonstrate the preparation of a carboxylic acid by the reaction of a Grignard reagent with carbon dioxide.

SAFETY ALERT

1. **Review the Safety Alert for Preparation of Grignard Reagents (Sec. 19.2).**

2. **Use tongs or thick gloves to handle the solid carbon dioxide (dry ice) because contact with skin can cause severe frostbite. Crush the dry ice by wrapping larger pieces in a clean, dry towel and striking them with a mallet or a wooden block. Use the dry ice *immediately* after crushing to minimize contact with atmospheric moisture.**

MINISCALE PROCEDURE

Preparation Refer to the online resources to answer Pre-Lab Exercises, access videos, and read the MSDSs for the chemicals used or produced in this procedure. Review Sections 2.17 and 2.21.

Apparatus Glass apparatus from the miniscale experimental procedure of Section 19.2, separatory funnel, ice-water bath, apparatus for vacuum filtration and *flameless* heating.

Setting Up After the reaction mixture for the preparation of phenylmagnesium bromide (Sec.19.2) has cooled to room temperature, place about 10 g of coarsely crushed dry ice in a 125-mL Erlenmeyer flask. The dry ice should be crushed *immediately* before use and protected as much as possible from atmospheric moisture.

Reaction Slowly pour the phenylmagnesium bromide solution onto the dry ice with vigorous swirling; the mixture normally becomes rather viscous. Rinse the original round-bottom reaction flask with 2–3 mL of *anhydrous* diethyl ether and add this wash to the Erlenmeyer flask. Continue swirling the contents of the flask for a few minutes. After the addition is complete, cover the flask with a small watchglass or piece of filter paper and allow the excess carbon dioxide to sublime by letting the flask stand in the hood, properly labeled with your name, until the next laboratory period.★

You may expedite the removal of the excess carbon dioxide by shaking or swirling the flask while warming it *very slightly* in a warm-water bath or by adding small amounts of warm water to the reaction mixture and then shaking or swirling the flask. Be careful, because both of these methods may cause a *sudden* evolution of carbon dioxide gas and splash the contents of the flask onto the floor or bench top. The flask should *never* be stoppered.

Work-Up, Isolation, and Purification After the excess dry ice has completely sublimed, most of the ether will have also evaporated, so add 20 mL of solvent-grade diethyl ether to the reaction flask. Combine about 5 g of ice and 10 mL of cold 3 *M* sulfuric acid in an Erlenmeyer flask, and add this cold acid to the reaction mixture *slowly* to avoid excessive foaming. If the ether evaporates appreciably during this operation, more solvent-grade diethyl ether should be added so that the total volume of ether is about 20–25 mL. Verify that the aqueous layer is acidic; if it is not, add cold 3 *M* sulfuric acid dropwise until the layer is acidic.

Swirl the mixture and transfer it to a separatory funnel. Rinse the flask with a small portion of solvent-grade diethyl ether and add the rinse to the separatory funnel. Shake the funnel cautiously with venting and separate the layers. Extract the aqueous layer with a 10-mL portion of solvent-grade diethyl ether and add this to the original ether layer. Extract the *combined* ethereal fractions sequentially with two 10-mL portions of a 1 *M* solution of sodium hydroxide; vent the funnel frequently during the extractions. Transfer the two alkaline extracts to an Erlenmeyer flask and slowly add 6 *M* hydrochloric acid until precipitation of the benzoic acid is complete and the aqueous mixture is *acidic.* Cool the solution in an ice-water bath, isolate the solid benzoic acid by vacuum filtration, wash it with water, and air-dry it.★ Recrystallize the benzoic acid from water or some other suitable solvent.

Analysis Weigh the benzoic acid and calculate the percent yield; determine its melting point. Obtain IR and ^{1}H NMR spectra of your starting material and product and compare them with those of authentic samples (Figs. 15.19, 15.20, 19.4, and 19.5).

Discovery Experiment   *Exploring the Influence of Mode of Addition*

Develop a protocol for investigating how using the "normal" mode addition of phenylmagnesium bromide and carbon dioxide affects the outcome of the reaction. You will need access to gaseous carbon dioxide to perform the reaction. Consult with your instructor before undertaking any experiments.

MICROSCALE PROCEDURE

Preparation Refer to the online resources to answer Pre-Lab Exercises, access videos, and read the MSDSs for the chemicals used or produced in this procedure. Review Sections 2.17 and 2.21.

Apparatus Glass apparatus from the microscale experimental procedure of Section 19.2, a 3-mL and a 5-mL conical vial, screw-cap centrifuge tube, Pasteur pipet with 0.5- and 1.0-mL calibration marks, ice-water bath, apparatus for vacuum filtration and Craig tube filtration.

Setting Up After the reaction mixture for the preparation of phenylmagnesium bromide (Sec.19.2) has cooled to room temperature, place about 1 g of coarsely crushed dry ice in a small beaker. The dry ice should be crushed immediately before use and protected as much as possible from atmospheric moisture.

Reaction Using a filter-tip pipet, slowly transfer the phenylmagnesium bromide solution onto the dry ice. Stir the mixture with a glass rod during the addition, as the mixture normally becomes rather viscous. Rinse the conical vial with 0.5 mL of *anhydrous* diethyl ether and add this wash to the beaker. After the addition is complete, cover the beaker with a small watchglass or piece of filter paper, and allow the excess carbon dioxide to sublime by letting the beaker stand in the hood, properly labeled with your name, until the next laboratory period.★

You may expedite the removal of the excess carbon dioxide by stirring the contents of the beaker while warming it *very slightly* in a warm-water bath or by adding small amounts of warm water to the reaction mixture with stirring. Because both of these methods may cause a *sudden* evolution of carbon dioxide gas and splash the contents of the beaker onto the floor or bench top, be careful using them.

Work-Up, Isolation, and Purification After the excess dry ice has completely sublimed, most of the ether will have also evaporated, so add 2.5 mL of solvent-grade diethyl ether. Then *slowly* add 1.5 mL of cold 3 *M* sulfuric acid to the reaction mixture to avoid excessive foaming. If the ether evaporates appreciably during this operation, more solvent-grade diethyl ether should be added so that the total volume of ether is about 2.0–2.5 mL. Verify that the aqueous layer is acidic; if it is not, add cold 3 *M* sulfuric acid dropwise until the layer is acidic.

Optional Measures If some solid material remains after the addition of aqueous acid is complete, the entire mixture may be transferred to a screw-cap centrifuge tube. Sequentially add solvent-grade diethyl ether and water in 0.5-mL portions to the centrifuge tube. After each addition, cap the tube and shake it to dissolve the solids. Continue the extraction as described using appropriately sized conical vials or screw-cap centrifuge tubes.

Using a Pasteur pipet, transfer the entire mixture to a 5-mL conical vial.★ Rinse the beaker with a small portion of solvent-grade diethyl ether and add the rinse to the conical vial. Cap the vial tightly, invert it gently several times with venting to mix the layers thoroughly, and allow the layers to separate. Transfer the aqueous layer to a 3-mL conical vial, extract the aqueous layer with a 0.5-mL portion of solvent-grade diethyl ether, and add this extract to the original ether layer. Extract the *combined* ethereal fractions sequentially with two 1-mL portions of a 1 *M* solution of sodium hydroxide, venting the vial frequently during the extractions. Transfer the two alkaline extracts to a small Erlenmeyer flask and warm the flask for about 5 min to remove the ether that is dissolved in the aqueous phase; a saturated solution of diethyl ether in water contains 7% diethyl ether. Cool the alkaline solution and slowly add 6 *M* hydrochloric acid until precipitation of the benzoic acid is complete and the aqueous mixture is acidic. Cool the solution in an ice-water bath, isolate the solid benzoic acid by vacuum filtration, and air-dry the sample.★ Recrystallize the benzoic acid from water or some other suitable solvent using a Craig tube.

Analysis Weigh the benzoic acid and calculate the percent yield; determine its melting point. Obtain IR and ^{1}H NMR spectra of your starting material and product and compare them with those of authentic samples (Figs. 15.19, 15.20, 19.4, and 19.5).

Discovery Experiment   *Exploring the Influence of Mode of Addition*

Develop a protocol for investigating how using the "normal" mode addition of phenylmagnesium bromide and carbon dioxide affects the outcome of the reaction. You will need access to gaseous carbon dioxide to perform the reaction. Consult with your instructor before undertaking any experiments.

WRAPPING IT UP

Dilute the *combined aqueous layers and washes* with water, neutralize the solution if necessary, and flush it down the drain with excess water. Place the *combined ethereal layers* in the container for nonhalogenated organic solvents. Spread the *calcium chloride* from the drying tube on a tray in the hood to allow the ether to evaporate; then place it and the *filter paper* in the container for nonhazardous solids.

C ▪ *Preparation of 2-Methyl-3-Heptanol*

Purpose To demonstrate the preparation of a secondary alcohol by the reaction of a Grignard reagent with an aldehyde.

SAFETY ALERT

Review the Safety Alert for Preparation of Grignard Reagents (Sec. 19.2).

MINISCALE PROCEDURE

Preparation Refer to the online resources to answer Pre-Lab Exercises, access videos, and read the MSDSs for the chemicals used or produced in this procedure. Review Sections 2.10, 2.11, 2.13, 2.17, 2.21, and 2.29.

Apparatus Glass apparatus from the miniscale experimental procedure of Section 19.2, a 25-mL round-bottom flask, separatory funnel, ice-water bath, apparatus for magnetic stirring, simple distillation, shortpath distillation, and *flameless* heating.

Setting Up While the reaction mixture for the preparation of 1-butylmagnesium bromide (Sec.19.2) is cooling to room temperature, dissolve 1.8 mL of *freshly distilled* 2-methylpropanal in 5 mL of *anhydrous* diethyl ether, and place this solution in the separatory funnel with the *stopcock closed*. Cool the reaction flask containing the 1-butylmagnesium bromide in an ice-water bath and begin stirring.

Reaction Over a period of about 5 min, add the solution of 2-methylpropanal *dropwise* to the stirred solution of 1-butylmagnesium bromide. Control the resulting

---

exothermic reaction by adjusting the rate of addition so that the ring of condensate rises no more than one-third of the way up the reflux condenser. After completing the addition, allow the reaction mixture to stand for about 15 min. The reaction mixture may be stored, loosely stoppered, *in the hood* until the next laboratory period.★

Work-Up, Isolation, and Purification Place about 10 mL of cold 6 *M* sulfuric acid and 5–10 g of crushed ice in a beaker. Pour the reaction mixture *slowly* with stirring into the ice-acid mixture. After the addition is complete, transfer the cold mixture, which may contain some precipitate, to a separatory funnel and shake it gently. The precipitate should dissolve. Separate the layers. Extract the aqueous layer sequentially with two 5-mL portions of solvent-grade diethyl ether, and add these extracts to the main ethereal layer. Transfer the combined ethereal extracts into the separatory funnel. Venting the funnel frequently to relieve pressure, wash the solution sequentially with 5 mL of *saturated* aqueous sodium bisulfite, two 5-mL portions of *saturated* aqueous sodium bicarbonate, and finally with 5 mL of *saturated* aqueous sodium chloride. Dry the ethereal solution over several spatula-tips full of *anhydrous* sodium sulfate.★ Swirl the flask occasionally for a period of 10–15 min to facilitate drying; add further small portions of *anhydrous* sodium sulfate if the solution remains cloudy.

Filter or carefully decant the dried organic solution into a round-bottom flask equipped for simple distillation, and remove the diethyl ether by simple distillation. Alternatively, use rotary evaporation or other techniques to concentrate the solution. The final traces of solvent may be removed by attaching the flask to a vacuum source and gently swirling the contents as the vacuum is applied.★

Using a Pasteur pipet, transfer the cooled pot residue to a 25-mL round-bottom flask containing a stirbar. Rinse the original stillpot with 1–2 mL of diethyl ether, and add the rinse to the 25-mL round-bottom flask. Remove the remaining diethyl ether by attaching the flask to a vacuum source and gently swirling the contents as the vacuum is applied. Fit this flask for a shortpath distillation, insulating the top of the distilling flask and the stillhead with cotton or glasswool wrapped with aluminum foil. Distill the 2-methyl- 3-heptanol into a *tared* receiver, collecting all distillate that boils above 130 °C (760 torr). Note that the head temperature may never reach the reported boiling point, owing to the high boiling temperature and the small amount of sample being distilled. Terminate heating when the dark pot residue becomes viscous and starts to evolve fumes.

Analysis Weigh the product and calculate the yield. Obtain IR and ^{1}H NMR spectra of your starting materials and product and compare them with those of authentic samples (Figs. 14.3, 14.4, 18.17, 18.18, 19.6, and 19.7).

Discovery Experiment | *Preparation and Characterization of a 3° Alcohol*

Following the general procedures outlined in Sections 19.2 and 19.4C for the preparation and reaction of 1-butylmagnesium bromide with 2-methylpropanal, perform a Grignard synthesis with 30 mmol of a 1° alkyl bromide and 28 mmol of a ketone, which are unknowns that you obtain from your instructor. Use 33 mmol of magnesium turnings. The possible alkyl halide and ketone unknowns are listed below in alphabetical order, as are the 3° alcohols that are products of the various combinations of reagents. Characterize your product by its boiling point, IR spectrum, and, if possible, ^{1}H NMR spectrum and use the data to determine which alcohol you have produced. Be certain to consult with your instructor before undertaking any procedures.

Alkyl Halide	Ketone	Alcohol
1-Bromobutane	Acetone	3-Ethyl-3-heptanol
Bromoethane	Butanone	4-Ethyl-4-heptanol
1-Bromopentane	2-Pentanone	2-Methyl-2-heptanol
1-Bromopropane	3-Pentanone	3-Methyl-3-heptanol
		4-Methyl-4-heptanol
		3-Ethyl-3-hexanol
		2-Methyl-2-hexanol
		3-Methyl-3-hexanol
		2-Methyl-2-octanol
		3-Methyl-3-octanol
		4-Methyl-4-octanol
		3-Ethyl-3-pentanol
		2-Methyl-2-pentanol
		3-Methyl-3-pentanol

WRAPPING IT UP

Dilute the *combined aqueous layers and washes* with water, neutralize the solution if necessary, and flush it down the drain with excess water. Place the *ether distillate* in the container for nonhalogenated organic solvents. Spread the *calcium chloride* from the drying tube and the *sodium sulfate* on a tray in the hood, and then place them in the container for nonhazardous solids after the volatiles have evaporated.

EXERCISES

General Questions

1. Arrange the following compounds in order of increasing reactivity toward attack of a Grignard reagent at the carbonyl carbon atom: methyl benzoate, benzoic acid, benzaldehyde, and benzophenone. Explain the basis for your decision, making use of mechanisms where needed.

2. What is (are) the product(s) of reaction of each of the carbonyl-containing compounds in Exercise 1 with *excess* Grignard reagent, RMgBr?

3. How might primary, secondary, and tertiary alcohols be prepared from a Grignard reagent and a suitable carbonyl-containing compound? Write chemical reactions for these preparations using any starting materials you wish; indicate stoichiometry where important.

Questions for Part A

4. Why is it unwise to begin addition of the solution of methyl benzoate to the Grignard reagent before the latter has cooled to room temperature and then been placed in an ice-water bath?

5. Why should anhydrous rather than solvent-grade diethyl ether be used to prepare the solution of methyl benzoate that is added to the Grignard reagent?

6. What is the solid that forms during the addition of the ester to the Grignard reagent?

7. Why is it necessary to acidify the mixture obtained after the reaction of methyl benzoate with phenylmagnesium bromide?

Refer to the online resources and use the spectra viewer and Tables 8.1–8.8 as needed to answer the blue-numbered questions on spectroscopy.

8. Cyclohexane is used as the recrystallization solvent to purify the triphenylmethanol by removing the biphenyl impurity. Why is this a better choice of solvent than a solvent such as isopropyl alcohol?

9. Comment on the use of steam distillation (Sec. 4.4) as a possible alternative procedure for purifying crude triphenylmethanol. Consider what possible starting materials, products, and by-products might be present, and indicate which of these should steam-distill and which should not. Would this method of purification yield pure triphenylmethanol? Give your reasoning.

10. Consider the spectral data for methyl benzoate (Figs. 8.48, 15.33, and 15.34).

   a. In the functional group region of the IR spectrum, identify the absorptions associated with the ester functional group and the aromatic ring.

   b. In the ^{1}H NMR spectrum, assign the various resonances to the hydrogen nuclei responsible for them.

   c. For the ^{13}C NMR data, assign the various resonances to the carbon nuclei responsible for them.

11. Consider the spectral data for triphenylmethanol (Figs. 19.2 and 19.3).

   a. In the functional group region of the IR spectrum, specify the absorptions due to the aromatic ring. There is a broad absorption at about 3450 cm^{-1}. What functional group is responsible for this absorption and why is the absorption broad?

   b. In the ^{1}H NMR spectrum, assign the various resonances to the hydrogen nuclei responsible for them.

   c. For the ^{13}C NMR data, assign the various resonances to the carbon nuclei responsible for them.

12. Discuss the differences observed in the IR and NMR spectra of methyl benzoate and triphenylmethanol that are consistent with the conversion of an ester into a tertiary alcohol in this experiment.

Questions for Part B

13. What molecular feature or features allow carbon dioxide to function as an electrophile?

14. Define the term *inverse addition* as it applies to a Grignard reaction and explain why this is the preferred mode of addition for the reaction between C_6H_5MgBr and CO_2.

15. Why is it necessary to acidify the mixture obtained after the reaction of phenylmagnesium bromide with carbon dioxide?

16. Consider the acid-base reaction that occurs when benzoic acid is extracted with aqueous sodium hydroxide.

   a. Write the equation for this reaction, labeling the acid, base, conjugate acid, and conjugate base.

   b. What is the anticipated relative solubility of benzoic acid and its conjugate base in diethyl ether?

   c. The pK_a of benzoic acid is 4.2, whereas that of carbonic acid, H_2CO_3, is 6.4. On the basis of this information, would aqueous sodium bicarbonate be sufficiently basic to deprotonate benzoic acid? Explain your reasoning.

   d. The pK_a of hydrochloric acid is about −6. On the basis of this information, would aqueous sodium chloride be sufficiently basic to deprotonate benzoic acid? Explain your reasoning.

17. What is the purpose of using diethyl ether to extract the aqueous mixture obtained after acidification of the reaction? What products are in the aqueous and organic layers?

18. What function does extracting the ethereal solution of organic products with aqueous base have in the purification of benzoic acid? What remains in the ether layer?

19. The yield of benzoic acid obtained when only enough acid is added to the aqueous solution of sodium benzoate to bring the pH to 7 is lower than that obtained if the pH is brought below 5. Explain.

20. Explain why neither triphenylmethanol, $(C_6H_5)_3COH$, nor benzophenone, $C_6H_5COC_6H_5$, was produced in the reaction between C_6H_5MgBr and CO_2.

21. Consider the spectral data for benzoic acid (Figs. 19.4 and 19.5).

Refer to the online resources and use the spectra viewer and Tables 8.1–8.8 as needed to answer the blue-numbered questions on spectroscopy.

   a. In the functional group region of the IR spectrum, specify the absorptions associated with the carbonyl component of the carboxyl group and the aromatic ring. There is a broad absorption in the region of about 2800–3050 cm^{-1}. What functional group is responsible for this absorption, and why is the absorption broad?

   b. In the ^1H NMR spectrum, assign the various resonances to the hydrogen nuclei responsible for them.

   c. For the ^{13}C NMR data, assign the various resonances to the carbon nuclei responsible for them.

22. Discuss the differences observed in the IR and NMR spectra of bromobenzene and benzoic acid that are consistent with the conversion of bromobenzene into benzoic acid in this experiment.

Questions for Part C

23. Which reagent functions as the electrophile in the reaction of 1-butylmagnesium bromide and 2-methylpropanal, and why should it be freshly distilled before use?

24. Why is it unwise to begin addition of the solution of 2-methylpropanal to the Grignard reagent before the latter has cooled to room temperature and then been placed in an ice-water bath?

25. What is the solid that forms upon reaction of 1-butylmagnesium bromide with 2-methylpropanal?

26. How may the exothermicity associated with the addition of the Grignard reagent to a solution of 2-methypropanal be controlled?

27. Why is it necessary to acidify the mixture obtained after the reaction of 1-butylmagnesium bromide with 2-methylpropanal?

28. Consumption of unreacted magnesium metal occurs when the reaction mixture is quenched with sulfuric acid. Write the balanced equation for the reaction between Mg^0 and sulfuric acid.

29. The work-up in this reaction calls for successive washes of an ethereal solution of the product with aqueous sodium bisulfite, sodium bicarbonate, and sodium chloride. What is the purpose of each of these steps?

30. Another organometallic reagent that reacts in a manner similar to the Grignard reagent is an alkyllithium, R–Li.

   a. Predict the addition product formed after acidification of the mixture resulting from reaction of $CH_2{=}CH{-}CH2{-}Li$ with 2-methylpropanal.

   b. Suppose $CH_3CHCH{=}CH{-}CH_2{-}Li$ had been used for the reaction with 2-methylpropanal. Write the addition product(s) expected after acidification of the reaction mixture.

31. Consider the spectral data for 2-methylpropanal (Figs. 18.17 and 18.18).

    a. In the functional group region of the IR spectrum, specify the absorptions associated with the carbonyl component and the carbon-hydrogen bond of the aldehyde group.

    b. In the ^{1}H NMR spectrum, assign the various resonances to the hydrogen nuclei responsible for them.

    c. For the ^{13}C NMR data, assign the various resonances to the carbon nuclei responsible for them.

32. Consider the spectral data for 2-methyl-3-heptanol (Figs. 19.6 and 19.7).

    a. In the IR spectrum, specify the absorptions associated with the hydroxyl group.

    b. In the ^{1}H NMR spectrum, assign the various resonances to the hydrogen nuclei responsible for them.

    c. For the ^{13}C NMR data, assign the various resonances to the carbon nuclei responsible for them. Explain why there are eight peaks in the ^{13}C NMR spectrum.

33. Discuss the differences observed in the IR and NMR spectra of 2-methyl-propanal and 2-methyl-3-heptanol that are consistent with the conversion of an aldehyde into a secondary alcohol in this experiment.

34. The IR spectrum of 2-methyl-3-heptanol prepared by the procedure of Part C sometimes shows a band at 1720 cm^{-1}. Give a possible source for this absorption.

Refer to the online resources and use the spectra viewer and Tables 8.1–8.8 as needed to answer the blue-numbered questions on spectroscopy.

SPECTRA

Starting Materials and Products

The ^{13}C NMR spectrum of methyl benzoate is shown in Figure 8.48; its IR and 1H NMR spectra are provided in Figures 15.33 and 15.34, respectively. The IR and NMR spectra for 1-bromobutane, bromobenzene, and 2-methylpropanal are presented in Figures 14.2, 14.4, 15.19, 15.20, 18.17, and 18.18, respectively.

Figure 19.2
IR spectrum of triphenylmethanol (IR card).

Figure 19.3

NMR data for triphenylmethanol (CDCl₃).

(a) ¹H NMR spectrum (300 MHz).
(b) ¹³C NMR data: δ 82.0, 127.2, 127.9, 146.9.

Figure 19.4

IR spectrum of benzoic acid (IR card).

Figure 19.5

NMR data for benzoic acid (CDCl₃).

(a) ¹H NMR spectrum (300 MHz).
(b) ¹³C NMR data: δ 128.5, 129.5, 130.3, 133.8, 172.7.

Figure 19.6
IR spectrum of 2-methyl-3-heptanol (neat).

Figure 19.7
NMR data for 2-methyl-3-heptanol (CDCl₃).

(a) ¹H NMR spectrum (300 MHz).
(b) ¹³C NMR data: δ 14.2, 17.5, 19.3, 23.1, 28.6, 33.8, 34.1, 76.6.

19.5 ORGANOZINC REAGENTS: PREPARATION AND REACTION

Organozinc compounds **1** (M = ZnX) represent another class of organometallic reagents. Similar to Grignard reagents with respect to their preparation and handling, organozinc reagents serve as nucleophiles in a number of carbon-carbon bond-forming reactions as shown in Equations 19.1 and 19.2 (M = ZnX). To illustrate their preparation and use, you will first generate allylzinc bromide (**16**) from the reaction of allyl bromide (**15**) and zinc metal (Eq. 19.14). This reaction is analogous to the reaction that produces Grignard reagents (Sec. 19.2), and the same precautions used to avoid side reactions when forming such reagents must also be applied when generating and handling organozinc reagents. Specifically, you must exclude moisture, carbon dioxide, and oxygen from reactions involving organozinc reagents to avoid side reactions related to those shown in Equations 19.6–19.8, for example. You will then treat **16** with 3-pentanone (**17**), the nucleophile and electrophile, respectively, in the reaction to produce the zinc alkoxide **18**. Subsequent protonation of **18** with aqueous acid provides the tertiary alcohol **19** (Eq. 19.15).

$$\text{(Br + Zn°)} \xrightarrow{\text{THF}} \text{(ZnBr)} \qquad (19.14)$$

15 Allyl bromide

16 Allylzinc bromide

$$\text{16} + \text{CH}_3 \overset{\delta^-}{\underset{\delta^+}{O}} \text{CH}_3 \xrightarrow{\text{THF}} \text{CH}_3 \overset{O^- \; ^+\text{ZnBr}}{\underset{\text{CH}_3}{}}$$

16   **17**   **18**

3-Pentanone   A zinc alkoxide

$$\xrightarrow{\text{H}_3\text{O}^+} \text{CH}_3 \overset{\text{OH}}{\underset{\text{CH}_3}{}} \qquad (19.15)$$

19

3-Ethylhex-5-en-3-ol

Preparation of 3-Ethylhex-5-En-3-Ol

Purpose To demonstrate the preparation and use of organozinc reagents for the synthesis of a tertiary alcohol.

SAFETY ALERT

1. **Wear safety glasses or goggles and suitable protective gloves while performing the experiment.**

2. **The *anhydrous* tetrahydrofuran used in this experiment is contained in capped bottles. To prevent evaporation, absorption of atmospheric moisture and oxygen, and accidental fires, the screw or plastic cap *must* always be in place when the bottle is not in use.**

3. **Open containers of tetrahydrofuran must *not* be kept at your laboratory bench or stored in your laboratory drawer. Estimate the total volume of tetrahydrofuran you will need and measure it in the hood into a container that is loosely stoppered.**

4. **You should use ovens to dry your glassware if possible. However, if drying ovens are not available and it is necessary to dry the glass apparatus with a**

flame or a heat gun, be certain that no one in the laboratory is working with tetrahydrofuran or diethyl ether, both of which are highly flammable. Consult with your instructor before using any open flame. Avoid excessive heating of the apparatus, especially the rubber septa.

5. Lubricate all ground-glass joints in the apparatus carefully and mate them tightly to prevent the escape of tetrahydrofuran during the reaction.

6. On the small scale of these experiments, the exothermic formation and reaction of organometallic reagents rarely causes a problem. Nevertheless, it is still good laboratory practice to also have an ice-water bath ready in the event the reaction proceeds too rapidly, as evidenced by refluxing of the solvent into the Claisen adapter.

MINISCALE PROCEDURE

Preparation Refer to the online resources to answer Pre-Lab Exercises, access videos, and read the MSDSs for the chemicals used or produced in this procedure. Review Sections 2.5, 2.9, 2.10, 2.11, 2.13, 2.21, 2.25, 2.27, 2.28, and 2.29.

Apparatus A 50-mL round-bottom flask, Claisen adapter, drying tube, two 10-mL syringes with needles, rubber septa, separatory funnel, two 10-mL screw-cap centrifuge tubes, apparatus for magnetic stirring and simple distillation.

Rubber septum

Setting Up Equip the 50-mL round-bottom flask with a stirbar and add 1.52 g of zinc granules to it. Place the flask and its contents, the Claisen adapter, and the drying tube in an oven at 110 °C for at least 30 min. Using gloves or tongs, remove the glassware from the oven and let it cool, preferably in a desiccator. After the glassware is cool enough to handle, add about 125 mg of iodine to the round-bottom flask. Lubricate all of the joints and quickly assemble the apparatus shown in Figure 10.46, positioning the flask in the room-temperature water bath located on the magnetic stirrer as shown. Add 5 mL of *anhydrous* tetrahydrofuran to the flask and begin stirring the contents of the flask. Dissolve 1.2 mL of allyl bromide in 5 mL of *anhydrous* tetrahydrofuran in a screw-cap centrifuge tube, cap, and set it aside. Dissolve 1.2 mL of 3-pentanone in 3mL of *anhydrous* tetrahydrofuran in a screw-cap centrifuge tube, cap, and set it aside as well.

Reaction Draw the solution of allyl bromide into a 10-mL syringe, pierce the rubber septum with the needle of the syringe, being careful not to press down on the plunger, and add the solution slowly over about 1 min to the stirred suspension of zinc granules. The reaction has started when the mixture becomes cloudy and bubbles form. When the reaction has subsided, the solution should be pale yellow and most of the zinc should have dissolved. Stir the mixture for 45 min.

Draw the solution of 3-pentanone into a 10-mL syringe, pierce the rubber septum with the needle of the syringe, being careful not to press down on the plunger as before, and add the solution directly to the stirred solution of allylzinc bromide contained in the round-bottom flask. After the addition is complete, continue to stir the mixture, at room temperature, for an additional 45 min.

Work-Up and Isolation *Slowly* add 3 mL of 1 *M* hydrochloric acid to the reaction mixture with stirring. Pour the mixture into the separatory funnel, rinse the round-bottom flask with 5 mL of diethyl ether and add the rinse to the separatory funnel. Shake the funnel vigorously with venting to relieve pressure; separate the aqueous layer. Wash the organic layer sequentially with 5 mL of saturated *aqueous* sodium thiosulfate solution, 5 mL of saturated *aqueous* sodium bicarbonate solution, and 5 mL of saturated *aqueous* sodium chloride solution. Transfer the organic layer to a dry flask. Dry the solution over several spatula-tips full of *anhydrous* sodium sulfate. Swirl the flask occasionally for a period of 10–15 min to facilitate drying; add further small portions of *anhydrous* sodium sulfate if the solution remains cloudy.★

Filter or carefully decant the dried organic solution into a round-bottom flask equipped for simple distillation, and remove the solvents by simple distillation. Alternatively, use rotary evaporation or other techniques to concentrate the solution. The final traces of solvent may be removed by attaching the flask to a vacuum source and gently swirling the contents as the vacuum is applied.

Analysis Weigh the product and calculate the percent yield. Obtain the IR and ^{1}H NMR spectra of your starting materials and products and compare them with those of authentic samples. (Figs. 19.8–19.13).

MICROSCALE PROCEDURE

Preparation Refer to the online resources to answer Pre-Lab Exercises, access videos, and read the MSDSs for the chemicals used or produced in this procedure. Review Sections 2.5, 2.9, 2.10, 2.11, 2.13, 2.21, 2.25, 2.27, 2.28, and 2.29.

Apparatus A 5-mL conical vial, Claisen adapter, drying tube, two 5-mL screw-cap centrifuge tubes and one 10-mL screw-cap centrifuge tube with caps, two 5-mL glass syringes with needles, water bath, apparatus for magnetic stirring and simple distillation.

Setting Up Equip the 5-mL conical vial with a spinvane and add 0.30 g of zinc granules to it. Place this vial and its contents, the Claisen adapter, and drying tube in an oven at 110 °C for at least 30 min. Using gloves or tongs, remove the glassware from the oven and let it cool, preferably in a desiccator. After the glassware is cool enough to handle, add about 25 mg of iodine. Lubricate all of the joints and quickly assemble the apparatus as shown in Figure 10.47, positioning the vial in the room-temperature water bath located on the magnetic stirrer as shown. Stir the solids and add 0.5 mL of *anhydrous* tetrahydrofuran to the vial with stirring. Dissolve 0.24 mL of allyl bromide in 1 mL of *anhydrous* tetrahydrofuran in a screw-cap centrifuge tube, cap, and set it aside. Dissolve 0.25 mL of 3-pentanone in 0.5 mL of *anhydrous* tetrahydrofuran in a screw-cap centrifuge tube, cap, and set it aside as well.

Reaction Draw the solution of allyl bromide into a 5-mL syringe, pierce the rubber septum with the needle of the syringe, being careful not to press down on the plunger, and add the solution slowly over about 15 sec to the stirred suspension of zinc granules. The reaction has started when the mixture becomes cloudy and bubbles form. When the reaction has subsided, the solution should be pale yellow and most of the zinc should have dissolved. Stir the mixture for 45 min.

Draw the solution of 3-pentanone into a syringe, pierce the rubber septum with the needle of the syringe, being careful not to press down on the plunger, and add the solution directly to the stirred solution of allylzinc bromide contained in the reaction vial. After the addition is complete, continue to stir the mixture, at room temperature, for an additional 30 min.

Work-Up and Isolation *Slowly* add 1 mL of 1 *M* hydrochloric acid to the reaction mixture with stirring. Stop the stirring, allow the layers to separate, and using a Pasteur pipet transfer the organic layer to a screw-cap centrifuge tube. Rinse the vial with 1 mL of diethyl ether and add the rinse to the screw-cap centrifuge tube. Cap the tube and shake it briefly with venting to relieve pressure. Separate the aqueous layer, and wash the organic layer sequentially with 1 mL of *saturated* aqueous sodium thiosulfate solution, 2 mL of *saturated* aqueous sodium bicarbonate solution, and 2 mL of saturated *aqueous* sodium chloride solution. Transfer the organic layer to a dry screw-cap centrifuge tube using a Pasteur pipet, and add several microspatula-tips full of *anhydrous* sodium sulfate. Allow the solution to stand for 15 min, with occasional shaking to hasten drying. Add additional portions of *anhydrous* sodium sulfate if the liquid remains cloudy.★

Using a Pasteur pipet, transfer the organic solution to a clean 5-mL conical vial, add a spinvane to the vial, and remove the volatile solvents by simple distillation. Alternatively, rotary evaporation or other techniques may be used to remove the solvent. The final traces of solvent may be removed by attaching the vial to a vacuum source and gently swirling the contents as the vacuum is applied.

Analysis Weigh the product and calculate the percent yield. Obtain the IR and ^{1}H NMR spectra of your starting materials and products and compare them with those of authentic samples. (Figs. 19.8–19.13).

WRAPPING IT UP

Dilute the *combined aqueous layers and washes* with water, neutralize the solution if necessary, and flush it down the drain with excess water. Place the *ether distillate* in the container for nonhalogenated organic solvents. Spread the *calcium chloride* from the drying tube and the *sodium sulfate* on a tray in the hood, and then place them in the container for nonhazardous solids after the volatiles have evaporated.

EXERCISES

1. How is the residual (unreacted) zinc removed from the reaction mixture?

2. The presence of water in reactions involving organozinc reagents is detrimental. Write the product of the reaction of **16** with water.

3. Write the structure of the product of the reaction of **16** with molecular oxygen.

4. Given the structure of your product **19**, would you expect it to be sensitive to basic conditions? Acidic conditions? Explain your answer.

5. When using Grignard reagents, a problematic side reaction is a Wurtz-type coupling (Eq. 19.9) between the reagent and the organohalide. Since organozinc reagents are less reactive than Grignard reagents, would you expect this side reaction to be a concern when using organozinc reagents?

6. Grignard and organolithium reagents will also react with esters. Would you expect organozinc reagents to add to esters? Explain your answer.

7. How is the unreacted 3-pentanone removed from the product mixture?

8. If there were an acidic hydrogen in another portion of the starting material, would you expect this reaction to be successful? Explain your answer.

9. Arrange the following compounds in order of increasing reactivity toward attack of an organozinc reagent at the carbonyl carbon atom: benzoic acid, benzaldehyde, and benzophenone. Explain the basis for your decision.

10. What is (are) the product(s) of reaction of benzoic acid, benzaldehyde, and benzophenone with excess organozinc reagent, RZnBr?

11. How might primary, secondary, and tertiary alcohols be prepared from an organozinc reagent and a suitable carbonyl-containing compound? Write chemical reactions for these reactions using any starting materials you wish; indicate stoichiometry where important.

Refer to the online resources and use the spectra viewer and Tables 8.1–8.8 as needed to answer the blue-numbered questions on spectroscopy.

12. Consider the spectral data for allyl bromide (Figs. 19.8 and 19.9).

  a. In the functional group region of the IR spectrum, identify any absorption bands associated with the carbon-carbon double bond.

  b. In the ^{1}H-NMR spectrum, assign the various resonances to the hydrogen nuclei responsible for them.

  c. For the ^{13}C-NMR data, assign the various resonances to the carbon nuclei responsible for them.

13. Consider the spectral data for 3-pentanone (Figs. 19.10 and 19.11).

  a. In the functional group region of the IR spectrum, identify any absorption bands associated with the carbonyl group.

  b. In the ^{1}H-NMR spectrum, assign the various resonances to the hydrogen nuclei responsible for them.

  c. For the ^{13}C-NMR data, assign the various resonances to the carbon nuclei responsible for them.

14. Consider the spectral data for 3-ethylhex-5-en-3-ol (Figs. 19.12 and 19.13).

  a. In the functional group region of the IR spectrum, identify any absorption bands associated with the carbon-carbon double bond or the alcohol.

  b. In the ^{1}H-NMR spectrum, assign the various resonances to the hydrogen nuclei responsible for them.

  c. For the ^{13}C-NMR data, assign the various resonances to the carbon nuclei responsible for them.

15. Discuss the differences observed in the IR and NMR spectra of 3-pentanone and 3-ethylhex-5-en-3-ol that are consistent with an addition reaction having occurred.

Starting Materials and Products

Figure 19.8
IR spectrum of allyl bromide (ATR).

Figure 19.9
NMR data for allyl bromide (CDCl₃).

(a) ¹H NMR spectrum (400 MHz).
(b) ¹³C-NMR data. Chemical shifts: δ 32.8, 119.0, 134.3.

Figure 19.10
IR spectrum of 3-pentanone (ATR).

Figure 19.11

NMR data for 3-pentanone (CDCl₃).

(a) ¹H NMR spectrum (400 MHz).
(b) ¹³C-NMR data. Chemical shifts: δ 7.9, 35.5, 212.1.

Figure 19.12

IR spectrum of 3-ethylhex-5-en-3-ol (ATR).

Figure 19.13

NMR data for 3-ethylhex-5-en-3-ol (CDCl₃).

(a) ¹H NMR spectrum (400 MHz).
(b) ¹³C-NMR data. Chemical shifts: δ 7.7, 30.9, 42.9, 74.3, 118.3, 134.0.

19.6 TRANSITION METAL CATALYZED REACTIONS AND PROCEDURE FOR SUZUKI REACTION

Organometallic compounds such as the Grignard and organozinc reagents discussed previously in this chapter (Secs. 19.1 and 19.5, respectively) are typically used in **stoichiometric** amounts, sometimes even more than a 1:1 ratio, with respect to other starting materials (Eq. 19.2) and are consumed as the reaction proceeds. However, many important industrial processes have been developed that use organometallic compounds as **catalysts**. As you recall, a catalyst is an additive that is used in a substoichiometric amount that increases the rate of a chemical reaction without itself being consumed or undergoing a net chemical change. In other words, a catalyst enables an otherwise slow or unselective reaction to proceed readily on a practical timescale by providing an alternate, lower-energy pathway without altering the relative energies of the reactant(s) or product(s). The active catalytic species is regenerated during the reaction and so only a small amount of the precatalyst (e.g., 0.01–0.05 equivalents) with respect to the starting materials is required.

Examples of important processes that use transition metals or their coordination complexes as catalysts include hydrogenation (Sec. 17.2), hydroformylation, alkene metathesis, and alkene polymerization (Scheme 19.2). The aerobic oxidation of alcohols using copper- or palladium-based catalysts is yet another example in which such catalysts are employed (Sec. 16.2).

Scheme 19.2

The **Suzuki-Miyaura** reaction that is the subject of the procedure in this section is another type of coupling that results in the formation of a new carbon-carbon bond. In a typical such reaction, an arylboronic acid reacts with an aryl halide under basic conditions in the presence of a Pd-derived catalyst to form a new single bond between two sp^2-hybridized carbon atoms (Eq. 19.16). If the two organic components are not identical (i.e., $R_1 \neq R_2$), the process is termed a cross-coupling reaction, which is the experiment you will perform in this section.

An arylboronic acid

An aryl halide
X = Cl, Br, I

(19.16)

new C-C σ bond

A biaryl compound

The mechanisms of stoichiometric organic reactions are typically shown in a stepwise manner. An alternate representation that is useful for analyzing catalytic processes is a **catalytic cycle**, which illustrates the key steps of the reaction and shows how the catalytically active species is regenerated. A generalized catalytic cycle for a Pd-catalyzed Suzuki-Miyaura cross-coupling reaction is exemplified in Scheme 19.3. The catalytic Pd-species is shown as L_nPd (L = ligand, typically a phosphine, PR_3), the stoichiometric coupling partners are bromobenzene and 4-methylphenylboronic acid, $ArB(OH)_2$, and potassium hydroxide is the added base. One complete catalytic cycle is termed a **turnover**, wherein one equivalent of reactant is converted to one equivalent of product per equivalent of catalyst. A high

Scheme 19.3

Simplified catalytic cycle for a Suzuki-Miyaura cross-coupling reaction turnover number is desirable for a catalytic process because smaller quantities of precatalyst, which are typically derived from expensive transition metals, are then required for complete conversion of starting materials into products.

The key steps in this cycle are common to many Pd-catalyzed coupling reactions and involve the following key transformations.

1. Insertion of the initial L_nPd^0 species into the C–Br bond of bromobenzene first occurs in a process termed **oxidative addition**. In this step, Pd^0 is oxidized to Pd^{11} to generate the complex **20**. The strength of the carbon-halogen bond, C–X, determines the ease with which the insertion occurs, with the general order of reactivity being aryl iodides > aryl bromides > aryl chlorides >> aryl fluorides. Accordingly, aryl bromides and iodides are most commonly used as reactants in such cross-coupling reactions.

2. Reaction of **20** with an activated form of 4-methylphenylboronic acid then occurs in a step termed **transmetallation**. This process involves the transfer of the aryl group from the boron atom of the arylboronic acid to the Pd atom to form the corresponding Pd^{II} species **21**. Activation of the boronic acid, which is required for a successful Suzuki-Miyaura coupling, weakens the carbon-boron bond and involves addition of hydroxide ion to the neutral boron atom, a Lewis acid, to form an anionic boronate. Transmetallation is crucial to the coupling process because the two aryl groups to be coupled are in close proximity on the palladium atom. The by-products of this step are potassium bromide and boric acid.

3. The final step of the catalytic cycle involves the formation of a new C–C σ-bond between the two aryl groups in **21** in a process termed **reductive elimination** to give the coupling product. This transformation is basically the microscopic reverse of oxidative addition. In addition to generating the biaryl product, the Pd^0 species is regenerated in this step, thereby completing the catalytic cycle.

It is important to note that organopalladium species are very different from Grignard and organozinc reagents with respect to their susceptibility to reacting with water. For example, a Grignard reagent has a highly polarized carbon-magnesium bond (Sec. 19.1), which is reflected in Pauling electronegativity values of 2.55 for carbon and 1.31 for magnesium. This difference imparts significant anionic character to the carbon atom, so Grignard reagents must be rigorously protected from moisture to prevent their protonation (Eq. 19.6). Because functional groups such as hydroxy or amino groups can also serve as proton donors, these functionalities cannot be present when generating or using Grignard reagents. In contrast, palladium has a Pauling electronegativity of 2.20, so the difference in electronegativities between carbon and palladium is much less, thereby decreasing the anionic character of the carbon atom and making carbonpalladium bonds more covalent. Consequently, the Suzuki-Miyaura coupling reaction can be performed in aqueous solution and in the presence of a wide range of organic functional groups. In addition, the arylboronic acid starting materials are non toxic, air- and moisture-stable, and readily prepared by the reaction of an trialkyl borate, $B(OR)_3$, with an aryl Grignard reagent (Eq. 19.17). These features, among others, make the Suzuki-Miyaura coupling a common reaction in industrial and pharmaceutical chemistry.

$$ArMgBr + B(OR^1)_3 \longrightarrow Ar-B(OR^1)_2 + R^1OMgBr \tag{19.17}$$

In the procedure that follows, a Suzuki-Miyaura cross-coupling reaction between 4-bromobenzyl alcohol (**22**) and 4-methylphenylboronic acid

(23) will be performed in ethanolic potassium hydroxide solution to produce 4'-methyl-(1,1'-biphenyl)-4-methanol (24), as shown in Equation 19.18 . The palladium catalyst in this case is ligand free, rather than a complex as shown in Scheme 19.3.

$$Br\!-\!\!\langle\;\rangle\!-\!CH_2OH + CH_3\!-\!\!\langle\;\rangle\!-\!B(OH)_2 \xrightarrow[\substack{KOH \\ 95\% \text{ ethanol}}]{\text{cat. Pd}^0} CH_3\!-\!\!\langle\;\rangle\!-\!\langle\;\rangle\!-\!CH_2OH \qquad (19.18)$$

22	**23**	**24**
4-Bromobenzyl alcohol	4-Methylphenyl-boronic acid	4'-Methyl-(1,1'-biphenyl)-4-methanol

EXPERIMENTAL PROCEDURES

Preparation of 4'-Methyl-(1,1'-Biphenyl)-4-Methanol

Purpose To demonstrate a palladium-catalyzed Suzuki-Miyaura cross-coupling reaction.

SAFETY ALERT

1. **Wear safety glasses or goggles and suitable protective gloves while performing the experiment.**
2. **Dispense dichloromethane in a well-ventilated work area.**

MINISCALE PROCEDURE

Preparation Refer to the online resources to answer Pre-Lab Exercises, access videos, and read the MSDSs for the chemicals used or produced in this procedure. Review Sections 2.11, 2.13, 2.17, and 2.29.

Apparatus A 125-mL round-bottom flask, apparatus for magnetic stirring, simple distillation, and vacuum filtration.

Setting Up Place 1.0 g of 4-bromobenzyl alcohol and a stirbar in the flask and add 10 mL of ethanolic 1 *M* potassium hydroxide solution. Stir the mixture until the solid is dissolved. Weigh 0.74 g of 4-methylphenylboronic acid into a small beaker and add 10 mL of 95% ethanol. Stir the mixture until the solid is dissolved. Add the solution of 4-methylphenylboronic acid to the *rapidly* stirred solution of 4-bromobenzyl alcohol. Rinse the beaker with 2 mL of 95% ethanol and add this wash to the reaction flask.

Reaction Stir the mixture for about 2 min and then add 1 mL of the stock aqueous solution of 1 mg of palladium/mL. Stir the reaction mixture rapidly for about 15 min.

Work-Up Add 25 mL of ice-cold water to the reaction mixture, continue stirring for 5 min, and collect the solid by vacuum filtration using a Büchner funnel. Rinse the isolated solid with 5 mL of ice-cold water on the filter and press on the crude product with a spatula while maintaining the vacuum to eliminate as much water as possible.

Isolation and Purification Dissolve the crude product in 25 mL of dichloromethane and dry the solution by stirring over *anhydrous* magnesium sulfate. Add further portions of *anhydrous* magnesium sulfate if the solution does not become clear. Remove the magnesium sulfate by gravity filtration, collecting the filtrate in a round-bottom flask. Equip the flask for simple distillation and remove the solvent by simple distillation. Alternatively, use rotary evaporation or other techniques to concentrate the solution. The final traces of solvent may be removed by attaching the flask to a vacuum source and gently swirling the contents as the vacuum is applied.

Analysis Weigh the isolated material and calculate the percent yield. Determine the melting point of the product. Obtain IR, ^{1}H-NMR, and ^{13}C-NMR spectra of your starting material and product, and compare the data with those from authentic samples (Figs. 19.14–19.19).

MICROSCALE PROCEDURE

Preparation Refer to the online resources to answer Pre-Lab Exercises, access videos, and read the MSDSs for the chemicals used or produced in this procedure. Review Sections 2.11, 2.13, 2.17, and 2.29.

Apparatus A 25-mL round-bottom flask, apparatus for magnetic stirring, simple distillation, and vacuum filtration.

Setting Up Place 0.20 g of 4-bromobenzyl alcohol and a stirbar in the flask and add 5 mL of 95% ethanol. Stir the mixture until the solid is dissolved. Weigh 0.15 g of 4-methylphenylboronic acid into a small vial. Add the solid 4-methylphenylboronic acid to the *rapidly* stirred solution of 4-bromobenzyl alcohol. Rinse the vial with 1 mL of 95% ethanol and add this wash to the reaction flask.

Reaction Stir the reaction mixture until the solids are fully dissolved. Add 2 mL of ethanolic 1 M potassium hydroxide solution and stir for about 1 min; then add 0.2 mL (8 drops) of the stock aqueous solution of 1 mg of palladium/mL. Stir the reaction mixture rapidly for about 10 min.

Work-Up Add 10 mL ice-cold water to the reaction mixture, continue stirring for 5 min, and collect the solid by vacuum filtration using a Hirsch funnel. Rinse the isolated solid with 2 mL of ice-cold water on the filter and press on the crude product with a spatula while maintaining the vacuum to eliminate as much water as possible.

Isolation and Purification Dissolve the crude product in 10 mL of dichloromethane and dry the solution by stirring over *anhydrous* magnesium sulfate. Add further portions of *anhydrous* magnesium sulfate if the solution does not become clear. Remove the magnesium sulfate from the solution by gravity filtration, collecting the filtrate in a round-bottom flask. Equip the flask for simple distillation and remove the solvent by simple distillation. Alternatively, use rotary evaporation or other techniques to concentrate the solution. The final traces of solvent may be removed by attaching the flask to a vacuum source and gently swirling the contents as the vacuum is applied.

Analysis Weigh the isolated material and calculate the percent yield. Determine the melting range of the product. Obtain IR, ^{1}H-NMR, and ^{13}C-NMR spectra of your starting material and product, and compare the data with those from authentic samples (Figs. 19.14–19.19).

WRAPPING IT UP

Flush all *aqueous solutions* down the drain. Spread the *magnesium sulfate* on a tray or filter paper in the fume hood to evaporate the solvent from it, and then place it in the container for non-hazardous solid waste. Transfer the recovered *dichloromethane* into the container for halogenated solvents and the ethanol into the container for nonhalogenated organic liquids.

EXERCISES

1. Draw the organic products obtained from the Suzuki-Miyaura coupling reactions shown below. Highlight the new C-C bond in each product.

2. Ethanol was used as the solvent for the cross-coupling reaction. Briefly explain why dichloromethane was added to the reaction mixture during the work-up process.

3. In addition to a white solid, you may have noticed a small amount of gray-black solid precipitate from the reaction mixture. What was this solid and how was it removed?

4. Briefly explain why the Suzuki-Miyaura coupling reaction can take place in aqueous solution whereas a Grignard reaction requires an *anhydrous* ethereal solvent.

5. Briefly explain why aryl bromides and iodides, rather than aryl chlorides or fluorides, are typically used as substrates in Pd-catalyzed coupling reactions.

6. Draw and label the full catalytic cycle for the Suzuki-Miyaura coupling of 4-bromobenzyl alcohol with 4-methylphenylboronic acid.

7. Construct a potential energy diagram that illustrates the difference between the Suzuki coupling reaction you performed (Eq. 19.18) in an uncatalyzed versus a catalyzed process. Assume that $\Delta H^{\ddagger}$ for the uncatalyzed reaction is 32 kcal/mol and that ΔH_{rxn} is −7 kcal/mol. For the catalyzed process, assume that the rate-determining step is the transmetallation of the complex analogous to **20** to produce the complex analogous to **21** in Scheme 19.3, labeling these intermediates as **20*** and **21***, respectively.

Refer to the online resources and use the spectra viewer and Tables 8.1–8.8 as needed to answer the blue-numbered questions on spectroscopy.

8. Consider the spectral data for 4-bromobenzyl alcohol (Figs. 19.14 and 19.15).

   a. In the functional group region of the IR spectrum, specify the absorptions associated with the alcohol group and the π-bonds of the aromatic ring.

   b. In the ^{1}H NMR spectrum, assign the various resonances to the hydrogen nuclei responsible for them.

   c. For the ^{13}C NMR data, assign the various resonances to the carbon nuclei responsible for them.

9. Consider the spectral data for 4-methylphenylboronic acid (Figs. 19.16 and 19.17).

   a. In the functional group region of the IR spectrum, specify the absorptions associated with the alcohol functional group and the π-bonds of the aromatic ring.

   b. In the ^{1}H NMR spectrum, assign the various resonances to the hydrogen nuclei responsible for them. Note that there is no resonance observed for the hydrogen nuclei of the OH groups.

   c. For the ^{13}C NMR data, assign the various resonances to the carbon nuclei responsible for them. Note that there is no resonance observed for the carbon atom bound to the boron atom.

10. Consider the spectral data for 4′-methyl-(1,1′-biphenyl)-4-methanol (Figs. 19.18 and 19.19).

    a. In the functional group region of the IR spectrum, specify the absorptions associated with the alcohol group and the π-bonds of the aromatic rings.

    b. In the ^{1}H NMR spectrum, assign the various resonances to the hydrogen nuclei responsible for them.

    c. For the ^{13}C NMR data, assign the various resonances to the carbon nuclei responsible for them.

11. Discuss the differences observed in the NMR spectra of your starting materials and product that are consistent with the coupling reaction having occurred.

SPECTRA

Starting Materials and Products

Figure 19.14

IR spectrum of 4-bromobenzyl alcohol (ATR).

Figure 19.15

NMR data for 4-bromobenzyl alcohol (CDCl₃).

(a) ¹H NMR spectrum (300 MHz).
(b) ¹³C NMR data: δ 64.7, 121.6, 128.8, 131.8, 139.9.

Figure 19.16

IR spectrum of 4-methylphenylboronic acid (ATR).

Figure 19.17

NMR data for 4-methylphenylboronic acid (CDCl₃).

(a) ¹H NMR spectrum (300 MHz).
(b) ¹³C NMR data: δ 22.1, 129.0, 135.9, 143.1 (signal for carbon atom bound to boron not detected).

Figure 19.18
*IR spectrum of 4'-methyl-
(1,1'-biphenyl)-4-methanol (ATR).*

Figure 19.19
*NMR data for 4'-methyl-
(1,1'-biphenyl)-4-methanol
(CDCI₃).*

(a) ¹H NMR spectrum (300 MHz).
(b) ¹³C NMR data: δ 21.3, 65.4, 127.1, 127.3, 127.7, 129.7, 137.3, 138.1, 139.8, 140.8.

CHAPTER 20

Carboxylic Acids and Their Derivatives

The chemistry of organic compounds containing the carbon-oxygen double bond is arguably richer than that of any other class of organic molecules. In the experiments in Chapter 18, there is an opportunity to learn that the chemistry of aldehydes and ketones is determined by the presence of the carbonyl functional group. In this chapter, you will be able to study the chemistry associated with carboxylic acids and some of their derivatives. These compounds contain a carbon-oxygen double bond, in which the carbonyl carbon atom bears an additional heteroatom such as oxygen, nitrogen, or halogen. The presence of this heteroatom dramatically changes the chemical properties of the carbonyl group, but in ways that you will learn are predictable. Derivatives of carboxylic acids are commonplace in materials we encounter daily and include fabrics, antibiotics, and insecticides, to name a few. In the experiments in this chapter, you will be introduced to the varied chemistry and the importance of carboxylic acid derivatives by preparing compounds having use as an anesthetic and as an insect repellent. You will also have the opportunity to study a fascinating chemical reaction that produces light, resulting from decomposition of a compound you have synthesized.

20.1 INTRODUCTION

Carboxylic acids are characterized by the presence of the **carboxyl functional group**, $-CO_2H$, which is composed of a **carb**onyl group and a hyd**roxyl** group, and have the general structural formula **1** in which R is a hydrogen or an alkyl or an aryl substituent. Their derivatives arise by replacing the hydroxyl group in **1** with other heteroatom substituents such as Cl in an **acid chloride**, $O=COR^1$ in an **acid anhydride**, OR^2 in an **ester**, and NR^3R^4 in an **amide**. These derivatives may be summarized by the general formula **2** in which the **acyl group**, $R-C(=O)-$, is bonded to an electronegative substituent Z, which may represent a single atom, as in acid chlorides, or a group of atoms, as in anhydrides, esters, and amides. Although they will not be included in the discussion that follows, **nitriles**, $R-C\equiv N$, are generally classified as derivatives of carboxylic acids because they may be hydrolyzed to produce these acids.

$$\underset{R}{\overset{\displaystyle O^{\delta^-}}{\underset{\delta^+}{\overset{\|}{C}}}}\, Z^{\delta^-}$$

1: Z = OH
2: Z = Cl, OCOR^1
OR^2, NR^{3}R^4

The acyl group defines the chemistry of carboxylic acids and their derivatives. Like the carbonyl group in aldehydes and ketones (Chap. 18), the carbon-oxygen π-bond in carboxylic acids and their derivatives is polarized because of the electronegativity of the oxygen atom. The *electrophilicity* of the **acyl carbon atom** induced by this polarization is further increased by the presence of the other electronegative substituent, Z. This combination of an electronegative substituent attached to the polarized carbon-oxygen double bond has important consequences on the reactivity of carboxylic acid derivatives toward nucleophiles. Indeed, the general type of reaction that is most characteristic of these compounds is **nucleophilic acyl substitution reaction** (Eq. 20.1).

$$\underset{R}{\overset{\displaystyle O}{\overset{\|}{C}}}\, Z \;+\; Nu^- \;\longrightarrow\; \underset{R}{\overset{\displaystyle O}{\overset{\|}{C}}}\, Nu \;+\; Z^- \tag{20.1}$$

A number of mechanistic pathways exist for nucleophilic acyl substitution. In the simplest of these, a negatively charged nucleophile, Nu$^-$, attacks the electrophilic acyl carbon atom of **3** to give the tetrahedral intermediate **4**. This then collapses to regenerate the carbon-oxygen double bond with loss of the **leaving group**, Z$^-$, to provide a substitution product **5**, which is also a carboxylic acid derivative (Eq. 20.2). The first step in this reaction may be considered to be a Lewis acid-Lewis base reaction in which the acyl carbon atom is the Lewis acid and the nucleophile is the Lewis base.

$$\underset{\substack{R \\ \mathbf{3}}}{\overset{\displaystyle O^{\delta^-}}{\underset{\delta^+}{\overset{\|}{C}}}}\, Z^{\delta^-} \;+\; Nu^- \;\longrightarrow\; \underset{\substack{\\ \mathbf{4}}}{R-\overset{\displaystyle O^-}{\underset{Nu}{\overset{|}{\underset{|}{C}}}}-Z} \;\longrightarrow\; \underset{\substack{R \\ \mathbf{5}}}{\overset{\displaystyle O}{\overset{\|}{C}}}\, Nu \;+\; Z^- \tag{20.2}$$

3
A carboxylic
acid derivative

5
A nucleophilic
substitution product

When weaker, typically neutral, nucleophiles or less reactive carboxylic acid derivatives are involved, the reaction may be catalyzed by acid as outlined in Scheme 20.1. The oxygen atom of the acyl group in **3** is first protonated to give the cation **6** in which the carbon-oxygen bond is even more highly polarized and the acyl carbon atom more electrophilic than in **3**. Nucleophilic attack by a neutral nucleophile, H–Nu, then gives the tetrahedral intermediate **7**, which will transfer a proton to a base, B$^-$, in the reaction mixture to give **8**. The details of how the tetrahedral intermediate **8** is converted to the substitution product **5** will depend upon the nature of the leaving group, Z$^-$, the nucleophile, Nu$^-$, and the reaction conditions. For example, if Z$^-$ is not a good leaving group relative to Nu$^-$, it will likely be protonated to give **9** prior to breaking of the C–Z bond; the cation **9** will then collapse to give **5**.

Scheme 20.1

Comparing the reactions of carboxylic acid derivatives with those of aldehydes and ketones (Chap. 18) reveals similarities as well as important differences. The reactions of aldehydes and ketones, whether they are acid-catalyzed or not, lead to the formation of *nucleophilic addition products* according to Equation 20.3, whereas carboxylic acid derivatives react with nucleophiles to give **substitution products** (Eq. 20.1). In both cases, the first step is nucleophilic attack on the electrophilic carbon atom of the carbon-oxygen double bond. However, aldehydes and ketones do not have an electronegative group, Z, that can form a stable anion and thus serve as a good leaving group. A good leaving group must be a relatively weak base such as Cl$^-$, RO$^-$, H$_2$O, and ROH, and not a very strong base like H$^-$ or R$^-$ (Sec. 14.1).

An aldehyde or a ketone   An addition product

$$ \tag{20.3} $$

The experiments described in this chapter illustrate some of the representative reactions of carboxylic acids and their derivatives. For example, you will perform a **Fischer esterification**, one of the classic reactions in organic chemistry that was discovered by the Nobel laureate Emil Fischer (see the Historical Highlight *Emil Fischer: Carbohydrate Chemist Extraordinaire,* which is available online, for an account of the life of this famous chemist), to prepare the anesthetic agent benzocaine from *p*-aminobenzoic acid (Sec. 20.2). You will also perform a Fischer esterification as part of an investigative procedure for identifying unknown esters. In another experiment, you will synthesize the mosquito repellent N,N-diethyl-*m*-toluamide (DEET) by a two-step process that involves the conversion of a carboxylic acid into an acid chloride and subsequent reaction with an amine to produce the desired amide (Sec. 20.3). Finally, you will prepare a hydrazide, a close relative of an amide, and study its chemical decomposition in a process that produces visible light.

20.2 ESTERS AND THE FISCHER ESTERIFICATION

The direct conversion of a carboxylic acid and an alcohol into an ester is usually catalyzed by a mineral acid such as *concentrated* sulfuric acid, as outlined in Scheme 20.2. The catalyst promotes the addition of the weakly nucleophilic alcohol to the carbon-oxygen double bond by protonating the carbonyl oxygen of the acid **10** to give **11**. Consequently, **11** is activated toward nucleophilic attack and hence reacts rapidly with an alcohol to give the tetrahedral intermediate **12**. Two proton transfer steps follow to give **13**, which collapses by loss of water, a good leaving group, to provide first **14** and then the ester **15**.

Scheme 20.2

As you can see from the multiple equilibria of Scheme 20.2, the acid-catalyzed hydrolysis of an ester **15** to the corresponding acid **10** is simply the reverse of the esterification. Special strategies are therefore required to effect esterification of a carboxylic acid in good yield. These may involve using a large excess of one of the reactants or removing one of the products as it is formed. Both of these experimental tactics make use of the Le Chatelier principle (Sec. 10.3). When the alcohol is inexpensive and volatile, it is frequently used as the solvent, thus forcing the equilibrium to the right and increasing the yield of the ester. In other cases, it may be possible to remove the product water from the reaction using a drying agent, like activated molecular sieves (Sec. 2.24), or by azeotropic distillation (Sec. 4.4). Often, ester formation by the Fischer esterification is inefficient because of the existence of the equilibration, and esters are more generally prepared by the reaction of alcohols with acid chlorides or acid anhydrides.

In one of the experiments described in this section, you will perform the Fischer esterification of *p*-aminobenzoic acid (**16**) with ethanol to give ethyl *p*-aminobenzoate (**17**), more commonly known as benzocaine, according to Equation 20.4. Benzocaine is a useful topical anesthetic, and the discovery of its anesthetic properties represents an interesting example of "rational drug design" as described in the Historical Highlight *Evolution of Synthetic Analgesics*, which is available online. In another experiment in this section, you will use a Fischer esterification procedure as part of a discovery process to identify unknown esters and the carboxylic acids and alcohols from which they are derived.

$$\underset{\substack{\textbf{16}\\ \text{\textit{p}-Aminobenzoic acid}}}{\underset{\text{NH}_2}{\text{CO}_2\text{H}}} + \text{CH}_3\text{CH}_2\text{OH} \underset{\text{(catalyst)}}{\overset{\text{conc. H}_2\text{SO}_4}{\rightleftharpoons}} \underset{\substack{\textbf{17}\\ \text{Ethyl \textit{p}-aminobenzoate}\\ \text{(benzocaine)}}}{\underset{\text{NH}_2}{\text{CO}_2\text{CH}_2\text{CH}_3}} + \text{H}_2\text{O} \qquad (20.4)$$

The equilibrium between the starting materials and products is readily established by heating a mixture of **16** and ethanol under reflux in the presence of a catalytic amount of sulfuric acid to give **17** and water (Eq. 20.4). However, if a microwave apparatus is available for heating (Sec. 2.9), the reaction can be effected in a pressurized vial. It is instructive to compare the reaction times required and the results obtained using conventional versus microwave heating for promoting the reaction. Your instructor may divide the class into two groups, so you can judge the merits of the two procedures.

Another experiment in this section is a Fischer esterification of either benzoic or propanoic acid and an *unknown* alcohol according to the reactions shown in Equations 20.5 and 20.6. You will use GC analysis (Sec. 6.4) to determine the identity of the ester thus formed, so you will be able to deduce the structures of the carboxylic acid and the alcohol.

There are two variations of the experimental procedure that can be used. One of these involves traditional heating and use of sulfuric acid, H_2SO_4, adsorbed on silica gel, which is a solid-supported catalyst that is easily removed upon completion of the reaction. The silica gel also functions as a desiccant (Sec. 2.24) to remove the water formed during the esterification. The other procedure involves use of a catalytic amount of sulfuric acid and heating in a microwave oven (Sec. 2.9). It is instructive to compare the reaction times required and the results obtained using the two procedures. Your instructor may divide the class into two groups, so you and your fellow students can judge the merits of the two techniques.

$$\underset{\text{Benzoic acid}}{\text{C}_6\text{H}_5\text{–CO–OH}} \xrightarrow[\text{catalyst}]{\text{ROH}} \underset{\text{An alkyl benzoate}}{\text{C}_6\text{H}_5\text{–CO–OR}} + \text{H}_2\text{O} \qquad (20.5)$$

$$R = CH_3, \; CH_2CH_3, \; CH_2CH_2CH_3, \; CH_2CH_2CH_2CH_3$$

$$\underset{\text{Propanoic acid}}{\text{CH}_3\text{CH}_2\text{–CO–OH}} \xrightarrow[\text{catalyst}]{\text{ROH}} \underset{\text{An alkyl propanoate}}{\text{CH}_3\text{CH}_2\text{–CO–OR}} + \text{H}_2\text{O} \qquad (20.6)$$

$$R = CH_3, \; CH_2CH_3, \; CH_2CH_2CH_3, \; CH_2CH_2CH_2CH_3$$

EXPERIMENTAL PROCEDURES

Fischer Esterification of Carboxylic Acids

A ▪ *Preparation of Benzocaine*

Purpose To demonstrate the acid-catalyzed esterification of a carboxylic acid with an alcohol.

SAFETY ALERT

1. **Wear safety glasses or goggles and suitable protective gloves while performing the experiment.**

2. *Concentrated* **sulfuric acid is** *very corrosive* **and may cause serious chemical burns if allowed to come into contact with your skin. When it is poured from the reagent bottle, some may run down the outside of the bottle; wipe the bottle clean with a sponge or paper towel. If any sulfuric acid comes into contact with your skin, immediately flood the affected area with cold water and then with 5% sodium bicarbonate solution.**

MINISCALE PROCEDURE

Preparation Refer to the online resources to answer Pre-Lab Exercises, access videos, and read the MSDSs for the chemicals used or produced in this procedure. Review Sections 2.9, 2.10, 2.11, 2.17, and 2.22.

Apparatus A 25-mL round-bottom flask, ice-water bath, apparatus for heating under reflux, magnetic stirring, vacuum filtration, and *flameless* heating.

Setting Up Add 1.0 g of *p*-aminobenzoic acid and 10 mL of *absolute* ethanol to the round-bottom flask containing a spinvane. Stir the mixture until the solid is *completely* dissolved. Add 1 mL of *concentrated* sulfuric acid *dropwise* to the ethanolic solution of *p*-aminobenzoic acid, equip the flask with a condenser, and set up the apparatus for heating under reflux.

Reaction Heat the mixture under gentle reflux for 30 min. If any solid remains in the flask at this time, remove the heat source and allow the mixture to cool for 2–3 min. Add 3 mL of ethanol and 0.5 mL of *concentrated* sulfuric acid to the reaction flask and resume heating under reflux. After the reaction mixture becomes homogeneous, continue heating it under gentle reflux for another 30 min. *For the experiment to succeed, it is important that all of the solids dissolve during the period of reflux.*★

Work-Up and Isolation Allow the reaction mixture to cool to room temperature, and then pour it into a beaker containing 30 mL of water. Bring the mixture to a pH of about 8 by *slowly* adding 10% aqueous sodium carbonate with stirring. Be careful in this step, as frothing occurs during the neutralization. Beforehand, you should calculate the approximate volume of 10% aqueous sodium carbonate that will be required to neutralize the *total* amount of sulfuric acid you used. By vacuum filtration, collect the crude benzocaine that precipitates. Use three 10-mL portions of cold water to rinse the solid from the beaker and wash the filter cake, and then air-dry the product.★

Purification Weigh the crude product and transfer it to a 50-mL Erlenmeyer flask containing a stirbar and 20 mL of water. Heat the mixture with stirring to about 60 °C and then add just enough methanol to dissolve the solid (5–10 mL); do *not* add more methanol than necessary. When the solid has dissolved, allow the solution to cool to room temperature and then cool it in an ice-water bath for 10–15 min to complete crystallization. Isolate the crystals by vacuum filtration, wash them with 5–10 mL of cold water, and allow them to air-dry.

Analysis Weigh the recrystallized product and calculate the percent yield; determine its melting point. Obtain IR and ^{1}H NMR spectra of your starting materials and product, and compare them with those of authentic samples (Figs. 20.1–20.6).

MINISCALE PROCEDURE FOR MICROWAVE OPTION

Preparation Refer to the online resources to answer Pre-Lab Exercises, access videos, and read the MSDSs for the chemicals used or produced in this procedure. Review Sections 2.9, 2.10, 2.11, 2.17, and 2.22.

Apparatus A 10-mL pressure-rated tube with cap, stirbar, ice-water bath, apparatus for microwave heating with magnetic stirring and vacuum filtration.

Setting Up Place 0.50 g of *p*-aminobenzoic acid and a stirbar in the 10-mL tube. Add 5 mL of *absolute* ethanol. Add about 0.3 mL (6 drops) of *concentrated* sulfuric acid, seal the vial with the cap, and place it in the cavity of the microwave apparatus.

Reaction Program the unit to heat the reaction mixture with stirring according to the directions provided by your instructor. Generally, the reaction temperature should be set at 140 °C and the power set at a maximum of 300 W with a 1-min ramp time and a 5-min hold time; the pressure limit should be set at 250 psi.

Work-Up and Isolation Allow the reaction mixture to cool to room temperature and then pour it into a beaker containing 15 mL of water. Bring the mixture to a pH of about 8 by *slowly* adding 10% aqueous sodium carbonate with stirring. Be careful in this step, as frothing occurs during the neutralization. Beforehand, you should calculate the approximate volume of 10% aqueous sodium carbonate that will be required to neutralize the amount of sulfuric acid that you used. By vacuum filtration, collect the crude benzocaine that precipitates. Use three 5-mL portions of cold water to rinse the solid from the beaker and wash the filter cake, and then air-dry product.★

Purification Weigh the crude product and transfer it to a 25-mL Erlenmeyer flask containing a stirbar and 20 mL of water. Heat the mixture with stirring to about 60°C and then add just enough methanol to dissolve the solid (3–5 mL); do *not* add more methanol than necessary. When the solid has dissolved, allow the solution to cool to room temperature and then cool it in an ice-water bath for 10–15 min to complete crystallization. Isolate the crystals by vacuum filtration, wash them with 3–5 mL of cold water, and allow them to air-dry.

Analysis Weigh the recrystallized product and calculate the percentage yield; determine the melting point. Obtain IR and ^{1}H NMR spectra of your starting materials and product, and compare them with those of authentic samples (Figs. 20.1–20.6).

MICROSCALE PROCEDURE

Preparation Refer to the online resources to answer Pre-Lab Exercises, access videos, and read the MSDSs for the chemicals used or produced in this procedure. Review Sections 2.9, 2.10, 2.11, 2.17, and 2.22.

Apparatus A 5-mL conical vial, ice-water bath, apparatus for heating under reflux, magnetic stirring, vacuum filtration, and *flameless* heating.

Setting Up Add 0.2 g of *p*-aminobenzoic acid and 2.0 mL of *absolute* ethanol to the conical vial containing a spinvane. Stir the mixture until the solid is *completely* dissolved. Add 0.2 mL of *concentrated* sulfuric acid *dropwise* to the ethanolic solution of *p*-aminobenzoic acid, equip the vial with a condenser, and set up the apparatus for heating under reflux.

Reaction Heat the mixture under gentle reflux for 20 min. If any solid remains in the vial at this time, remove the heat source and allow the mixture to cool for 2–3 min. Then add 0.5 mL of *absolute* ethanol and 0.1 mL of *concentrated* sulfuric acid to the reaction vial and resume heating under gentle reflux. After the reaction mixture becomes homogeneous, continue heating it under gentle reflux for another 30 min. *For the experiment to succeed, it is important that all of the solids dissolve during the period of reflux.*★

Work-Up and Isolation Allow the reaction mixture to cool to room temperature and then pour it into a beaker containing 5 mL of water. Bring the mixture to a pH of about 8 by *slowly* adding 10% aqueous sodium carbonate with stirring. Be careful in this step, as frothing occurs during the neutralization. Beforehand, you should calculate the approximate volume of 10% aqueous sodium carbonate that will be required to neutralize the *total* amount of sulfuric acid you used. By vacuum filtration, collect the crude benzocaine that precipitates. Use three 2-mL portions of cold water to rinse the solid from the beaker and wash the filter cake, and then air-dry the product.★

Purification Weigh the crude product and transfer it to a 10-mL Erlenmeyer flask containing a magnetic stirbar and 4 mL of water. Heat the mixture with stirring to about 60 °C and then add just enough methanol to dissolve the solid (1–2 mL); do *not* add more methanol than necessary. When the solid has dissolved, allow the solution to cool to room temperature and then cool it in an ice-water bath for 10–15 min to complete crystallization. Isolate the crystals by vacuum filtration, wash them with 1–2 mL of cold water, and allow them to air-dry.

Analysis Weigh the recrystallized product and calculate the percent yield; determine its melting point. Obtain IR and ¹H NMR spectra of your starting materials and product, and compare them with those of authentic samples (Figs. 20.1–20.6).

WRAPPING IT UP

Flush all *filtrates* down the drain.

EXERCISES

1. Why is absolute ethanol rather than 95% ethanol used in this experiment?

2. Propose a structure for the solid that separates when *concentrated* sulfuric acid is added to the solution of *p*-aminobenzoic acid in ethanol.

3. Why is it important that the solid referred to in Exercise 2 dissolve during the reaction to obtain a good yield of ethyl *p*-aminobenzoate?

4. Sodium carbonate is used to adjust the pH to 8 during the work-up of the reaction.

   a. Why is it necessary to adjust to this pH prior to extracting the aqueous mixture with diethyl ether?

   b. What undesired reaction might occur if you made the solution strongly basic with aqueous sodium hydroxide?

   c. Using curved arrows to symbolize the flow of electrons, write the stepwise mechanism for the reaction of Part **b**.

5. Consider the equilibrium for the esterification of *p*-aminobenzoic acid (**16**) to give ethyl *p*-aminobenzoate (**17**) shown in Equation 20.4.

   a. Write the mathematical expression for the equilibrium constant, K_{eq}, as a function of the concentrations of the products and reactants for this reaction.

   b. Provide two specific ways whereby the Le Chatelier principle may be applied to this equilibrium to drive the reaction completely to the ester **17**.

   c. Which one of the techniques given in Part b is used in this experiment?

6. Using curved arrows to symbolize the flow of electrons, write the stepwise mechanism for the acid-catalyzed esterification of **16** to give **17**.

7. A strong acid is generally used to catalyze the Fischer esterification of carboxylic acids. What *two* steps in the reaction are accelerated by the presence of strong acid, and what function does the acid play in each of these steps?

8. Consider the spectral data for *p*-aminobenzoic acid (Figs. 20.1 and 20.2).

   a. In the functional group region of the IR spectrum, specify the absorption associated with the carbonyl component of the carboxyl group. What functional group is responsible for the broad absorption in the region of about 2800 cm⁻¹, and why is the absorption broad?

   b. In the ¹H NMR spectrum, assign the various resonances to the hydrogen nuclei responsible for them.

   c. For the ¹³C NMR data, assign the various resonances to the carbon nuclei responsible for them.

Refer to the online resources and use the spectra viewer and Tables 8.1–8.8 as needed to answer the blue-numbered questions on spectroscopy.

9. Consider the spectral data for ethanol (Figs. 20.3 and 20.4).

   a. What functional group accounts for the broad absorption centered at about 3350 cm^{-1} in the IR spectrum, and why it is broad?

   b. In the ^{1}H NMR spectrum, assign the various resonances to the hydrogen nuclei responsible for them.

   c. In the ^{13}C NMR spectrum, assign the various resonances to the carbon nuclei responsible for them.

10. Consider the spectral data for ethyl *p*-aminobenzoate (Figs. 20.5 and 20.6).

   a. In the functional group region of the IR spectrum, specify the absorption associated with the carbonyl component of the ester group. What functional group is responsible for the broad absorption in the region of about 3200 cm^{-1}, and why is the absorption broad?

   b. In the ^{1}H NMR spectrum, assign the various resonances to the hydrogen nuclei responsible for them.

   c. For the ^{13}C NMR data, assign the various resonances to the carbon nuclei responsible for them.

11. Discuss the differences observed in the IR and NMR spectra of *p*-aminobenzoic acid and ethyl *p*-aminobenzoate that are consistent with the formation of the latter in this procedure.

SPECTRA

Starting Materials and Products

Figure 20.1
IR spectrum of p-*aminobenzoic acid (IR card).*

Figure 20.2

NMR data for p-aminobenzoic acid (DMSO-d_6).

(a) ^{1}H NMR spectrum (300 MHz).
(b) ^{13}C NMR data: δ 112.6, 116.9, 131.2, 153.1, 167.5.

Figure 20.3

IR spectrum of ethanol (neat).

Figure 20.4

NMR data for ethanol (CDCl$_3$).

(a) ^{1}H NMR spectrum (300 MHz).
(b) ^{13}C NMR data: 18.1, 57.8.

Figure 20.5
*IR spectrum of ethyl
p-aminobenzoate (IR card).*

Figure 20.6
*NMR data for ethyl
p-aminobenzoate (CDCl₃).*

(a) ^{1}H NMR spectrum (300 MHz).
(b) ^{13}C NMR data: δ 14.4, 60.9, 113.8, 120.2, 131.5, 150.7, 166.7.

B ▪ *Identifying Unknown Esters Produced by Fischer Esterification*

Discovery Experiment   **Purpose** Using GC to identify unknown esters prepared by the Fischer esterification of an unknown acid and an unknown alcohol.

SAFETY ALERT

1. **Wear safety glasses or goggles and suitable protective gloves while performing the experiment.**

2. *Concentrated* **sulfuric acid is very corrosive and may cause serious chemical burns if allowed to come into contact with your skin. When it is poured from the reagent bottle, some may run down the outside of the bottle; wipe the bottle clean with a sponge or paper towel. If any** *concentrated* **sulfuric acid comes into contact with your skin, immediately flood the affected area with cold water and then with 5% sodium bicarbonate solution.**

---

3. *Concentrated* sulfuric acid adsorbed on silica gel is corrosive and may cause serious chemical burns if allowed to come into contact with your skin. If any of this solid catalyst comes into contact with your skin, immediately flood the affected area with cold water and then with 5% sodium bicarbonate solution.

4. If using a microwave oven, be sure to read and carefully follow the manufacturer's procedures for safe operation of that particular oven. In general, be sure that the reaction vessel is tightly capped prior to heating the reaction mixture. Before opening the reaction vessel, be careful to cool the mixture to room temperature and *slowly* open the container.

MINISCALE PROCEDURE

Preparation Refer to the online resources to answer Pre-Lab Exercises, access videos, and read the MSDSs for the chemicals used or produced in this procedure. Review Sections 2.9, 2.10, 2.11, 2.17, 2.19, 2.21, 2.22, and 6.4.

Apparatus A 10-mL conical vial, apparatus for heating under reflux, magnetic stirring, and *flameless* heating.

Setting Up Add 3 mL of the unknown stock solution, a spinvane, and about 100 mg of the silica-gel sulfuric acid catalyst to the 10-mL conical vial. Attach the condenser to the vial and set up the apparatus for heating under reflux.

Reaction Heat the reaction under gentle reflux for 30 min.

Work-Up and Isolation Remove the heating source and allow the reaction mixture to cool to room temperature. *Slowly* decant the mixture into a beaker containing 3 mL of water; the solid silica gel should remain in the conical vial. Bring the mixture to a pH of about 8 by *slowly* adding 10% *aqueous* sodium carbonate solution. Be careful because some frothing may occur during neutralization. Add about 3 mL of diethyl ether to the beaker and transfer the liquids to a 15-mL conical vial. Using a pipet, remove the lower layer, and wash the organic layer with 3 mL of *saturated* sodium chloride solution. Using a pipet, remove the lower layer again, add about 0.3 g of *anhydrous* sodium sulfate, and cap the vial loosely.

Analysis Following the directions of your instructor, perform a gas chromatographic analysis of your sample and compare your result to the known standards. Identify the ester you made and deduce the structures of the carboxylic acid and alcohol that were in your unknown mixture.

MINISCALE PROCEDURE FOR MICROWAVE OPTION

Preparation Refer to the online resources to answer Pre-Lab Exercises, access videos, and read the MSDSs for the chemicals used or produced in this procedure. Review Sections 2.9, 2.10, 2.11, 2.17, 2.19, 2.21, 2.22, and 6.4.

Apparatus A 10-mL conical vial, apparatus for microwave heating.

Setting Up Add 3 mL of the unknown stock solution and a stirbar to the 10-mL vial. Add about 6 drops of *concentrated* sulfuric acid, seal the vial with the cap, and place it in the cavity of the microwave apparatus.

Reaction Consult with your instructor for specific directions on the use of your microwave heating apparatus. Generally, the unit should be programmed to heat the reaction mixture to 140 °C with a maximum power setting to 250 W. The temperature program is a 1-min ramp, with a 5-min hold time; the internal pressure limit should be set to 250 psi.

Work-Up and Isolation Remove the vial from the microwave apparatus, and allow the reaction mixture to cool to room temperature. *Slowly* pour the mixture into a beaker containing 3 mL of water. Bring the mixture to a pH of about 8 by *slowly* adding 10% *aqueous* sodium carbonate solution with stirring. Be careful, as some frothing may occur during neutralization. Add about 3 mL of diethyl ether to the beaker and transfer the liquids to a 15-mL conical vial. Using a pipet, remove the lower layer, and wash the organic layer with 3 mL of *saturated* sodium chloride solution. Using a pipet, remove the lower layer again, add about 0.3 g of *anhydrous* sodium sulfate, and cap the vial loosely.

Analysis Following the directions of your instructor, perform a gas chromatographic analysis of your sample and compare your result to the known standards. Identify the ester you made and deduce the structures of the carboxylic acid and alcohol that were in your unknown mixture.

WRAPPING IT UP

Dilute the *combined aqueous layers* with water and flush them down the drain. Place the *ethereal solution* of the ester in the container for nonhalogenated organic solvents.

EXERCISES

1. What would happen if the *concentrated* sulfuric acid were omitted from these reactions?

2. Why are the alcohols used as the solvents in these reactions?

3. How is the excess carboxylic acid (if any) removed from the reaction mixture?

4. What undesired reaction might occur if you made the solution strongly basic with aqueous sodium hydroxide?

5. Consider the equilibrium for the esterification of benzoic acid to give methyl benzoate.

   a. Write the mathematical expression for the equilibrium constant, K_{eq}, as a function of the concentrations of the products and reactants for this reaction.

   b. Provide two specific ways whereby the Le Châtelier principle may be applied to this equilibrium to drive the reaction completely to the ester.

   c. Which one of the techniques is used in this experiment?

6. A strong acid is generally used to catalyze the Fischer esterification of carboxylic acids. What *two* steps in the reaction are accelerated by the presence of strong acid, and what function does the acid play in each of these steps?

7. Using curved arrows to symbolize the flow of electrons, write the stepwise mechanism for the reaction below.

$$\text{C}_6\text{H}_5\text{CO}_2\text{H} + \text{CH}_3\text{OH} \underset{}{\overset{\text{H}_2\text{SO}_4}{\rightleftharpoons}} \text{C}_6\text{H}_5\text{CO}_2\text{CH}_3 + \text{H}_2\text{O}$$

Refer to the online resources and use the spectra viewer and Tables 8.1–8.8 as needed to answer the blue-numbered questions on spectroscopy.

8. Consider the spectral data for benzoic acid (Figs. 19.6 and 19.7).

   a. In the functional group region of the IR spectrum, specify the absorptions associated with the carbonyl component of the carboxyl group and the aromatic ring. There is a broad absorption in the region of about 2800–3050 cm^{-1}. What functional group is responsible for this absorption, and why is the absorption broad?

   b. In the ^{1}H NMR spectrum, assign the various resonances to the hydrogen nuclei responsible for them.

   c. For the ^{13}C NMR data, assign the various resonances to the carbon nuclei responsible for them.

9. Consider the spectral data for ethanol (Figs. 20.3 and 20.4).

   a. What functional group accounts for the broad absorption centered at about 3350 cm^{-1} in the IR spectrum, and why it is broad?

   b. In the ^{1}H NMR spectrum, assign the various resonances to the hydrogen nuclei responsible for them.

   c. For the ^{13}C NMR data, assign the various resonances to the carbon nuclei responsible for them.

10. Consider the spectral data for methyl benzoate (Figs. 19.2, 19.3, and 8.48).

   a. What functional group accounts for the sharp absorption centered at about 1750 cm^{-1} in the IR spectrum?

   b. In the ^{1}H NMR spectrum, assign the various resonances to the hydrogen nuclei responsible for them.

   c. For the ^{13}C NMR data, assign the various resonances to the carbon nuclei responsible for them.

20.3 AMIDES AND INSECT REPELLENTS

Amides cannot be easily prepared by the direct reaction of ammonia or amines with carboxylic acids because amines function as bases to convert carboxylic acids into their unreactive ammonium carboxylate salts (Eq. 20.7). Although these salts may be converted into amides by heating to high temperatures (>200 °C), the forcing nature of these conditions severely limits the applicability of this method (see Sec. 20.4).

$$\underset{\substack{\text{A carboxylic} \\ \text{acid}}}{\text{R}-\overset{\text{O}}{\underset{\text{OH}}{\text{C}}}} + \underset{\substack{\text{An amine}}}{\text{R}^1\text{NH}_2} \longrightarrow \underset{\substack{\text{An ammonium} \\ \text{carboxylate}}}{\text{R}-\overset{\text{O}}{\underset{\text{O}^-}{\text{C}}} \ \text{R}^1\overset{+}{\text{N}}\text{H}_3} \overset{200\ °\text{C}}{\longrightarrow} \underset{\substack{\text{An amide}}}{\text{R}-\overset{\text{O}}{\underset{\text{NHR}^1}{\text{C}}}} + \text{H}_2\text{O} \qquad (20.7)$$

A more general method for preparing amides involves a two-step procedure in which the carboxylic acid is first converted to an acid chloride that is then treated with an amine to give the desired amide according to Equation 20.8. Acid chlorides are among the most reactive derivatives of carboxylic acids and thus are often used to prepare other organic compounds having an acyl group, including esters and aryl alkyl ketones (Sec. 15.3).

$$
\underset{\substack{\text{A carboxylic acid}}}{R-\overset{\displaystyle O}{\underset{\displaystyle OH}{C}}} \xrightarrow{\text{SOCl}_2} \underset{\substack{\text{An acid chloride}}}{R-\overset{\displaystyle O}{\underset{\displaystyle Cl}{C}}} \xrightarrow{\text{2 R}^1\text{NH}_2} \underset{\substack{\text{An amide}}}{R-\overset{\displaystyle O}{\underset{\displaystyle NHR^1}{C}}} + \underset{\substack{\text{An ammonium}\\\text{salt}}}{R^1\overset{+}{N}H_3\ Cl^-} \tag{20.8}
$$

The reagent commonly used to prepare an acid chloride from a carboxylic acid is thionyl chloride, $SOCl_2$. The reaction probably proceeds via the intermediate *mixed anhydride* **18** (Eq. 20.9), which undergoes rapid attack by chloride ion via a nucleophilic acyl substitution; sulfur dioxide and chloride ion are lost, and the acid chloride results.

$$\tag{20.9}$$

18

A carboxylic acid	Thionyl chloride	An acid chloride

The acid chlorides prepared in this fashion are often not isolated and purified, but rather are used directly in a subsequent reaction. This is possible because both of the side products, HCl and SO_2, are gases and readily lost or removed from the mixture. For example, if an amide is the desired product, then the crude acid chloride is simply allowed to react with an excess of an amine. An excess of the amine is used to neutralize the HCl that is generated during the preparation of the acid chloride (Eq. 20.9) *and* by the reaction of the acid chloride with the amine (Eq. 20.8). Like **18**, acid chlorides are highly reactive, and they tend to undergo nucleophilic acyl substitution according to the general mechanism shown in Equation 20.2.

The experimental procedure for the synthesis of *N,N*-diethyl-*m*-toluamide (**21**) nicely illustrates a common method for preparing amides. In the first step, *m*-toluic acid (**19**) reacts with thionyl chloride to give the acid chloride **20** (Eq. 20.10). The acid chloride is treated directly with an excess of diethylamine to give *N,N*-diethyl-*m*-toluamide and diethylammonium hydrochloride (Eqs. 20.11 and 20.12).

$$\tag{20.10}$$

19

m-Toluic acid

20

m-Toluoyl chloride

$$\underset{\substack{\textbf{20}\\ m\text{-Toluoyl chloride}}}{\text{(m-methylbenzoyl chloride)}} + (CH_3CH_2)_2NH \longrightarrow \underset{\substack{\textbf{21}\\ N,N\text{-Diethyl-}m\text{-toluamide}}}{\text{N(CH}_2\text{CH}_3)_2} + HCl \quad (20.11)$$

$$HCl + (CH_3CH_2)_2NH \longrightarrow \underset{\text{Diethylammonium chloride}}{(CH_3CH_2)_2\overset{+}{N}H_2 \; Cl^-} \quad (20.12)$$

N,N-Diethyl-*m*-toluamide (**21**), commonly known as DEET, is widely used as an insect repellent and is the active ingredient in the commercial repellent OFF, for example. Whereas other repellents are useful toward only one or two types of insects, DEET is unusual in that it is effective against a broad spectrum of insects, including mosquitoes, fleas, chiggers, ticks, deerflies, and gnats. Although the exact mode of action of *N,N*-diethyl-*m*-toluamide as a repellent is not known, at least for mosquitoes, it appears to interfere with the receptors or sensors that enable the insect to detect moisture gradients. Hence, the female mosquito is unable to find a suitable host or target, like you! So this is an experiment you can do without fear of being bit by a stray mosquito that might wander into the lab, no matter where you live.

EXPERIMENTAL PROCEDURES

Preparation of N,N-Diethyl-m-Toluamide

Purpose To demonstrate the synthesis of an amide from a carboxylic acid via the corresponding acid chloride.

Note: The entire procedure up to the first stopping point should be completed in one laboratory period to obtain a good yield of product.

SAFETY ALERT

1. **Wear safety glasses or goggles and suitable protective gloves while performing the experiment.**

2. **Thionyl chloride is a volatile and *corrosive* chemical that undergoes rapid reaction with water and atmospheric moisture to produce sulfur dioxide and hydrogen chloride, both of which are also corrosive. Do not inhale this chemical or allow it to come in contact with your skin. If contact occurs, immediately flood the affected area with cold water and then rinse with 5% sodium bicarbonate solution.**

3. **Dispense thionyl chloride only at a hood and minimize its exposure to atmospheric moisture by keeping the bottle tightly closed when not in use.**

4. Diethyl ether is *extremely* flammable and volatile, and its vapors can easily travel several feet along the bench top or the floor and then be ignited. Consequently, *be certain there are no open flames anywhere in the laboratory* whenever you are working with it. Use a *flameless* heating source *whenever* heating is required.

5. Diethylamine is a noxious-smelling and corrosive chemical. Dispense this chemical only at a hood.

6. Make sure your glassware is dry, lubricate the ground-glass joints of the apparatus carefully, and be certain that the joints are intimately mated. Otherwise, noxious gases will escape.

7. If possible, this experiment should be performed in a hood.

8. Heptane is extremely flammable; use *flameless* heating to concentrate solutions of heptane.

MINISCALE PROCEDURE

Preparation Refer to the online resources to answer Pre-Lab Exercises, access videos, and read the MSDSs for the chemicals used or produced in this procedure. Review Sections 2.9, 2.10, 2.11, 2.13, 2.21, 2.22, 2.23, 2.25, 2.29, and 6.3.

Apparatus A 100-mL round-bottom flask, a 50-mL Erlenmeyer flask, Claisen adapter, gas trap, separatory funnel, separatory funnel with standard-taper glass joint, ice-water bath, reflux condenser, apparatus for magnetic stirring, column chromatography, simple distillation, and *flameless* heating.

Setting Up Oven-dry the apparatus prior to assembling it. Because an *airtight* seal is required for all connections, *carefully grease all joints* when assembling the apparatus. Place 2.0 g of *dry m*-toluic acid in the round-bottom flask and equip the flask with a stirbar and the Claisen adapter. Fit the *sidearm* of the adapter with the reflux condenser and fit the top of the condenser with the gas trap specified by your instructor; one example is illustrated in Figure 20.7. *Working at the hood* and using a graduated pipet, measure 2.2 mL of thionyl chloride into the separatory funnel; be certain that the stopcock of the funnel is firmly seated and closed prior to this transfer. Stopper the funnel and place it on the straight arm of the Claisen adapter above the reaction flask (Fig. 20.7). Prepare a solution of 5.0 mL of diethylamine in 10 mL of *anhydrous* diethyl ether in a small Erlenmeyer flask, stopper the flask, and place it in an ice-water bath. *Be sure the flask is secured so it does not tip over while you are making the acid chloride.*

Reaction Open the stopcock of the funnel fully so that the thionyl chloride is added as rapidly as possible to the flask containing the *m*-toluic acid; it may be necessary to lift the stopper on the funnel *briefly* to equalize the pressure in the system if the flow of thionyl chloride becomes erratic or slow. When the addition has been completed, *close the stopcock* and heat the mixture under gentle reflux for 15 min. The bubbling that is initially observed should slow during the reflux period and the reaction mixture should become clear.

Cool the mixture to room temperature with continued stirring and add 30 mL of *anhydrous* diethyl ether to the separatory funnel. Open the stopcock of the funnel fully so that the ether drains as rapidly as possible into the flask. *Close the stopcock,* transfer the *cold* ethereal solution of diethylamine to the funnel, and stopper

Figure 20.7

Miniscale apparatus for heating reaction under reflux, equipped with a gas trap and a separatory funnel for adding a solution.

the funnel. Some white smoke forms in the funnel during this transfer. Add the ethereal solution of diethylamine dropwise to the stirred solution of acid chloride over a period of about 10–15 min. While this solution is being added to the flask, quantities of white smoke will be produced. After the addition is complete, continue stirring for 15 min to ensure completion of the reaction.

Work-Up and Isolation Remove the gas trap, separatory funnel, condenser, and Claisen adapter from the reaction flask. Add 15 mL of 2.5 *M* aqueous sodium hydroxide to the stirred solution and continue stirring for 15 min; swirl the flask if necessary to dissolve all of the white solid. Transfer the reaction mixture to a separatory funnel, separate the layers, and remove the lower aqueous layer. Wash the organic layer with 15 mL of *cold* 3 *M* hydrochloric acid, *venting as necessary,* and then with 15 mL of cold water. Dry the organic layer over several spatula-tips full of *anhydrous* sodium sulfate.★ Filter or carefully decant the solution into a *tared* round-bottom flask and remove most of the ether by simple distillation. Alternatively, use rotary evaporation or other techniques to concentrate the solution. The final traces of solvent may be removed by connecting the cool stillpot to a vacuum source. Determine the yield of crude product.★

Purification The crude product may be purified by column chromatography (Sec. 6.3). To do this, prepare an alumina column using heptane and 20 g of alumina in a column about 1.5 cm in diameter. Drain the heptane from the column until it is level with the layer of sand on the top of the column. Add about 2 mL of heptane to the flask containing the crude *N,N*-diethyl-*m*-toluamide and swirl the flask to effect mixing. Using a Pasteur or filter-tip pipet, transfer this solution to the top of the column. Open the stopcock of the column and allow the liquid to drain until the level of solvent in the column is the same as the layer of sand; close the stopcock. Wash the flask with 2–3 mL of heptane, transfer this rinse to the top of the column, and drain the column as before. Elute the column with about 50 mL of heptane, collecting the eluant as a single fraction. To determine whether all of the product has eluted, collect several drops on a watchglass and evaporate the liquid. If there is no residue, the amide is no longer eluting.★ Transfer the eluant into a *tared* round-bottom flask, and remove most of the heptane by simple distillation. Alternatively, use rotary evaporation or other techniques to concentrate the solution. The final traces of heptane may be removed by connecting the cool stillpot to a vacuum source.

Analysis Weigh the product and calculate the percent yield. Obtain IR and ^{1}H NMR spectra of your starting material and product, and compare them with those of authentic samples (Figs. 20.9–20.12).

MICROSCALE PROCEDURE

Preparation Refer to the online resources to answer Pre-Lab Exercises, access videos, and read the MSDSs for the chemicals used or produced in this procedure. Review Sections 2.5, 2.9, 2.10, 2.11, 2.13, 2.21, 2.22, 2.23, 2.25, 2.29, and 6.3.

Apparatus A 3- and 5-mL conical vial, 2-mL glass syringe, Claisen adapter, gas trap, ice-water bath, reflux condenser, a screw-cap centrifuge tube, apparatus for magnetic stirring, column chromatography, simple distillation, and *flameless* heating.

Setting Up This reaction should be set up at a hood. Oven-dry the apparatus prior to assembling it. Because an *airtight* seal is required for all connections, *carefully grease all joints* when assembling the apparatus. Place 0.2 g of *dry m*-toluic acid in the 5-mL conical vial containing a spinvane. Equip the vial with a Claisen adapter, and fit the *sidearm* of the adapter with the reflux condenser. Fit the top of the condenser with the gas trap specified by your instructor; one example is illustrated in Figure 20.8. If you use the gas trap shown, be sure that no water from the moistened cotton runs into the condenser. Prepare a solution of 0.5 mL of diethylamine in 2.0 mL of *anhydrous* diethyl ether in a screw-cap centrifuge tube, screw on the cap, and place the tube in an ice-water bath. *Be sure the tube is secured so it does not tip over while you are making the acid chloride.*

Reaction *Working at the hood,* transfer 0.2 mL of thionyl chloride into the vial using a graduated pipet. Seal the joint of the *straight portion* of the adapter with a rubber septum. The apparatus may now be moved to your desk if desired. Heat the mixture under gentle reflux for 15 min. The bubbling that is initially observed should slow during the reflux period and the reaction mixture should become clear.

Cool the mixture to room temperature with continued stirring. Using the syringe, add 1.0 mL of *anhydrous* diethyl ether through the rubber septum to the vial. Add the ethereal solution of diethylamine dropwise to the stirred solution of acid chloride over a period of about 2–3 min. While this solution is being added to the flask, quantities of white smoke will be produced. After the addition is complete, continue stirring for 15 min to ensure completion of the reaction.

Work-Up and Isolation Remove the rubber septum from the Claisen adapter, add 1.5 mL of 2.5 *M* aqueous sodium hydroxide to the stirred solution, and continue stirring for 15 min. If necessary, screw a cap onto the vial and shake it briefly to dissolve all of the white solid. Remove the lower aqueous layer with a Pasteur pipet and wash

Moistened cotton or
glasswool

Drying tube

Syringe

Water out

Condenser

Water in

Clamps

Claisen adapter

Thermometer

Conical reaction vial

Spinvane

Heating source

Magnetic stirrer

Laboratory jack

Figure 20.8
*Microscale apparatus for heating
under reflux, equipped with a gas
trap and a syringe for adding a
solution.*

the organic layer with 1.5 mL of *cold 3 M* hydrochloric acid, *venting as necessary,*
and then with 1.5 mL of cold water. Transfer the organic layer to a dry test tube
using a Pasteur or filter-tip pipet and add several microspatula-tips full of *anhydrous*
sodium sulfate. Allow the solution to stand for 10–15 min, with occasional swirling
to hasten drying.★ Using a Pasteur or filter-tip pipet, transfer the organic solution to
a *tared* dry 5-mL conical vial, and add a spinvane to the vial. Remove most of the
ether by simple distillation. Alternatively, use rotary evaporation or other techniques
to concentrate the solution. The final traces of solvent may be removed by connecting the cool stillpot to a vacuum source. Determine the yield of crude product.★

Purification The crude product may be purified by column chromatography
(Sec. 6.3). To do this, prepare an alumina column using heptane and 2 g of alumina in a Pasteur filtering pipet (see Fig. 2.9b). Dissolve your crude product in about
0.2 mL of heptane and transfer the solution to the top of the alumina column. Wash
the column with about 5 mL of heptane, collecting the eluant as a single fraction. To
determine whether all of the product has eluted, collect several drops on a watchglass and evaporate the liquid. If there is no residue, the amide is no longer eluting.★
Transfer the eluant into a *tared* round-bottom flask and remove most of the heptane
by simple distillation. Alternatively, use rotary evaporation or other techniques to
concentrate the solution. The final traces of heptane may be removed by connecting
the cool stillpot to a vacuum source. The pure product should be a colorless liquid.

Analysis Weigh the product and calculate the percent yield. Obtain IR and ^{1}H NMR
spectra of your starting material and product and compare them with those of
authentic samples (Figs. 20.9–20.12).

WRAPPING IT UP

Combine all of the *aqueous layers* and determine the pH with pHydrion paper. Adjust the pH to approximately 7 with either solid sodium carbonate or 3 *M* HCl. Flush all *filtrates* down the drain. Pour the recovered *diethyl ether* and the *heptane* in a container for nonhalogenated organic liquids. Spread the *sodium sulfate* and the *alumina* on a tray in the hood to evaporate residual solvent, and then put them in the container for nonhazardous solids.

EXERCISES

1. Write the balanced equation for the reaction of thionyl chloride with water.

2. Write the balanced equation for the reaction of *m*-toluoyl chloride with water.

3. What is the white solid that is formed when diethylamine is added to the reaction mixture containing *m*-toluoyl chloride?

4. What is the white smoke that is formed when diethylamine is added to the reaction mixture containing *m*-toluoyl chloride?

5. The thionyl chloride remaining after the formation of the acid chloride also reacts with diethylamine. Write a balanced equation for this reaction.

6. In the work-up of the reaction, the organic layer is extracted with 2.5 *M* sodium hydroxide solution. What is removed from the organic layer in this extraction?

7. What is the purpose of the next step of the work-up in which the organic layer is washed with 3 *M* hydrochloric acid?

8. Using curved arrows to symbolize the flow of electrons, write a stepwise mechanism for the formation of *m*-toluoyl chloride from *m*-toluic acid using thionyl chloride according to Equation 20.10.

9. Using curved arrows to symbolize the flow of electrons, write a stepwise mechanism for the formation of *N,N*-diethyl-*m*-toluamide from *m*-toluoyl chloride with diethylamine according to Equation 20.11.

Refer to the online resources and use the spectra viewer and Tables 8.1–8.8 as needed to answer the blue-numbered questions on spectroscopy.

10. Consider the spectral data for *m*-toluic acid (Figs. 20.9 and 20.10).

    a. In the functional group region of the IR spectrum, specify the absorption associated with the carbonyl component of the carboxyl group and with the π-bonds of the aromatic ring.

    b. In the ^{1}H NMR spectrum, assign the various resonances to the hydrogen nuclei responsible for them.

    c. For the ^{13}C NMR data, assign the various resonances to the carbon nuclei responsible for them.

11. Consider the spectral data for *N,N*-diethyl-*m*-toluamide (Figs. 20.11 and 20.12).

    a. In the functional group region of the IR spectrum, specify the absorption associated with the carbonyl component of the amide group and with the π-bonds of the aromatic ring.

    b. In the ^{1}H NMR spectrum, assign the various resonances to the hydrogen nuclei responsible for them.

    c. For the ^{13}C NMR data, assign the various resonances to the carbon nuclei responsible for them.

12. Discuss the differences observed in the IR and NMR spectra of *m*-toluic acid and *N,N*-diethyl-*m*-toluamide that are consistent with the formation of the latter in this procedure.

SPECTRA

Starting Materials and Products

Figure 20.9
IR spectrum of m-*toluic acid (IR card).*

Figure 20.10
NMR data for m-*toluic acid (CDCl$_3$).*

(a) ^{1}H NMR spectrum (300 MHz).
(b) ^{13}C NMR data: δ 21.2, 127.4, 128.4, 129.2, 130.7, 134.6, 138.3, 172.8.

Figure 20.11

IR spectrum of N,N-diethyl-m-toluamide (neat).

Figure 20.12

NMR data for N,N-diethyl-m-toluamide (CDCl₃).

(a) ¹H NMR spectrum (300 MHz).
(b) ¹³C NMR data: δ 14.2, 21.4, 39.1, 123.1, 126.9, 128.2, 129.8, 137.2, 138.2, 171.5.

20.4 AMIDES AND CHEMILUMINESCENCE

A phenomenon related to the fundamental principles associated with UV-Vis spectroscopy (Sec. 8.4) is that of **chemiluminescence**, which is a process whereby light is generated by a chemical reaction. This occurs when the reaction produces a product in an *excited* electronic state that ultimately relaxes to the *ground* electronic state, with concomitant emission of energy in the form of a photon (light). In UV-Vis spectroscopy (Sec. 8.4), of course, exposing a sample to light of the proper wavelength fosters just the opposite behavior, namely, transforming a ground electronic state to an excited electronic state by promotion of an electron from one orbital to another (Fig. 8.49). In any case, the excited state may relax to the ground state through **radiationless** loss of energy or through emission of visible light, as seen when **fluorescence** and **phosphorescence** occur (Fig. 20.13). These last two modes of relaxation involve the **singlet** and **triplet** electronic states, respectively (see below).

Each electronic state comprises a number of vibrational energy states, labeled as v_0–v_5 (Fig. 20.13). For simplicity's sake, we've shown the relaxation phenomena

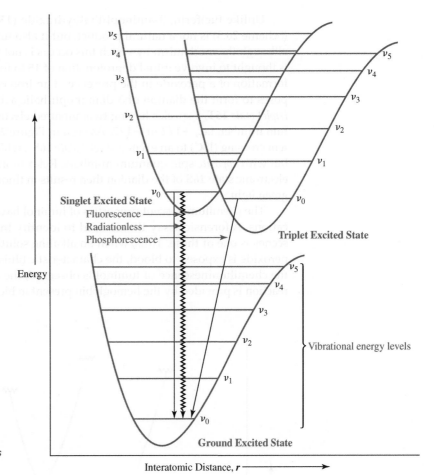

Figure 20.13
Potential energy diagram for singlet and triplet electronic states of a molecule.

as originating solely from the lowest vibrational energy state, v_0, of the electronically excited states. In this simplified diagram, you can see that the energy released upon conversion of a triplet excited state to the ground electronic state is less than that for corresponding transformation of a singlet excited state. A consequence of this difference is that light of longer wavelength, and thus lower in energy (Eq. 8.1), results from phosphorescence as compared to fluorescence.

A number of living organisms emit light through chemiluminescent reactions involving naturally occurring compounds. If you have ever watched fireflies on a summer evening, you have witnessed one of the more familiar examples of **bioluminescence**, an example of chemiluminescence in which a biochemical process results in the emission of light. When in search of a mate, the male firefly emits flashes of visible light that result from the reaction of *luciferin* with molecular oxygen in a process catalyzed by the enzyme luciferase (Eq. 20.13).

$$\text{Luciferin} \xrightarrow[-CO_2]{\text{Luciferase/O}_2} \text{Decarboxyketoluciferin} + \text{Light} \tag{20.13}$$

Unlike luciferin, 3-aminophthalhydrazide (**15**), commonly called luminol (Scheme 20.3) is not a natural product, but it also undergoes chemiluminescence, although the mechanism by which this occurs is not fully understood. However, it is thought to involve initial deprotonation of **15** to form a dianion that leads to the formation of a peroxide in the presence of an iron catalyst. The peroxide decomposes to form the dianion of 3-diaminophthalic acid in an electronically excited *triplet state* **16T**, a species having two unpaired electrons with the same spin quantum number, i.e., +1/2 or −1/2. As seen in Figure 20.14, **16T** undergoes intersystem crossing (ISC) to an excited *singlet state* **16S***, which has two unpaired electrons but with *opposite* spin quantum numbers. Relaxation (decay) to the *singlet* ground electronic state **16S** of the dianion then results in fluorescence in the form of bluish-green light.

The chemiluminescent properties of luminol have a number of practical applications. A forensic assay that is used to identify the presence of blood at crime scenes is one of these. Thus, when an alkaline solution of luminol and hydrogen peroxide is exposed to blood, the characteristic bluish-green light associated with the chemiluminescence of luminol is observed; the iron catalyst required for the reaction is provided by the hemoglobin present in blood.

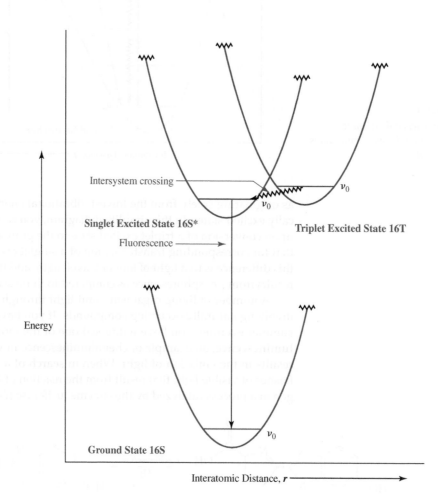

Figure 20.14
*Potential energy diagram
illustrating intersystem crossing
and fluorescence.*

Scheme 20.3

15
Luminol   Luminol dianion   Proposed peroxide

16T   **16S***   **16S**
Triplet dianion   Excited-state singlet dianion   Ground-state singlet dianion

Note that the half-headed arrows (⚊) represent the orientation of the spin of an electron.

In the procedure described below, luminol is synthesized by a sequence of reactions that involves the condensation of 3-nitrophthalic acid (**16**) with hydrazine (**18**) to produce 3-nitrophthalhydrazide (**20**), which is then reduced by sodium dithionite to form luminol (Eq. 20.14). The chemiluminescent reaction will then be observable by mixing luminol with hydrogen peroxide and potassium ferricyanide, $K_2Fe(CN)_6$ (Scheme 20.3).

(20.14)

17   **19**   **20**   **15**
3-Nitrophthalic acid   Hydrazine   3-Nitrophthalhydrazide   Luminol

The formation of **20** is basically a reaction in which a 1,2-dicarboxylic acid and a 1,2-diamine react to form a cyclic diamide or hydrazide. The mechanistic steps by which an acid and amine are converted to an amide are portrayed in Scheme 20.4. The acid and amine are in rapid equilibrium with the corresponding ammonium salt and the carboxylate ion, the latter of which is not subject to nucleophilic attack at the carbonyl carbon atom. The acid itself may undergo nucleophilic attack, however, so the small amount of it that exists at equilibrium (see Exercise 4) reacts with free amine to produce the tetrahedral zwitterionic intermediate **21**. Subsequent proton transfers and loss of water afford the amide **22**. Repetition of this series of steps generates **20**.

Scheme 20.4

The reduction of the nitro group of 3-nitrophthalhydrazide (**20**) with dithionate is mechanistically complex, but a possible sequence of events is illustrated in Scheme 20.5, with $ArNO_2$ representing **20**. The reaction presumably is initiated by transfer of an electron from the SO_2 anion radical **23**, which is derived from dithionite, to produce the anion radical **23**. Subsequent steps involving protonation and addition of a second electron from **23** afford **25**, which can lose hydroxide to produce the nitroso compound **26**. Addition of three more electrons from **23** and protonations ultimately lead to the reduced product **15**.

Scheme 20.5

EXPERIMENTAL PROCEDURES

Preparation and Chemiluminescence of Luminol

Purpose To demonstrate the preparation of luminol and its chemiluminescence.

SAFETY ALERT

1. **Wear safety glasses or goggles and suitable protective gloves while performing the experiment.**
2. **If hydrazine comes into contact with your skin, immediately wash the area with soap and water and flush it with copious amounts of water.**

A ■ *Preparation of Luminol*

MINISCALE PROCEDURE

Preparation Refer to the online resources to answer Pre-Lab Exercises, access videos, and read the MSDSs for the chemicals used or produced in this procedure. Review Sections 2.9, 2.10, and 2.17.

Apparatus A 20-mm × 150-mm test tube, 25-mL Erlenmeyer flask, thermometer, apparatus for vacuum filtration and *flameless* heating.

Setting Up Combine 1 g of 3-nitrophthalic acid and 2 mL of an 8% aqueous solution of hydrazine in the test tube and carefully heat the mixture until the solid dissolves. Add 3 mL of triethylene glycol and a boiling stone and clamp the vial in a vertical position. Insert a thermometer into the solution, securing the thermometer with a clamp.

Reaction Bring the solution to a vigorous boil to remove excess water. During this time, the temperature should be around 110–120 °C. After the water has evaporated, the temperature should rise to 215 °C in a 3–4-min period. Maintain the temperature at 215–220 °C for 2 min and then remove the test tube from the heat source and allow the solution to cool to about 100 °C. While the test tube and its contents are cooling, bring about 15 mL of water to boiling in the Erlenmeyer flask. Slowly add the hot water to the test tube, stir the contents with a glass stirring rod, cool the mixture to room temperature, and collect the solid nitrohydrazide by vacuum filtration.

Return the damp solid to the uncleaned test tube, add 5 mL of 3 *M* aqueous sodium hydroxide, and mix the contents until the solid is dissolved. Add 3 g of fresh sodium hydrosulfite dihydrate to the solution, rinsing any solid adhering to the walls of the test tube into the solution with a few drops of water, and heat the resulting mixture to just below boiling for 5 min, taking care to avoid bumping.

Work-Up and Isolation Add 2 mL of *glacial* acetic acid to the reaction mixture, cool the test tube in a beaker of cold water, and collect the crude luminol by vacuum filtration. A second crop of product may separate from the filtrate upon standing, but do not combine it with the first crop, which is to be used in the chemiluminescence experiment.

Analysis If instructed to do so, recrystallize the 3-nitrophthalhydrazide and luminol, and obtain their IR and ^{1}H NMR spectra and those of 3-nitrophthalic acid; compare these spectra with those of authentic samples (Figs. 20.15–20.20).

MICROSCALE PROCEDURE

Preparation Refer to the online resources to answer Pre-Lab Exercises, access videos, and read the MSDSs for the chemicals used or produced in this procedure. Review Sections 2.9, 2.10, and 2.17.

Apparatus A 5-mL conical vial, 25-mL Erlenmeyer flask, thermometer, apparatus for vacuum filtration and *flameless* heating.

Setting Up Combine 200 mg of 3-nitrophthalic acid and 0.4 mL of an 8% aqueous solution of hydrazine in the conical vial and heat the mixture until the solid dissolves. Add 0.6 mL of triethylene glycol and a boiling stone, and clamp the vial in a vertical position. Insert a thermometer into the solution, securing the thermometer with a clamp.

Reaction Bring the solution to a vigorous boil to remove excess water. During this time, the temperature should be around 110–120 °C. After the water has evaporated, the temperature should rise to 215 °C in a 3–4-min period. Maintain the temperature at 215–220 °C for about 2 min, and then remove the vial from the heat source and allow the solution to cool to about 100 °C. While the tube and its contents are cooling, bring about 10 mL of water to boiling in the Erlenmeyer flask. Slowly add 3 mL of boiling water to the reaction mixture, stir the contents of the conical vial with a glass stirring rod, cool the resulting mixture to room temperature, and collect the solid nitrohydrazide by vacuum filtration.

Return the damp solid to the conical vial, add 1 mL of 3 *M* aqueous sodium hydroxide, and mix the contents until the solid is dissolved. Add 0.6 g of fresh sodium hydrosulfite dihydrate to the solution, and heat the resulting mixture to just below boiling for 5 min, taking care not to cause bumping.

Work-Up and Isolation Add 0.4 mL of *glacial* acetic acid to the reaction mixture, cool the vial in a beaker of cold water, and collect the luminol by vacuum filtration.

Analysis If instructed to do so, recrystallize the 3-nitrophthalhydrazide and luminol, and obtain their IR and ^{1}H NMR spectra and those of 3-nitrophthalic acid; compare these spectra with those of authentic samples (Figs. 20.15–20.20).

WRAPPING IT UP

Flush the *aqueous filtrates* down the drain with excess water. Put the *filter papers* in the container for nontoxic waste.

B ▪ *Chemiluminescence*

MINISCALE PROCEDURE

Preparation Refer to the online resources to answer Pre-Lab Exercises, access videos, and read the MSDSs for the chemicals used or produced in this procedure. Review Section 2.11.

Apparatus Two 50-mL, one 125-mL, one 250-mL Erlenmeyer flask.

Setting Up Prepare stock solution A by placing the crude luminol (40–60 mg dry weight, but which need not be dry) in a 50-mL Erlenmeyer flask and dissolving it in 2 mL of 3 *M* aqueous sodium hydroxide and 18 mL of water. Prepare solution B in the second 50-mL Erlenmeyer flask by combining 4 mL of 3% aqueous potassium ferricyanide, 4 mL of 3% aqueous hydrogen peroxide, and 32 mL of water. Combine 5 mL of solution A with 35 mL of water in the 125-mL Erlenmeyer flask.

Reaction Working in a low-light environment, simultaneously pour the diluted portion of solution A and all of solution B into the 250-mL Erlenmeyer flask. Swirl the flask and record what you observe. Test what happens if you add additional small portions of the aqueous NaOH solution and crystals of potassium ferricyanide to the reaction mixture.

MICROSCALE PROCEDURE

Preparation Refer to the online resources to answer Pre-Lab Exercises, access videos, and read the MSDSs for the chemicals used or produced in this procedure. Review Section 2.11.

Apparatus Two 50-mL and one 125-mL Erlenmeyer flask.

Setting Up Working with a partner, prepare solution A by combining two samples of crude luminol, which need not be dry, in a 50-mL Erlenmeyer flask and dissolving the solids in 2 mL of 3 *M* aqueous sodium hydroxide and 18 mL of water. Prepare solution B in the second 50-mL Erlenmeyer flask by combining 4 mL of 3% aqueous potassium ferricyanide, 4 mL of 3% aqueous hydrogen peroxide, and 32 mL of water in the other small Erlenmeyer flask.

Reaction Working in a low-light environment, simultaneously pour solution A and solution B into the 125-mL Erlenmeyer flask. Swirl the flask and record what you observe.

WRAPPING IT UP

Neutralize the *luminol solution* and flush it down the drain with excess water.

EXERCISES

1. Define the following terms.
   a. fluorescence
   b. phosphorescence
   c. chemiluminescence
   d. intersystem crossing
2. What is the difference in electronic configuration of a singlet and a triplet state?

3. There are two carboxylic acid functions in 3-nitrophthalic acid (17). The pK_a of the acid function at the 2-position is approximately 2, whereas that at the 1-position is about 3. Which of the two is the more acidic and why?

4. Given the approximate pK_as provided, compute K_{eq} for the reaction below.

$$pK_a < 2 \qquad\qquad\qquad pK_a < 9$$

5. Using curved arrows to symbolize the flow of electrons, propose a mechanism for the transformation shown below.

20

6. What factor(s) make the hydrogen atoms of the hydrazide moiety sufficiently acidic so that hydroxide can convert luminol to the dianion portrayed in Scheme 20.3?

Refer to the online resources and use the spectra viewer and Tables 8.1–8.8 as needed to answer the blue-numbered questions on spectroscopy.

7. Consider the spectral data for 3-nitrophthalic acid (Figs. 20.15 and 20.16).

   a. In the functional group region of the IR spectrum, specify the absorptions associated with the O–H bonds and the carbonyl and nitro functions of the molecule.

   b. In the ^{1}H NMR spectrum, assign the various resonances to the hydrogen nuclei responsible for them.

   c. In the ^{13}C NMR spectrum, assign the various resonances to the carbon nuclei responsible for them.

8. Consider the spectral data for 3-nitrophthalhydrazide (Figs. 20.17 and 20.18).

   a. In the functional group region of the IR spectrum, specify the absorptions associated with the N–H bonds and carbonyl and nitro functions of the molecule.

   b. In the ^{1}H NMR spectrum, assign the various resonances to the hydrogen nuclei responsible for them.

   c. In the ^{13}C NMR spectrum, assign the various resonances to the carbon nuclei responsible for them.

9. Consider the spectral data for luminol (3-aminophthalhydrazide) (Figs. 20.19 and 20.20).

   a. In the functional group region of the IR spectrum, specify the absorptions associated with the N–H bonds of the hydrazide and of the amino group and that of the carbonyl functions of the molecule.

b.   In the ^{1}H NMR spectrum, assign the various resonances to the hydrogen nuclei responsible for them.

c.   In the ^{13}C NMR spectrum, assign the various resonances to the carbon nuclei responsible for them.

10. Discuss the differences observed in the IR and NMR spectra of 3-nitrophthalic acid and 3-nitrophthalhydrazide that are consistent with replacement of the carboxylic acid functions with the hydrazide moiety in this experiment.

11. Discuss the differences observed in the IR and NMR spectra of 3-nitrophthalhydrazide and luminol that are consistent with reduction of the nitro function to an amino group in this experiment.

SPECTRA

Starting Material and Products

Figure 20.15

IR spectrum of 3-nitrophthalic acid (IR card).

Figure 20.16

NMR data for 3-nitrophthalic acid (DMSO-d$_6$).

(a) ^{1}H NMR spectrum (300 MHz).

(b) ^{13}C NMR data: δ127.4, 130.4, 130.7, 131.3, 134.9, 146.5, 165.7, 165.8.

Figure 20.17

IR spectrum of 3-nitrophthalhy-drazide (IR card).

Figure 20.18

NMR data for 3-nitrophthalhy-drazide (DMSO-d₆).

(a) ¹H NMR spectrum (300 MHz).
(b) ¹³C NMR data: δ 118.4, 126.1, 127.6, 127.8, 133.9, 147.7, 151.9, 152.7.

Figure 20.19

IR spectrum of luminol (3-aminophthalhydrazide) (KBr pellet).

Figure 20.20

NMR data for luminol (3-amin-ophthalhydrazide) (DMSO-d₆).

(a) ¹H NMR spectrum (300 MHz).

(b) ¹³C NMR data: δ 109.8, 110.6, 116.7, 126.5, 133.9, 150.2, 151.5, 161.2.

CHAPTER 21

Multistep Organic Synthesis

An everyday application of organic chemistry is the synthesis of compounds that enable us to lead better and more enjoyable lives. Such compounds may be drugs, pesticides, insecticides, or polymers, or they may be used as additives in fragrances, flavors, foods, cosmetics, adhesives, or detergents. The syntheses of these substances from simple starting materials typically involve numerous chemical steps and hence require careful planning and extensive experimentation to optimize the overall processes. It's usually necessary to marshal the combined skills of expert synthetic and mechanistic organic chemists to solve the problems that arise. In the experiments in this chapter, you will become better acquainted with the practical aspects of organic chemistry by preparing a useful compound by a multistep sequence. In so doing, perhaps you will further understand the importance of studying organic chemistry.

21.1 INTRODUCTION

Many organic compounds that we use daily are prepared in large quantities, especially by companies in the pharmaceutical and agricultural industries. Hence, solving the problems associated with the efficient and cost-effective syntheses of complex, multifunctional molecules from simple starting materials is a challenge that must be routinely addressed by organic chemists. In these synthetic processes, it is necessary to perform each transformation selectively and to avoid all side reactions that may lead to reduced yields and impure products. It is perhaps equally important to minimize the number of individual steps that are required because of the "arithmetic demon" that haunts the overall yield in sequential reactions that are not quantitative. For example, even a five-step synthesis in which each step occurs in a respectable 80% yield gives an overall yield of only about 33% $[(0.8)^5 \times 100\%]$. In view of the manifold problems, it is noteworthy that there are presently economical routes to biologically important molecules that require as many as 20 steps.

The *optimal route* to a particular compound *is not always the shortest*, as is illustrated by the simple problem of preparing 2-chloropropane on an industrial scale, shown in Scheme 21.1. Although 2-chloropropane is produced from propane in one step by direct chlorination, it must be separated from the 1-chloropropane formed concurrently; this is *not* an easy separation. However, if the mixture of 1- and 2-chloropropane is treated with a base, both isomers undergo dehydrochlorination to give propane, and subsequent addition of hydrogen chloride produces 2-chloropropane as the *sole* product. In this case, a three-step process is the preferred way to obtain pure 2-chloropropane. As this simple example demonstrates, an important consideration in designing a synthesis is the ease of separating the desired product from unwanted by-products and purifying it. Other considerations include the availability and cost of starting materials, simplicity of equipment and instrumentation, energy costs, activity of catalysts, selectivity of reactions in polyfunctional molecules, and stereochemical control.

Scheme 21.1

$$CH_3\overset{\overset{\displaystyle H}{|}}{CH}-CH_2-H + Cl_2 \xrightarrow{\text{heat or } h\nu} \left\{ \begin{array}{c} \overset{\overset{\displaystyle Cl}{|}}{CH_3CHCH_3} \\ \text{2-Chloropropane} \\ + \\ CH_3CH_2CH_2-Cl \\ \text{1-Chloropropane} \end{array} \right\} \xrightarrow[\text{base}]{-HCl}$$

$$CH_3CH{=}CH_2 \xrightarrow{\text{HCl}} CH_3\overset{\overset{\displaystyle Cl}{|}}{CH}CH_3$$
2-Chloropropane

Because so many variables must be considered in planning the synthesis of complex molecules, it is not surprising that organic chemists sometimes use computers to help design and analyze multistep syntheses. As you know, computers can handle an enormous amount of information, and it may eventually be possible to design optimal sequences of reaction steps using sophisticated software. In considering this fascinating prospect, however, it must be remembered that the information is provided to the computer by humans, specifically chemists, who will be vital to the success of such a project. Chemists still have at least one significant competitive advantage over computers—they are more creative!

Appreciation of and insight into some of the problems that may be encountered in planning a multistep synthesis may be gained from preparing a compound that requires a fairly small number of steps. In this chapter, the syntheses of sulfanilamide (Sec. 21.2), 1-bromo-3-chloro-5-iodobenzene (Sec. 21.3), and lidocaine (Sec. 21.4) are used to illustrate the fundamental principles. The product of one reaction is used in a subsequent step in these experiments, and if you are to be successful, you *must* use good experimental technique.

You may either isolate and purify the intermediates by distillation or recrystallization, or you may use the crude material directly in the next step. This is a choice the practicing synthetic organic chemist constantly faces. In general, extensive purification of each intermediate in a sequence is avoided *if* the impurities can eventually be removed *and* their presence does not interfere with the course of the desired reactions. However, even if the entire quantity of an intermediate is not purified, it is good scientific practice to purify a small sample of it for *complete* characterization by spectroscopic (Chap. 8) and physical (bp, mp, etc.) methods.

21.2 SULFANILAMIDE: DISCOVERY AND SYNTHESIS OF THE FIRST ANTIBIOTIC

An important area of research in the pharmaceutical industry is the discovery and development of new orally active **antibiotics** to treat bacterial infections. Commonly used antibiotics fall into several important classes, including **β-lactams**, the

Sulfamethoxazole
A sulfa drug

Terramycin
A tetracycline antibiotic

Ampicillin
A β-lactam antibiotic

Figure 21.1
Classes of antibiotics.

Erythromycin A
A macrolide antibiotic

Spectinomycin
An aminocyclitol antibiotic

macrolides, the **tetracyclines**, the **aminocyclitols**, and the **sulfa drugs** (Fig. 21.1). Each antibiotic has a different profile of biological activity, and all of them are used in modern medicine to treat various infections.

Background

The broad-spectrum antibacterial activity of sulfanilamide (**1**) was first revealed in the mid-1930s by serendipity, as are many important discoveries in science; an interesting account of this discovery can be found in the Historical Highlight *Discovery of Sulfa Drugs*, which is available online. Over 1000 derivatives of sulfanilamide have been synthesized and tested as potential antibiotics, and some of these are still used today. The mode of action of the sulfa drugs is another interesting story because it provides some insights into strategies that might be generally exploited to design biologically active compounds.

Early in the development of sulfa drugs as antibiotics, it was found that *p*-aminobenzoic acid (PABA, **2**), which is now used in sunscreens and sunblocks, inhibits the antibacterial action of sulfanilamide. Since *p*-aminobenzoic acid and sulfanilamide are structurally similar, this discovery led to the speculation that the two compounds competed with each other in some biological process that was essential for bacterial growth. This speculation was eventually supported by experimentation. *p*-Aminobenzoic acid is used by bacteria in the synthesis of the essential enzyme cofactor folic acid (**3**). When sulfanilamide is present, it successfully competes with *p*-aminobenzoic acid for the active site in the enzyme that incorporates *p*-aminobenzoic acid into folic acid. By functioning as a competitive inhibitor of this enzyme, sulfanilamide blocks the biosynthesis of folic acid, and without folic acid, the bacteria cannot grow. On the other hand, animal cells cannot synthesize folic acid, which is an **essential vitamin**, and it therefore must be part of the diet. Because only bacteria rely on the biosynthesis of folic acid from *p*-aminobenzoic acid, the sulfa compounds are "ideal" drugs, as they kill only the bacteria and not the animal host. Of course, they are not truly ideal, since some people have allergic reactions to sulfa drugs and because bacteria develop resistance to them over time.

1
Sulfanilamide

2
p-Aminobenzoic
acid (PABA)

3
Folic acid

The preparation of sulfanilamide (**1**) from benzene (**4**) (Scheme 21.2) serves as an excellent example of a multistep synthesis that produces a biologically active, nonnatural molecule. Owing to the toxicity of benzene, the experimental sequence that follows will commence with the second step, the reduction of nitrobenzene (**5**), which is considerably less toxic. Each of the intermediate compounds in the synthesis may either be isolated or used directly in the subsequent step without further purification. A discussion of the reactions involved in each step follows.

Scheme 21.2

4
Benzene

5
Nitrobenzene

6
Aniline

7
Acetanilide

8
4-Acetamidobenzene-
sulfonyl chloride

9
4-Acetamidobenzene-
sulfonamide

1
Sulfanilamide

Reduction of Aromatic Nitro Compounds: Preparation of Aniline (6)

Aromatic nitro compounds, which are readily prepared by electrophilic aromatic nitration (Sec. 15.1), are reduced to the corresponding aromatic amines by a variety of methods. Chemical reduction is commonly used in the research laboratory, but the most important commercial method is catalytic hydrogenation (Chap. 17). Although the precise mechanism of the reduction of nitro compounds is not fully understood, various intermediates in this stepwise process have been isolated, as shown in Equation 21.1. The symbol [H] over the arrows represents reduction without specifying the reagent. The product that is ultimately isolated from the reduction of an aromatic nitro compound can be controlled to a significant degree by the nature of the reducing agent. For example, the reduction of nitrobenzene (**5**) with zinc metal and ammonium chloride gives only *N*-phenylhydroxylamine (**11**), whereas the use of tin metal and hydrochloric acid gives aniline (**6**).

$$\tag{21.1}$$

5   **10**   **11**   **6**

Nitrobenzene   Nitrosobenzene   *N*-Phenylhydroxylamine   Aniline

The reduction of nitrobenzene to aniline is a typical oxidation-reduction reaction in which tin metal, Sn°, is oxidized to stannic ion, Sn^{4+}, in the form of stannic chloride, SnCl$_4$; hydrochloric acid serves as the source of protons. A plausible mechanism of this reaction is outlined in Scheme 21.3. Generally, the reduction occurs by a sequence of steps in which an electron is first transferred from a tin atom to the organic substrate to give an intermediate radical ion that is then protonated. The oxygen atoms on the nitro group are eventually removed as water molecules. It is left as an exercise to write a *balanced* equation for the overall reaction and to provide a mechanism for the reduction of *N*-phenylhydroxylamine (**11**) into aniline (see Exercises 13 and 14 at the end of this section).

Scheme 21.3

The procedure for isolating and purifying the aniline formed in this experiment represents an excellent example of how the physical and chemical properties of a component in a mixture of organic substances can be exploited to isolate it *without* using chromatographic techniques. For example, after completing the procedure for the chemical reduction of nitrobenzene, it is necessary to seperate aniline from its principal impurities, which are unchanged nitrobenzene and two by-products, benzidine (**12**) and 4-aminophenol (**13**).

The purification can be achieved using steam distillation (Secs. 2.16 and 4.5), a good technique for separating volatile organic compounds from nonvolatile organic and inorganic substances. Because the reaction mixture in the procedure is *acidic*, it must first be made *basic* prior to performing the steam distillation so aniline is present as the free base rather than the anilinium salt **14**, and the 4-aminophenol is converted to its *water-soluble* sodium phenoxide salt. The aniline and nitrobenzene are then removed from the reaction mixture by steam distillation. The phenoxide salt and benzidine are both nonvolatile and remain in the aqueous phase.

| **12** | **13** | **14** |
| Benzidine | 4-Aminophenol | Anilinium hydrochloride |

Aniline may then be separated from nitrobenzene by first acidifying the steam distillate with hydrochloric acid to convert aniline to its water-soluble hydrochloride salt **14**. The acidic solution is then extracted with diethyl ether to remove nitrobenzene. The aqueous layer, which contains **14**, is made basic, and the aniline that is formed is isolated from the mixture by extraction with diethyl ether.

Protecting Groups: Preparation of Acetanilide (7)

The amino function on aniline is an electron-donating group that *activates* the aromatic ring toward electrophilic aromatic substitution reactions (Chap. 15). However, if an aryl amino group is protonated or complexed with a Lewis acid, the resulting ammonium group *deactivates* the ring toward electrophilic aromatic substitution. Electrophilic reagents or other functional groups that may be present on the ring can also react directly with the amino group. Thus, because of the basic and nucleophilic properties of the amino group, it is necessary to protect it to ensure that the desired reaction will take place. In general, placing a **protecting group** on a reactive functional group alters the chemical properties of that group, thereby rendering it unreactive toward certain reagents. The protecting group must be carefully selected, however, because it must be stable to the reaction conditions employed in the various transformations, *and* it must be removable under conditions that do not adversely affect other functional groups in the molecule.

In our synthesis of sulfanilamide (**1**), adding a sulfonamido, –SO$_2$NH$_2$, group to the aromatic ring requires the use of a protecting group for the amino function. This is necessary because introducing a sulfonamido function onto an arene, Ar–H, commonly involves sequential chlorosulfonation, an electrophilic aromatic substitution reaction, followed by treatment of the intermediate aryl sulfonyl chloride, Ar–SO₂Cl, with ammonia to give the sulfonamide, Ar–SO₂NH₂ (Eq. 21.2). If aniline itself were treated with chlorosulfonic acid, the amino group could react with either hydrogen chloride or SO₃, both of which are generated *in situ*, to provide ammonium salts such as **14** or **15**; these groups would then deactivate the ring toward electrophilic aromatic substitution. Moreover, a chlorosulfonyl group cannot be generated in the presence of an amino group since the chlorosulfonyl group of one molecule would react with the amino group of another, thereby forming a *polymeric* material containing sulfonamide linkages.

$$\text{Ar}{-}\text{H} \xrightarrow[\text{(2) NH}_3]{\text{(1) ClSO}_3\text{H}} \text{Ar}{-}\text{SO}_2\text{NH}_2 \tag{21.2}$$

An arene   An aryl sulfonamide

15

In the experiment you will perform, the amino group is protected as an amide by acetylation. The free amine is later regenerated by removing the acetyl group via controlled hydrolysis *after* the chlorosulfonyl moiety has been introduced onto the aromatic ring and converted into the sulfonamido group. The selective hydrolysis of the amido group of **9** in the presence of the sulfonamido group is possible because the sulfur atom of the sulfonamido group is *tetrahedral* and hence more sterically hindered toward nucleophilic attack by water than the *trigonal* carbonyl carbon atom (Scheme 21.2).

Aniline (**6**) is converted to acetanilide (**7**) by acetylation according to the sequence shown in Equation 21.3. In this reaction, aniline is first converted to its hydrochloride salt **14**. Acetic anhydride and sodium acetate are then added to give a mixture in which sodium acetate and **14** are in equilibrium with acetic acid and aniline. As free aniline is produced by this acid-base reaction, it rapidly undergoes acetylation by acetic anhydride to give acetanilide. This particular method for acetylating amines is general, and the yields are usually high.

$$\text{(6)} \xrightarrow{\text{HCl}} \text{(14)} \xrightarrow[\text{CH}_3\text{CO}_2\text{Na}]{\text{(CH}_3\text{CO)}_2\text{O}} \text{(7)} + \text{CH}_3\text{CO}_2\text{H} + \text{NaCl} \tag{21.3}$$

It is always important to consider possible side reactions that may occur. One example in this procedure is diacetylation (Eq. 21.4). Although the acetylation of aniline in pure acetic anhydride does give substantial quantities of the diacetyl compound **16**, this side reaction is minimized in aqueous solution.

$$\text{7} \xrightarrow{\text{(CH}_3\text{CO)}_2\text{O}} \text{16} + CH_3CO_2H \qquad (21.4)$$

Chlorosulfonation: Preparation of 4-Acetamidobenzenesulfonyl Chloride (8)

The sulfonyl group, $-SO_2Cl$, can be introduced *para* to the *N*-acetyl group of acetanilide by a one-step process called chlorosulfonation (Eq. 21.5). Although an electron-donating group can direct an incoming electrophile to either the *ortho* or *para* position, the acetamido group orients the incoming group predominantly *para*, presumably as a consequence of the steric bulk of the acetamido group; virtually none of the *ortho* isomer is observed. The reaction is known to proceed through the intermediate sulfonic acid **17**, which is converted to the sulfonyl chloride **8** by reaction with chlorosulfonic acid. At least two equivalents of chlorosulfonic acid are thus required per equivalent of acetanilide (**7**). The attacking electrophile is probably SO_3, which is generated *in situ* from chlorosulfonic acid according to Equation 21.6.

$$\text{7} \xrightarrow{\text{ClSO}_3\text{H (1 mol)}} \text{17} + HCl \xrightarrow{\text{ClSO}_3\text{H (1 mol)}} \text{8} + H_2SO_4 \qquad (21.5)$$

4-Acetamidobenzene-
sulfonic acid

$$ClSO_3H \rightleftharpoons SO_3 + HCl \qquad (21.6)$$

The reaction mixture is worked up by pouring it into ice water, whereupon the 4-acetamidobenzenesulfonyl chloride (**8**) separates as a white solid. The excess chlorosulfonic acid is hydrolyzed by water (Eq. 21.7). In general, sulfonyl chlorides are much less reactive toward water than are carboxylic acid chlorides, but you should not expose them to water for extended periods of time because they hydrolyze slowly to give the corresponding sulfonic acids according to Equation 21.8. It is unnecessary to dry or purify the 4-acetamidobenzenesulfonyl chloride (**8**) in this sequence; rather, **8** is simply treated directly with aqueous ammonia.

$$ClSO_3H + H_2O \longrightarrow HCl + H_2SO_4 \qquad (21.7)$$

$$\text{8} + H_2O \longrightarrow \text{17} + HCl \qquad (21.8)$$

Ammonolysis and Hydrolysis: Preparation of Sulfanilamide (1)

The final steps in the preparation of sulfanilamide (1) involve treating 4-acetamidobenzenesulfonyl chloride (8) with an excess of aqueous ammonia to give 4-acetamidobenzenesulfonamide (9), followed by selective hydrolytic removal of the *N*-acetyl protecting group (Eq. 21.9). To avoid hydrolysis of the sulfonyl chloride group (Eq. 21.8) before its reaction with ammonia, it is *imperative* to treat 8 with ammonia *immediately* after its isolation in the previous experiment. If the conversion of acetanilide (7) to 9 is not completed within a *single* laboratory period, the overall yield for the sequence may be significantly reduced.

Hydrolysis of the acetamido moiety of 9 may be effected using either aqueous acid or aqueous base without affecting the sulfonamido group, which hydrolyzes slowly. An acid-catalyzed hydrolysis is performed in this experiment. Because the amine group will form a hydrochloride salt under the conditions of the hydrolysis, it is necessary to neutralize the solution with a base, such as sodium carbonate, to isolate sulfanilamide, which may then be purified by recrystallization from water.

$$
\underset{\substack{\text{SO}_2\text{Cl} \\ \mathbf{8}}}{\underset{}{\text{NHCOCH}_3}} \xrightarrow{\text{NH}_4\text{OH}} \underset{\substack{\text{SO}_2\text{NH}_2 \\ \mathbf{9}}}{\underset{}{\text{NHCOCH}_3}} \xrightarrow[\Delta]{\text{HCl/H}_2\text{O}} \underset{\substack{\text{SO}_2\text{NH}_2}}{\underset{}{\overset{+}{\text{NH}_3}\,\text{Cl}^-}} \xrightarrow[\text{H}_2\text{O}]{\text{Na}_2\text{CO}_3} \underset{\substack{\text{SO}_2\text{NH}_2 \\ \mathbf{1}}}{\underset{}{\text{NH}_2}} \qquad (21.9)
$$

EXPERIMENTAL PROCEDURES

Synthesis of Sulfanilamide

Purpose To demonstrate principles of multistep synthesis, protecting groups, and electrophilic aromatic substitution.

Note Throughout the sequence of reactions, quantities of reagents must be adjusted according to the amount of starting material that is available from the previous reaction, but you should *not* run the reactions on a larger scale than is indicated. If the amount of your starting material varies from that indicated, do *not* change the reaction times unless directed to do so by your instructor. You should be able to obtain sufficient quantities of product from each reaction to complete the entire sequence starting with the amount of acetanilide provided in the first step.

A ■ *Preparation of Aniline*

SAFETY ALERT

1. **Wear safety glasses or goggles and suitable protective gloves while performing the experiment.**
2. **Nitrobenzene is toxic; avoid inhaling its vapors or allowing it to come into contact with your skin.**

3. *Concentrated* hydrochloric acid can cause burns if it comes in contact with the skin. Should contact occur, flood the affected area with water and rinse it thoroughly with dilute aqueous sodium bicarbonate solution.

4. The solution of 12 *M* sodium hydroxide is highly caustic and can cause burns and loss of hair; avoid contact of this solution with your skin. Should contact occur, flood the affected area with water and rinse it thoroughly with dilute aqueous acetic acid solution.

5. Use *flameless* heating when concentrating ethereal solutions.

MINISCALE PROCEDURE

Preparation Refer to the online resources to answer Pre-Lab Exercises, access videos, and read the MSDSs for the chemicals used or produced in this procedure. Review Sections 2.9, 2.10, 2.11, 2.13, 2.16, 2.21, 2.22, 2.25, and 2.29.

Apparatus A 250-mL round-bottom flask, thermometer, ice-water bath, separatory funnel, apparatus for heating under reflux, magnetic stirring, steam distillation, simple distillation, and *flameless* heating.

Setting Up Add 13.1 g of finely divided tin powder and 5.2 mL of nitrobenzene to the round-bottom flask equipped with a stirbar.

Reduction Add 28 mL of *concentrated* hydrochloric acid to the flask, insert a thermometer, and stir the contents of the flask to mix the three-phase system. Monitor the temperature and cool the flask as necessary in an ice-water bath to maintain the temperature below 60 °C. The initial exothermic portion of the reaction should be complete after about 15 min. Attach a reflux condenser and heat the contents of the flask under reflux for about 20 min with continued stirring. During this time, the color, which is due to the intermediate reduction product, and any droplets of nitrobenzene in the condenser should disappear.

Work-Up When the reflux period is complete, cool the acidic solution in an ice-water bath and slowly add 50 mL of 12 *M* sodium hydroxide solution directly to the reaction mixture. Check the pH of the mixture to ensure that it is basic; if it is not, add additional base. Equip the flask for steam distillation and steam-distill the mixture until the condensate no longer appears oily, although it may still be slightly cloudy; about 35–40 mL of distillate should be collected.★

Add 4.5 mL of *concentrated* hydrochloric acid to the distillate and transfer the mixture to a separatory funnel. Check the pH of the mixture to be certain it is acidic, and then extract it sequentially with two 15-mL portions of diethyl ether to remove any unreacted nitrobenzene. Transfer the *aqueous* layer to an Erlenmeyer flask. Cool the contents of the flask in an ice-water bath and make the aqueous solution *basic* by slowly adding a *minimum* volume of 12 *M* sodium hydroxide solution. Saturate the aqueous solution with solid sodium chloride, cool the mixture to room temperature, and transfer it to a *clean* separatory funnel. Extract the aqueous mixture sequentially with two 15-mL portions of diethyl ether, using the first portion to rinse the flask in which the neutralization was done. Separate the aqueous layer from the organic layer as thoroughly as possible each time. Dry the *combined* organic extracts over several spatula-tips full *anhydrous* sodium sulfate.★ Decant the dried solution into a round-bottom flask, and remove most of the diethyl ether by simple distillation. Alternatively, use rotary evaporation or other techniques to concentrate the solution. Remove the final traces of solvents by connecting the cool stillpot to a vacuum source.

The aniline remaining in the stillpot may be used directly in the next step of the sequence. Alternatively, you may purify the aniline by shortpath distillation. Wrap the stillhead with a layer of glasswool and then aluminum foil to minimize heat losses, thereby facilitating the distillation. Collect three fractions having approximate boiling ranges (760 torr) of 35–90 °C, 90–180 °C, and 180–185 °C, respectively.

Analysis Weigh the product and calculate its yield. Obtain IR and ^{1}H NMR spectra of your starting material and product, and compare them with those of authentic samples (Figs. 21.2–21.5). Pure aniline is colorless but may darken soon following distillation, owing to air-oxidation.

WRAPPING IT UP

Vacuum-filter the *pot residue from the steam distillation* through a bed of filter-aid, and place the *tin salts* together with the *filter-aid* and *filter paper* in the container for nonhazardous solids. Neutralize the *filtrate* and the other *aqueous layers* with dilute hydrochloric acid and flush them down the drain with excess water. Pour the combined *ethereal extracts* and *distillate* into a container for nonhalogenated organic liquids. Put the *sodium sulfate* on a tray in the hood and, after the ether evaporates, pour it into the container for nonhazardous solids.

B ■ *Preparation of Acetanilide*

SAFETY ALERT

1. **Wear safety glasses or goggles and suitable protective gloves while performing the experiment.**

2. **Acetic anhydride is a lachrymator. Transfer this material in a hood and do not inhale its vapors.**

MINISCALE PROCEDURE

Preparation Refer to the online resources to answer Pre-Lab Exercises, access videos, and read the MSDSs for the chemicals used or produced in this procedure. Review Sections 2.9, 2.10, 2.11, 2.17, and 2.26.

Apparatus A 250-mL Erlenmeyer flask, thermometer, ice-water bath, apparatus for magnetic stirring, vacuum filtration, and *flameless* heating.

Setting Up Add 3.6 mL of aniline to the Erlenmeyer flask containing 100 mL of 0.4 *M* hydrochloric acid and equipped with a stirbar. Stir the mixture to effect solution and warm it to about 50 °C. Prepare a solution containing 6.0 g of sodium acetate trihydrate dissolved in 20 mL of water, and, in a *separate* container, measure out 4.4 mL of acetic anhydride.

Reaction and Isolation Add the acetic anhydride in *one* portion to the warm solution of anilinium hydrochloride. Stir the solution *vigorously* and then add the solution of sodium acetate *immediately and in one portion.* Cool the reaction mixture to 5 °C in an ice-water bath, and continue to stir the mixture until the crystalline product completely precipitates. Collect the acetanilide by vacuum filtration, wash it with a small portion of ice-cold water, and air-dry it.★ If impure or slightly colored acetani-lide is obtained, recrystallize it from a minimum volume of hot water, using decolor-izing carbon to give a colorless product.

Analysis Weigh the product and calculate its yield. Determine its melting point. Obtain IR and ^{1}H NMR spectra of your starting material and product and compare them with those of an authentic sample (Figs. 8.7 and 15.31).

WRAPPING IT UP

After neutralizing the *aqueous filtrate* with sodium carbonate, flush it and the *mother liquor* from crystallization down the drain with a large excess of water.

C ■ *Preparation of 4-Acetamidobenzenesulfonyl Chloride*

Note This entire procedure as well as the next one, up to the first stopping point, must be completed in a *single* laboratory period.

SAFETY ALERT

1. **Wear safety glasses or goggles and suitable protective gloves while perform-ing the experiment.**

2. **This experiment should be performed in a hood, if possible. Otherwise, use the gas trap prescribed by your instructor to prevent escape of hydrogen chloride, SO_2, and SO_3 into the laboratory.**

3. **Chlorosulfonic acid is *highly corrosive* and may cause serious burns if it is allowed to come in contact with your skin. Use *extreme* care when working with this chemical. Should any chlorosulfonic acid come in contact with your skin, *immediately* flood the affected area with cold water and then rinse it with 5% sodium bicarbonate solution.**

4. **Chlorosulfonic acid reacts *violently* with water. Open containers of chloro-sulfonic acid will fume upon exposure to atmospheric moisture owing to the reaction of chlorosulfonic acid with water to give HCl and SO_3, both of which are noxious and corrosive gases. Measure and transfer this acid only *in the hood.* Several graduated cylinders should be kept in the hood for all students to use for measuring the volume of chlorosulfonic acid needed.**

5. **To destroy residual chlorosulfonic acid in graduated cylinders and other glassware that has contained it, add cracked ice to the glassware *in the hood* and let the glassware remain there until the ice has melted. Then rinse the apparatus with large amounts of cold water.**

6. **Make sure your glassware is dry, lubricate the ground-glass joints of the appa-ratus carefully, and be certain that the joints are intimately mated. Otherwise, noxious gases will escape.**

MINISCALE PROCEDURE

Preparation Refer to the online resources to answer Pre-Lab Exercises, access videos, and read the MSDSs for the chemicals used or produced in this procedure. Review Sections 2.9, 2.10, 2.11, 2.17, 2.23, and 2.28.

Apparatus A 50-mL round-bottom flask, Claisen adapter, gas trap, separatory funnel with ground-glass joint, thermometer, ice-water bath, apparatus for magnetic stirring, vacuum filtration, and *flameless* heating.

Setting Up Oven-dry the glass apparatus prior to assembling it. Because an *airtight* seal is required for all connections, *carefully grease all joints* when assembling the apparatus. Place 2.7 g of *dry* acetanilide in the round-bottom flask and equip the flask with a stirbar and the Claisen adapter. Fit the *sidearm* of the adapter with the gas trap specified by your instructor; some examples are illustrated in Figure 2.68. *Working at the hood,* measure 8.0 mL of chlorosulfonic acid into a small separatory funnel having a standard-taper ground-glass joint; be certain that the stopcock of the funnel is firmly seated and closed *prior* to this transfer. Stopper the funnel and place it on the straight arm of the Claisen adapter above the reaction flask so that the chlorosulfonic acid will drip *directly* onto the acetanilide. Cool the flask in a water bath that is maintained at 10–15 °C; do *not* allow the bath temperature to drop below 10 °C.

Reaction Open the stopcock of the funnel fully so that the chlorosulfonic acid is added as rapidly as possible to the flask containing the acetanilide. It may be necessary to *briefly* lift the stopper on the funnel to equalize the pressure in the system if the flow of chlorosulfonic acid becomes erratic or slow. When the addition has been completed, stir the mixture to hasten dissolution of the acetanilide; maintain the temperature of the water bath below 20 °C. After most of the solid has dissolved and the initial reaction has subsided, remove the cooling bath, and allow the reaction mixture to warm to room temperature with stirring. To complete the reaction, warm the mixture at about 70–80 °C with continued stirring until there is no longer any increase in the rate of gas evolution; about 10–20 min of heating will be required.

Work-Up and Isolation Cool the mixture to room temperature or slightly below using an ice-water bath. *Working at the hood,* place about 150 g of crushed ice in a beaker and pour the reaction mixture slowly onto the ice with stirring; stir the mixture thoroughly with a glass rod during the addition. *Be careful not to add the mixture too quickly to avoid splattering the chlorosulfonic acid.* Rinse the reaction flask with a little ice water and transfer this to the beaker. The remainder of the work-up procedure may be performed at the bench. The precipitate that forms is crude 4-acetamidobenzenesulfonyl chloride and may be white to pink in color. If it is allowed to stand without stirring, a hard mass may form. Any lumps that form should be thoroughly broken up with a stirring rod. Collect the crude material by vacuum filtration. Wash the solid thoroughly with 15-mL portions of ice water *until the filtrate is neutral;* press the filter cake as dry as possible with a clean cork. The crude sulfonyl chloride should be used *immediately* in the following procedure.

Purification Purify about 50 mg of the product by dissolving it in a minimum amount of boiling dichloromethane in a test tube. Using a Pasteur pipet, remove the upper, aqueous layer as quickly as possible. Be careful when separating the layers. Cool the organic layer, collect the purified product by vacuum filtration, and air-dry it.

Analysis   Determine the melting point of the recrystallized solid. Obtain its IR and ¹H NMR spectra and compare them with those of an authentic sample (Figs. 21.6 and 21.7).

WRAPPING IT UP

Neutralize the combined *aqueous solutions* with sodium carbonate and flush them down the drain with a large excess of water. Cover any spilled droplets of *chlorosulfonic acid* with sodium carbonate and flush the resulting powder down the drain. If a gas trap that contains a drying agent has been used, put the *drying agent* in the container for nonhazardous solids.

D ■ *Preparation of 4-Acetamidobenzenesulfonamide*

SAFETY ALERT

1. **Wear safety glasses or goggles and suitable protective gloves while performing the experiment.**

2. **Ammonia is evolved in this procedure. If possible, perform this reaction in a hood. Otherwise, use an inverted funnel attached to a vacuum source (Fig. 2.71b) to minimize escape of vapors of ammonia into the laboratory.**

3. ***Concentrated* ammonium hydroxide is caustic and may cause burns if it is allowed to come in contact with your skin. Use care when working with this reagent. Should any of the liquid come in contact with your skin, *immediately* flood the affected area with cold water.**

MINISCALE PROCEDURE

Preparation   Refer to the online resources to answer Pre-Lab Exercises, access videos, and read the MSDSs for the chemicals used or produced in this procedure. Review Sections 2.9, 2.10, 2.11, and 2.17.

Apparatus   A 125-mL Erlenmeyer flask, ice-water bath, apparatus for vacuum filtration and *flameless* heating.

Reaction, Work-Up, and Isolation   Transfer the crude 4-acetamidobenzenesulfonyl chloride obtained in the previous experiment to the Erlenmeyer flask, and add 15 mL of *concentrated* (28%) ammonium hydroxide. A rapid exothermic reaction may occur if the crude 4-acetamidobenzenesulfonyl chloride contains acidic contaminants that were not removed by the aqueous washings in the previous step. Use a stirring rod to break up any lumps of solid that may remain; the reaction mixture should be thick but homogeneous. Heat the mixture at 70–80 °C for about 0.5 h. Cool the reaction mixture in an ice-water bath and collect the product by vacuum filtration. Wash the crystals with cold water and air-dry them.★

Analysis Weigh the product and calculate its yield. Determine its melting point. Obtain IR and ¹H NMR spectra of your product and compare them with those of an authentic sample (Figs. 21.8 and 21.9).

WRAPPING IT UP

Neutralize the combined *aqueous solutions* and *filtrates* with 10% hydrochloric acid and flush them down the drain with a large excess of water.

E ▪ *Preparation of Sulfanilamide*

MINISCALE PROCEDURE

Preparation Refer to the online resources to answer Pre-Lab Exercises, access videos, and read the MSDSs for the chemicals used or produced in this procedure. Review Sections 2.9, 2.10, 2.11, 2.17, and 2.22.

Apparatus A 50-mL round-bottom flask, ice-water bath, apparatus for heating under reflux, magnetic stirring, vacuum filtration, and *flameless* heating.

Setting Up Prepare 20 mL of a solution of *dilute* hydrochloric acid by adding 10 mL of *concentrated* hydrochloric acid to 10 mL of water; swirl to mix. Equip the round-bottom flask with a stirbar and add the 4-acetamidobenzenesulfonamide and an amount of *dilute* hydrochloric acid that is about *twice* the weight of the crude 4-acetamidobenzenesulfonamide. Assemble the apparatus for heating under reflux.

Reaction, Work-Up, and Isolation Heat the reaction mixture with stirring under gentle reflux for 45 min and then allow the mixture to cool to room temperature. If any solid, which is unreacted starting material, appears on cooling, reheat the mixture under reflux for an additional 15 min. Add an equal volume of water to the cooled solution and transfer the resulting mixture to a beaker. Neutralize the excess hydrochloric acid by adding *small* quantities of *solid* sodium carbonate until the solution is slightly alkaline.

A precipitate should form during neutralization; scratching at the air-liquid interface may be necessary to initiate crystallization. Cool the mixture in an ice-water bath to complete the precipitation of product.* Collect the crystals by vacuum filtration, wash them with a small amount of cold water, and air-dry them.

Purification Purify the crude product by recrystallization from the minimum volume of hot water. If necessary, decolorize the hot solution and use a preheated funnel for the hot-filtration step to prevent premature crystallization of the product in the funnel. Cool the aqueous solution of sulfanilamide in an ice-water bath, isolate the crystalline product by vacuum filtration, and air-dry it.[★]

Analysis Weigh the product and calculate its yield. Determine its melting point. Test the solubility of sulfanilamide in 1.5 M hydrochloric acid solution and in 1.5 M sodium hydroxide solution. Obtain IR and ^{1}H NMR spectra of your product and compare them with those of an authentic sample (Figs. 21.10 and 21.11).

WRAPPING IT UP

Neutralize the combined *aqueous solutions and filtrates* with sodium carbonate and flush them down the drain with a large excess of water.

EXERCISES

General Questions

1. What biological properties of sulfanilamide enable it to interfere with bacterial but not human cell growth?

2. If each step in the synthesis of sulfanilamide from nitrobenzene proceeded in 90% yield, how much nitrobenzene would be needed to obtain 1.0 kg of sulfanilamide according to the sequence of reactions in Scheme 21.2?

3. Using benzene as the starting material together with any organic or inorganic reagents you desire, outline a synthesis of *p*-aminobenzoic acid (PABA) (2).

4. Outline a possible synthesis of the sulfanilamide derivative, **18**, using benzene as the only source of an aromatic ring. Use any needed aliphatic or inorganic reagents.

$$H_2N-\!\!\!\bigcirc\!\!\!-SO_2NH-\!\!\!\bigcirc$$

18

5. Two hypothetical sequences have been developed for converting an arbitrary compound A into E in the same overall yield.

$$A \xrightarrow{25\%} B \xrightarrow{49\%} C \xrightarrow{60\%} D \xrightarrow{57\%} E$$

$$A \xrightarrow{58\%} B \xrightarrow{57\%} C \xrightarrow{51\%} D \xrightarrow{30\%} E$$

   a. What is the overall yield for each sequence?

   b. As the production manager responsible for selecting the more economical of the two routes, what factor(s) would you consider in reaching a decision? Assume that no new capital investment would be required for either sequence.

6. A student proposed the following alternative sequence of reactions for preparing **1** from sulfanilic acid. Discuss this modified approach, commenting on any possible side reactions that might be encountered in either of the two steps. Propose a modification of this sequence that could be used to prepare sulfanilamide.

Sulfanilic acid   *p*-Aminosulfonyl
chloride

Questions for Part A

7. Why must an organic compound be *immiscible* with water in order to be purified by steam distillation?

8. Use Dalton's law and Raoult's law to explain why the steam distillation of aniline occurs at a lower temperature than the boiling point of aniline (Sec. 4.5).

9. Outline in a flow diagram the procedure for purifying aniline. Indicate the importance of each step in the procedure, and give reasons for doing the steam distillation with a *basic* solution. What is the purpose of performing the subsequent extractions from aqueous acid and then base? Write the equation(s) for the reactions that occur when base and acid are added.

10. What gas is evolved upon adding *concentrated* hydrochloric acid to tin powder in this reduction?

11. Suppose a student decided that a suitable stopping point in the procedure might be immediately after the initially exothermic reaction between tin and hydrochloric acid. What is the potential problem with storing the reaction mixture at this stage?

12. Why is sodium chloride added to the steam distillate *after* it has been made basic but *before* extraction with diethyl ether?

13. Write the balanced half-reactions for the reduction of nitrobenzene (**5**) into aniline (**6**) and for the oxidation of tin metal into stannic ion, Sn^{4+}. Sum these equations and write the balanced equation for the overall process.

14. Write a mechanism for the reduction of *N*-phenylhydroxylamine (**11**) into aniline (**6**). Show all electron and proton transfer steps, using curved arrows to symbolize the flow of electrons.

15. Protection and deprotection of the aryl amino group might be avoided if it were possible to reduce the nitro group as the last step in the synthesis. Explain why this is not an available option.

Refer to the online resources and use the spectra viewer and Tables 8.1–8.8 as needed to answer the blue-numbered questions on spectroscopy.

16. Consider the spectral data for nitrobenzene (Figs. 21.2 and 21.3).

    a. In the functional group region of the IR spectrum, specify the absorptions associated with the nitro group and with the aromatic ring.

    b. In the ^1H NMR spectrum, assign the various resonances to the hydrogen nuclei responsible for them.

    c. For the ^{13}C NMR data, assign the various resonances to the carbon nuclei responsible for them.

17. Consider the spectral data for aniline (Figs. 21.4 and 21.5).

    a. In the functional group region of the IR spectrum, specify the absorptions associated with the amino group and with the aromatic ring.

b. In the ^{1}H NMR spectrum, assign the various resonances to the hydrogen nuclei responsible for them.

c. For the ^{13}C NMR data, assign the various resonances to the carbon nuclei responsible for them.

18. Discuss the differences observed in the IR and NMR spectra of nitrobenzene and aniline that are consistent with the formation of the latter in this procedure.

Questions for Part B

19. Why is it necessary to *N*-acetylate aniline to give acetanilide (Part B) when the *N*-acetyl group (CH$_3$CO–) will be removed at the end of the synthesis (Part E)?

20. Why should the acetic anhydride not be allowed to stay in contact with the aqueous solution of anilinium hydrochloride (**14**) for an extended period of time before the solution of sodium acetate is added?

21. Why is aqueous sodium acetate preferred to aqueous sodium hydroxide for the conversion of anilinium hydrochloride (**14**) to aniline?

22. Why is aniline (**6**) soluble in aqueous hydrochloric acid, whereas acetanilide (**7**) is not?

23. Using curved arrows to symbolize the flow of electrons, give a stepwise reaction mechanism for the reaction of aniline (**6**) with acetic anhydride.

Refer to the online resources and use the spectra viewer and Tables 8.1–8.8 as needed to answer the blue-numbered questions on spectroscopy.

24. Consider the spectral data for acetanilide (Figs. 8.7 and 15.31).

a. In the functional group region of the IR spectrum, specify the absorptions associated with the nitrogen-hydrogen bonds, with the carbon-oxygen double bond, and with the aromatic ring.

b. In the ^{1}H NMR spectrum, assign the various resonances to the hydrogen nuclei responsible for them.

c. For the ^{13}C NMR data, assign the various resonances to the carbon nuclei responsible for them.

25. Discuss the differences observed in the IR and NMR spectra of aniline and acetanilide that are consistent with the formation of the latter in this procedure.

Questions for Part C

26. Why does the sulfonation of acetanilide (**7**) only occur in the *para* position?

27. Explain why 4-acetamidobenzenesulfonyl chloride (**8**) is much *less* susceptible to hydrolysis of the acid chloride function than is 4-acetamidobenzoyl chloride.

4-Acetamidobenzoyl chloride

28. What materials, organic or inorganic, may contaminate the crude sulfonyl chloride **8** prepared in this reaction? Which of them are likely to react with the ammonia used in the next reaction step of the sequence?

29. Using curved arrows to symbolize the flow of electrons, write a stepwise mechanism for the conversion of acetanilide (**7**) into 4-acetamidobenzenesulfonic acid (**17**).

30. Write the structure of the material containing sulfonamide linkages that might be obtained upon polymerization of 4-aminobenzenesulfonyl chloride.

$$H_2N-\text{〈benzene ring〉}-SO_2Cl$$

4-Aminobenzenesulfonyl chloride

31. Predict the product of monochlorosulfonation of the following compounds.

a. [structure: H₃CO-substituted benzene ring with NO₂]

b. [structure: phenyl benzoate]

Refer to the online resources and use the spectra viewer and Tables 8.1–8.8 as needed to answer the blue-numbered questions on spectroscopy.

32. Consider the spectral data for 4-acetamidobenzenesulfonyl chloride (Figs. 21.6 and 21.7).

   a. In the functional group region of the IR spectrum, specify the absorptions associated with the nitrogen-hydrogen bond, with the carbon-oxygen double bond, and with the aromatic ring.

   b. In the ^{1}H NMR spectrum, assign the various resonances to the hydrogen nuclei responsible for them.

   c. For the ^{13}C NMR data, assign the various resonances to the carbon nuclei responsible for them.

33. Discuss the differences observed in the IR and NMR spectra of acetanilide and 4-acetamidobenzenesulfonyl chloride that are consistent with the formation of the latter in this procedure.

Questions for Part D

34. What acids might be present in the crude sulfonyl chloride that would cause an exothermic reaction with ammonia?

35. Why would you expect the yield of 4-acetamidobenzenesulfonamide (**9**) to be lowered if the crude sulfonyl chloride **8** were not combined with ammonia until the laboratory period following its preparation?

36. *Concentrated* ammonia contains 72% water, yet its reaction with 4-acetamidobenzenesulfonyl chloride (**8**) produces 4-acetamidobenzenesulfonamide (**9**) rather than 4-acetamidobenzenesulfonic acid (**17**). Why is ammonia more reactive than water toward **9**?

37. Using curved arrows to symbolize the flow of electrons, write a stepwise mechanism for the aminolysis of 4-acetamidobenzenesulfonyl chloride (**8**) to give 4-acetamidobenzenesulfonamide (**9**).

Refer to the online resources and use the spectra viewer and Tables 8.1–8.8 as needed to answer the blue-numbered questions on spectroscopy.

38. Consider the spectral data for 4-acetamidobenzenesulfonamide (Figs. 21.8 and 21.9).

   a. In the functional group region of the IR spectrum, specify the absorptions associated with the nitrogen-hydrogen bonds, with the carbon-oxygen double bond, and with the aromatic ring.

b. In the ^{1}H NMR spectrum, assign the various resonances to the hydrogen nuclei responsible for them.

c. For the ^{13}C NMR data, assign the various resonances to the carbon nuclei responsible for them.

39. Discuss the differences observed in the IR and NMR spectra of 4-acetamido-benzenesulfonyl chloride and 4-acetamidobenzenesulfonamide that are consistent with the formation of the latter in this procedure.

Questions for Part E

40. In the preparation of sulfanilamide (**1**) from 4-acetamidobenzenesulfonamide (**9**), only the acetamido group is hydrolyzed. Explain this difference in reactivity of the acetamido and sulfonamido groups toward aqueous acid.

41. Following hydrolysis of 4-acetamidobenzenesulfonamide (**9**) with aqueous acid, the reaction mixture is homogeneous, whereas 4-acetamidobenzenesulfonamide is insoluble in aqueous acid. Explain the change in solubility that occurs as a result of the hydrolysis.

42. Explain the results obtained when the solubility of sulfanilamide was determined in 1.5 *M* hydrochloric acid and in 1.5 *M* sodium hydroxide. Write equations for any reaction(s) that occurred.

43. Why might there be a preference for using solid sodium carbonate instead of solid sodium hydroxide for basifying the acidic hydrolysis solution obtained in this experiment?

44. What would be observed if 4-acetamidobenzenesulfonamide were subjected to vigorous hydrolysis conditions, such as *concentrated* hydrochloric acid and heat, for a long period of time? Write an equation for the reaction that might occur.

45. Calculate the overall yield of sulfanilamide obtained in the sequence of reactions that you performed.

46. An amide derivative of a carboxylic acid is hydrolyzed in aqueous acid or base more slowly than an ester, anhydride, or acid chloride of the same acid.

a. Provide a rationale for this observation, using derivatives of acetic acid to illustrate your answer.

b. Would you expect the amide function in 4-acetamidobenzenesulfonamide to be more or less readily hydrolyzed than the amide group of an aliphatic amide? Explain your answer.

47. Using curved arrows to symbolize the flow of electrons, write a stepwise mechanism for the hydrolysis of 4-acetamidobenzenesulfonamide (**9**) in aqueous acid to give sulfanilamide (**1**).

Refer to the online resources and use the spectra viewer and Tables 8.1–8.8 as needed to answer the blue-numbered questions on spectroscopy.

48. Consider the spectral data for sulfanilamide (Figs. 21.10 and 21.11).

a. In the functional group region of the IR spectrum, specify the absorptions associated with the nitrogen-hydrogen bonds and with the aromatic ring.

b. In the ^{1}H NMR spectrum, assign the various resonances to the hydrogen nuclei responsible for them.

c. For the ^{13}C NMR data, assign the various resonances to the carbon nuclei responsible for them.

49. Discuss the differences observed in the IR and NMR spectra of 4-acetamido-benzenesulfonyl chloride and 4-acetamidobenzenesulfonamide that are consistent with the formation of the latter in this procedure.

Starting Materials and Products

The IR and NMR spectra of acetanilide are provided in Figures 8.7 and 15.31, respectively.

Figure 21.2
IR spectrum of nitrobenzene (neat).

Figure 21.3
NMR data for nitrobenzene (CDCl₃).

(a) ¹H NMR spectrum (300 MHz).
(b) ¹³C NMR data: δ 123.5, 129.5, 134.8, 148.3.

Figure 21.4
IR spectrum of aniline (neat).

Figure 21.5
NMR data for aniline (CDCl₃).

(a) ¹H NMR spectrum (300 MHz).
(b) ¹³C NMR data: δ 115.1, 118.2, 129.2, 146.7.

Figure 21.6
IR spectrum of 4-acetamidoben-zenesulfonyl chloride (IR card).

Figure 21.7

NMR data for 4-acetamidobenz-enesulfonyl chloride (DMSO-d₆).

(a) ¹H NMR spectrum (300 MHz).
(b) ¹³C NMR data: δ 25.3, 119.7, 127.5, 141.6, 141.7, 170.3.

Figure 21.8

IR spectrum of 4-acetamidobenz-enesulfonamide (IR card).

Figure 21.9

NMR data for 4-acetamidobenz-enesulfonamide (DMSO-d₆).

(a) ¹H NMR spectrum (300 MHz).
(b) ¹C NMR data: δ 24.2, 118.5, 126.7, 138.1, 142.3, 169.0.

Figure 21.10
IR spectrum of sulfanilamide (IR card).

Figure 21.11
NMR data for sulfanilamide (DMSO-d₆).

(a) ^{1}H NMR spectrum (300 MHz).
(b) ^{13}C NMR data: δ 113.1, 127.6, 130.2, 151.6.

21.3 SYNTHESIS OF 1-BROMO-3-CHLORO-5-IODOBENZENE

The preparation of 1-bromo-3-chloro-5-iodobenzene (**24**), as illustrated in Scheme 21.4, is a fascinating excursion into the role of **protecting groups** and **substituent effects** as important factors in multistep synthesis. The approach used here involves a sequence of **electrophilic aromatic substitutions** and illustrates two important general strategies in synthesis: (1) protecting a reactive functionality to prevent undesired side reactions and (2) using substituents to control the regiochemical course of a reaction. In the present example, the first principle is represented by the conversion of the *amino* group of **6** into the *amido* group of **7**, and the second is illustrated by the directing and ring-activating properties of the amino and amido functions. The yields of the individual steps are good to high, ranging from 60% to 96%, an important consideration in designing multistep syntheses, as was noted in Section 21.1.

Although benzene (**4**) is a potential starting point of our synthesis (Scheme 21.4), it is a potential carcinogen, so the sequence of reactions begins with nitrobenzene (**5**). Discussion of and experimental procedures for the reduction of nitrobenzene and the conversion of aniline (**6**) to acetanilide (**7**) are found in Section 21.2, so the present description starts with the bromination of acetanilide.

Scheme 21.4

Acetylation and Bromination: Preparation of 4-Bromoacetanilide (19)

The introduction of bromine by electrophilic aromatic substitution is not performed directly on aniline (**6**) because its amino group so greatly activates the ring toward electrophiles that even mild brominating agents such as an aqueous solution of bromine lead to trisubstitution (Eq. 21.10). Acetylating the amino group of aniline attentuates the activating effect of the nonbonded pair of electrons on the nitrogen atom. Like the amino group, the acetamido group, CH_3CONH-, of acetanilide (**7**) is an activating group, so selective monobromination of **7** to give **19** occurs under mild conditions, such as bromine in acetic acid *without* a Lewis acid catalyst. The acetamido group is *ortho-* and *para-directing*, so both 2-bromo- and 4-bromoacetanilide are produced. However, the steric bulk of the acetamido substituent hinders attack at the 2-position, thereby strongly favoring formation of the 4-bromo isomer **19** (Eq. 21.11).

(21.10)

6   2,4,6-Tribromoaniline

(21.11)

7   **19**   2-Bromoacetanilide
          major   minor

As a general rule, separating *ortho*- and *para*-isomers from a mixture containing both is fairly simple. Because *para*-isomers are more symmetrical than *ortho*-isomers, they are *less* soluble in a given solvent, as noted in Section 15.4. 4-Bromoacetanilide (**19**), for example, is separable from the 2-bromo isomer by a single recrystallization of the crude mixture. A second consequence of the greater molecular symmetry of *para*-isomers is that they typically have higher melting points than their *ortho* relatives. In the case of the *ortho*- and *para*-isomers of bromonitrobenzene, this difference is approximately 70 °C.

Chlorination: Preparation of 4-Bromo-2-Chloroacetanilide (20)

4-Bromoacetanilide is slightly less reactive toward further electrophilic substitution than acetanilide, but it is still sufficiently reactive to be chlorinated under mild conditions *without* a strong Lewis acid catalyst. Thus, molecular chlorine dissolved in acetic acid chlorinates 4-bromoacetanilide to produce 4-bromo-2-chloroacetanilide (**20**) as shown in Equation 21.12. The formation of the 2- rather than the 3-chloro isomer demonstrates the more powerful directing effect of the acetamido group as compared to the bromo substituent.

(21.12)

19   **20**

To avoid the hazards associated with using chlorine gas, chlorine may be easily generated *in situ* by reaction of *concentrated* hydrochloric acid and sodium chlorate ($NaClO_3$). In this reaction, an oxidation-reduction process occurs by which chloride ion is oxidized to chlorine (Eq. 21.13) and chlorate ion is reduced to chlorine (Eq. 21.14); balancing of these equations is left to you. Not only is this method for generating chlorine easy and safe, but it also allows for the preparation of precise amounts of molecular chlorine.

$$Cl^- \rightarrow 1/2\ Cl_2 + e^- \tag{21.13}$$

$$6H^+ + ClO_3^- + 5e^- \rightarrow 1/2\ Cl_2 + 3H_2O \tag{21.14}$$

Hydrolysis: Preparation of 4-Bromo-2-Chloro-6-Iodoaniline (22)

The acetamido group of 4-bromo-2-chloroacetanilide (**20**) is the most powerful directing group of the three substituents on the aromatic ring, so further electrophilic substitution will preferentially occur at the 6-position of the ring. However, attempted iodination of **20** fails because of the deactivating effects of the halogen substituents and because the iodonium ion, I^+, is a weak and unreactive electrophile compared to bromonium, Br^+, or chloronium, Cl^+, ions. The solution to this problem is to regenerate the amino group, a more powerful ring activator, by hydrolyzing the amide (Eq. 21.15). This reaction is effected using a solution of *concentrated* hydrochloric acid in ethanol.

$$\tag{21.15}$$

The iodinating agent in our procedure is iodine monochloride, ICl. Because iodine is the less electronegative of the two halogens, this reagent is polarized to make the iodine atom electrophilic, and it may dissociate to give Cl^- and I^+. The electrophile for the reaction is thus either polarized molecular I–Cl or iodonium ion (Eq. 21.16).

$$\tag{21.16}$$

Diazotization and Protonolysis: Preparation of 1-Bromo-3-Chloro-5-Iodobenzene (24)

The final step in the synthesis is the removal of the amino group from 2-chloro-4-bromo-6-iodoaniline (**22**). After the amino group has been diazotized with nitrous acid at 0 °C to produce **23**, the diazo group may be replaced by a hydrogen atom upon treatment with ethanol or hypophosphorous acid, H_3PO_2, as depicted in Equation 21.17. Although the latter is the more general reagent for replacing a diazonium group by a hydrogen atom, ethanol works well with certain types of aromatic compounds, in particular those containing halogen, and it will be used in our procedure. The transformation presumably involves an oxidation-reduction reaction in which ethanol is oxidized to acetaldehyde or acetic acid. The process is mechanistically ill-defined, but donation of a hydride ion to reduce the diazo compound is a likely possibility (Eq. 21.18).

(21.17)

(21.18)

A potentially important side reaction that might occur is dissociation of the diazonium salt **23** to a carbocation and molecular nitrogen. The carbocation could then react with ethanol to give an aryl ethyl ether as a by-product (Eq. 21.19). Fortunately, ether formation is a minor process in the present reaction, and any of the ether that is produced is removed by recrystallization of the product.

(21.19)

Overview

The logic underlying the specific sequence of steps for making **24** is summarized as follows. Introducing the amino group as a director on the benzene ring provides a way to control the locations at which the various halogen substituents are introduced. When the directive effects of the amino group are no longer needed, this functionality is replaced with a hydrogen atom, the final step of the present synthesis. Before introducing a halogen atom onto the ring, the ring-activating power of the amino group is lessened by converting it to an acetamido group, which makes monohalogenation possible. The reason the bromo substituent is added first is that the acetamido group preferentially directs an electrophile to the *para* position. The chloro substituent was introduced next because chlorination of 4-bromoacetanilide is possible, whereas iodination is not. If the amino function had been regenerated *prior* to chlorinating and iodinating the ring, undesired polyhalogenation would have occurred. The iodo group is added last because the aromatic ring must be highly activated for this reaction to succeed, and there is only one activated position remaining where substitution can occur, making poly-iodination improbable.

EXPERIMENTAL PROCEDURES

Synthesis of 1-Bromo-3-Chloro-5-Iodobenzene

Purpose To demonstrate principles of multistep synthesis, protecting groups, and electrophilic aromatic substitution.

Note Throughout the sequence of reactions, quantities of reagents must be adjusted according to the amount of starting material that is available from the previous reaction, but you should *not* run the reactions on a larger scale than is indicated. If the amount of your starting material varies from that indicated, do *not* change the reaction *times* unless directed to do so by your instructor. You should be able to obtain sufficient quantities of product from each reaction to complete the entire sequence starting with the amount of acetanilide used in Step B.

A ■ *Preparation of Aniline and Acetanilide*

See Section 21.2 for Pre-Lab Exercises and Experimental Procedures for the preparation of aniline and acetanilide and Figures 21.4, 21.5, 8.7, and 15.31, respectively, for their spectra.

B ■ *Preparation of 4-Bromoacetanilide*

SAFETY ALERT

1. **Wear safety glasses or goggles and suitable protective gloves while performing the experiment.**

2. **Bromine is a hazardous chemical that may cause serious chemical burns. Do not breathe its vapors or allow it to come into contact with the skin. Perform all operations involving the transfer of solutions of bromine at a hood. If you get bromine on your skin, immediately wash the affected area with soap and warm water and soak the skin in 0.6 *M* sodium thiosulfate solution for up to 3 h if the burn is particularly serious.**

3. **Do *not* use acetone to rinse glassware containing residual bromine! This prevents formation of α-bromoacetone, a severe *lachrymator*. First rinse the glassware with aqueous sodium bisulfite.**

4. **Glacial acetic acid is a corrosive liquid. If this reagent comes in contact with your skin, thoroughly wash the area with large amounts of water.**

MINISCALE PROCEDURE

Preparation Refer to the online resources to answer Pre-Lab Exercises, access videos, and read the MSDSs for the chemicals used or produced in this procedure. Review Sections 2.9, 2.11, and 2.17.

Apparatus A 250-mL round-bottom flask, ice-water bath, apparatus for magnetic stirring, vacuum filtration, and *flameless* heating.

Setting Up Working at the hood, prepare a solution of 3.2 mL of bromine in 6 mL of *glacial* acetic acid in an Erlenmeyer flask. Dissolve 8.1 g of acetanilide in 30 mL of *glacial* acetic acid in the round-bottom flask containing a stirbar.

Bromination, Work-Up, and Isolation Add the bromine solution to the rapidly stirred solution of acetanilide over a period of 1–2 min; continue stirring the reaction mixture for about 10 min after completing the addition. Slowly add 100 mL of ice-cold water with stirring and then add just enough ice-cold saturated aqueous sodium bisulfite to discharge the color of the mixture that results. Cool the mixture in an ice-water bath and collect the product by vacuum filtration. If the crude product appears yellow, wash it with aqueous sodium bisulfite. In any case, wash it well with cold water and press it as dry as possible on the filter. The product may be recrystallized from methanol.★

Analysis Weigh the product and calculate its yield. Determine its melting point. Obtain IR and ¹H NMR spectra of your starting material and product and compare them with those of authentic samples (Figs. 8.7, 15.31, 21.12, and 21.13).

MICROSCALE PROCEDURE

Preparation Refer to the online resources to answer Pre-Lab Exercises, access videos, and read the MSDSs for the chemicals used or produced in this procedure. Review Sections 2.9, 2.11, and 2.17.

Apparatus A 5-mL conical vial, ice-water bath, apparatus for magnetic stirring, vacuum filtration, and *flameless* heating.

Setting Up Working at the hood, prepare a solution of 0.15 mL of bromine and 0.3 mL of *glacial* acetic acid in a test tube. Prepare a solution of 375 mg of acetanilide and 3 mL of *glacial* acetic acid in the conical vial equipped with a spinvane.

Bromination, Work-Up, and Isolation In one portion, add the bromine solution to the rapidly stirred solution of acetanilide and continue stirring for about 10 min at room temperature. Pour the reaction mixture into 25 mL of ice-cold water contained in a beaker that is equipped with a stirbar. Add just enough ice-cold saturated aqueous sodium bisulfite to discharge the color of the mixture. Stir the resulting mixture in an ice-water bath for 15 min and isolate the precipitated product by vacuum filtration.★ If the crude product appears yellow, wash it with aqueous sodium bisulfite. In any case, wash it well with cold water and press it as dry as possible on the filter. The product may be recrystallized from methanol.★

Analysis Weigh the product and calculate its yield. Determine its melting point. Obtain IR and ¹H NMR spectra of your starting material and product and compare them with those of authentic samples (Figs. 8.7, 15.31, 21.12, and 21.13).

WRAPPING IT UP

If the *filtrate* and *washings* contain residual bromine, as evidenced by a yellow color, add saturated aqueous sodium bisulfite until the solution is colorless. Flush the *decolorized solution* down the drain.

C ▪ *Preparation of 4-Bromo-2-Chloroacetanilide*

SAFETY ALERT

1. **Wear safety glasses or goggles and suitable protective gloves while performing the experiment.**

2. **Chlorine gas is evolved in this procedure. To minimize exposure to it, perform this reaction in the hood, if possible. Alternatively, invert a funnel over the reaction flask and attach the funnel to the water aspirator to remove the gaseous chlorine (Fig. 2.71b).**

3. *Concentrated* **hydrochloric acid and** *glacial* **acetic acid are corrosive liquids. Thoroughly wash any areas of your skin that may come in contact with them.**

MINISCALE PROCEDURE

Preparation Refer to the online resources to answer Pre-Lab Exercises, access videos, and read the MSDSs for the chemicals used or produced in this procedure. Review Sections 2.9, 2.10, 2.11, and 2.17.

Apparatus A 250-mL round-bottom flask, ice-water bath, apparatus for magnetic stirring, vacuum filtration, and *flameless* heating.

Setting Up Suspend 10.7 g of 4-bromoacetanilide in a stirred solution of 23 mL of *concentrated* hydrochloric acid and 28 mL of *glacial* acetic acid contained in the round-bottom flask. Heat the mixture gently with stirring until it becomes homogeneous and then cool the solution to 0 °C. Prepare a solution of 2.8 g of sodium chlorate in about 7 mL of water.

Chlorination, Work-Up, and Isolation Add the aqueous sodium chlorate solution dropwise to the stirred cold mixture over a period of about 5 min; some chlorine gas is evolved. After completing the addition, stir the reaction mixture at room temperature for 1 h and then collect the precipitate by vacuum filtration.★ Wash the crude product *thoroughly* with ice-cold water until the washings are neutral or nearly so. The product may be recrystallized from methanol.

Analysis Weigh the product and calculate its yield. Determine its melting point. Obtain its IR and ^{1}H NMR spectra and compare them with those of an authentic sample (Figs. 21.14 and 21.15).

MICROSCALE PROCEDURE

Preparation Refer to the online resources to answer Pre-Lab Exercises, access videos, and read the MSDSs for the chemicals used or produced in this procedure. Review Sections 2.9, 2.10, 2.11, and 2.17.

Apparatus A 5-mL conical vial, ice-water bath, apparatus for magnetic stirring, vacuum filtration, and *flameless* heating.

Setting Up Suspend 500 mg of 4-bromoacetanilide in a stirred solution of 1 mL of *concentrated* hydrochloric acid and 1.3 mL of *glacial* acetic acid contained in the conical vial. If necessary, heat the mixture gently with stirring until it becomes homogeneous and then cool the solution to 0 °C. Prepare a solution of 150 mg of sodium chlorate in about 0.5 mL of water.

Chlorination, Work-Up, and Isolation Add the aqueous sodium chlorate solution dropwise to the stirred cold mixture over a period of about 2 min; some chlorine gas is evolved. After completing the addition, stir the reaction mixture at room temperature for 1 h and then collect the precipitate by vacuum filtration.★ Wash the crude product *thoroughly* with ice-cold water until the washings are neutral or nearly so. The product may be recrystallized from methanol.

Analysis Weigh the product and calculate its yield. Determine its melting point. Obtain its IR and ^{1}H NMR spectra and compare them with those of an authentic sample (Figs. 21.14 and 21.15).

WRAPPING IT UP

If the *filtrate* and *washings* contain residual chlorine, as evidenced by formation of a blue-black color when a drop of the liquid is placed on starch/iodide paper, add saturated aqueous sodium bisulfite until the solution gives a negative test to starch/iodide paper. Flush the *decolorized solution* down the drain.

D ▪ *Preparation of 4-Bromo-2-Chloroaniline*

SAFETY ALERT

1. **Wear safety glasses or goggles and suitable protective gloves while performing the experiment.**

2. **Concentrated hydrochloric acid is a corrosive liquid, and 14 *N* aqueous sodium hydroxide is highly caustic. Thoroughly wash any areas of your skin that may come in contact with them.**

MINISCALE PROCEDURE

Preparation Refer to the online resources to answer Pre-Lab Exercises, access videos, and read the MSDSs for the chemicals used or produced in this procedure. Review Sections 2.9, 2.11, 2.17, and 2.22.

Apparatus A 250-mL round-bottom flask, ice-water bath, apparatus for heating under reflux, magnetic stirring, vacuum filtration, and *flameless* heating.

Setting Up Mix 11.2 g of crude 4-bromo-2-chloroacetanilide with 20 mL of 95% ethanol and 13 mL of *concentrated* hydrochloric acid in the flask. Add a stirbar and assemble the apparatus for heating under reflux.

Hydrolysis and Isolation Heat the mixture under reflux for *ca.* 0.5 h. At the end of this period, add 90 mL of *hot* water, swirl the mixture to dissolve any solids completely, and pour the hot solution onto about 150 g of ice contained in a beaker. Slowly add 12 mL of 14 *M* sodium hydroxide solution to the resulting mixture, stirring the mixture well during the addition. Add more ice if it has all melted before completing addition of the aqueous base. Check the pH of the reaction mixture to ensure that it is basic; if not, add more aqueous base. Isolate the solid by vacuum filtration, wash it thoroughly with ice-cold water, and press it as dry as possible on the filter.★

Purification The crude product may be recrystallized by first dissolving it in boiling methanol, allowing the solution to cool to below 40 °C, and adding cold water dropwise until crystallization begins. Should the product begin to oil out rather than crystallize, add a few drops of methanol to redissolve the oil. Cool the mixture thoroughly in an ice-water bath and collect the product by vacuum filtration, washing it with a *small* volume of ice-cold 1:1 methanol-water.

Analysis Weigh the product and calculate its yield. Determine its melting point. Obtain its IR and ^{1}H NMR spectra and compare them with those of an authentic sample (Figs. 21.16 and 21.17).

MICROSCALE PROCEDURE

Preparation Refer to the online resources to answer Pre-Lab Exercises, access videos, and read the MSDSs for the chemicals used or produced in this procedure. Review Sections 2.9, 2.11, 2.17, and 2.22.

Apparatus A 5-mL conical vial, ice-water bath, apparatus for heating under reflux, magnetic stirring, vacuum filtration, Craig tube filtration, and *flameless* heating.

Setting Up Mix 450 mg of crude 4-bromo-2-chloroacetanilide with 1.5–2 mL of 95% ethanol and 0.5 mL of *concentrated* hydrochloric acid in the conical vial. Add a spinvane and assemble the apparatus for heating under reflux.

Hydrolysis and Isolation Heat the mixture under reflux for *ca.* 0.5 h. At the end of this period, add 2 mL of *hot* water, stir the mixture to dissolve any solids completely, and pour the hot solution onto 3–4 g of ice contained in a beaker. Slowly add 0.5 mL of 14 *M* sodium hydroxide solution to the resulting mixture, stirring the mixture well during the addition. Add more ice if it has all melted before completing addition of the aqueous base. Check the pH of the reaction mixture to ensure that it is basic; if not, add more aqueous base. Isolate the solid by vacuum filtration, and thoroughly dry it.★

Purification The crude product may be recrystallized by first dissolving it in boiling methanol, allowing the solution to cool to below 40 °C, and adding cold water dropwise until crystallization begins. Should the product begin to oil out rather than crystallize, add a few drops of methanol to redissolve the oil. Cool the mixture thoroughly in an ice-water bath and collect the product by Craig tube or vacuum filtration, washing it with a *few drops* of ice-cold 1:1 methanol-water.

Analysis Weigh the product and calculate its yield. Determine its melting point. Obtain its IR and ^{1}H NMR spectra and compare them with those of an authentic sample (Figs. 21.16 and 21.17).

WRAPPING IT UP

Neutralize the *filtrate* and flush it down the drain.

E ▪ *Preparation of 4-Bromo-2-Chloro-6-Iodoaniline*

SAFETY ALERT

1. **Wear safety glasses or goggles and suitable protective gloves while performing the experiment.**
2. *Glacial* **acetic acid is a corrosive liquid, can cause burns, and is harmful if inhaled. Thoroughly wash with water any areas of your skin that may come in contact with it.**

MINISCALE PROCEDURE

Preparation Refer to the online resources to answer Pre-Lab Exercises, access videos, and read the MSDSs for the chemicals used or produced in this procedure. Review Sections 2.9, 2.10, 2.11, and 2.17.

Apparatus A 250-mL round-bottom flask, thermometer, ice-water bath, apparatus for magnetic stirring, vacuum filtration, and *flameless* heating.

Setting Up With magnetic stirring, dissolve 2.5 g of recrystallized 4-bromo-2-chloroaniline in 40 mL of *glacial* acetic acid in the round-bottom flask and add 10 mL of water to the mixture. In an Erlenmeyer flask, prepare a solution of 2.5 g of technical iodine monochloride in 10 mL of *glacial* acetic acid.

Iodination, Work-Up, and Isolation Over a period of 8–10 min, add the iodine monochloride solution to the stirred solution of the aniline. Heat the resulting mixture to 90 °C with continued stirring and then allow the solution to cool to about 50 °C. Add just enough saturated aqueous sodium bisulfite to turn the color of the mixture bright yellow; record the volume of sodium bisulfite solution required. Dilute the reaction mixture with enough extra water such that the *combined* volume of the aqueous sodium bisulfite *plus* the volume of added water is about 15 mL. Cool the reaction mixture in an ice-water bath and isolate the crude product that results by vacuum filtration. Wash the crystals with a small amount of ice-cold 33% aqueous acetic acid and then at least 50 mL of ice-cold water. Press the filter cake as dry as possible.★

Purification Recrystallize the crude product from acetic acid-water as follows. Mix the product with *glacial* acetic acid in the ratio of about 20 mL of *glacial* acetic acid per gram of product. Heat the mixture and slowly add 5 mL of water per gram of product to the mixture as it is heating. Terminate heating once dissolution has occurred and allow the solution to cool slowly to room temperature. Complete the crystallization by cooling the mixture in an ice-water bath and isolate the product by vacuum filtration. Wash the product with cold water and air-dry it.

Analysis Weigh the product and calculate its yield. Determine its melting point. Obtain its IR and ^{1}H NMR spectra and compare them with those of an authentic sample (Figs. 21.18 and 21.19).

MICROSCALE PROCEDURE

Preparation Refer to the online resources to answer Pre-Lab Exercises, access videos, and read the MSDSs for the chemicals used or produced in this procedure. Review Sections 2.9, 2.10, 2.11, and 2.17.

Apparatus A 25-mL Erlenmeyer flask, thermometer, ice-water bath, apparatus for magnetic stirring, vacuum filtration, and *flameless* heating.

Setting Up With magnetic stirring, dissolve 250 mg of recrystallized 4-bromo-2-chloroaniline in 4 mL of *glacial* acetic acid in the Erlenmeyer flask, and add 1 mL of water to the mixture. In a test tube, prepare a solution of 250 mg of technical iodine monochloride in 1 mL of *glacial* acetic acid.

Iodination, Work-Up, and Isolation Over a period of 3–5 min, add the iodine monochloride solution to the stirred solution of the aniline. Heat the resulting mixture to 90 °C with continued stirring and then allow the solution to cool to about 50 °C. Add just enough saturated aqueous sodium bisulfite to turn the color of the mixture bright yellow; record the volume of sodium bisulfite solution required. Dilute the reaction mixture with enough extra water such that the *combined* volume of the aqueous sodium bisulfite *plus* the volume of added water is about 3 mL. Cool the reaction mixture in an ice-water bath and isolate the crude product that results by vacuum filtration. Wash the crystals with a small amount of ice-cold 33% aqueous acetic acid and then with at least 5 mL of ice-cold water. Press the filter cake as dry as possible.★

Purification Using a test tube, recrystallize the crude product from acetic acid-water as follows. Mix the product with *glacial* acetic acid in the ratio of about 2 mL of *glacial* acetic acid per 100 mg of product. Heat the mixture and slowly add 0.5 mL of water per 100 mg of product to the mixture as it is heating. Terminate heating once dissolution has occurred and allow the solution to cool slowly to room temperature. Complete the crystallization by cooling the mixture in an ice-water bath and isolate the product by vacuum filtration. Wash the product with cold water and air-dry it.

Analysis Weigh the product and calculate its yield. Determine its melting point. Obtain its IR and ^{1}H NMR spectra and compare them with those of an authentic sample (Figs. 21.18 and 21.19).

WRAPPING IT UP

Flush the *aqueous filtrates* down the drain. Use solid sodium carbonate to neutralize the *filtrate* from the recrystallization before flushing it down the drain.

F ■ *Preparation of 1-Bromo-3-Chloro-5-Iodobenzene*

MINISCALE PROCEDURE

Preparation Refer to the online resources to answer Pre-Lab Exercises, access videos, and read the MSDSs for the chemicals used or produced in this procedure. Review Sections 2.9, 2.10, 2.11, 2.13, 2.16, 2.17, 2.22, and 2.29.

Apparatus A 250-mL and a 50-mL round-bottom flask, thermometer, ice-water bath, apparatus for heating under reflux, magnetic stirring, steam distillation, vacuum filtration, simple distillation, and *flameless* heating.

Setting Up Suspend 2.0 g of 4-bromo-2-chloro-6-iodoaniline in about 10 mL of *absolute* ethanol in the 250-mL round-bottom flask containing a stirbar. While stirring the mixture, add 4.0 mL of *concentrated* sulfuric acid dropwise. Equip the flask for heating under reflux.

Diazotization, Isolation, and Purification Add 0.7 g of powdered sodium nitrite in small portions through the top of the condenser. When the addition is complete, use a few drops of *absolute* ethanol to rinse any solid that may stick to the walls of the condenser into the flask and then heat the mixture under gentle reflux for about 10 min. Add 125 mL of hot water to the flask and equip it for steam distillation. Steam-distill the mixture, collecting about 50 mL of distillate, which should be clear and not contain any organic product. The desired product should form as a solid in the condenser, so monitor the condenser closely to ensure that it does not become plugged during the distillation. Should the condenser become clogged with solid, open it by running hot water or a slow stream of steam through the condenser jacket.

Upon completing the distillation, extract the distillate with two 15-mL portions of dichloromethane and put the extract in the 50-mL round-bottom flask. Use a few milliliters of dichloromethane to rinse any solid in the condenser into the flask containing the dichloromethane extract.* Remove the dichloromethane by simple distillation. Alternatively, use rotary evaporation or other techniques to remove the solvent. Recrystallize the residue from methanol or some other suitable solvent, and isolate the product by vacuum filtration.

Analysis Weigh the product and calculate its yield. Determine its melting point. Obtain its IR and ¹H NMR spectra, and compare them with those of an authentic sample (Figs. 21.20 and 21.21).

MICROSCALE PROCEDURE

Preparation Refer to the online resources to answer Pre-Lab Exercises, access videos, and read the MSDSs for the chemicals used or produced in this procedure. Review Sections 2.9, 2.10, 2.11, 2.13, 2.16, 2.17, 2.22, and 2.29.

Apparatus A 5-mL conical vial, ice-water bath, apparatus for heating under reflux, magnetic stirring, simple distillation, Craig tube filtration, and *flameless* heating.

Setting Up Suspend 200 mg of 4-bromo-2-chloro-6-iodoaniline in about 1.5 mL of absolute ethanol in the conical vial containing a spinvane. While stirring the mixture, add 0.4 mL of *concentrated* sulfuric acid dropwise.

Diazotization, Isolation, and Purification Add 0.07 g of powdered sodium nitrite in small portions to the vial. When the addition is complete, use a drop or two of *absolute* ethanol to rinse any solid that may stick to the sides of the vial into the solution. Equip the vial with a condenser and heat the mixture under gentle reflux for about 10 min. Add 2 mL of hot water to the vial and equip it for simple distillation. Steam-distill the mixture until about 1 mL of liquid remains in the vial. Remove distillate from the Hickman stillhead as necessary and place it in a conical test tube or vial.

Upon completing the distillation, combine all distillates in the conical vessel. Rinse any solid contained in the Hickman stillhead into the conical vessel with 1–2 mL of dichloromethane and separate the layers. Extract the aqueous layer once with 2 mL of dichloromethane and combine the dichloromethane solutions.★ Remove the dichloromethane by simple distillation. Alternatively, use rotary evaporation or other techniques to remove the solvent. Recrystallize the residue from methanol or some other suitable solvent in a Craig tube and isolate the product by Craig tube filtration.

Analysis Weigh the product and calculate its yield. Determine its melting point. Obtain its IR and ^{1}H NMR spectra and compare them with those of authentic samples (Figs. 21.20 and 21.21).

WRAPPING IT UP

Neutralize the *pot residue* from the steam distillation of the *reaction mixture* and flush the resulting *aqueous layer* and that obtained from the *steam distillate* down the drain. Pour the *organic layer* from the *pot residue,* the recovered *dichloromethane,* and the *filtrate* from recrystallization into the container for halogenated compounds.

EXERCISES

General Questions

1. Polyhalogenation is a potential problem at several stages of the synthesis of 1-bromo-3-chloro-5-iodobenzene. Describe what tactics are used to avoid polyhalogenation at each of these stages.

2. Which step(s) in the multistep synthesis of 1-bromo-3-chloro-5-iodobenzene from benzene is/are:

   a. a protection?

   b. a deprotection?

   c. a reduction?

   d. an electrophilic aromatic substitution reaction?

3. Using the concepts developed in the synthesis of 1-bromo-3-chloro-5-iodobenzene, outline synthetic procedures, starting from acetanilide, for preparing each of the following compounds:

   a. 1,3,5-tribromobenzene

   b. 2-bromo-4-chloro-6-iodophenol

   c. 2-bromo-4,6-dichloroaniline

   d. 3-bromotoluene

4. Why is the amino group a more powerful ring activator than the amido group?

5. In the sequence used, which steps demonstrate that the amido group is a more powerful *ortho-, para*-director than a bromo or chloro substituent?

6. Hydroxylic solvents, such as water or low-molar-mass alcohols, are often used to purify amides such as the acetanilides produced in this experiment. Briefly explain why these are used in preference to hydrocarbon solvents like petroleum ether.

Questions for Part A

See Questions 9–25 of Section 21.2.

Questions for Part B

7. Sodium bisulfite is used to discharge the yellow color produced in the electrophilic aromatic bromination of acetanilide.

   a. What is the source of the yellow color?

   b. Write the chemical equation for the reaction in which this yellow color is discharged.

8. Explain why electrophilic bromination of acetanilide produces 4- rather than 2-bromoacetanilide as the major product.

9. Compared to $Br_2/HOAc$, $Br_2/FeBr_3$ is a more powerful combination of reagents for effecting electrophilic aromatic bromination. What difficulty might arise if this combination were used for brominating acetanilide?

Refer to the online resources and use the spectra viewer and Tables 8.1–8.8 as needed to answer the blue-numbered questions on spectroscopy.

10. Account for the color change you observed upon reaction of bromine with acetanilide.

11. Using curved arrows to symbolize the flow of electrons, provide a stepwise mechanism for the bromination of acetanilide with $Br_2/HOAc$.

12. Consider the spectral data for 4-bromoacetanilide (Figs. 21.12 and 21.13).

    a. In the functional group region of the IR spectrum, specify the absorptions associated with the nitrogen-hydrogen bond, carbon-oxygen double bond, and the aromatic ring.

    b. In the ^1H NMR spectrum, assign the various resonances to the hydrogen nuclei responsible for them.

    c. For the ^{13}C NMR data, assign the various resonances to the carbon nuclei responsible for them.

13. How many resonances would you expect for the ^1H-decoupled ^{13}C spectrum of 2-bromoacetanilide?

14. Discuss the differences observed in the IR and NMR spectra of acetanilide and 4-bromoacetanilide that are consistent with the formation of the latter in this procedure.

Questions for Part C

15. What is the oxidation number of the chlorine atom in the chlorate ion, ClO_3, and in molecular chlorine, Cl_2, in Equation 21.14?

16. Why does an amide group activate an aromatic ring toward electrophilic aromatic substitution more than a halogen atom?

17. How might acetic acid interact with molecular chlorine to make it more electrophilic in character?

18. Would 4-bromoacetanilide or 4-bromo-2-chloroacetanilide be more reactive toward electrophilic substitution? Explain.

19. 4-Bromo-3-chloroacetanilide is a possible product from chlorination of 4-bromoacetanilide. Why is this isomer *not* produced to any significant extent?

Refer to the online resources and use the spectra viewer and Tables 8.1–8.8 as needed to answer the blue-numbered questions on spectroscopy.

20. Consider the spectral data for 4-bromo-2-chloroacetanilide (Figs. 21.14 and 21.15).

    a. In the functional group region of the IR spectrum, specify the absorptions associated with the nitrogen-hydrogen bond, the carbon-oxygen double bond, and the aromatic ring.

    b. In the ^1H NMR spectrum, assign the various resonances to the hydrogen nuclei responsible for them.

    c. For the ^{13}C NMR data, assign the various resonances to the carbon nuclei responsible for them.

21. Discuss the differences observed in the IR and NMR spectra of 4-bromoacetanilide and 4-bromo-2-chloroacetanilide that are consistent with the formation of the latter in this procedure.

Questions for Part D

22. What is the white precipitate that forms during the hydrolysis of 4-bromo-2-chloroacetanilide?

23. Why must aqueous base be added to isolate 4-bromo-2-chloroaniline from the aqueous acidic mixture?

24. Using curved arrows to symbolize the flow of electrons, provide a stepwise mechanism for the acid-catalyzed hydrolysis of 4-bromo-2-chloroacetanilide.

Refer to the online resources and use the spectra viewer and Tables 8.1– 8.8 as needed to answer the blue-numbered questions on spectroscopy.

25. Why is it necessary to remove the protecting group on nitrogen at this stage of the synthesis rather than as the last step?

26. Consider the spectral data for 4-bromo-2-chloroaniline (Figs. 21.16 and 21.17).

    a. In the functional group region of the IR spectrum, specify the absorptions associated with the amino group and the aromatic ring.

    b. In the ^{1}H NMR spectrum, assign the various resonances to the hydrogen nuclei responsible for them.

    c. For the ^{13}C NMR data, assign the various resonances to the carbon nuclei responsible for them.

27. Discuss the differences observed in the IR and NMR spectra of 4-bromo-2-chloroacetanilide and 4-bromo-2-chloroaniline that are consistent with the formation of the latter in this procedure.

Questions for Part E

28. The combination of reagents used in this experiment to iodinate 4-bromo-2-chloroaniline (**21**) is I–Cl/CH$_3$CO$_2$H because I$_2$/CH$_3$CO$_2$H is not as effective. Provide a mechanistic rationale to explain this fact.

29. a. Using curved arrows to symbolize the flow of electrons, provide a step-wise mechanism for the reaction that occurs between iodine monochloride and 4-bromo-2-chloroaniline.

    b. Suppose that this same reaction were carried out using bromine monochloride, BrCl. What electrophilic substitution reaction might occur were this reagent, rather than iodine monochloride, allowed to react with 4-bromo-2-chloroaniline? Explain.

30. Iodonium ion, I$^+$, is a less reactive electrophile than bromonium ion, Br$^+$. Explain why.

31. Halogen atoms deactivate the aromatic ring toward electrophilic substitution.

    a. Provide a mechanistic rationale to explain this observation.

    b. Which halogen is the most deactivating? Explain your answer.

32. Explain why halogen atoms direct electrophilic aromatic substitution to the *ortho*- and *para*-positions rather than the *meta*-position.

33. Why is iodine the last halogen substituent introduced onto the aromatic ring?

34. The pK_a of acetic acid is about 4.8, whereas that of 4-bromo-2-chloro-6-iodoanilinium ion is estimated to be about 3.

    a. Using these values, calculate K_{eq} for the acid-base reaction given below.

    b. Given that the anilinium ion should be soluble in polar media such as the combination of HOAc and H$_2$O used in recrystallizing 4-bromo-2-chloro-6-iodoaniline (**22**), explain how your calculated K_{eq} is consistent with the fact that **22** precipitates from aqueous acetic acid.

22

4-Bromo-2-chloro-6-iodoanilinium ion

Refer to the online resources and use the spectra viewer and Tables 8.1–8.8 as needed to answer the blue-numbered questions on spectroscopy.

35. Consider the spectral data for 4-bromo-2-chloro-6-iodoaniline (Figs. 21.18 and 21.19).

    a. In the functional group region of the IR spectrum, specify the absorptions associated with the amino group and the aromatic ring.

    b. In the ^{1}H NMR spectrum, assign the various resonances to the hydrogen nuclei responsible for them.

    c. For the ^{13}C NMR data, assign the various resonances to the carbon nuclei responsible for them.

36. Discuss the differences observed in the IR and NMR spectra of 4-bromo-2-chloroaniline and 4-bromo-2-chloro-6-iodoaniline that are consistent with the formation of the latter in this procedure.

Questions for Part F

37. Using curved arrows to symbolize the flow of electrons, provide a stepwise mechanism for the conversion of 4-bromo-2-chloro-6-iodoaniline (**22**) to the diazonium salt **23**.

38. The reduction of the diazonium salt **23** may involve transfer of hydrogen from ethanol, which is oxidized to acetaldehyde. Propose an experiment that would allow you to determine whether the hydrogen comes from the hydroxyl group or from Cl of ethanol.

39. Calculate the overall yield of 1-bromo-3-chloro-5-iodobenzene obtained in the sequence of reactions that you performed.

40. Arene diazonium salts are relatively stable in aqueous solutions below about 10 °C. At higher temperatures, or if allowed to dry, they may decompose explosively.

    a. Write the products of this decomposition reaction, using **23** as an example.

    b. What is the driving force for decomposition?

41. In addition to reduction, diazonium salts can be converted directly to various other useful aromatic functional groups. Predict the product formed when the diazonium salt **23** is treated with the following reagents:

    a. H_2O

    b. CuCl

    c. CuCN

    d. HBF_4

Refer to the online resources and use the spectra viewer and Tables 8.1–8.8 as needed to answer the blue-numbered questions on spectroscopy.

42. Consider the spectral data for 1-bromo-3-chloro-5-iodobenzene (Figs. 21.20 and 21.21).

    a. In the functional group region of the IR spectrum, specify the absorptions associated with the aromatic ring.

    b. In the ^{1}H NMR spectrum, assign as many as possible of the various resonances to the hydrogen nuclei responsible for them.

    c. For the ^{13}C NMR data, assign the various resonances to the carbon nuclei responsible for them.

43. Discuss the differences observed in the IR and NMR spectra of 1-bromo-3-chloro-5-iodobenzene and 4-bromo-2-chloro-6-iodobenzene that are consistent with the formation of the latter in this procedure.

SPECTRA

Starting Materials and Products

The IR and NMR spectra of acetanilide are provided in Figures 8.7 and 15.31, respectively.

Figure 21.12

IR spectrum of 4-bromoacetanilide (IR card).

Figure 21.13

NMR data for 4-bromoacetanilide (DMSO-d₆).

(a) ¹H NMR spectrum (300 MHz).
(b) ¹³C NMR data: δ 25.1, 115.5, 121.9, 132.5, 139.7, 169.5.

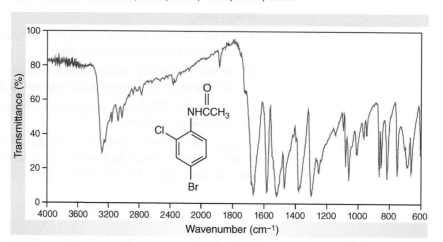

Figure 21.14

IR spectrum of 4-bromo-2-chloroacetanilide (IR card).

Figure 21.15

NMR data for 4-bromo-2-chloroacetanilide (CDCl₃).

(a) ¹H NMR spectrum (300 MHz).
(b) ¹³C NMR data: δ 24.8, 116.2, 122.6, 123.2, 130.7, 131.3, 133.7, 168.2.

Figure 21.16

IR spectrum of 4-bromo-2-chloroaniline (IR card).

Figure 21.17

NMR data for 4-bromo-2-chloroaniline (CDCl₃).

(a) ¹H NMR spectrum (300 MHz).
(b) ¹³C NMR data: δ 109.2, 116.8, 119.8, 130.4, 131.5, 142.1.

Figure 21.18

IR spectrum of 4-bromo-2-chloro-6-iodoaniline (IR card).

Figure 21.19

NMR spectrum of 4-bromo-2-chloro-6-iodoaniline (CDCl₃).

(a) ¹H NMR spectrum (300 MHz).
(b) ¹³C NMR data: δ 84.0, 109.7, 118.6, 132.4, 139.7, 143.2.

Figure 21.20

IR data for 1-bromo-3-chloro-5-iodobenzene (IR card).

Figure 21.21

NMR data for 1-bromo-3-chloro-5-iodobenzene (CDCl₃).

(a) ¹H NMR spectrum (300 MHz).

(b) ¹³C NMR data: δ 94.7, 123.7, 131.6, 136.3, 136.5, 138.7.

21.4 LIDOCAINE: SYNTHESIS OF AN ANESTHETIC AGENT

Lidocaine (**25**) is the *generic* name of an important member of a category of useful drugs that exhibit local anesthetic activity. One of the common trade names for this compound is Xylocaine, but its chemical name is 2-diethylamino-*N*-2′,6′-dimethylphenylacetamide. Two other prominent members of this family of anesthetic agents are procaine (**26**), known more commonly by the trade name Novocain, and benzocaine (**27**) (Sec. 20.2). In addition to having applications as topical anesthetics, lidocaine and procaine are effective in treating arrhythmia, a condition involving erratic beating of the heart. Procaine has a very short half-life in the body because the ester group is readily hydrolyzed by enzymes called lipases. However, the amide derivative procainamide (**28**) is more stable *in vivo* and serves as a useful cardiac depressant and antiarrhythmic agent. Interestingly, the antiarrhythmic properties of these compounds were discovered accidentally by cardiologists who used these drugs as anesthetics during surgical procedures—a nice example of serendipitous discovery in science.

25

Lidocaine

26

Procaine

27

Benzocaine

28

Procainamide

Examination of biologically active compounds often reveals certain structural or functional subunits that are common to other naturally and nonnaturally occurring substances having a broad spectrum of important medicinal applications. For example, compounds **25**, **26**, and **28** each possess a dialkylamino moiety, a group that is also found in the *tranquilizer* perazine (**29**), the *antidepressant* imipramine (**30**), the *antihistamine* chlorpheniramine (**31**), the *psychotomimetic* psilocybin (**32**), and the *antimalarial* chloroquine (**33**).

29
Perazine

30
Imipramine

31
Chlorpheniramine

32
Psilocybin

33
Chloroquine

34
Cocaine

With the exception of psilocybin, compounds **25–33** are all synthetic, or man-made. Indeed, many of the drugs commonly used today to treat diseases are the result of the ingenuity and synthetic skills of organic chemists who work in the pharmaceutical industry. Sometimes the inspiration for a new drug is based upon analogy with other known drugs, but new drugs are frequently developed from lead compounds found in nature. For example, compounds with beneficial pharmacological and medicinal properties are often found in folk medicines that are used by native populations in various parts of the world. Indeed, the anesthetic agents **25–27** were discovered during the course of developing structural mimics of cocaine (**34**), an alkaloid found in the coca plant indigenous to South America. Cocaine has addictive as well as anesthetic properties, so there were compelling reasons to develop a synthetic analog lacking addicting potential. An interesting account of the evolution of **25–27** from cocaine can be found in the Historical Highlight *Evolution of Synthetic Analgesics,* which is available online.

The present synthesis of lidocaine (**25**) from 2,6-dimethylnitrobenzene (**35**) requires three steps, as shown in Scheme 21.5. Some of the details of each step in the sequence are discussed in the following paragraphs.

Scheme 21.5

35
2,6-Dimethylnitrobenzene

36
2,6-Dimethylaniline

37
α-Chloro-2,6-dimethylacetanilide

25
Lidocaine

Reduction: Preparation of 2,6-Dimethylaniline (36)

The preparation of lidocaine starts with the chemical reduction of 2,6-dimethylnitrobenzene (**35**) to give 2,6-dimethylaniline (**36**) using stannous chloride as a reducing agent (Eq. 21.20). In this reaction, stannous chloride, $SnCl_2$, is oxidized to stannic chloride, $SnCl_4$. This reduction is similar to the reduction of nitrobenzene to aniline, which was discussed in detail in Section 21.2 (Scheme 21.3), and the side reactions associated with this process are discussed in that section.

35
2,6-Dimethyl-
nitrobenzene

38
2,6-Dimethylanilinium
hydrochloride

36
2,6-Dimethylaniline

(21.20)

The hydrochloride salt **38** that is formed upon reduction of the nitro group precipitates from the reaction mixture. Isolating this salt by filtration makes purification easy, as this operation frees **38** from contaminants such as unchanged **35** and all by-products that are soluble in the reaction medium. The **36** that is obtained after treating **38** with aqueous potassium hydroxide is sufficiently pure for use in the next step of the sequence. On the other hand, purification of aniline (**6**) from the chemical reduction of nitrobenzene is more tedious (Sec. 21.2, Part A).

Acylation: Preparation of α-Chloro-2,6-Dimethylacetanilide (37)

The substituted aniline **36** is next converted into the amide **37** by reaction with the difunctional reagent α-chloroacetyl chloride, $ClCH_2COCl$ (Eq. 21.21). Selective substitution at the acyl carbon atom in this step reflects the substantially greater reactivity of nucleophiles with acid chlorides as compared to alkyl chlorides. This results from the difference in electrophilicities and steric environments of the two possible sites for nucleophilic attack. Therefore, reaction at the carbon atom alpha to the carbonyl carbon atom to give **39** or the disubstitution product **40** is a minor competing reaction at best.

$$\text{(21.21)}$$

36
2,6-Dimethylaniline

37
α-Chloro-2,6-dimethylacetanilide

39

40

This reaction is performed in *glacial* acetic acid, so **36** will be in equilibrium with the salt **41** (Eq. 21.22). As the reaction proceeds, hydrochloric acid is liberated, so **36** is also partially converted to its hydrochloride salt **38**. Any **38** remaining at the end of the reaction period would contaminate the precipitated **37** because **38** is also insoluble in cold acetic acid. Hence, to avoid coprecipitation of the desired **37** and the hydrochloride acid salt **38**, aqueous sodium acetate is added to the warm reaction mixture to reestablish the process shown in Equation 21.22 as the only significant acid-base equilibrium involving **36**. The acetate salt **41** *is* soluble in cold aqueous acetic acid, so that filtration allows isolation of crystalline **37**. Any unreacted **36**, or its acetate salt **41**, will remain in solution and appear in the filtrate.

$$\text{(21.22)}$$

36
2,6-Dimethylaniline

41

Nucleophilic Substitution: Preparation of Lidocaine (25)

The reaction of the alkyl chloride **37** with diethylamine, which is an S_N2 process (Sec. 14.2), completes the synthetic sequence and is another example of a selective reaction. In this case, nucleophilic attack at the carbonyl function of the amido group is disfavored relative to reaction at the α-carbon atom because of the disruption of amide resonance that would accompany attack at the carbonyl group. The diethylamine serves the dual roles of acting as a nucleophile and as a base in this step. Namely, it displaces a chloride ion from **37**, and it reacts with the hydrogen chloride formed in the reaction (Eq. 21.23). Because of this latter reaction, it is necessary to use an excess of diethylamine to obtain high yields of lidocaine (**25**).

$$\text{(21.23)}$$

37
α-Chloro-2,6-dimethylacetanilide

25
Lidocaine

Diethylammonium hydrochloride

Isolation of **25** involves filtering the reaction mixture to remove the diethyl-ammonium hydrochloride, followed by extraction of the basic **25** into aqueous hydrochloride. All nonbasic contaminants, such as unchanged **37**, remain in the toluene solution. Lidocaine is liberated from its hydrochloride acid salt with aqueous base and then extracted into diethyl ether. After removal of the diethyl ether, the lidocaine is isolated as a low-melting solid or oil, depending upon its purity. To facilitate purification of lidocaine in this procedure, it is converted into its solid bisulfate salt **42** by reaction with sulfuric acid (Eq. 21.24). Many drugs that contain a basic site, like lidocaine, are sold in the form of their hydrochloride or sulfate salts because these salts are typically more stable and more soluble in water than the free base.

$$\text{(21.24)}$$

25
Lidocaine

42
Lidocaine bisulfate

EXPERIMENTAL PROCEDURES

Synthesis of Lidocaine

Purpose To demonstrate the principles of multistep synthesis and substitution reactions.

Note Quantities of reagents must be adjusted according to the amount of starting material that is available from the previous reaction, but you should *not* run the reactions on a larger scale than is indicated. If the amount of your starting material varies from that indicated, do *not* change the reaction *times* unless directed to do so by your instructor. You should be able to obtain sufficient quantities of product from each reaction to complete the entire sequence starting with the amount of 2,6-dimethylnitrobenzene used in the first step.

A ▪ *Preparation of 2,6-Dimethylaniline*

SAFETY ALERT

1. **Wear safety glasses or goggles and suitable protective gloves while performing the experiment.**

2. ***Concentrated* hydrochloric acid and *glacial* acetic acid are corrosive liquids that may cause burns if allowed to come in contact with your skin. Should contact occur, flood the affected area with water and then rinse it thoroughly with dilute aqueous sodium bicarbonate solution.**

3. **The solution of 8 *M* sodium hydroxide is highly caustic and can cause burns and loss of hair; avoid contact of this solution with your skin. Should contact occur, flood the affected area with water and then rinse it thoroughly with dilute aqueous acetic acid solution.**

4. **Be certain there are no open flames in the vicinity when you are working with diethyl ether. Use *flameless* heating when concentrating ethereal solutions.**

MINISCALE PROCEDURE

Preparation Refer to the online resources to answer Pre-Lab Exercises, access videos, and read the MSDSs for the chemicals used or produced in this procedure. Review Sections 2.9, 2.11, 2.13, 2.17, 2.21, 2.25, and 2.29.

Apparatus A 125- and 250-mL Erlenmeyer flask, 50-mL round-bottom flask, water bath, separatory funnel, apparatus for vacuum filtration, simple distillation, and *flameless* heating.

Setting Up Dissolve 33.9 g of stannous chloride dehydrate in 40 mL of *concentrated* hydrochloric acid contained in the 125-mL Erlenmeyer flask. Heating may be required for complete dissolution and the solution may appear milky. Dissolve 5.0 g of 2,6-dimethylnitrobenzene in 50 mL of *glacial* acetic acid contained in the 250-mL Erlenmeyer flask.

Reduction Add the solution stannous chloride dehydrate in *concentrated* hydrochloric acid *in one portion* to the solution of 2,6-dimethylnitrobenzene in *glacial* acetic acid. Swirl the resulting mixture briefly and let the resulting warm solution stand for 15 min.

Work-Up and Isolation Cool the reaction mixture to room temperature using a water bath and collect the precipitate that has formed by vacuum filtration.* Transfer the damp solid to an Erlenmeyer flask containing 25 mL of water. Add 40–50 mL of 8 *M* aqueous *potassium hydroxide* solution to the flask and check the pH of the mixture to ensure that it is strongly basic (pH >10). Cool the warm mixture to room temperature using a water bath and transfer it to a separatory funnel. Extract the aqueous mixture sequentially with two 15-mL portions of diethyl ether, using the first portion to rinse the Erlenmeyer flask in which the aqueous mixture was made basic. Wash the *combined* organic extracts sequentially with two 10-mL portions of water and then dry the organic layer over several spatula-tips full of *anhydrous* sodium sulfate.* Filter or carefully decant the solution into the *tared* round-bottom flask and remove most of the diethyl ether by simple distillation. Alternatively, use rotary evaporation or other techniques to concentrate the solution. Remove the final traces of diethyl ether by connecting the cool stillpot to a vacuum source.

Analysis Weigh the product and calculate its yield. Obtain IR and ^{1}H NMR spectra of your starting material and product, and compare them with those of authentic samples (Figs. 21.22–21.25).

MICROSCALE PROCEDURE

Preparation Refer to the online resources to answer Pre-Lab Exercises, access videos, and read the MSDSs for the chemicals used or produced in this procedure. Review Sections 2.9, 2.11, 2.13, 2.17, 2.21, 2.25, and 2.29.

Apparatus A 25-mL Erlenmeyer flask, a 5-mL conical vial, a 13-mm × 100-mm test tube, water bath, separatory funnel, apparatus for vacuum filtration, simple distillation, and *flameless* heating.

Setting Up Dissolve 3.4 g of stannous chloride dehydrate in 4 mL of *concentrated* hydrochloric acid contained in the test tube. Heating may be required for complete dissolution and the solution may appear milky. Dissolve 0.5 g of 2,6-dimethylnitrobenzene in 5 mL of *glacial* acetic acid contained in the Erlenmeyer flask.

Reduction Add the solution of stannous chloride dehydrate in *concentrated* hydrochloric acid *in one portion* to the solution of 2,6-dimethylnitrobenzene in acetic acid. Swirl the resulting mixture briefly and let the resulting warm solution stand for 15 min.

Work-Up and Isolation Cool the reaction mixture to room temperature using a water bath and collect the precipitate that has formed by vacuum filtration.★ Transfer the damp solid to an Erlenmeyer flask containing 2.5 mL of water. Add 4–5 mL of 8 *M* aqueous potassium hydroxide solution to the flask and check the pH of the mixture to ensure that it is strongly basic (pH >10). Cool the warm mixture to room temperature using a water bath and transfer the mixture to a separatory funnel. Extract the aqueous mixture sequentially with two 1.5-mL portions of diethyl ether, using the first portion to rinse the Erlenmeyer flask in which the aqueous mixture was made basic. Wash the *combined* organic extracts sequentially with two 1-mL portions of water and then dry the organic layer over several microspatula-tips full of *anhydrous* sodium sulfate.★ Filter or carefully decant the solution into the *tared* conical vial, and remove most of the diethyl ether by simple distillation. Alternatively, use rotary evaporation or other techniques to concentrate the solution. Remove the final traces of diethyl ether by connecting the cool stillpot to a vacuum source.

Analysis Weigh the product and calculate its yield. Obtain IR and ¹H NMR spectra of your starting material and product and compare them with those of authentic samples (Figs. 21.22–21.25).

WRAPPING IT UP

Combine all of the *aqueous solutions* and neutralize them with either dilute sodium hydroxide or dilute hydrochloric acid. Flush the combined solutions down the drain with excess water. Place the *ethereal distillate* in a container for nonhalogenated organic solvents. Put the *sodium sulfate* on a tray in the hood, and after the diethyl ether evaporates, pour it into the container for nonhazardous solids.

B ■ *Preparation of α-Chloro-2,6-Dimethylacetanilide*

SAFETY ALERT

1. Wear safety glasses or goggles and suitable protective gloves while performing the experiment.

2. *Glacial* acetic acid is a corrosive liquid. Should contact with your skin occur, flood the affected area with water and rinse it thoroughly with dilute aqueous sodium bicarbonate solution.

3. *α*-Chloroacetyl chloride is irritating to the mucous membranes and to the skin. Work at the ventilation hood when handling and transferring containers of this chemical. Should contact with your skin occur, flood the affected area with water and rinse it thoroughly with dilute aqueous sodium bicarbonate solution.

MINISCALE PROCEDURE

Preparation Refer to the online resources to answer Pre-Lab Exercises, access videos, and read the MSDSs for the chemicals used or produced in this procedure. Review Sections 2.9, 2.10, 2.11, 2.17, and 2.26.

Apparatus Two 250-mL Erlenmeyer flasks, ice-water bath, apparatus for vacuum filtration, and *flameless* heating.

Setting Up Dissolve 3.0 g of 2,6-dimethylaniline in 20 mL of *glacial* acetic acid contained in the Erlenmeyer flask. Dissolve 4.3 g of sodium acetate trihydrate in 80 mL of water in the other Erlenmeyer flask.

Reaction Add 2.8 g of *α*-chloroacetyl chloride to the solution of 2,6-dimethylaniline in acetic acid and swirl the solution to mix the reactants. Allow the reaction to proceed for 15 min with occasional swirling. Add the aqueous solution of sodium acetate in one portion with swirling.

Work-Up and Isolation Cool the resulting mixture in an ice-water bath and collect the product by vacuum filtration. Rinse the filter cake with water until the odor of acetic acid can no longer be detected. Press the solid with a clean spatula while the vacuum source is still attached to dry it as completely as possible, and then air-dry it for at least 24 h.

Analysis Weigh the product and calculate its yield. Determine its melting point (reported mp, 145–146 °C). The product may be recrystallized from a small volume of toluene. Obtain IR and ^{1}H NMR spectra of your product and compare them with those of an authentic sample (Figs. 21.26 and 21.27).

MICROSCALE PROCEDURE

Preparation Refer to the online resources to answer Pre-Lab Exercises, access videos, and read the MSDSs for the chemicals used or produced in this procedure. Review Sections 2.9, 2.10, 2.11, 2.17, and 2.26.

Apparatus Two 25-mL Erlenmeyer flasks, ice-water bath, apparatus for vacuum filtration, and *flameless* heating.

Setting Up Dissolve 0.3 g of 2,6-dimethylaniline in 2 mL of *glacial* acetic acid contained in an Erlenmeyer flask. Dissolve 0.45 g of sodium acetate trihydrate in 8 mL of water in the other Erlenmeyer flask.

Reaction Add 0.28 g of α-chloroacetyl chloride to the solution of 2,6-dimethylaniline in acetic acid and swirl the solution to mix the reactants. Allow the reaction to proceed for 15 min with occasional swirling. Add the aqueous solution of sodium acetate in one portion with swirling.

Work-Up and Isolation Cool the resulting mixture in an ice-water bath and collect the product by vacuum filtration. Rinse the filter cake with water until the odor of acetic acid can no longer be detected. Press the solid with a clean spatula while the vacuum source is still attached to dry it as completely as possible, and then air-dry it for at least 24 h.

Analysis Weigh the product and calculate its yield. Determine its melting point (reported mp, 145–146 °C). Obtain IR and ^{1}H NMR spectra of your product and compare them with those of an authentic sample (Figs. 21.26 and 21.27).

WRAPPING IT UP

Neutralize the *filtrate* with dilute sodium hydroxide and flush it down the drain with excess water.

C ■ *Preparation of Lidocaine*

SAFETY ALERT

1. **Wear safety glasses or goggles and suitable protective gloves while performing the experiment.**
2. **Diethylamine is an unpleasant-smelling and corrosive liquid. Measure it out in a hood. Should any of this chemical come in contact with the skin, flood the affected area with cold water.**
3. **The solution of 8 *M* potassium hydroxide is highly caustic and can cause burns and loss of hair; avoid contact of this solution with your skin. Should contact occur, flood the affected area with water and rinse it thoroughly with dilute aqueous acetic acid solution.**
4. **Be certain there are no open flames in the vicinity when you are working with diethyl ether. Use *flameless* heating when concentrating diethyl ether solutions.**

MINISCALE PROCEDURE

Preparation Refer to the online resources to answer Pre-Lab Exercises, access videos, and read the MSDSs for the chemicals used or produced in this procedure. Review Sections 2.7, 2.9, 2.10, 2.11, 2.17, 2.21, 2.22, 2.28, and 2.29.

Apparatus A 125-mL round-bottom flask, 250-mL Erlenmeyer flask, an ice-water bath, separatory funnel, apparatus for heating under reflux, magnetic stirring, vacuum filtration, simple distillation, and *flameless* heating.

Setting Up Oven-dry the glass apparatus for heating under reflux. Dissolve 2.2 g of dry α-chloro-2,6-dimethylacetanilide in 30 mL of toluene in the dry round-bottom flask containing a stirbar. Assemble the apparatus for heating under reflux.

Reaction Add 2.4 g of diethylamine to the solution of α-chloro-2,6-dimethylacet-anilide with stirring, and then heat the reaction mixture under reflux for 90 min. Allow the mixture to cool to room temperature, and then cool it briefly in an ice-water bath.

Work-Up and Isolation Filter the reaction mixture by vacuum filtration and rinse the filter cake with about 5 mL of toluene. Transfer the combined filtrate and washing to a separatory funnel and extract it with two 20-mL portions of 3 *M* hydrochloric acid. Shake the funnel vigorously with venting. Combine the two acidic aqueous extracts in an Erlenmeyer flask and add about 25 mL of 8 *M* potassium hydroxide solution to make the mixture strongly basic to pHydrion paper. Cool the alkaline mixture thoroughly in an ice-water bath. Transfer the chilled alkaline solution to a separatory funnel. Extract the aqueous mixture sequentially with two 15-mL portions of diethyl ether, using the first portion to rinse the Erlenmeyer flask in which the aqueous mixture was basified. The extractions should be carried out with vigorous shaking and frequent venting. Wash the *combined* organic extracts with 25 mL of water and dry the organic layer over several spatula-tips full of *anhydrous* sodium sulfate.★ Filter or carefully decant the solution into a dry, *tared* round-bottom flask, and remove most of the diethyl ether by simple distillation. Alternatively, use rotary evaporation or other techniques to concentrate the solution. Remove the final traces of solvent by connecting the cool stillpot to a vacuum source.

Salt Formation Dissolve the crude lidocaine in diethyl ether (10 mL of solvent per g of solute; *no flames!*) and then add a solution of 2 mL of 2.2 *M* sulfuric acid in ethanol per g of solute. Mix the solutions thoroughly and scratch at the air-liquid interface with a glass rod to induce crystallization. Dilute the mixture with an equal volume of reagent-grade acetone and isolate the precipitated salt by vacuum filtration. Rinse the filter cake with a few milliliters of cold reagent-grade acetone and then air-dry and weigh the product. The lidocaine bisulfate may be recrystallized by dissolving it in an equal weight of hot water and then adding *in one portion* a volume of reagent-grade acetone equal to 20 times the volume of this aqueous solution. Swirl the mixture briefly and then allow the solution to stand until crystallization is complete; you may need to scratch at the air-liquid interface with a glass rod to induce crystallization. Cool the solution in an ice-water bath for 15 min to complete crystallization.

Analysis Weigh the lidocaine bisulfate and calculate its yield. Determine its melting point (reported mp, 210–212 °C). (*Caution:* Do *not* use a liquid-bath melting-point apparatus to determine the melting point; use a metal block apparatus, as illustrated in Fig. 2.19, Sec. 2.7.) Obtain IR and ^{1}H NMR spectra of your product and compare them with those of an authentic sample (Figs. 21.28 and 21.29).

MICROSCALE PROCEDURE

Preparation Refer to the online resources to answer Pre-Lab Exercises, access videos, and read the MSDSs for the chemicals used or produced in this procedure. Review Sections 2.7, 2.9, 2.10, 2.11, 2.17, 2.21, 2.22, 2.28, and 2.29.

Apparatus A 5-mL conical vial, 25-mL Erlenmeyer flask, two screw-cap centrifuge tubes, an ice-water bath, apparatus for heating under reflux, magnetic stirring, vacuum filtration, simple distillation, and *flameless* heating.

Setting Up Oven-dry the glass apparatus for heating under reflux. Dissolve 0.2 g of dry α-chloro-2,6-dimethylacetanilide in 2.5 mL of toluene in the dry conical vial containing a spinvane. Assemble the apparatus for heating under reflux.

Reaction Add 0.22 g of diethylamine to the solution of α-chloro-2,6-dimethylacetanilide with stirring, and then heat the reaction mixture under reflux for 90 min. Allow the mixture to cool to room temperature and then cool it briefly in an ice-water bath.

Work-Up and Isolation Filter the reaction mixture by vacuum filtration and rinse the filter cake with about 0.5 mL of toluene. Transfer the combined filtrate and washing to a screw-cap centrifuge tube and extract it with two 1.5-mL portions of 3 *M* hydrochloric acid. Shake the tube vigorously with venting. Combine the two acidic aqueous extracts in the Erlenmeyer flask and add about 2.0 mL of 8 *M* potassium hydroxide solution to make the mixture strongly basic to pHydrion paper. Cool the alkaline mixture thoroughly in an ice-water bath. Transfer the chilled alkaline solution to a screw-cap centrifuge tube. Extract the aqueous mixture sequentially with two 1.5-mL portions of diethyl ether, using the first portion to rinse the Erlenmeyer flask in which the aqueous mixture was basified. The extractions should be carried out with vigorous shaking and frequent venting. Wash the *combined* organic extracts with 2 mL of water and dry the organic layer over several microspatula-tips full of *anhydrous* sodium sulfate.★ Filter or carefully decant the solution into a dry, *tared* conical vial, and remove most of the diethyl ether by simple distillation. Alternatively, use rotary evaporation or other techniques to concentrate the solution. Remove the final traces of solvent by connecting the cool stillpot to a vacuum source.

Salt Formation Dissolve the crude lidocaine in diethyl ether (1 mL of solvent per 0.1 g of solute; *no flames!*) and then add a solution of 0.2 mL of 2.2 *M* sulfuric acid in ethanol per 0.1 g of solute. Mix the solutions thoroughly and scratch at the air-liquid interface with a glass rod to induce crystallization. Dilute the mixture with an equal volume of reagent-grade acetone and isolate the precipitated salt by vacuum filtration. Rinse the filter cake with a few milliliters of cold reagent-grade acetone and then air-dry and weigh the product. The lidocaine bisulfate may be recrystallized by dissolving it in an equal weight of hot water and then adding *in one portion* a volume of reagent-grade acetone equal to 20 times the volume of this aqueous solution. Swirl the mixture briefly and then allow the solution to stand until crystallization is complete; you may need to scratch at the air-liquid interface with a glass rod to induce crystallization. Cool in an ice-water bath for 15 min to complete crystallization.

Analysis Weigh the lidocaine bisulfate and calculate its yield. Determine its melting point (reported mp, 210–212 °C). (*Caution:* Do *not* use a liquid-bath melting-point apparatus to determine the melting point; use a metal block apparatus, as illustrated in Fig. 2.19, Sec. 2.7.) Obtain IR and ^{1}H NMR spectra of your product and compare them with those of an authentic sample (Figs. 21.28 and 21.29).

WRAPPING IT UP

Neutralize the *aqueous solution* with *dilute* hydrochloric acid and flush it down the drain with excess water. Place the *combined diethyl ether distillate and filtrates* in a container for nonhalogenated organic solvents. Put the *sodium sulfate* on a tray in the hood, and after the diethyl ether evaporates, pour it and the *diethylammonium hydrochloride* into the container for nonhazardous solids.

EXERCISES

General Questions

1. 2,6-Dimethylnitrobenzene (**35**) may be prepared in one step from 1,3-dimethylbenzene. What reagents would be necessary to effect this transformation, and would this reaction give **35** as the exclusive product? Explain your answer.

2. Given the yields that you obtained in the laboratory, calculate the number of moles of 2,6-dimethylnitrobenzene (**35**) that would be needed to prepare one mole of lidocaine bisulfate (**42**).

Questions for Part A

3. Write the balanced equation for the reduction of 2,6-dimethylnitrobenzene (**35**) to give **36** using stannous chloride, SnCl$_2$, in *concentrated* hydrochloric acid.

4. What is the precipitate that is originally collected in the reduction of 2,6-dimethylnitrobenzene (**35**) by stannous chloride?

5. What would be the consequence of failing to make the aqueous solution basic prior to extracting the 2,6-dimethylaniline (**36**) into ether?

6. Write structures for at least two organic by-products expected to be produced by the reduction of 2,6-dimethylnitrobenzene (**35**) with stannous chloride.

7. Write a mechanism for the reduction of **35** into **36**. Show all electron and proton transfer steps using curved arrows to symbolize the flow of electrons.

8. Consider the spectral data for 2,6-dimethylnitrobenzene (Figs. 21.22 and 21.23).

    a. In the IR spectrum, specify the absorptions in the functional group region associated with the nitro group and with the aromatic ring.

    b. In the ^{1}H NMR spectrum, assign the various resonances to the hydrogen nuclei responsible for them.

    c. For the ^{13}C NMR data, assign the various resonances to the carbon nuclei responsible for them.

9. Consider the spectral data for 2,6-dimethylaniline (Figs. 21.24 and 21.25).

    a. In the IR spectrum, specify the absorptions in the functional group region associated with the nitrogen-hydrogen bonds and with the aromatic ring.

    b. In the ^{1}H NMR spectrum, assign the various resonances to the hydrogen nuclei responsible for them.

Refer to the online resources and use the spectra viewer and Tables 8.1–8.8 as needed to answer the blue-numbered questions on spectroscopy.

c. For the ^{13}C NMR data, assign the various resonances to the carbon nuclei responsible for them.

10. Discuss the differences observed in the IR and NMR spectra of 2,6-dimethyl-nitrobenzene and 2,6-dimethylaniline that are consistent with the formation of the latter in this procedure.

Questions for Part B

11. Why would ethanol be a poor choice of a solvent for the reaction between 2,6-dimethylaniline (**36**) and α-chloroacetyl chloride?

12. Why is the anilide **37** much less basic than 2,6-dimethylaniline (**36**)?

13. What organic by-product(s) might be formed if **36** were allowed to react with two moles of α-chloroacetyl chloride?

14. What factor(s) favor attack of 2,6-dimethylaniline (**36**) at the carbonyl carbon atom rather than the α-carbon atom of α-chloroacetyl chloride?

15. Using curved arrows to symbolize the flow of electrons, write a stepwise mechanism for the reaction of **36** with α-chloroacetyl chloride.

Refer to the online resources and use the spectra viewer and Tables 8.1–8.8 as needed to answer the blue-numbered questions on spectroscopy.

16. Consider the spectral data for α-chloro-2,6-dimethylacetanilide (Figs. 21.26 and 21.27).

a. In the IR spectrum, specify the absorptions in the functional group region associated with the nitrogen-hydrogen bond, with the carbon-oxygen double bond, and with the aromatic ring.

b. In the ^{1}H NMR spectrum, assign the various resonances to the hydrogen nuclei responsible for them.

c. For the ^{13}C NMR data, assign the various resonances to the carbon nuclei responsible for them.

17. Discuss the differences observed in the IR and NMR spectra of 2,6-dimethylaniline and α-chloro-2,6-dimethylacetanilide that are consistent with the formation of the latter in this procedure.

Questions for Part C

18. What would be the expected effect on yield if only one mole of diethylamine per mole of **37** were used in the final step of the preparation of lidocaine? Explain.

19. What side reaction(s) would be expected if **37** were wet when the reaction with diethylamine was performed?

20. What is the solid that is isolated initially in the reaction between **37** and diethylamine?

21. Why is **37** less susceptible to nucleophilic attack at the carbonyl carbon atom than is α-chloroacetyl chloride?

22. Why does sulfuric acid protonate the nitrogen atom of the diethylamine group of lidocaine preferentially to that of the amido group, as shown by formation of **42**?

Refer to the online resources and use the spectra viewer and Tables 8.1–8.8 as needed to answer the blue-numbered questions on spectroscopy.

23. Consider the spectral data for lidocaine bisulfate (Figs. 21.28 and 21.29).

a. In the IR spectrum, specify the absorptions in the functional group region associated with the nitrogen-hydrogen bond, with the carbon-oxygen double bond, and with the aromatic ring.

b. In the ^{1}H NMR spectrum, assign the various resonances to the hydrogen nuclei responsible for them.

c. For the ^{13}C NMR data, assign the various resonances to the carbon nuclei responsible for them.

24. Discuss the differences observed in the IR and NMR spectra of α-chloro-2,6-dimethylacetanilide and lidocaine bisulfate that are consistent with the formation of the latter in this procedure.

SPECTRA

Starting Materials and Products

Figure 21.22
IR spectrum of 2,6-dimethylnitro-benzene (neat).

Figure 21.23
NMR data for 2,6-dimethylnitro-benzene (CDCl₃).

(a) ^{1}H NMR spectrum (300 MHz).
(b) ^{13}C NMR data: δ 17.3, 128.8, 129.4, 129.9, 153.4.

Figure 21.24
IR spectrum of 2,6-dimethylaniline (IR card).

Figure 21.25
NMR data for 2,6-dimethylaniline (CDCl₃).

(a) ¹H NMR spectrum (300 MHz).
(b) ¹³C NMR data: δ 17.3, 117.6, 121.3, 128.0, 142.5.

Figure 21.26
IR spectrum of α-chloro-2,6-dimethylacetanilide (IR card).

Figure 21.27

NMR data for α-chloro-2,6-dimethylacetanilide (CDCl₃).

(a) ¹H NMR spectrum (300 MHz).
(b) ¹³C NMR data: δ 18.3, 42.8, 127.9, 128.3, 132.6, 135.3, 164.3.

Figure 21.28

IR spectrum of lidocaine bisulfate (IR card).

Figure 21.29

NMR data for lidocaine bisulfate (DMSO-d₆).

(a) ¹H NMR spectrum (300 MHz).
(b) ¹³C NMR data: δ 8.8, 18.1, 48.5, 52.7, 127.1, 129.9, 133.6, 135.0, 163.3.

CHAPTER 22

Polymers

Bigger is better? You may think so after performing experiments in this chapter because they will give you the opportunity to explore various aspects of the chemistry of **polymers**, molecules having molar masses in the tens of thousands. In addition to your everyday use of polymers in a variety of forms, including the fibers in the clothing you wear and the adhesive on Post-it notes that you stick on everything, you yourself are constantly manufacturing **biopolymers**, in the form of hair, skin, DNA, and RNA, for example. Although in this chapter we can provide only a small sampling of the wealth of chemistry surrounding polymers, it should give you a sense of this fascinating branch of organic chemistry.

22.1 INTRODUCTION

Polymers are a class of molecules characterized by their high molar masses, which range from the thousands to the hundreds of thousands, and by the presence of simple repeating structural units called **monomers**. Because of their large size, polymers are often referred to as **macromolecules**. A polymer comprised of a *single* recurring monomer, M, is termed a **homopolymer**, whereas one containing at least two structurally distinct monomeric units, M_1 and M_2, distributed at random in the molecule is called a **heteropolymer** or **copolymer**. These two classes of macromolecules are represented by **1** and **2**, respectively.

$$R-M-(M)_n-M-R \qquad R-M_1-(M_1-M_2)_n-M_2-R$$

1   **2**

Homopolymer   Copolymer

Polymers are found in nature and may also be produced by laboratory synthesis. Important examples of naturally occurring macromolecules, or **biopolymers**, are **proteins**, **polysaccharides**, **terpenes**, and **nucleic acids**. General representations of these substances are provided by structures **3–6**, respectively, in which the monomeric subunits of an α-amino acid, **3**, a pyranose, **4**, an isoprene, **5**, and a ribonucleotide phosphate, **6**, are seen. Synthetic, or man-made, polymers are represented by the myriad **plastics**, **elastomers**, and **fibers** that are commonplace in contemporary society.

Two primary methods are commonly used to convert monomers into synthetic polymers. In the older literature, these techniques are referred to as **addition** and **condensation polymerization**, but because of ambiguities in these terms, the preferred names now are **chain-reaction polymerization** and **step-growth**

3

4

5

6

(Base = purine or pyrimidine)

polymerization, respectively. As discussed in Sections 22.2 and 22.3, the major distinction between these two types of polymerization is the general mechanism by which the polymer forms.

The term *addition polymerization* arose because such polymers are produced by combining a large number of monomer molecules through addition reactions. For example, the self-addition of thousands of ethylene molecules yields polyethylene (**7**; see the Historical Highlight *Discovery of Polyethylene and Nylon*, which is available online), a homopolymer (Eq. 22.1). Another homopolymer is polystyrene (**9**), which is formed by self-addition of styrene (**8**) as shown in Equation 22.2. We note that representing the molecular formula of the polymer as essentially n times that of the monomer, as shown in these two equations, is only a slight oversimplification of the actual formula. Because n is such a large number, inclusion of the elemental compositions of the end-groups, R, that appear at the termini of the polymeric chain makes an insignificant change in the molecular formula in comparison with the rest of the molecule.

$$n\ CH_2{=}CH_2 \xrightarrow[\text{and/or heat}]{\text{catalyst}} \left(CH_2{-}CH_2\right)_n \qquad (22.1)$$

Ethylene
7
$n = 10{,}000{-}30{,}000$
Polyethylene

$$\qquad (22.2)$$

8
Styrene

9
Polystyrene

Some other common addition polymers have trade names that do not indicate their structure. For example, Teflon and Plexiglas are homopolymers of tetrafluoroethylene (**10**), and methyl methacrylate (**11**), respectively.

$$CF_2{=}CF_2 \qquad\qquad CH_2{=}C\begin{smallmatrix} CO_2CH_3 \\ \\ CH_3 \end{smallmatrix}$$

10
Tetrafluoroethylene

11
Methyl methacrylate

Copolymers are produced from a *mixture* of monomers, as noted earlier. For example, Saran (**14**), a widely used plastic film, is made by polymerizing a mixture of vinyl chloride (**12**) and vinylidene chloride (**13**), as depicted in Equation 22.3. You should be aware that the abbreviated formula of **14** is not meant to imply that the two monomeric units appear as a sequence of two distinct blocks, each of which individually represents a homopolymer. Although such **block copolymers** can be produced by special techniques, copolymers usually have the two monomers distributed randomly along the chain, as in **15.**

$$n\ CH_2{=}CH{-}Cl\ +\ m\ CH_2{=}CCl_2\ \longrightarrow\ {-}(CH_2{-}CHCl)_n(CH_2{-}CCl_2)_m{-} \quad (22.3)$$

12	**13**	**14**
Vinyl chloride	Vinylidene chloride	Saran

$$R{-}M_1{-}(M_2{-}M_2{-}M_1{-}M_2{-}M_1{-}M_1{-}M_1)_n{-}M_2{-}R$$
15

The term *condensation polymerization* was originally used because this form of polymerization involves condensation reactions. Such transformations normally involve combining two functionalized molecules through an addition-elimination process, termed nucleophilic acyl substitution (Sec. 20.1), that results in the loss of a small molecule, H–L, such as water (Eq. 22.4). The aldol condensation (Sec. 18.3) and esterification (Sec. 20.2) are examples of such reactions. In the case of condensation or step-growth polymerization, *di*functionalized substrates are required as monomers, as illustrated in Equation 22.5.

$$R{-}C({=}O){-}L\ +\ HNu\ \longrightarrow\ R{-}C({=}O){-}Nu\ +\ HL \quad (22.4)$$

$$L{-}C({=}O){\sim}C({=}O){-}L\ +\ HNu{\sim}NuH\ \longrightarrow\ {-}(C({=}O){\sim}C({=}O){-}Nu{\sim}Nu){-}_n \quad (22.5)$$

The following sections contain discussions of the mechanisms of chain-reaction and step-growth polymerization. The associated experimental procedures illustrate the preparation of polymers derived from each type of polymerization.

22.2 CHAIN-REACTION POLYMERIZATION

As the name implies, *chain-reaction polymerization* is a chain reaction in which the initiator may be a cation, anion, or free radical. An example of **cationic polymerization** is found in the polymerization of isobutylene (2-methylpropene) in the presence of protic or Lewis acid catalysts to give poly(isobutylene) (**16**), as depicted in Equation 22.6. The conversion of acrylonitrile to poly(acrylonitrile) (**17**) using sodium amide, a strong base, represents **anionic polymerization** (Eq. 22.7).

$$(CH_3)_2C\!=\!CH_2 \xrightarrow{\;H^+\;} (CH_3)_2\overset{+}{C}\!-\!CH_3 \xrightarrow{\;H_2C\!=\!C(CH_3)_2\;} (CH_3)_3C\!-\!CH_2\overset{CH_3}{\underset{CH_3}{\overset{\displaystyle|}{\underset{\displaystyle|}{\overset{+}{C}}}}}$$

Isobutylene

$$n\, H_2C\!=\!C(CH_3)_2 \qquad\qquad (22.6)$$

$$(CH_3)_3C\!\left(\!CH_2\overset{CH_3}{\underset{CH_3}{\overset{\displaystyle|}{\underset{\displaystyle|}{C}}}}\!\right)_{\!n}\!\!CH_2\overset{\displaystyle|}{\underset{\displaystyle|}{C}}\!-\!R$$

16

Poly(isobutylene)

$$CH_2\!=\!CHCN \xrightarrow{\;H_2N^-\;} \overset{NH_2}{\underset{}{\overset{\displaystyle|}{CH_2}}}\!-\!\overset{-}{C}HCN \xrightarrow{\;CH_2\!=\!CHCN\;} H_2NCH_2\overset{CN}{\overset{\displaystyle|}{CH}}\!-\!CH_2\overset{CN}{\overset{\displaystyle|}{\overset{-}{C}}}\!-\!H$$

Acrylonitrile

$$n\, CH_2\!=\!CHCN \qquad\qquad (22.7)$$

$$H_2N\!\left(\!CH_2\overset{CN}{\overset{\displaystyle|}{CH}}\!\right)_{\!n}\!\!CH_2\overset{CN}{\overset{\displaystyle|}{CH}}\!-\!R$$

17

Poly(acrylonitrile)

Free-radical polymerization is a widely used method to induce chain-reaction polymerization, and its mechanistic course is parallel to that of the free-radical halogenation of hydrocarbons (Sec. 9.2). In the experiment in this section, you will perform the free-radical polymerization of styrene to give polystyrene (**24**, Eq. 22.12).

The reaction is started by the thermal decomposition of an *initiator*, which in our experiment is *tert*-butyl peroxybenzoate (**18**), a compound that produces the free radicals **19** and **20** when heated (Eq. 22.8). If In• represents one or both of these free radicals, the course of the polymerization may be illustrated as shown in Equations 22.9–22.12. Equation 22.9 indicates the function of the free radicals in **initiating** the polymerization. Equations 22.10a and 22.10b represent the **propagation** of the growing polymer chain. Equations 22.11 and 22.12 show possible **termination** processes. In Equation 22.11, the free-radical end of one growing polymer chain abstracts a hydrogen atom from the carbon atom next to the end of another polymer radical to produce the unsaturated and saturated polymer molecules **22** and **23**, respectively, in a process termed **disproportionation**. For the termination reaction illustrated by Equation 22.12, Rad• may be one of the initiating radicals, In•, or another growing polymer chain.

$$\text{CH}_3-\underset{\underset{\text{CH}_3}{|}}{\overset{\overset{\text{CH}_3}{|}}{\text{C}}}-\text{O}-\text{O}-\overset{\overset{}{}}{\underset{\underset{\text{O}}{\|}}{\text{C}}}-\text{Ph} \quad \xrightarrow{\text{heat}} \quad (\text{CH}_3)_3\text{C}-\text{O}^{\bullet} \; + \; {}^{\bullet}\underset{\underset{\text{O}}{\|}}{\overset{\overset{\text{O}}{\|}}{\text{C}}}-\text{Ph} \qquad (22.8)$$

18   **19**   **20**

tert-Butyl peroxybenzoate

$$\text{In}^{\bullet} + \text{CH}_2{=}\text{CHC}_6\text{H}_5 \longrightarrow \text{In}-\text{CH}_2-\underset{\underset{\text{C}_6\text{H}_5}{|}}{\overset{\bullet}{\text{CH}}} \qquad (22.9)$$

8

$$\text{InCH}_2-\underset{\underset{\text{C}_6\text{H}_5}{|}}{\overset{\bullet}{\text{CH}}} + \text{CH}_2{=}\text{CHC}_6\text{H}_5 \longrightarrow \text{InCH}_2\text{CH}-\text{CH}_2\underset{\underset{\text{C}_6\text{H}_5}{|}}{\overset{\bullet}{\text{CH}}} \qquad (22.10a)$$

$$\underset{\underset{\text{C}_6\text{H}_5}{|}}{}$$

21

$$\text{InCH}_2\text{CH}-\text{CH}_2\underset{\underset{\text{C}_6\text{H}_5}{|}}{\overset{\bullet}{\text{CH}}} + n\,\text{CH}_2{=}\text{CHC}_6\text{H}_5 \longrightarrow \text{InCH}_2\text{CH}-(\text{CH}_2\text{CH})_n-\text{CH}_2\underset{\underset{\text{C}_6\text{H}_5}{|}}{\overset{\bullet}{\text{CH}}} \qquad (22.10b)$$

$$\underset{\underset{\text{C}_6\text{H}_5}{|}}{} \quad \underset{\underset{\text{C}_6\text{H}_5}{|}}{} \qquad\qquad \underset{\underset{\text{C}_6\text{H}_5}{|}}{} \quad \underset{\underset{\text{C}_6\text{H}_5}{|}}{}$$

21

$$\text{C}_6\text{H}_5-\overset{\bullet}{\text{CH}}-\overset{\overset{\text{H}}{|}}{\text{CHR}} + \overset{\bullet}{\text{CH}}-\text{CH}_2\text{R} \longrightarrow \text{C}_6\text{H}_5-\text{CH}{=}\text{CHR} + \text{C}_6\text{H}_5-\overset{\overset{\text{H}}{|}}{\text{CH}}-\text{CH}_2\text{R} \qquad (22.11)$$

$$\underset{\underset{\text{C}_6\text{H}_5}{|}}{} \qquad\qquad\qquad \textbf{22} \qquad\qquad\qquad \textbf{23}$$

$$\text{R} = -(\text{CH}_2\text{CH})_n-$$
$$\underset{\underset{\text{C}_6\text{H}_5}{|}}{}$$

$$\text{InCH}_2\text{CH}-(\text{CH}_2\text{CH})_n-\text{CH}_2\overset{\bullet}{\text{CH}} + \text{Rad}^{\bullet} \longrightarrow \text{InCH}_2\text{CH}-(\text{CH}_2\text{CH})_n-\text{CH}_2\text{CH}-\text{Rad} \qquad (22.12)$$

$$\underset{\underset{\text{C}_6\text{H}_5}{|}}{} \quad \underset{\underset{\text{C}_6\text{H}_5}{|}}{} \quad \underset{\underset{\text{C}_6\text{H}_5}{|}}{} \qquad\qquad \underset{\underset{\text{C}_6\text{H}_5}{|}}{} \quad \underset{\underset{\text{C}_6\text{H}_5}{|}}{} \quad \underset{\underset{\text{C}_6\text{H}_5}{|}}{}$$

24

The commercially available styrene used in our experiments contains *tert*-butylcatechol (**25**), a phenol that stabilizes styrene by functioning as a **radical scavenger**. The catechol does this by donating a hydrogen atom to reactive free radicals to convert them into nonradical products as shown in Equation 22.13. The resulting phenoxy radical **26** is relatively unreactive as an initiator of a free-radical chain reaction. The stabilizer is necessary to prevent premature polymerization of styrene during storage or shipment because it is so readily polymerized by traces of substances such as atmospheric oxygen. Thus, the stabilizer must first be removed before polymerizing styrene.

$$(22.13)$$

25
tert-Butycatechol   **26**

In the discovery experiment that follows, you will explore an important aspect of polymer chemistry by preparing polystyrene (**24**) under different reaction conditions and test whether it, like many polymers, may be produced in a variety of physical forms such as an amorphous solid, a film, and a clear glass.

EXPERIMENTAL PROCEDURES

Preparation of Polystyrene

 Discovery Experiment   **Purpose** To demonstrate the synthesis of polystyrene by free-radical polymerization under different conditions.

SAFETY ALERT

1. **Wear safety glasses or goggles and suitable protective gloves while performing the experiment.**

2. **The free-radical initiator *tert*-butyl peroxybenzoate is a safe material to use in this experiment because it decomposes at a moderate rate when heated. Nonetheless, do not heat this catalyst excessively when performing the polymerization.**

MINISCALE PROCEDURE

Preparation Refer to the online resources to answer Pre-Lab Exercises, access videos, and read the MSDSs for the chemicals used or produced in this procedure. Review Sections 2.9, 2.17, 2.19, 2.21, and 2.22.

Apparatus A separatory funnel, small soft-glass test tube, 25-mL round-bottom flask, microburner, apparatus for magnetic stirring, heating under reflux, and *flameless* heating.

A ▪ *Removal of the Inhibitor from Commercial Styrene*

Place about 10 mL of commercial styrene in a small separatory funnel and add 4 mL of 3 *M* sodium hydroxide and 15 mL of water. Shake the mixture thoroughly, allow the layers to separate, and withdraw the aqueous layer. Wash the organic layer sequentially with two 8-mL portions of water, carefully separating the aqueous layers after each wash. Dry the styrene by pouring it into a small Erlenmeyer flask containing a little *anhydrous* calcium chloride and then swirling the flask. Allow the mixture to stand for 5–10 min, decant the liquid from the drying agent, and use the dried styrene in the following experiments.

Analysis Obtain IR and NMR spectra of styrene and compare them with those of an authentic sample (Figs. 22.1 and 22.2).

WRAPPING IT UP

Allow the volatiles to evaporate from the *calcium chloride* by placing it on a tray in the hood; then discard it in the container for nontoxic solids. Neutralize the **aqueous layers** before flushing them down the drain.

B ■ *Polymerization of Pure Styrene*

Place about 2–3 mL of dry styrene in a small soft-glass test tube, and add 2 or 3 drops of *tert*-butyl peroxybenzoate. Clamp the test tube in a *vertical* position over a wire gauze, insert a thermometer so that its bulb is in the liquid, and heat the styrene and catalyst with a *small* burner flame. When the temperature reaches 140 °C, temporarily remove the flame. If boiling stops, resume heating to maintain gentle boiling. The exothermicity of the polymerization increases the rate of formation of free radicals by thermal decomposition of the initiator, and this in turn increases the rate of polymerization. Thus be watchful for a rapid increase in the rate of boiling and remove the flame if the refluxing liquid nears the top of the test tube.

After the onset of polymerization, the temperature should rise to 180–190 °C, much above the boiling point of styrene. The viscosity of the liquid will increase rapidly during this time. As soon as the temperature begins to decrease, remove the thermometer and pour the polystyrene onto a watchglass. Do not touch the thermometer *before* the temperature decreases, because moving it in the boiling liquid might cause a sudden "bump," which could throw hot liquid out of the tube. Note the formation of fibers as the thermometer is pulled out of the polymer. The rate of solidification of the polystyrene depends on the amount of catalyst used, the temperature, and the length of time the mixture is heated.

C ■ *Solution Polymerization of Styrene*

Place about 2 mL of dry styrene and 5 mL of xylene in a 25-mL round-bottom flask and add 7 drops of *tert*-butyl peroxybenzoate from a Pasteur pipet. Assemble the apparatus for heating under reflux and heat the mixture under reflux for 20 min. Cool the solution to room temperature and then pour about *half* of it into 25 mL of methanol. Collect the white precipitate of polystyrene that forms by decantation or by vacuum filtration if decantation is not practical. Resuspend the polystyrene in fresh methanol and stir it vigorously. Collect the polystyrene by filtration and allow it to dry in the hood.

Pour the remaining *half* of the polystyrene solution onto a watchglass or the bottom of a large inverted beaker and allow the solvent to evaporate. A clear film of polystyrene should form.

WRAPPING IT UP

Place the *filtrate* containing a mixture of xylene and methanol in the container for nonhalogenated organic liquids. Flush the *methanolic filtrate* obtained after resuspension of the polystyrene down the drain.

Optional Discovery Experiments   Two of the following experiments allow you to explore whether polystyrene is stable toward different organic solvents and to assess the interaction between an ionic polymer, sodium polyacrylate (**27**), and water. The third lets you investigate the change in the properties of a polymer when it is cross-linked with other strands of itself or other polymers.

Discovery Experiment   *Stability of Polystyrene toward Organic Solvents*

Styrofoam is a "puffed-up" form of **24** that is produced by polymerizing styrene in the presence of a "blowing agent" like pentane. The heat of the polymerization causes the agent to vaporize, and it is temporarily trapped in the polymerizing material, forming a bubble or cell if the viscosity of the material is high enough to prevent the gases from escaping. The final polymer has many of these cells, resulting in a foam-like material. The foam is a good semirigid shock absorber and also serves as an insulator because air is not very effective at transferring heat or cold. As a consequence of these physical properties, Styrofoam may be used in a variety of ways, ranging from packaging materials to ice chests.

Work in groups of at least two on this experiment and carefully observe what happens as you perform it. All of the results should be compiled and separately interpreted by each member of the team.

Working at the fume hood and away from any flames and hot plates, fill a petri dish about half-full with an organic solvent. Place a Styrofoam cup bottom-down in the solvent and record what happens. Use a glass stirring rod to explore what remains of the cup. Some solvents that might be tested are acetone, 95% ethanol, dichloromethane, and toluene. Properly dispose of all liquids and solids remaining at the completion of this experiment.

Discovery Experiment   *Polymers and Water*

Most polymers are water-insoluble even though they may contain polar, but neutral, substituents along the carbon backbone. Sodium polyacrylate, however, is a cross-linked ionic polymer that is water-soluble or, equivalently, is a solid polymer in which water is soluble. It is used in products as diverse as disposable diapers and as a replacement for plant soil. Just what happens when this polymer and water are combined is the subject of this experiment.

$$-(CH_2-CH)_n-$$
$$|$$
$$CO_2^- \ Na^+$$

Sodium poly(acrylate)

Weigh out approximately 0.5 g of sodium polyacrylate and transfer it to a *dry* 250-mL Erlenmeyer flask. Add 100 mL of water and immediately stir the mixture. Record your observations.

Discovery Experiment   *Cross-Linking of Polymers*

Poly(vinyl alcohol), like sodium poly(vinyl acrylate), is another water-soluble polymer and is composed of repeating units of vinyl alcohol. In contrast to the acrylate, however, the solubility is not due to a charged polar functionality; rather, extensive hydrogen-bonding involving the hydroxyl groups accounts for the water solubility.

$$\text{--(CH}_2\text{--CH)}_{\overline{n}}\text{--}$$
$$\underset{\text{OH}}{|}$$

Poly(vinyl alcohol)

When sodium tetraborate is dissolved in water, an equilibrium is established according to the equation below, resulting in the formation of a buffer having a pH of about 9. The protonated borate ion reacts with the hydroxyl groups of one strand of the poly(vinyl alcohol) and subsequently with the hydroxyl groups of another strand, possibly with the elimination of water. The resulting array may be represented in a general way as shown below. By joining strands of the polymer through such cross-links, a measure of rigidity is imparted to the molecular array that has been formed. In addition to the cross-links, the molecules have extensive intra- and intermolecular hydrogen-bonding, which also effects a form of cross-linking, albeit a weak one. Breaking and reforming of the hydrogen bonds presumably accounts for the viscoelastic properties of the material that is produced. This material can be formed into a ball, but you should see what happens when it is left untouched.

$$B_4O_7{}^{2-}{}_{(aq)} + H_2O \; \rightleftharpoons \; HB_4O_7{}^-{}_{(aq)} + HO^-{}_{(l)}$$

$$\text{--(CH}_2\text{--CH)}_{\overline{n}}\text{--}$$
$$\underset{\underset{\displaystyle\ \ \overset{}{B}}{O\diagdown\ \diagup O}}{|}$$
$$\text{(CH--CH}_2)_{\overline{n}}\text{--}$$

Apparatus A 100-mL and a 250-mL beaker, thermometer, glass stirring rod, and apparatus for magnetic stirring and *flameless* heating.

Setting Up Add 50 mL of distilled water into the 250-mL beaker equipped with a stirbar. Heat the water with stirring but do *not* exceed a temperature of 90 °C.

Gel Formation With continued stirring and warming, slowly sprinkle 2 g of poly(vinyl alcohol) having an average molar mass of at least 10^5 g/mol onto the surface of the water; this procedure prevents the formation of a sticky mass of polymer that is difficult to dissolve. Combine the poly(vinyl alcohol) solution and 5 mL of a 4% (by mass) aqueous solution of sodium tetraborate, $Na_2B_4O_7$, in the 100-mL beaker, and stir the mixture vigorously with the glass rod. A material that you may consider to be "slime" should form almost immediately.

This cross-linked polymer is a gel that has interesting physical properties. You may explore some of them by forming your material into a ball and then seeing what happens when you carefully tip it in the palm of your hand. If a long column of polymer forms, jerk it abruptly and record the result. Be creative and test the properties of the gel in other ways!

WRAPPING IT UP

Clean the *beaker* and *stirring rod* used to prepare the gel with soap and water. Mix the *solution of polymer* with a copious amount of water and flush the *mixture* down the drain. Discard the *gel* itself in the container for nonhazardous waste.

EXERCISES

1. *tert*-Butylcatechol (**25**) is capable of reacting with *two* equivalents of a radical, R•, to produce two moles of RH and a stable nonradical oxidation product of **25**. Propose a structure for this product and write a stepwise reaction mechanism for its formation. Use curved "fish-hook" arrows to symbolize the flow of electrons.

2. The use of phenols such as *tert*-butylcatechol as free-radical scavengers is based on the fact that phenolic hydrogens are readily abstracted by radicals, producing relatively stable phenoxyl radicals that interrupt chain processes of oxidation and polymerization. Alcohols such as cyclohexanol, on the other hand, do *not* function as radical scavengers. Explain why the two types of molecules differ in their abilities to donate a hydrogen atom to a radical, R•.

3. Write an equation for the reaction involved in the removal of *tert*-butylcatechol from styrene by extraction with sodium hydroxide.

4. Why is it necessary to remove *tert*-butylcatechol from commercially available styrene prior to preparing polystyrene?

5. Why is *tert*-butyl peroxybenzoate a good radical initiator?

6. Explain why only a catalytic amount of the radical initiator is required in a free-radical-chain polymerization reaction.

7. Why is the polymerization of styrene an exothermic reaction? Explain in terms of a calculation based on the following equation using these bond dissociation energies: PhCH(R)–H, 83 kcal/mol; CH$_2$=CHPh, 53 kcal/mol (π-bond only); PhCH$_2$CH$_2$–CH(CH$_3$)Ph, 73 kcal/mol

$$\text{PhCH—H} + \text{CH}_2\text{=CH} \longrightarrow \text{PhCH—CH}_2\text{CHPh}$$
$$\quad\ |\qquad\qquad\quad |\qquad\qquad\qquad |\qquad\quad |$$
$$\ \ \text{CH}_3\qquad\qquad\ \text{Ph}\qquad\qquad\qquad \text{CH}_3\quad\ \text{H}$$

8. Explain why polystyrene is soluble in xylene but insoluble in methanol.

9. What effect would using a smaller proportion of catalyst to styrene have on the average molar mass of polystyrene?

10. In principle, radicals could add to styrene at the carbon atom bearing the phenyl group rather than the other one, yet they do not. Explain the basis of this selectivity for the addition reaction.

11. Specify whether polystyrene is a condensation polymer, a homopolymer, a copolymer, or a block polymer.

12. Some monomers polymerize to produce a polymer having centers of chirality (Chap. 7). If there is no preference for one configuration over another, the

configuration of the centers will be random throughout the polymer. This type of polymer is called *atactic*. When the stereocenters are nonrandom, the polymer may be either *syndiotactic* or *isotactic* (see your lecture textbook for a definition of these terms).

 a. Write a portion of polystyrene containing two monomeric units and circle any stereocenters that are present.

 b. Would you expect the polystyrene generated in this experiment to be *atactic*, *syndiotactic*, or *isotactic*?

13. Circle the monomeric unit in the polysaccharide shown below.

14. Teflon is produced from the polymerization of tetrafluoroethene. Write the structure of Teflon showing the monomeric unit in parentheses.

15. Why does the nucleophilic attack of isobutylene on the $(CH_3)_3C^+$ cation in Equation 22.6 form a new carbon-carbon bond at C(2) of isobutylene instead of C(1)?

16. Super Glue is a polymer formed via an anionic polymerization of methyl cyanoacrylate, $CH_2={=}(C{\equiv}N)CO_2CH_3$. Predict the structure of this glue.

Refer to the online resources and use the spectra viewer and Tables 8.1-8.8 as needed to answer the blue-numbered questions on spectroscopy.

17. Consider the spectral data for styrene (Figs. 22.1 and 22.2).

 a. In the functional group region of the IR spectrum, specify the absorptions associated with the carbon-carbon double bond and the aromatic ring. Also specify the absorptions in the fingerprint region that are characteristic for the terminal vinyl group.

 b. In the ^1H NMR spectrum, assign the various resonances to the hydrogen nuclei responsible for them.

 c. For the ^{13}C NMR data, assign the various resonances to the carbon nuclei responsible for them.

18. Consider the IR spectrum of polystyrene (Fig. 8.10). In the functional group region of the spectrum, specify the range of absorptions associated with C–H stretching vibrations for the hydrogen atoms attached to the aromatic ring and for those of the methylene groups. Also specify the absorption associated with the carbon-carbon double bond of the aromatic rings.

19. Discuss the differences in the IR spectrum of styrene and polystyrene that are consistent with the loss of the vinyl group during the polymerization of styrene in this experiment. The IR spectrum of polystyrene is presented in Figure 8.10.

SPECTRA

Starting Material and Product

The IR spectrum of polystyrene is provided in Figure 8.10.

Figure 22.1

IR spectrum of styrene (neat).

Figure 22.2

NMR data for styrene (CDCl₃).

(a) ¹H NMR spectrum (300 MHz)

(b) ¹³C NMR data: δ 113.5, 126.2, 127.8, 128.5, 137.0, 137.7.

22.3 STEP-GROWTH POLYMERIZATION

Step-growth polymerization typically involves the reaction between two different *difunctionalized* monomers. Both functionalities of each monomer react and this leads to the formation of polymers. For example, a diacid such as terephthalic acid (**27**) can react with a diol such as ethylene glycol (**28**) in the presence of an acid catalyst to produce a polyester, as shown in Equation 22.14.

$$n \; HO_2C\!-\!\!\left\langle \right\rangle\!\!-\!CO_2H \xrightarrow[n \; HOCH_2CH_2OH]{H^+} \quad (22.14)$$

27
Terephthalic acid

28
Ethylene glycol

$$H\!\!\left(\!O\!-\!\overset{O}{\overset{\|}{C}}\!-\!\!\left\langle \right\rangle\!\!-\!\overset{O}{\overset{\|}{C}}\!-\!O\!-\!CH_2CH_2\!\right)_{\!\!n}\!OH \; + \; (2n-1)\,H_2O$$

A polyester

Chain growth is initiated by the acid-catalyzed reaction of a carboxyl group of the diacid with a hydroxy function of the diol to provide an ester and a molecule of water. The free carboxyl group or hydroxyl group of the resulting dimer then reacts with an appropriate functional group in another monomer or dimer, and the process is repeated by such *steps* until all of the monomers are converted into dimers, trimers, tetramers, and, eventually, polymers.

Step-growth polymerization processes are much slower than chain-reaction processes. Because they also typically have higher activation enthalpies, heating is often required to attain satisfactory rates of polymerization. Finally, step-growth polymers generally have lower average molar masses than polymers formed by chain-reaction polymerization.

Polyamides are one type of useful polymer that is produced by a step-growth process, and a variety of such polymers are preparable from various diacids and diamines. Nylon-6,6 was the first commercially successful polyamide and is derived from the two monomers, hexanedioic acid (**29**) and 1,6-hexanediamine (**30**), as shown in Equation 22.15. Its trade name reflects the presence of six carbon atoms in each of the monomeric units that comprise the polymer. Of interest regarding the commercial importance of the nylons is the fact that the tremendous financial success enjoyed by E. I. Du Pont and Company from these types of polymers stems from the firm's patent on a method to draw the substance into fibers rather than from a patent on the molecular composition of the polymer itself. (See the Historical Highlight *Discovery of Polyethylene and Nylon,* which is available online.)

$$n \; HO\!-\!\overset{O}{\overset{\|}{C}}(CH_2)_4\overset{O}{\overset{\|}{C}}\!-\!OH \; + \; n \; H_2N(CH_2)_6NH_2 \longrightarrow$$

29
Hexanedioic acid

30
1,6-Hexanediamine

$$(22.15)$$

$$H\!\!\left(\!O\!-\!\overset{O}{\overset{\|}{C}}(CH_2)_4\overset{O}{\overset{\|}{C}}\!-\!HN(CH_2)_6NH\!\right)_{\!\!n}\!H \; + \; (2n-1)\,H_2O$$

Nylon-6,6

In the typical industrial process for preparing polyamides, equimolar amounts of the diacid and diamine are mixed to give a salt that is then heated to high temperature under vacuum to eliminate the water. The resulting polymer has a molar mass of about 10,000 and a melting point of about 250 °C. Fibers may be spun from melted polymer, and if the fibers are stretched to several times their original length, they become very strong. This "cold-drawing" orients the polymer molecules parallel to one another so that hydrogen bonds form between C–O and N–H groups on adjacent polymer chains, as shown in **31**, greatly increasing the strength of the fibers. The strength of the fibers of silk, a well-known biopolymer involving protein molecules, is ascribed to the same stabilizing factor.

31

The preparation of Nylon-6,10, rather than Nylon-6,6, has been chosen to illustrate step-growth polymerization for the present experiment. This polyamide is commercially produced from decanedioic acid (**32**) and 1,6-hexanediamine (**30**), as shown in Equation 22.16. To facilitate forming the polyamide under simple laboratory conditions, however, the diacid dichloride of **32** is used because it is more reactive toward diamine **30**. Using the diacid dichloride means that the small molecule eliminated in this step-growth polymerization is hydrogen chloride rather than water (Eq. 22.17). Sodium carbonate is added to neutralize the acid formed to prevent consumption of the expensive diamine via an acid-base reaction. If the base were not added, an excess of diamine would be required for complete polymerization.

$$n \; HO-\overset{\overset{\displaystyle O}{\|}}{C}(CH_2)_8\overset{\overset{\displaystyle O}{\|}}{C}-OH \; + \; n \; H_2N(CH_2)_6NH_2 \longrightarrow$$

32   **30**

Decanedioic acid

$$H\!\!\left(\!O-\overset{\overset{\displaystyle O}{\|}}{C}(CH_2)_8\overset{\overset{\displaystyle O}{\|}}{C}-HN(CH_2)_6NH\!\right)_{\!\!n}\!\!H \; + \; (2n-1)\,H_2O \qquad (22.16)$$

Nylon-6,10

$$Cl\overset{\overset{\displaystyle O}{\|}}{C}(CH_2)_8\overset{\overset{\displaystyle O}{\|}}{C}Cl \; + \; H_2NCH_2-R \longrightarrow Cl\overset{\overset{\displaystyle O}{\|}}{C}(CH_2)_8\overset{\overset{\displaystyle O}{\|}}{C}-NHCH_2-R \qquad (22.17)$$

Decanedioyl dichloride
(Sebacoyl chloride)

The reactivity of the diacid chloride toward nucleophilic acyl substitution (Eq. 22.17) allows this polymerization to be performed under mild conditions. When a solution of the diacid chloride in a water-immiscible solvent is brought

Figure 22.3
Apparatus for the "Nylon Rope Trick."

Aqueous solution

Dichloromethane solution

into contact with an aqueous solution of the aliphatic diamine, a film of polymer of high molar mass instantly forms at the interface between the two solutions. The film is thin but strong, and can be pulled from the interface, where it is immediately and continuously replaced by additional polymer. In this way a long cord or rope of polyamide can be produced, much as a magician pulls a string of silk handkerchiefs out of a top hat. When this experiment was first described by two Du Pont chemists, they characterized it as the "Nylon Rope Trick." It does seem to be almost magic that a polymer can attain an average molar mass in the range 5000–20,000 in a fraction of a second!

To perform this experiment properly, the necessary equipment must be assembled so the polymer rope can be pulled from the reaction zone rapidly. A convenient way to do this is illustrated in Figure 22.3. A can, preferably with a diameter of 10 cm or more, makes a good drum on which to wind the polymer. After puncturing the can in the center of each end, a wooden or metal rod is passed through the center holes to make an axle for the drum. The rod is supported horizontally by clamps attached to ring stands in the usual way.

The circumference of the drum should be measured and a reference point should be marked on the drum so that an estimate of the length of the nylon rope can be made by counting the revolutions made as the rope is produced. A length of 6 m or more can usually be obtained with the procedure described here.

If instructed to do so, a much simpler procedure may be used to draw a polymeric fiber, but it will not allow you to determine the length of the fiber easily. Thus, you may dip a glass stirring rod or copper wire into the polymerizing mixture, lift the rod or wire out, and wind the fiber around it as the polymerization proceeds.

The dry polymer obtained in this experiment does not appear to have the properties expected of a nylon; it is fragile and of low density. However, fibers produced from it are much more dense and have the appearance and strength more characteristic of a typical polyamide. You will make fibers and films from the dry polymer so you can compare the properties of the different forms of the polymer.

EXPERIMENTAL PROCEDURES

Preparation of Nylon-6,10

Discovery Experiment   **Purpose** To demonstrate step-growth polymerization by the synthesis of Nylon-6,10.

SAFETY ALERT

1. **Wear safety glasses or goggles and suitable protective gloves while performing the experiment.**

2. **If Pasteur pipets are used instead of syringes to measure the reactants, use a rubber bulb to draw up the liquid.**

3. **Do not handle the polymer rope with your bare hands more than is necessary until it has been washed free of solvent and reagents. Use tongs or forceps to manipulate it. If the crude polymer comes in contact with your skin, immediately wash the area with soap and warm water.**

4. **If you use formic acid to form a film, do not let it get on your skin, because it causes deep skin burns that are not immediately obvious. If the acid does accidentally come in contact with your skin, wash the affected area immediately with 5% sodium bicarbonate solution and then with copious amounts of water.**

MINISCALE PROCEDURE

Preparation Refer to the online resources to answer Pre-Lab Exercises, access videos, and read the MSDSs for the chemicals used or produced in this procedure. Discuss the experiment with your lab partner.

Apparatus A 250-mL beaker, a 5-mL syringe, separatory funnel, apparatus for the "Nylon Rope Trick."

Setting Up Measure 2 mL of decanedioyl dichloride into a 250-mL beaker using the syringe. *The size of the beaker is important:* In smaller beakers, the polymer tends to stick to the walls, whereas in larger beakers, poor "ropes" are obtained unless larger amounts of reagents are used. Dissolve the decanedioyl dichloride in 100 mL of dichloromethane. Place 1.1 g of crystalline 1,6-hexanediamine or 1.3 mL of its commercially available 80–95% aqueous solution in a separatory funnel, and add 2.0 g of sodium carbonate and 50 mL of water. Gently shake the mixture to dissolve both reactants. Arrange the drum onto which the polymer is to be wound at a height such that the beaker containing the decanedioyl dichloride solution can be placed on the lab bench about 40 cm beneath and slightly in front of the drum (Fig. 22.3). Support the separatory funnel containing the other reagents so the lower tip of the funnel is centered no more than a centimeter above the surface of the dichloromethane solution of the decanedioyl dichloride.

Reaction Open the stopcock of the separatory funnel slightly so the aqueous solution runs *slowly and gently* onto the surface of the organic solution. A film of polymer will form immediately at the interface of the two solutions. Use long forceps or tongs to grasp the *center* of the polymer film and pull the rope that forms up to the front of the drum, loop it over the drum, and rotate the drum so as to wind the rope onto the drum. For the first turn or two it may be necessary to use your fingers to secure the rope to the drum. Continue to rotate the drum and rapidly wind the nylon rope onto the drum until the reactants are consumed, remembering to count the revolutions of the drum as you wind.

Work-Up and Isolation Replace the beaker with a large dish or pan containing about 200 mL of 50% aqueous ethanol and unwind the nylon rope into the wash solution. After stirring the mixture gently, decant the wash solution, and transfer the polymer to a filter on a Büchner funnel. Press the polymer as dry as possible, and then place it in your desk until the next laboratory period. When the nylon is thoroughly dry, weigh it and calculate the yield. Note how the bulk of polymer is affected upon drying.

Film Formation To produce a film of Nylon-6,10, dissolve the dry polymer in about 10 times its weight of 90–100% formic acid (*Caution:* See entry 4 of the Safety Alert) by stirring the mixture at *room temperature;* heating to achieve dissolution degrades the polymer. Spread the viscous solution on a glass plate. Leave the plate *in a hood* until the next laboratory period to allow evaporation of the formic acid.

Alternative Procedure

Apparatus A 100-mL beaker, 10-mL graduated cylinder, Pasteur pipet, glass rod or 15-cm length of copper wire.

Setting Up Pour 10 mL of a 5% aqueous solution of 1,6-hexanediamine into the beaker and add 10 drops of 20% aqueous sodium hydroxide solution. Stir the solution to ensure homogeneity.

Reaction Carefully pour 10 mL of a 5% solution of decanedioyl dichloride into the beaker to produce a biphasic mixture. Touch the tip of the glass rod or end of the copper wire to the interface between the two layers and gently remove the rod or wire from the mixture, pulling the fiber of polymer along with it. Twist the rod or wire to spin the fiber around it.

Work-Up and Isolation Follow the procedure described for the "Nylon Rope Trick."

Fiber Formation Form fibers by carefully melting the dry polymer in a metal spoon or spatula with gentle heating over a very small burner flame or a hot plate, and then drawing fibers from the melt with a small glass rod. If necessary, combine your polymer with that of several students to provide enough material to be melted and drawn successfully. Do not heat the polymer much above the melting temperature because it becomes discolored and charred.

Obtain IR and NMR spectra of your starting materials and compare them to those of authentic specimens (Figs. 22.4–22.7).

WRAPPING IT UP

After the rope has been drawn, stir the remaining reaction mixture thoroughly until no more *polymer* forms. Isolate any additional polymer and, after thoroughly washing it with water, put it in the container for nonhazardous organic solids. Separate the *dichloromethane and aqueous layers* of the reaction mixture. Pour the *dichloromethane* into the container for halogenated organic liquids. Flush the *aqueous layer and all aqueous solutions* down the drain.

EXERCISES

1. Explain why the preparation of Nylon-6,10 occurs under milder conditions when decanedioyl dichloride is used instead of decanedioic acid.

2. Using curved arrows to symbolize the flow of electrons, write the stepwise mechanism for the condensation reaction between decandioyl dichloride and 1,6-hexanediamine.

3. Write an equation for the formation of the salt produced from one molecule of hexanedioic acid and two molecules of 1,6-hexanediamine.

4. Why is sodium carbonate used in the reaction to prepare Nylon-6,10?

5. Explain the large decrease in the bulk of the rope of Nylon-6,10 upon drying.

6. Using full structural formulas, draw a typical portion of a Nylon-6,6 molecule; that is, expand a portion of the formula given in Equation 22.15. Show at least two hexanedioic acid units and two 1,6-hexanediamine units.

7. Draw formulas that illustrate the hydrogen bonding that may exist between two polyamide molecules after fibers have been "cold-drawn."

8. Nylons undergo *de*polymerization when heated in aqueous acid. Propose a reaction mechanism that accounts for this fact, using curved arrows to symbolize the flow of electrons.

9. Nylon-6 is produced from caprolactam by adding a small amount of aqueous base and then heating the mixture to about 270 °C.

Caprolactam

a. Draw a representative portion of the polyamide molecule.

b. Suggest a mechanism for the polymerization, using curved arrows to symbolize the flow of electrons, and indicate whether it is of the chain-reaction or step-growth type.

10. Would you expect polyesters to be stabilized by hydrogen bonding? Explain.

11. Proteins are polyamides formed from α-amino acids.

R

H₂N⟍⟍OH

O

An α-amino acid

    a. Write a partial structure of a protein by drawing three monomeric units of it.

    b. Are proteins *chain-reaction* or *step-growth* polymers?

Refer to the online resources and use the spectra viewer and Tables 8.1-8.8 as needed to answer the blue-numbered questions on spectroscopy

12. Consider the spectral data for decanedioyl dichloride (Figs. 22.4 and 22.5).

    a. In the functional group region of the IR spectrum, specify the absorptions associated with the carbonyl component of the acid chloride.

    b. In the ¹H NMR spectrum, assign the various resonances to the hydrogen nuclei responsible for them.

    c. For the ¹³C NMR data, assign the various resonances to the carbon nuclei responsible for them.

13. Consider the spectral data for 1,6-hexanediamine (Figs. 22.6 and 22.7).

    a. In the IR spectrum, specify the absorptions associated with the amino group.

    b. In the ¹H NMR spectrum, assign the various resonances to the hydrogen nuclei responsible for them.

    c. For the ¹³C NMR data, assign the various resonances to the carbon nuclei responsible for them.

SPECTRA

Starting Materials

Figure 22.4

IR spectrum of decanedioyl dichloride (neat).

Figure 22.5
NMR data for decanedioyl dichloride (CDCl₃).

(a) ¹H NMR spectrum (300 MHz).
(b) ¹³C NMR data: δ 24.7, 27.9, 28.4, 46.7, 173.2.

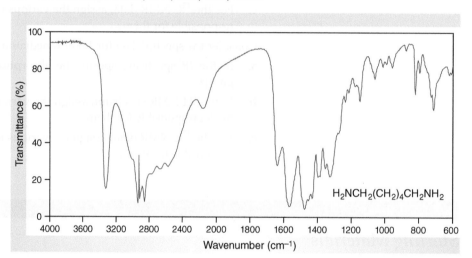

Figure 22.6
IR spectrum of 1,6-hexanediamine (neat).

Figure 22.7
NMR data for 1,6-hexanediamine (CDCl₃).

(a) ¹H NMR spectrum (300 MHz).
(b) ¹³C NMR data: δ 26.0, 33.0, 41.4.

Carbohydrates

Carbohydrates are one of the important classes of organic biomolecules that are essential to life as we know it and are found in every living organism. For example, carbohydrates comprise an essential part of the backbone of the nucleic acids that carry the genetic information of living cells. The sugar and starch in the food you eat are carbohydrates, as is the cellulose in wood and paper. Modified carbohydrates, especially those linked to proteins, are found in the membranes on the surfaces of many cells, and they play an important role in mediating interactions between cells in a number of signaling processes. Attractions between carbohydrate groups on the surface of white blood cells, or leukocytes, and carbohydrates on the surface of an injured cell are involved in the inflammatory response, and entry of the AIDS virus into a target cell depends upon interactions between carbohydrates on the viral coat and the cell receptor.

In this chapter we'll introduce you to some of the rich chemistry of carbohydrates. For example, you may already know that invert sugar is sweeter than sucrose, or table sugar, but do you know how it's formed? In one of the experiments in this chapter, you will learn how to make invert sugar. In others, you will learn about some of the chemical properties of simple carbohydrates.

23.1 INTRODUCTION

The term **carbohydrate** originates from the fact that many, but not all, members of this class have the general molecular formula $C_nH_{2n}O_n$ and thus were once considered to be "hydrates of carbon." Examination of the structures of carbohydrates reveals that this view is inaccurate, but use of the term persists. The simplest carbohydrates are either polyhydroxy aldehydes, which have the general structure **1** and are referred to as **aldoses**, or polyhydroxy ketones, which have the general structure **2** and are called **ketoses**. Such simple carbohydrates are termed **monosaccharides**.

$$
\begin{array}{cc}
\text{CHO} & \text{CH}_2\text{OH} \\
| & | \\
(\text{CHOH})_n & \text{C}=\text{O} \\
| & | \\
\text{CH}_2\text{OH} & (\text{CHOH})_n \\
& | \\
& \text{CH}_2\text{OH} \\
\mathbf{1} & \mathbf{2} \\
\text{An aldose} & \text{A ketose}
\end{array}
$$

Two monosaccharides may be combined with the loss of one molecule of water to form a **disaccharide**. **Polysaccharides** and other complex carbohydrates are then produced by the condensation of more monosaccharide subunits with the loss of a molecule of water for each additional monosaccharide. Depending upon the constitution of a polysaccharide, its hydrolysis yields either a single monosaccharide or a mixture of monosaccharides. In this chapter, some of the fundamental chemical and physical properties of carbohydrates will be investigated using simple mono- and disaccharides; several classic qualitative tests to characterize and classify carbohydrates will also be performed.

23.2 MONOSACCHARIDES: GENERAL PRINCIPLES

The monosaccharide D-glucose, whose chemistry is representative of all aldoses containing four or more carbon atoms, exists predominantly in the two **pyranose** forms **4** and **5**. These are six-membered **hemiacetals** formed by the reversible cyclization of the acyclic polyhydroxy aldehyde **3** (Eq. 23-1). In the cyclic forms **4** and **5**, the ring carbon that is derived from the carbonyl group is referred to as the **anomeric carbon** atom. The **specific rotation**, $[\alpha]_D^{25}$ (Sec. 7.5), of α-D-(+)-glucose (**4**) is $+112°$ whereas that of the β-anomer **5** is $+19°$. When crystals of either pure **4** or pure **5** are dissolved in water, the $[\alpha]_D^{25}$ changes to an equilibrium value of $+52.7°$. This process is termed **mutarotation**. At equilibrium in water, the α- and β-forms are present in the ratio of 36:64; only about 0.03% of D-glucose is in the acyclic form **3**.

4	**3**	**5**
α-D-Glucopyranose	D-Glucose	β-D-Glucopyranose

$$(23.1)$$

Another common monosaccharide is D-fructose (**6**), a ketose. In aqueous solution, D-fructose also undergoes mutarotation to produce a complex equilibrium mixture of the acyclic form **6** ($< 1\%$), the five-membered hemiacetals **7** (31%) and **8** (9%), which are called **furanoses**, and the cyclic pyranoses **9** (57%) and **10** (3%); this mixture exhibits an $[\alpha]_D^{25} = -92°$ (Scheme 23.1).

Scheme 23.1

The ready oxidation of the aldehyde group of an aldose to a carboxylic acid function forms the basis of a number of useful qualitative tests for classifying a carbohydrate as a **reducing sugar**. Ketoses also yield positive tests as reducing sugars even though no aldehyde group is present. This positive test for ketoses arises because the α-hydroxy keto group of the open form of a ketose undergoes base-catalyzed

CH₂OH

7

β-D-Fructofuranose

CH₂OH
C=O
HO—C—H
H—C—OH
H—C—OH
CH₂OH

6

D-Fructose

9

β-D-Fructopyranose

CH₂OH   CH₂OH

8

α-D-Fructofuranose

CH₂OH

10

α-D-Fructopyranose

tautomerization to give an **enediol** that is protonated to provide a pair of epimeric aldehydes differing in configuration at C2. These transformations are outlined in the form of partial structures in Equation 23.2. Since the tests for reducing sugars are performed under basic conditions that allow the equilibria of Equation 23.2 to be established, *all* known *monosaccharides* are reducing sugars.

$$
\begin{array}{ccc}
\text{CH}_2\text{OH} & \text{CHOH} & \text{CHO} \qquad\qquad \text{CHO} \\
\text{C=O} \underset{\text{H}_2\text{O/HO}^-}{\rightleftarrows} \text{C—OH} \underset{\text{H}_2\text{O/HO}^-}{\rightleftarrows} \text{H—C—OH} + \text{HO—C—H} \\
\text{HO—C—H} & \text{HO—C—H} & \text{HO—C—H} \qquad \text{HO—C—H}
\end{array}
\tag{23.2}
$$

A ketose   An enediol   An aldose   An aldose

23.3 DISACCHARIDES: HYDROLYSIS OF SUCROSE

Disaccharides are *O*-glycosides in which the hydroxyl group attached to the anomeric carbon atom of one monosaccharide is replaced with a hydroxyl group of another monosaccharide, thereby forming a cyclic **acetal**. For example, maltose (**11**) is a disaccharide in which the anomeric hydroxyl group of one *α*-D-glucopyranose has been replaced with the C(4)-hydroxyl group of another D-glucopyranose subunit. Because the other D-glucopyranose ring of **11** is in equilibrium with the open-chain isomer **12** having a free aldehyde function (Eq. 23.3), maltose is a reducing sugar. Indeed, any disaccharide in which *one* of the rings is a hemiacetal or hemiketal is a *reducing sugar* (Sec. 23.2), because the cyclic hemiacetal or hemiketal moiety is in equilibrium with the open-chain isomer in which the aldehyde or the α-hydroxyketone function can be oxidized.

$$(23.3)$$

11
Maltose

12

Disaccharides in which *both* rings are in the **acetal** or **ketal** form are *not* reducing sugars because they *cannot* be in equilibrium with an aldehydo or keto form under neutral or basic conditions. For example, sucrose (**13**), a common foodstuff, is a **nonreducing** disaccharide because the glycosidic linkage between the two monosaccharide subunits is formed between the two anomeric carbon atoms, thereby incorporating the potential aldehyde and α-hydroxyketone functions into cyclic acetal and ketal groups. Thus sucrose, $[\alpha]_D^{25} = 166.5°$, does not undergo mutarotation under either neutral or alkaline conditions. However, sucrose hydrolyzes under acidic conditions to give the reducing sugars D-glucose and D-fructose (Eq. 23.4). The 1:1 mixture of D-glucose and D-fructose that is produced upon hydrolysis of sucrose is called **invert sugar** because the specific rotation of the mixture is dominated by the negative rotation of D-fructose.

Fructose is the sweetest common sugar, being about twice as sweet as sucrose; consequently, invert sugar is sweeter than sucrose. The enzyme **invertase**, which bees use in making honey, accomplishes the same chemical result as does the acid-catalyzed hydrolysis of sucrose.

In the experiment in this section, you will examine the acid-catalyzed hydrolysis of sucrose to give a mixture of D-glucose and D-fructose and monitor the consequent change in the specific rotation that occurs using a polarimeter.

$$\xrightarrow{H_3O^+} \text{D-Glucose} + \text{D-Fructose} \qquad (23.4)$$

13
Sucrose

EXPERIMENTAL PROCEDURE

Hydrolysis of Sucrose

Purpose To demonstrate the acid-catalyzed hydrolysis of a disaccharide into its component monosaccharides.

SAFETY ALERT

1. **Wear safety glasses or goggles and suitable protective gloves while performing the experiment.**

2. **If any *concentrated* hydrochloric acid comes in contact with your skin, wash the area with large volumes of water and then with *dilute* sodium bicarbonate solution to neutralize any residual acid.**

MINISCALE PROCEDURE

Preparation Refer to the online resources to answer Pre-Lab Exercises, access videos, and read the MSDSs for the chemicals used or produced in this procedure. Review Sections 2.9, 2.11, and 2.22.

Apparatus A 100-mL round-bottom flask, separatory funnel, 50-mL volumetric flask, polarimeter, apparatus for magnetic stirring, heating under reflux, and *flameless* heating.

Setting Up and Reaction Accurately weigh about 7.5 g of sucrose and place it in the round-bottom flask containing a stirbar. Add about 40 mL of water, swirl the contents of the flask to effect solution, and add about 0.5 mL of *concentrated* hydrochloric acid. Assemble the apparatus for heating under reflux. Heat the solution under reflux for about 2 h.

Determining the Specific Rotation of Sucrose While the solution of sucrose in aqueous acid is being heated under reflux, determine the specific rotation of sucrose according to the general directions that follow, together with any specific directions of your instructor. Carefully fill the sample tube for the polarimeter with water; *be certain that no air bubbles are trapped in it.* Place the sample tube in the polarimeter and determine the blank reading for the solvent. Record the blank rotation, the temperature, and the length of the tube. Empty and carefully dry the tube. Accurately weigh 5–10 g of sucrose to the nearest 0.05 g, and transfer the sample *quantitatively* to the volumetric flask. Fill the flask to within several milliliters of the volumetric mark, tightly stopper the flask, and shake the flask thoroughly. Carefully fill the flask to the mark and again shake the flask thoroughly. Transfer a portion of this solution to the polarimeter tube, measure the rotation, and record this value.

Analysis At the end of the period of reflux, cool the reaction mixture to room temperature and transfer the solution to a clean volumetric flask. Use small amounts of water to rinse the round-bottom flask and add the rinses to the flask to effect quantitative transfer; fill the flask to the volumetric mark. Using the polarimeter as before, determine the specific rotation of the product mixture from the hydrolysis of sucrose. Compare this value with the specific rotation of pure sucrose determined earlier.

WRAPPING IT UP

Neutralize the *aqueous solutions* with sodium carbonate and flush them down the drain.

EXERCISES

1. Explain the change in sign of the optical rotation that occurs when sucrose undergoes hydrolysis.

2. When determining the rotation, why is it important that no air bubbles be present in the polarimeter tube?

3. Calculate the specific rotation of *invert* sugar from the known rotations of the equilibrium mixtures of the anomers of D-glucose and D-fructose. How does this number compare with that determined experimentally?

4. The specific rotation of invert sugar and of a racemic mixture represent average rotations produced by the molecules in solution. Why is the rotation of a racemic mixture 0°, whereas that of invert sugar is not?

5. Using curved arrows to symbolize the flow of electrons, write the stepwise mechanism for the intramolecular cyclization of D-glucose (3) into β-D-glucopyranose (5).

6. Using curved arrows to symbolize the flow of electrons, write a stepwise reaction mechanism for the acid-catalyzed hydrolysis of sucrose to D-glucose and D-fructose.

7. Specify which of the structures in Scheme 23.1 are α-anomers and which are β-anomers.

8. Using curved arrows to symbolize the flow of electrons, write a stepwise reaction mechanism for acid-catalyzed isomerization of 9 to 6, followed by cyclization to produce a five-membered ring as in 7.

9. Which hydroxyl group (i.e., α-, β-, γ-) of an aldoheptose adds to the aldehyde group to produce a furanose? A pyranose?

10. Would you classify the stereochemical relationship between the α- and β-anomers of a cyclic sugar as being enantiomeric, diastereomeric, *meso*, or none of these? Explain your answer.

11. Explain why D-fructose is considered a D-sugar even though its optical rotation is levorotatory.

12. The Fischer projection of D-galactose is shown below. Write the structure of α-D-galactopyranose and explain how it differs from α-D-glucopyranose.

D-Galactose

13. Explain what is meant by the term *reducing sugar*.

14. What functional group must be present or formed under the reaction conditions for a sugar to be classified as a *reducing sugar*?

15. Using curved arrows to symbolize the flow of electrons, write a stepwise mechanism for the base-catalyzed conversion of a ketose into an aldose.

16. Lactose is a disaccharide in which the anomeric hydroxyl group of β-D-galactopyranose has been replaced with the C(4)-hydroxyl group of one α-D-glucopyranose unit.

    a. Write the structure of lactose.

    b. Is lactose a reducing sugar? Explain your answer.

    c. Would you expect lactose to undergo mutarotation? Explain your answer.

17. Consider the structure of raffinose, a trisaccharide, shown below.

Raffinose

    a. Label each of the anomeric carbon atoms with an asterisk.

    b. Circle all the α-glycosidic linkages.

    c. Why is raffinose considered a *trisaccharide*?

    d. Is raffinose a reducing sugar?

    e. Write the structure(s) of the monosaccharide(s) that would be produced when raffinose is heated with aqueous acid.

    f. Write the structures and names of the monosaccharide and the disaccharide that would be produced when raffinose is incubated with *invertase*, the enzyme used by bees to make honey from the carbohydrates contained in nectar.

    g. Why is the disaccharide in Part **f** not hydrolyzed further into monosaccharides by invertase?

18. The specific rotations for α-D-mannose and β-D-mannose are +29.3° and –17.0°, respectively. When either of these monosaccharides is dissolved in water, the rotation reaches an equilibrium value of +14.2. Calculate the percentage of the α-anomer present at equilibrium. Show your calculations.

23.4 CARBOHYDRATES: THEIR CHARACTERIZATION AND IDENTIFICATION

Determining the complete structure of an unknown monosaccharide was a formidable challenge to early organic chemists (see the Historical Highlight *Emil Fischer: Carbohydrate Chemist Extraordinaire,* which is available online). Many carbohydrates, particularly when impure, have a tendency to form syrups rather than crystallize from solution, and this sometimes makes it difficult to obtain pure compounds for characterization and identification. The number and nature of functional groups and the number of stereocenters present in carbohydrates further exacerbates the problem of assigning their structures.

Some of these difficulties are illustrated by considering the example of D-(+)-glucose, whose nature in solution is shown in Equation 23.1. The first step toward solving the structure of D-(+)-glucose was to identify D-(+)-glucose as a derivative of a 2,3,4,5,6-pentahydroxyhexanal 14. Because D-(+)-glucose exists primarily as a cyclic hemiacetal at equilibrium, another issue involved determining the ring size. Namely, did D-(+)-glucose exist in the form of a *furanose,* 15, or a *pyranose,* 16? Finally, it was necessary to assign the relative and absolute configuration to each of the centers of chirality (Sec. 7.1), including the anomeric carbon atom (Sec. 23.2). Despite these tremendous experimental challenges, the complete structures of many monosaccharides, some containing up to nine carbon atoms, are now known. It is amazing that much of this structural work was performed *prior* to the availability of the modern spectroscopic methods discussed in Chapter 8.

14   15   16

Elucidation the structures of polysaccharides represents an even greater challenge. The individual monosaccharide subunits that constitute the unknown polysaccharide must first be identified. Then the ring size and position in the polysaccharide sequence must be elucidated for each monosaccharide. Finally, the nature of the glycosidic linkages that form the polysaccharide backbone must be defined. Toward this end, it is necessary to establish which hydroxyl group on one monosaccharide is involved in the formation of the acetal or ketal that forms the glycosidic bond to the adjoining monosaccharide; the stereochemistry at this anomeric center must also be determined.

Classification Tests

A number of useful qualitative chemical tests have been devised to obtain information for unknown carbohydrates. Some of these tests are used to classify such molecules according to their structural type. In the experiments that follow, you will use such tests to identify certain structural features that are found in mono- and polysaccharides. Data derived from these experiments provide information that may be used to prove the structure of an unknown carbohydrate.

Tollens's Test

Tollens's test is a method for distinguishing between aldehydes and ketones. A positive test indicates the presence of an aldehyde functional group, whereas ketones do not undergo reaction with the reagent. Tollens's reagent consists of silver-ammonia complex, $Ag(NH_3)_2^+$, in an ammonia solution. This reagent oxidizes both aliphatic and aromatic aldehydes to the corresponding carboxylic acids; silver ion is reduced to elemental silver, which is deposited as a silver mirror on the glass wall of a *clean* test tube. Thus, the formation of the silver mirror or of a precipitate is considered a positive test. Equation 23.5 shows the reaction that occurs.

$$RCHO + 2Ag(NH_3)_2^+ + 2\,HO^- \longrightarrow 2\,\underline{Ag^\circ} + RCO_2^-\,NH_4^+ + H_2O + 3\,NH_3 \qquad (23.5)$$

An aldehyde   (metallic silver)

Benedict's Test

This test is based on the fact that cupric ion will oxidize aliphatic aldehydes, including α-hydroxyaldehydes, but not aromatic aldehydes. The reagent used in this test is a solution of cupric sulfate, sodium citrate, and sodium carbonate. The citrate ion forms a complex with the cupric ion, Cu^{2+}, so that cuprous hydroxide, $Cu(OH)_2$, does not precipitate from the basic solution. A positive test for the presence of the aliphatic aldehyde group in an aldose is evidenced by the formation of a red precipitate of cuprous oxide, Cu_2O (Eq. 23.6). A yellow precipitate is occasionally observed as a positive test. This yellow solid has not been characterized, but its formation seems to depend on the amount of oxidizing agent present.

$$RCHO + 2\,Cu^{2+} + 5\,HO^- \xrightarrow{\text{Citrate}} RCO_2^- + \underline{Cu_2O} + 3\,H_2O \qquad (23.6)$$

(brick red)

Barfoed's Test for Monosaccharides

Like Tollens's and Benedict's tests, this test depends on the *reducing* properties of the saccharides. However, because of the specific conditions employed for the test, it is possible to distinguish between monosaccharides and disaccharides. The test reagent consists of an aqueous solution of cupric acetate and acetic acid. Thus, in contrast to the previous tests, the reaction is carried out under *acidic* conditions. A positive test for monosaccharides is indicated by the formation of the brick-red precipitate of cuprous oxide within *two* or *three* minutes. For reasons that are not fully understood, disaccharides require a longer time, providing the precipitate only after about 10 minutes or more. Nonreducing sugars such as sucrose undergo slow hydrolysis under the acidic aqueous conditions of the test, and therefore they give a precipitate after *extended* time.

EXPERIMENTAL PROCEDURES

Classification Tests for Carbohydrates

 Discovery Experiment   **Purpose** To demonstrate several classification tests for carbohydrates.

 SAFETY ALERT

Wear safety glasses or goggles and suitable protective gloves while performing the tests. Read the additional cautions associated with each test.

MICROSCALE PROCEDURES

Preparation Refer the online resources to answer Pre-Lab Exercises, access videos, and read the MSDSs for the chemicals used or produced in this procedure.

Apparatus A number of 13-mm × 100-mm test tubes.

Setting Up Obtain the carbohydrates that you will use in the following classification tests from your instructor. Perform each test on each carbohydrate, and enter your observations and conclusions in your notebook.

EXPERIMENTAL PROCEDURE

Tollens's Test

SAFETY ALERT

1. **Wear safety glasses or goggles and suitable protective gloves while performing the tests.**
2. **Avoid spilling Tollens's reagent on your skin, because staining may result. If the reagent comes into contact with your skin, immediately flood the affected area with water.**
3. **Do not store unused Tollens's reagent, because it decomposes on standing and yields silver azide, AgN$_3$, an explosive precipitate. Follow the directions under Wrapping It Up for proper disposal of the solution.**

Preparation Refer to the online resources to read the MSDSs for the chemicals used or produced in this procedure.

Setting Up Tollens's reagent must be prepared at the time of use by combining two other solutions, A and B, which should be available as stock solutions. If not, first make up solution A by dissolving 0.25 g of silver nitrate in 4.3 mL of distilled water and solution B by dissolving 0.3 g of potassium hydroxide in 4.2 mL of distilled water. Then prepare the test reagent according to the following directions. Add *concentrated* ammonium hydroxide solution dropwise to 3 mL of solution A until the initial brown precipitate begins to clear. The solution should be grayish and almost clear. Finally, add 3 mL of solution B to this mixture, and again add *concentrated* ammonium hydroxide dropwise until the solution is almost clear.

Reaction To carry out the test, add 0.5–1 mL of the reagent to 3 drops or 50–100 mg of the carbohydrate; the formation of a silver mirror or black precipitate constitutes a positive test. The deposition as a silver mirror requires a *clean* glass surface. A black precipitate, although not as visually pleasing, still constitutes a positive test. If no reaction occurs at room temperature, warm the solution slightly in a beaker of warm water.

For comparative purposes, perform the test simultaneously on glucose, sucrose, and an unknown if one is assigned by your instructor. Record the results.

WRAPPING IT UP

Remove the *silver mirrors* from the test tube with *concentrated* nitric acid, and pour the resulting *solution* into a beaker containing unused *Tollens's reagent, ammonium hydroxide,* and *sodium hydroxide.* Acidify this mixture with *concentrated* nitric acid to destroy any excess Tollens's reagent. Neutralize the solution with sodium carbonate and add saturated sodium chloride solution to precipitate silver chloride. Collect the silver chloride by vacuum filtration and place the *filter cake* in a container for recovered silver halides; flush the filtrate down the drain.

EXPERIMENTAL PROCEDURE

Benedict's Test

SAFETY ALERT

1. **Wear safety glasses or goggles and suitable protective gloves while performing the tests.**
2. **If Benedict's reagent comes in contact with your skin, wash the area thoroughly with soap and water.**

Preparation Refer to the online resources to read the MSDSs for the chemicals used or produced in this procedure.

Apparatus Test tube, graduated cylinder, apparatus for *flameless* heating.

Setting Up Place about 0.1 g of the carbohydrate in the test tube and add 2–3 mL of the stock solution of Benedict's reagent followed by 2 mL of water.

Reaction Stir the mixture to effect solution and then heat the solution to boiling. The formation of a yellow to red precipitate is a positive test for aliphatic aldehydes and α-hydroxyaldehydes; the precipitate appears green when viewed in the blue solution of the reagent.

For comparative purposes, perform the test simultaneously on glucose, sucrose, and an unknown if one is assigned by your instructor. Record the results.

EXPERIMENTAL PROCEDURE

Barfoed's Test

SAFETY ALERT

1. **Wear safety glasses or goggles and suitable protective gloves while performing the tests.**
2. **If Barfoed's reagent comes in contact with your skin, wash the area thoroughly with soap and water.**

Preparation Refer to the online resources to read the MSDSs for the chemicals used or produced in this procedure.

Apparatus Test tube, graduated cylinder, water bath, 25- or 50-mL beaker, apparatus for *flameless* heating.

Setting Up Place 3 mL of Barfoed's reagent and 1 mL of a 1% aqueous solution of the carbohydrate in a test tube.

Reaction Place the test tube in a beaker of boiling water for 5 min. Remove the test tube and cool it under running water. A red precipitate of cuprous oxide is a positive test. To see the precipitate, it may be necessary to view the tube against a dark background in good light. For comparison, run the test simultaneously on the unknown, on glucose, and on lactose, and record the results.

For comparative purposes, perform the test simultaneously on glucose, sucrose, and an unknown if one is assigned by your instructor. Record the results.

WRAPPING IT UP

Remove any precipitated *cuprous oxide* by vacuum filtration, and place it in the container for nonhazardous solids. Flush the *filtrate* down the drain with an excess of water.

FORMATION OF OSAZONES

Carbohydrates that can exist in solution in an acyclic form with a free aldehyde or ketone function react with three equivalents of phenylhydrazine to form bright yellow crystalline derivatives called **phenylosazones**, together with aniline and ammonia as the other products of the reaction. For example, α- and β-D-glucopyranoses **4** and **5** are in dynamic equilibrium in aqueous solution with the ring-opened form **3** (Sec. 23.2), which, because of its free aldehyde group, will react with phenylhydrazine to form the phenylosazone **17**, as shown in Equation 23.7. Since phenylosazones are readily identified by either their melting points or temperatures of thermal decomposition, these derivatives may be used to characterize carbohydrates.

The accepted mechanism for the formation of phenylosazones is presented in Scheme 23.2. Following initial formation of the phenylhydrazone **18** from the aldose, a double tautomerization leads to the ketone **19**. The newly formed carbonyl group then condenses with a second equivalent of phenylhydrazine to give **20**, which tautomerizes to **21**. After 1,4-elimination of aniline to produce either **22** or **23**, a third equivalent of phenylhydrazine condenses with the imine group of **22** or **23** to yield the phenylosazone **24** and ammonia. Although it may appear that **24** should undergo further reaction with phenylhydrazine via tautomerization involving the secondary alcohol group at C3 and the hydrazone group at C2, the reaction stops at this point so that only two phenylhydrazine units are introduced. The formation of the intramolecular hydrogen bond shown by the dashed line in **24** apparently limits the reaction to the first two carbon atoms of the chain. In support of this hypothesis, when 1-methylphenylhydrazine [H$_2$NNMeC$_6$H$_5$] is used in place of phenylhydrazine, the reaction readily proceeds down the chain at least as far as C5.

$$\text{D-Glucose (3)} + 3\ C_6H_5NHNH_2 \xrightarrow{\ H_3O^+\ } \text{Phenylosazone of D-glucose (17)} + C_6H_5NH_2 + NH_3 \qquad (23.7)$$

Scheme 23.2

EXPERIMENTAL PROCEDURE

Formation of Osazones

Purpose   To demonstrate the preparation of osazones.

SAFETY ALERT

1. **Wear safety glasses or goggles and suitable protective gloves while performing the experiment.**

2. **Phenylhydrazine is toxic; avoid contact with it. If you spill some on your skin, rinse it off thoroughly with household bleach, *dilute* acetic acid, and then water.**

MICROSCALE PROCEDURE

Preparation   Refer to the online resources to answer Pre-Lab Exercises, access videos, and read the MSDSs for the chemicals used or produced in this procedure. Review Sections 2.9 and 2.17.

Apparatus   Test tubes, hot-water bath, apparatus for vacuum filtration.

Setting Up   Heat a large beaker of water to boiling. In each of three test tubes, *separately* dissolve 0.2-g portions of D-glucose, D-fructose, and sucrose in 4 mL of water. In a fourth test tube, place 4 mL of the solution that was used for rotation experiments for the hydrolysis of sucrose (Sec. 23.3). Add 0.5 mL of saturated sodium bisulfite solution to each test tube to suppress oxidation of the phenylhydrazine during the reaction; this avoids contaminating the phenylosazone with tarry by-products.

Reactions   To each of the test tubes, add *either* 0.6 mL of *glacial* acetic acid, 0.6 g of sodium acetate, and 0.4 g of phenylhydrazine *or* 0.6 g of sodium acetate and 0.6 g of phenylhydrazine hydrochloride. Stir the solutions thoroughly, place the test tubes in the beaker of boiling water, and discontinue heating. Allow the test tubes to remain in the hot-water bath for 30 min. Remove the test tubes from the hot-water bath and cool the contents to room temperature.★

Isolation and Purification   Collect any precipitates by vacuum filtration.★ Using ethanol-water, recrystallize each of the phenylosazones formed.

Analysis   Determine the melting points of the purified phenylosazones. Make a mixture containing equal parts of the two phenylosazones obtained from the pure carbohydrates, and determine the melting point of this mixture. Because the melting points of phenylosazones may depend on the rate of heating, perform these determinations simultaneously. Make a mixture containing equal parts of the phenylosazone obtained from the hydrolysis of sucrose and each of the phenylosazones obtained from the pure carbohydrates and determine their melting points as well. Record your observations.

WRAPPING IT UP

Neutralize the *filtrates* obtained from isolating the crude phenylosazones with a small amount of sodium carbonate, and add 2 mL of laundry bleach (5.3% aqueous sodium hypochlorite) for every 1 mL of the filtrate. Heat the mixture at 45–50 °C for 2 h to oxidize and decompose any phenylhydrazine that is present. Cool the mixture and then flush it down the drain with excess water. Flush the *filtrates from the recrystallizations* down the drain.

EXERCISES

1. Consider Tollens's test.
   a.   What causes the formation of a silver mirror?
   b.   Write the half reaction for silver in the test and indicate whether silver undergoes oxidation or reduction.
   c.   Write the half reaction for the sugar in the test and indicate whether the sugar undergoes oxidation or reduction.

2. List the classification tests that may be used to determine whether a sugar is a reducing sugar.

3. What ion is responsible for the blue color of Benedict's reagent?

4. Why does a solution of D-glucose and Benedict's reagent turn brick red when heated?

5. Using curved arrows to symbolize the flow of electrons, write a stepwise mechanism for the acid-catalyzed reaction of an aldose with phenylhydrazine to give a phenylhydrazone like **18**.

6. Which of the pure carbohydrates formed phenylosazones?

7. What conclusions can be drawn from the results of the mixed melting-point determination on the phenylosazones formed from the pure carbohydrates?

8. What conclusions can be drawn from the results of the mixed melting-point determinations on the phenylosazones formed from the pure carbohydrates and that obtained from the hydrolysis product of sucrose?

9. D-Glucose and D-mannose give the same phenylosazone. Explain.

10. Draw the structure of the product that would be expected upon treatment of D-glucose with a large excess of 1-methylphenylhydrazine.

11. In what way does the intramolecular hydrogen bond in the phenylosazone **24** prevent further reaction with phenylhydrazine?

α-Amino Acids and Peptides

α-Amino acids are organic molecules that contain both an amine and a carboxylic acid functional group on the same carbon atom. α-Amino acids thus undergo some reactions that are typical of both of these functionalities, but they also exhibit special chemical and physical properties that are a consequence of having these two functional groups attached to one carbon atom. You have an opportunity to learn about some of the chemistry of amines and carboxylic acids in Chapters 17, 20, and 21. In this chapter, you will study several aspects of the chemistry of α-amino acids and how these compounds can be combined to make peptides, substances that comprise two or more amino acids joined by an amide bond between the nitrogen atom of one amino acid and the carboxyl carbon atom of another.

α-Amino acids are the essential building blocks of proteins, which are high-molar-mass biopolymers comprising many α-amino acids linked by peptide bonds. Examples of proteins in living cells include the enzymes that catalyze essential biochemical transformations and the receptors in cell membranes that bind smaller molecules such as neurotransmitters, hormones, and pharmaceuticals. Useful materials like wool or silk are also constituted of proteins. In the experiments in this chapter, you will learn how to combine individual α-amino acids to form peptides in exactly the same way such compounds are prepared in research laboratories.

24.1 INTRODUCTION

Amino acids are generally characterized by the presence of an **amine functional group**, $-NH_2$, and a **carboxylic acid functional group**, $-CO_2H$. Although the amino and carboxyl groups may be located anywhere in the molecule, the most important amino acids in biological systems are **α-amino acids**, as these are the monomeric building blocks for forming **proteins** in living organisms. These compounds are commonly represented as **1**, a general structural formula in which R^1 is H or variously substituted alkyl groups. The R^2 substituent on the nitrogen atom is usually an H. Whenever R^1 is an alkyl group, the carbon atom bearing the amino and carboxyl groups is a *center of chirality* or *stereocenter* (see Sec. 7.1), and the vast majority of amino acids found in mammalian proteins have the absolute configuration shown, which may be either *S*- or *R*- depending on the functionality in R^1, and are classified as L-amino acids because of their stereochemical relationship with L-glyceraldehyde.

$$R^1 \diagup CO_2H \qquad HOCH_2 \diagup CHO$$
$$H \diagup NHR^2 \qquad H \diagup OH$$

1

An α-amino acid   L-Glyceraldehyde

The R^1 substituent in **1** is commonly referred to as the **side chain**, and the variations in the structure and functionality of these groups play an important role in determining the chemical and physical properties of the individual amino acids. Based upon the nature of the side chain, the 20 common L-amino acids found in proteins may be grouped into four general categories as shown in Table 24.1, in which the structures, names, and three-letter abbreviations are listed. The largest single class consists of those amino acids with nonpolar side chains containing simple alkyl and various aromatic and heteroaromatic groups. A second category comprises polar but neutral or uncharged functional groups such as amides and alcohols. The third and fourth groups possess either acidic or basic side chains that will be at least partially ionized at the physiological pH of about 7.4. Because of the manner in which these side chains interact with water, those amino acids with nonpolar side chains are sometimes referred to as *hydrophobic,* or "water-fearing," whereas those with polar side chains are considered *hydrophilic,* or "water-loving." Although it is beyond the scope of this discussion, those of you who study biochemistry will learn that the amino acid side chains in proteins play important roles in protein folding, in substrate recognition, and in catalyzing biological reactions.

Because amino acids have both a basic (the amine) and an acidic (the carboxylic acid) functional group, the state of protonation of the molecule varies with pH as shown in Equation 24.1. The presence of a functional group in the side chain may have considerable influence upon the position of this equilibrium at different pHs. The pH at which the vast majority of the molecules are in the **zwitterionic** form, and therefore have a net charge of zero, is referred to as the **isoelectric point, pI**. An amino acid is *least* soluble in water at its isoelectric point, which is different for each amino acid. If the side chain bears an ionizable group such as a carboxylic acid or an amine, the state of protonation of that functional group will also vary as a function of pH. However, the amino acids used in the experiments in this chapter are nonpolar, so we will not concern ourselves with the complexities associated with ionizable side chains.

$$R^1 \diagup CO_2^- \quad \underset{-H^+}{\overset{+H^+}{\rightleftharpoons}} \quad R^1 \diagup CO_2^- \quad \underset{-H^+}{\overset{+H^+}{\rightleftharpoons}} \quad R^1 \diagup CO_2H \qquad (24.1)$$
$$NHR^2 \qquad\qquad +NH_2R^2 \qquad\qquad +NH_2R^2$$

Species present   Zwitterionic form   Species present
in strong base   in strong acid

An amide bond formed between the nitrogen atom of one amino acid and the acyl carbon atom of another is referred to as a **peptide bond**, a term coined in 1902 by the Nobel laureate Emil Fischer (see the Historical Highlight *Emil Fischer: Carbohydrate Chemist Extraordinaire*, which is available online, for an account of the life of this famous chemist). For example, **aspartame (2)**, the active ingredient in the artificial sweetener NutraSweet, is the methyl ester of a dipeptide derived from joining a molecule of L-aspartic acid (Asp) and

Table 24.1 *The 20 Amino Acids Commonly Found in Proteins*

Nonpolar Side Chains

L-Glycine (Gly)

L-Methionine (Met)

L-Alanine (Ala)

L-Proline (Pro)

L-Valine (Val)

L-Phenylalanine (Phe)

L-Isoleucine (Ile)

L-Tryptophan (Trp)

L-Leucine (Leu)

Polar Side Chains

L-Asparagine (Asn)

L-Serine (Ser)

L-Glutamine (Gln)

L-Threonine (Thr)

Acidic Side Chains   *Basic Side Chains*

L-Tyrosine (Tyr)

L-Arginine (Arg)

L-Cysteine (Cys)

L-Histidine (His)

L-Aspartic acid (Asp)

L-Glutamic acid (Glu)

L-Lysine (Lys)

L-phenylalanine (Phe) by a peptide bond. By convention, **peptides**, the general name given to chains comprising up to approximately 30–40 amino acid residues, are written from the left, beginning with the amino acid having a free $-NH_3^+$ group, the **N-terminus**, and progressing to the amino acid with the free COO^- group, the **C-terminus**. **Proteins** are biopolymers that are typically formed by joining more than about 100 amino acids in a single polypeptide chain. The definitions of *N*- and *C*-terminus and peptide bonds are illustrated in the structure of **2**.

2

Aspartame

24.2 SYNTHESIS OF PEPTIDES AND POLYPEPTIDES

The synthesis of peptides and proteins in nature is a complex process that you may study in detail in a biochemistry course. However, because short peptides and their derivatives may exhibit useful properties and potent biological activities, chemists in academic and pharmaceutical laboratories have developed efficient methods for their synthesis.

Short peptides consisting of four or five amino acids may be easily prepared in solution, but problems associated with the insolubility of larger peptide derivatives cause difficulties with purification and isolation of intermediates as well as of the desired polypeptides. A major advance in polypeptide synthesis was reported by R. Bruce Merrifield in 1962. He discovered that all the steps in the synthesis of polypeptides could be performed on the surface of an insoluble polymer rather than in solution; you can read some of the details of this approach in the Historical Highlight *Invention of a Method for Solid-Phase Peptide Synthesis*, which is available online. This method for preparing polypeptides revolutionized the field, and it enabled Merrifield to complete the synthesis of ribonuclease, an enzyme containing 124 amino acids, in only six weeks! The significance of Merrifield's work earned him the Nobel Prize in Chemistry in 1984. You will not have enough time and the necessary apparatus to complete the synthesis of such a large protein as ribonuclease, of course, but the experiments in this chapter allow you to learn the essential principles of peptide synthesis by preparing a simple dipeptide derivative.

The fundamental problem in the synthesis of peptides is that amino acids must be connected in a defined sequence by specifically forming a peptide bond between the carboxylic acid group of one amino acid and the amino group of another. The potential difficulties arising in such endeavors may be illustrated by considering the synthesis of the simple dipeptide Ala–Phe (**5**) from the individual amino acids L-alanine (Ala) (**3**) and L-phenylalanine (Phe) (**4**), as seen in Equation 24.2. Formation of peptide bonds in a random manner could lead to four different dipeptides. This mixture arises because the carboxylic acid of Ala may react with either the amino group of another molecule of Ala or with the amino group of a Phe. Similarly, the carboxyl group of the Phe may react with the amino group of a molecule of Phe or Ala. Obtaining such a mixture is clearly undesirable, as the desired Ala–Phe would be difficult to isolate in pure form and high yield. Hence, the problem facing the early peptide chemist was determining how to make the amino group of one amino acid react selectively with the carboxyl group of another.

3
L-Alanine (Ala)

4
L-Phenylalanine (Phe)

5
Ala–Phe

(24.2)

Phe–Ala

Ala–Ala

Phe–Phe

24.3 SYNTHESIS OF THE PROTECTED DIPEPTIDE ALA–PHE–OME

By completing the experimental procedures of this chapter, you will prepare the methyl ester of Ala–Phe (Ala–Phe–OMe). You will be using a general strategy that has been developed to direct the synthesis of polypeptides so that only one specific peptide bond can be formed in the coupling step, thereby avoiding the production of mixtures (Eq. 24.2). The first stage of this process involves protecting (see Sec. 21.2 for more on protecting groups) the amino group of one amino acid so it cannot serve as a *nucleophile,* and also protecting the carboxylic acid group of the other amino acid so it cannot serve as an *electrophile* or an acylating agent. Coupling two such protected α-amino acids furnishes a diprotected dipeptide exclusively, as shown in Equation 24.3.

(24.3)

An *N*-protected
α-amino acid

A *C*-protected
α-amino acid

A diprotected dipeptide

In developing a strategy for selecting amine and carboxylic acid protecting groups for the synthesis of peptides, it is essential that each protecting group may be *selectively* removed in the presence of the other; such protecting groups are often referred to as being *orthogonal*. For example, selective removal of the nitrogen-protecting group P^1 from the diprotected dipeptide will give a *C*-protected dipeptide that may then be coupled with another *N*-protected amino acid. The selective deprotection and coupling process may be repeated to continue chain growth from the *N*-terminal end of the peptide. Alternatively, the protecting group P^2 on the carboxylic acid can be removed first to give a *N*-protected dipeptide that may be coupled with a *C*-protected amino acid to propagate chain growth from the *C*-terminal end. For reasons that are beyond the scope of this discussion, peptide synthesis in the laboratory is normally performed by extending the growing chain from the *N*-terminus. On the other hand, the enzyme-catalyzed biological synthesis of peptides *in vivo* proceeds by addition of new amino acids to the *C*-terminus of the growing chain.

A *C*-protected dipeptide   An *N*-protected dipeptide

There are a variety of protecting groups for an amino group that may be used, but you will protect this functionality in L-alanine (**3**) with a *tert*-butoxycarbonyl (Boc) group by treating **3** with di-*tert*-butyl dicarbonate (**6**) in aqueous base to give *N*-*tert*-butoxycarbonyl-L-alanine (Boc–Ala) (**7**) (Eq. 24.4). The carboxylic acid group of an amino acid is typically protected as an ester, and you will protect L-phenylalanine (**4**) by forming a methyl ester using thionyl chloride, $SOCl_2$, in methanol (Eq. 24.5) to give methyl L-phenylalaninate hydrochloride (**8**). It is left as an exercise to provide mechanisms for the reactions depicted in Equations 24.4 and 24.5 (see Exercises 17 and 25 at the end of the chapter). The relative solubilities of **7** and **8** in water or organic solvents depend upon whether or not they are in their ionized or neutral forms.

3   **6**   **7**

        Di-*tert*-butyl dicarbonate   *N*-*tert*-Butoxycarbonyl-L-alanine
                                                        (Boc–Ala)

$$(24.4)$$

4   **8**

                                            Methyl L-phenylalaninate
                                                    hydrochloride
                                                (Phe–OMe • HCl)

$$(24.5)$$

The next step in our synthesis of Ala–Phe requires activating the carboxyl group of *N-tert*-butoxy-L-carbonylalanine (**7**) so it can serve as an acylating agent for the amino group of methyl L-phenylalaninate to give the protected dipeptide, methyl *N-tert*-butoxycarbonyl-L-alanyl-L-phenylalaninate (**11**). The acid chlorides of carboxylic acids are often used as intermediates for forming amides (Sec. 20.3), but such derivatives are unsuitable for use in procedures for coupling peptides because their high reactivity leads to side reactions and undesirable by-products.

Although a variety of specialized methods and reagents for forming peptide bonds have been developed, the simple and effective method used in this experiment involves converting **7** to the *mixed anhydride* **10** by reaction with isobutyl chloroformate (**9**) in the presence of a tertiary amine such as *N*-methylmorpholine (Eq. 24.6). This mixed anhydride is easily hydrolyzed, so it is not isolated. Rather, a solution of methyl L-phenylalaninate hydrochloride (**8**) and *N*-methylmorpholine is added to a solution of **10** to give **11** (Eq. 24.7). It is left as an exercise to provide mechanisms for the reactions depicted in Equations 24.6 and 24.7 (see Exercises **34** and **35** at the end of the chapter). Unlike free amino acids and peptides, protected peptides are not zwitterionic, so they are more soluble in organic solvents than in water.

$$(CH_3)_3CO\overset{O}{\overset{\|}{C}}\underset{H}{\overset{CH_3}{N}}\overset{CH_3}{\underset{}{CH}}COOH \ + \ Cl\overset{O}{\overset{\|}{C}}O\text{-}i\text{-}C_4H_9 \quad \xrightarrow[\text{DMF}]{O\bigcirc N\text{–Me}} \quad (CH_3)_3CO\overset{O}{\overset{\|}{C}}\underset{H}{N}\overset{CH_3}{\underset{}{CH}}\overset{O}{\underset{\|}{C}}O\overset{O}{\underset{\|}{C}}O\text{-}i\text{-}C_4H_9 \quad (24.6)$$

7   **9**   **10**

Isobutyl chloroformate

$$(CH_3)_3CO\overset{O}{\overset{\|}{C}}\underset{H}{N}\overset{CH_3}{\underset{}{CH}}\overset{O}{\underset{\|}{C}}O\overset{O}{\underset{\|}{C}}O\text{-}i\text{-}C_4H_9 \ + \ Cl^-\overset{+}{H_3}N\overset{}{CH}COOCH_3 \quad \xrightarrow[\text{DMF}]{O\bigcirc N\text{–Me}}$$

10   **8**

$$(CH_3)_3CO\overset{O}{\overset{\|}{C}}\underset{H}{N}\overset{CH_3}{\underset{}{CH}}\overset{}{\underset{\underset{O}{\|}}{C}}\underset{H}{N}\overset{}{CH}COOCH_3 \quad (24.7)$$

11

Methyl *N-tert*-butoxycarbonyl-L-alanyl-L-phenylalaninate
(Boc–Ala–Phe–OMe)

To convert the diprotected dipeptide Boc–Ala–Phe–OMe (**11**) into Ala–Phe, it is necessary to remove the protecting groups from the nitrogen atom of the *N*-terminal L-alanine and the carboxylic acid function of the *C*-terminal L-phenylalanine. This might be accomplished by hydrolysis of both groups under acidic conditions. However, the zwitterionic Ala–Phe is difficult to isolate, so in this experiment you will only selectively remove the Boc group from the L-alanine residue to give methyl

L-alanyl-L-phenylalaninate trifluoroacetate (**12**). This involves an acid-catalyzed reaction in which carbon dioxide and isobutylene are also produced (Eq. 24.8). It is left as an exercise to provide a mechanism for this transformation (see Exercise 38 at the end of the chapter).

11

$$(24.8)$$

12

Methyl L-alanyl-L-phenylalaninate trifluoroacetate
(Ala–Phe–OMe • CF₃CO₂H)

The conversion of **12** into Ala–Phe (**5**) requires hydrolysis of the methyl ester, a reaction that may be accomplished by hydrolysis of **12** in aqueous acid or base followed by neutralization (Eq. 24.9). You may have an opportunity to develop a procedure to effect this transformation and isolate **5** if you perform the suggested Discovery Experiment.

$$(24.9)$$

12   **5**

L-Alanyl-L-phenylalanine
(Ala–Phe)

Dipeptides are rarely the final products of peptide synthesis, but the principles described for the preparation of **5** may be readily applied to the synthesis of peptides containing more amino acids. In practice, the stepwise extension is performed at the *N*-terminus of a growing peptide chain that is protected with an ester group at the *C*-terminus. *N*-Protected amino acids are sequentially added by a coupling reaction followed by removal of the protecting group on the *N*-terminus until the desired peptide, in its diprotected form, has been prepared. The protecting groups at the *N*- and *C*-termini are then removed to provide the peptide of interest. Indeed, this is precisely the strategy that is used in the **solid-phase synthesis** of large peptides (see the Historical Highlight *Invention of a Method for Solid-Phase Peptide Synthesis*, which is available online). It is left as an exercise for you to apply these principles to preparing the tripeptide Val–Ala–Phe from the monoprotected dipeptide **12** (see Exercise 10 at the end of the chapter).

EXPERIMENTAL PROCEDURES

A ■ *Preparation of* N-tert-*Butoxycarbonyl* L-*Alanine*

Purpose To demonstrate the protection of the amine group of an α-amino acid by forming a *tert*-butyl carbamate.

MINISCALE PROCEDURE

Preparation Refer to the online resources to answer Pre-Lab Exercises, access videos, and read the MSDSs for the chemicals used or produced in this procedure. Review Sections 2.5, 2.10, 2.11, 2.13, 2.17, 2.21, and 2.29.

Apparatus A 50-mL round-bottom flask, 5-mL glass syringe with a needle, separatory funnel, ice-water bath, apparatus for magnetic stirring, simple distillation, vacuum filtration, and *flameless* heating.

Set-Up and Reaction Place 0.90 g of L-alanine in a 50-mL round-bottom flask containing a stirbar. Add 5 mL of a 3 *M* solution of aqueous sodium hydroxide and 5 mL of *tert*-butyl alcohol to the round-bottom flask and stir the mixture until all of the solids have dissolved. With continued stirring, add 2.5 mL of di-*tert*-butyl dicarbonate and then stir the reaction mixture for 45 min at room temperature.

Work-Up Add 10 mL of water to the reaction mixture and transfer the solution to a separatory funnel. Rinse the round-bottom flask with 1–2 mL of water and transfer the rinse to the separatory funnel. Add 25 mL of diethyl ether to the separatory funnel and shake the funnel gently, *venting as necessary*. Separate the layers and acidify the aqueous layer to pH 2 with 3 *M* HCl. Extract the aqueous layer with two 10-mL portions of diethyl ether and then wash the combined organic layers with

10 mL of brine. Dry the organic layer over several spatula-tips full of *anhydrous* sodium sulfate. Occasionally swirl the mixture over a period of 10–15 min to facilitate drying; add further small portions of *anhydrous* sodium sulfate if the solution remains cloudy.

Isolation and Purification Filter or carefully decant the dried organic solution into a round-bottom flask equipped for simple distillation and concentrate the solution to a volume of approximately 2 mL. Alternatively, use rotary evaporation or other techniques to concentrate the solution. Allow the solution to cool to room temperature and transfer it to a 50-mL Erlenmeyer flask. Rinse the round-bottom flask with 0.5–1 mL of diethyl ether and transfer this rinse to the Erlenmeyer flask. Add approximately 25 mL of hexanes to the ethereal solution, stir the solution with a glass stirring rod, and place the flask in an ice-water bath for 10 min. If no solid precipitates, use one of the methods described in Chapter 3 to induce crystallization. Isolate the white solid by vacuum filtration and air-dry it. Recrystallize the crude product by a mixed solvent recrystallization according to the second option for dissolution (Sec. 3.2) by suspending the solid in about 20 mL of boiling hexanes and adding ethyl acetate dropwise until all the solids have dissolved.

Analysis Weigh the recrystallized product and calculate the percent yield; determine its melting point. Obtain IR and ^{1}H NMR spectra of your starting materials and product, and compare them with those of authentic samples (Figs. 24.1–24.6).

MICROSCALE PROCEDURE

Preparation Refer to the online resources to answer Pre-Lab Exercises, access videos, and read the MSDSs for the chemicals used or produced in this procedure. Review Sections 2.5, 2.10, 2.11, 2.17, 2.21, and 2.29.

Apparatus A 5-mL conical vial, 1-mL glass syringe with a needle, a screw-cap centrifuge tube, ice-water bath, apparatus for magnetic stirring, simple distillation, vaccum filtration, and *flameless* heating.

Set-Up and Reaction Place 0.20 g of L-alanine in a 5-mL conical vial containing a spinvane. Add 1 mL of a 3 *M* solution of aqueous sodium hydroxide and 1 mL of *tert*-butyl alcohol to the conical vial, and stir the mixture until all of the solids have dissolved. With continued stirring, add 0.5 mL of di-*tert*-butyl dicarbonate and then stir the reaction mixture for 45 min at room temperature.

Work-Up Add 2 mL of water to the reaction mixture and transfer the solution to a screw-cap centrifuge tube. Rinse the conical vial with about 5 drops of water and transfer the rinse to the screw-cap centrifuge tube. Add 3 mL of diethyl ether to the screw-cap centrifuge tube, screw the cap onto the centrifuge tube, and shake the tube gently, *venting as necessary*. Separate the layers and acidify the aqueous layer to pH 2 with 3 *M* HCl. Extract the aqueous layer with two 1.5-mL portions of diethyl ether and then wash the combined organic layers with 2 mL of brine. Transfer the organic layer to a dry test tube and add several microspatula-tips full of *anhydrous* sodium sulfate. Occasionally swirl the mixture over a period of 10–15 min to facilitate drying; add further small portions of *anhydrous* sodium sulfate if the solution remains cloudy.

Isolation and Purification Transfer the organic solution to a conical vial equipped for simple distillation and remove most of the solvent. Alternatively, use rotary evaporation or other techniques to concentrate the solution. Allow the vial to cool to room temperature. Add about 2 mL of hexanes to the residue in the vial, stir the solution with a glass stirring rod, and place the flask in an ice-water bath for 10 min. If no solid precipitates, use one of the methods described in Chapter 3 to induce crystallization. Isolate the white solid by vacuum filtration and air-dry it. Recrystallize the crude product by a mixed solvent recrystallization according to the second option for dissolution (Sec. 3.2) by suspending the solid in about 4 mL of boiling hexanes and adding ethyl acetate dropwise until all the solids have dissolved.

Analysis Weigh the recrystallized product and calculate the percent yield. Determine its melting point. Obtain IR and ^{1}H NMR spectra of your starting materials and product and compare them with those of authentic samples (Figs 24.1–24.6).

WRAPPING IT UP

Combine all of the *aqueous layers,* adjust the pH of the solution to approximately 7 with solid sodium carbonate, and flush it down the drain with excess water. Pour the *organic layer from the first extraction* and the recovered *diethyl ether* and *recrystallization solvent* into a container for nonhalogenated organic liquids. Spread the *sodium sulfate* on a tray in the hood to evaporate residual solvent, and then put it in the container for nonhazardous solids.

B ■ *Preparation of Methyl L-Phenylalaninate Hydrochloride*

Purpose To demonstrate the protection of the carboxylic acid group of an α-amino acid by forming a methyl ester.

SAFETY ALERT

1. **Wear safety glasses or goggles and suitable protective gloves while performing the experiment.**

2. **Thionyl chloride is a volatile and *corrosive* chemical that undergoes rapid reaction with water and atmospheric moisture to produce sulfur dioxide and hydrogen chloride, both of which are also corrosive. Do not breathe this chemical or allow it to come in contact with your skin. If contact occurs, immediately flood the affected area with cold water and then rinse with 5% sodium bicarbonate solution. Dispense thionyl chloride only at a hood and minimize its exposure to atmospheric moisture by keeping the bottle tightly closed when not in use.**

3. **This experiment should be performed in a hood if possible.**

4. **Diethyl ether is *extremely* flammable and volatile, and its vapors can easily travel several feet along the bench top or the floor and then be ignited. Consequently, *be certain there are no open flames anywhere in the laboratory* whenever you are working with it. Use a flameless heating source *whenever* heating is required.**

MINISCALE PROCEDURE

Preparation Refer to the online resources to answer Pre-Lab Exercises, access videos, and read the MSDSs for the chemicals used or produced in this procedure. Review Sections 2.10, 2.11, 2.17, and 2.22.

Apparatus A 25-mL round-bottom flask, 1-mL glass syringe with needle, ice-water bath, apparatus for heating under reflux, magnetic stirring, vacuum filtration, and *flameless* heating.

Set-Up Place 1.0 g of L-phenylalanine in a 25-mL round-bottom flask containing a stirbar, add 5 mL of MeOH, and begin stirring the mixture.

Reaction Add 0.5 mL of thionyl chloride dropwise using a syringe. After the addition of thionyl chloride is completed, the solution should be homogeneous. Equip the round-bottom flask with a reflux condenser and heat the mixture under gentle reflux for 45 min.

Isolation and Purification Allow the reaction mixture to cool to room temperature and then transfer the solution to an Erlenmeyer flask. Rinse the round-bottom flask with a 0.5–1-mL portion of methanol and transfer this rinse to the Erlenmeyer flask. Place the Erlenmeyer flask in an ice-water bath and add 25 mL of diethyl ether. Using a glass stirring rod, scratch the inside of the flask at the air-liquid interface to induce crystallization. Isolate the white solid by vacuum filtration and air-dry it. Recrystallize the crude product by a mixed solvent recrystallization according to the second option for dissolution (Sec. 3.2) by suspending the solid in about 20 mL of boiling diethyl ether and adding methanol dropwise until all the solids have dissolved.

Analysis Weigh the recrystallized product and calculate the percent yield. Determine its melting point. Obtain IR and ^{1}H NMR spectra of your starting material and product, and compare them with those of authentic samples (Figs. 24.7–24.10).

MICROSCALE PROCEDURE

Preparation Refer to the online resources to answer Pre-Lab Exercises, access videos, and read the MSDSs for the chemicals used or produced in this procedure. Review Sections 2.10, 2.11, 2.17, and 2.22.

Apparatus A 3-mL conical vial, 1-mL glass syringe with needle, ice-water bath, apparatus for heating under reflux, magnetic stirring, vacuum filtration, and *flameless* heating.

Set-Up Place 0.2 g of L-phenylalanine in a 3-mL conical vial containing a spinvane, add 1 mL of MeOH, and begin stirring the mixture.

Reaction Add 0.1 mL of thionyl chloride dropwise using a syringe. After the addition of thionyl chloride is completed, the solution should be homogeneous. Equip the conical vial with a reflux condenser, and heat the mixture under gentle reflux for 45 min.

Isolation and Purification Allow the reaction mixture to cool to room temperature and then transfer the solution to a small Erlenmeyer flask. Rinse the round-bottom flask with a few drops of methanol and transfer this rinse to the Erlenmeyer flask. Place the Erlenmeyer flask into an ice-water bath and add 5 mL of diethyl ether. Use a glass stirring rod to scratch the inside of the flask at the air-liquid interface to induce crystallization. Isolate the white solid by vacuum filtration and air-dry it. Recrystallize the crude product by a mixed solvent recrystallization according to the second option for dissolution (Sec. 3.2) by suspending the solid in about 4 mL of boiling diethyl ether and adding methanol dropwise until all the solids have dissolved.

Analysis Weigh the recrystallized product and calculate the percent yield. Determine its melting point. Obtain IR and ¹H NMR spectra of your material and product and compare them with those of authentic samples (Figs. 24.7–24.10).

WRAPPING IT UP

Combine all the *organic filtrates* and pour them into a container for nonhalogenated organic liquids.

C ▪ *Preparation of Methyl N-tert-Butoxycarbonyl L-Alanyl-L-Phenylalaninate*

Purpose To demonstrate the formation of a diprotected dipeptide by the reaction of the mixed anhydride of an *N*-protected amino acid with the ester of a second amino acid.

SAFETY ALERT

1. **Wear safety glasses or goggles and suitable protective gloves while performing the experiment.**

2. **Isobutyl chloroformate is a lachrymator and a volatile, *corrosive* chemical that undergoes rapid reaction with water and atmospheric moisture to produce carbon dioxide, isobutyl alcohol, and hydrogen chloride, which is also corrosive. Do not breathe this chemical or allow it to come in contact with your skin. If contact occurs, immediately flood the affected area with cold water and then rinse with 5% sodium bicarbonate solution. Dispense isobutyl chloroformate only at a hood and minimize its exposure to atmospheric moisture by keeping the bottle tightly closed when not in use.**

3. **This experiment should be performed in a hood if possible.**

4. **Diethyl ether is *extremely* flammable and volatile, and its vapors can easily travel several feet along the bench top or the floor and then be ignited. Consequently, *be certain there are no open flames anywhere in the laboratory* whenever you are working with it. Use a *flameless* heating source *whenever* heating is required.**

MINISCALE PROCEDURE

Preparation Refer to the online resources to answer Pre-Lab Exercises, access videos, and read the MSDSs for the chemicals used or produced in this procedure. Review Sections 2.5, 2.10, 2.11, 2.13, 2.17, 2.21, and 2.29.

Apparatus A 100-mL round-bottom flask, 50-mL Erlenmeyer flask, two 1-mL glass syringes with needles, ice-water bath, separatory funnel, apparatus for magnetic stirring, simple distillation, vacuum filtration, and *flameless* heating.

Set-Up and Reaction Place 0.63 g of methyl L-phenylalaninate hydrochloride in a 50-mL Erlenmeyer flask. Add 10 mL of dimethylformamide and 0.3 mL of *N*-methylmorpholine to the flask, swirl the flask to mix the contents, and place the flask in an ice-water bath. Label the flask as Solution A.

Place 0.50 g of *N-tert*-butoxycarbonyl-L-alanine in a 100-mL round-bottom flask containing a stirbar. Add 10 mL of dimethylformamide and 0.3 mL of *N*-methylmorpholine to the flask. Place the flask in an ice-water bath and stir the solution for 5 min. Add 0.4 mL of isobutyl chloroformate dropwise and continue stirring the reaction mixture in the ice-water bath for 5–10 min. Transfer Solution A into the round-bottom flask and continue to stir the reaction with cooling in the ice-water bath for 45 min.

Work-Up Add 20 mL of water to the reaction mixture and transfer it to a separatory funnel. Rinse the round-bottom flask with a 30-mL portion of diethyl ether and transfer the rinse to the separatory funnel. Shake the funnel gently, *venting as necessary*. Separate the layers and wash the organic layer with two 25-mL portions of 1 *M* HCl, a 25-mL portion of saturated sodium bicarbonate, and a 25-mL portion of brine. Transfer the organic layer to a 125-mL Erlenmeyer flask and add several spatula-tips full of *anhydrous* sodium sulfate. Occasionally swirl the mixture over a period of 10–15 min to facilitate drying; add further small portions of *anhydrous* sodium sulfate if the solution remains cloudy.

Isolation and Purification Filter or carefully decant the dried organic solution into a round-bottom flask equipped for simple distillation, and concentrate to a volume of approximately 5 mL. Alternatively, use rotary evaporation or other techniques to concentrate the solution. Allow the solution to cool to room temperature. Transfer this solution to a 50-mL Erlenmeyer flask. Rinse the round-bottom flask with 0.5–1 mL of diethyl ether and transfer this rinse to the Erlenmeyer flask. Add approximately 15 mL of hexanes to the flask and place the flask in an ice-water bath. If crystals do not form within 5–10 min, use one of the methods described in Chapter 3 to induce crystallization. Isolate the crystals by vacuum filtration and air-dry them. Recrystallize the product from diethyl ether/hexanes (2:1).

Analysis Weigh the recrystallized product and calculate the percent yield. Determine its melting point. Obtain IR and ¹H NMR spectra of your product and compare them with those of authentic samples (Figs. 24.11 and 24.12).

MICROSCALE PROCEDURE

Preparation Refer to the online resources to answer Pre-Lab Exercises, access videos, and read the MSDSs for the chemicals used or produced in this procedure. Review Sections 2.5, 2.10, 2.11, 2.13, 2.17, 2.21, and 2.29.

Apparatus A 5-mL conical vial, 13-mm × 100-mm test tube, two 1-mL glass syringes with needles, ice-water bath, screw-cap centrifuge tube, apparatus for magnetic stirring, simple distillation, vacuum filtration, and *flameless* heating.

Set-Up and Reaction Place 0.12 g of methyl L-phenylalaninate hydrochloride in a 13-mm × 100-mm test tube. Add 1.0 mL dimethylformamide and 0.06 mL of

N-methylmorpholine to the test tube, swirl the tube to mix the contents, and place it in an ice-water bath. Label the test tube as Solution A.

Place 0.10 g of *N-tert*-butoxycarbonyl-L-alanine in a 5-mL conical vial containing a spinvane. Add 1.0 mL of dimethylformamide and 0.06 mL of *N*-methylmorpholine to the vial. Place the vial in an ice-water bath and stir the solution for 5 min. Add 0.08 mL of isobutyl chloroformate dropwise and continue stirring the reaction mixture in the ice-water bath for 5–10 min. Transfer Solution A into the conical vial and continue to stir the reaction with cooling in the ice-water bath for 45 min.

Work-Up Add 20 mL of water to the reaction mixture and transfer it to a screw-cap centrifuge tube. Rinse the conical vial with a 4-mL portion of diethyl ether and transfer the rinse to the screw-cap centrifuge tube, screw the cap onto the centrifuge tube, and shake the tube gently, *venting as necessary*. Separate the layers and wash the organic layer with two 4-mL portions of 1 *M* HCl, a 4-mL portion of saturated sodium bicarbonate, and a 4-mL portion of brine. Transfer the organic layer to a dry test tube and add several spatula-tips full of *anhydrous* sodium sulfate. Occasionally swirl the mixture over a period of 10–15 min to facilitate drying; add further small portions of *anhydrous* sodium sulfate if the solution remains cloudy.

Isolation and Purification Transfer the organic solution to a conical vial equipped for simple distillation and distill most of the solvent. Alternatively, use rotary evaporation or other techniques to concentrate the solution. Allow the vial and its contents to cool to room temperature, add approximately 3 mL of hexanes to the vial, and place it in an ice-water bath. If crystals do not form within 5–10 min, use one of the methods described in Chapter 3 to induce crystallization. Isolate the crystals by vacuum filtration and air-dry them. Recrystallize the product from diethyl ether/hexanes (2:1).

Analysis Weigh the recrystallized product and calculate the percent yield. Determine its melting point. Obtain IR and ¹H NMR spectra of your product and compare them with those of authentic samples (Figs. 24.11 and 24.12).

WRAPPING IT UP

Combine all of the *aqueous layers*, adjust the pH of the solution to approximately 7 with either solid sodium carbonate or 3 *M* HCl, and flush it down the drain with excess water. Pour the recovered *diethyl ether* and *recrystallization solvent* into a container for nonhalogenated organic liquids. Spread the *sodium sulfate* on a tray in the hood to evaporate residual solvent, and then put it in the container for nonhazardous solids.

D ▪ *Preparation of Methyl L-Alanylphenyl-L-Alaninate Trifluoroacetate*

Purpose To demonstrate the selective removal of a *tert*-butylcarbamate of a diprotected dipeptide in the presence of a methyl ester.

SAFETY ALERT

1. **Wear safety glasses or goggles and suitable protective gloves while performing the experiment.**

2. **Trifluoroacetic acid is a volatile and *corrosive* chemical. Do not breathe this chemical or allow it to come in contact with your skin. If contact occurs, immediately**

flood the affected area with cold water and then rinse with 5% sodium bicarbonate solution. Dispense trifluoroacetic acid only at a hood.

3. **This experiment should be performed in a hood if possible.**

4. **Diethyl ether is *extremely* flammable and volatile, and its vapors can easily travel several feet along the bench top or the floor and then be ignited. Consequently, *be certain there are no open flames anywhere in the laboratory* whenever you are working with it. Use a *flameless* heating source *whenever* heating is required.**

MINISCALE PROCEDURE

Preparation Refer to the online resources to answer Pre-Lab Exercises, access videos, and read the MSDSs for the chemicals used or produced in this procedure. Review Sections 2.5, 2.10, 2.11, 2.13, 2.17, and 2.29.

Apparatus A 10-mL round-bottom flask, 1-mL glass syringe with needle, ice-water bath, apparatus for magnetic stirring, simple distillation, vacuum filtration, and *flameless* heating.

Set-Up and Reaction Place 300 mg of methyl *N-tert*-butoxycarbonyl-L-alanyl-L-phenylalaninate in the 10-mL round-bottom flask containing a stirbar, and add 6 mL of dichloromethane. Begin stirring the solution, add 1.5 mL of trifluoroacetic acid to the flask, and continue stirring the mixture for 30 min at room temperature.

Isolation and Purification Equip the flask for simple distillation and distill the solvent. Alternatively, use rotary evaporation or other techniques to concentrate the solution. Allow the flask to cool to room temperature and then add 4 mL of diethyl ether to the flask. Using a spatula, thoroughly mix the solvent with the residue to induce the formation of the solid product. Place the flask in an ice-water bath for 5–10 min to complete the precipitation. Isolate the white solid by vacuum filtration and air-dry it. Recrystallize the crude product from ethyl acetate.

Analysis Weigh the recrystallized product and calculate the percent yield. Determine its melting point. Obtain IR and ^{1}H NMR spectra of your product and compare them with those of authentic samples (Figs. 24.13 and 24.14).

MICROSCALE PROCEDURE

Preparation Refer to the online resources to answer Pre-Lab Exercises, access videos, and read the MSDSs for the chemicals used or produced in this procedure. Review Sections 2.5, 2.10, 2.11, 2.13, 2.17, and 2.29.

Apparatus A 3-mL conical vial, 1-mL glass syringe with needle, apparatus for magnetic stirring, simple distillation, vacuum filtration, and *flameless* heating.

Set-Up and Reaction Place 50 mg of methyl *N-tert*-butoxycarbonyl-L-alanyl-L-phenylalaninate in the 3-mL conical vial containing a spinvane and add 1 mL of dichloromethane. Begin stirring the solution, add 0.2 mL of trifluoroacetic acid to the conical vial, and continue stirring the mixture for 30 min at room temperature.

Isolation and Purification Equip the conical vial for simple distillation and distill the solvent. Alternatively, use rotary evaporation or other techniques to concentrate the solution. Allow the conical vial to cool to room temperature and then add 1 mL of diethyl ether to the conical vial. Using a spatula, thoroughly mix the solvent with the residue to induce formation of the solid product. Place the conical vial in an ice-water bath for 5–10 min to complete the precipitation. Isolate the white solid by vacuum filtration and air-dry it. Recrystallize the crude product from ethyl acetate.

Analysis Weigh the recrystallized product and calculate the percent yield. Determine its melting point. Obtain IR and ¹H NMR spectra of your product and compare them with those of authentic samples (Figs. 24.13 and 24.14).

WRAPPING IT UP

Transfer the *distillate* to a separatory funnel and wash it *carefully with venting* with a small amount of *saturated* aqueous sodium bicarbonate. Dilute the *aqueous layer* with water, and flush it down the drain with excess water. Pour the *organic layer* into a container for halogenated organic liquids. Pour the *ethereal filtrate* into a container for nonhalogenated organic liquids.

Discovery Experiment   *Synthesis of L-Alanyl-L-Phenylalanine*

Consult with your instructor before performing this experiment, in which you will prepare the dipeptide L-alanyl-L-phenylalanine by hydrolysis of the methyl ester of the monoprotected dipeptide Ala–Phe–OMe (**12**). The ester may be hydrolyzed in either aqueous base or acid and you should design procedures for both. You should consider the chemical and physical properties of **12** and develop a reaction and a work-up procedure that will most easily enable you to isolate the product. Obtain the IR, ¹H, and ¹³C NMR spectra of the purified product for characterization and compare these with the literature data.

EXERCISES

General Questions

1. Of the 19 amino acids with a center of chirality in Table 24.1, identify the one that has the *R*-configuration rather than the *S*-configuration.

2. Amino acids are least soluble in water when the pH of the solution is the same as the isoelectric point for the amino acid. Explain this observation.

3. Write the structures of all the possible products that would be obtained by random peptide bond formation involving L-isoleucine and L-asparagine.

4. Another common protecting group for the amine function of amino acids is the benzyloxycarbonyl (Cbz) group, as exemplified by the protection of L-valine by reaction with benzyloxycarbonyl chloride in the presence of an aqueous base to give *N*-benzyloxycarbonyl-L-valine (Cbz–Val). Using curved arrows to symbolize the flow of electrons, write a stepwise mechanism for this reaction.

L-Valine   Benzyloxycarbonyl chloride (Cbz–Cl)

(1) aq. NaOH
(2) H⁺

N-Benzyloxycarbonyl-L-valine (Cbz–Val)

5. One method for removing an *N*-benzyloxycarbonyl group is by catalytic *hydrogenolysis* using dihydrogen in the presence of a palladium catalyst. Another method involves use of hydrogen bromide in *glacial* acetic acid. Using curved arrows to symbolize the flow of electrons, write a stepwise mechanism for the deprotection of *N*-benzyloxycarbonyl-L-valine by this latter technique.

N-Benzyloxycarbonyl-L-valine   L-Valine

6. Another common ester for protecting amino acids is the *tert*-butyl ester, which may be prepared as exemplified by the reaction of L-valine with isobutylene in the presence of acid to give *tert*-butyl L-valinate. Using curved arrows to symbolize the flow of electrons, write a stepwise mechanism for this reaction.

L-Valine   Isobutylene   *tert*-Butyl L-valinate
                                                        (Val–O–*t*-Bu)

7. It is possible to hydrolyze the methyl ester group of **11** selectively using lithium hydroxide in methanol, followed by acidification to give *N-tert*-butoxycarbonyl-L-alanyl-L-phenylalanine.

11   *N-tert*-Butoxycarbony-L-alanyl-L-phenylalanine
                                                        (Boc–Ala–Phe)

  a. Using curved arrows to symbolize the flow of electrons, write a stepwise mechanism for the hydrolysis.

  b. There are three different carbonyl groups in **11** that could undergo hydrolysis, two of which are associated with esters, yet the methyl ester is hydrolyzed preferentially. Provide an explanation for this observation.

8. When exposed to solutions of aqueous base, the *tert*-butyl esters of amino acids and peptides are much more stable toward hydrolysis than the corresponding methyl esters. Provide an explanation for this observation.

9. The *tert*-butyl esters of amino acids and peptides may be readily removed upon treatment with trifluoroacetic acid in dichloromethane. Using curved arrows to symbolize the flow of electrons, write a stepwise mechanism for this reaction.

10. Starting with L-aspartic acid and L-phenylalanine and, using any of the protecting groups presented in this chapter, including those in Exercises 3 and 5, outline a synthesis of the artificial sweetener aspartame (**2**).

11. Starting with the monoprotected dipeptide **12** and the unprotected amino acid L-valine, outline a sequence of reactions for the preparation of the monoprotected tripeptide Val–Ala–Phe.

Questions for Part A

12. Write the balanced equation for the reaction of di-*tert*-butyl dicarbonate with water.

13. Using curved arrows to symbolize the flow of electrons, write a stepwise mechanism for the reaction of di-*tert*-butyl dicarbonate with water in Exercise 12.

14. Write the structure of the charged form of L-alanine in aqueous base.

15. In the work-up of the reaction, the diethyl ether layer and aqueous layers are separated. What compound(s) remain in the diethyl ether layer? The aqueous layer?

16. Why is it necessary to acidify the aqueous layer before extracting it with diethyl ether?

17. What compound(s) might remain in the aqueous layer after this extraction with diethyl ether?

18. Using curved arrows to symbolize the flow of electrons, write a stepwise mechanism for the formation of *N-tert*-butoxycarbonyl-L-alanine from L-alanine and di-*tert*-butyl dicarbonate according to Equation 24.4.

Refer to the online resources and use the spectra viewer and Tables 8.1–8.8 as needed to answer the blue-numbered questions on spectroscopy.

19. Consider the spectral data for L-alanine (Figs. 24.1 and 24.2).

    a. In the functional group region of the IR spectrum, specify the absorptions associated with the nitrogen-hydrogen bonds and carbonyl component of the carboxyl group.

    b. In the ^{1}H NMR spectrum, assign as many as possible of the various resonances to the hydrogen nuclei responsible for them.

    c. For the ^{13}C NMR data, assign the various resonances to the carbon nuclei responsible for them.

20. Consider the spectral data for di-*tert*-butyl dicarbonate (Figs. 24.3 and 24.4).

    a. In the functional group region of the IR spectrum, specify the absorption associated with the carbonyl component of the dicarbonate group.

    b. In the ^{1}H NMR spectrum, assign the resonances to the hydrogen nuclei responsible for them.

    c. For the ^{13}C NMR data, assign the various resonances to the carbon nuclei responsible for them.

21. Consider the spectral data for *N-tert*-butoxycarbonyl-L-alanine (Figs. 24.5 and 24.6).

    a. In the functional group region of the IR spectrum, specify the absorptions associated with the nitrogen-hydrogen bond and the carbonyl components of the carboxyl and the carbamate groups.

    b. In the ^{1}H NMR spectrum, assign as many as possible of the various resonances to the hydrogen nuclei responsible for them.

    c. For the ^{13}C NMR data, assign the various resonances to the carbon nuclei responsible for them.

22. Discuss the differences observed in the IR and NMR spectra of L-alanine, di-*tert*-butyl dicarbonate, and *N-tert*-butoxycarbonyl-L-alanine that are consistent with the formation of the latter in this procedure.

Questions for Part B

23. Write the balanced equation for the reaction of thionyl chloride with an excess of methanol.

24. Using curved arrows to symbolize the flow of electrons, write a stepwise mechanism for the reaction of thionyl chloride with methanol.

25. Write the structure of the charged form of L-phenylalanine in a solution of methanolic hydrogen chloride.

26. Assume that the HCl generated by the reaction of thionyl chloride with methanol catalyzes the esterification of L-phenylalanine. Using curved arrows to symbolize the flow of electrons, write a stepwise mechanism for the acid-catalyzed esterification of phenylalanine with methanol according to Equation 24.5.

Refer to the online resources and use the spectra viewer and Tables 8.1–8.8 as needed to answer the blue-numbered questions on spectroscopy.

27. Consider the spectral data for L-phenylalanine (Figs. 24.7 and 24.8).

    a. In the functional group region of the IR spectrum, specify the absorptions associated with the nitrogen-hydrogen bonds and the carbonyl component of the carboxyl group.

    b. In the ^{1}H NMR spectrum, assign as many as possible of the various resonances to the hydrogen nuclei responsible for them.

    c. For the ^{13}C NMR data, assign the various resonances to the carbon nuclei responsible for them.

28. Consider the spectral data for methyl L-phenylalaninate hydrochloride (Figs. 24.9 and 24.10).

    a. In the functional group region of the IR spectrum, specify the absorptions associated with the nitrogen-hydrogen bonds and the carbonyl component of the ester group.

    b. In the ^{1}H NMR spectrum, assign as many as possible of the various resonances to the hydrogen nuclei responsible for them.

    c. For the ^{13}C NMR data, assign the various resonances to the carbon nuclei responsible for them.

29. Discuss the differences observed in the IR and NMR spectra of L-phenylalanine and methyl L-phenylalaninate hydrochloride that are consistent with the formation of the latter in this procedure.

Questions for Part C

30. Write the balanced equation for the reaction of isobutyl chloroformate with water.

31. Why is *N*-methylmorpholine required in the conversion of **7** and **9** into **10** (Eq. 24.6) in the first step of the procedure used in this experiment?

32. Why is *N*-methylmorpholine required in the conversion of **8** and **10** into **11** (Eq. 24.7) in the second step of the procedure used in this experiment?

33. In the work-up of the reaction, the organic layer is washed with dilute aqueous acid. What is the purpose of this step?

34. In the work-up of the reaction, the organic layer is washed with dilute aqueous base. What is the purpose of this step?

35. Using curved arrows to symbolize the flow of electrons, write a stepwise mechanism for the formation of **10** from *N-tert*-butoxycarbonyl-L-alanine (**7**) according to Equation 24.6.

36. Using curved arrows to symbolize the flow of electrons, write a stepwise mechanism for the formation of methyl *N-tert*-butoxycarbonyl-L-alanylphenyl-L-alaninate (**11**) from **8** and **10** according to Equation 24.7.

Refer to the online resources and use the spectra viewer and Tables 8.1–8.8 as needed to answer the blue-numbered questions on spectroscopy.

37. Consider the spectral data for methyl *N-tert*-butoxycarbonyl-L-alanyl-L-phenyl-alaninate (Figs. 24.11 and 24.12).

    a. In the functional group region of the IR spectrum, specify the absorptions associated with the nitrogen-hydrogen bonds and the carbonyl components of the ester, amide, and carbamate groups.

    b. In the ^{1}H NMR spectrum, assign as many as possible of the various resonances to the hydrogen nuclei responsible for them.

    c. For the ^{13}C NMR data, assign the various resonances to the carbon nuclei responsible for them.

38. Discuss the differences observed in the IR and NMR spectra of *N-tert*-butoxycarbonyl-L-alanine, methyl L-phenylalaninate hydrochloride, and methyl *N-tert*-butoxycarbonyl-L-alanyl-L-phenylalaninate that are consistent with the formation of the latter in this procedure.

Questions for Part D

39. Using curved arrows to symbolize the flow of electrons, write a stepwise mechanism for the formation of methyl L-alanyl-L-phenylalaninate (**12**) from **11** according to Equation 24.8.

40. Consider the spectral data for methyl L-alanyl-L-phenylalaninate (Figs. 24.13 and 24.14).

    a. In the functional group region of the IR spectrum, specify the absorptions associated with the nitrogen-hydrogen bonds and the carbonyl components of the ester and amide groups.

    b. In the ^{1}H NMR spectrum, assign as many as possible of the various resonances to the hydrogen nuclei responsible for them.

    c. For the ^{13}C NMR data, assign the various resonances to the carbon nuclei responsible for them.

41. Discuss the differences observed in the IR and NMR spectra of methyl *N-tert*-butoxycarbonyl-L-alanyl-L-phenylalanine and methyl L-alanyl-L-phenylalaninate that are consistent with the formation of the latter in this procedure.

SPECTRA

Starting Materials and Products

Figure 24.1

IR spectrum of L-alanine (IR card).

Figure 24.2
NMR data for L-alanine (D₂O).

(a) ¹H NMR spectrum (300 MHz).
(b) ¹³C NMR data: δ 17.3, 51.7, 176.6.

Figure 24.3
IR spectrum of di-tert-butyl dicarbonate (neat).

Figure 24.4
NMR data for di-tert-butyl dicar-bonate (CDCl₃).

(a) ¹H NMR spectrum (300 MHz).
(b) ¹³C NMR data: δ 27.5, 85.2, 146.9.

Figure 24.5
IR spectrum of N-tert-butoxycarbonyl-L-alanine (IR card).

Figure 24.6
NMR data for N-tert-butoxycarbonyl-L-alanine (DMSO-d_6).

(a) ^{1}H NMR spectrum (300 MHz).
(b) ^{13}C NMR data: δ 17.7, 28.9, 49.5, 78.6, 155.9, 175.4.

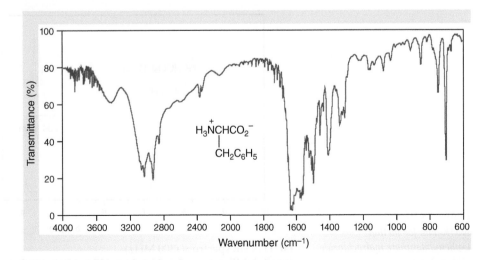

Figure 24.7
IR spectrum of L-phenylalanine (IR card).

Figure 24.8

NMR data for L-phenylalanine (D₂O).

(a) ¹H NMR spectrum (300 MHz).
(b) ¹³C NMR data: δ 37.4, 56.9, 128.4, 129.9, 130.2, 136.3, 174.2.

Figure 24.9

IR spectrum of methyl L-phenylalaninate hydrochloride (IR card).

Figure 24.10

NMR data for methyl L-phenylalaninate hydrochloride (DMSO-d₆).

(a) ¹H NMR spectrum (300 MHz).
(b) ¹³C NMR data: δ 35.7, 52.4, 53.2, 127.2, 128.5, 129.3, 134.7, 169.3.

Figure 24.11
IR spectrum of methyl N-tert-butoxycarbonyl-L-alanyl-L-phenylalaninate (IR card).

Figure 24.12
NMR data for methyl N-tert-butoxycarbonyl-L-alanyl-L-phenylalaninate (DMSO-d$_6$).

(a) ^{1}H NMR spectrum (300 MHz).
(b) ^{13}C NMR data: δ 18.8, 28.9, 37.5, 50.2, 52.5, 54.1, 78.7, 127.2, 128.9, 129.8, 129.9, 137.7, 155.6, 172.5, 173.4.

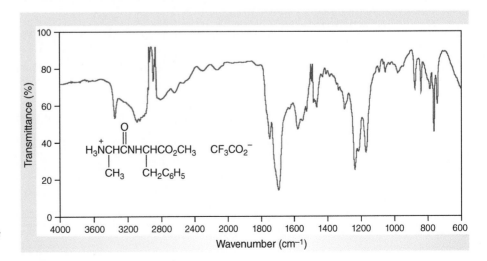

Figure 24.13
IR spectrum of methyl L-alanyl-L-phenylalaninate trifluoroacetate (IR card).

Figure 24.14
NMR data for methyl L-alanyl-L-phenylalaninate trifluoroacetate (DMSO-d₆).

(a) ¹H NMR spectrum (300 MHz).
(b) ¹³C NMR data: δ 17.8, 39.3, 48.6, 52.7, 54.6, 127.4, 129.0, 129.8, 137.6, 170.6, 172.1.

Index

Note: Boldface numbers indicate a figure.

Periodic Table of the Elements

1 1A	Alkaline earth metals 2 2A											metals ← → nonmetals 13 3A	14 4A	15 5A	16 6A	Halogens 17 7A	Noble gases 18 8A
1 **H** 1.008																	2 **He** 4.003
3 **Li** 6.941	4 **Be** 9.012	3	4	5	6	7	8	9	10	11	12	5 **B** 10.81	6 **C** 12.01	7 **N** 14.01	8 **O** 16.00	9 **F** 19.00	10 **Ne** 20.18
11 **Na** 22.99	12 **Mg** 24.31				Transition metals							13 **Al** 26.98	14 **Si** 28.09	15 **P** 30.97	16 **S** 32.07	17 **Cl** 35.45	18 **Ar** 39.95
19 **K** 39.10	20 **Ca** 40.08	21 **Sc** 44.96	22 **Ti** 47.88	23 **V** 50.94	24 **Cr** 52.00	25 **Mn** 54.94	26 **Fe** 55.85	27 **Co** 58.93	28 **Ni** 58.69	29 **Cu** 63.55	30 **Zn** 65.38	31 **Ga** 69.72	32 **Ge** 72.59	33 **As** 74.92	34 **Se** 78.96	35 **Br** 79.90	36 **Kr** 83.80
37 **Rb** 85.47	38 **Sr** 87.62	39 **Y** 88.91	40 **Zr** 91.22	41 **Nb** 92.91	42 **Mo** 95.94	43 **Tc** (98)	44 **Ru** 101.1	45 **Rh** 102.9	46 **Pd** 106.4	47 **Ag** 107.9	48 **Cd** 112.4	49 **In** 114.8	50 **Sn** 118.7	51 **Sb** 121.8	52 **Te** 127.6	53 **I** 126.9	54 **Xe** 131.3
55 **Cs** 132.9	56 **Ba** 137.3	57 **La*** 138.9	72 **Hf** 178.5	73 **Ta** 180.9	74 **W** 183.9	75 **Re** 186.2	76 **Os** 190.2	77 **Ir** 192.2	78 **Pt** 195.1	79 **Au** 197.0	80 **Hg** 200.6	81 **Tl** 204.4	82 **Pb** 207.2	83 **Bi** 209.0	84 **Po** (209)	85 **At** (210)	86 **Rn** (222)
87 **Fr** (223)	88 **Ra** 226	89 **Ac†** (227)	104 **Rf** (261)	105 **Db** (262)	106 **Sg** (263)	107 **Bh** (264)	108 **Hs** (265)	109 **Mt** (268)	110 **Ds** (271)	111 **Rg** (272)	112 **Cn** (285)	113 **Uut**	114 **Fl** (289)	115 **Uup**	116 **Lv** (293)	117 **Uus**	118 **Uuo**

Alkali metals (column 1A)

*Lanthanides	58 **Ce** 140.1	59 **Pr** 140.9	60 **Nd** 144.2	61 **Pm** (145)	62 **Sm** 150.4	63 **Eu** 152.0	64 **Gd** 157.3	65 **Tb** 158.9	66 **Dy** 162.5	67 **Ho** 164.9	68 **Er** 167.3	69 **Tm** 168.9	70 **Yb** 173.0	71 **Lu** 175.0
†Actinides	90 **Th** 232.0	91 **Pa** (231)	92 **U** 238.0	93 **Np** (237)	94 **Pu** (244)	95 **Am** (243)	96 **Cm** (247)	97 **Bk** (247)	98 **Cf** (251)	99 **Es** (252)	100 **Fm** (257)	101 **Md** (258)	102 **No** (259)	103 **Lr** (262)

Group numbers 1–18 represent the system recommended by the International Union of Pure and Applied Chemistry.